Y0-CAO-056

BOTANY

BOTANY

A TEXTBOOK FOR COLLEGES

FOURTH EDITION

J. BEN HILL
Deceased

HENRY W. POPP

ALVIN R. GROVE, JR.

Professors of Botany
The Pennsylvania State University

ROBERT E. KRIEGER PUBLISHING COMPANY
HUNTINGTON, NEW YORK
1977

Original Edition 1936
Reprint 1977 *(from Fourth Edition)*

Printed and Published by
ROBERT E. KRIEGER PUBLISHING CO., INC.
645 NEW YORK AVENUE
HUNTINGTON, NEW YORK 11743

Printed in The United States of America

Library of Congress Cataloging in Publication Data

Hill, John Benjamin, 1879-1961.
Botany, a textbook for colleges.

Reprint of the 4th edition published by McGraw-Hill, New York.
Includes bibliographies and index.
1. Botany. I. Popp, Henry William, 1892-joint author. II. Grove, Alvin Russell, 1914-joint author. III. Title.
[QK47.H65 1977] 581 76-57931
ISBN 0-88275-516-1

PREFACE

Because of the many advances in the science of botany in recent years, the changing emphasis in the various branches of the subject, and the continued use of this textbook (including editions in Spain, India, and Japan), another revision has become necessary.

Electron-microscope studies have greatly increased the botanist's knowledge of the cell. Biochemical investigations have expanded his understanding of photosynthesis, protein synthesis, respiration, and other metabolic processes. The rapidly growing research on DNA has greatly influenced his understanding of heredity and biological activity in general. In fact, so great has been the impact of these findings that there has been a tendency to minimize or neglect the other basic physiological and morphological features of plants. In this textbook we have tried to remember that we are still dealing with plants. We have attempted to give a well-balanced account of the structure, biochemistry, physiology, genetics, and ecology of plants, together with a survey of the plant kingdom, and to emphasize the importance of plants to man.

Acting on suggestions made by colleagues who have used the textbook in the past, we have made a number of changes in the sequence of topics and in the relative emphasis of different subjects. A brief summary has been added at the beginning of each chapter in addition to questions and general references that have been added at the end of each

chapter. Chapters 2, 13, 14, 15, 16, and 18 have been rewritten, and we have added a chapter on ecology without eliminating the ecological treatment of other sections of the book. Cell division has been moved to the chapter on the cell. The discussion of metabolism—including photosynthesis, protein synthesis, and respiration, and the nature and functions of DNA and RNA—has been brought up to date. All other sections have been revised in the light of recent advances in the general botanical field. A substantial addition of terms has been made to the glossary.

We have eliminated some of the illustrations of the previous edition and have improved others. A substantial number of new figures has been added. We are greatly indebted to the various persons who have assisted in the preparation of the illustrations or have furnished photographs. Acknowledgment has been made individually in connection with each figure contributed. In addition, we wish to express our appreciation particularly to the following persons who have each made a considerable number of drawings: Mrs. Elsie M. McDougle, Mrs. Edna Stamy Fox, Dr. Helen D. Hill, Miss Florence Brown, Ernest Geisweite, Christian Hildebrandt, and Otto E. Schultz. We are likewise indebted to Dr. D. A. Kribs for the many excellent photomicrographs he made for the book and to Dr. F. D. Kern for many borrowed figures.

We are especially indebted to Dr. Helen D. Hill for her valuable criticism and her untiring assistance in the preparation of the manuscript; to Miss Florence Brown for her criticism and constant help in the preparation of the first edition; to Dr. Charles L. Fergus for his valuable criticism and assistance in rewriting the chapter on fungi; to Dr. John S. Boyle for his help with the viruses; to Dr. Robert H. Hamilton for his assistance in the revision of the metabolism chapters; to Dr. Paul Grun for his criticism of the chapters on cells and on heredity; to Dr. Charles J. Hillson for his suggestions on the chapter on algae; to Dr. Ronald A. Pursell for his criticism of the chapter on bryophytes; to Dr. Richard D. Schein for his help with the chapter on ecology; to Dr. Herbert A. Wahl for his criticism of the chapters on taxonomy and the vascular plants and other sections; to Dr. Eugene S. Lindstrom for his criticism of the section on bacteria; and to the many members of the various departments of The Pennsylvania State University who have made suggestions or have assisted in one way or another.

The authors wish to acknowledge their indebtedness to the late Dr. L. O. Overholts, coauthor of the first edition, for which he provided many of the illustrations, especially in the chapter on the fungi.

We wish especially to express our appreciation to Dr. O. A. Stevens of the Department of Botany of North Dakota State University, who examined critically and in detail the third edition and made numerous valuable constructive suggestions, and to Dr. H. W. Rickett of the New York Botanical Garden for his suggestions on inflorescences and nomenclature.

While these colleagues have all generously offered their criticism and help, the authors alone are responsible for any shortcomings and errors in the text.

Henry W. Popp
Alvin R. Grove, Jr.

CONTENTS

BOTANY

1 THE PLANT KINGDOM

Spice plants. Top, allspice, native of South America and the West Indies; bottom, nutmeg, native of the Moluccas and South India. The fruit of the allspice plant is the spice of commerce. The seed of the nutmeg plant and its covering, the aril, constitute the spices. The aril furnishes mace, and the seed, ground nutmeg.

"The Persians, Arabians and Egyptians formerly brought cloves and nutmegs to the ports in the Mediterranean, and hither the Venetians and Genoese resorted to buy the spices until the Portuguese, in 1551, discovered the country of their production. The Dutch drove the Portuguese from the Moluccas and for a long time maintained a strict monopoly over the production of these islands." (From William Rhind, A History of the Vegetable Kingdom. Blackie and Son, Glasgow, Edinburgh, and London, 1857.)

The role of plants in man's history, development, activities, and progress; the value of a study of botany; the branches of botany; the kinds of plants and their parts are considered briefly in this chapter.

Why study botany? Botany is a science that deals with plants: what kinds of plants there are, how they live and grow, how they respond to their surroundings, to what diseases they are susceptible, and how these are prevented and cured, and, among other things, how plants influence our everyday lives. Further study reveals the broad scope of the subject and gives a clearer insight into man's dependence upon plants and the great influence they have had in the origin and progress of civilization.

Not only do plants furnish us food, clothing, and shelter, but they keep the air we breathe enriched with oxygen, without which life would be impossible. Some of them, like the bacteria, cause serious diseases of man and the lower animals, but at the same time, antibiotics, such as penicillin, and other drugs obtained from plants help to prevent and cure these diseases. Plants furnish power to operate mills and factories and, in many cases, the raw materials such as cotton, oils, fats, waxes, rubber, and timber that go into the manufacture of their products. A majority of the world's workers earn a living by working with plants or plant products.

From the dawn of history plants have ministered to the needs of man and determined his progress. Probably one of the most significant landmarks in the history of civilization was the discovery that seeds dropped into the earth would grow and produce food-giving plants. This required man to remain in one place long enough to harvest his crops and influenced the formation of social groups, which in turn led to the division of labor and to the origin of trades and professions. Some of these were already well developed hundreds of years before Christ. The early white civilization that arose in the Tigris and Euphrates Valleys was centered around the native home of wheat. The ancient Chinese civilization depended upon rice, and the ancient Mayan civilization of the Western Hemisphere arose where corn (maize) had its native home.

The search for food-producing and other useful plants led to exploration and discovery of new lands. Mention need be made only of the search for a new route to the land where spices grew, which led Columbus to the discovery of America, or the search for the Northwest Passage, which led Henry Hudson and others to the exploration of the Hudson River basin and much of eastern Canada, to indicate the importance of plants in the discovery and development of our own country. In these explorations many new and useful plants were found, which in turn greatly influenced the history of mankind.

Exploration led to settlement and to the establishment of new colonies and countries. The struggle among nations for these new lands led to wars. The history of the world reveals a continuous struggle among nations for lands where useful plants were indigenous or could be grown. In the winning of these wars food obtained from plants has been, and still is, as important as ammunition, but even much of the ammunition is obtained from plant products. Thus plants have played a dominant role in the history of every nation.

Furthermore plants beautify our surroundings and furnish the setting or the inspiration for some of the world's best literature, art, and music. The Bible abounds in references to them. Such poems as Bryant's "Thanatopsis," Wordsworth's "Daffodils," Dickens's "The Ivy Green," and such well-known songs as Moore's "The Last Rose of Summer," Goethe's "Heidenröslein," and MacDowell's "To a Wild Rose," to mention but a few examples, were inspired by plants. Down through the years the beauty of plants has influenced the composers of music and the authors of lasting prose and verse. Plant structures have likewise been used extensively in art. We see them in book designs, in wallpaper, in rugs, in clothing, in famous paintings, and in architectural designs.

The study of plants by some of the world's most accomplished scientists has added immeasurably to man's general enlightenment and understanding of the universe. It has resulted in vast improvement in the production of farm crops, in manufacturing processes, in medicine, and in all things that assuage the rigor and augment the comforts of life and raise man's standard of living.

It should be obvious, in consideration of

these facts, that a general knowledge of objects which constitute so large a part of our environment and play so prominent a part in our lives is essential to a broad education. For students in agriculture, biology, forestry, and the natural sciences in general, botany is the foundation upon which their more specialized knowledge is built.

Aside from the purely utilitarian value of botany, the study of plants in the past has proved to be a most pleasing and gratifying avocation for lawyers, bankers, physicians, and other professional men. Many of our best early publications in botany were provided by such amateurs. These men discovered that an interest in plants opens up new viewpoints of life. It takes one out into the open fields, meadows, and forests where fresh air and beauty prevail and where the "quiet stillness" of Nature furnishes the best sedative for the tensions, fears, and struggles of man in the modern world.

Branches of botany Botany is one of the main divisions of biology, the science of life. Other branches of biology include zoology, biochemistry, biophysics, psychology, and the medical sciences. The element common to all of these fields of learning is that they deal with living organisms. This emphasizes the fact that plants are living organisms and as such have many things in common with other forms of life.

For convenience of study, the subject of botany has been divided into several important branches. Among these are taxonomy, morphology, physiology, pathology, ecology, paleobotany, and plant genetics. **Taxonomy,** or systematic botany, deals with nomenclature and the classification of plants. **Morphology** considers the form and structure of plants, together with the relationships of the parts of plants to each other, and comprises a study of anatomy, cytology (study of the cell), and embryology. **Physiology** is concerned with the life processes of plants and the functions of the different organs and tissues. **Pathology** deals with diseases of plants; **ecology,** with the relations of plants to their surroundings; **paleobotany,** or fossil botany, with the plants of past geological periods; and **plant genetics,** with the study of heredity in plants.

Other branches of the subject are concerned with an intensive study of separate groups of plants. Thus **bacteriology** is confined to the study of bacteria, **mycology** to the study of fungi, **algology,** or **phycology,** to the study of the algae, and **bryology** to the study of mosses and liverworts.

In addition to these definite branches of botany, many of the agricultural and other sciences either have had their origin in botany or may be considered as resting on a foundation of botany. Among these may be mentioned horticulture, forestry, landscape gardening, floriculture, plant breeding, and to some extent agronomy and pharmacy.

Different kinds of plants An excursion into the woods or fields during the summer or autumn reveals a wide diversity of form and structure in the plants encountered. Some are tall trees; others are low-growing herbs or shrubs; some have beautiful flowers and produce seed, while others, like the ferns, produce no flowers at all but reproduce by means of tiny structures called spores; some live on land, others in water; some are giants, while others are so small as to be seen only with a microscope. These wide differences led botanists years ago to attempt to arrange plants into different groups for the convenience of study and discussion. Many different systems of classification have been used, but the ones most generally accepted are based upon fundamental similarities or relationships and are called natural systems. According to these systems, the entire plant kingdom is usually subdivided into several large main divisions. Each of these is then subdivided

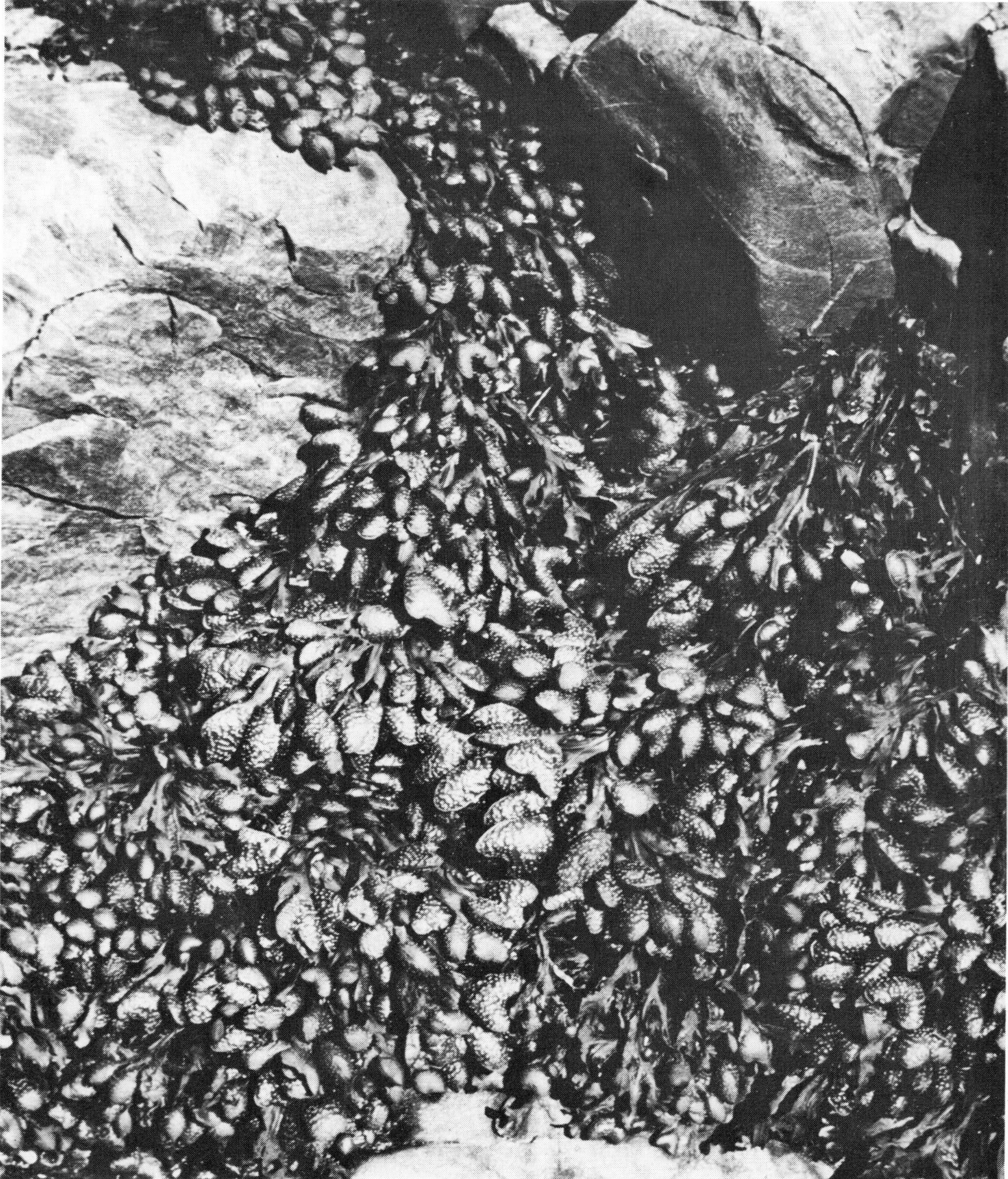

Fig. 1-1 Fucus, *a brown alga commonly called rockweed. Bar Harbor, Maine. (Photograph by Dr. Samuel Chan.)*

into appropriate groups. A discussion of the systems of classification may be found in Chap. 14. It will suffice here to call attention merely to some of the different kinds of plants that make up the plant kingdom.

Among the plants with the least structural differentiation are the bacteria, the fungi, and the algae. These plants have no true roots, stems, or leaves. A plant body of this type with relatively simple construction is called a **thallus.** The simplest of these plants are the bacteria, most of which are unicellular; i.e., the entire plant consists of but a single cell. The individual plants are visible only under the high magnification of a microscope. They are commonly called microbes. These organisms are universally present in the soil, in air, in water, on decaying animal and vegetable matter, or in other living organisms where they sometimes cause serious diseases. They are of vast importance to man in many ways. Pneumonia, tuberculosis, influenza, various infections, and many other diseases of human beings are caused by bacteria. On the other hand, many of the bacteria are extremely beneficial to man. This is especially true of soil bacteria.

Molds, mildews, yeasts, smuts, rusts, mushrooms, and toadstools (Fig. 1-2) are examples of fungi. Like the bacteria, the fungi all lack the green chlorophyll pigments essential for independent existence, and hence they must get their food from other living organisms or from dead or decaying organic matter. Some of the fungi, like the rusts and smuts, cause serious diseases of crop plants. Others cause diseases of human beings and of other animals. Many, however, are of great benefit to man. Thus mushrooms are used as food; yeasts are highly important in the baking, brewing, and wine-making industries; and many of the soil fungi are important in crop production. Penicillin, aureomycin, and other antibiotics first obtained from fungi are extremely important in medicine.

The algae (Fig. 1-1), examples of which are the common pond scums and seaweeds, usually grow submerged in water or in moist situations. Some are found in soil, and others grow on the bark of trees. They are of various colors—green, brown, red, and blue-green, but they contain the green chlorophyll pigments and hence are independent plants. Many of the algae either are unicellular or consist of a colony of cells. Most of them are relatively simple in structure, although some of the larger seaweeds show considerable tissue differentiation. Algae are of economic importance as the food basis for all aquatic animal life. They also make up an important part of the soil flora essential in soil productivity.

The mosses and the liverworts make up a group of plants commonly found growing in moist places throughout the world. These plants also lack roots, stems, and leaves, in the sense that we think of these structures in higher plants. The liverworts (Fig. 1-3) and mosses (Fig. 1-4) differ from algae not so much in size or appearance as in their reproductive features. Whereas the reproductive organs of algae lack

Fig. 1-2 Amanita muscaria, *a fungus (small sector of a fairy ring that was 27 ft in diameter and contained 144 sporophores).*

Fig. 1-3 Marchantia polymorpha, *a liverwort. (Photograph courtesy of F. D. Kern from "The Essentials of Plant Biology," Harper & Row, Publishers, Incorporated, New York, 1947.)*

Fig. 1-4 Polytrichum commune, *a moss.*

a sterile jacket of cells, those of the liverworts and mosses have this structure. Although liverworts and especially the mosses add considerable beauty and interest to the wild landscapes of the world, compared with the algae and fungi, they are of only slight economic importance.

The common ferns and the horsetails or scouring rushes, together with the lycopodiums and selaginellas, or club mosses differ from the algae, fungi, liverworts, and mosses in that they have more highly differentiated plant bodies (Fig. 1-5). They have true roots, stems, and leaves and a well-defined vascular, or conducting, system. They differ from the higher plants in that they do not produce flowers, fruits, or seeds. Some of these plants, especially the ferns, are ornamental plants and as such have some economic value.

The plants with the highest development and greatest differentiation are the seed plants. They have true roots, stems, and leaves and a highly developed vascular, or conducting, system. The most important feature about them, however, is the fact that they produce seeds. It is to this group that most crop plants belong, as well as all the common trees, shrubs, and flowering

Fig. 1-5 Ferns. (Photograph by Robert S. Beese.)

plants with which everyone is familiar. Any plant that forms true seeds belongs to this group.

There are two principal groups of seed plants, namely, the **gymnosperms** and the **angiosperms.** The gymnosperms are characterized by producing their seeds exposed, i.e., not enclosed in a fruit. The term gymnosperm means "naked seed." The gymnosperms are represented in the north temperate regions by the pines, spruces, hemlocks, cedars, and other evergreens (Fig. 1-6). Many of them bear their seeds in cones, and none of them has flowers. The angiosperms, on the other hand, have well-developed flowers and produce their seeds in an enclosed structure which is called the fruit. The term angiosperm means "enclosed or hidden seed." The members of this group are very numerous and embrace all the well-known flowering plants. The angiosperms are further subdivided into two coordinate groups, the **monocotyledons** and the **dicotyledons.** The monocotyledons (Fig. 1-7) are represented by the grasslike plants, such as corn, wheat, oats, and other cereals, and by such plants as the lilies, orchids, bamboo, and banana. To the dicotyledons (Figs. 1-8 to 1-10) belong all the broad-leaved forest trees and many ornamental and crop plants, such as clover, beans, peas, buckwheat, cotton, and geranium. The essential differences between these two groups can best be considered later in the text.

Fig. 1-6 Douglas fir (Pseudotsuga taxifolia), *a gymnosperm at Fairchild Ranger Station, Lincoln National Forest, N.M. (Photograph by U.S. Forest Service.)*

The parts of a seed plant Because practically all the first half of this textbook deals with the seed plants, it may be well to consider briefly the main parts of such a plant as they develop during its life cycle. These parts are roots, stems, leaves, flowers, fruits, and seeds (Fig. 1-11).

Seeds consist of a partially developed young plant, the **embryo,** and various types of stored food enclosed in an outer protective structure called the **seed coat.** In some cases the food is stored in the embryo itself, and in others it is stored around the embryo in a form called **endosperm.** When a seed is placed under suitable conditions of air, temperature, and moisture, as when we plant it in the soil, it may germinate; that is, the embryo, or young plant within the seed, begins to grow. In so doing, it ruptures the seed coat and emerges from the seed as a small plant called a **seedling.** During this development the stored foods are digested and utilized. The first part of the embryo to emerge from the seed usually grows downward into the soil to form the first **root** of the plant.

Fig. 1-7 Sugarcane (Saccharum officinarum), *a monocotyledonous plant. (Photograph courtesy of F. D. Kern.)*

This primary root may branch repeatedly and thus spread out into the soil to form the **root system,** which anchors the plant to the soil. The finer roots of this system remain in intimate contact with the soil and absorb from it water and inorganic substances in solution.

Another part of the embryo grows upward out of the soil to form the first **stem** of the plant. This stem may also develop many branches bearing the **leaves** of the plant. The place on the stem where a leaf originates is called a **node,** and the region between two adjacent nodes is called an **internode.** There may be one or more leaves at a node. The stem is made up partly of tough fibers which are effective in giving it strength to support the leaves and later the flowers and fruits. It also contains conducting elements which function in the transport of water and inorganic substances from the roots to the leaves and in the movement of foods and other organic substances from the leaves to the roots and to other parts of the plant. Usually

Fig. 1-8 Hepatica acutiloba, *a dicotyledonous plant.*

Fig. 1-9 Cotton plant (Gossypium *sp.*), *a member of the dicotyledons of considerable economic importance. (Photograph courtesy of B. M. Waddle, U.S.D.A., Agricultural Research Service, Cotton and Cordage Fibers Research Branch, Beltsville, Md.)*

considerable amounts of food are also stored in stems.

A **leaf** usually consists of an expanded portion called the **blade,** a stemlike part called the **petiole,** and often two small leaflike structures at the base called **stipules.** The blade, when held toward light, can be seen to have many **veins.** These veins usually have many branches, forming a regular network which extends to all parts of the blade. The veins are branches of the vascular system and are continuous with the vascular structures in the petiole and stem. Through them water and minerals are obtained from the stem, and out of them move the foods and other organic substances made in the leaf. There are many pores, called **stomata** (singular, stoma), especially on the lower surfaces of the blades. These pores facilitate gaseous exchange with the outer air.

Leaves are usually green in color because of the presence of green pigments called **chlorophyll.** By means of this chlorophyll, leaves are able to synthesize carbohydrates, like sugar, out of carbon dioxide, obtained from the air through the stomata, and water, obtained from the soil through the roots. In this process light energy is stored as chemical energy and oxygen gas is liberated. With the exception of a few microscopic organisms, only plants containing chlorophyll can carry on this process, which is ultimately the source of all organic matter. It is this fact that gives green plants so prominent a place in the world, since upon them depend all forms of life, including man. Green plants, unlike animals, are able to make all the foods and other organic substances they need, being dependent upon an outside source only for inorganic substances and light. The outstanding physiological activity of leaves is synthesizing carbohydrates. Out of these carbohydrates all other organic substances of the plant are ultimately made.

As the plant matures, the stem also produces **flowers.** Flowers may be produced singly or in clusters of various types called **inflorescences.** The apex of the stemlike structure bearing the flower is the **receptacle.** The various parts of the flower are borne on it. There is usually an outer whorl of leaflike, often green parts, the **sepals,** collectively called the **calyx,** and a whorl of often showy, colored, leaflike parts called **petals,** collectively called the **corolla.** The corolla surrounds one or more whorls of structures called **stamens.** Each stamen consists of a stemlike part, the **filament,** having at its summit a pollen-bearing structure, the **anther.** At the center of the flower are one or more **pistils,** each of which may consist of a swollen basal part, the **ovary,** and an elongated portion, the **style,** the tip of which usually has a sticky surface and is called the **stigma.** The ovary contains within it one or more smaller units, the **ovules.**

When the anthers of the flower are mature, they open and shed the **pollen.** The pollen often looks like a yellow dust. By means of the wind, insects, gravity, or other forces, the pollen may be carried to the stigma of the same or another flower, where it adheres to the sticky surface. Here, when conditions are favorable, the pollen grain germinates and sends a tube down through the inside of the style to the ovary, ultimately entering one of the ovules. Meanwhile there have developed within the pollen grain two minute male cells called **sperms,** or **male gametes.** Similarly, there has developed within each ovule a structure containing a female cell, called the **egg,** or **female gamete.** The sperms move down the pollen tube and enter the ovule, where one of them fuses with the egg to form a cell, the fertilized egg, from which develops the **embryo,** or young plant. The entire ovary usually grows larger at this time and develops into the **fruit.** The ovules, each containing an embryo, become **seeds,** and the cycle is ready to begin again. There are many different kinds of fruits, but essentially they are all ripened ovaries containing seeds, which are ripened ovules.

In general the roots, stems, and leaves rep-

Fig. 1-10 Hops, a dicotyledonous plant growing in Alberta, Canada.

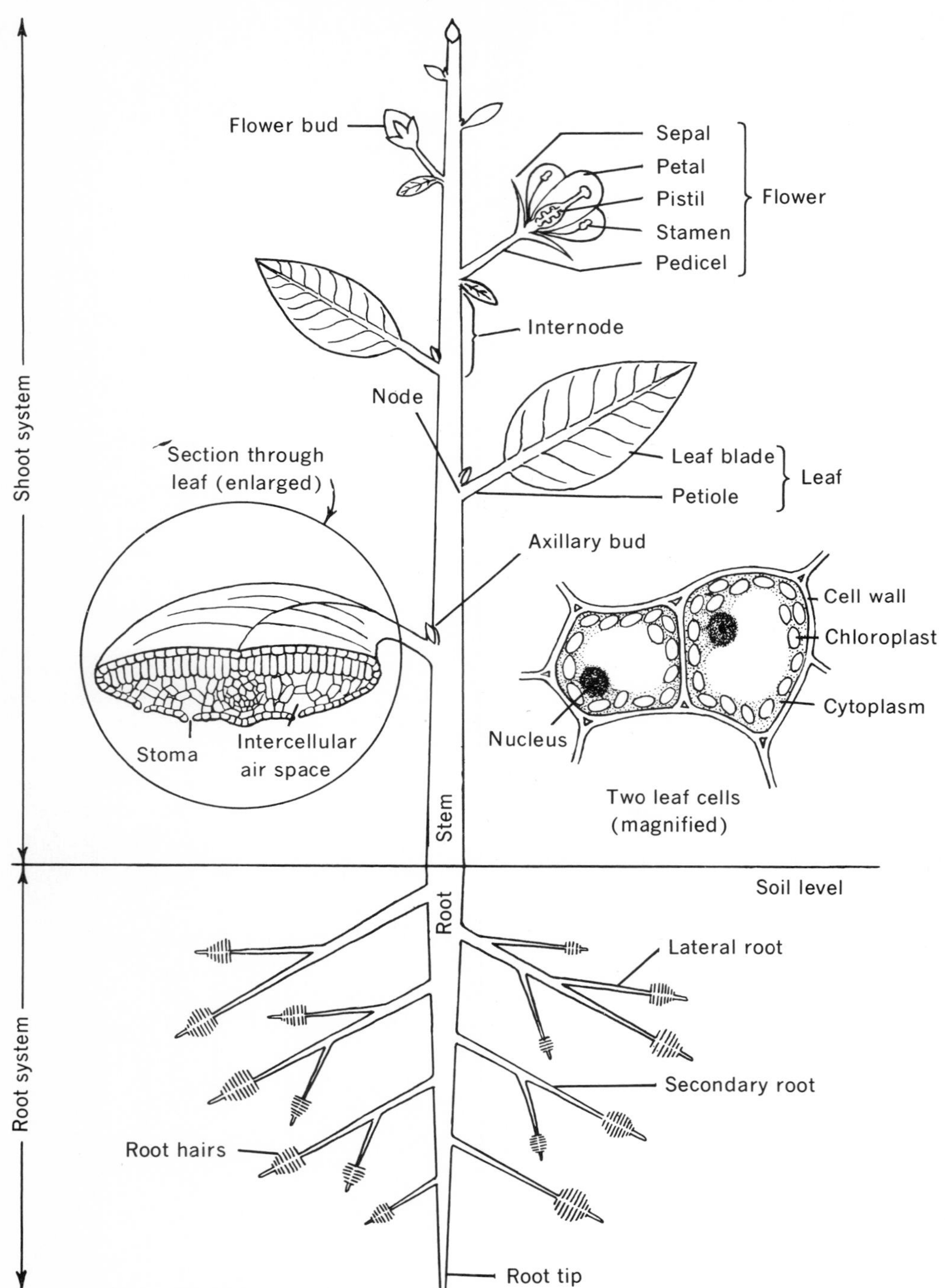

Fig. 1-11 The parts of a seed plant (diagrammatic). (Courtesy of F. D. Kern from "The Essentials of Plant Biology," Harper & Row, Publishers, Incorporated, New York, 1947.)

resent the vegetative stage of the life cycle. The flowers, fruits, and seeds are involved in gametic reproduction.

REFERENCES

CRAWFORD, M. D. C. 1948. The Heritage of Cotton. Fairchild Publications, Inc., New York. The history of cotton and its influence as a factor in the artistic, economic and social, technological and commercial life of the last 30 centuries.

DORRANCE, ANNE. 1945. Green Cargoes. Doubleday & Company, Inc., Garden City, N.Y. The story of how seeds and plants have been transported across vast seas and continents is a story of early exploration, of famous plant hunters, scientists, and such colorful amateurs as Johnny Appleseed. Gives brief histories of discovery and transport of plant materials from the early Phoenicians to the present time.

DURAN-REYNALS, M. L. 1946. The Fever Bark Tree, The Pageant of Quinine. Doubleday & Company, Inc., Garden City, N.Y. The story of man's fight against malaria, from the time when it killed Alexander of Macedon to World War II, and the history of quinine.

HATTON, RICHARD G. 1909. The Craftsman's Plant Book. Chapman & Hall Ltd., London. Takes up the use of plant materials in design for decorative purposes in art and for other uses and is illustrated with woodcuts from the old herbals.

PEATTIE, DONALD CULROSS. 1936. Cargoes and Harvests. Appleton-Century-Crofts, Inc., New York. A popular account of exploration and discovery of plants important to man and the struggles of nations over these plants; includes the history of spices, coffee, rubber, opium, dyes, the potato, and other economically important plants and plant products.

VERRILL, A. HYATT. 1937. Foods America Gave the World. L. C. Page & Company, Boston. The strange, fascinating, and often romantic histories of many native American food plants, their origin, and other interesting and curious facts concerning them.

VERRILL, A. HYATT. 1940. Perfumes and Spices, Including an Account of Soaps and Cosmetics. L. C. Page & Company, Boston. Interesting, little-known facts about spices, perfumes, soaps, and cosmetics, including their origins and histories and the part they have played in international trade, in exploration, and in advancing the spread of civilization.

2

CELLS AND CELL CHEMISTRY

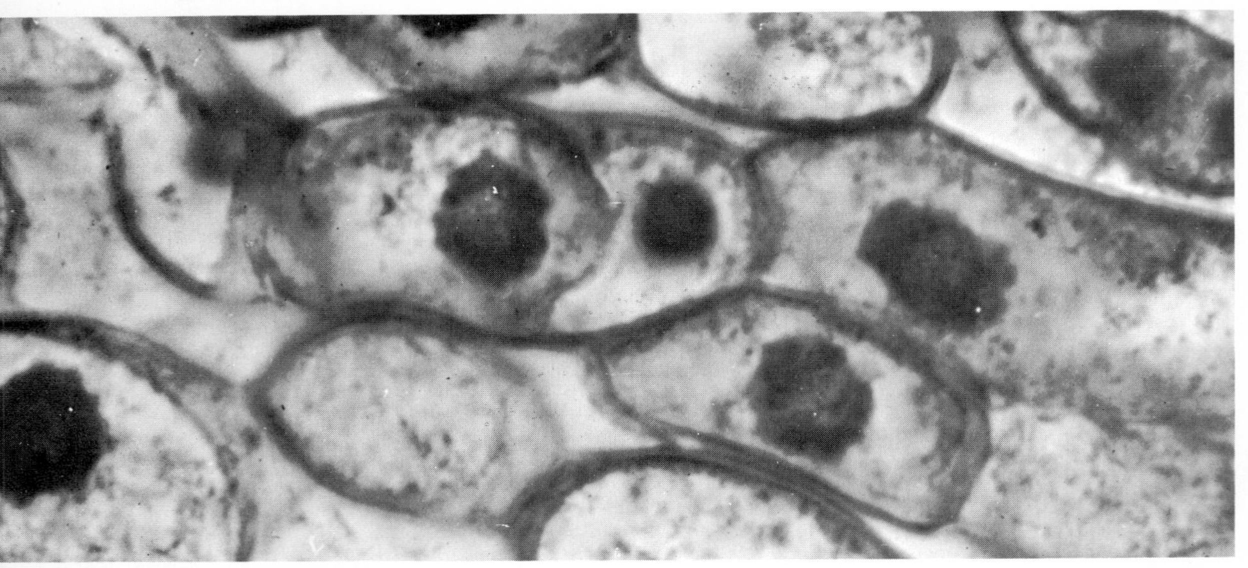

Cortical cells of magnolia. (Photomicrograph by Dr. D. A. Kribs.)

Knowledge of the cell, both of its fine structure and its chemical behavior, is increasing at a rapid rate. This chapter briefly reviews significant historical highlights, especially the cell theory and concepts of the cell as the physical basis of life. The size and shape of cells, the parts of a cell and their location, and the behavior of the cell during nuclear and cellular division are discussed. Protein synthesis and the genetic code, cell chemistry, and the organization of cells into tissues are presented.

The body of a seed plant is made up of roots, stems, leaves, flowers, fruits, and seeds (see also Chap. 1). These structures are composed of various kinds of tissues, such as storage, conducting, supporting, and protective tissues. These in turn are composed of structural and physiological units called **cells.** The study of the cell is of great importance because the cell is the physical unit of structure, the seat of vital physiological processes of the organism, and the bearer of the hereditary material.

The cell as a unit of structure was first seen in slices of cork and named by Robert Hooke, an Englishman, in 1665. The cell theory, which postulates that all plants and animals are composed of similar fundamental structural and physiological units, was an idea synthesized by Schleiden and Schwann, two Germans, in 1838–1839, from their own observations and from those of many other scientists. Hooke regarded the compartment and the cell wall as the fundamental structure, and although he observed the protoplasmic content, he thought of it only as a "nourishing juice." It is now known that although the cell wall is characteristic of plant cells and a significant structural feature in plant tissues, it is a product of the protoplast. Physiologically the wall is of less importance than the living portion of the cell.

THE PROTOPLAST

The **protoplast** is made up of **protoplasm,** which is the living material of the cell. Although protoplasm is composed of proteins, nucleic acids, carbohydrates, fats, minerals, water, and other compounds, it is more than a mere mixture of chemicals. It is an organized substance that Huxley referred to as the "physical basis of life." According to the modern concept of the physical basis of life, the whole mass of protoplasm is the seat of life. The organization, cooperation, and adjustment of one cell part to another constitute life. From this viewpoint the seat of life is not to be sought in any ultimate living particle.

The most outstanding features of protoplasm are its physiological properties because these characteristics distinguish living from nonliving things. They include metabolism, regulation, irritability, growth, differentiation, and reproduction. **Metabolism** is the sum of the physiological processes in the building up and tearing down of protoplasm. It includes all of the chemical changes that take place in the normal behavior of living matter, including the synthesis of organic substances, digestion, respiration, and finally **assimilation,** which is the conversion of nonliving material into protoplasm. **Regulation** refers to the ability of protoplasm to regulate the speed of its own physiological processes; i.e., there is within the protoplasm the element of regulation whereby the various chemical processes proceed in orderly fashion and at ordinary temperatures. **Irritability** implies the ability of protoplasm to respond to external stimuli, such as light, temperature, gravity, and chemicals. The response may be a single movement, or it may involve more complicated changes in the protoplasm. **Growth** involves not only an increase in mass or volume but also development, i.e., **differentiation** or irreversible change in form that only living protoplasm can achieve. Finally, **reproduction** refers to the ability of protoplasm to produce new individuals.

The shape and size of cells When a cell is grown as an isolated unit, it often assumes a roughly spherical shape or a pancake shape as it rests on the substratum. However, when a cell is surrounded by other cells of a tissue, it is polyhedral in shape. In plants, cells of the fundamental tissues are more spherical in general appearance than cells of elongating stem and root tips. The latter cells are elongated and appear more rectangular. In the case of large cells, where the surface-volume ratio is smaller, cell shape may be more threadlike, resulting in an increased surface area. Most cells vary in size between 0.5 and 20 μ in diameter, but some bacterial cells are as small as 250 mμ and plant cells of several millimeters in length are not unusual.

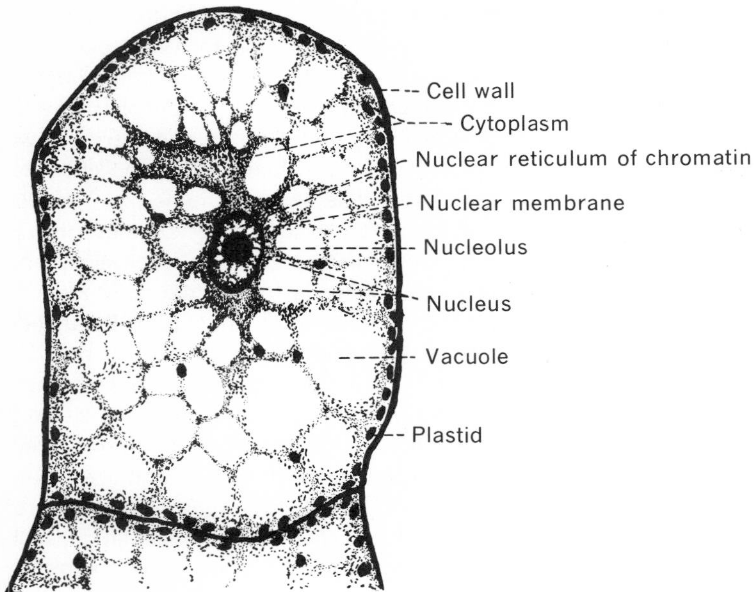

Fig. 2-1 Apical cell, showing protoplasmic structure, from Stypocaulon, *a brown alga, as seen with a light microscope.*

The structure of the protoplast Biologists are reasonably certain of the presence, structure, and function of the constituents of the protoplast. Over the past several decades the use of the electron and phase-contrast microscopes, the more sophisticated techniques in cytochemistry, and the use of the tagged radioactive atoms have opened up a new area of discovery in cytology. The increase in resolving power, as a result of the development and use of new equipment, has made visible for study protoplasmic structures that cannot be seen with the light microscope.

Fig. 2-2 Cell of moss leaf with chloroplasts, as seen with a light microscope.

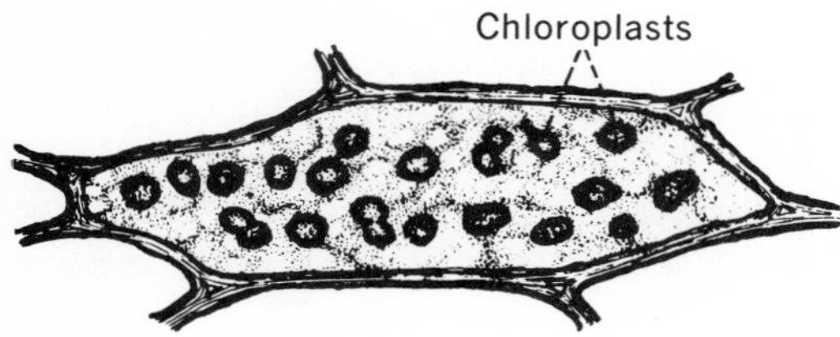

Generally protoplasts are differentiated into **nucleus** and **cytoplasm,** which are recognizable as definite cell parts (Figs. 2-1 to 2-3). The term nucleus figuratively means the center, and it is in a sense the center of activity within the cell. Structurally the cytoplasm has greater mass than the nucleus and occupies a larger part of the cell. The nucleus is located within the mass of the cytoplasm as are **ribosomes, mitochondria, plastids,** and other **organelles** (Fig. 2-4). Chemical reactions occur in the protoplasm, and in unicellular organisms it must be assumed that the functional aspects of the protoplast are broad and inclusive. In some multicellular organisms particular cells become more specialized, performing only some of the chemical functions found in unicellular organisms. The protoplast is surrounded by a living, selectively permeable **plasma membrane** (Fig. 2-10), and included within the protoplast are **vacuoles** (Figs. 2-3, 2-10) of many sizes.

Nucleus Each typical protoplast contains a single, conspicuous nucleus (Figs. 2-1, 2-3). However, in certain of the lower forms of plant life, diversity occurs in this regard. In the blue-green algae the protoplast is without a definite nucleus although nuclear material including chromatin is present. In certain other species of algae and in some fungi, a multinucleate (coenocytic) condition is found (Fig. 15-22). In seed plants certain cells rendered abnormal by age, disease, or injury and other cells, such as some reproductive cells and nutritive cells, are multinucleate.

The nucleus is typically spherical or ovoid, but its shape may vary in relation to the shape of the cell. In general there may be a direct relationship between the size of the nucleus and the size of the cell, but this relationship is not always found. As the plant cell increases in size, in most instances as a result of increasing vacuole size, there is no corresponding increase in the size of the nucleus. The nucleus therefore occupies relatively less space in mature cells. Although nuclei vary in size from 1 μ (1/25,000 in.) to 660 μ in diameter, their average size in plant cells is 15 μ in diameter.

The nucleus is surrounded by a double membrane which structurally separates it from other parts of the cytoplasm (Fig. 2-4). It is likely that each membrane is a "unit membrane," similar to the plasma membrane of the cell (Fig. 2-10) (see also discussion of plasma membrane). Projections of the outer nuclear membrane contribute to the endoplasmic reticulum (Fig. 2-4) (see also discussion of plasma membrane). The function of the pores seen in the nuclear membranes is still under investigation, but the pores would obviously provide the simplest and most direct route between nucleus and cytoplasm (Fig. 2-4). The membranes that surround the nucleus during the interphase disappear at the time of nuclear division, and new membranes are formed when the new daughter nuclei are organized (Figs. 2-18, 2-19).

The body of the nucleus, within the nuclear membranes, is made up of the **nuclear gel,** or **karyolymph,** the **chromatin,** and a **nucleolus.** The chromatin material appears to be dispersed through the nuclear gel in the form of a much-distended network, or **reticulum** (Fig. 2-18*A*), and the nucleolus is organized at a particular locus of one of the chromosomes. The chromatin is regarded by biologists as the most important constituent of the nucleus because of its importance in the modern concept of the physical basis of heredity.

At the time of nuclear division the chromatin is recognizable as definite rod-shaped bodies called **chromosomes** (Fig. 2-18). The behavior of the chromosomes and of the genes which are located on the chromosomes constitutes the

Fig. 2-3 Structure of the cell, as seen with a light microscope. Cell from a stamen hair of Tradescantia, *the arrows indicating direction of streaming cytoplasm; the granules in the cytoplasm were observed to move at the rate of 10 μ per second.*

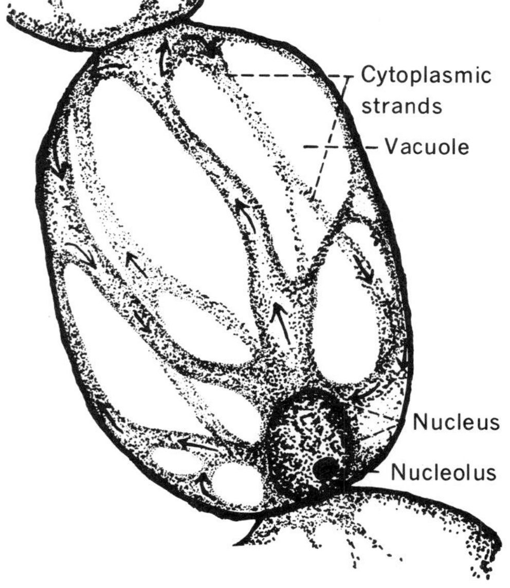

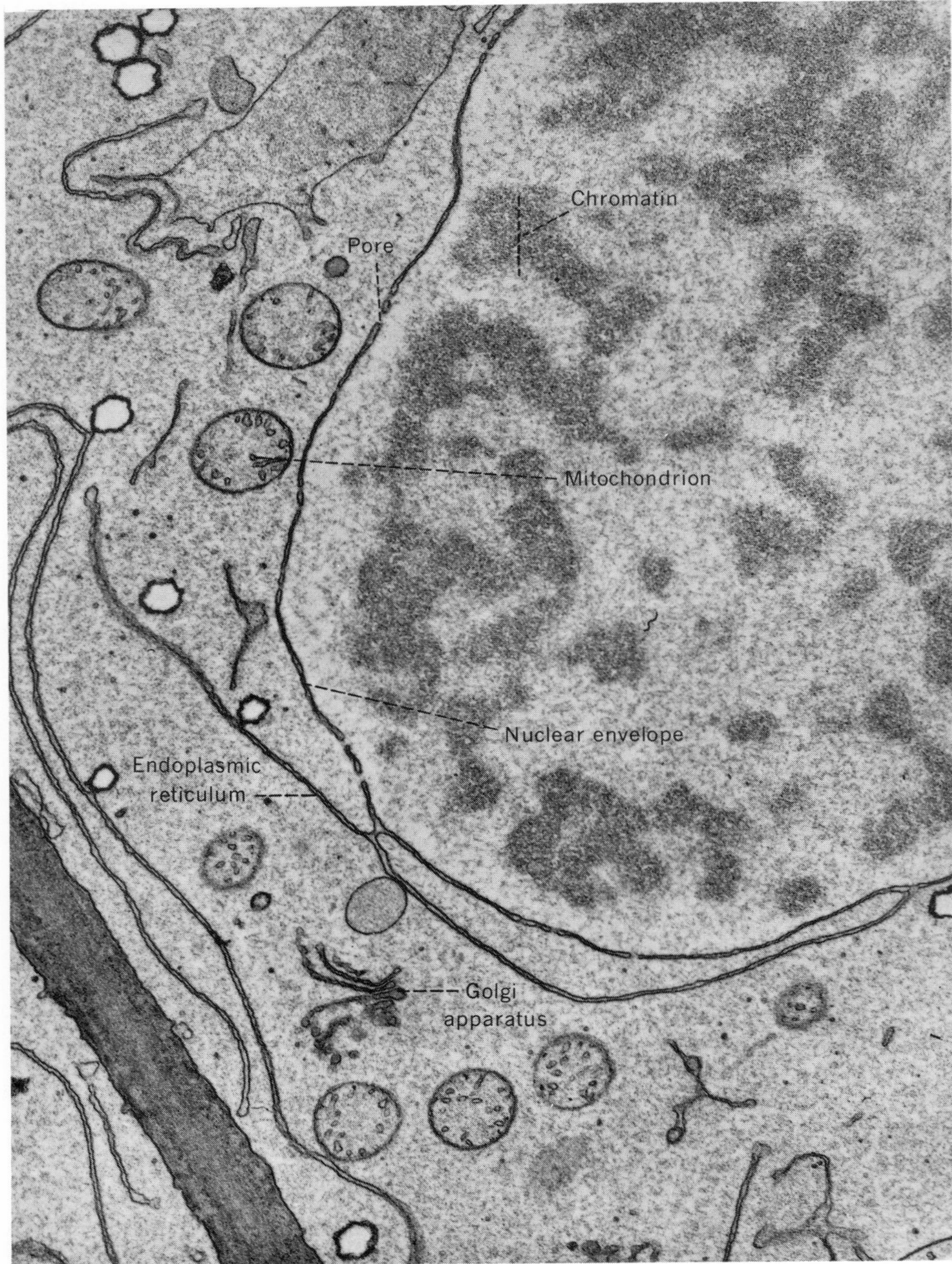

Fig. 2-4 Electron micrograph of a portion of a meristematic root-cap cell of corn. The double nature of the nuclear envelope, the pores, and the connection of the outer portion of the nuclear envelope with the endoplasmic reticulum can be seen. Fixation in 2.0 percent aqueous solution of $KMnO_4$. Approximately × 15,000. (Photograph courtesy of Dr. W. Gordon Whaley.)

essentials of nuclear division. The **genes** are the determiners of what is inherited from one cell generation to the next. The cell, in large measure, functions on the basis of information provided by the nucleus, and when cells are enucleated, the activities of cytoplasm may continue for a short period of time but the cytoplasm will die unless a nucleus is put back into the cell to supply such direction.

Chromatin consists of long strands of a complex chemical material of large molecular size, including proteins and nucleoproteins (see also Chap. 5 and later discussion in this chapter). This chemical complex (chromatin) contains deoxyribonucleic acid, or DNA. In 1953 Watson and Crick, collaborators at Cavendish Laboratory, Cambridge, England, proposed a structural model for DNA. The model illustrates a double helix of strands of DNA (Fig. 2-5).

Nucleolus Most often there is a single nucleolus (plural, nucleoli) in a nucleus, but the number varies in different cells. The nucleolus may best be described as a dense globular body of generally homogeneous appearance (Figs. 2-1, 2-3). Chemically it is composed of ribonucleic acid (RNA) "complexed" with proteins (see also Chap. 5 and later discussion in this chapter).

Cytoplasm and organelles Cytoplasm varies in density; it is fluid, semifluid, or viscous in nature and frequently can be seen to be in an active state of streaming. Cytoplasm is a very fine colloidal emulsion, may be hyaline, and is generally proteinaceous.

Plastids Plastids are prominent, important structures which, to a considerable degree, have an independent existence in the mass of the cytoplasm (Fig. 2-2). They are of several different types, each associated with a definite physiological function and often distinguished by a specific color. Emphasis should be placed upon the plastid as a protoplasmic body rather than upon its color. Plastids arise by division of preexisting proplastids, just as nuclei arise from nuclei and are never made *de novo* (anew) in the cytoplasm. Dividing plastids can easily be seen in the thin moss leaves and in algae. Plastids are of several kinds, viz., **leucoplasts** (Fig. 2-6) which are colorless, **amyloplasts** (Fig. 2-7) which are leucoplasts containing starch, **chloroplasts** (Fig. 2-8) which contain chlorophyll and carotenoid pigments, and **chromoplasts** which contain carotenoid pigments in the absence of chlorophyll.

The chloroplasts originate from smaller proplastids and are the most important of the plastids because of the presence of the green chlorophyll pigments, which are associated with the process by which carbohydrates are formed (see also Chap. 5). The chloroplasts of the higher plants are small, spherical, ovoid, or disk-shaped bodies. In the algae they are often larger and are of unusual shape in some groups, such as the radiate plastids of *Zygnema* and the elaborate spiral bands of *Spirogyra* (Fig. 15-13).

The chloroplast consists of lamellar areas of pigment concentration called the grana, clear nonlamellar areas, in which pigments are absent, called the stroma, and two delimiting selectively permeable membranes (Fig. 2-8).

Fig. 2-5 Model of DNA.

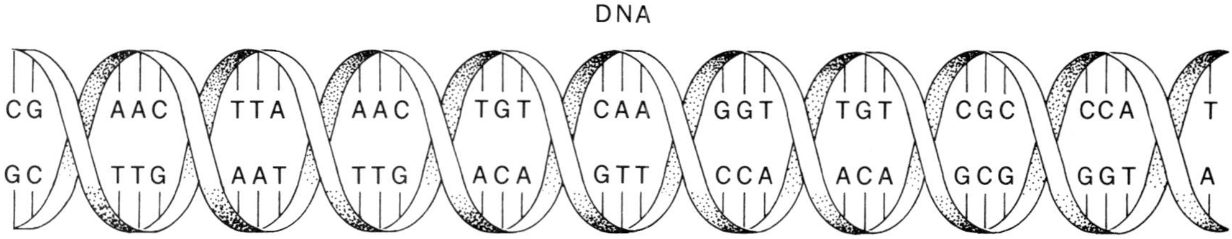

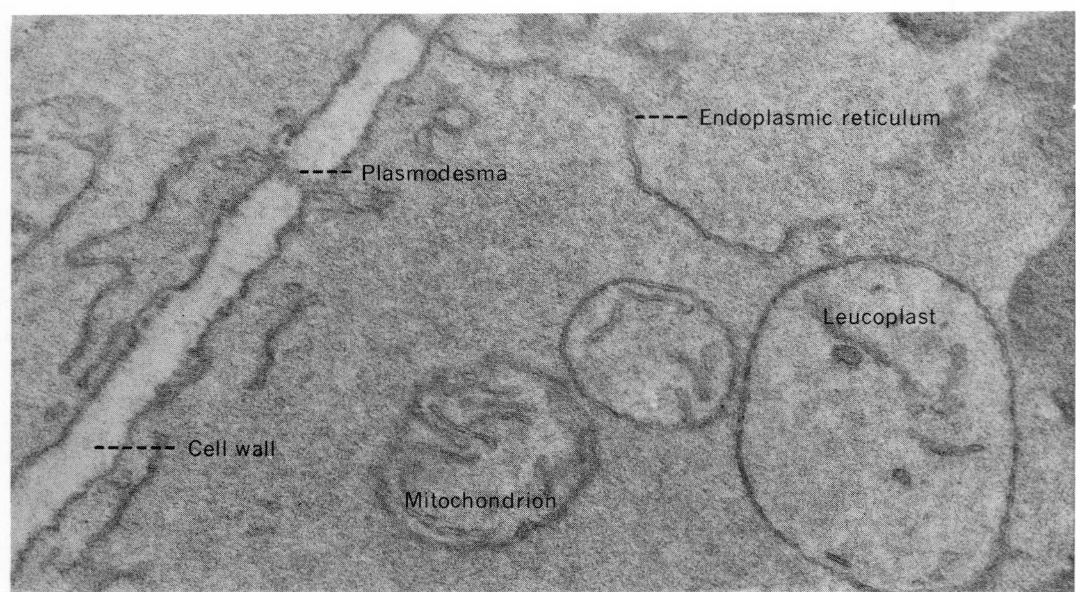

Fig. 2-6 Electron micrograph of root-tip cell from species of Solanum. *Fixed in* $KMnO_4$. × *47,200. (Photograph by Dr. Paul Grun.)*

Fig. 2-7 Electron micrograph of cell from tip of stolon of Solanum chacoense. *Fixed with* OsO_4 *and stained with lead acetate.* × *39,000. (Photograph by Dr. Paul Grun.)*

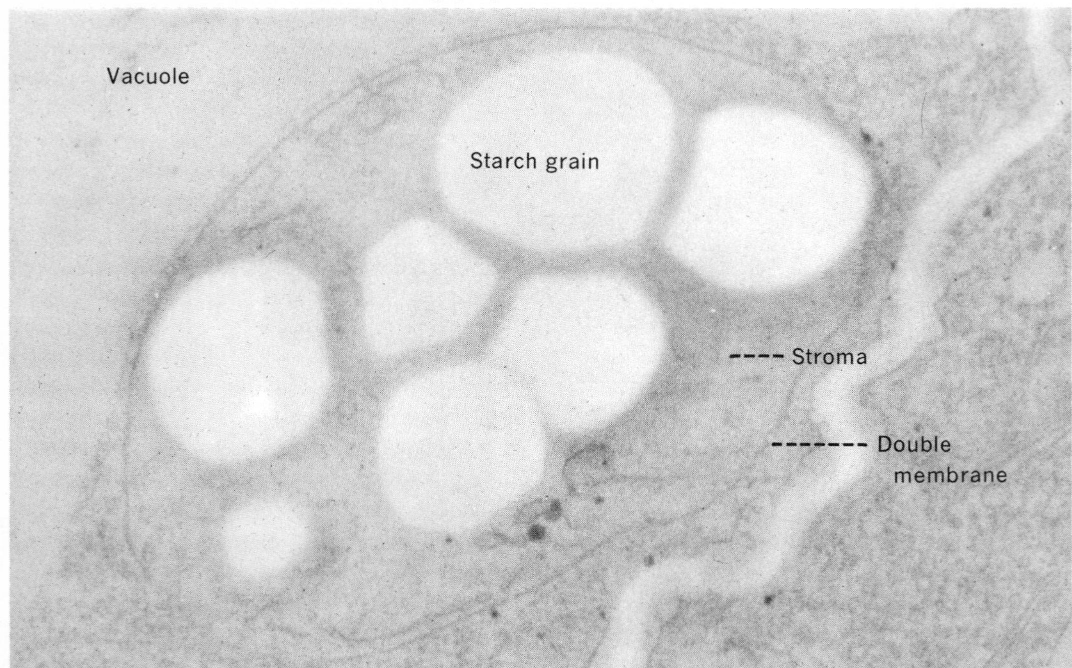

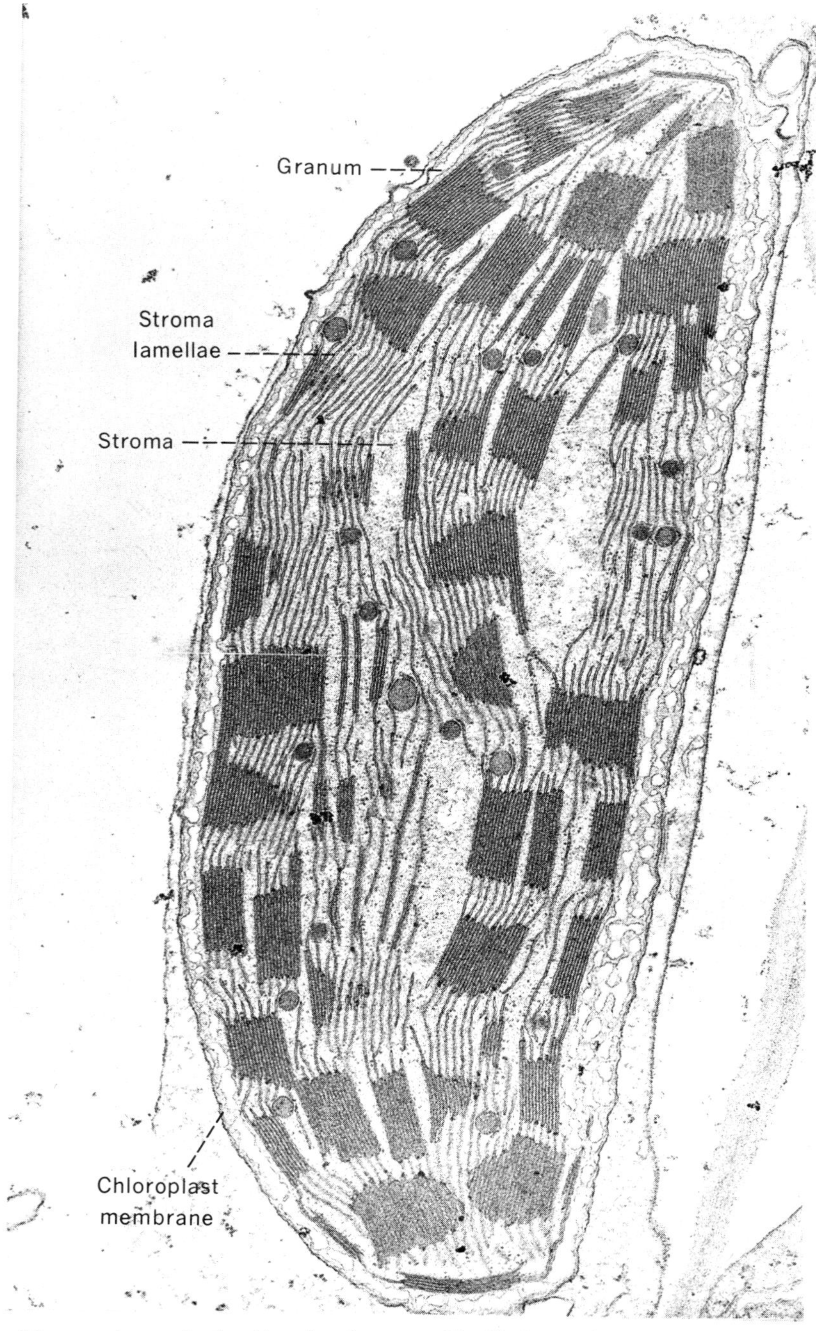

Fig. 2-8 Electron micrograph of a chloroplast from corn. The black spots have the characteristics of ribosomes and the light areas show the delicate fibrils of DNA. (Photograph courtesy of Dr. K. L. Shumway.)

The **grana** that make up the lamellar areas have been compared with pieces of multilayered plywood within which a layer of chlorophyll molecules associated with phospholipids and carotenoids is oriented between layers of protein like a sandwich. This lamellar organization is necessary for photosynthesis and any disturbance in structure leads to a reduction in its efficiency. Such disturbance is seen when *Euglena,* a unicellular green alga, is grown in the dark with the concomitant destruction of the chlorophyll and disorganization of the lamellae. When it is returned to the light, the reconstruction of new lamellae starts within several hours, and at the end of several days the structural development of the chloroplast has been completed.

The orange, yellow, and red plastids, called chromoplasts, owe their color to the presence of carotenoid pigments and are commonly found in many fruits and flowers (Fig. 3-1).

The brightly colored **eyespot** of the lower plants seems to be associated with a plastid (Fig. 15-1). It is regularly present in some single-celled algae and in the reproductive cells of other algae. In some cases it certainly has a stroma and pigments and is also sensitive to light. In other instances the evidence for the plastid nature of the eyespot is not so clear.

Fig. 2-9 Diagram of a mitochondrion.

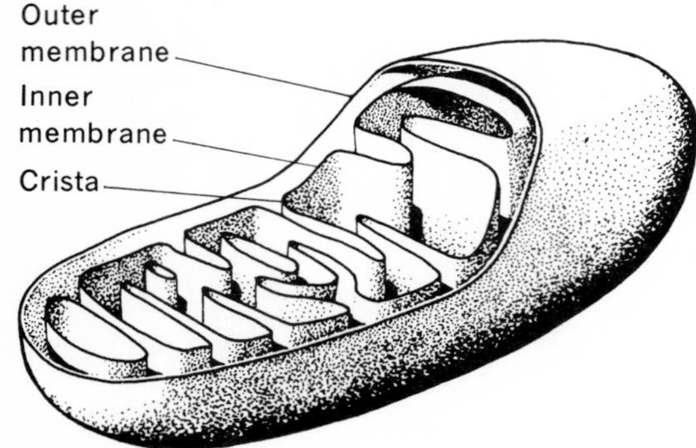

Mitochondria Mitochondria, like the chloroplasts, are involved in the conversion of energy within the cell. They are usually short, fat rods varying in length from 1 to 2 μ and their diameter is about half their length. They are surrounded by two membranes similar to the membranes of the chloroplast. The inner membrane is much convoluted into cristae which provide an increased surface (Figs. 2-6, 2-9). For some time after the discovery of mitochondria, their function was not well known, but it is now established that they contain enzymes and coenzymes that take part in respiration. They have been isolated from other cell contents by centrifugation, and they can carry on the oxidation of organic molecules if fed the necessary raw materials. There is evidence that mitochondria originate by division of preexisting mitochondria, as in the case of proplastids.

Vacuoles Within the body of the cytoplasm, there are usually one or more sap cavities, or vacuoles. Vacuoles are of two kinds, water vacuoles (Figs. 2-1, 2-3) and oil vacuoles. The water vacuoles contain a solution of organic and inorganic substances, collectively called the **cell sap.** The materials in solution in the cell sap may include anthocyanins, flavones, organic acids and their salts, inorganic compounds, and sugars. As the cells increase in size, the vacuoles continue to enlarge until finally they become confluent. In mature cells, in all probability, there will be one large vacuole in the center of the cell with the cytoplasm occupying a peripheral position against the cell wall (Fig. 7-2). Each water vacuole is surrounded by a specialized membrane called the **tonoplast** (Fig. 2-10). The oil vacuoles are droplets of oil and never attain the large dimension of water vacuoles.

Golgi bodies The Golgi body, or apparatus, is a complex of cytoplasmic membranes similar in many ways to the endoplasmic reticulum,

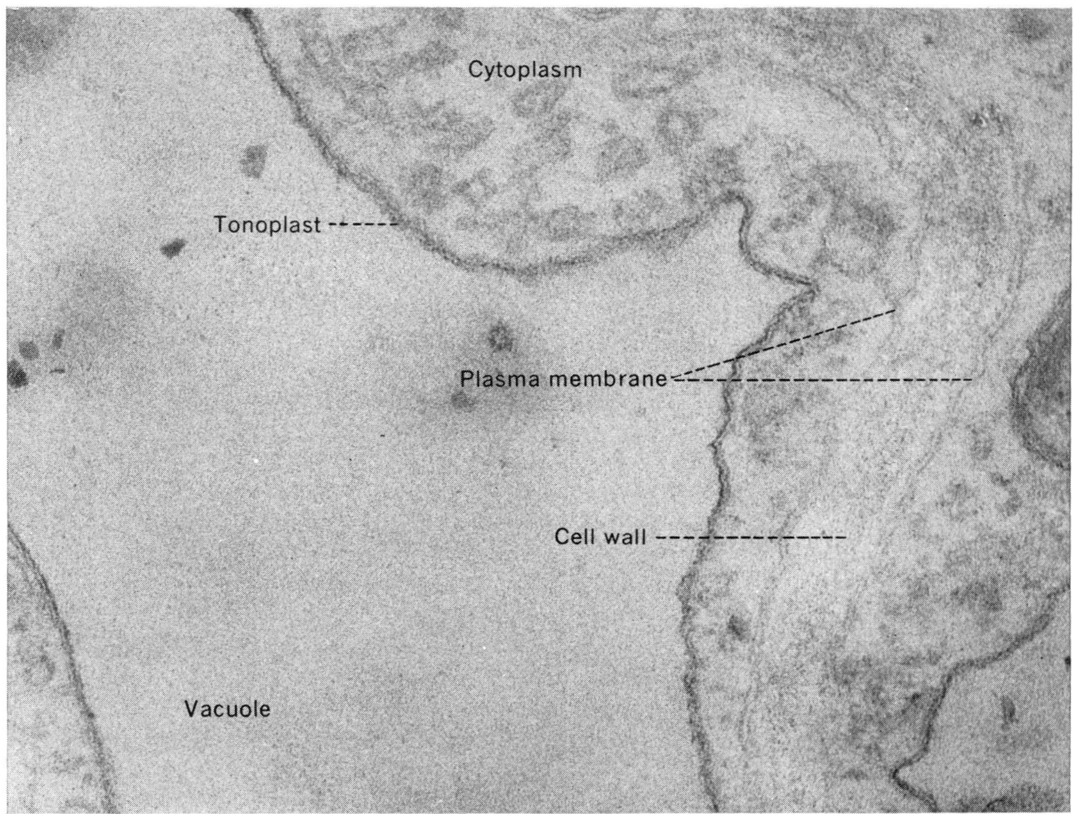

Fig. 2-10 Electron micrograph of root-tip cell from a species of Solanum. *Fixed in OsO_4 and stained in lead acetate.* × *131,000. (Photograph by Dr. Paul Grun.)*

but an actual morphological connection between the endoplasmic reticulum and the Golgi complex is debatable. The function of the Golgi apparatus is still under investigation, but it would seem to be involved in the formation of vesicles which serve to store certain cell secretory products. The general absence of RNA would indicate that these complexes are not involved in protein synthesis (Fig. 2-14).

Centrosome The centrosome is a small area of the cytoplasm near the nucleus. It is regularly present in animal cells and in the cells of some plants. In nondividing cells it appears as a clear area and often has a centrally located, deep-staining body, the centriole. With the electron microscope it has been demonstrated that the internal structure of the centriole consists of nine double tubules arranged in a circle and similar to such other cytoplasmic extensions as cilia and flagella. At the time of cell division a conspicuous system of cytoplasmic radiations, collectively called the **aster,** forms around it.

Blepharoplast The term blepharoplast has been quite generally applied to the structure from which flagella arise in the motile male cells of plants. The manner of origin and the

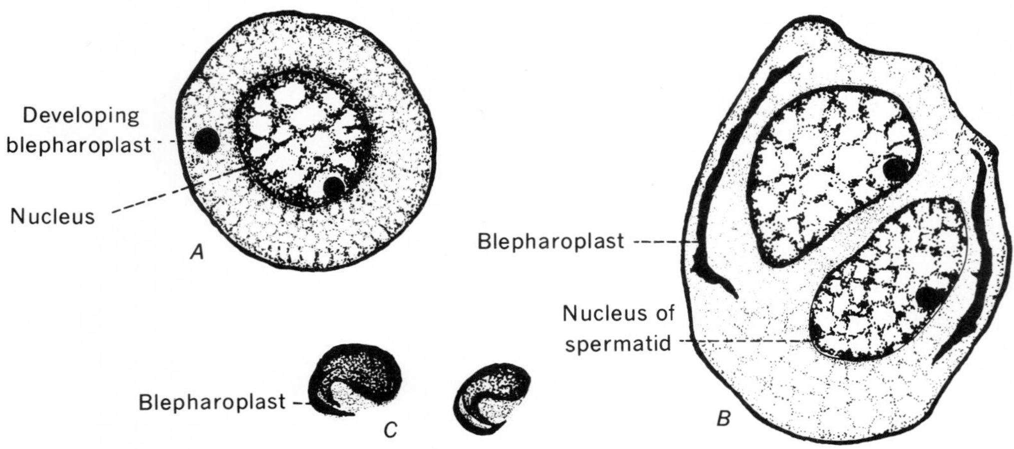

Fig. 2-11 A, B, C, development of the blepharoplast in Ceratozamia *sperm; A, young sperm mother cell; B, older stage with larger blepharoplasts and sperm; C, sperm of* Equisetum *with mature blepharoplasts.*

ultimate structure of the blepharoplast vary greatly. In most of the plants investigated, it originates as a centriolelike body (Fig. 2-11). Ultimately the blepharoplast assumes a form and position related to the cilia or flagella which at maturity appear to be attached to it.

Ergastic substances Various nonliving inclusions called ergastic substances are found in the cytoplasm. Some of these are carried in the cytoplasm, whereas others are localized in the vacuoles. Most of these inclusions are food, either reserve or in transit from one part of the cell or organism to another. Fat and oil globules, starch, protein, crystals of calcium oxalate, calcium carbonate, and silica are common inclusions.

Fig. 2-12 Portion of Fig. 2-10 enlarged to show unit membranes of two adjacent cells and location of the cell wall.

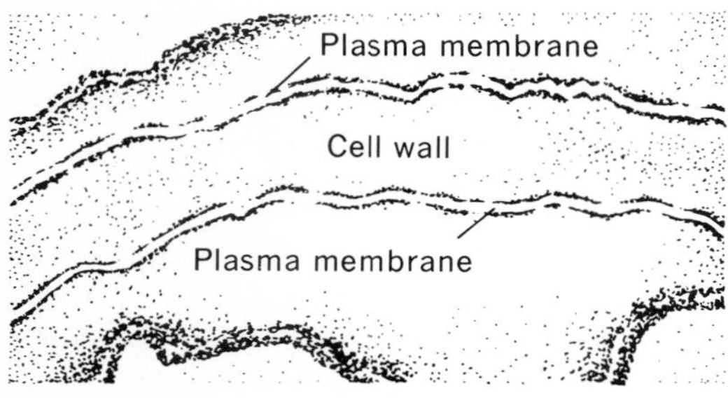

Plasma membrane The plasma membrane is a part of the living cell and limits the extent of the cell's content (Fig. 2-12). Its presence has been universally accepted for many years and electron micrographs have added to our knowledge of its physical and chemical structure. The plasma membrane has been described as a "unit membrane," a single membrane with three layers. The outer and inner layers are proteinaceous. The middle layer, seen as a single layer in electron micrographs but described as bimolecular in structure, is made up of leaflets of lipid (Figs. 2-12, 2-13). The lipoprotein nature of the plasma membrane was recognized prior to examination with the electron microscope. Permeability studies, involving especially the easy passage of fat solvents, suggested a lipid content, and surface-tension studies, along with the obvious plasticity of the membrane, suggested a protein constituent (see also Chap. 7). The selectively permeable nature of the plasma membrane establishes it as an exceedingly important structure. All osmotic action, as well as the absorption of inorganic and organic substances by the cell, depends upon the permeability of the plasma membrane.

Electron-microscope studies of the cytoplasm have forced the abandonment of the simple concept that cytoplasm is a viscous liquid without any structure. Electron micrographs show that internal projections of the plasma membrane and projections of the nuclear membrane contribute to a more or less continuous but labile system of convoluted membranes called the **endoplasmic reticulum** (Figs. 2-4, 2-6, 2-14). The ribosomes (Fig. 2-15) (small granules rich in RNA) present along the surfaces of parts of the endoplasmic reticulum provide a roughness which has been identified as an area of protein synthesis. It should be recognized that ribosomes also occur free in the cytoplasm and are not exclusively associated with the endoplasmic reticulum. The surface formed by the endoplasmic reticulum probably provides a suitable place for a variety of chemical reactions to take place.

Gene action and protein synthesis Genes are the determiners of what is inherited from one cell generation to the next (see mitosis), and, through a system of chemical control, they establish and maintain the pattern of cell behavior. The cell is a minute chemical factory in which many reactions take place, and the most important reaction involves protein synthesis (see also Chap. 5). A significant part of the gene consists of **deoxyribonucleic acid (DNA),** a nucleic acid made up of other chemical units called **nucleotides.** Each nucleotide consists of a molecule of **deoxyribose** (ribose in which hydrogen has been substituted for a hydroxyl group), combined with a molecule of phosphoric acid (to form a phosphorylated sugar), and a molecule of one of four organic bases. These organic bases are **adenine** (A), **guanine** (G), **cytosine** (C), and **thymine** (T). Adenine and guanine are **purines;** thymine and cytosine are **pyrimidines** (Fig. 2-16). In the formation of the double helical form of nucleic acid, DNA, the purines can cross-link only with the pyrimidines and, in fact, the purine adenine links only with the pyrimidine thymine and guanine with cytosine (Fig. 2-5).

Biologists agree that duplication of units of the living cell and consequently cell activity or behavior are the result of a code or template. The chemical template is called **ribonucleic acid (RNA),** and the kind of RNA that carries the message or code from the gene DNA to centers of protein synthesis in the cytoplasm is called **messenger RNA.** Centers of protein synthesis are ribosomes, which are often associated with the endoplasmic reticulum, forming its rough

Fig. 2-13 Diagrammatic representation of a membrane as conceived by J. F. Danielli and H. J. Davson.

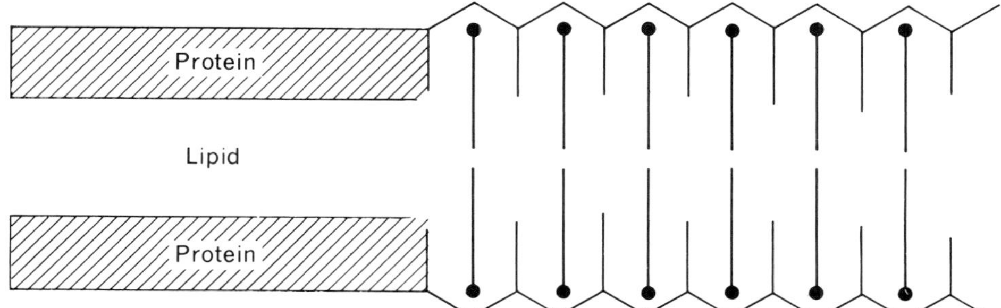

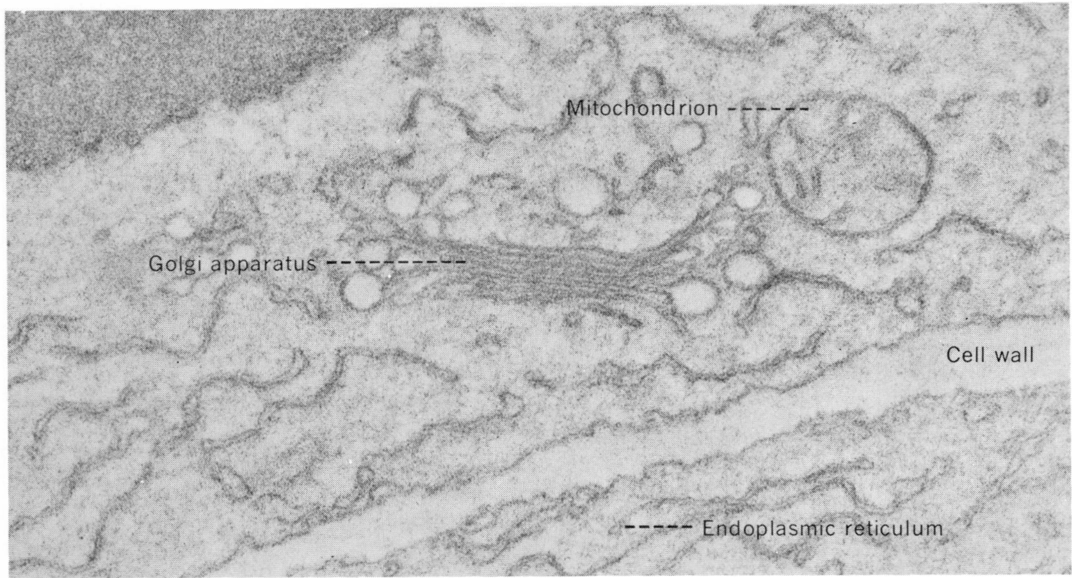

Fig. 2-14 Electron micrograph of a root-tip cell from a species of Solanum. *Fixed in $KMnO_4$. × 47,000. (Photograph by Dr. Paul Grun.)*

Fig. 2-15 Electron micrograph of root-tip cell from a species of Solanum. *Fixed in OsO_4 and stained with lead acetate. × 160,000. The small dark bodies are ribosomes. (Photograph by Dr. Paul Grun.)*

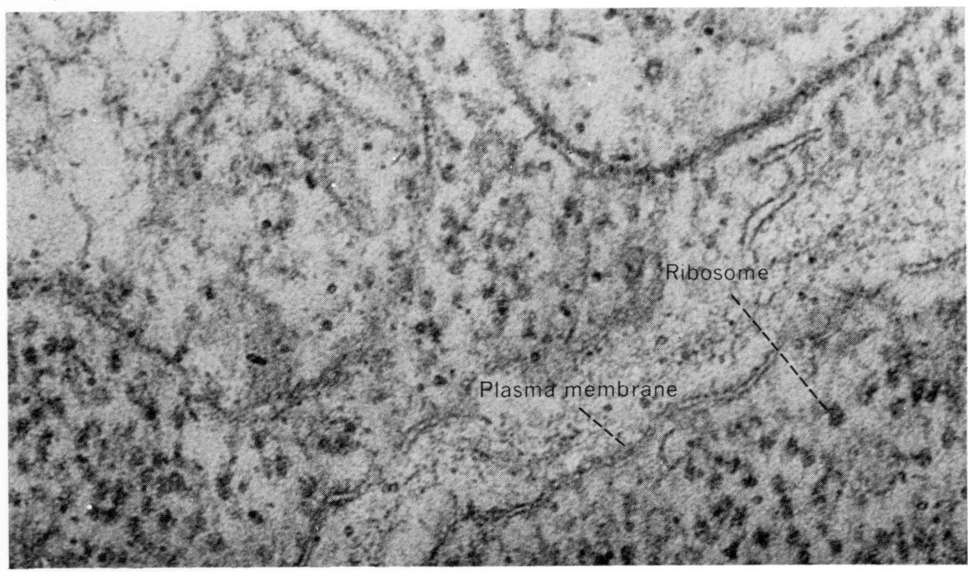

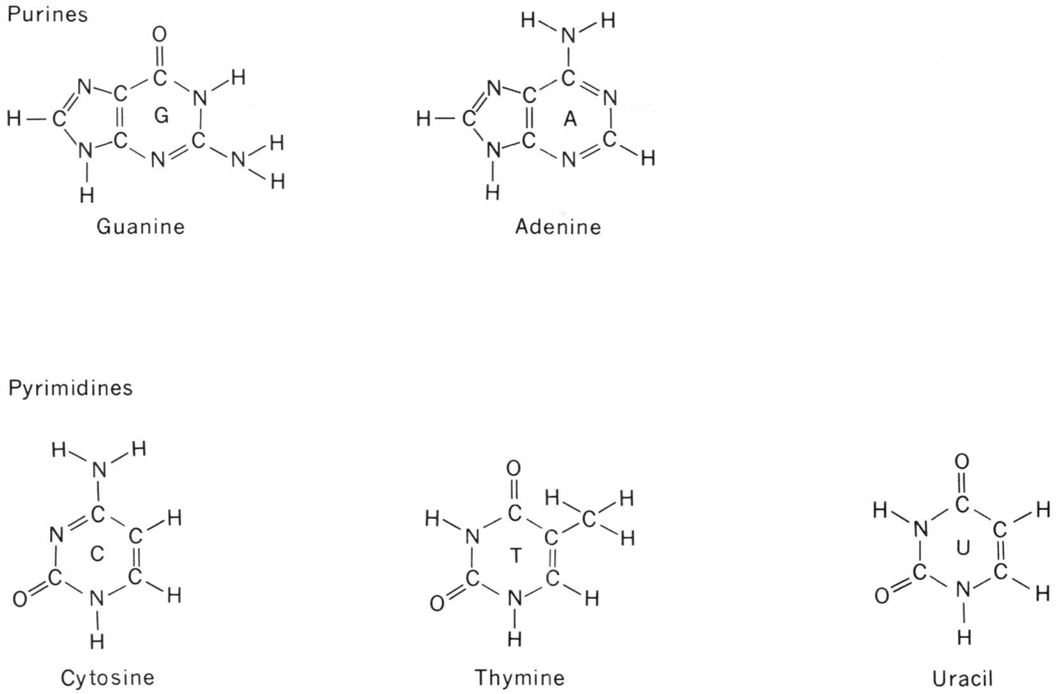

Purines

Guanine

Adenine

Pyrimidines

Cytosine

Thymine

Uracil

Fig. 2-16 Organic bases.

surface, and frequently are found distributed in the cytoplasm (Fig. 2-15). There is evidence that protein synthesis also occurs in mitochondria and chloroplasts.

Ribonucleic acid differs from deoxyribonucleic acid in that each nucleotide molecule of RNA contains the sugar ribose instead of deoxyribose, which is found in DNA, and a pyrimidine base called **uracil** (Fig. 2-16), which replaces the thymine of DNA. Probably RNA is arranged in long single chains. There is evidence suggesting that RNA of the nucleolus finds its way into the cytoplasm to form the ribosomes.

If a cell is disrupted and the ribosomes are removed by centrifuging the cellular components, it is then possible to use the removed ribosomes to synthesize proteins outside the living cell, provided the necessary messenger RNA and transfer RNA are supplied, along with the needed amino acids.

Duplication of a specific protein is accomplished by limiting new chemical unions to specific chemical unions determined by the arrangement of the four nucleotides of DNA. If adenine is bound to thymine (uracil of RNA) and guanine to cytosine, then an impression of AGC of DNA on messenger RNA serves as the template for the arrangement UCG, the only one allowable within the system. The synthesis of a specific protein, determined by messenger RNA codes, is accomplished when amino acids, transferred to the surface of the ribosomes from a supply located in the cytoplasm by the appropriate transfer RNA, react and unite in chemical synthesis (Fig. 2-17). Of the thousands of proteins synthesized within a cell, at the surface

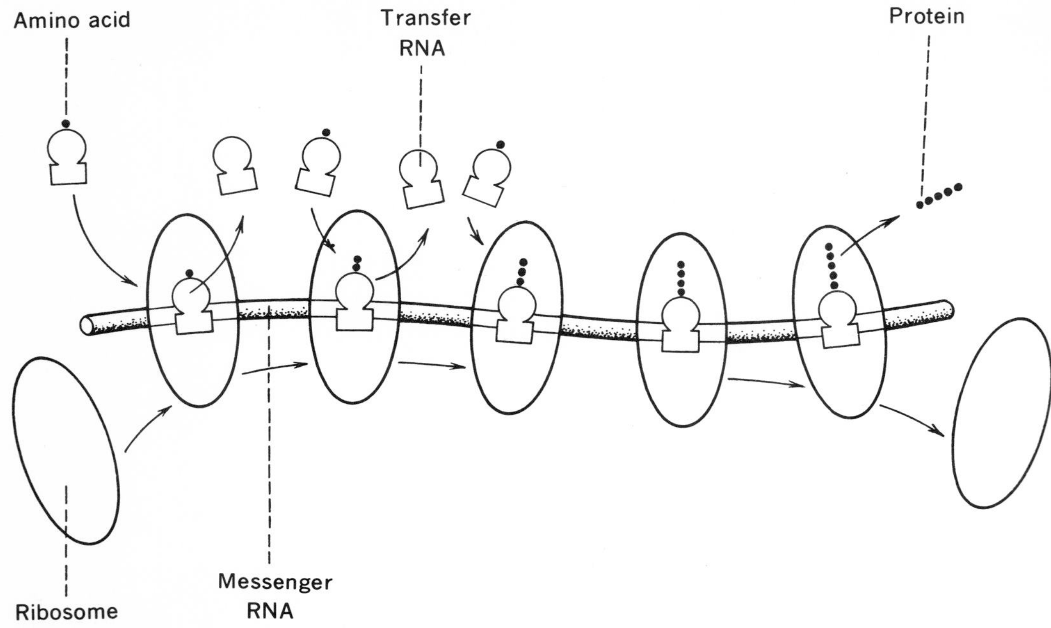

Fig. 2-17 Model of polyribosome mechanism. The long strand represents messenger RNA, and transfer RNA carries amino acids. As each ribosome travels along the messenger RNA, it provides the surface on which transfer RNA and messenger RNA interact. The result of this interaction is the production of a protein from amino acids supplied by the transfer RNA.

of ribosomes, and at other locations, many are enzymes that exert their influence on cellular reactions involving synthesis, degradation, oxidation, energy transfer, and other processes of a determinate or specific nature.

NUCLEAR AND CELL DIVISION

Cell division is an important, fundamental part of the processes of growth and reproduction in living organisms. In the development of an organism, plant or animal, the cells in a part or an organ increase in size and after a time cell division occurs. In most plant cells, division may be considered to involve both the nucleus and the remainder of the cell. First the nucleus divides by **mitosis.** Generally, but not always, this is accompanied by division of the cytoplasm, followed by formation of a cell wall separating the two new protoplasts.

Mitosis The term mitosis refers exclusively to the process of division of the nucleus in growing cells of living organisms. The word comes from a combination of two Greek words, *mitos,* "thread," and *osis,* "state" or "condition." It thus refers to the slender, threadlike condition of the chromosomes during the division process (Figs. 2-18 to 2-20). The study of the structure and behavior of chromosomes is of great significance because the chromosomes carry the hereditary units called genes from cell to cell in plants and animals. Although mitosis is a continuous process, certain stages or phases in the division of the nucleus are recognized. They are **prophase, metaphase, anaphase, telophase,** and **interphase** (Figs. 2-18 to 2-20). These are not separate and distinct periods in the process but are stages that have been recognized and named.

Prophase (Greek *pro,* "before" or "early," and *phasis,* "appearance") The term prophase refers to the earliest recognizable stage of mitosis. During the prophase the nuclear membrane and the nucleoli disappear. The chromosomes appear as a tangled mass of long, slender, threadlike structures in the nucleus. Each of the chromosomes is actually divided lengthwise, forming two parts twisted around each other and attached together at one place. The replication is preliminary to the separation of the two half chromosomes at a later stage in nuclear division. Each of the halves is a **chromatid.** The region of attachment of the chromatids, the **centromere,** may be at any position within the chromatid except at the exact end. The position of the centromere usually is constant for each chromosome.

Careful examination shows at least one inner strand in each chromatid. This is the **chromonema** (plural, chromonemata). There are two strands, or chromonemata, recognizable at some time in each chromatid (Fig. 2-21). Just as the term chromosome means colored body, a chro-

Fig. 2-18 Mitosis in cells of root tip. A–D, prophase stages; E–F, metaphase stages; A, early prophase; B, medium prophase with chromosomes forming; C, later prophase, nuclear membrane disappearing and spindle forming; D, later prophase, spindle becoming distinct and chromosomes grouping toward equator of spindle, nuclear membrane not evident; E, early metaphase, chromosomes definitely at equator of spindle; duplication of chromosomes can be seen in some instances; F, later metaphase, chromosomes grouped at equator of spindle.

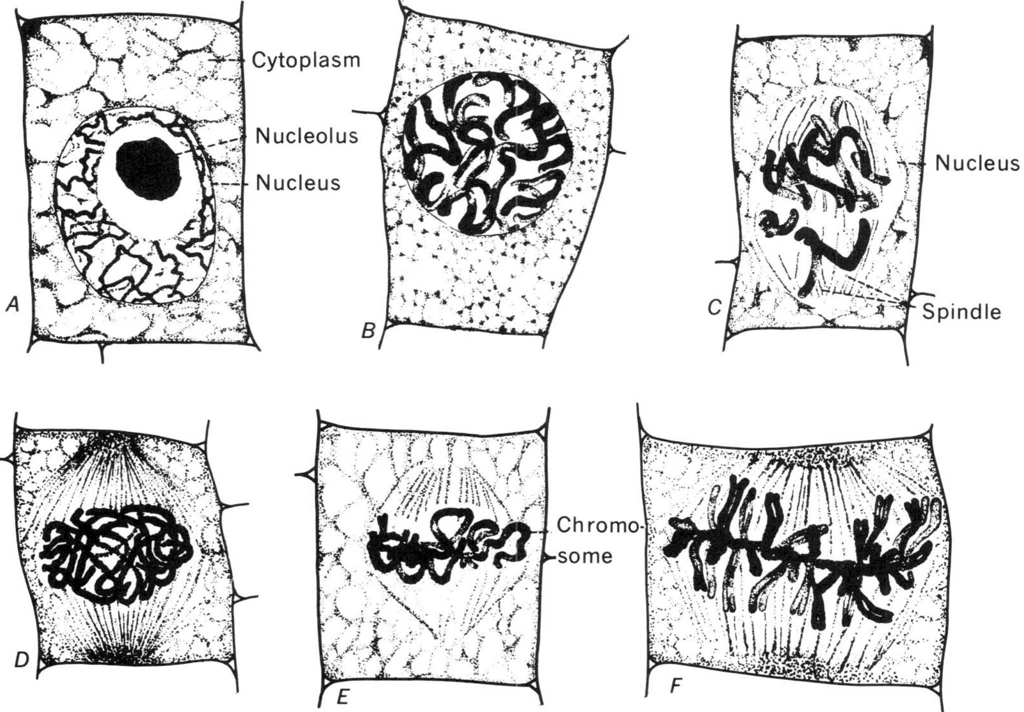

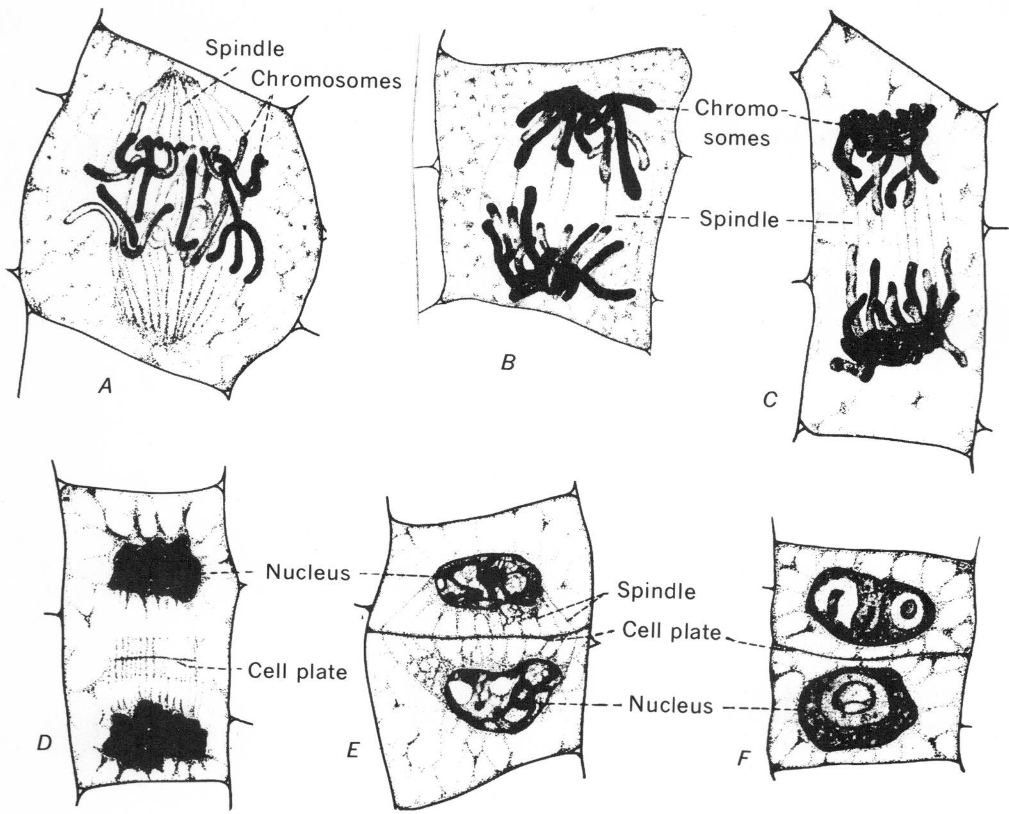

Fig. 2-19 Mitosis (continued). A–B, anaphase; C–F, telophase; A, early anaphase; the duplicated chromosomes are passing from the equator toward the poles of the spindle; B, late anaphase; most of the chromosomes have reached the poles of the spindle; C, early telophase; chromosomes grouping at each pole of the spindle; D, somewhat later telophase; chromosomes rounding up at poles, preceding the formation of nuclear membrane; cell plate forming on spindle at equator; E, late telophase; nuclear membrane formed around new nuclei; spindle becoming indistinct; cell division is completed by formation of new wall from the cell plate; F, two new cells formed.

monema means colored thread, from the Greek words *chromat,* "color," and *nemat,* "thread." In both cases the term *chrom* refers to the affinity of these structures for the nuclear dyes used in cell studies. The chromonemata contain the genes, or hereditary units, and their behavior during all nuclear divisions is important in the distribution of hereditary material to the cells.

Later in prophase, the chromosomes with their chromatids become shorter and thicker, as a result of a tight coiling of the chromonemata. Because the matrix of the chromosomes in late prophase and following phases has an affinity for certain stains, called nuclear stains, the characteristic features of the chromosomes can then be seen readily under the light microscope. As the division proceeds, the chromosomes tend to move toward the center of the

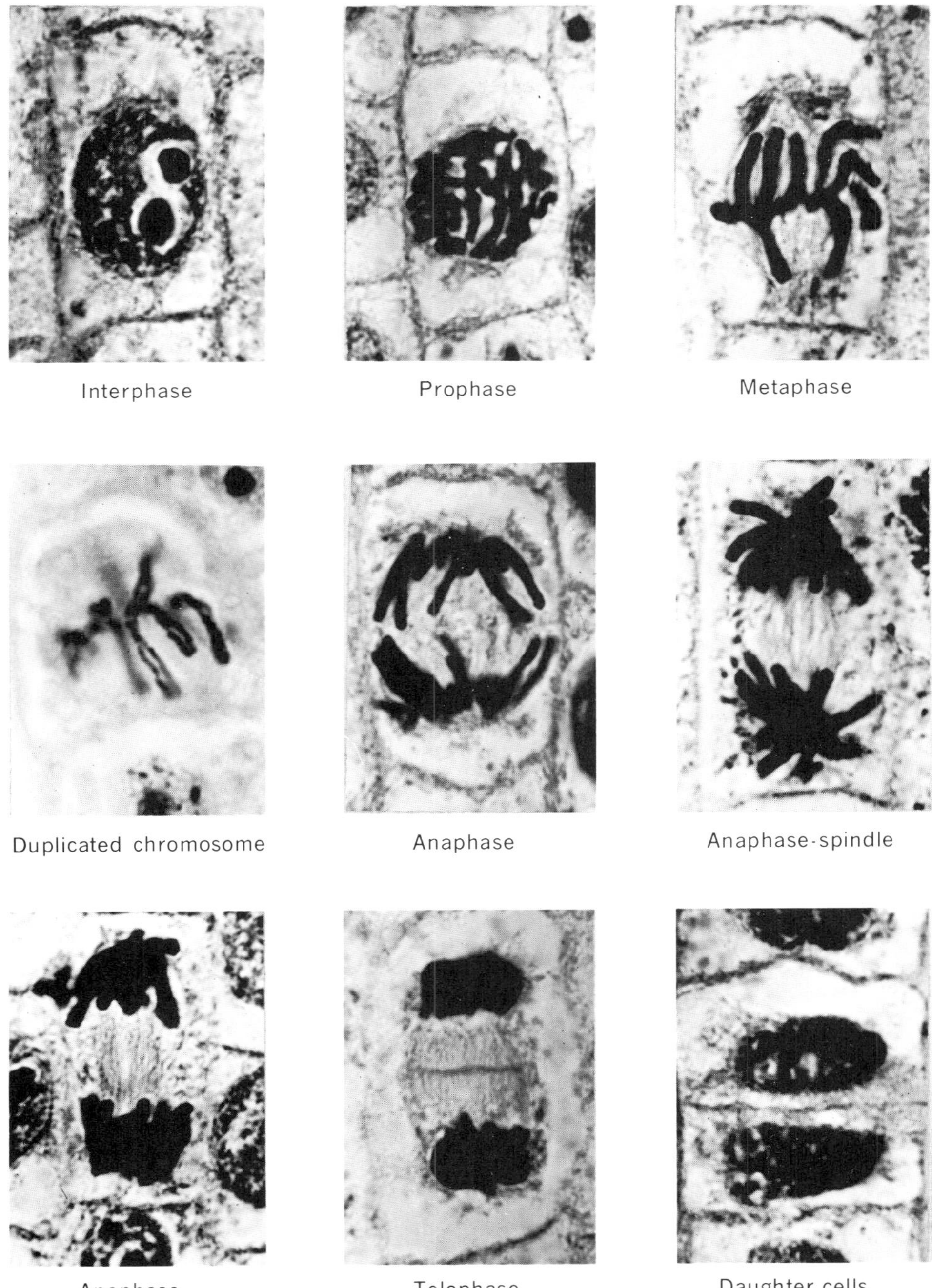

Fig. 2-20 Mitosis in hyacinth root. (Photomicrograph by Dr. D. A. Kribs.)

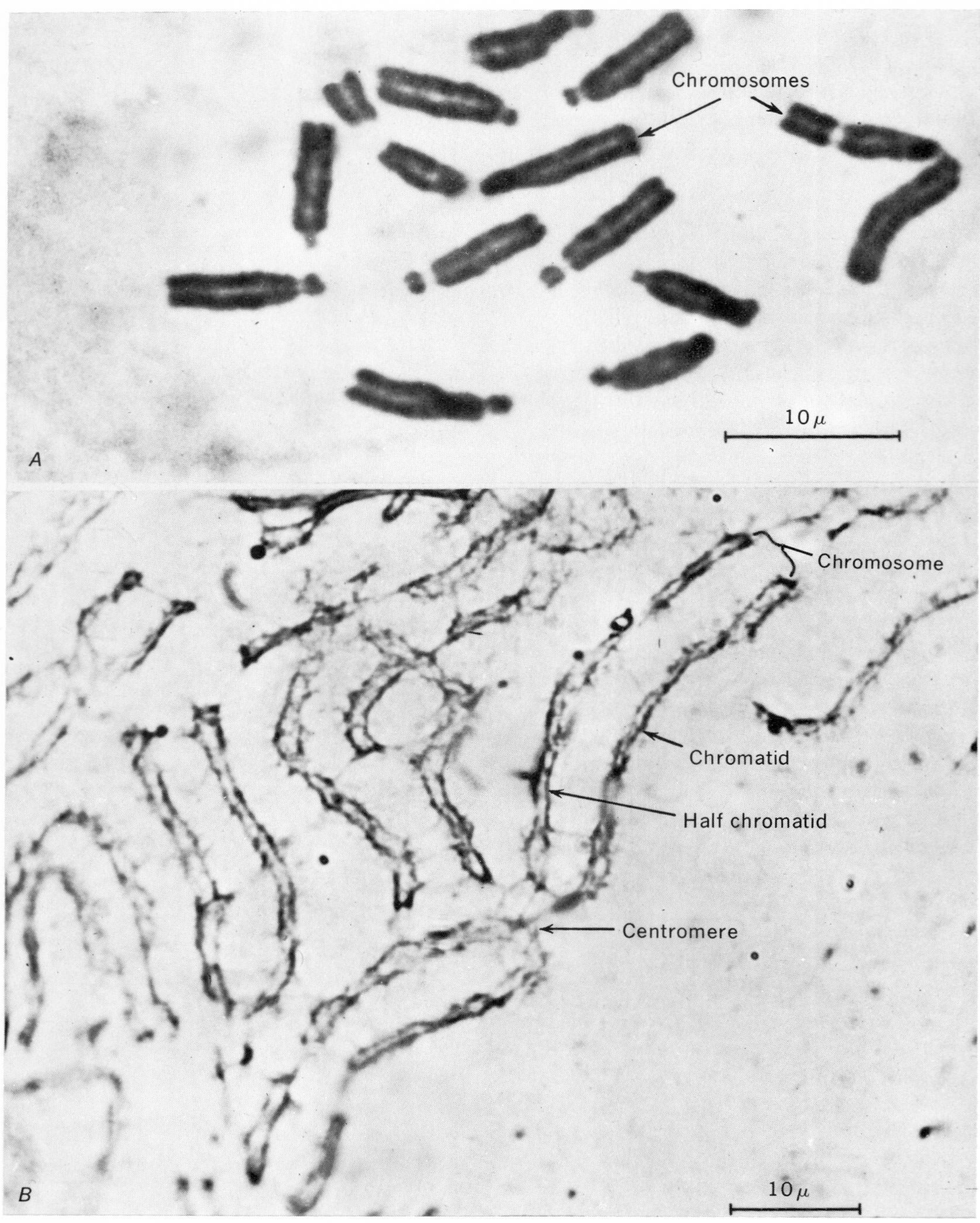

Fig. 2-21 A, normal Vicia Faba *chromosome complement at metaphase. × 2,200. B, Feulgen-stained, trypsin-treated isolated metaphase chromosomes. × 1,800. The chromosome is seen with each of its chromatids split into two half chromatids. (Photograph courtesy of Dr. James E. Trosko.)*

developing spindle (Fig. 2-18*C, D*). This event marks the approximate end of prophase, or first stage of nuclear division.

Metaphase (Greek *meta,* "later," and *phasis,* "appearance") Metaphase means, literally, a later appearance, i.e., following the prophase which has preceded it in nuclear division. During early metaphase the division spindle with its poles becomes strikingly evident, apparently floating free in the cytoplasm of the cell. The spindle has many conspicuous spindle fibers, some of which appear to extend from pole to pole. Others are more radiating and appear to end in the adjacent cytoplasm (Figs. 2-18*F,* 2-19*A*). In some cases, notably the cells of certain animals, definite asters, each with its centriole, or central body, and many prominent radiating strands may be seen at the poles of the spindle. Asters are generally absent in the dividing cells of the higher plants, but they do occur at the time of division of reproductive cells of some of the lower plants.

As metaphase proceeds, the chromosomes move to the center of the spindle, with the centromeres approximately at the equator and the arms of the chromosomes radiating outward in all directions. At this stage the divided nature of the chromosomes may be conspicuous (Figs. 2-18*F;* 2-21*A, B*). Metaphase ends and anaphase begins when the two chromatids of each chromosome separate.

Anaphase (Greek *ana,* "movement," possibly "return," and *phasis*) The anaphase is thus the stage of movement, emphasizing especially the movement of the separated chromatids (chromosomes) to opposite poles of the division spindle. The equal distribution of the daughter chromosomes to the two poles maintains in the new cells the chromosome number characteristic of the organism and assures equal distribution of the genetic material.

At about this stage certain spindle fibers are attached to each of the chromatids in the region of the centromere. The parts of the chromosome on each side of the centromere are called the arms. The spindle fibers seem to assist or guide the chromatids to the poles of the spindle after their separation at the equator. As they move toward the poles, generally the part of the chromatid containing the centromere is nearest to the pole and directed toward it, with the arms extending backward toward the equator. If the centromere is near the middle of the chromatid, the structure forms a V shape. In other cases the migrating chromatid assumes a J, or hook, shape. When the chromatids, which enlarge to the size of chromosomes, reach the poles, the anaphase may be regarded as ended.

Telophase (Greek *telos,* "last" or "final," and *phasis*) Telophase thus refers to the last recognizable stage of mitosis. The events of telophase have been described as the reverse of the happenings of prophase. During early telophase the chromosomes that have moved from the equator to the two poles of the division spindle tend to come together at the poles, forming a spherical or oval mass of tangled threads. The chromosomes gradually lose their visible identity. As telophase progresses, nuclear membranes form around each of the two nuclei, and the chromatin net, or reticulum, characteristic of the interphase nucleus appears. Meanwhile the number of nucleoli typical of the organism again becomes visible. With the reorganization of the two new nuclei, telophase, or the last stage of mitosis, is completed (Fig. 2-19*F*).

Cytokinesis The term **cytokinesis** refers to the division of the cytoplasm of the cell; generally it is associated with nuclear division. During telophase, evidence of the development of a cell plate, which will divide the original cell into two, may be seen within the confines of the equatorial region of the division spindle. The cell plate, started at the center, develops until the structure reaches the side walls of the mother cell, effectively cutting it into two approximately equal parts.

If a fixed amount of chromosomal material

within a nucleus is divided equally between two daughter nuclei and if this process is repeated some indefinite number of times, it is immediately evident that the amount of chemical material (DNA) would be reduced by half in each instance. With the constant reduction of chemical material, it is reasonable to assume that the physiological behavior of the nucleus and, in consequence, the cell, tissue, and organism would be seriously disrupted. It should be remembered that not only is the division of a nucleus and its chromosomes a qualitative process but it is also a quantitative process. Duplication or replenishment of DNA, as well as other proteins within the nucleus, is an important, necessary prerequisite to nuclear division.

It is generally accepted by biologists that some uncoiling of the helically arranged molecules of DNA occurs at some time during interphase and there is breaking of the hydrogen bonds between the organic bases. One half of the original DNA chain subsequently passes to each of the daughter nuclei. To each of these halves, however, are added, from a reservoir of necessary chemical materials in the cell, the organic bases and the phosphorylated deoxyribose molecules that have split from each of the remaining halves of the original DNA chain (Fig. 2-22). Variation exists in the exact time of organization of the organic bases (A, T, C, G) with the phosphorylated deoxyribose molecules during interphase. But regardless of this, there results a qualitative, as well as a quantitative, replacement of DNA that duplicates the original.

THE CELL WALL

The cell wall is the result of the activity of the protoplast; i.e., it is built up by the protoplast. The presence of rigid cell walls is the primary structural characteristic of plant tissues. Animal cells have no such walls—generally only delicate, limiting membranes surrounding the protoplasts.

In young tissues the cell wall consists of a single layer, whereas in older tissues it is composed of two or more layers. The first wall of a cell is called the **primary wall** or layer. It is often thin and flexible and capable of considerable extension. The additional thickenings laid down between the primary wall and the plasma membrane are known as the secondary or tertiary layers. Collectively they are called **secondary thickenings.** At the time of division of a plant cell, an intercellular material, the **middle lamella,** is formed and is shared by adjacent cells. It serves as an intercellular cementing substance. Secondary thickenings are often laid down on the primary cell wall in an uneven manner,

Fig. 2-22 Diagram to show DNA duplication.

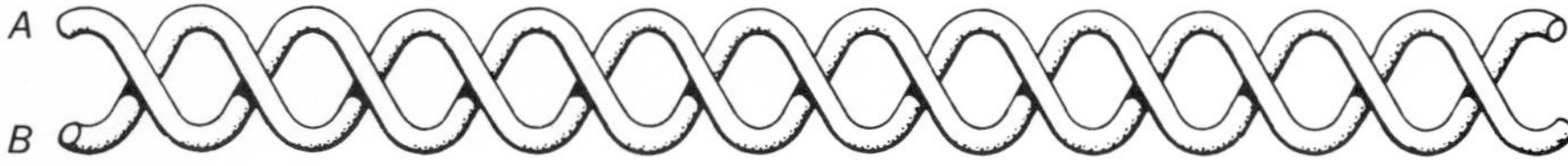

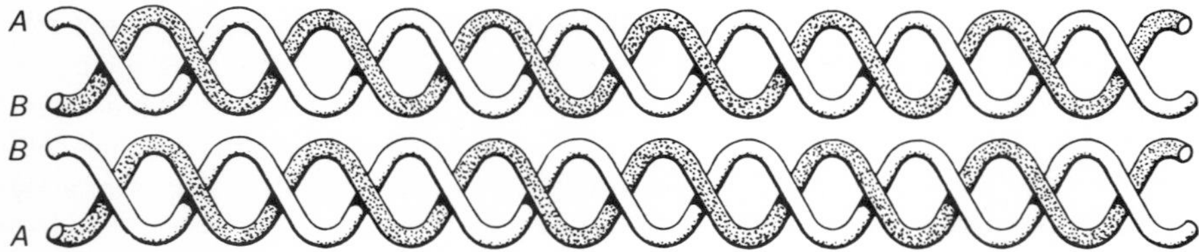

especially in walls that become greatly thickened. Interruptions, known as **pits,** are left in such walls. These are sometimes distributed in a uniform pattern over the wall of the cell. In some cells, if the later thickening overhangs the border of these places, the pits are known as **bordered pits** (Fig. 2-25*B*). Pits without such overhanging borders are known as **simple pits.**

Cell walls are composed of various substances. The middle lamella usually consists of **pectic materials,** such as calcium pectate. Pectic materials are hydrophilic colloids, i.e., colloids that are able to absorb and hold water. The outer walls of root hairs also consist of these materials. Of the substances found in the primary and secondary walls, **cellulose** is by far the most common. Cellulose is so universally found in the cell walls of plants that it is considered to be a distinguishing feature of plants. It is largely cellulose that makes plant-cell walls more or less rigid, as opposed to the pliable limiting membranes of animal cells. Cellulose walls are readily permeable to water and to most of the substances found dissolved in the water in plant cells. Cellulose is usually found in plant-cell walls, no matter what other substances may be present. Cellulose, a large molecule built up of many sugar units, is, with other cellulose molecules, organized into **microfibrils,** which are in turn organized into still larger bundles called **fibrils.** The arrangement of fibrils within the cell wall may vary from one of little organization to a parallel or spiral arrangement. Cell walls sometimes contain **hemicelluloses,** which are complex organic compounds related to cellulose but more soluble. They make the walls tough and are found in the walls of cells of some seeds and in other parts of the plant.

The cell walls of the woody parts of plants are strengthened by a complex organic substance called **lignin.** Lignin makes walls tough and hard, but it does not prevent the passage of water and the substances dissolved in water. Cells that have been strengthened by the addition of lignin are said to be lignified. Lignin replaces the pectic substances in the middle lamellae of wood (xylem) cells and to some extent is added to the cellulose content of the secondary thickenings. It is the combination of lignin and cellulose (lignocellulose) that gives wood many of its properties.

Cutin and **suberin** are closely related, waxy substances found in the walls of cells that are exposed to the atmosphere. Cutin is restricted to epidermal tissues, such as those found in leaves, young stems, flowers, and fruits, whereas suberin is found only in cork cells of the stem, root, or another plant organ. Both cutin and suberin render cells more or less waterproof, thereby checking the rate of loss of water from the interior of the organ in which they are found. In addition to the substances already mentioned, many different kinds of inorganic materials are sometimes found in plant-cell walls. Some of these cause the walls to become very hard and brittle.

Plasmodesmata (plasmodesmi) The protoplasts of adjacent cells are sometimes connected by extremely fine strands of cytoplasm extending through the cell walls. These strands are known as plasmodesmata (Fig. 2-23). In most cells they are usually so fine as to be entirely invisible under the ordinary microscope unless a special technique has been used to bring them out. They are almost universally seen in electron micrographs.

TYPES OF CELLS: TISSUES AND TISSUE SYSTEMS

In the simplest plants the whole plant body consists of either a single cell or a colony of cells all essentially alike. Since plants of this kind usually are aquatic, each cell is in direct contact with its supply of minerals and water. There is little difference among the cells as to form or function. In the more complex plants, particularly in plants growing on land, a division of labor

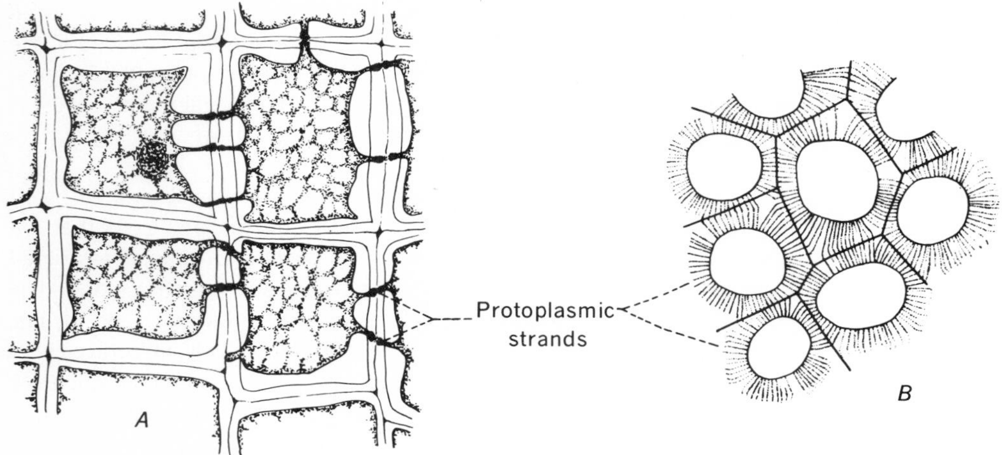

Fig. 2-23 Plasmodesmata (plasmodesmi). A, in walls of parenchyma cells in stem of Cryptomeria; *B, in walls of endosperm of* Diospyros. *(Drawings by Helen D. Hill.)*

occurs among cells. Specialized tissues develop and organs are made up of many different kinds of cells showing considerable variation in size, shape, and general characteristics.

It is in the higher plants that the greatest diversity of form and structure of the plant body occurs. In such plants no one organ is independent of the others but all work together in performing the life functions of the plant as a whole. This is accomplished by a high degree of specialization in the cells and by the development of a highly complex system of tissues.

A **tissue,** in the most general use of the term, may be looked upon as a group of similar cells performing a common function. Such a tissue is called a **simple tissue.** Often, however, the term is applied to a more complex group of cells, some of which may differ markedly from others but all of which work together as a unit. These might be referred to as **complex tissues.** This is true of such tissues as the phloem and the xylem. The name of a simple tissue is often applied to the type of cells of which the tissue is composed. This is true of parenchyma, collenchyma, and sclerenchyma, all of which are simple tissues.

Parenchyma Of the usual types of cells and tissues found in plants, parenchyma is the most common. Parenchyma tissue (Fig. 2-24*A, B*) consists typically of thin-walled cells that are approximately isodiametric, i.e., not much longer than they are wide. The individual cells may be spherical, cubical, many sided, frequently with 14 faces, or irregular in shape. They usually contain a living protoplasmic content and retain their capacity to divide even though division may never occur after the cells are mature. The parenchyma cells of the growing tips of roots and stems are known as **meristematic tissue.** These cells are in an active state of division, giving rise to the primary tissues of the plant. Parenchyma cells are found in all organs of the plant. When parenchyma tissue contains chloroplasts, it is called **chlorenchyma.** The epidermis, which is the outermost layer of cells of young stems, young roots, leaves, flower parts, and fruits, consists of somewhat specialized parenchyma cells.

Collenchyma Collenchyma tissue (Fig. 2-24*C, D*) is made up of somewhat elongated cells with cellulose thickening occurring as longitudinal

sieve tubes. The phloem is the chief food-conducting tissue of the plant.

The tissue systems All the tissues found in plants may be grouped into three systems: the epidermal tissue system, the fundamental tissue system, and the vascular tissue system.

THE EPIDERMAL TISSUE SYSTEM The epidermal tissue system consists of the outer protective layer, or epidermis, covering leaves, young stems, young roots, and the parts of flowers and fruits. The cells comprising the epidermis are specialized parenchyma cells. Epidermal cells may be of very irregular shapes because of the relative plasticity of the cell walls and the tension of the whole tissue. Stomata, which are pores through which gases are exchanged, occur regularly in the epidermis of leaves and young stems. The epidermis generally consists of a single layer of cells; rarely there are several layers. In this case it is called a multiple epidermis. Epidermal cells frequently show considerable thickening of the walls. On the outer cell wall there usually is a cuticle, which is a deposit of waxy material called cutin. This cutin helps to prevent the loss of water from the plant body. The epidermis of stems that increase in diameter from season to season is broken and lost, and a corky tissue develops beneath the epidermal layer.

THE FUNDAMENTAL TISSUE SYSTEM The fundamental, or ground, tissue system, making up a major portion of the young root, the young stem, and the leaf, as well as the flower and fruit, is composed chiefly of parenchyma cells. The internal tissues of the leaf belong almost entirely to this system. In the young root the fundamental tissue system consists of the cortex, and in the stem it is made up of the same tissue together with the pith. Although the major part of these tissues consists of parenchyma cells, there are also found in them collenchyma and sclerenchyma cells.

THE VASCULAR TISSUE SYSTEM The third of the tissue systems is the vascular. This is a complex system, consisting chiefly of phloem and xylem which facilitate conduction of water, mineral salts, and foods, as well as providing strength and support. In older perennial plants the vascular tissue system makes up the major portion of the stem and root. The arrangement of the vascular tissues varies in different organs of the plant, as well as in different kinds of plants.

QUESTIONS

1. What is a light microscope?

2. What is the importance of the cell theory to an understanding of the basic biological phenomena of all living things?

3. What relationships exist between surfaces, such as endoplasmic reticulum, mitochondria, etc., and physiological reactions?

4. To what do the terms cellular and subcellular refer?

5. When can chromosomes be counted most easily during mitosis?

6. How is protein synthesis in the cell related to nuclear division and the transmission of genetic material?

7. How are the structural features of cells related to their functions?

8. Diagram your own model of DNA.

9. How did botanists study cells before the invention of the electron microscope?

REFERENCES

CRICK, F. H. C. 1954. The Structure of the Hereditary Material. *Sci. Am.,* **191**(4):54–61.

ESAU, KATHERINE. 1965. Plant Anatomy, 2d ed. John Wiley & Sons, Inc., New York. An excellent reference and textbook on all parts of the plant.

MCELROY, WILLIAM D. 1964. Cell Physiology and

Biochemistry, 2d ed. Foundations of Modern Biology Series. Prentice-Hall, Inc., Englewood Cliffs, N.J. Concise review with adequate depth.

PETERMANN, MARY L. 1964. The Physical and Chemical Properties of Ribosomes. American Elsevier Publishing Company of New York. Ribosomes in protein synthesis.

RICH, ALEXANDER. 1963. Polyribosomes. *Sci. Am.,* **209**(6):44–53. Polyribosomes and protein synthesis.

SWANSON, CARL P. 1964. The Cell, 2d ed. Foundations of Modern Biology Series. Prentice-Hall, Inc., Englewood Cliffs, N.J. Excellent treatment of cells, their structure, and behavior in genetics, development, and division.

TAYLOR, J. H. 1958. The Duplication of Chromosomes. *Sci. Am.,* **198**(6):37–42.

WARDROP, A. B. 1962. Cell Wall Organization in Higher Plants. I. The Primary Wall. *Botan. Rev.,* **28**:241–285.

WATSON, J. 1963. Involvement of RNA in the Synthesis of Protein. *Science,* **140**:17–25.

WHITE, PHILIP R. 1963. The Cultivation of Animal and Plant Cells, 2d ed. The Ronald Press Company, New York. A concise, but inclusive, treatment of the methods used in cultivating cells outside the body.

3

COLORATION IN PLANTS; PLANT PIGMENTS

Autumn coloration.

In this chapter are considered the principal colors found in plants and some of the pigments responsible for them; the general properties, methods of extraction, and possible functions of the chlorophylls, the carotenoids, and the anthocyanins, and a brief statement on the flavones.

Of the common external characteristics of plants, the most impressive and distinctive is probably color. Color is not only a conspicuous feature of vegetation, but some of the pigments responsible for color are closely tied up with the physiological activities of the plant itself. Hence

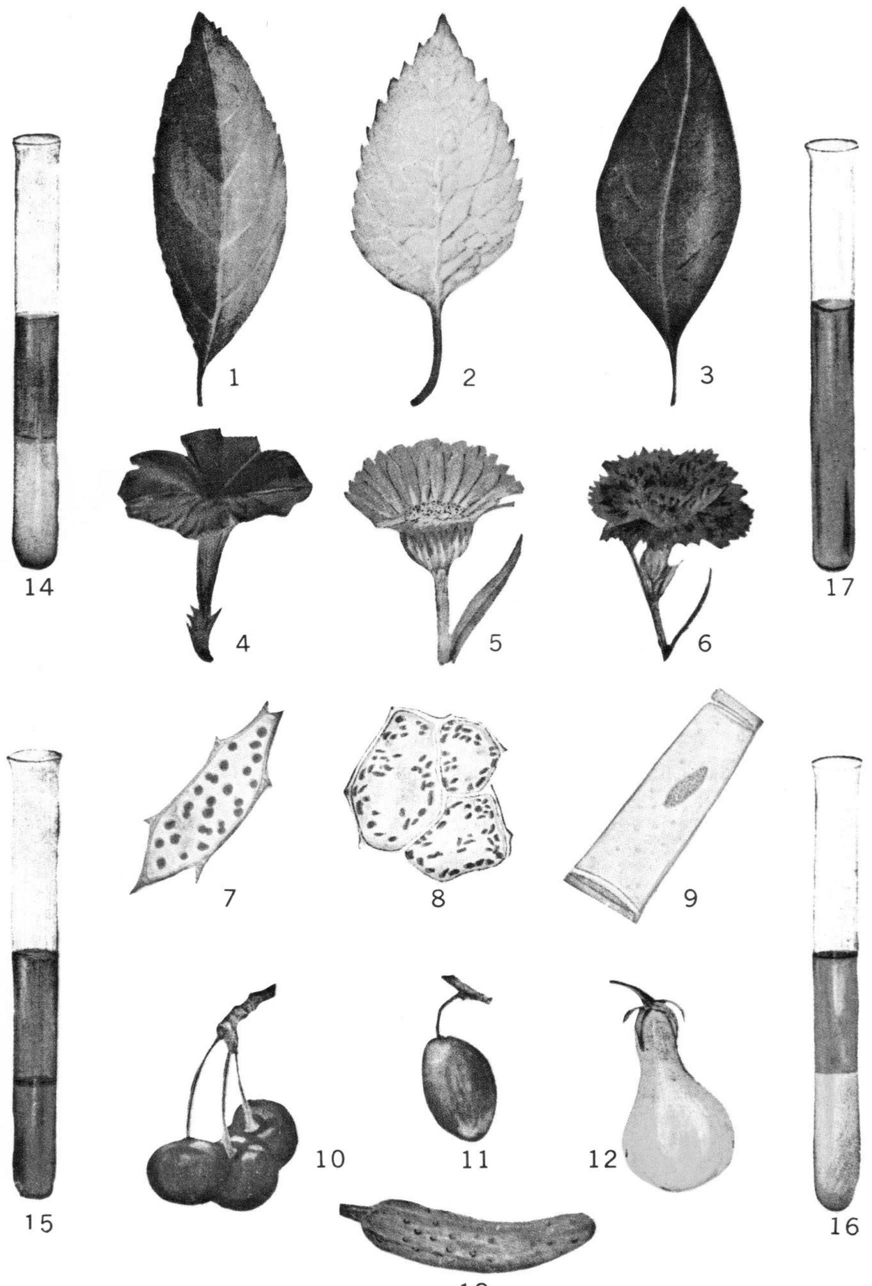

14
1
2
3
17
4
5
6
7
8
9
15
10
11
12
16
13

a knowledge of plant pigments is essential to an understanding of how plants live and grow.

Although it is possible to find in the plant kingdom as a whole all shades and combinations of the colors of the spectrum, there is in general a predominance of the colors green, yellow, red, and blue (Fig. 3-1). These colors are imparted to the plant by definite chemical compounds, or pigments, each of which has its own characteristic color. The particular color which a plant organ assumes is usually caused by the predominance of one or another of these pigments in a combination of several of them. When plant parts appear white, it is because of the absence of pigments. Sunlight falling on such parts is not so selectively absorbed as it is in colored parts but is transmitted or reflected practically as received, and hence such parts appear white or colorless. The opaqueness of this white is intensified by the refractive powers of cell walls and often by the presence of air spaces in the tissues.

The chlorophylls The green color so uniformly present in plants is caused by the presence of two closely related pigments which have been designated as **chlorophyll** *a* and **chlorophyll** *b* (Fig. 3-2). These two pigments are commonly referred to simply as chlorophyll. They occur in practically all seed plants, ferns, mosses, and algae. They may develop in roots, stems, leaves, and fruits, provided that these organs are above ground and exposed to light (Fig. 3-1, 1, 13). They are not normally present in internal tissues where light does not penetrate, such as the wood of a tree or the flesh of an apple, nor are they usually found in underground structures like roots and tubers. Though apparently absent in red or yellow leaves, when the other

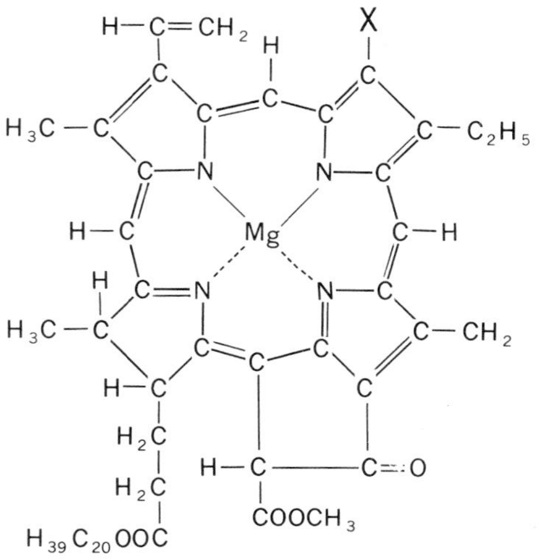

Chlorophyll

Fig. 3-2 Structure of chlorophylls a and b. In chlorophyll a, $X = -CH_3$*; in chlorophyll b,* $X = -C{=}O(H)$*.*

Fig. 3-1 Coloration in plants. (Painting by Elsie M. McDougle.)

1. *Cherry leaf.*
2. *Leaf of yellow coleus.*
3. *Leaf of* Achyranthes.
4. Petunia *flower.*
5. *Flower of pot marigold* (Calendula).
6. *Flower of (pink)* Dianthus.
7. *Enlarged view of cell of a moss leaf showing chloroplasts.*
8. *Enlarged view of cell of rose fruit showing chromoplasts.*
9. *Leaf hair of velvet plant* (Gynura aurantiaca).
10. *Cherry.*
11. *Plum.*
12. *Yellow tomato.*
13. *Cucumber.*

14–16. *Alcoholic extracts of leaves to which benzol has been added, shaken, and allowed to settle.*

The upper layer in each case is the benzol layer containing the green pigments, chlorophyll a and chlorophyll b; the lower layer is the alcohol layer containing in each case yellow carotenoid pigments and, in addition, in 15, anthocyanin. The yellow pigments in 15 are masked by the red anthocyanin. Green leaves were used in 14, red leaves in 15, and yellow leaves in 16.

17. *A water extract of red leaves of* Achyranthes *containing anthocyanin.*

coloring matters are extracted from such leaves, the chlorophylls may be found to be present even there, having been obscured by other pigments. Thus red *Coleus* or red *Achyranthes* leaves when boiled in water become green after the red has thus been extracted. The chlorophyll pigments are actually absent, however, in such lower plants as the bacteria, mushrooms, toadstools, and other fungi, and in such flowering plants as the Indian pipe and beechdrops. The absence of the chlorophylls has a profound effect on the method by which plants obtain their carbohydrate food. Green plants, by means of their chlorophyll, are able to manufacture their own foods, whereas plants lacking chlorophyll are dependent for their food on other plants or animals.

CONDITION WITHIN THE CELL Within the cells in which they are found, the chlorophyll pigments are contained in **chloroplasts.** Chloroplasts are plastids containing chlorophyll pigments (Fig. 3-1, 7). When the pigments are extracted, the plastid still remains. When a plant grows in the dark and is more or less colorless, proplastids may be present in the cells. If such a plant is brought out into light, chlorophyll pigments may be formed in these plastids, whereupon they become chloroplasts. Associated with the two chlorophylls, there are also in the chloroplast two kinds of yellow pigments, viz., carotenes and xanthophylls, or carotenols, to be discussed later. Thus a chloroplast contains four kinds of pigments, two that are green and two that are yellow.

EXTRACTION AND PROPERTIES The chlorophyll pigments are not soluble in water even under prolonged boiling but may be readily extracted from green tissues by the use of ethyl, or grain alcohol, methyl, or wood alcohol, acetone, ether, chloroform, or benzene (benzol). When these solvents are used, other pigments are extracted simultaneously. An alcoholic extract of green leaves is a solution consisting of several pigments. When an equal quantity of benzene is added to such a solution and the mixture is gently inverted several times, it separates into two layers, an upper green layer and a lower yellow layer (Fig. 3-1, 14). The upper layer contains the two green pigments dissolved in benzene, viz., chlorophyll *a*, $C_{55}H_{72}O_5N_4Mg$, and chlorophyll *b*, $C_{55}H_{70}O_6N_4Mg$. The lower alcoholic layer is a solution of yellow carotenoid pigments. When red leaves are used for extraction, the lower alcoholic layer appears pink to red (Fig. 3-1, 15) owing to a red pigment present in red leaves; with yellow leaves the yellow of the lower layer is intensified (Fig. 3-1, 16) because of the relatively higher percentage of carotenoid pigments present in such leaves.

Alcoholic extracts of green leaves are fluorescent; i.e., they appear green by transmitted light and brownish red by reflected light. A pure alcoholic solution of chlorophyll *a* is blue-green by transmitted light and bloodred by reflected light. An alcoholic solution of chlorophyll *b* is yellow-green by transmitted light and brownish red by reflected light. If the alcoholic extract of green leaves is allowed to stand in bright sunlight for an hour or two, the solution becomes a dirty brown color, indicating the destruction of the chlorophyll pigments by sunlight. A similar destruction probably occurs in living leaves, but since the chlorophylls in this case are bound up with the colloids of the plastid, the destruction is not nearly so rapid, and in addition new chlorophyll pigments are constantly being manufactured. Hence there is no change in color of green tissues in light so long as they remain in a healthy and vigorous condition.

A more refined separation of the chloroplast pigments can be made by means of **chromatography,** a technique which takes advantage of differential adsorption and solubility of various substances, including pigments. An inert carrier such as filter paper or a column of a finely powdered substance is employed together with a wide variety of solvents. The inert carrier acts as an adsorbant.

The chloroplasts of all higher plants contain both chlorophyll *a* and chlorophyll *b*, but in some of the lower plants (diatoms and some of the red, brown, and blue-green algae) chlorophyll *b* may be absent altogether or be present in very minute quantities. Two additional chlorophylls, chlorophyll *c* and chlorophyll *d*, have been reported to be present in some of these plants. Of the total green pigment present in leaves, usually about three fourths is chlorophyll *a* and one fourth is chlorophyll *b*. Apparently the proportion of chlorophyll *b* is larger in plants grown in the shade than in plants exposed to direct sunlight.

CONDITIONS NECESSARY FOR CHLOROPHYLL FORMATION It is well known that certain conditions and substances are necessary for chlorophyll formation. There must be light of proper intensity. A medium light intensity is most favorable. Plants grown in total darkness do not become green. This is well illustrated by potatoes growing in cellars or by grass growing under boards. Frequently, when plants that are already green are placed in the dark, they lose their green color, as occurs in the blanching of celery. With most plants, however, a long period of darkness is required for complete blanching.

A favorable temperature is also necessary for chlorophyll formation, medium temperatures being best. There is little, if any, greening at temperatures lower than 2 to 4°C or higher than 38 to 40°C. Most rapid formation occurs at about 20 to 30°C. In addition, there must be an available supply of oxygen and of salts containing iron, nitrogen, magnesium, and possibly other mineral elements. Nitrogen and magnesium occur in the molecules of both chlorophylls. A trace of iron salts is necessary even though iron is not a part of the molecules of these pigments. Finally, plants cannot synthesize chlorophyll pigments unless they have a supply of sugar, such as sucrose or glucose. When any of these substances or conditions are deficient, plants tend to assume a white or yellow appearance, a condition which is called **chlorosis.** Chlorotic plants make poor growth and may die unless the condition is corrected.

IMPORTANCE OF CHLOROPHYLL PIGMENTS The details of the function of chlorophyll pigments in the plant are taken up later in the text, but it may be stated here that they are necessary in the plant for the manufacture of carbohydrates such as glucose, fructose, sucrose, and starch, substances which form the basis of all foods for both plants and animals and which enter into the composition of all plant tissues. It is for this reason that the chlorophyll pigments are regarded among the most important chemical substances in nature.

The chlorophyll pigments are used to some extent in medicines; as a coloring for waxes, candles, resins, soaps, foods, and oils; and as deodorants. A pound (453.59 g) of fresh leaves yields about 0.5 to 1 g of pure chlorophyll. Although this seems a small amount, it has been estimated that in the United States alone more than 6 million tons of chlorophyll are produced yearly by corn and the small grain crops alone.

The carotenoids A number of yellow, orange, or sometimes red-colored pigments, collectively known as the **carotenoids,** are found in plants. More than 60 different kinds of these pigments have been isolated thus far. They may be divided into two principal classes, viz., the **carotenes** and the **xanthophylls,** or **carotenols.** Often these two classes of pigments are found together in the same cell. As has been stated in a previous paragraph, they are both present with the chlorophylls in the chloroplasts, often giving to healthy green leaves a decidedly yellow tinge. They are not restricted to the chloroplasts, however, but are also found in the absence of the chlorophylls, in the yellow- to orange-colored **chromoplasts** (Fig. 3-1, 8). They are always found in plastids and never dissolved in the cell sap. Chromoplasts are found in many flowers, fruits, seeds, and roots as well as in leaves (Fig.

3-1, 2, 5, 12). While the colors of the carotenoid pigments may range from yellow through orange to red, the commonest color is yellow.

When the chlorophylls are extracted from green leaves with alcohol, some of the carotenoid pigments are removed at the same time. As previously stated, some of them can be separated from the chlorophylls by adding benzene to the extract, in which case the carotenoid pigments remain in the lower alcoholic layer (Fig. 3-1, 14, 16). Carotenoids are not readily decomposed by light or heat and hence, when associated with the chlorophylls, they become prominent after the leaf vitality begins to wane in the autumn. The carotenoids are also widespread in lower plants and in animal tissue such as the fat of many animals, egg yolk, and butterfat.

THE CAROTENES Chemically the carotenes are hydrocarbons (compounds consisting of only carbon and hydrogen), most of which have the general formula $C_{40}H_{56}$. They are insoluble in water or aqueous alcohol but can be extracted readily with petroleum ether, ethyl ether, chloroform, or carbon disulfide. It is the occurrence of carotenes in the roots of the carrot that gives these roots their bright yellow color and from its presence in this plant originated the name of this class of pigments. The commonest one found in green plants is β-**carotene** (Fig. 3-3), but α-**carotene** and γ-**carotene** are also common in green and yellow leaves, in carrots, in pumpkins, and in other parts of plants. Another well-known carotene is **lycopene,** found in the fruits of the tomato, the pepper and other members of the nightshade family, in rose hips, apricots, citrus fruits, marigold flowers, orange blossoms, and in many other seed plants and even in some of the bacteria.

β-Carotene and a few other carotenes are converted into fat-soluble vitamin A in the animal body. It is for this reason that carrots and leafy vegetables in general are important dietary sources of this vitamin.

THE XANTHOPHYLLS, OR CAROTENOLS **Xanthophylls** (meaning leaf yellow) were first isolated from leaves. Most of those who work with carotenoid pigments prefer to call them **carotenols.** They differ from the carotenes in that they all contain oxygen in addition to carbon and hydrogen and hence are not hydrocarbons. They also differ in solubility, being insoluble in petroleum ether and only slightly soluble in carbon disulfide, but readily soluble in ethyl alcohol. Like the carotenes they are insoluble in water but readily soluble in ethyl ether. The different kinds of xanthophylls differ in chemical constitution and properties, but many of them are alcohols. They are a much larger group than the carotenes. The most abundant xanthophyll in leaves is **luteol** (lutein), $C_{40}H_{54}(OH)_2$. It also occurs in many yellow flowers, such as those of the sunflower and the dandelion, and in egg yolks. **Zeaxanthol** (zeaxanthin), $C_{40}H_{54}(OH)_2$, is the chief yellow pigment of corn (maize). **Fucoxanthol** (fucoxanthin), $C_{40}H_{60}O_6$, is one of the principal pigments of the brown algae.

Fig. 3-3 Structure of β-carotene.

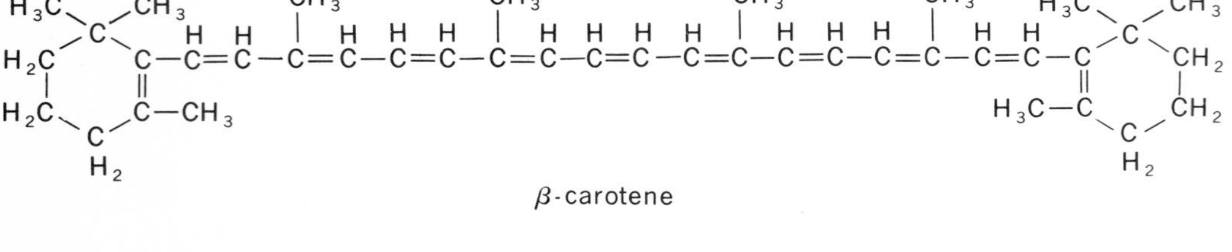

```
H3C   CH3        CH3         CH3            CH3          CH3      H3C   CH3
    C        H H  |  H H H    |  H H H H     |  H H H     |  H H       C
H2C   C—C=C—C=C—C=C—C=C—C=C—C=C—C=C—C=C—C=C—C            CH2
 |    ||                                                 ||      |
H2C   C—CH3                                          H3C—C     CH2
    C                                                       C
    H2                                                      H2
```

β-carotene

POSSIBLE FUNCTIONS OF CAROTENOID PIGMENTS Not all the functions of the carotenoid pigments in the plant are known. Some of them, as already mentioned, are involved in vitamin A production, which may be as important to plants as it is to animals. Their presence in flowers is thought to attract insects and thereby to secure cross-pollination. Their presence in fruits may render fruits more attractive and thus secure for them a wider distribution by animals. The carotenoid pigments probably function in absorbing light energy and transferring it to chlorophyll *a* in photosynthesis. The relation of the carotenoids to vitamin A renders these pigments of great importance to both plants and animals and has stimulated a vast amount of research on them.

The anthocyanins Most of the red, blue, and violet colors of plants ranging from the mosses through the seed plants are caused by a class of chemical compounds known as **anthocyanins** (Fig. 3-4). These pigments are not contained in plastids as are the chlorophylls and carotenoids but are found dissolved in the cell sap of the cell. They therefore usually appear uniformly distributed throughout the cell (Fig. 3-1, 9). In some cases they occur as crystals. The red or purplish leaf hairs of the velvet plant (*Gynura aurantiaca*), the red roots of the common beet, the red lower surface of the leaves of wandering Jew and other plants, many red, blue, or violet flowers and fruits all owe their color to anthocyanins (Fig. 3-1, 4, 6, 10, 11). In leaves the anthocyanins sometimes completely obscure the chlorophyll pigments, as in red cabbage or in *Achyranthes* (Fig. 3-1, 3).

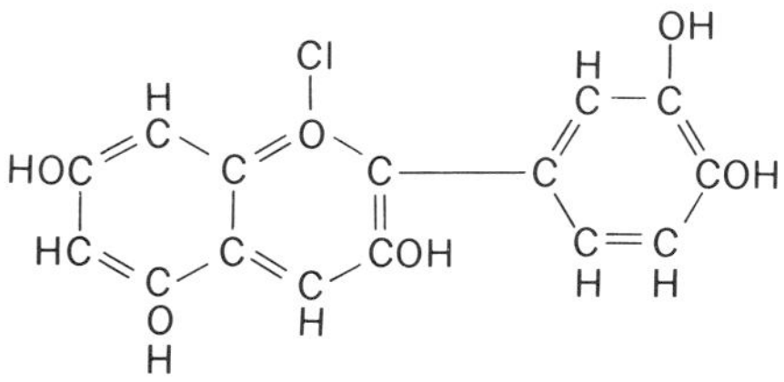

EXTRACTION AND PROPERTIES Chemically the anthocyanins are glycosides, i.e., compounds which, on being broken down into less complex components, yield, among other substances, a sugar. The nonsugar part of a glycoside is called an **aglycon.** The aglycons of the anthocyanins are called **anthocyanidins** and are rather complex organic compounds with a similar organic nucleus. The commonest sugar component is glucose, but several other sugars may be found. Some anthocyanins contain one molecule of sugar, and some contain two. There are many different kinds of anthocyanins in different species of plants. Two or more kinds are often present in the same species.

Anthocyanin (cyanidin chloride)

Fig. 3-4 Structure of anthocyanins (cyanidin chloride).

The anthocyanins are all soluble in water and can also be extracted with alcohol. They are not soluble in ether, chloroform, or benzene. The red color obtained when beets or red leaves are boiled in water (Fig. 3-1, 17) is caused by anthocyanins.

Their color in the plant depends upon a number of factors among which may be mentioned the concentration of the anthocyanin, the simultaneous presence of several anthocyanins and other pigments, the ash content of the cell sap, the colloidal condition of the cell content, and the hydrogen-ion concentration of the cell sap. Of these, the last has been most emphasized. When the cell sap is acid, the color is often red (Fig. 3-1, 3) and when the cell sap is alkaline, it is usually blue (Fig. 3-1, 4, 11). This can be demonstrated in the laboratory by placing blue flowers in a weak acid solution. After enough time has elapsed to permit penetration, they become red in color. Similarly, pink or red flowers will often turn blue in weak ammonia. The same thing may be demonstrated by extracting the pigments of blue flowers with water and adding acid to the extract. It immediately changes to a cherry red. Adding an alkali to the extract usually does not give a blue but a yellowish-green color because of the presence of yellow flavone pigments and other impurities.

FACTORS AFFECTING ANTHOCYANIN FORMATION In many species of plants the production of anthocyanins is a hereditary character; i.e., such plants come true to color from seed. This fact is utilized by plant breeders. The intensity of color, however, is often greatly influenced by environmental conditions and by internal factors. Purple-leaved types of lettuce, for example, fail to develop anthocyanin when deprived of blue-violet and ultraviolet radiation, and many highly colored flowers and leaves become much paler under these conditions or under weak light. Some plants do not develop anthocyanin in darkness; others, like beets, do. Temperature, the internal food supply, the supply of inorganic substances, disease and injury, and perhaps other factors may affect the formation of these pigments. Often anthocyanin does not appear until late in the season as a result of ordinary ripening processes, as in apples and other fruits. The exact mechanism of the formation of anthocyanins in the plant is not known, but it is likely that sugars are involved in their synthesis, not only as the sugar component but also in the formation of the aglycons.

POSSIBLE FUNCTIONS OF ANTHOCYANIN PIGMENTS The function of the anthocyanin pigments has been much in dispute. Where they occur in flowers, the colors have been thought to guide or attract insects or hummingbirds and thus facilitate cross-pollination. By some it is thought that anthocyanins in some plants, as in the tender opening foliage of maples, act as a protective screen against intense sunlight and especially ultraviolet radiation, which may be injurious to living protoplasm. Another theory is that anthocyanins absorb certain of the sun's rays and transform them into heat, which raises the internal temperature of the parts in which the pigments occur and thus protects them against low temperatures.

In favor of both of these ideas, it may be stated that plants growing in high altitudes, where it is cooler and where the intensity of the ultraviolet portion of sunlight may be greater, usually develop much more anthocyanin than do plants in lowland regions. However, in the present state of experimental evidence, it must be admitted that neither of these theories rests on a firm foundation. The same may be said of attempts to link the anthocyanins with various physiological processes such as respiration and food synthesis or to consider them as reserve food substances or waste products of metabolism. The varied distribution of these pigments, their presence in so many different organs of the plant, the different conditions under which they are formed, and other factors will always make it difficult to assign to them specific functions.

The flavones Another group of pigments fairly widespread in plants is the **flavones.** Their color is yellow and is often masked by other pigments. Water extracts of green leaves often have a yellowish tinge caused by these pigments. The flavones, like the anthocyanins, are water soluble and hence occur in the cell sap and not in plastids. Chemically they are related to the anthocyanins. One of the most common of them is **quercetin,** $C_{15}H_{10}O_7$, which has been isolated from such plants as sumach, horse chestnut, tea, hops, onions, and the North American dyer's oak (*Quercus tinctoria*). Many of the flavone pigments were formerly used as mordant dyes.

Many other pigments are found in plants, especially in the red algae and in the fungi. Some of these are considered in later chapters.

Autumn coloration A consideration of color in plants would not be complete without some reference to the brilliant display of color manifested each autumn, particularly in the leaves of trees and shrubs. Our knowledge of the underlying causes of autumn coloration is far from complete. Two types of factors are probably involved. On the one hand are the internal physiological conditions of the leaf itself, particularly those concerned with pigment development, and on the other the changing external conditions of the environment.

Primarily, autumn coloration is associated with the waning vitality of the leaf. As the growth season draws to a close, a special layer of cells known as the abscission layer is formed through the base of the leaf (Fig. 4-19). The cells behind this layer become corky and impervious to water, and the water-conducting vessels themselves often become clogged. The transport of materials to and from the leaf is thus seriously checked. This reduces the activities of the leaf and interferes with the development of chlorophyll. Since chlorophyll is being continually destroyed by sunlight, the leaves soon lose their green color when more chlorophyll cannot be made to replace this loss. This allows any other coloring matter that may be present to show up and also is conducive to the formation of other coloring substances, although probably these new ones do not originate from the decomposition of the chlorophyll, as formerly believed. Thus the principal foundation on which autumn coloration rests is the destruction of the chlorophyll, thereby allowing other pigments formerly masked to become prominent, and the formation of new coloring matters not previously present in the leaf.

As previously mentioned, the yellow carotenoid pigments are constantly associated with chlorophyll. Because of the greater stability of these pigments in sunlight, they persist in the leaves after the chlorophyll is destroyed. In the absence of other coloring matters, therefore, autumn leaves are yellow in color, and in some trees, like the ash, walnut, sycamore, poplar, and some birches, this is the predominating if not the only color shown. In others the yellow, while always present, becomes masked by secondary coloring compounds that are formed. Yellow must be regarded as the most commonly present, though perhaps not the most conspicuous, autumn color.

The red colors that appear in autumn vegetation are caused by pigments often not previously present in the leaf but formed after the vitality of the leaf begins to decline. The exact manner of their origin is not known. That light is influential is shown by the fact that leaves of such trees as maples and oaks, if kept covered or well shaded, do not become red but show only yellow. The disappearance of the chlorophyll tends to admit more light to the leaf, which may affect the development of the red pigments. It is perhaps partly for this reason that the brightest colors are often found in vegetation when the autumn season is bright and sunny rather than dull and cloudy.

It is often found that leaves which develop the greatest intensity of red, like the maples, are rich in sugar. The presence of sugar in a leaf is probably essential for anthocyanin formation. Furthermore coloration is usually richest in seasons when there is an abrupt change from high summer temperatures to low autumn temperatures. Under such conditions the movement of sugars and other materials out of the leaf may be retarded and thus may provide the conditions conducive to pigment formation.

Further evidence of the possible influence of sugars and other substances of the leaf on the development of anthocyanin is furnished by leaves in which one or more of the veins have been cut or injured (Fig. 3-5). It is often found in such cases that the portion of the leaf formerly supplied by the cut vein turns a bright red, while the rest of the leaf may turn yellow in autumn. The cut or injury probably occurs while there is still considerable sugar in the leaf, and since the sugar cannot be translocated from the severed section, it remains available

Fig. 3-5 Autumn maple leaf in which the central vein has been injured by insect or fungus attack. The dark portion above the injury was bright red, while the rest of the leaf was yellow.

for anthocyanin formation. The portion above the cut in such leaves usually remains green longer than the other portions of the leaf but finally turns red in leaves that make anthocyanin.

Frost may play a part in autumn coloration, but that it is not essential is indicated by the fact that leaves frequently turn red or yellow before the first frost occurs.

The discussion thus far has been concerned with the three predominating colors of autumn vegetation. The various shades, ranging from brilliant red to orange and golden yellow, usually result from combinations of these three colors. These brilliant colors usually last for only a comparatively short time. As the season advances and the nights become colder, the colors gradually fade, through internal disorganization of the pigments and other leaf compounds, and the leaf dies. Decomposition products accumulate, and the leaf assumes a brown color and falls to the ground.

QUESTIONS

1. What is the principal color found in natural vegetation and what causes it?

2. What parts of a plant are not green in color? Why?

3. Why do green apples often become red or yellow on ripening?

4. When McIntosh apples are heavily shaded on the tree, they seldom turn red on ripening. Why?

5. What influence has the sun on the color of plants?

6. When leaves are boiled in water, what pigments are extracted?

7. What is the color of a red leaf after it has been boiled in water?

8. Chlorophyll *a* is found in all chlorophyll-containing plants. Do they all also contain chlorophylls *b*, *c*, and *d*?

9. Why do chlorotic plants fail to thrive?

10. Why is chlorophyll considered to be one of the most important chemical substances in nature?

11. Of what importance to man are the carotenoid pigments?

12. What causes leaves to change color in autumn?

REFERENCES

ARONOFF, S. 1950. Chlorophyll. *Botan. Rev.*, **16:** 525–588. A detailed review of the literature on the chlorophylls up to 1950 with a bibliography of 243 titles.

BLANK, F. 1947. The Anthocyanin Pigments of Plants. *Botan. Rev.,* **13**:241–317. A detailed review of the chemistry, distribution, physiology, and genetics of the anthocyanins, with a bibliography of 594 titles.

FRANK, SYLVIA. 1956. Carotenoids. *Sci. Am.,* **194** (1):80–86.

MEYER, B. S., D. B. ANDERSON, and R. H. BÖHNING. 1960. Introduction to Plant Physiology. D. Van Nostrand Company, Inc., Princeton, N.J.

STRAIN, HAROLD H. 1958. Chloroplast Pigments and Chromatographic Analysis. Thirty-second annual Priestley lectures, The Pennsylvania State University, University Park, Pa. (obtainable from Phi Lambda Upsilon, Department of Chemistry, Pennsylvania State University, University Park, Pa.). A series of lectures on the chloroplast pigments of higher plants and algae with particular emphasis on methods of separation of the pigments.

THOMPSON, J. F., S. I. HONDA, et al. 1959. Partition Chromatography and Its Use in the Plant Sciences. *Botan. Rev.,* **25**:1–263. An extensive review of all types of chromatography, with a section on plant pigments.

WILLSTÄTTER, R., and A. STOLL. 1928. Investigations on Chlorophyll. Methods and Results. English translation by F. M. Schertz and A. R. Merz. Science Press, Lancaster, Pa. A translation of the original work of the Nobel Prize winner, Willstätter, on the chemistry of chlorophyll.

4

LEAVES

Leaves of Coleus.

The external and internal structure of leaves, stomata and their operation, abscission, general physiological processes of leaves, transpiration and guttation, and specialized leaves are taken up in this chapter.

ORIGIN AND EXTERNAL FORM

Leaves are perhaps the most conspicuous parts of plants. Being rich in chlorophyll, they are responsible for the common green color of forests and fields. They are always borne on stems. The part of the stem to which a leaf is attached is called a **node.** The upper angle the leaf makes with the stem at its point of attachment is called the **axil** of the leaf (Fig. 4-1). Invariably a bud is found in this axil, although the bud may be so immature as to be invisible to the unaided eye or it may be covered by a sheathing leaf base.

Leaves in general are characterized by their thin and expanded form. Though usually very thin, they are able to maintain their shape because of an internal framework, or skeleton, of more or less rigid **veins.** In size, leaves vary from tiny, almost microscopic structures to forms 10 to 20 ft and more long. As will be seen later, leaves are the outstanding organs in which carbohydrate and other foods are made and hence are very important to the life of the plant.

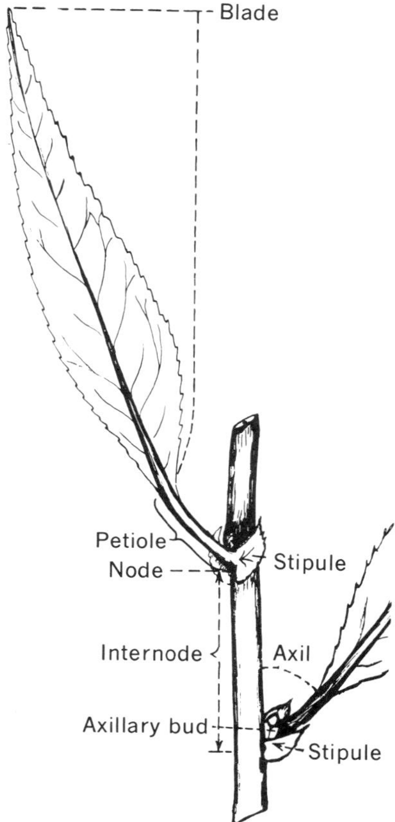

Fig. 4-1 Portion of a branch of willow showing parts of a leaf and the relation of the leaf to the stem. (Drawing by Florence Brown.)

Origin and development of leaves In general, leaves arise in buds that contain the undeveloped growing points of stems. In perennial plants these growing points are often protected by a series of scalelike leaves that make the bud more or less conspicuous. The growing point itself is made up of cells that are all essentially alike. These cells are characterized by being thin walled and isodiametric and by consisting of rather dense protoplasm with relatively large nuclei. During the growing season they are practically all in an active state of division, forming new cells and causing the growing point to expand and move forward. Cells with these characteristics are known as **meristematic cells.** At regular intervals around the sides of the growing tip, certain cells of the outermost layers, by a series of divisions, form small outgrowths which are known as **leaf primordia** (Figs. 4-2, 9-1) and by the continued growth of these primordia, the leaves are developed. At first the growth is chiefly at the tip or apex of the primordium (apical growth), but later, especially in pinnately veined leaves, the sides of the primordium begin to enlarge rapidly by intercalary growth (i.e., growth not restricted to the apex) to form the expanded portion of the leaf. Sooner or later growth at the apex ceases, and further expansion of the leaf takes place only at the base.

All this development occurs while the leaf is still in the bud and before it has reached a length of more than a few millimeters. When a leaf is large enough to be handled, the growing region can be demonstrated easily by marking the leaf into small squares with India ink and observing how these squares enlarge as the leaf grows (Fig. 4-3). The full development of the leaf often takes place in a comparatively short time. When the leaf has reached mature size,

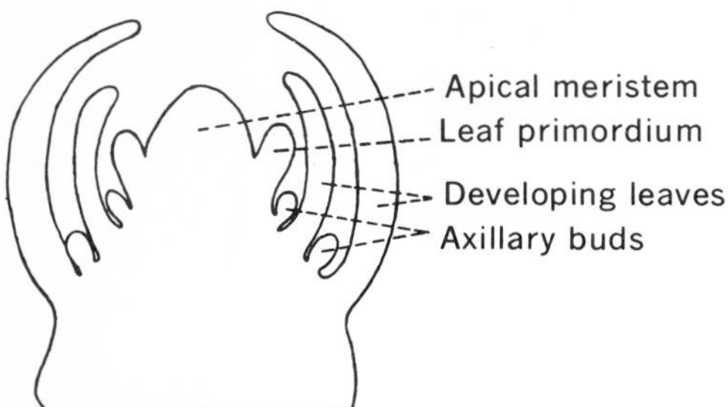

Fig. 4-2 Diagrammatic representation of formation of leaves at the stem apex.

all growth ceases. In this respect leaves differ from roots and stems, which continue to grow as long as the plant is living.

In many perennial plants the new leaves are differentiated in the buds during the summer and go into a dormant state from which they emerge the following spring. In many of the monocotyledonous plants, basal growth of the leaves continues for some time and results in a long narrow form such as is common in the grasses. In the ferns apical growth proceeds after basal growth has ceased. Such leaves keep unrolling at the tips until mature (Fig. 4-7*D*).

The parts of a leaf A leaf may consist of an expanded portion known as the **lamina,** or **blade,** a **leafstalk,** or **petiole,** by which the blade is attached to the stem, and two small appendages at the base of the petiole, called **stipules** (Fig. 4-1).

THE LEAF BLADE, OR LAMINA The leaf blade is the most conspicuous and, in general, the most important part of the leaf. It is usually thin, and its construction is such as to expose the greatest number of its chlorophyll-containing cells to light. The form or outline of the leaf blade, while constant for any particular species of plant, presents wide variations. The different forms are usually described by such terms as linear, lanceolate, elliptic, ovate, orbicular, cordate, and reniform (Fig. 4-4, I). Likewise the margin of the leaf blade may vary from an entire, or even, condition, through all gradations, to deep sinuses extending almost to the middle of the blade. The various types of margins are designated by such terms as entire, undulate, crenate, dentate, serrate, and lobed (Fig. 4-4, II).

THE PETIOLE The petiole is usually a stem-like structure which connects the leaf blade with the stem (Fig. 4-1). Largely through the growth in length or the bending and twisting of the petiole, the leaf blade is brought into a favorable position with respect to light. This position is usually broadside toward the light, regardless of what may be the position of the stem from which the leaf originates. It is best seen in climbing plants that grow over the sides of stone buildings. All the leaves of such plants usually lie in a plane nearly parallel to the side of the building. Usually the entire available space is covered with leaves in such a manner that they do not shade each other to any great extent. This is brought about chiefly by the wide differences in the lengths and positions of the petioles of the different leaves. For the same reason, the leaves of a tree are all directed outward and present an entirely different aspect when viewed from beneath the tree from what they do when viewed at a distance.

The shape and form of the petiole vary in different species of plants. In some leaves petioles are absent. Such leaves are said to be **sessile** (Fig. 4-8*A*). In many of the monocotyledonous plants, and especially in the grasses, the base of the leaf forms a sheath around the stem, usually extending from one node to the next lower one (Fig. 4-4, III).

STIPULES There are usually two stipules present at the base of the petiole (Figs. 4-1, 4-6*B*), but some leaves, like those of the maple, lack them altogether. In some leaves the stipules drop off shortly after the leaf matures. Stipules vary widely in shape and form. Ordinarily they

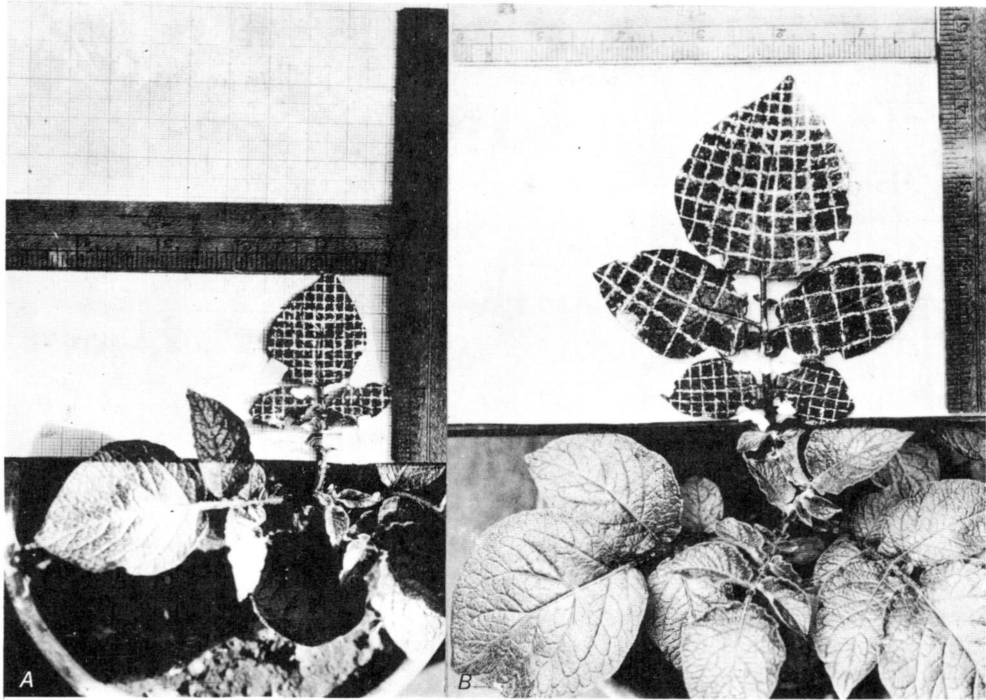

Fig. 4-3 Region of growth in the leaf of a dicotyledonous plant (potato). A, young leaf marked into ⅛-in. squares; B, the same leaf as it appeared 10 days later. (Photographs courtesy of E. L. Nixon.)

are tiny leaflike organs, but sometimes, as in the leaves of the garden pea, they are quite large and assist materially in photosynthesis. In the black locust and a few other plants, the stipules are thorns, and in some species of *Smilax* they are tendrils.

Venation The blade of the leaf is strengthened by the presence of veins. These veins are made up chiefly of vascular, or conducting, tissue which is continuous with that of the petiole and stem. Hence the veins serve to distribute water and dissolved inorganic salts throughout the different parts of the blade and to carry away elaborated foods as they are made.

There are two principal types of venation, or veining, viz., **parallel venation** and **netted venation** (Fig. 4-5). In parallel-veined leaves at least the principal veins all run parallel to each other from the base to the tip of the leaf or they may run parallel at right angles to a main central vein, as in the banana. This type of venation is characteristic of the monocotyledonous plants and is especially well exemplified in the grasses. In most leaves of this type there are tiny veins, invisible to the unaided eye, that connect the principal veins.

In netted-veined leaves the veins branch again and again, forming a complete network through the leaf (Fig. 4-5*D*). The extreme tips of these branches usually end in the blade tissue. The nature of this type of venation can best be seen by holding a leaf like that of the

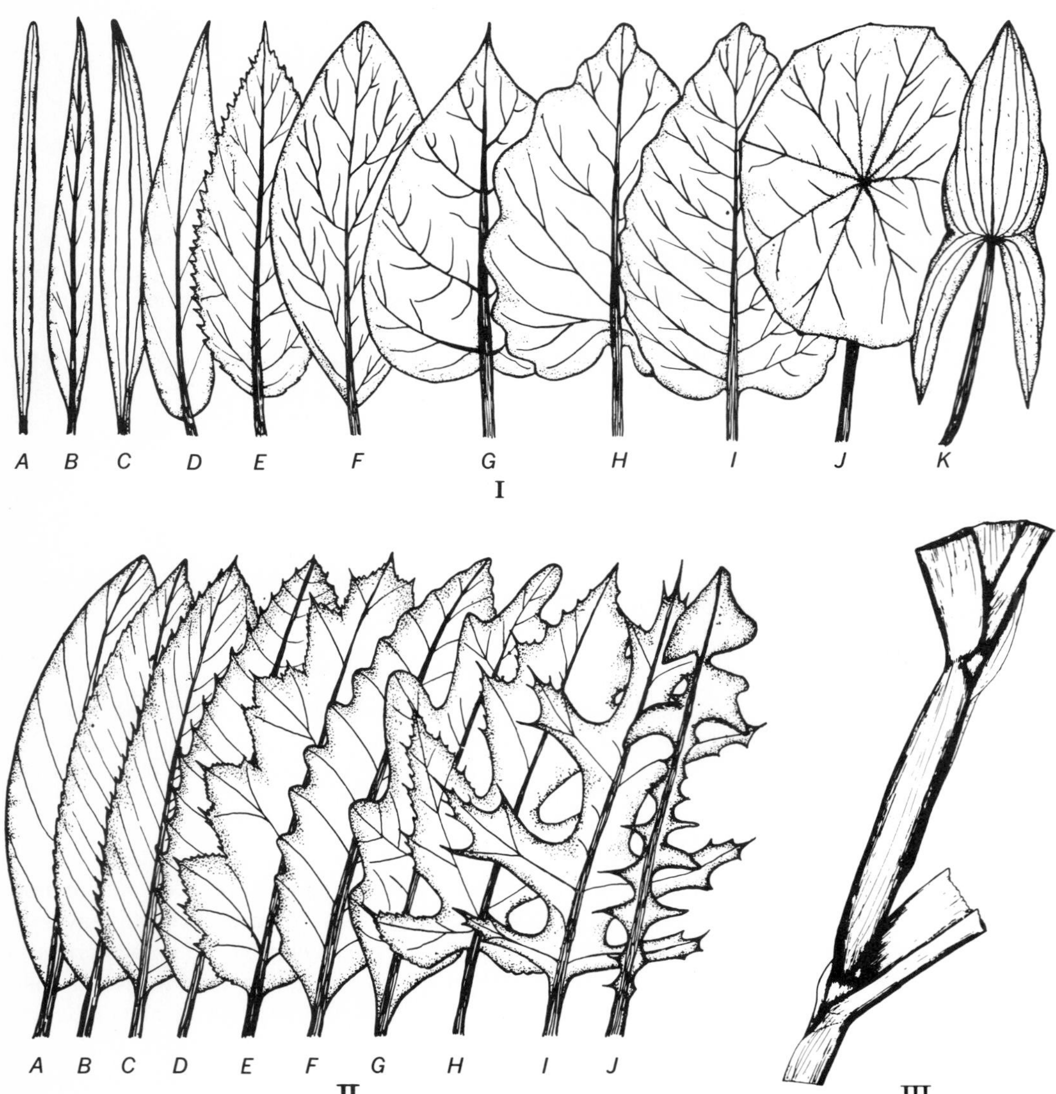

Fig. 4-4 I, *forms of leaves. A, linear; B–D, lanceolate; E, elongate ovate; F, elliptical; G, H, cordate; I, ovate; J, peltate; K, sagittate. II, types of leaf margins. A, entire; B, finely serrate; C, sharply serrate; D, coarsely serrate; E, dentate; F, coarsely crenate; G–I, deeply lobed; J, pinnatifid. III, portion of stem of corn, showing leaf sheaths extending from one node to the next lower one. (I and II drawn by Edna S. Fox; III drawn by Elsie M. McDougle.)*

maple, the catalpa, or the oak toward light. Netted venation is characteristic of dicotyledonous plants. In some netted-veined leaves, like those of the elm or the oak, there is a principal central vein, called the **midrib,** from which all other veins branch out. From its featherlike appearance, this type has been called **pinnate venation** (Fig. 4-5*B*). In others there are several large veins of equal size, all arising at a common point at the tip of the petiole and spreading out fanlike through the blade. This type of netted venation is called **palmate venation** and occurs in leaves of mallow, geranium, maple, and many other plants (Fig. 4-5*A*).

Simple and compound leaves The blades of some leaves are deeply indented at the margins. Others are completely separated into individual parts called **leaflets** (Fig. 4-6). In both instances, regardless of the degree of dissection of the leaf blade, a bud will be located in the axil of the leaf. As long as the blade is in one piece, even though deeply lobed, the leaf is said to be **simple** (Fig. 4-4, II*G–J*). When the blade is completely dissected into leaflets, the leaf is said to be **compound.** There are two types of compound leaves: **palmately compound** leaves, in which the leaflets are all attached at a common point at the tip of the petiole (Fig. 4-6*A*) as in clover and horse chestnut, and **pinnately compound** leaves (Fig. 4-6*B*), in which there is a principal central axis, called the **rachis,** to which all the leaflets are attached. This type occurs in the rose and in the ash. The relation of these types of leaves to palmate and pinnate venation is obvious. The rachis of the pinnately compound leaf is homologous with the midrib of a pinnately veined leaf. The leaflets of a compound leaf may be either sessile or stalked and may have any of the types of margins characteristic of simple leaves. The leaflets may themselves be compound as in honey locust or in baneberry.

Fig. 4-5 Types of venation. A, B, netted venation; C, parallel venation; A shows palmate type of netted venation, and B shows pinnate type; D, leaf skeleton showing the connecting network of veins. (Drawings by Edna S. Fox.)

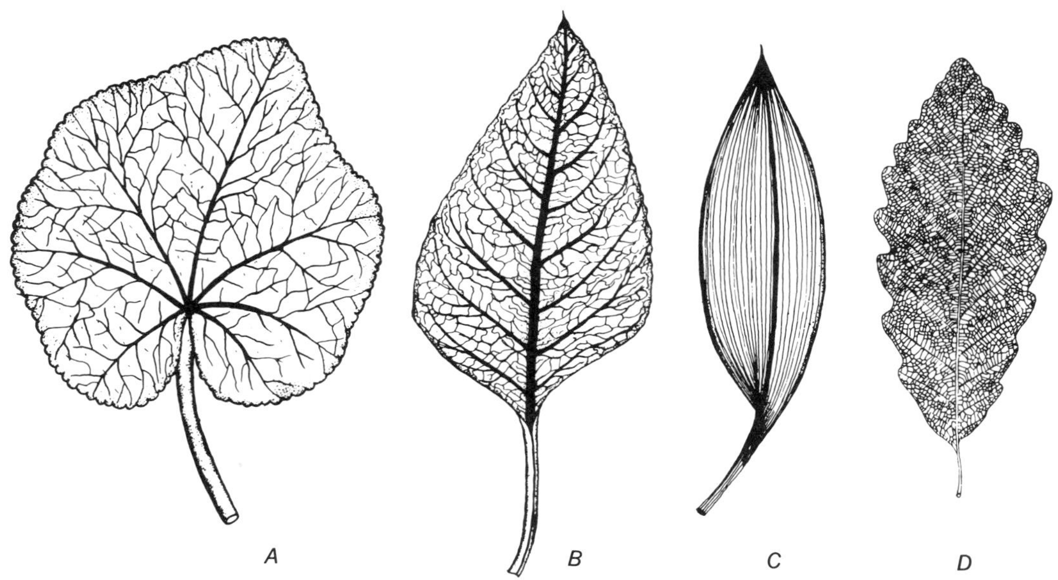

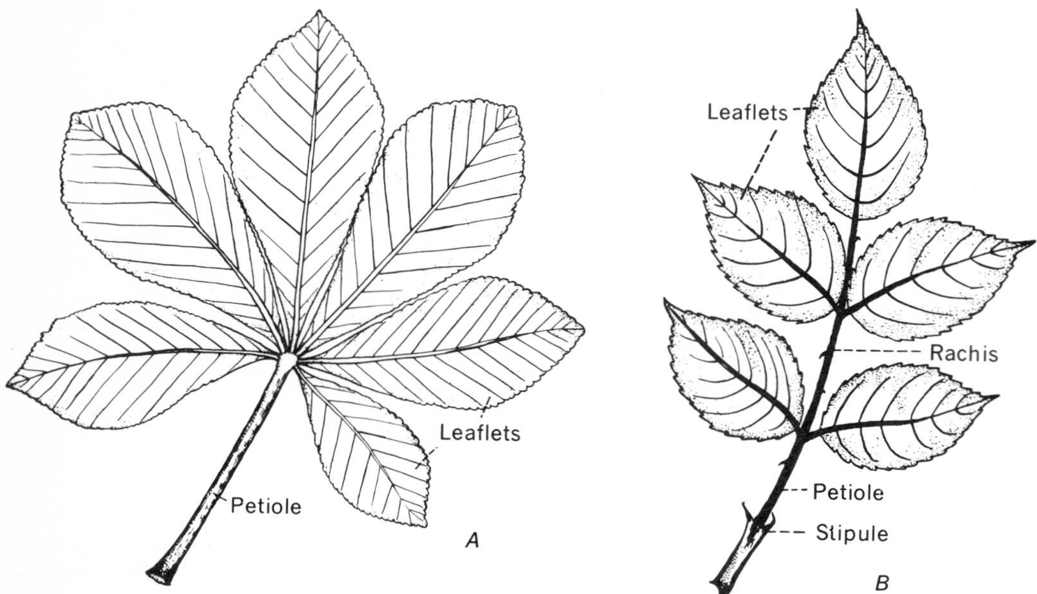

Fig. 4-6 Compound leaves. A, palmately compound; B, pinnately compound. (Drawings by Edna S. Fox.)

In this case the leaf is said to be twice or bipinnately compound. In a few cases leaves are three times compound as in some of the acacias.

Vernation The manner in which the leaves are folded or rolled up in the bud is called **vernation.** There are several distinct types of vernation, each characteristic of particular kinds of plants (Fig. 4-7). They are (1) *reclinate,* or *inflexed,* when the upper part of the leaf is bent down on the lower part as in the tulip tree; (2) *conduplicate,* when the leaf is folded lengthwise along the midrib, bringing the two halves of the blade face to face, usually with the lower surfaces outermost, as in the leaves of the cherry, the oak, and members of the magnolia family; (3) *plicate,* when the blade is folded back and forth along the main veins like a closed fan, as in palmately veined leaves like those of geranium, mallow, and maple; (4) *circinate,* when the leaf is rolled from the tip downward to the base, as in ferns and in sundew; (5) *convolute,* when the leaf is rolled lengthwise from side to side, scroll like, as in many members of the rose family and in lily of the valley; (6) *involute,* when both edges of the leaf are inrolled lengthwise on the upper surface toward the midrib, as in violets and water lilies; and (7) *revolute,* when both edges of the leaf are inrolled lengthwise on the lower surface toward the midrib, as in azalea and dock.

Leaf arrangement—phyllotaxy The actual positions occupied by leaves on the stem are determined (1) by inherent characteristics of the plant and (2) by environmental conditions. It is rather interesting that leaves are not haphazardly placed on the stem but arise with mathematical precision at definite and regular intervals. The arrangement of leaves on the stem is referred to as **phyllotaxy,** a term meaning "leaf order."

Leaves may be *opposite* (Fig. 4-8*B*), in which case two leaves occur at a node, as in maples and mints; *whorled* (Fig. 4-8*A*), when more than two leaves occur at a node, as in bedstraw, loosestrife, and some lilies; or *spiral* (Fig. 4-8*C*, *D*), in which case only one leaf occurs at a node. When two leaves occur at a node, each pair usually stands at right angles to the next pair above or below, when the stem is in a vertical position, thus forming four rows or ranks around the stem as seen from above. This type of arrangement is also called *decussate* (Fig. 4-8*B*). If the stem is horizontally placed, the petioles of the leaves of alternate nodes are usually bent around so as to bring the leaf blades into two horizontal rows or ranks.

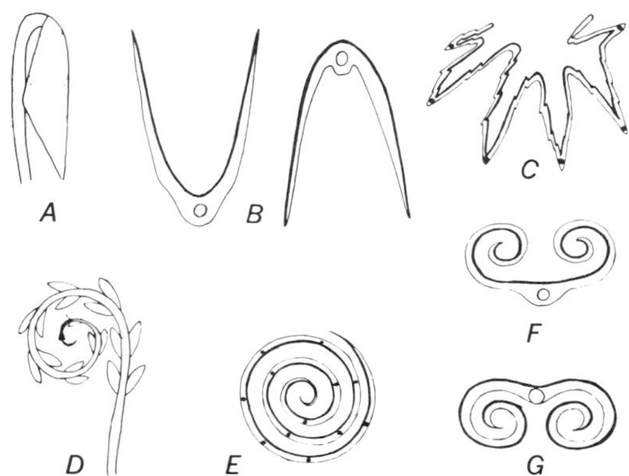

Fig. 4-7 Types of vernation illustrated diagrammatically. A, reclinate or inflexed; B, conduplicate; C, plicate; D, circinate; E, convolute; F, involute; G, revolute. (Drawn by Florence Brown.)

In the simplest case of spiral phyllotaxy, every third leaf is directly over the first one of the cycle (Fig. 4-8*C*), thus forming two vertical rows around the stem, as in elm and in corn and many other grasses. (This type is correctly referred to as *alternate,* but the term alternate has come to be used generally by botanists to refer to any spiral arrangement, i.e., to any type in which there is only one leaf at a node.)

In the next type every fourth leaf is over the first, forming three vertical rows, as in the beech, white hellebore, and the sedges. Perhaps the commonest spiral arrangement is the next type, in which every sixth leaf is directly over the first, forming five vertical rows or ranks. This is the type found in oaks, cherries, apples, and poplars. In the remaining types, the ninth, fourteenth, twenty-second, thirty-fifth, or fifty-sixth leaf, respectively, is over the first.

It is customary to designate spiral phyllotaxy by fractions representing the portion of the circumference of the stem between any two suc-

Fig. 4-8 Types of phyllotaxy. A, whorled; B, opposite; C, spiral, showing the ½ fractional arrangement; D, spiral, with the ⅖ arrangement. (Drawings by Elsie M. McDougle.)

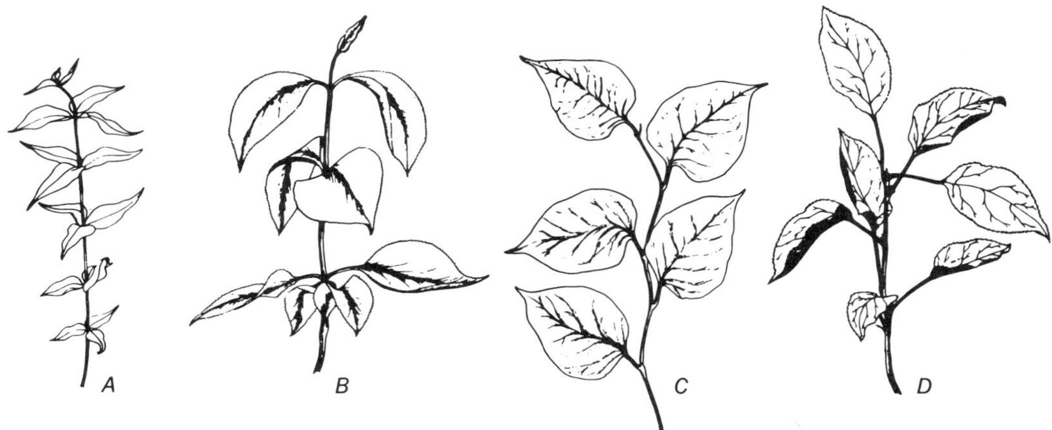

ceeding leaves in the spiral. These fractions may be determined by tying a string to the base of the petiole of any leaf and proceeding with it, by the shortest route, to the base of the petiole of the next leaf, and so on, until a leaf is reached that arises directly over the first one. The numerator of the fraction would then be the number of complete turns made with the string around the stem, and the denominator the number of leaves touched, not including the first. In an apple twig, for example (Fig. 4-8 *D*), in which the sixth leaf is over the first, the string would pass twice around the stem and would touch five leaves, not counting the first, making the fraction $2/5$. The actual fractions found in this manner are $1/2$, $1/3$, $2/5$, $3/8$, $5/13$, $8/21$, $13/34$, and $21/55$ and are known as the Fibonacci series. The first three represent the first three types mentioned in the two preceding paragraphs. The higher fractions are usually found in plants with greatly shortened stems, like the common rosette plants, and in pine cones.

Fig. 4-9 Leaf mosaic of maple showing variations in lengths of petioles which result in the orientation of the leaf blades more or less in a single plane facing the direction of the incident light. (Drawn by Elsie M. McDougle.)

The arrangement of leaves on a stem usually falls into one or another of these fractions. There are seldom any intermediates or variations from them, except that stems are sometimes twisted enough to obscure the actual type of phyllotaxy. Even in this case the change will usually be to another of the fractions already mentioned. It is also interesting to note, on examining these fractions, that in all cases after the first two the numerators and denominators are each the sum of the two preceding and that the numerators are the same as the denominators of the second preceding fractions. The denominators also represent the number of rows or ranks of leaves one would see on looking vertically down on the stem producing them.

Phyllotaxy deals only with the arrangement of leaves as determined by their points of origin on the stem. The actual arrangement of the leaves is greatly influenced by the position, with respect to light, of the leaf, the stem, or the plant as a whole. Leaves are usually found in positions that enable them to receive a maximum amount of the incident light. Thus on vertical stems, and especially in rosette plants, the petioles of the leaves are usually increasingly longer from the tip to the base of the stem. One-sided illumination may cause the true arrangement of the leaves to be obscured, as in a leafy vine growing on a brick wall. In this case the leaves are all directed outward, forming a **mosaic** (Fig. 4-9). In the compass plant (*Silphium laciniatum*) and in wild lettuce (*Lactuca scariola*) the leaves all face east or west regardless of their phyllotaxy. The crowding of plants in dense colonies also greatly affects the arrangement of their leaves. Sometimes the leaves of such plants are long and narrow and are placed in a position parallel to the rays of the sun. This is the situation with many reeds and sedges and in cattail colonies. Such arrangements cut down shading to a minimum.

Other arrangements result from the water relations of the plant. In the water lilies, the leaves float on the surface of the water, appar-

Fig. 4-10 Floating leaves of water lilies (Victoria regia). *(Photograph by Paul Shope, taken at Tower Grove Park, St. Louis, Mo.)*

ently without any definite arrangement, the rest of the plant being submerged, with the roots in the soil, under the water (Fig. 4-10).

ANATOMY OF THE LEAF

The internal arrangement of the tissues of the leaf can be understood best by a microscopic examination of very thin slices or cross sections cut at right angles to the broad surface of the blade. Sections cut in this manner will show several well-defined regions in the leaf blade (Fig. 4-11). The outermost layer of cells, which extends all over the surface of the leaf, is called the **epidermis.** The interior of the leaf, between the upper epidermis and the lower epidermis, is called the **mesophyll,** a term meaning "middle of the leaf." The mesophyll is further differentiated into an upper region, the **palisade mesophyll,** and a lower region, the **spongy mesophyll.** The **veins,** or **vascular tissues,** extend throughout the spongy mesophyll.

The epidermis The epidermis usually consists of a single layer of cells that appear more or less rectangular in shape as seen in cross section (Fig. 4-11). In surface view they are often very irregular in shape. On the upper surface of the leaf the exterior or upper walls of the epidermal cells are thickened by a special colorless layer known as the **cuticle** (Fig. 4-11). This cuticle is made up of a waxy, waterproof material, called cutin, which prevents the drying out of the underlying tissues and protects them to some extent against mechanical and other

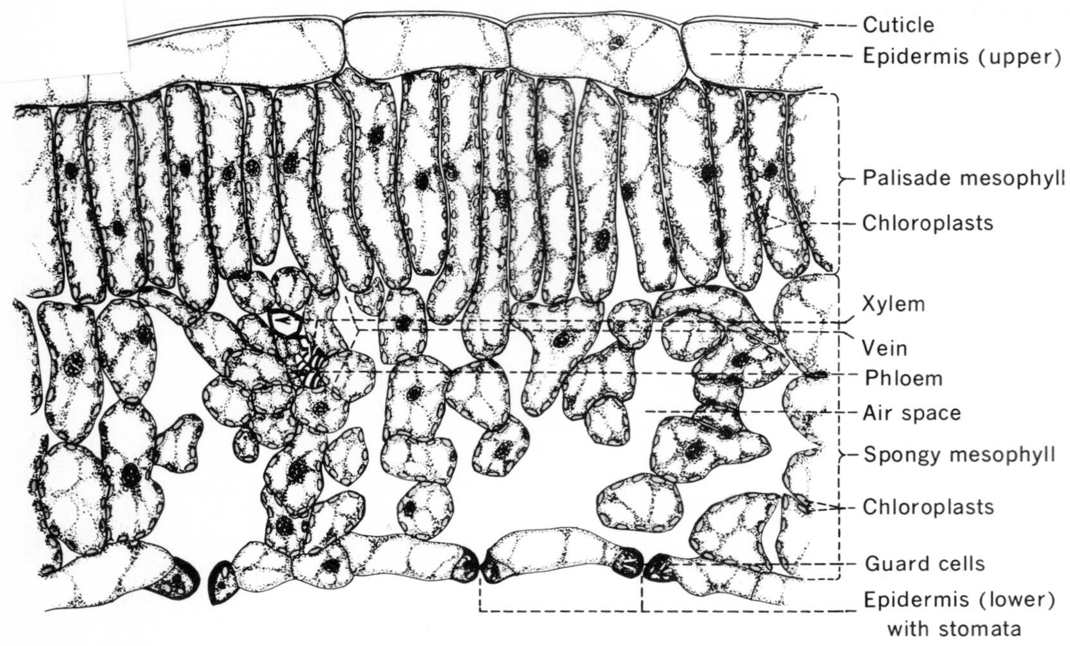

Fig. 4-11 Cross section of a portion of a leaf of Lonicera tatarica, *showing internal structure. (Drawing by Helen D. Hill from a slide furnished by J. W. Sinden.)*

injuries. The mature cells of the epidermis are usually colorless throughout and contain a comparatively small amount of protoplasm lying against the cell walls and a large central vacuole filled with cell sap. In some seed plants the cell sap of the epidermal cells may be colored with anthocyanin, but chlorophyll is never found except in guard cells and in a few submersed plants.

Depending upon the species and upon the environmental conditions under which a plant grows, there may be one or several layers of epidermal cells. The epidermis is more likely to consist of several layers in plants growing under intense sunlight and in dry situations. The epidermal cells of submersed plants and of those growing in moist places are usually larger than those of plants growing in dry habitats. The epidermis of the lower surface of the leaf usually consists of a single layer of cells, similar in structure to those of the upper surface but less regular, and with thinner outer cell walls, covered with less cutin.

EPIDERMAL HAIRS As previously stated, the outer surface of the epidermis consists of a cuticle. In addition to this cuticle there may be other waxy or resinous coverings, sometimes forming a "bloom" over the leaf surface. In many plants the epidermis is covered with various kinds of hairs (Fig. 4-12). These hairs may be restricted to the lower or to the upper epidermis or may occur on both upper and lower surfaces. They may be simple, unicellular epidermal outgrowths or complex, multicellular structures. They are often much branched. Sometimes they consist of living cells. Many of these are glandular, secreting oils and other substances. The stinging hairs of nettle come under this category. They are elongated, tapering, unicellular structures, containing a fluid which is held under considerable turgor pressure. The tips of these hairs are enlarged somewhat and are turned slightly to one side. When touched, the end usually breaks off, leaving a point like a hypodermic needle. This easily pierces the skin and injects the fluid, which causes considerable irritation.

The epidermal leaf hairs of other plants are dead and often filled with air. In the common mullein (*Verbascum Thapsus*) these dead hairs are branched (Fig. 4-12*C*) and are produced in such profusion as to give the leaf a woolly or feltlike appearance. In many members of the mustard and mallow families the hairs are star shaped (stellate). In other plants they may be scalelike. Sometimes they are soft, filamentous structures giving the leaf a velvety appearance, while at other times they are stiff and hard, like the leaf spines of thistles.

While most of the different types of epidermal hairs are characteristic of particular species of plants, their development is greatly influenced, at least in some plants, by environmental conditions. The production of leaf hairs is usually most pronounced in plants living under dry conditions. Young leaves are usually more densely covered with hair than older leaves. This is especially true of many perennial plants with resting buds that open during cool weather in the spring. The young leaves in the buds of horse chestnut are so hairy as to be hardly recognizable as leaves; but as the leaves emerge from the bud and expand, this hairiness gradually disappears.

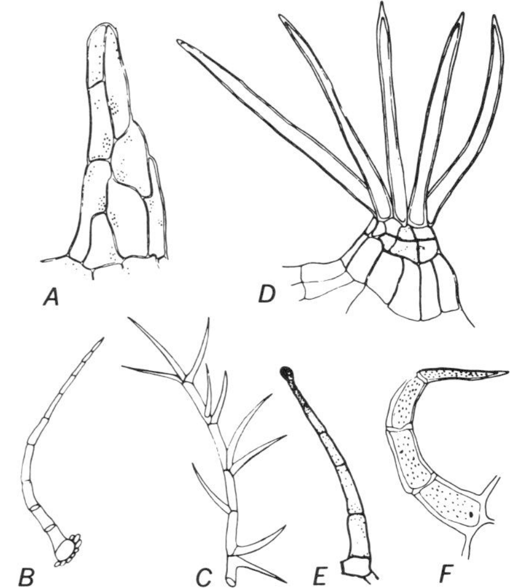

Fig. 4-12 Types of epidermal hairs. A, Begonia; *B,* Coleus; *C,* Verbascum; *D,* Malva; *E,* Petunia; *F,* Pleroma. *(A, B, D, and F drawn by L. J. McConnel and Christian Hildebrandt.)*

Stomata The lower epidermis, and sometimes the upper, is perforated by numerous pores known as stomata (singular, stoma, a Greek term meaning "mouth") (Figs. 4-11, 4-13 to 4-15). Each stoma is a minute opening between two highly specialized epidermal cells called **guard cells.** While the term stoma is applied, strictly, only to the opening between the guard cells, it is sometimes used also to refer to the whole structure comprising the guard cells.

Stomata can best be observed by examining the epidermis in surface view. The lower epidermis of many leaves can be peeled off readily, especially after the leaf has been rubbed for a while between the fingers and thumb. When a section of the epidermis, peeled off in this manner, is examined under a microscope, the stomata stand out in striking contrast to the rest of the epidermis. The two guard cells appear bean or crescent shaped and are attached to each other by the curved ends of their concave sides, leaving a slitlike opening, the stoma proper, between them and presenting together a circular or oval form (Fig. 4-13). In its most expanded condition, the opening between the guard cells is considerably smaller than a pinhole. The smallest visible hole that one could make with an ordinary laboratory dissecting needle in a sheet of paper of the thickness of this page would be about 100 μ ($1\ \mu = 0.001$ mm). In comparison with this, the average stoma is only about 18 μ long and 6 μ wide. By actual measurement, the average area of the opening in 37 kinds of cultivated plants has been found to be 92 sq μ.

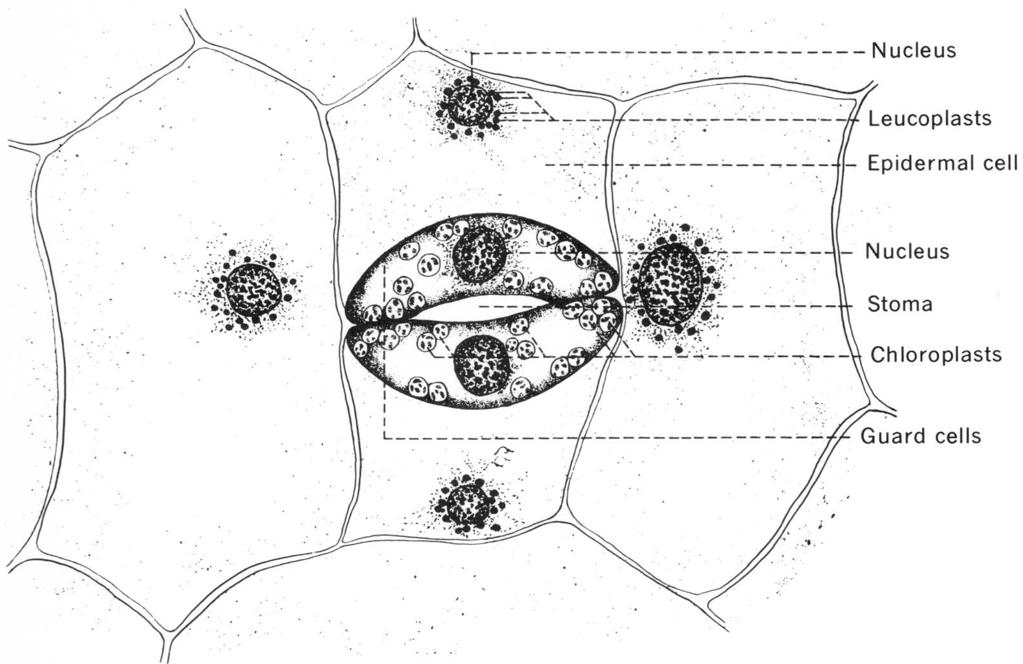

Fig. 4-13 Surface view of a stoma from a leaf of Tradescantia. *(Drawn by Elsie M. McDougle.)*

The walls of the guard cells that border on the stoma are usually thicker than the outer walls (Fig. 4-11). The protoplasm is usually denser than that of the ordinary epidermal cells, and there is always present a rather conspicuous nucleus. Starch is usually found in the protoplasm of guard cells even though it may not normally be present in the other cells of the leaf, as is the case in many monocotyledonous plants. Unlike the ordinary epidermal cells, the guard cells contain chloroplasts and hence appear green.

OPENING AND CLOSING OF STOMATA The stomata are not merely pores in the epidermis but pores capable of being opened and closed. It has already been mentioned in connection with guard cells that the walls that border on the stoma proper are usually thicker than the outer walls. This unequal thickness of the opposite walls, together with the shape of the guard cells and their position with respect to each other, is responsible for the fact that when the guard cells are turgid, i.e., bulged out by the internal pressure of their liquid content, the opening between them is greatest or, in other words, the stoma is wide open. When the guard cells are limp (flaccid), the stoma is closed. The opening and closing of the stomata are thus governed by changes in the turgor of the guard cells.

Several kinds of explanations have been advanced to account for the turgor changes in the guard cells. The oldest of these deals with osmotic changes resulting from carbohydrate transformations in the guard cells. Another deals with hydration and dehydration of the colloids in the guard cells. In these explanations light, the presence of chloroplasts, the concentration of CO_2, and changes in acidity of the guard cells are considered as contributing factors.

The opening and closing of the stomata regulate the exchange of gases between the interior tissues of the leaf and the surrounding atmosphere, this exchange being necessary for respiration and for photosynthesis. Under these conditions the escape of water vapor also takes place, even though it may prove detrimental to the plant. It is important to note that the plant is wholly incapable of closing its stomata in anticipation of wilting. The opening and closing of stomata are entirely chemical and physical processes to which the plant is entirely passive. The diffusion of gases and water vapor through the stomata is further facilitated by the invariable presence of an air chamber in the spongy mesophyll immediately beyond the stoma and communicating with it.

In most thin-leaved dicotyledonous plants of humid regions, the stomata are open all day and closed all night. In the potato the stomata are open continuously except for about 3 hr following sundown. The stomata of cabbage, pumpkin, squash, tulip, and onion also tend to remain open continuously. The stomata of cereals like wheat, oats, and barley are closed all night and only partially open during the day.

Fig. 4-14 Cross section of a portion of a barberry leaf showing a stoma in communication with an air space. (Drawn by J. G. Bechtel.)

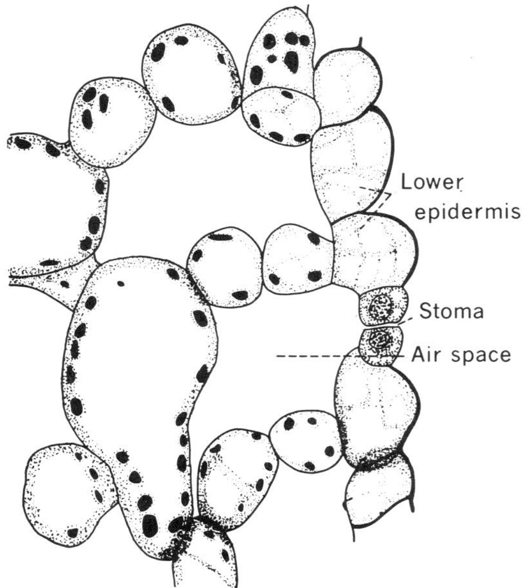

DISTRIBUTION OF STOMATA In many leaves the stomata are confined to the lower surface. In others they occur on both surfaces, and in the floating leaves of water plants they may be present only on the upper surface. In number, they average around 100 to 300 per square millimeter of leaf surface (Fig. 4-15), but as many as 1,400 per square millimeter have been found in mature leaves of oak. When it is realized that a square millimeter is an extremely small area, slightly less than the area of the end of the lead in a lead pencil, it becomes obvious that stomata are numerous. The actual numbers of stomata found in the leaves of some common plants are given in the accompanying table.

Average Number of Stomata per Square Millimeter of Leaf Surface

plant	*upper surface*	*lower surface*
Common pea	101	216
White water lily	460	0
Olive	0	625
Sunflower	175	325
Norway maple	0	400
Black walnut	0	461
Tiger lily	62	62
White pine	142	0
Pumpkin	28	269
Wheat	33	14

It should perhaps be mentioned that stomata are not confined to leaves but occur on all parts of the plant that have a functioning epidermis, except roots.

The mesophyll Between the upper and the lower epidermal layers is the important region of the leaf known as the mesophyll. The cells lying next to the upper epidermis are roughly

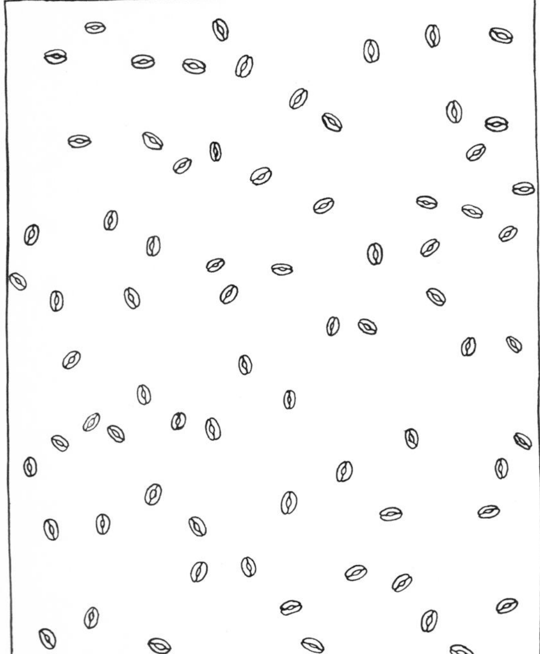

Fig. 4-15 Diagram illustrating distribution of stomata on a geranium leaf; dimensions of the section, 576 by 800 μ, or about 0.46 sq mm.

rectangular in shape as seen in the cross section of the leaf, with the long axes of the cells at right angles to the surface of the leaf. These cells are arranged, palisadelike, in one or more rather compact layers or rows; hence they have received the name **palisade mesophyll** (Fig. 4-11). What air spaces there are between these cells are usually very small. The cells are filled with chloroplasts, which give the upper surface of the leaf a much darker green color than the lower.

The cells of the lower part of the mesophyll are very irregular in shape and much less compact. There are large and numerous intercellular spaces extending throughout this region, giving it a rather spongy appearance. Hence it has been called the **spongy mesophyll** (Fig. 4-11). The intercellular spaces of the spongy mesophyll are in direct communication with the stomata of the lower leaf surface and hence with the surrounding air. By this means almost every cell of the mesophyll can receive carbon dioxide and oxygen directly from the atmosphere or pass off these gases to the atmosphere. The cells of the spongy mesophyll also contain chloroplasts, but these are seldom as numerous as they are in the palisade mesophyll.

All the ordinary cells of the mesophyll belong to the type of tissue known as parenchyma. Since the cells of the mesophyll also contain chloroplasts, they are sometimes called **chlorenchyma.** The chlorenchyma cells of the mesophyll are the chief carbohydrate-manufacturing cells of the plant.

The description thus far given for the mesophyll applies chiefly to ordinary dicotyledonous plants. Even in these the structure is sometimes quite different from the one described. Some of these differences are characteristic of particular species of plants, and others are caused by the environment in which the plant grows. In general the mesophyll tissues are much more compact, with fewer air spaces, when plants are grown in direct sunlight. Under these conditions there may be two or three palisade layers. In the shade, on the other hand, the palisade mesophyll may appear more like the spongy mesophyll. The monocotyledonous plants as a whole, and the grasses in particular, often present an entirely different appearance in cross section (Fig. 4-16). In the grasses there is no well-defined differentiation of the mesophyll into palisade and spongy tissue. The cells are more or less compact and angular throughout, with air spaces occurring only next to the stomata, but these occur on both the upper and the lower surfaces of the leaves. Many other variations occur.

The veins Standing out rather conspicuously in the spongy mesophyll tissue, as seen in cross section, are the veins of the leaf (Figs. 4-11, 4-17*A*). The veins are the terminals of a vascular system which extends throughout the plant from root to leaf and functions in the conduction of water, minerals, and organic substances and in strengthening and support. The veins, or vascular bundles, of the leaf blade are, in general, restricted to the spongy mesophyll, but the largest ones, like the midribs, may extend from the upper to the lower epidermis. In netted-veined leaves the veins branch again and again and in all directions. Consequently, in a cross section of such a leaf, some of the veins will be cut lengthwise and others crosswise, with all gradations between. Since the veins become progressively smaller as they branch out in the mesophyll tissue, the size of the different sections will also vary. Cross sections of leaves of grasses, in which the veins are parallel, will show only cross sections of the veins (Fig. 4-16*A*).

The largest veins usually consist of three dis-

Fig. 4-16 Leaves of monocotyledonous plants in cross section; A, corn (Zea mays); *B, Kentucky bluegrass* (Poa pratensis). *(Photomicrographs by Dr. D. A. Kribs.)*

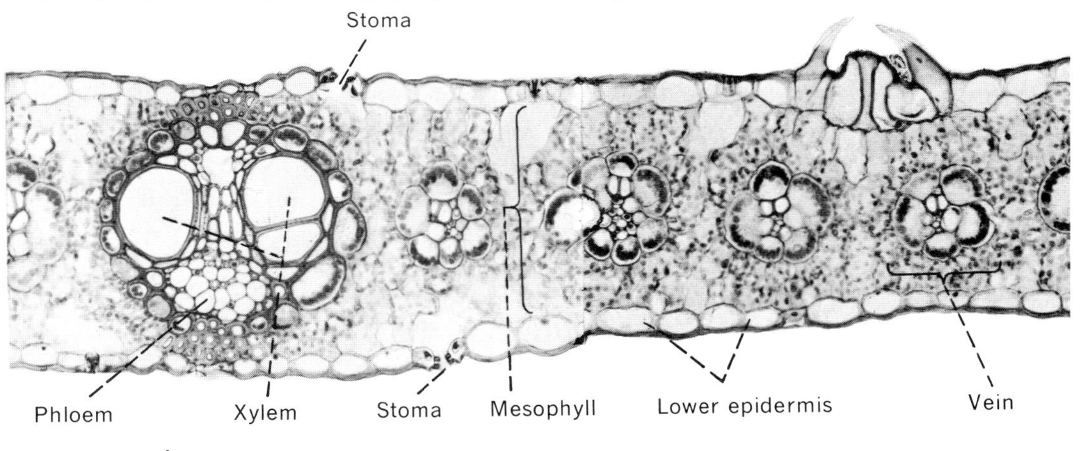

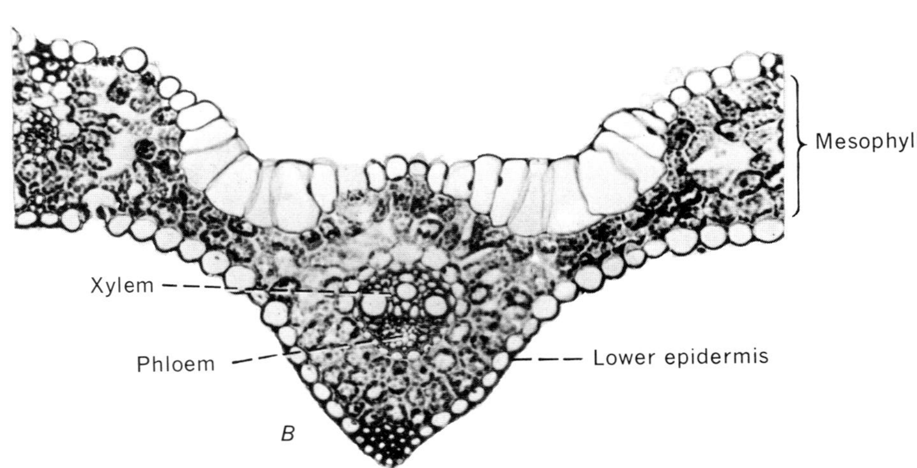

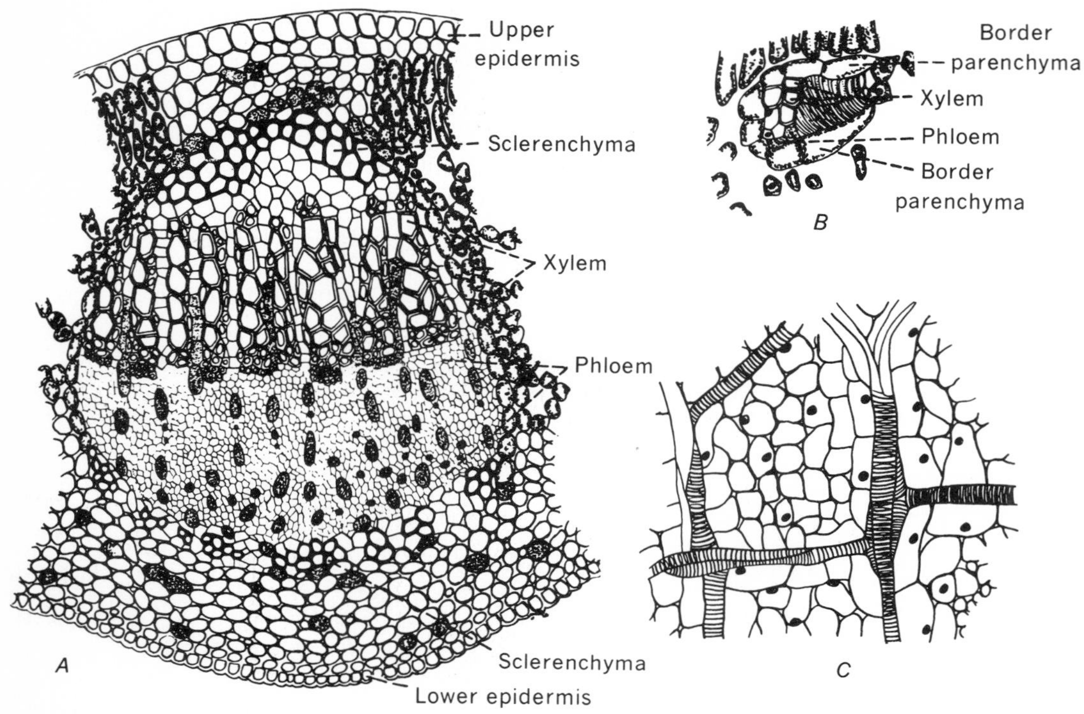

Fig. 4-17 A, cross section through one of the larger veins of a pear leaf; B, cross section through a smaller vein of the same leaf; C, section through a leaf of tobacco cut parallel to the surface showing veins in lengthwise view. (Drawings by Helen D. Hill.)

tinct tissues, an outer **sclerenchyma sheath;** a group of water-conducting cells, called **xylem,** lying toward the upper surface of the leaf; and **phloem,** or food-conducting tissue, lying toward the lower surface (Fig. 4-17*A–C*).

The **sclerenchyma sheath** consists of rather thick-walled, fibrous cells that give strength to the vascular bundle. This sheath may completely surround the bundle or it may be restricted to the upper and lower sides of the bundle. It is only the main veins that have a sclerenchyma sheath. In the smaller veins the sheath consists of colorless parenchyma cells usually referred to as **border parenchyma** (Fig. 4-17*B*).

The **xylem** of the vein consists of elongated cells with internal, often spiral or reticulate, thickening (Fig. 4-17*C*). These are the principal water-conducting cells and have much the same structure as they do in other parts of the plant. The individual cells are joined end to end, to form long tubes called tracheae or vessels. The xylem of the main veins also consists of fibers and parenchyma cells. As the veins become smaller by branching, the size of the xylem decreases. Finally, a single vessel, or tracheid, may be found ending in the mesophyll tissue.

The **phloem** of the bundle, like that found in other parts of plants, usually consists of sieve tubes, companion cells, and parenchyma cells. It is the sieve tubes that transport the foods out of the leaf as they are manufactured. As the veins become progressively smaller, the phloem tissues are reduced so as to consist ultimately of only parenchyma cells.

The veins are so thoroughly distributed throughout the spongy mesophyll tissue that any one manufacturing cell is, at most, only a few cells away from its water supply and is also in more or less direct communication with the phloem.

The petiole The internal structure of the petiole appears, in cross section, more like that of a young stem (Fig. 4-18). There is an epidermis around the outer portion and a large mass of fundamental or parenchyma tissue within, traversed by one or more strands of vascular tissue. Usually there are several separate vascular bundles arranged in a circle or in a semicircle as seen in cross section. These bundles consist of a sclerenchyma sheath, xylem, and phloem. The xylem is situated toward the upper side of the petiole and the phloem toward the lower side. The vascular bundles are continuous with those of the stem and also join those of the leaf blade. Often there are groups of sclerenchyma cells just inside the epidermis. These, together with the sclerenchyma around the bundles, serve to strengthen the petiole.

Abscission and leaf fall In temperate regions the leaves of plants sooner or later cease functioning and die. In some plants the dead leaves remain on the plant until they decay and disintegrate, but in many woody plants, like the common broad-leaved trees, all the leaves are regularly shed in autumn. This is brought about by the development, at the base of the petiole where it is attached to the stem, of a special layer of cells called the **abscission layer** (Fig. 4-19). This layer usually starts to develop long before the leaves actually fall. It consists of a band of parenchyma cells extending all through the basal region of the petiole except through the vascular tissues.

As the season progresses, the walls of the cells of the abscission layer become softened and gelatinous by chemical action. This causes them to separate from one another until finally the leaf is held only by the vascular tissue. Sooner or later, the vascular tissue also becomes ruptured by the swaying of the leaf in the wind or by the action of frost, and the leaf falls. The action of frost in this connection may be observed on bright, sunny mornings following cold, frosty nights. As the sun melts the ice formed where the abscission takes place, the leaves often fall in showers. In some oaks and in other plants it is not uncommon to find some of the dead leaves remaining on the tree over winter. This may be caused by the failure of the vascular strands to break or by the failure of the abscission layer to develop fully.

Along with the abscission layer there is usually developed, between it and the stem, a protective layer consisting of corky (lignosuberized) cells. This layer may develop before the leaf falls or later. In some cases it is formed by transformations in the walls of cells already present and in others by cell division, giving rise to a corky tissue similar to that found in the outer portion of the stem. In any case this corky layer tends to heal over the scar left by the falling leaf.

Plants that regularly shed all their leaves at the end of the growing season are known as deciduous plants. Those that retain them throughout the year are known as evergreen plants. The leaves of such evergreens as the rhododendrons, some laurels, pines, spruces,

Fig. 4-18 Diagram of a cross section through a petiole of a potato leaf. (Drawn by Helen D. Hill.)

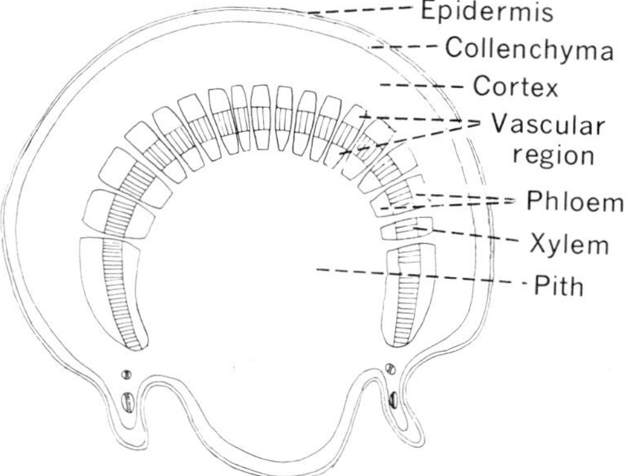

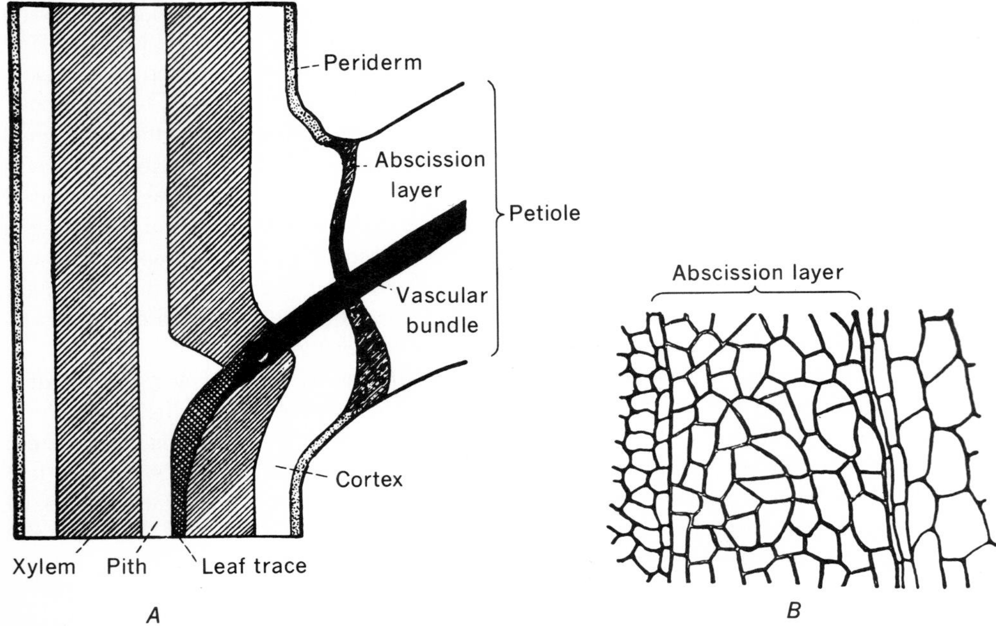

Fig. 4-19 A, diagram to show leaf abscission layer; radial section through twig and leaf base in Juglans cinerea. *The abscission layer extends through the vascular bundle only in parenchyma cells; all other cells are broken mechanically. B, detail of cellular structure of a small part of the layer three weeks before leaf fall. (From A. J. Eames and L. H. MacDaniels, "Introduction to Plant Anatomy," McGraw-Hill Book Company, New York, 1925.)*

and most other conifers often persist from three to four years or more. Even in these plants there is a more or less periodic fall of the older leaves. The white pine, for example, sheds rather abundantly in late autumn, but the leaves shed are usually those which were produced the second or third year preceding.

PHYSIOLOGICAL PROCESSES OF LEAVES

Certain physiological processes are common to all living cells wherever they occur in the plant. Among these may be mentioned digestion, respiration, and assimilation. Since the leaf consists almost entirely of living cells, it carries on all these processes. Because of its particular structure and position, however, the ordinary leaf carries out certain physiological processes more intensively than others. Most prominent among these are photosynthesis, or the manufacture of carbohydrates, and transpiration, which is the giving off of water vapor by the plant. Photosynthesis is taken up in the following chapter. It may be emphasized here, however, that the leaf is the outstanding center of this important process in the plant. In the present chapter only transpiration is considered.

TRANSPIRATION

Transpiration may be defined as the giving off of water vapor from the internal tissues of living plants. This type of water loss is common to all terrestrial plants. It may take place from any exposed part of the plant, but the structure and the position of the leaves are such that the greatest loss of water usually occurs through them. That such loss actually occurs can be readily demonstrated by placing a potted plant under a dry bell jar. Even when the soil is sealed off with paraffin, so that no water can escape except through the plant, moisture will soon collect on the sides of the bell jar as the inside air becomes saturated and will ultimately run down the walls in rivulets. A bell jar without a plant will not collect condensed moisture in this manner under the same conditions.

The actual amount of water lost by transpiration can be determined quantitatively by weighing at regular intervals a potted plant, the soil of which has been sealed over with paraffin or rubber dam in such a manner that no water can escape except from the plant (Fig. 4-20). When this is done, it will be found that the plant continues to lose weight. It can easily be proved that this loss in weight is almost entirely caused by transpiration. A striking visual method of determining the rate of transpiration is by means of a potometer (Fig. 4-21), in which the loss of water causes a bubble of air to move across a scale.

Cuticular and stomatal transpiration Two types of transpiration are recognized, viz., cuticular transpiration, or the diffusion of water vapor directly through the cuticle of the epidermis, and stomatal transpiration, which takes place through the stomata. The former is usually of less importance, being responsible for less than 10 percent of the total amount of transpiration from leaves under ordinary conditions.

By far the greatest loss of water from the plant takes place through the stomata of the leaves. The structure and mode of operation of the stomata have already been considered. It was stated that the opening of the stomata is necessary for the exchange of gases in photosynthesis and that, when they are open, water is unavoidably lost through them. The loss of water may be so great that transpiration exceeds the rate of water absorption and hence results in wilting. It is for this reason that the broad leaves of plants like the common pumpkin sometimes wilt partially on a hot, sunny day. The plant has no power to stop this loss. Only after wilting has begun do the stomata close, and then only because the guard cells, along with the other leaf cells, have lost their turgidity.

Relation to evaporation Transpiration might also be defined as evaporation from living plant tissues. One essential part of the process is the same as in evaporation, viz., the transformation of water into water vapor. The same factors that influence the rate of evaporation from a nonliving surface have similar effects on transpiration. The difference is that transpiration takes place from living tissues and is therefore influenced by the structure and physiology of the plant, while evaporation is controlled entirely by environmental factors. It is for this reason

Fig. 4-20 Apparatus for determining weight loss resulting from transpiration. (Photograph courtesy of F. D. Kern from "The Essentials of Plant Biology," Harper & Row, Publishers, Incorporated, New York, 1947.)

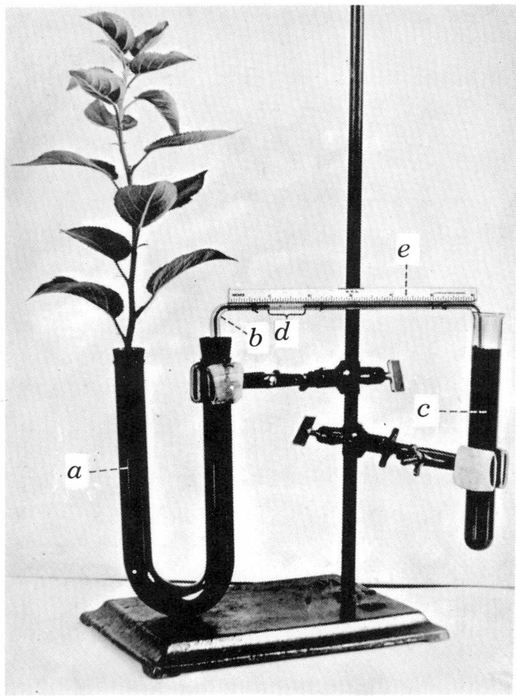

Fig. 4-21 Potometer for studying rate of transpiration. A leafy twig is fastened, by means of a perforated and cut rubber stopper, into one arm of a U-tube, a, filled with boiled and cooled water to which a little thionin has been added. A bent capillary tube, b, passing through a one-hole stopper is forced down into the other arm of the U-tube until the liquid has been forced out of the free end of the capillary tube. As water is lost by transpiration from the leafy shoot, air is drawn into the free end of the capillary tube. As soon as a bubble of air has been drawn into the tube, the free end is lowered into a test tube, c, also filled with boiled water colored with thionin. The rate of transpiration is determined by noting the time required for the bubble, d, to move across the scale, e, fastened to the back of the capillary tube. To force the bubble back to the zero mark of the scale, the capillary tube is pushed down lower into the U-tube. The comparative influence of light, darkness, air currents, and various temperatures can be determined by this method.

that transpiration is considered to be a physiological process.

When considered on the basis of leaf area involved, transpiration is always considerably less than evaporation from an equal area. Thus it has been found that under the same conditions the rate of transpiration from a sunflower plant at night is only about one fourth of the rate of evaporation from a free surface of water and during the day about three fourths. These values vary considerably with different plants.

Conditions affecting the rate of transpiration The rate at which a plant gives off water vapor is not constant but fluctuates with the condition of the surrounding atmosphere as well as with a number of conditions operating within the plant itself. These two types of conditions may be referred to as external factors and internal, or plant, factors, respectively.

External factors The most important of the external factors affecting the rate of transpiration are radiant energy, humidity, temperature, air currents, atmospheric pressure, soil factors, and films of dust and other materials on leaves.

RADIANT ENERGY The source of radiation for plants growing out of doors consists of direct and reflected radiation from the sun. This radiation consists not only of visible rays, or light, but also of invisible ultraviolet and infrared rays. It is the visible radiation, or light, that causes the stomata to open, thereby exposing the saturated interior cells of the leaf to the atmosphere and greatly increasing the rate of transpiration. But in addition to this, transpiration is still further increased because many of the cell constituents absorb radiant energy and transform it into heat, which is effective in increasing the vaporization of water. Such absorption is not restricted to visible radiation but takes place also in the ultraviolet and especially in the infrared.

Of the total radiant energy that falls on a

leaf, about 10 percent is reflected, 10 percent is transmitted through the leaf, and about 80 percent is absorbed. Some of the absorbed energy is used in photosynthesis and perhaps in other ways, and some is reradiated from the leaf, but a large amount is dissipated in transpiration. It is for this reason that the rate of transpiration is closely correlated with changes in intensity of the incident radiation. Since the intensity of daylight gradually increases from morning to noon and falls again from noon to night, we should expect a corresponding increase in transpiration rate from morning to noon and a decrease from noon to night. Such is actually found to be the case. The maximum rate, however, usually occurs between 1:00 and 2:00 PM and the minimum rate at night. Transpiration is never so rapid at night as it is on a bright, sunny day. In the cereals the average rate of transpiration at night has been found to be only 3 to 5 percent of the day rate, and in many plants the rate is at least five times as great during the day. The wide variations in intensity of the sun's radiation with cloudiness, altitude, latitude, and season cause corresponding changes in the rate of transpiration. The combined effect of radiation in causing the stomata to open and in increasing the vaporization of water makes it the most important single external factor affecting transpiration.

HUMIDITY The amount of water vapor in the atmosphere immediately surrounding the plant is also an important factor influencing water loss. Transpiration is to be regarded as a diffusion process in which water vapor passes from a region of high concentration (the intercellular spaces of the leaf) to a region of lower concentration (the outside air). The rate of such diffusion depends directly upon the relative concentrations of the diffusing material. The air in the intercellular spaces of the leaf is usually near the saturation point. If the external atmosphere is nearly or quite saturated, as on damp, foggy days, little transpiration will occur. In general the rate of transpiration varies inversely as the humidity of the atmosphere, the effect of humidity in this case being the same as it is with evaporation. It has been found, however, that some transpiration still occurs even in a saturated atmosphere. This is probably because the internal temperature of the leaf may be a few degrees higher than that of the surrounding atmosphere.

TEMPERATURE The effect of increased temperature is to increase transpiration, not only because it hastens the purely physical process of transforming water into water vapor but also because air at a high temperature is capable of holding more moisture than cold air. When the temperature of the atmosphere is increased, therefore, its water-holding capacity is increased and thus transpiration increases. Low temperatures, conversely, lower the amount of transpiration, other factors being constant. It is thus obvious that temperature effects are closely correlated with effects of humidity. The importance of radiant energy in tending to increase the internal temperature of the leaf has already been mentioned. Plants, unlike higher animals, do not maintain a constant temperature but tend to acquire the temperature of their surroundings. Any increase in the temperature of the atmosphere, therefore, tends to increase the temperature of the plant and hence increases the rate of transpiration.

AIR CURRENTS Air currents have the same significance in transpiration as in evaporation from a free water surface. When the air immediately surrounding the transpiring surface is being constantly renewed by air currents, transpiration will be increased. If not so removed, this air becomes more saturated with water vapor, the difference in humidity between the external and the internal air becomes less, and transpiration is lessened. It has been found, however, that moderate air currents are more effective than a strong wind.

ATMOSPHERIC PRESSURE The barometric pressure of the atmosphere affects the rate of transpiration because it influences the rate of vaporization of water. For this reason evaporation of water is more rapid in a vacuum than it is in air. In general the higher the pressure, the lower will be the rate of transpiration. This factor is of minor importance in any given locality but is quite effective in a comparison of transpiration rate at high altitudes with the rate at low altitudes. In high altitudes the pressure is low. Since there is likely to be also a low humidity and a high radiation intensity at high altitudes, the rate of transpiration under such conditions is often high. It is interesting to note that plants growing in high altitudes are usually dwarfed and often have special structures that tend to reduce water loss.

SOIL FACTORS Certain soil conditions, such as water content, concentration of the soil solution, composition, and temperature, indirectly influence the rate of transpiration because they affect the rate of water absorption by the plant. If the water content is very low or if the concentration of the soil solution is too high, water cannot be so readily absorbed by the plant. This will ultimately result in decreased transpiration. On the other hand, transpiration rate reaches a maximum value in a soil having only enough water present for good tilth. A greater amount of water in the soil has little if any effect. When the water content of the soil is so low as to cause the plant to wilt, the guard cells of the stomata, along with the other cells of the leaf, lose their turgidity and the stomata close. Once the leaf has started to wilt, therefore, the transpiration rate falls.

Films of dust or of spray materials on leaves in general tend to increase the amount of transpiration. Often the increase is greater at night than in the daytime.

Internal or plant factors As previously stated, the rate of transpiration is also influenced by structural and other features of the plant. It is this fact that makes transpiration different from ordinary evaporation. Because of the great differences in general structure among different species of plants, wide differences in the rate of transpiration between species may occur, even when they are under identical external conditions. Structures that are effective in reducing water loss are usually found best developed in plants growing in dry habitats. Such plants are called **xerophytes.** Many swamp plants also have xerophytic structures because of the fact that the water in a swamp, for various reasons, is not readily available to the plant. Xerophytic structures, while often characteristic of a species, are at least partly a result of the external conditions under which a plant grows. The plant has no ability to develop such structures in anticipation of their need.

SIZE, ARRANGEMENT, AND POSITION OF LEAVES The size, arrangement, and position of the leaves of a plant have much to do with the rate of water loss from the plant. Shade plants usually have broad, thin leaves. The rate of transpiration from such leaves is usually high, but this does not cause wilting because of the abundant supply of water ordinarily available in shaded places. The same species of plant growing in direct sunlight often develops a much-reduced leaf surface from which the absolute amount of water lost in a given time is considerably lower. Many shade plants, however, are not able to survive in direct sunlight. Plants that are commonly found in dry situations (xerophytes) practically all have a reduced leaf surface. In many of the cacti the leaves are mere spines. A less modified type of xerophytic leaf occurs in pines and spruces. The leaves of these plants are needlelike in form. A similar reduction in exposed leaf surface is often brought about by the rolling or folding of the leaves. This is frequently seen in the corn plant on hot, windy days or in the folding together of the leaflets of the clover plant under similar conditions. Such positions of leaves tend to reduce transpiration. The same thing results from the edgewise position of leaves toward the sun that is found in the so-called compass plants.

LEAF STRUCTURE The general structure of the leaf in particular, and of other plant organs as well, influences the rate of transpiration. The epidermis is particularly effective in cutting down transpiration, since, with its cuticle, it presents an almost impassable barrier against water loss. Removal of the epidermis results in rapid wilting. In plants exposed to conditions conducive to excessive transpiration, a double layer of epidermal cells, one or more layers of hypodermal cells, or a thicker cuticle is often developed, particularly on the upper leaf surface. Evergreen plants usually have highly cutinized leaves, and the same is true of arctic and alpine plants. Submersed water plants, on the other hand, develop a very thin epidermis with no cuticle at all. Such plants wilt immediately on being exposed to air. In some cases waxy, resinous, or hairy coverings are found over the epidermis. These also tend to cut down transpiration. In stems waterproof suberin takes the place of cutin in the outer corky layers of the bark.

The number of air spaces in the leaf and the general compactness of the mesophyll are also important structural features. In general the looser the structure, the more readily is water lost from the leaf because of the greater transpiring surface. Xerophytic leaves usually have a very compact mesophyll.

STOMATA The importance of stomata in transpiration has already been indicated. The number, distribution, structure, and condition of the stomata all have their effects on the rate of water loss through them.

The fact that the stomatal openings are slit-shaped and relatively small and numerous increases the rate of diffusion through them. This has been demonstrated in a purely physical manner by the use of artificial, perforated membranes. When the pores are very minute and sufficiently close together, it has been found that the velocity of flow of a gas through them varies inversely as the radius or diameter and not as the area of the pore. In other words, the velocity of flow of water vapor through such pores increases as the diameter of the pore decreases. For this reason, it is conceivable that, with the size and distribution of stomata often found in leaves, the rate of water loss, when the stomata are open, might be almost as great as though there were no epidermis at all present. In any case the actual losses that have been found are much greater than the areas of the epidermis occupied by stomata (1 to 3 percent) would lead one to expect to find.

The actual number of stomata per unit of leaf area, as has been mentioned on a previous page, varies in different plants. Obviously, the greater this number, the more readily can water vapor escape when the stomata are open. Leaves with stomata on both the upper and the lower surfaces will lose water more rapidly than those in which the stomata are restricted to the lower surface. Frequently the stomata are found in furrows or depressions in the leaf surface as in the pines and other plants (Fig. 4-22). This allows a pocket of saturated air to be retained just outside the pore and thus checks transpiration.

The condition of the stomata, i.e., whether they are open or closed, is obviously very effec-

Fig. 4-22 Stomata sunken in cavity in leaf of Nerium oleander. *(Drawn by W. L. Dennis.)*

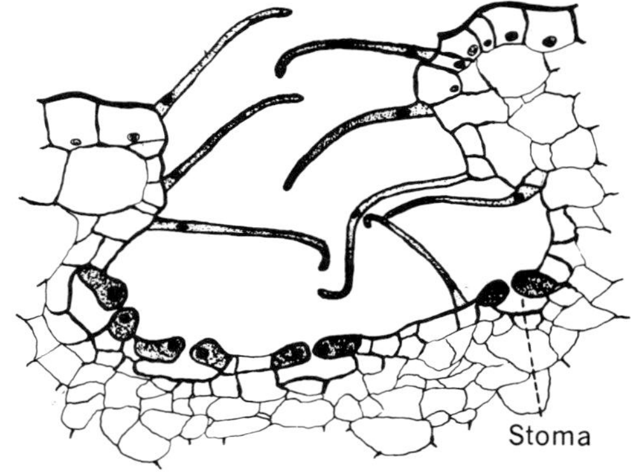

tive. It is chiefly the closing of the stomata in darkness that causes the transpiration rate to be less at night. The closing of the stomata when wilting begins also checks further water loss. It should be emphasized again that the opening and closing of stomata are governed entirely by physical and chemical conditions and that the plant has no control over them. In fact, it frequently happens that at times when the loss of water might prove most harmful to the plant, the stomata are wide open.

WATER-RETAINING ABILITY OF TRANSPIRING TISSUES The water-retaining ability of the tissues of the leaf and of other transpiring organs is a very important internal factor which influences water loss. In all living plant tissues there are found hydrophilic colloids, i.e., substances that absorb and hold water by imbibitional force. Some of the water held by these colloids is bound water, which, unlike the free water in the tissues, cannot easily be removed. The degree to which such colloids develop in the plant has much to do with the rate and ease of water loss from the tissues and hence with the ability of the plant to withstand drought. The development of hydrophilic colloids is determined partly by inherent characteristics of the plant and partly by external conditions. Some plants fail to develop them under any conditions. Such plants are usually killed by drought, through excessive transpiration. Other plants will develop them if gradually subjected to drier and drier conditions and hence are drought resistant.

The selection of drought-resistant varieties of crop plants is very important in semiarid regions. The best development of hydrophilic colloids is found in many desert plants, such as the cacti. It is this fact that enables such plants to survive under xerophytic conditions that would prove fatal to ordinary plants.

A similar relationship exists between the water-holding ability of plants and their winter hardiness. The effect of freezing is somewhat similar to that of drying, since in both cases water is removed from the living tissues to the detriment of the plant. The presence of sufficient bound water is probably one of the most important factors that enable plants like wheat, planted in late autumn, to survive the winter and enable other species of plants to live in arctic and alpine regions.

AGE AND MATURITY OF THE PLANT The rate of transpiration varies as the plant matures. In annual crop plants it begins at a slow rate in the seedling, gradually rises to a maximum a little beyond the middle of the growth period, and then gradually decreases until the plants are harvested. Probably one half of the total water absorbed by a plant during its growing season is transpired during the development of its maximum leaf area. In corn this occurs during the tasseling and earing period. Corn loses half of the total water lost during the life of the plant in a 5-week period embracing this time of tasseling and earing. It has been estimated that during a 10-day period of maximum transpiration (usually in July) our annual crop plants lose about one fourth of the total water lost during the whole season. Naturally this is a very critical period in the life of the plant. If the weather is dry during this period of maximum transpiration, it greatly reduces growth and seriously curtails yields.

DISEASE AND INJURY Finally it may be mentioned that disease and injury often increase the rate of transpiration by exposing, to drying out, tissues that normally are protected. For this reason leaves attacked by fungi or insects frequently wither and die. In some diseases, bacteria or fungi may clog the vascular tissues and thereby interfere with the water supply to the leaves. In other cases chemicals and toxic substances may be produced by a parasite, or structural changes may occur in the parasitized tissues, which result in a changed rate of transpiration.

Amount of transpiration From the previous paragraphs it is clear that the amount of transpiration varies greatly not only in different kinds of plants but also in the same plant at different times and under different conditions. It is therefore impossible to state precisely how much water a plant will lose unless we actually determine the loss under controlled conditions. We can, however, get a general idea of the magnitude of the loss by examining the data obtained by different workers with different plants. In the table below are given the rates of transpiration of fruit trees and several other species, as determined by the potometer method, using only leafy shoots.

The figures in this table seem small because the water loss is figured on the basis of a square inch of leaf surface. It is only when the figures are calculated on the basis of the loss per plant or per acre of plants that the great magnitude of the water loss becomes evident. Thus, if we assume that a mature apple tree has 100,000 leaves, that each leaf has an area of 5 sq in., and that there are 40 trees to an acre, the average rate of 0.0306 cc per square inch of leaf surface per hour, given in the table for the Grimes variety, amounts to 15,300 cc, or about 4 gal per tree per hour, 96 gal per tree per day, 2,880 gal per tree per month, and 480 tons of water per acre per month. It should be borne in mind, however, that these figures are only approximate, since the rates were determined for leafy shoots removed from the tree. In any case the loss is very great. Similar large losses have been reported for other plants. Thus grass plants have been found to lose 6.5

Transpiration Rates of Trees and Shrubs with Edible Fruits or Nuts

common name	*genus and species*	*variety*	*transpiration rate, cubic centimeters per square inch per hour*		
			day	*night*	*average day and night*
Apple	*Pyrus malus*	Grimes	0.0341	0.0256	0.0306
Pear	*Pyrus communis*	Kieffer	0.0320	0.0066	0.0217
Peach	*Prunus persica*	Elberta	0.0048	0.0042	0.0045
Cherry	*Prunus cerasus*	Montmorency	0.0183	0.0081	0.0141
Quince	*Cydonia oblonga*	Champion	0.0350	0.0186	0.0242
Gooseberry	*Ribes grossularia*	Poorman	0.0161	0.0094	0.0117
Currant	*Ribes vulgare*	Perfection	0.0452	0.0233	0.0308
Blackberry	*Rubus nigrobacchus*	Taylor	0.0434	0.0253	0.0315
Raspberry	*Rubus occidentalis*	Quillen	0.0325	0.0212	0.0252
Pecan	*Carya pecan*	Warrick	0.0069	0.0064	0.0066
Hazelnut	*Corylus maxima*	Daviana	0.0347	0.0237	0.0275
Black walnut	*Juglans nigra*	Ten Eyck	0.0057	0.0049	0.0052

NOTE. For first four, day = 5:00 AM to 7:00 PM; night = 7:15 PM to 5:00 AM. For last eight, day = 2:00 PM to 7:30 PM; night = 7:30 PM to 6:00 AM. For last eight, relative humidity = 45 to 60 percent during day and 61 to 65 percent at night and temperature 75°F during day and 68 to 73°F at night. Measurements made in July, 1927, weather "clear." Data from Victor W. Kelley, Univ. Illinois Agr. Expt. Sta. Bull. 341, 1930.

tons of water per acre daily during the summer. A single sunflower or corn plant has been found to transpire 440 lb of water during the total vegetative period.

Sometimes a comparison is made between the total amount of water transpired and the total dry weight of the plant. The value obtained by dividing the weight of the water transpired by the dry weight produced is called the **water requirement** of the plant. Stated in other words, it is the number of pounds of water used by the plant in producing 1 lb of dry matter. This value varies greatly in different plants and in the same plant under different conditions but usually has a value of 200 to 500 for most crop plants growing in humid regions. The accompanying table gives some of the values actually found for different plants.

Water Requirements of Plants

(In units of water transpired for each unit of dry matter produced)			
Millet	310	Pumpkin	834
Corn	368	Cotton	646
Wheat	513	Alfalfa	831
Oats	557	Red clover	453
Potatoes	636	Pigweed	287
Cucumbers	713	Ragweed	948

Is transpiration useful or harmful to the plant? The enormous loss of water resulting from transpiration has led many plant physiologists to doubt whether this process is at all beneficial to the plant. Certainly many plants perish from excessive water loss by transpiration during hot, dry periods. It is true that transpiration has a slight cooling effect, but actual experimentation has demonstrated that such cooling is insignificant in preventing damage from excessive radiation. Transpiration during hot, sunny days causes a more rapid upward movement of sap in the xylem. Since this is a mass movement, it carries with it inorganic substances in solution, which may facilitate the transport of these substances to parts of the plant, and particularly to the leaves, where they are utilized in organic synthesis. This may explain the fact that leaves usually contain higher concentrations of inorganic substances than do other parts of the plant. While transpiration helps to create the pull that lifts the sap in the xylem, the processes of growth and photosynthesis and other physiological processes that consume water would probably create sufficient pull to satisfy the needs of the plant for water.

In summary it may be stated that transpiration is certainly sometimes detrimental to the plant. That it may be of some use to the plant is also likely. As long as the stomata of the leaves are open during the exchange of gases in photosynthesis, the escape of water vapor from the moist exposed cells of the mesophyll must necessarily occur. The rate at which this occurs and the conditions resulting from it in the plant certainly influence the growth and development of the plant, regardless of what interpretation is placed on the process.

Guttation Transpiration is the giving off of water in vapor form. Water is sometimes exuded from uninjured plants in liquid form. This process is called guttation. Guttation occurs under atmospheric conditions that normally check transpiration. It is particularly pronounced on cool, humid nights following hot days. The water usually forms in drops along the edges or tips of leaves (Fig. 4-23), where it is frequently mistaken for dew. Usually the water of guttation is exuded through special structures of the leaves known as **hydathodes.** Hydathodes may be simple epidermal outgrowths consisting of one or at most a few cells or they may be rather complex, multicellular structures connecting with the vascular system of the leaf and terminating in a so-called water pore resembling a fixed stoma. Guttation also occurs from fungi and other lower plants.

The amount of water given off by plants in

Fig. 4-23 Guttation from leaves of a strawberry plant. (Photograph by R. S. Beese.)

this manner varies greatly with different plants and under different conditions. In some species of *Colocasia* more than 100 drops per minute have been observed. In most plants, however, the rate is much lower. The water exuded contains small amounts of organic and inorganic salts.

SPECIALIZED LEAVES

Some leaves or parts of leaves are so highly specialized and so greatly altered in appearance that upon superficial examination they are not recognized as leaves. Although the best criterion for the identification of a leaf as such is its morphological origin, specialized leaves when mature can usually be recognized by their positions at stem nodes, the presence of axillary buds, or the presence of recognizable leaf parts such as stipules.

Storage organs In certain perennial monocotyledonous plants such as the lily and the onion, food manufactured during one growing season is stored up for use during the next growing season in the leaf bases. The concentric layers of the onion bulb are the greatly swollen, sheathing leaf bases of the long, green cylindrical leaves, attached to a much-shortened stem axis. The green portion of the leaf carries on photosynthesis, and some of the manufactured food is stored in the base of the leaf, which persists as a storage organ after the green portion dies. These fleshy, sheathing leaf bases together with the short stem axis to which they are attached form the onion bulb.

The seed leaves, or cotyledons, of many plants such as the garden pea and the bean also serve as food-storage organs. These are described in more detail in the sections on seeds and seedlings.

Much-thickened green leaves such as those of *Haworthia* and species of *Sedum* have water-storage regions consisting of comparatively large parenchyma cells with big central vacuoles containing hydrophilic colloids.

Bud scales Leaves or parts of leaves in the form of bud scales which protect the developing shoot ensheathed by them are present on many of our woody perennials (Figs. 8-2, 8-4*B*). Bud scales are small, thick, corky structures, often resinous and sometimes hairy. They are usually modified leaf blades, but in some cases petioles, and in others stipules.

Spines or thorns In the common barberry some of the leaves are thorns or spines. They are recognizable as leaves by their position on the stem and by the presence of axillary buds. Moreover, on almost every leafy branch all gradations may be seen from ordinary leaves with bristly teeth along the margins to thorns (Fig. 4-24); this is further evidence that the thorns are really leaves. In some cacti all the leaves are spines and the photosynthetic function is carried on entirely by the stems. In other plants, as on the black locust (Fig. 4-25*A*), the spines represent only the stipules of the leaf.

Fig. 4-24 Leaves of common barberry (Berberis vulgare), *all taken from the same plant, showing various degrees of specialization. Some are ordinary leaves with spiny margins; others are spines.*

Fig. 4-25 Leaves of unusual form. A, pinnately compound leaf of black locust of which the stipules are sharp, tough, persistent spines; B, branch of Smilax *with leaves, the stipules of which are tendrils; C, leaf of garden pea* (Pisum sativum) *with large stipules, s, that carry on much of the photosynthetic activity of the leaf; the tendrils are leaflets.*

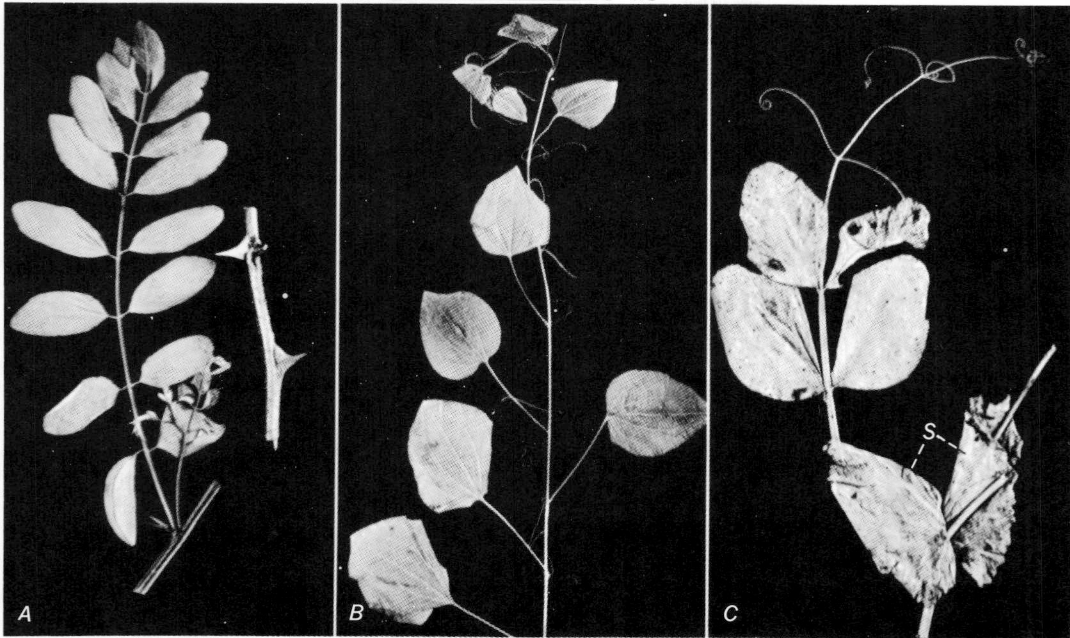

Not all spines are leaves. Those of roses are merely epidermal outgrowths, and those of the honey locust are stems.

Tendrils Tendrils are slender, elongated threadlike structures which either twist around various supporting structures or are provided with small disks by means of which they adhere to supports. Many tendrils are transformed stems, but some are leaves or leaf parts. The terminal leaflets of the garden pea (Fig. 4-25*C*), for instance, are tendrils, while the lower ones are typical foliage leaflets. In a certain European vetch all the leaflets are missing and the end of the petiole develops into a tendril. The stipules on this plant become enlarged and look like leaf blades. In some species of *Smilax*, on the contrary, the stipules are tendrils (Fig. 4-25*B*) and the other parts of the leaf are not modified.

Vegetative reproductive organs The leaves of *Bryophyllum* (Fig. 4-26), if removed from the parent plant and placed on moist soil, are capable of giving rise to buds in the leaf notches. These buds develop into new plants. Certain leaves of the walking fern will take root at their tips and form new plants (Fig. 19-35). These leaves perform the function of vegetative propagation.

Insect traps In conclusion, a few words may be said concerning a group of plants highly specialized in the manner in which they obtain a part of their food. These have been called **insectivorous plants** because they are able, by one means or another, to trap insects. The insects are usually caught in a special structure which is usually a specialized leaf or part of a leaf. By means of enzymes secreted by the specialized leaf structure, the bodies of the insects are digested and the digested products absorbed by the plant.

The so-called pitcher plants (Fig. 4-27*A*) commonly found in bogs belong to this group. These plants are provided with cuplike or pitcherlike leaves containing water, into which insects wander and are prevented, by inwardly directed spines and other devices, from getting out again. The insects drown and gradually disintegrate, and the digested constituents of their bodies are absorbed by the plant tissues.

The common sundew (*Drosera*) (Fig. 4-28*B*) and the Venus's-flytrap (*Dionaea muscipula*) (Figs. 4-27*B*, 4-28*A*) have unique trapping devices. The more or less circular leaf of the sundew is covered with long glandular hairs or "tentacles," at the tips of which is secreted a sticky substance which holds small insects that alight on it. The tentacles then all bend over toward the center of the leaf and hold the insect fast. The outer ends of the leaves of the Venus's-flytrap resemble somewhat a spring trap. Along the

Fig. 4-26 Leaf of Bryophyllum, *showing new shoots developed from adventitious buds in the notches of the margin. (Drawn by Florence Brown.)*

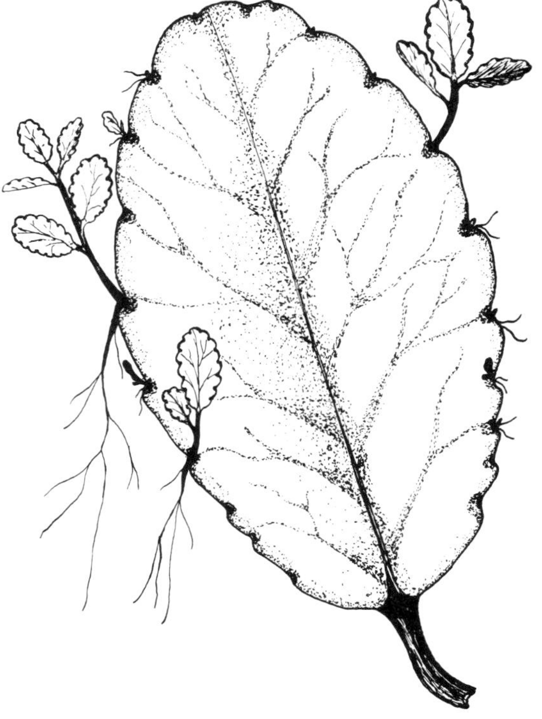

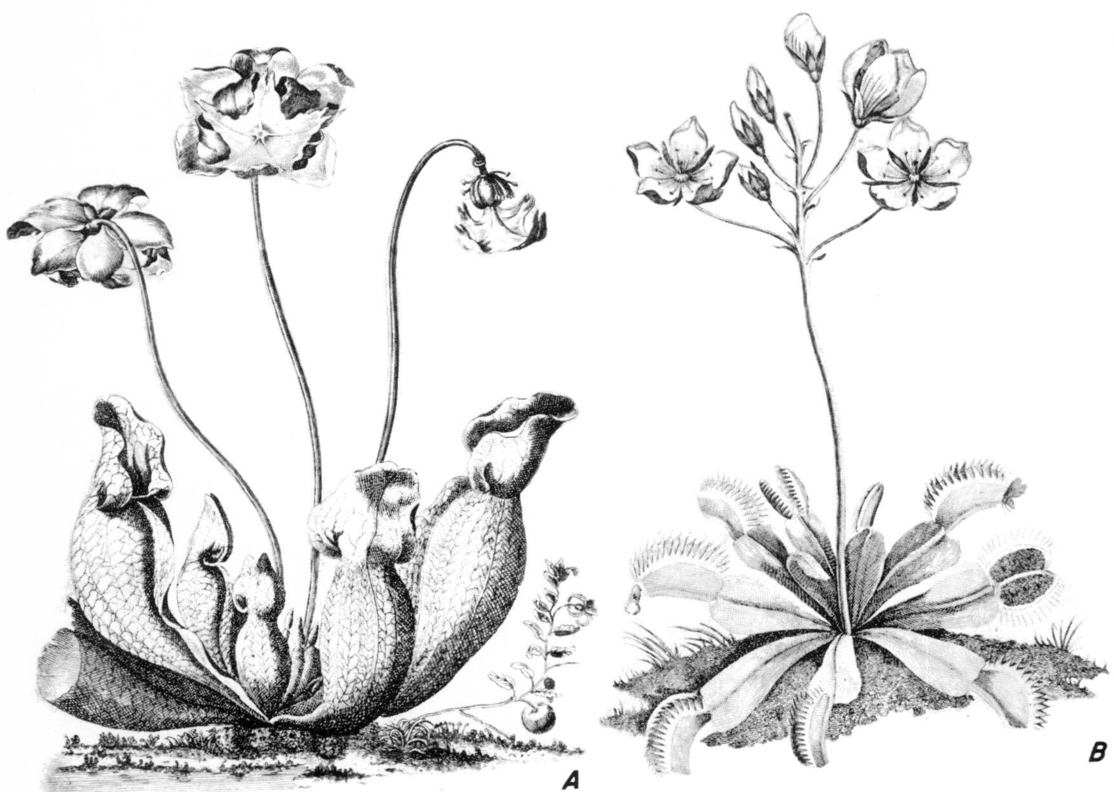

Fig. 4-27 A, pitcher plant (Sarracenia purpurea), *an insectivorous plant found growing in bogs. The leaves consist of hollow receptacles, or "pitchers," which are usually partly filled with water in which insects are trapped and drown. B, Venus's-flytrap* (Dionaea muscipula), *an insectivorous plant commonly found in southern bogs. (These two illustrations were taken from the book "Elements of Botany," by Benjamin Smith Barton, who was one of the early botanists of the United States, a physician, and professor of materia medica, natural history, and botany in the University of Pennsylvania. This book, first published in 1803, was one of the first textbooks of general botany, if not the very first, published in the United States. The illustrations in it were almost all taken from drawings by William Bartram, son of John Bartram, who was one of the most prominent botanists of America during Revolutionary times.)*

outer margin of each half of the leaf blade there is a row of stout teeth, and in the center of each half, on the upper surface, there are three sensitive hairs. When an insect alights on the leaf and touches the sensitive hairs, the two halves of the leaf spring together, folding along the midrib, and hold the insect fast. In both of these plants the digestion of the tissues of the insect and the absorption of the digested products by the leaf proceed apparently as in the pitcher plants.

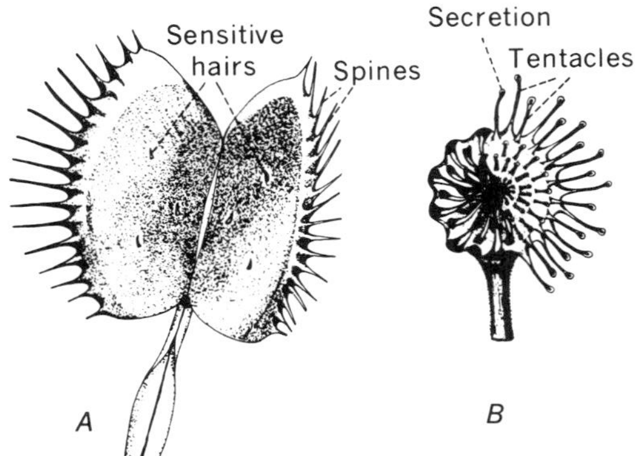

Fig. 4-28 A, leaf of Venus's-flytrap (Dionaea muscipula) *enlarged, showing sensitive hairs and marginal spines; B, leaf of sundew* (Drosera rotundifolia) *enlarged, showing tentacles with a liquid secretion at their tips. The tentacles on the left side of the blade have closed over a small insect. (Drawings by Florence Brown.)*

QUESTIONS

1. How do you distinguish between a simple leaf and a compound leaf?

2. Of what advantage to the plant is it to have a thinner cuticle on the lower surface of the leaves than on the upper surface?

3. Of what advantage to the plant is it to have no chloroplasts in the ordinary epidermal cells?

4. If you were to study the development of leaf primordia in an apple tree, what part of the tree would you examine?

5. Of what advantage is the compact arrangement of the cells in the palisade mesophyll of a leaf?

6. What is the advantage of having stomata only on the lower surface of a leaf?

7. What causes leaves to fall in autumn?

8. What is the difference between transpiration and evaporation?

9. At what seasons of the year is it best to transplant deciduous perennials?

10. What causes the leaves of squash plants to wilt in the hot sun even though there is an abundant supply of water in the soil?

11. What disadvantage is there in growing pumpkins in a cornfield?

REFERENCES

HARDER, RICHARD, WALTER SCHUMACHER, FRANZ FIRBAS, and DIETRICH VON DENFFER. 1965. Strasburger's Textbook of Botany. Translated from the 28th German edition (1962) by Peter Bell and David Coombe. Longmans, Green & Co., Inc., New York. A very comprehensive general textbook of botany which has been a standard reference work through the years, now brought up to date.

KETELLAPPER, H. S. 1963. Stomatal Physiology. *Ann. Rev. Plant Physiol.,* **14**:249–270. A review of recent work on stomata and their operation.

MILLER, EDWIN C. 1938. Plant Physiology, 2d ed., chap. VII, The Loss of Water from Plants, pp. 407–522. McGraw-Hill Book Company, New York. A thorough discussion on stomata and on all the literature on transpiration up to 1938 including the author's own research; includes an extensive bibliography.

ZELITCH, ISRAEL (ed.). 1963. Stomata and Water Relations in Plants. Conn. Agr. Expt. Sta., New Haven, Bull. 664. Papers and discussions given July 1–12, 1963, as part of the Advanced Seminar on the physiology and biochemistry of leaf stomata.

5

METABOLISM; THE ANABOLIC PHASE; FOOD SYNTHESIS

Continuous-algae-culture apparatus used in the study of photosynthesis. (Photograph by Jon Brenneis.)

The nature of metabolism; what constitutes plant food; general characteristics of carbohydrates, fats, and proteins; the anabolic phase of metabolism; photosynthesis; fat or oil synthesis; protein synthesis; and the food habits of parasitic and saprophytic plants are the topics of this chapter.

In all living organisms there is an unceasing procession of chemical reactions and processes which lead to fundamental internal changes. Living substance is constantly being torn down and built up; the general activities of protoplasm are being maintained; waste products are being eliminated, and a vast amount of energy is constantly being expended. The sum total of these processes and changes is called **metabolism.** Since the ultimate seat of this activity is the cell, we may define metabolism as the sum of the processes concerned in the building up and tearing down of protoplasm. This capacity for metabolic activity is one of the features distinguishing living protoplasm from all nonliving substances.

PLANT FOODS

General It is obvious that if an organism is to make new cells or to expend energy in the maintenance of life, it must be supplied with working material. This material is obtained in the form of food. Foods are substances which can be used by the organism either as a source of energy or directly for the building of protoplasm and other structural components of the cell. It is important for the student to realize that the food of the higher plants is exactly the same as the food of animals and consists in every case of organic compounds. The strictly inorganic mineral salts which a plant obtains from the soil and which are added to the soil in the form of fertilizers are often incorrectly called foods. Such substances, while very important to the life of the plant, are used neither as a source of energy nor directly for the building of tissues and therefore are in no sense to be considered as foods. The actual foods of the plant, like those of animals, can be divided into three groups of compounds, viz., **carbohydrates, fats,** and **proteins.** As regards foods, chlorophyll-containing plants differ from animals in that they manufacture their own food within their tissues, while animals obtain their food either directly from plants or from other animals which feed on plants. A knowledge of some of the characteristics of the principal plant foods is essential to an understanding of how these substances are made and used by the plant.

The carbohydrates Carbohydrates are compounds consisting of carbon, hydrogen, and oxygen, the hydrogen and oxygen being usually in the same proportions as found in water. Most of the carbohydrates commonly found in plants contain six, or multiples of six, carbon atoms and a like number of oxygen atoms less one for each group of six carbon atoms above the first. Those that contain only six carbon atoms are called **hexoses** and have the general formula $C_6H_{12}O_6$. Of less common occurrence in plants in uncombined form, and derived usually by the conversion of more complex compounds, are the **pentoses,** which are carbohydrates consisting of five carbon atoms and having the general formula $C_5H_{10}O_5$. Two common pentoses found in plants are *arabinose,* obtained from the gum of cherry trees and peach trees, and *xylose,* obtained from other wood gums and from grains, fruits, and straws. The 5-carbon sugars occurring in combined form, such as *ribose,* found combined in RNA, *2-deoxyribose,* found in DNA, and *ribulose,* combined with phosphate and important in photosynthesis, are of great importance in metabolism.

A carbohydrate containing five or six carbon atoms or fewer is known as a **monosaccharide.** When the molecule is made up of two monosaccharide units, the carbohydrate is called a **disaccharide;** when it consists of three monosaccharide units, a **trisaccharide;** and when it consists of many monosaccharide units, a **polysaccharide.** The monosaccharides and disaccharides are readily soluble in water, have a sweet taste, and are known as **sugars;** the polysaccharides are mostly insoluble in water, are tasteless, and include such compounds as *starch, glycogen, inulin, cellulose,* and *hemicelluloses.* Trisaccharides are less common in plants than other types of carbohydrates. The trisaccharide *raffinose,* $C_{18}H_{32}O_{16}$, occurs in cottonseeds, barley, beet roots, and in other plants.

Of all the monosaccharides found in plants, *glucose* (*dextrose*) and *fructose* (fruit sugar) are the most common. Glucose, $C_6H_{12}O_6$, is probably found in every living plant cell. Many of the other carbohydrates are probably changed to glucose before they are transported from one part of the plant to another or before they are broken down in respiration or built into protoplasm. Glucose and fructose are probably the most common substances used in respiration as well as the basic material out of which many of the more complex chemicals in the plant are manufactured. Fructose, $C_6H_{12}O_6$, has the same empirical formula as glucose but differs from the latter structurally and chemically. It is often found with glucose and is particularly abundant in sweet fruits. It is used by the plant in much the same way that glucose is used.

Sucrose, or cane sugar, $C_{12}H_{22}O_{11}$, is the most widely distributed disaccharide found in plants. It occurs particularly in sweet fruits, in stems, seeds, roots, bulbs, and in the sap of many trees like the maple. It is the common table sugar, the commercial product being obtained almost exclusively from sugarcane and from sugar beets. In the plant it is usually a storage form of carbohydrate. It is readily soluble in water. On being hydrolyzed (i.e., broken down by the chemical addition of water, as in digestion), it yields one molecule of glucose and one of fructose for every molecule of sucrose.

Other disaccharides of less common occurrence in plants include *maltose,* obtained chiefly by the digestion of starch but also found as such in leaves, rhizomes, or roots of some plants, *cellobiose,* resulting from the hydrolysis of cellulose, and *trehalose,* which is found widely distributed in fungi. Each of these disaccharides, though structurally different, yields two molecules of glucose on hydrolysis. The general formula $C_{12}H_{22}O_{11}$ applies to all of them.

The polysaccharide *starch,* $(C_6H_{10}O_5)n$, is the commonest storage form of carbohydrate found in plants. It may be found in any part of the plant but is found in greatest abundance in such storage organs as roots, tubers, and seeds. In the grains of cereals it may form from 50 to 70 percent of the dry weight, and in potato tubers from 15 to 30 percent of the dry weight. It always occurs in the form of small white grains, the starch grains. In storage tissues of the plant these grains begin as small leucoplasts, which increase in size as more starch is laid down in them. The full-sized grain displays the original core, called the hilum, surrounded by rings or striations of different densities showing where additional layers of starch were laid down. In the leaf, starch is deposited in the chloroplasts. Starch grains have characteristic shapes and markings which differ with the species of plant in which they are found (Fig. 12-2).

Starch is made by the plant from glucose, which alone it yields on complete hydrolysis. When heated in water, it forms a translucent paste. It gives a characteristic blue color with iodine solution. Since it is insoluble in water, it must always be changed back to sugar before it can be moved out of the cell or used in any way by the plant. Some plants, and particularly some of the monocotyledons, do not make starch but store their carbohydrates in the form of sugar or inulin.

Glycogen, $(C_6H_{10}O_5)n$, sometimes called animal starch, resembles starch in many of its properties. Like starch, it yields only glucose on hydrolysis. Unlike starch, it gives a red to brown color with iodine solution. It is widely distributed in the animal kingdom but has a restricted distribution in plants, being found chiefly in fungi and especially in yeast. It also occurs in some of the algae.

Inulin, $(C_6H_{10}O_5)n$, differs from starch and glycogen in that it yields only fructose on hydrolysis. It is found in many kinds of plants but is especially abundant in the tubers of dahlia and artichoke and in other composites. It is a common reserve food in these plants.

Cellulose, $(C_6H_{10}O_5)n$, is another polysaccharide of great importance, often comprising about one half of the dry matter of certain parts of plants. Cotton often consists of over 90 percent cellulose. It is the principal material used in building the framework of plants and is found in the walls of all plant cells, giving them a rigidity not found in animal cells. Cellulose has the same basic or general formula as starch but is very much more complex. It is probably the most insoluble of all the carbohydrates. It is not only insoluble in the plant but also more or less indigestible and therefore, except in the case of certain bacteria and fungi, is seldom used as a food. It is probably made from glucose, which alone it yields on hydrolysis. Cellulose is frequently found combined with other substances and especially with lignin, with which it forms *lignocellulose*, the common strengthening material of the cells of wood.

The *hemicelluloses* are a group of compounds of varying composition, resembling true cellulose in their physical properties but unlike cellulose in their chemical properties. Although insoluble in water, the hemicelluloses are readily soluble in alkali. Like cellulose, they are frequently found in the walls of cells, imparting to these walls strength and toughness. Unlike cellulose, they are at times digested by appropriate enzymes in the plant and hence may be used to some extent as reserve foods. On hydrolysis they yield one or more monosaccharides which may be hexoses or pentoses. Hemicelluloses are found in many seeds, especially those of members of the palm and lily families, in the wood and leaves of many trees, and in some fungi. There are several different kinds of hemicelluloses, the distribution of each of which varies considerably in the plant kingdom. Vegetable ivory, obtained from the seeds of a palm and used in making buttons, umbrella handles, and other articles, owes its properties largely to the hemicelluloses of which it consists.

In addition to the carbohydrates already mentioned, many other substances that are commonly classed with the carbohydrates or yield carbohydrates on hydrolysis are found in plants. Common among these are *gums, mucilages, pectins,* and *glycosides.* As these compounds are not extensively used as foods by the plant, they need not be considered here.

Fats and oils Fats and oils, like carbohydrates, consist of carbon, hydrogen, and oxygen, but these elements never occur in the same relative proportions as found in the carbohydrates. There is always less oxygen and relatively more carbon and hydrogen in the fats. The chemical properties are in no way similar. Fats that are liquids at ordinary temperatures are known as oils. In plants the fats commonly exist in the form of oils. Two kinds of oils may be distinguished, viz., *fixed oils* and *volatile oils.* The fixed oils have no particular odor or taste and are the only kinds classed as foods. The volatile oils, as the name implies, are volatile when exposed to air and have very characteristic odors. Common examples of volatile oils are oil of wintergreen, lemon oil, mustard oil, oil of turpentine, and oil of peppermint. The volatile oils have a varied composition which always differs from that of fixed oils. Since the volatile oils are probably never used as plant foods, we need not consider them here.

The *fixed oils* and *fats* have very characteristic properties. They always leave a permanent grease spot or translucent mark on paper. They are insoluble in water, but easily soluble in ether and in chloroform. They are all lighter than water, their specific gravity ranging from 0.875 to 0.970. When treated with sodium hydroxide, NaOH, or potassium hydroxide, KOH, they saponify; i.e., they form glycerol and soap, which is the sodium or potassium salt of a fatty acid. They are all made up of glycerol, CH_2OH—$CHOH$—CH_2OH, and fatty acids. The fatty acids are organic acids having the general formula C_xH_y—COOH. Two of the commonest fatty acids found in vegetable oils are palmitic acid, $C_{15}H_{31}$—COOH, and oleic acid, $C_{17}H_{33}$—COOH. The manner in which these fatty acids unite with glycerol to form fats is taken up later under fat synthesis.

Fats and oils are found in greatest quantity in storage regions of plants and particularly in seeds but are also a necessary part of every living cell. They are much more efficient storage foods than carbohydrates, a given amount of fat yielding 2¼ times as much energy as the same quantity of carbohydrate. This is because of their relatively higher carbon content and relatively lower oxygen content. For the same reason they require a relatively larger volume of oxygen for their complete oxidation than do carbohydrates. They are readily converted into carbohydrates in the plant and are used chiefly as a source of energy in respiration.

The proteins The proteins are among the most complex and most important organic compounds found in plants. Besides being classed as foods, they form an integral part of protoplasm and hence are present in every living cell. The nucleus of the cell contains protein. Proteins are always found in abundance where there is active cell division and growth. When stored as reserve food in such organs as seeds, they serve chiefly as tissue or protoplasm builders in the developing embryo. Besides carbon, hydrogen, and oxygen, they all contain nitrogen and are thus classed as nitrogenous foods. All of them contain, in addition, sulfur and many contain phosphorus. The protein molecules are always very large as seen in the following examples: *gliadin* (a protein in wheat), $C_{685}H_{1068}N_{196}O_{211}S_5$; *zein* (from corn), $C_{736}H_{1161}N_{184}O_{208}S_3$; and *casein* (from milk), $C_{708}H_{1130}N_{180}O_{224}S_4P_4$.

Many different kinds of proteins with various properties are found in plants, but all are composed of chains of simpler compounds, known as *amino acids*. The complete digestion of a protein, therefore, always yields amino acids. The amino acids are organic acids containing one or more basic NH_2 groups, usually in a very definite position in the molecule and one or more acidic COOH groups. A general formula for an amino acid is R—$CHNH_2$—COOH. Examples of those composing the proteins of plants are *glycine*, H—$CHNH_2$—COOH, and *alanine*, CH_3—$CHNH_2$—COOH. More than 25 different amino acids have been isolated. It will be noticed that they all contain the basic NH_2 group and the acid COOH group. Amino acids can react with each other, the NH_2 group of one combining with the COOH group of another, with the loss of water, to form what is called the **peptide linkage** (—CONH—). The proteins are basically long chains of amino acids combined in this manner, although other substances often enter into their composition.

Proteins can be detected in plants by means of various color reactions, common among which are the *biuret test* and the *xanthoproteic acid reaction*. In the biuret test a solution of copper sulfate is added to an alkaline solution of the material to be tested. When protein is present, this gives a bluish-violet color. In the xanthoproteic acid reaction a little strong nitric acid is added, which gives a yellow color to protein material. This color changes to orange on the addition of ammonia. It is this reaction that causes the fingers to turn yellow when they are brought in contact with nitric acid.

The proteins of plants are most often found in colloidal solution in water. Excessive heat or cold, acids, heavy metals, and other agents may cause them to coagulate as the albumen of an egg coagulates on boiling; from this condition a return to the normal is impossible. In storage regions they may exist in small grains, the **aleurone grains,** somewhat comparable with starch grains, though usually much smaller. Aleurone grains are common in seeds. In the cereals, such as wheat, oats, and corn, they are localized in a single layer of cells next to the seed coat. Proteins as such are not able to pass through plant-cell membranes and hence must be converted into amino acids before they can be removed from storage regions to other parts of the plant where they are needed.

The student should remember that the proteins are preeminently the protoplasm- and tissue-building materials of both plants and animals. They may also be used in the synthesis of many other complex substances such as enzymes and secretions.

THE PHASES OF METABOLISM

It has already been stated that metabolism includes all the processes concerned in the building up and tearing down of protoplasm. The building-up processes comprise the synthesis of all the foods as well as protoplasm and are collectively referred to as the **anabolic phase** of metabolism. The tearing-down processes consist chiefly in the conversion of the foods to soluble and usable forms or in the oxidation of these foods, thereby releasing and making available to the plant their contained energy. This is the **catabolic phase** of metabolism. Naturally it is impossible to have the one phase without the other. Both are in operation simultaneously as long as the plant is living. That is, the two phases are absolutely interdependent. For convenience of study, the two phases are considered separately. Under the anabolic phase are considered the synthesis of carbohydrates (photosynthesis); the synthesis of fats and oils; the synthesis of proteins; and, finally, the conversion of these foods into protoplasm itself (assimilation). The catabolic phase is discussed in Chap. 12, under the topics of digestion, respiration, and fermentation.

Types of plants in relation to metabolism According to the method by which they obtain their foods, all plants can be divided into two classes, viz., **autotrophic plants** and **heterotrophic plants.** The former are the self-nourishing or independent plants; i.e., they make within their tissues all the foods they need, obtaining only inorganic materials and some form of energy from their environment. There are two kinds of autotrophic plants, the **photosynthetic** group and the **chemosynthetic** group. The former obtain materials from the soil, carbon dioxide from the air, and energy from sunlight. They are all plants that contain chlorophyll and hence can carry on photosynthesis. All the higher green plants as well as the lower vascular plants, bryophytes, and algae belong to this group.

The chemosynthetic group differs from the former in that the energy they store in foods is obtained from some chemical source instead of light. They need, therefore, to be supplied only with inorganic materials. The carbon they use is obtained from the carbon dioxide of the air as in the former group. The most important members of this group are certain bacteria. None of them contains chlorophyll. Examples are the hydrogen bacteria, some sulfur bacteria, the iron bacteria, and the nitrifying bacteria. The substances which these bacteria oxidize as a source of chemical energy are, respectively, hydrogen, hydrogen sulfide, ferrous iron, and ammonia.

The heterotrophic plants lack the power to synthesize carbohydrates out of CO_2 and water, although most of them can make other foods. They are the dependent plants. In general they must rely on other plants or on dead or nonliving organic material for their organic carbon and must be supplied also with inorganic substances. To this group belong all the fungi and such higher plants as dodder and Indian pipe (Fig. 5-1). The heterotrophic plants in general are plants lacking chlorophyll.

Significance of the synthetic ability of plants It is the almost unlimited power which plants have of making their own foods that gives them so prominent a place in the universe. While it is true that animals can, to a limited extent, synthesize organic compounds from CO_2, utilizing the energy required by oxidizing organic compounds, only the green plants can use the energy from an outside source, the sun, and thus build up the stored-up energy available to all organisms. They do this by making carbohydrates in photosynthesis. Out of these carbohydrates is ultimately built all the other organic material used by both plants and animals. That is, it is the building source of all organic structure, including our own bodies. Likewise, these carbohydrates are our chief source of energy, not only animal, plant, and human energy but also that energy, obtained directly from coal, petroleum, and wood, with which we operate our many industries.

Animals can synthesize some proteins but not all the kinds they need. They must depend upon plants for several of the most important amino acids used in protein building. Plants here again are practically unrestricted. Similarly, animals are very greatly limited in the synthesis of vitamins, the majority of which they must obtain ready-made either directly or indirectly from plants.

This ability of plants to synthesize foods out of inorganic materials makes them the connecting link between the mineral and the organic worlds.

PHOTOSYNTHESIS

In the preceding section it was stated that the food of plants consists of carbohydrates, fats, and proteins and that these foods are manufactured by the plant within its tissues. The purpose of the present section is to consider food synthesis in plants. Since the carbohydrates are the materials out of which fats and proteins are made, their synthesis is considered first. Because light is used by the plant in this process, the name **photosynthesis** (a synthesis utilizing light) has been given to it.

Location of the process Carbohydrates are made by the plant in the chloroplasts. Any part of the plant that contains chlorophyll and is exposed to light can carry on the process. The outstanding organs of photosynthesis, however, are the leaves. All features of leaves seem to be directed toward their efficiency in carrying on this process. The position and the arrangement of leaves on the stem are such as most effectively to enable them to receive light and air. In form and internal structure leaves are ideal for the process. They are relatively broad and thin, thus enabling them efficiently to intercept large quantities of light. On the upper surface they have a transparent cuticle and a transparent epidermis which prevent water loss but permit light to pass. Beneath the epidermis of leaves of dicotyledonous plants is the compact palisade mesophyll filled with chloroplasts and so placed as to absorb a maximum amount of the available light. Then the spongy mesophyll, loosely constructed with many air chambers which communicate through the stomata with the outside air, permits an orderly exchange of gases in the process. The stomata themselves are so constructed as to open in the daytime when photosynthesis can be carried on. The veins, ending in the mesophyll, bring in an abundant supply of water for the process and carry off the food as it is made. It would be difficult to imagine a more ideal structure for photosynthesis. Young stems and other green parts, however, also carry on the process, though not so extensively.

The raw materials From the chemical composition of carbohydrates, it is evident that there must be a supply of compounds containing carbon, hydrogen, and oxygen for their synthesis. All these elements are obtained from just two ordinary compounds, viz., water, H_2O, and carbon dioxide, CO_2. No other compounds obtained from the exterior are used directly in photosynthesis. Ordinarily, terrestrial plants obtain their water from the soil. It is absorbed by the roots and conducted from them through the stem to the leaf. A special tissue, the xylem, is used for this purpose. The xylem vessels are essentially a system of very fine tubes beginning in the roots and ending in the leaves and containing branches to all parts of the plant, so that every cell is in close proximity to its water supply.

The carbon dioxide enters in gaseous form from the atmosphere, gaining entrance to the leaf by diffusion through the stomata. Soils ordinarily contain more carbon dioxide than is found in the atmosphere. Whereas the average amount of carbon dioxide in the atmosphere is 3 parts in 10,000 parts of air, or 0.03 percent, soils may contain as much as 5 percent or more of the gas. The high amounts in soils result chiefly from the respiration of roots and of soil organisms. This gas may be displaced to the surface during heavy rains and thereby increase the concentration of carbon dioxide of the atmosphere near the ground level and in the vicinity of growing plants. It is very unlikely that carbon dioxide can be absorbed by roots and carried up internally to other parts of the plant in sufficient quantity to be of material use in photosynthesis. This is shown by the fact that when carbon dioxide gas is withheld from the aboveground parts of plants, so that the sole source is that absorbed by roots, photosynthesis does not take place in measurable quantity, regardless of how much of the gas is available to the roots.

The method by which carbon dioxide enters the leaf is one of simple diffusion. Air consists of a mixture of several gases. It is one of the laws of diffusion that the rate at which a gas travels from one region to another depends upon the difference in concentration of that particular gas in the two regions, regardless of the percentage of other gases present. Hence the movement of carbon dioxide is independent of that of other gases, or, in other words, it is not necessary for large volumes of air to be drawn into the leaf in order that the necessary amount of carbon dioxide be obtained. The rate of diffusion is always greater from a region of higher concentration toward one of lower concentration. Such being the case, it is obvious that if the interior of the leaf is to obtain a continuous supply of carbon dioxide from the air, the outside concentration must be greater than that inside the leaf. As long as photosynthesis is going on, this condition is met because the carbon dioxide within the leaf is constantly being consumed.

After the carbon dioxide has passed through the stomata and into the air spaces of the spongy mesophyll, it goes into solution in the water in the walls of the chlorophyll-containing cells and then diffuses into the cell cavity where it is immediately taken up by the chloroplasts and used in photosynthesis.

Chlorophyll and photosynthesis The general properties of chlorophyll have already been taken up in Chap. 3. There remains to be considered the part that this pigment plays in photosynthesis. From the fact that no photosynthesis can take place in the absence of chlorophyll, it is obvious that its role is an essential one. Chlorophyll probably takes part in the process in several ways. One of the most important of these is through its absorption of light. White light, as it comes from the sun, consists of a series of rays of different wavelengths. When a beam of light is passed through a prism, each of the different rays suffers a different amount of refraction or bending, the longer (red) rays being refracted least and the shorter (blue-violet) rays most. The result is that a spectrum is formed, part of which is visible to the eye and appears as a band of colors consisting of red, orange, yellow, green, blue, and violet, in that sequence. At the red end of this spectrum, but beyond the visible red rays, there is a series of longer invisible rays called the infrared. Similarly, beyond the violet there is a series of shorter invisible rays called the ultraviolet.

If now we place a tube of chlorophyll in the path of the beam of light before it reaches the prism, certain parts of the spectrum are absorbed by the chlorophyll and a continuous band of colors is no longer obtained, but the absorbed portions appear as dark bands. These dark bands are called the **absorption bands** of chlorophyll. The darkest of them will be in the visible red. Others appear in the yellow and in the blue-violet and ultraviolet. Green is least absorbed and hence chlorophyll has a green color. It is significant to note that this part of the spectrum which is absorbed by the chlorophyll is probably the principal light used in photosynthesis. In other words, chlorophyll enables the plant to absorb light for photosynthesis. Since no decrease has been observed in the amount of chlorophyll in leaves during intense photosynthesis, chlorophyll also functions as a photocatalyst, i.e., a substance which accelerates a reaction in light, appearing unchanged at the end of the reaction.

Relation of light to the process Photosynthesis is strictly an endothermic process, i.e., an energy-storing process. The energy stored is the radiant energy of sunlight which is transformed into a potential form in the resulting carbohydrates. Not nearly all of the radiant energy that falls on plants is utilized in the process. To begin with, only the part that is absorbed can be used. In the previous paragraph it was stated that only certain parts of the spectrum are absorbed. These parts contain only a relatively small part of the total energy contained in the incident light. According to the physicist Langley, who has made a very careful study of the energy of sunlight, the distribution of the energy in the spectrum is as follows:

spectral region	*percent of total energy*
Infrared	62 to 63
Visible	37.0
Ultraviolet	0.6

Since none of the infrared is used in photosynthesis, it is clear that 62 to 63 percent of the total energy has no value in the process. Maximum absorption occurs in the red part of the spectrum, and the next largest amount in the blue-violet region. Green is not absorbed to any great extent. The total amount of energy utilized varies with different plants. The thicker the leaf, the more light it can absorb, but in general only about 1 percent of the total energy incident on the leaf is used in photosynthesis. This seems an insignificant amount; yet it is sufficient to supply practically all of the energy used by man, animals, and plants.

The student should keep in mind that it is this storing of energy from an outside source, thereby making it available for use by all forms of life, that makes photosynthesis stand out as the leading synthetic process in nature. All the net activities of animals result in the dissipation or loss of energy. Green plants alone, through photosynthesis, are able to add to and build up the biological supply of energy. The storing of energy in photosynthesis is therefore the most significant part of the entire process.

Mechanism of photosynthesis The well-known facts about photosynthesis in green plants can be stated as follows: (1) water and carbon dioxide are the raw materials used; (2) chloroplasts are necessary; (3) light energy is stored; (4) oxygen is liberated; and (5) carbohydrates are formed. Most of these facts are suggested in what is usually called the photosynthetic equation:

$$6CO_2 + 6H_2O + \text{light energy} \rightarrow \underset{\text{sugar}}{C_6H_{12}O_6} + 6O_2$$

This is a purely arbitrary equation which merely indicates the raw materials and the end products of the process. It is correct in one detail; viz., it indicates that for every volume of CO_2 consumed, an equal volume of oxygen is liberated. This can be verified by experiment.

The problem of how the plant is able to achieve this synthesis of carbohydrates has attracted the attention of some of the world's best chemists.

Aristotle (384–322 B.C.) proclaimed that plants obtain their foods already elaborated in pure form from the soil. This idea held sway for over seventeen centuries. One of the earliest men to question it was J. B. van Helmont (1577–1644). He grew a willow tree in a covered pot of soil to which nothing was added but water. After five years of growth the willow tree had gained 164 lb in weight, while the dry weight of the soil in which it was grown had decreased only 2 oz. Van Helmont naturally but erroneously concluded that the entire substance of plants was formed from water. This was, however, the first experimental evidence that water enters into the composition of plants. That gaseous constituents of the air also contribute to the building up of the body of the plant was first proved by Stephen Hales (1677–1761) in his "Vegetable Staticks" (1727). As early as 1671 Marcello Malpighi, in his "Anatome Plantarum Idea," had indicated that the green leaves are the organs which prepare the food of plants. Priestley discovered oxygen in 1774, and in 1779 he found that the green parts of plants occasionally exhale oxygen. In the same year Jan Ingenhousz showed that this takes place only under the influence of light. Jean Senebier, in his "Physiologie Vegetale," published in 1800, stated that the oxygen given off from plants in light comes from the carbon dioxide which has been absorbed. This idea was accepted generally by plant scientists until the comparatively recent use of "tagged" elements proved it to be erroneous. Although it had been indicated earlier that the carbon in the composition of plants was derived from CO_2, it was Theodore de Saussure (1767–1845) who in 1804, by careful experiments, proved that the elements of water, as well as carbon, are fixed in the plant. He also proved that nitrates and mineral matter enter into the composition of plants. It remained for Julius von Sachs, the father of plant physiology, to show that chlorophyll is important in the assimilation of carbon dioxide and that carbohydrates appear as end products.

In 1870 von Baeyer, a German chemist, suggested that formaldehyde (CH_2O) may be formed during the process as an intermediate product and that it could be converted into glucose. A union of six molecules of formaldehyde would give one molecule of glucose ($C_6H_{12}O_6$). This idea had a wide influence on investigators, with the result that many of the earlier theories considered the formation of formaldehyde as a part of the process. More recent work, however, which has done much to clear up the nature of photosynthesis, indicates that it is very unlikely that formaldehyde plays any part in the process.

A detailed discussion of the possible mechanism of photosynthesis is beyond the scope of a general textbook of botany. It might be said, however, that evidence is accumulating to indicate that the process is fundamentally a series of oxidation-reduction reactions involving a number of steps, including both chemical and photochemical reactions by which, in the presence of light and chloroplasts, the carbon dioxide is reduced with hydrogen obtained from water. The oxygen liberated in the process is dehydrogenated water. Many different enzymes are involved.

According to this interpretation, molecules of chlorophyll *a* in the chloroplasts are activated or excited by light, thereby becoming ionized by the loss of electrons ($e-$). Chlorophyll *b*, carotenoids, and perhaps other pigments present, by absorbing parts of the spectrum not equally absorbed by chlorophyll *a*, may contribute by transferring some of the absorbed energy to chlorophyll *a*. Through the action of the activated chlorophyll, a series of energy or electron transfers take place involving a number of enzymes and coenzymes (see page 293): cytochromes, flavin mononucleotide, ferredoxin, nicotinamide adenine dinucleotide **(NAD)**,[1] nicotinamide adenine dinucleotide phosphate **(NADP)**,[1] nicotinamide adenine dinucleotide-H_2 **($NADH_2$)**, or nicotinamide adenine dinucleotide phosphate-H_2 **($NADPH_2$)** (see page 293).

[1] NAD was formerly called diphosphopyridine nucleotide (DPN) and NADP, triphosphopyridine nucleotide (TPN).

Through this action of activated chlorophyll, water is split into hydrogen (H)+ and hydroxyl (OH)− ions. Following extraction of an electron (e^-), the OH radicals may combine to re-form water and liberate oxygen. The electrons (e^-) are ultimately transferred, through chlorophyll and the above-mentioned intermediates, to NAD or NADP to form $NADH_2$ or $NADPH_2$. During this process adenosine diphosphate (ADP) (see page 293) combines with phosphate to form adenosine triphosphate (ATP) containing additional stored energy. So far, these reactions constitute the photochemical phase or light reactions of photosynthesis. The remaining changes are referred to as dark reactions, although darkness is not necessary.

The CO_2 which has been absorbed by the chloroplasts has been shown, by means of tracer techniques and by chromatography, to combine with a phosphorylated carbohydrate called 1,5-diphosphoribulose. The result is an unstable 6-carbon compound which is split into two molecules of phosphoglyceric acid (PGA). By the addition of hydrogen the PGA is converted into triose phosphate, the first sugar formed in photosynthesis. The hydrogen for this reaction is supplied by $NADPH_2$, which thereby becomes converted back to NADP. ATP is used to form the PGA and is converted into ADP (compare Fig. 12-4). Some of the triose phosphate is converted into hexose phosphate and some into ribulose diphosphate which is free to combine with more CO_2. The hexose phosphate is ultimately converted into glucose and other sugars.

These steps may be summarized as follows:

Light-dependent phase:

1. **4Chl** (inactive chlorophyll) + light → **4Chl**+ (activated chlorophyll) + $4e^-$
2. $4\mathbf{H_2O} \rightarrow 4\mathbf{H}^+ + 4\mathbf{(OH)}^-$
3. $4\mathbf{(OH)}^- + 4\mathbf{Chl}^+ \rightarrow 2\mathbf{H_2O} + \mathbf{O_2} +$ 4**Chl**
4. $4e^- + 2\mathbf{NADP} + 4\mathbf{H}^+ \rightarrow 2\mathbf{NADPH_2}$; **ADP** + phosphate → **ATP**

Dark phase:

5. $\mathbf{CO_2}$ + ribulose diphosphate → unstable 6-carbon compound
6. Unstable 6-carbon compound → 2**PGA**
7. 2**PGA** + 2$\mathbf{NADPH_2}$ + 2**ATP** → triose phosphate + 2**NADP** + 2**ADP**
8. Triose phosphate → hexose phosphate + ribulose phosphate
9. Ribulose phosphate + **ATP** → ribulose diphosphate + **ADP**
10. Hexose phosphate → glucose

That the oxygen liberated in step 3 comes from the water and not from the CO_2, as formerly believed, has been demonstrated by the use of water or carbon dioxide containing heavy oxygen, i.e., an isotope of oxygen having an atomic weight of 18 instead of the usual 16. By thus tagging the oxygen, it could be traced through the end products of the process. Thus, when water containing heavy oxygen (O^{18}) was supplied to plants carrying on photosynthesis, the molecular oxygen liberated was found to contain heavy oxygen. When carbon dioxide containing O^{18} was supplied, the molecular oxygen liberated was O^{16}, or ordinary oxygen, proving that the liberated oxygen was derived from water. Steps 1 and 3 indicate that the chlorophyll behaves as a light-energy absorber and appears unchanged at the ends of the reactions.

Since water is a product of photosynthesis, as well as a raw material, a better representation of the so-called photosynthetic equation would be the following:

$$6\mathbf{CO_2} + 12\mathbf{H_2O} + \text{light} + \text{chlorophyll } a \rightarrow \mathbf{C_6H_{12}O_6} + 6\mathbf{O_2} + 6\mathbf{H_2O}$$

The steps in the process as given above are an abbreviated or summarized account which does not include some of the energy or electron transfers which probably take place. Furthermore the light-phase steps do not necessarily take place in the order given. Some of them probably occur simultaneously.

Products of the process The hexose phosphate of step 10 may be converted into fructose, glucose, sucrose, starch, and other carbohydrates. Fatty acids, amino acids, and other plant constituents may also be formed from the phosphoglyceric acid. Such transformations are not confined to the chloroplasts but may take place in the living cells of any part of the plant and hence are not solely a part of the photosynthetic process.

During photosynthesis, in many plants, glucose is rapidly transformed into starch. The presence of starch can be verified by a simple iodine test. An advantage of this change is that starch is insoluble in the water content of the cell and therefore does not affect the osmotic properties of the cell. Starch also occupies less volume and does not retard photosynthesis so rapidly as would an accumulation of a soluble sugar. At night the starch is converted back to sugar, which is free to move out of the leaf through the veins to other parts of the plant. The cells of the leaf are thus usually free of starch and sugars in the morning and therefore in better condition to continue to carry on photosynthesis. Some plants, particularly many of the monocotyledons, like onions and lilies, do not make starch. In that case the carbohydrates may remain in the form of glucose or be converted into fructose, sucrose, and other forms. Sucrose (cane sugar) is formed in the leaves of many plants.

Oxygen is always liberated as a by product in photosynthesis. The net result of gaseous exchange in the leaf in daytime, therefore, is that CO_2 is taken in and O_2 given off. The CO_2 liberated by respiration in the leaf is also used in photosynthesis. The liberation of oxygen by plants in photosynthesis is the chief method by which the supply of oxygen in the atmosphere is kept constant. Here again plants make life possible for animals and man, since all animals constantly use up oxygen and must have a new supply.

Rate of photosynthesis and factors affecting it The rate at which carbohydrates are made by the plant depends upon the combined action of external and internal factors. The most important external factors are **temperature, carbon dioxide supply,** the **kind and amount of light,** and the **water supply.** Two internal factors are of importance, viz., the **chlorophyll content** and a **protoplasmic factor.**

TEMPERATURE In general, as the temperature rises above the minimum, the rate of photosynthesis rises in a geometrical way. That is, for every 10° rise in temperature the rate of photosynthesis increases 2.2 to 2.6 times, until a temperature of 30 to 35°C has been reached, beyond which no further increase in the rate occurs. If the temperature rises very far above 30 to 35°C, the rate may decrease. As will be seen later, temperature effects are closely tied up with the internal factors limiting the process.

CARBON DIOXIDE SUPPLY Probably no other single external factor under natural conditions has a greater influence on the rate of photosynthesis than the carbon dioxide supply of the atmosphere. As previously stated, there is an average of only 0.03 percent of this gas in the air. Careful experiments have shown that plants could use much higher percentages if they were available. When greater amounts are supplied artificially to plants, the rate of photosynthesis increases until a maximum point is reached, beyond which no increase occurs. For many of the common plants this point is reached at a concentration of about 0.5 to 1 percent of the gas. Some plants probably could utilize still higher percentages of carbon dioxide. When it is borne in mind that this carbon dioxide is the only source of carbon for the plant and that carbon makes up about 50 percent of the dry weight of plants, the importance of an ample supply of the gas becomes apparent.

Much has been done to increase yields of farm crops by adding mineral fertilizers to the soil; yet all the minerals in plants combined make up only from 1 to 10 percent of the dry weight of plants. The application of additional amounts of carbon dioxide to crop plants has been tried by many different investigators and the results have generally been very promising. Increased yields of from 30 to 300 percent have been reported with such crops as potatoes, tomatoes, beets, carrots, and barley. The small percentage of carbon dioxide in air makes it likely that it is this factor that ultimately determines how fast photosynthesis can go on in the plant under natural conditions, although it is likely that the concentration of carbon dioxide in the air near the ground level is always higher because of its escape from the soil.

THE KIND AND AMOUNT OF LIGHT Light is one of the most variable factors in nature. It may vary in quality, in intensity, and in duration, and each of these will affect the rate of photosynthesis. While some photosynthesis can go on under all parts of the visible spectrum and to some extent in the ultraviolet, not all parts of the spectrum are of equal value. If the spectrum of sunlight is divided at the middle point of the visible region and the product of photosynthesis in the red half is represented by 100, then the value of the blue half would be about 54, or roughly half that of the red. The lowest rate occurs in the green region. Infrared radiation is not used in photosynthesis, and the intensity of the ultraviolet portion of sunlight is so low as to make this region also more or less insignificant in the process. While the rate is highest in the red end, because of its greater energy value in sunlight and the stronger absorption of light by chlorophyll in this region, absence of the blue-violet end of the spectrum causes a marked lowering of the rate of photosynthesis. It is partly for this reason that photosynthesis proceeds at a lower rate under artificial light, most artificial lights, such as tungsten incandescent lamps, being deficient in blue-violet rays.

As to the intensity of light, if we begin with darkness or a very low light intensity and gradually increase it, the rate of photosynthesis increases as the intensity increases, up to a maximum point. Plants vary considerably as to the location of the maximum point, but for most plants it is far below the intensity of daylight at noon. It has been found that noon daylight intensity during the summer can be reduced to one-twelfth of its value before any decrease in the rate of photosynthesis occurs in individual leaves. This means that there is ordinarily much more light available in nature than plants can use, as long as the other factors such as carbon dioxide supply remain the same.

The duration of the light or the length of time the plant is in the light will obviously affect the amount of carbohydrate that can be made. This factor becomes important during the short days of autumn and winter. It is of considerable practical importance in the production of greenhouse crops in winter.

WATER SUPPLY Water being one of the raw materials out of which the carbohydrates are made, it is easily seen that a deficiency of water might check the rate of photosynthesis. Only when water becomes so low as to cause wilting, however, does this factor become important. Under ordinary growing conditions of plants the water supply seldom becomes a limiting factor in the process.

THE INTERNAL FACTORS Of the internal factors that affect the rate of photosynthesis only two need be considered here, viz., the chlorophyll content of the leaf and a protoplasmic factor. The protoplasmic factor involves a number of features definitely tied up with the activities of protoplasm and not well understood. The importance of a protoplasmic factor is indicated by the fact that attempts to induce photosynthesis *in vitro* with chlorophyll extracts have generally proved unsuccessful although photosynthesis has been accomplished with isolated chloroplasts. Part of the protoplasmic factor involves the action of several enzymes associated with the process.

The two internal factors mentioned are best considered together. Both of these factors are closely tied up with temperature relations. The action of enzymes is always accelerated as temperature rises. It may thus be that the increased rate of photosynthesis which results from increased temperature is due partly, at least, to the speeding up of the action of enzymes associated with the process. Acceleration of the enzymes can lead to increased photosynthesis, however, only when there is an abundant supply of chlorophyll present. In general the greater the amount of chlorophyll present, the higher will be the rate of photosynthesis because more light can then be absorbed. When the chlorophyll content is low, an increase in temperature has little effect on the rate of photosynthesis because, even though this increase accelerates the enzyme activity, the process cannot go on more rapidly because the chlorophyll is probably already working at maximum capacity and cannot, therefore, absorb the additional light which would be needed to make use of the increased enzyme action. On the other hand, when the chlorophyll content is high, an increase in temperature will have a marked effect on the rate of photosynthesis because there will then probably be a sufficient absorption of light to make use of the increased activity of the enzymes. It is thus possible that in plants high in chlorophyll content the activity of the enzymes may become the factor that limits the rate of photosynthesis, while in plants low in chlorophyll it is the absorption of light that limits the rate.

LIMITING FACTORS It should be kept in mind that all these factors are operating simultaneously and that the amount of carbohydrate made will depend upon their joint action. Yet, as a chain is no stronger than its weakest link, so the rate of photosynthesis in the last analysis is probably determined by that factor which occurs in minimum. For example, if there is a very favorable temperature, a sufficiently intense light of the proper quality, an abundance of water and chlorophyll, but a very low supply of carbon dioxide, then the rate of the process will be determined by the carbon dioxide supply or, in other words, the carbon dioxide supply operates as the **limiting factor.** Any one of the external or internal factors that have just been discussed may become a limiting factor in photosynthesis.

Quantity of carbohydrates made in photosynthesis In view of what has been said concerning the effect of different factors on the rate of photosynthesis, it is obvious that the amount of carbohydrate made will vary greatly in different plants as well as in the same plant under different conditions. It is therefore impossible to state precisely how much carbohydrate a plant can make in a given time unless we measure it. Such measurements have often been made and average figures have been obtained for general average conditions. Ganong gives as an average for many common plants 1 g of carbohydrate per square meter of leaf surface per hour. Some plants will make much more than this and some less. A gram an hour produced by a square meter of leaf surface seems like an insignificant amount; yet when all the plants that carry on the process are considered, this soon adds up to tremendous figures. To illustrate: for the crop season 1964, the United States produced 2,224,000 tons of sugar from sugarcane and 3,037,000 tons from sugar beets, or a total of 5,261,000 tons of refined sugar. These figures merely represent the output of two cultivated species of plants, sugarcane and the sugar beet, and in reality are the surplus produced by these plants over and above what they used themselves.

The United States produces per year about 3.5 billion bu of corn, 1.5 billion bu of wheat, 1.5 billion bu of oats, over 300 million bu of barley, 25 million bu of rye, and over 400 million bu of potatoes. These crops consist mostly of carbohydrates made by photosynthesis. The total value of these and other crops averages between 10 and 20 billion dollars per year. When it is remembered that crop plants make up only a small part of the total vegetation of the earth, the vast magnitude of the product of photosynthesis in the aggregate becomes apparent.

Fate of the products of photosynthesis The carbohydrates made in photosynthesis are used by the plant in various ways. If starch is made in the leaf, it is first changed to the soluble carbohydrate glucose and then it is carried to other parts of the plant through the phloem. There it may be oxidized at once for its contained energy or it may be transformed into other carbohydrates and related compounds or to fats and proteins. Some of it is used in building new protoplasm and new tissues. After all these needs have been satisfied, there is always a surplus which the plant stores as reserves in such organs as roots, stems, and seeds. It is these stored reserves that are made use of by man in many of his crop plants. These uses may be summarized as follows:

1. Changed to soluble forms (digestion)
2. Carried to other parts of the plant (translocation)
3. Oxidized with the liberation of energy (respiration or fermentation)
4. Used in the synthesis of fats, proteins, and other compounds
5. Used in the building of new tissues (assimilation and growth)
6. Carried to storage organs as reserves (storage)

Summary Photosynthesis is a synthetic plant process by which chlorophyll-containing cells convert the radiant energy of sunlight into chemical energy and liberate oxygen in making carbohydrates out of water obtained from the soil and carbon dioxide gas obtained from the atmosphere. Fundamentally the process is one of oxidation reduction in which carbon dioxide is reduced with hydrogen obtained from water, the oxygen liberated in the process being dehydrogenated water. The rate at which the process goes on is dependent upon such external factors as temperature, light, carbon dioxide supply, and water supply and the internal factors, chlorophyll content, and a protoplasmic factor, any one of which may operate as a limiting factor. The average plant makes about 1 g of carbohydrate per square meter of leaf surface per hour. This is sufficient to supply all forms of plant and animal life with food. The carbohydrates so made are used by the plant in digestion, translocation, respiration, synthesis of other substances, assimilation, and storage.

THE SYNTHESIS OF FATS AND PROTEINS

Fat and oil synthesis The general properties of the fats have already been considered. It will be recalled that the fixed oils or fats, which alone are used as food, consist of glycerol and fatty acids; i.e., they are glycerol esters of fatty acids. An ester is an organic salt derived from an organic acid in much the same way that a mineral salt is derived from an inorganic acid and a metallic base. The formation of an inorganic salt is illustrated in the following equation:

$NaOH$	+	HCl	→	$NaCl$	+	H_2O
sodium hydroxide		hydrochloric acid		sodium chloride (a salt)		water

In the same way the following equation illustrates the formation of an organic salt or ester:

C_2H_5OH	+	$HOOC \cdot H$	→	$C_2H_5 \cdot OOC \cdot H$	+	H_2O
ethyl alcohol		formic acid		ethyl formate (an ester)		water

In general,

$$\underset{\text{any alcohol}}{R_1 \cdot OH} + \underset{\text{any fatty acid}}{HOOC \cdot R_2} \rightarrow \underset{\text{an ester}}{R_1 \cdot OOC \cdot R_2} + \underset{\text{water}}{H_2O}$$

Glycerol is an alcohol with three OH groups, and therefore three molecules of a monobasic acid are required to react with all the OH groups. When this happens, a fat or oil is formed as follows:

$$\underset{\text{glycerol}}{\begin{array}{c} H_2COH \\ | \\ HCOH \\ | \\ H_2COH \end{array}} + \underset{\text{fatty acid}}{\begin{array}{c} HOOC \cdot R \\ \\ HOOC \cdot R \\ \\ HOOC \cdot R \end{array}} \rightarrow \underset{\text{fat}}{\begin{array}{c} H_2COOC \cdot R \\ | \\ HCOOC \cdot R \\ | \\ H_2COOC \cdot R \end{array}} + 3H_2O$$

Many of the oils found in plants contain stearic, palmitic, or oleic acid. It is possible, and in fact usually happens, that two or three different fatty acids may react with a single glycerol molecule in the above reaction.

It is evident from the last equation given that if a plant is to manufacture a fat, it must first have glycerol and fatty acids. Hence both of these substances must be made first. There is reason to believe that they are both made out of carbohydrates through a series of intermediate steps. Glycerol is probably made from glucose. The synthesis of fatty acids probably begins with the fermentation of sugars. Neither light nor chlorophyll is necessary. The condensation of the glycerol and fatty acids to form the fat is brought about by an enzyme called *lipase.* This phase of the process occurs particularly in those parts of the plant where the water content is diminishing, as in developing seeds. It is part of the ripening process in seeds. If the water content remains fairly high, fatty acids and glycerol may remain together as such for a long time without forming fat. The steps in fat synthesis in plants may be summarized as follows:

1. Formation of glycerol by the splitting of glucose or other hexose sugars or directly from the triose which is one of the intermediate products of photosynthesis
2. Formation of fatty acids, probably beginning with the fermentation of sugars
3. Condensation of glycerol and fatty acids to form fat, brought about by the enzyme lipase

From the facts that fats are a part of all living protoplasm and that fats as such cannot move from cell to cell, it is likely that all protoplasm is able to synthesize fats or, at least, to carry out the last step in the process.

Fats are used chiefly as a source of energy, their energy value, as explained before, being 2¼ times that of carbohydrates. Besides making up a part of all protoplasm, they are commonly found stored in seeds such as flax, peanuts, castor bean, and sunflowers and in the embryo of corn and other cereals. Commercial oils are obtained from each of these types of seed.

Protein synthesis The protein molecule always contains nitrogen and usually sulfur, in addition to carbon, hydrogen, and oxygen. Phosphorus may also be present. It is in the making of the proteins that the plant uses many of the minerals which it absorbs from the soil, particularly nitrates, sulfates, and phosphates.

The method by which proteins are made probably varies in different plants. Carbohydrates from the photosynthetic process serve as a source of the carbon, hydrogen, and oxygen. The higher plants generally use nitrates obtained from the soil as their source of nitrogen for the process. Some, however, are able to use ammonium salts. A wide variation exists in the lower plants as to what form of nitrogen they can use. Some of the bacteria can use atmospheric nitrogen; some use ammonium salts; some nitrites and some nitrates. Still others use organic salts of nitrogen. The steps in the process of protein synthesis will therefore depend upon what form of nitrogen is utilized. In higher plants there are probably at least three steps, viz.,

1. The reduction of nitrates

2. The synthesis of amino acids
3. The linking together of the amino acids to form protein

The reduction of the nitrates to nitrites and to ammonia is necessary before the second step can take place. This reduction probably occurs in different places in the plant. There is evidence that in apple trees it occurs in the roots, while in plants such as the tomato it has been found to occur in stem tips, in leaf cells, in the cortex of leaf petioles and stems, and near the food-conducting tissues of the stem. It also takes place in many of the lower plants. In all cases it probably involves the action of reductases coupled with the oxidation of pyridine nucleotides, with nitrite, hyponitrite, and hydroxylamine as intermediates before reduction to ammonia occurs.

There are several ways by which amino acids may be synthesized out of the product of the reduction of nitrates. Carbohydrates are the basic substances from which the intermediates arise. Sulfates, phosphates, and possibly other substances obtained from the soil are also involved in the synthesis of some of the amino acids. Some of them undoubtedly are made from some of the intermediates of photosynthesis. Amino acids can be moved as such from one part of a plant to another, and hence their presence in a given region does not necessarily mean that they were synthesized there. Furthermore they may be formed in digestion by the splitting of proteins already existing in the plant. This fact makes it difficult to determine where and how amino acids are made in plants. As previously mentioned, there are about 20 different amino acids occurring in proteins, the synthesis of each of which involves a separate problem.

The linking together of the amino acids to form protein takes place on the ribosomes of the cytoplasm. The ribosomes contain ribonucleic acid (RNA) (see Chap. 2). The amino acids are first activated by combining with ATP. Each activated amino acid then unites with an RNA molecule, called the **transfer RNA,** thereby releasing AMP. For each amino acid there is a separate type of transfer RNA. The amino acids are next joined together, the NH_2 group of one of them uniting with the COOH group of another, and thus a whole chain of amino acids results, forming the protein molecule. Each protein molecule is made up of sometimes hundreds of these amino acids arranged, not in a haphazard fashion, but in a precise order. The precise order for a particular protein is determined by a special type of RNA called **messenger RNA,** so called because it is obtained from the DNA of the nucleus and carries the message as to the sequence of amino acids in the protein molecule from the nucleus to the ribosomes of the cytoplasm which contain additional RNA. The ribosomes occur in clusters called **polysomes,** each of which has a strand of messenger RNA.

The essential features of protein synthesis in plants are that they are built up of amino acids, which in turn are made out of the carbohydrates produced in photosynthesis and such salts obtained from the soil as nitrates, sulfates, and phosphates. Proteins are an essential part of every living cell, being one of the chief materials out of which protoplasm itself is made.

SUMMARY OF THE ANABOLIC PHASE OF METABOLISM IN GREEN PLANTS

An attempt has been made in the present chapter to show how the plant builds up the various substances that are needed for its existence. Beginning with the inorganic materials, water and carbon dioxide, it first manufactures carbohydrates, a process in which the energy of light is stored for future use. This process, called photosynthesis, is carried on solely by chlorophyll-bearing plants. Out of the carbohydrates thus made are formed most of the other compounds found in plants. Starch, cellulose, and other carbohydrates are made directly out of the original glucose resulting from photosynthesis. Some of the glucose is converted into glycerol and fatty acids from which fats are made. By the addition to the breakdown products of glucose of such minerals as nitrates, phosphates, and sulfates, amino acids are made. These in turn are linked together to form proteins. Out of the proteins may be made enzymes, secretions, other complex organic compounds, and protoplasm itself.

The making of the living substance protoplasm is called **assimilation.** Assimilation is the climax of all the anabolic processes. Almost nothing is known as to how the nonliving substances become living in assimilation except that it can take place only where life already exists. Before any substance can be assimilated, it must be in a soluble form and must be carried to those regions where assimilation is in progress, such as the growing parts of the plant. Since much of the food that is used is stored in an insoluble form, this necessitates that certain breaking-down processes precede assimilation. Such processes constitute the catabolic phase of metabolism. Anabolism and catabolism thus go on hand in hand, simultaneously.

FOOD HABITS OF PLANTS LACKING CHLOROPHYLL

The foregoing discussion has been concerned chiefly with plants that contain chlorophyll, such plants being able to carry on photosynthesis. Many plants, however, do not contain this pigment. Prominent among these are most bacteria, the fungi, and some flowering plants such as the Indian pipe. With the exception of certain bacteria, such plants cannot carry on photosynthesis. Hence they must obtain their carbohydrates in some other manner. Since chlorophyll is not needed in the synthesis of fats and proteins, these foods can be made by plants lacking the pigment and are made in the same manner as described previously. As stated in the previous section, we may subdivide plants that lack chlorophyll into two groups, viz., the autotrophic plants, consisting of certain bacteria that use energy obtained from some chemical source instead of sunlight to form carbohydrates, and the heterotrophic plants, which cannot carry on photosynthesis but must obtain an organic source of carbon from other organisms, living or nonliving. A heterotrophic plant that obtains its food from another living organism is called a **parasite,** the organism from which the food is taken being called the **host.** If the food is obtained from a dead or decaying organism, the heterotrophic plant is called a **saprophyte.** The association of the two organisms in the first instance is called **parasitism,** and in the second instance **saprophytism.** A mushroom or a toadstool growing on the ground or on an old log is a saprophyte. Such a fungus as the one causing the late blight of potatoes is a parasite.

The nutrition of saprophytes In the case of saprophytes, the problem of obtaining food is relatively a simple one. Such plants are largely confined to the groups of the fungi and the bacteria. These grow directly on the nonliving substance and absorb their food in solution from it. The fungi usually develop an extensive system of vegetative filaments collectively termed the **mycelium,** which grows through the substratum and absorbs the food from it. The bacteria, on the other hand, are one-celled plants, and each cell must absorb its material directly through its own cell membrane.

The nutrition of parasites Among the parasitic fungi there occurs the same formation of a more or less extensive mycelium noted above for the saprophytic species. Any part of the mycelium may function in absorption of foods. More often, however, these species have special structures which penetrate the host cells and absorb food directly from these cells. Penetration, particularly in the case of leaf parasites, is often directly through the outer protective cells such as the epidermis. This is secured by the activity of special enzymes secreted by the tips of the mycelium and capable of dissolving out the substances of the cell walls so as to provide a place of entry for the mycelium. Once within the tissues of the host, these parasites often develop special organs for absorbing the food from the living cells, for many parasites enter the living cells and continue absorbing the foods without killing the cells. For this purpose the tips of the mycelium are specialized and become more or less rounded or pouchlike sacs known as **haustoria.**

In such parasitic flowering plants as the dodder (Fig. 5-1*B*) and the mistletoe, multicellular, greatly modified roots penetrate the host tissues and do the work of absorption. The conducting cells of these parasites unite directly with those of the host so that the relationship is extremely intimate. The Indian pipe (Fig. 5-1*A*), though usually referred to as a saprophyte, actually gets its nourishment from fungi which surround its roots. It is a seed plant belonging to the wintergreen family.

It is a well-known fact that parasites are often highly injurious to their hosts: (1) They may extract so much of the food as to leave the host in a starving condition without enough food to carry on its growth. As a result the host may

Fig. 5-1 Dependent seed plants. A, Indian pipe (Monotropa uniflora), *a dependent flowering plant; B, common dodder* (Cuscuta *sp.*), *a twining seed plant parasitizing water willow* (Justicia americana).

become considerably dwarfed or may even die. (2) The presence of the parasite may stimulate to extraordinary growth local regions of the host, thus giving rise to various kinds of deformities, such as swellings, galls, tumors, and witches' brooms. (3) They may throw off highly injurious waste products often known as toxins, which may poison the host and thereby cause its death or at least cause the death of some of its organs. (4) They may penetrate and disorganize the cells or their content directly, resulting in more or less extended death of tissues. In general the effects of plant parasites on their plant hosts are quite comparable with the effects of plant or animal parasites on animal hosts. It follows that plants may become diseased even as animals and man do. On this basis a specialized branch of botany known as plant pathology has been developed. Every state university in the United States now has with it one or more specialists whose duties are to investigate these diseases and devise methods of control.

QUESTIONS

1. Why are commercial fertilizers not plant foods?

2. Of what value are green plants in a fishbowl?

3. Why do lawns often fail to develop well under trees?

4. How do roots obtain food?

5. Cherry trees are sometimes killed by the attack of fungi which cause most of the leaves to fall in midsummer. Why are cherry trees not killed when they normally lose all their leaves in the fall?

6. Plants grown in the shade often have larger leaves than the same plants grown in direct sunlight. Of what advantage is this to the plant?

7. Our knowledge of the mechanism of photosynthesis has improved greatly in the last few years. How did the use of radioactive elements contribute to this advance?

8. Why is the catabolic phase of metabolism necessary for the proper operation of the anabolic phase?

REFERENCES

ACKLEY, MEREDITH E., and PHILIP B. WHITFORD. 1965. The Chemistry of Photosynthesis. Appleton-Century-Crofts, Inc., New York.

ARNON, D. I. 1965. Ferredoxin and Photosynthesis. *Science,* **149**(3691):1460–1470.

ARNON, D. I. 1960. The Role of Light in Photosynthesis. *Sci. Am.,* **203**(5):104.

BASSHAM, J. A. 1962. The Path of Carbon in Photosynthesis. *Sci. Am.,* **206**(6):88.

CHELDELIN, V. H., and R. W. NEWBURGH. 1964. The Chemistry of Some Life Processes. Reinhold Publishing Corporation, New York.

MEYER, B. S., D. B. ANDERSON, and R. H. BÖHNING. 1960. Introduction to Plant Physiology. D. Van Nostrand Company, Inc., Princeton, N.J.

YEMM, E. W., and B. F. FOLKES. 1958. The Metabolism of Amino Acids and Proteins in Plants. *Ann. Rev. Plant Physiol.,* **9:**245–280.

6

ROOTS

Roots. In this chapter the general features of roots, their internal structure, root hairs, specialized roots, and the functions of roots are considered.

GENERAL FEATURES

Distinguishing features The root is the part of the plant body which ordinarily grows downward into the soil, anchoring the plant and absorbing water and inorganic salts in solution. Yet roots are not necessarily underground structures. The prop roots of corn and the air roots of orchids are examples of roots that normally remain partly or wholly aboveground. On the other hand, some recognized stem structures such as tubers and rhizomes normally grow underground. These exceptions emphasize the fact that roots cannot be distinguished from stems on the basis of their position with respect to the soil. Such distinction is based rather upon external and internal structural differences. Unlike stems, roots ordinarily do not bear leaves and buds and are not divided into nodes and internodes. Roots do, however, under certain conditions, develop adventitious buds which give rise to leafy shoots. Such shoots are produced very irregularly and often in profusion when the roots of plants like poplars, black locust, Osage orange, and apple are exposed or are near the surface of the ground. The presence of a protective structure over the end of the root, known as the root cap, is distinctly a root characteristic found on no other part of the plant. The arrangement of the internal tissues of the young root is also an important distinguishing feature of roots.

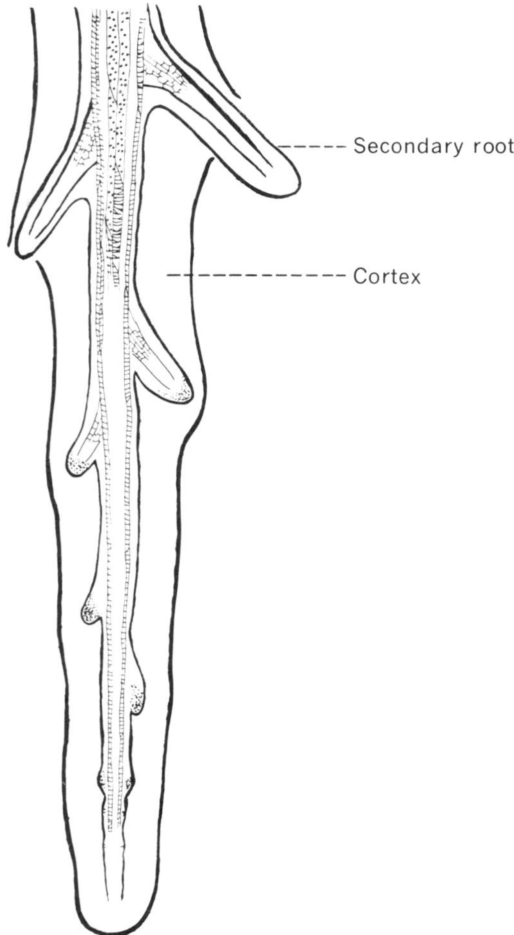

Fig. 6-1 Diagrammatic representation of origin of secondary roots in water lettuce (Pistia *sp.*)*; branches toward top of figure are older and show slight differentiation of cells which will lead to development of conducting tissues; cortex of main root broken at older branches.*

Kinds of roots As to size and order of development, roots can be classified as **primary roots, secondary,** or **lateral, roots,** and **adventitious roots.** They may also be classified as **taproots, fibrous roots,** and **fleshy roots.**

PRIMARY AND SECONDARY ROOTS The first root of a germinating seed usually grows directly downward and is known as the **primary root** (Fig. 6-6). The branches from this primary root are called **secondary,** or **lateral, roots.** They originate from within the vascular cylinder, in a parenchyma tissue called the **pericycle** (Figs. 6-1, 6-2). Here a region of growth is initiated which develops into a young root tip with its point directed outward. This secondary root forces its way through the surrounding cortical tissues and thus reaches and penetrates the soil.

ADVENTITIOUS ROOTS Many grass stems root at the nodes if the stems become prostrate; cut stems of such herbaceous plants as geranium, coleus, and carnation, and one-year-old stems of willow (Fig. 6-3*A*), poplar, and rose will develop roots if placed in moist sand or in water,

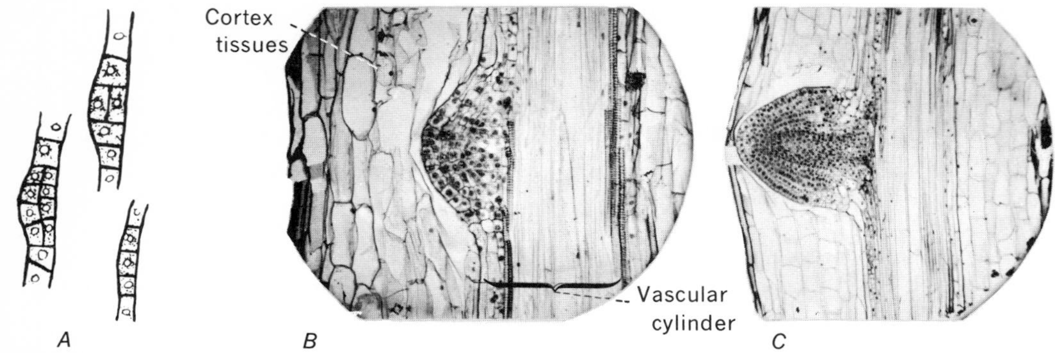

Fig. 6-2 Successive changes in the development of a secondary root. A, origin of secondary root primordium in the pericycle of the primary root; B, C, development of the young root tip, as it forces its way through the surrounding cortical tissue of the primary root.

and even the leaves of such plants as *Begonia* and *Bryophyllum* (Fig. 4-26) will, in a like manner, strike root under suitable conditions. All roots so developed are to be classed as **adventitious roots.** In the cereals adventitious roots developed at the junction of the root with the stem

Fig. 6-3 A, willow cutting, showing development of adventitious roots from the stem; B, fibrous roots of the broad-leaved plantain (Plantago major); *C, a young ragweed plant showing a taproot. (Drawings by Elsie M. McDougle.)*

of the seedling and sometimes at the first nodes of the stem become the principal roots of the plant.

TAPROOTS, FIBROUS ROOTS, AND FLESHY ROOTS If the primary root remains the largest root of the plant and continues its growth in a downward direction, so as to become the main root, with all other roots of the plant branching off from it, it is known as a **taproot** (Fig. 6-3*C*). Ragweeds, burdock, dandelions, and oak and hickory trees have well-developed taproots. Monocotyledonous plants usually do not develop taproots. When numerous long, slender roots of about equal size are developed, they are known as **fibrous roots** (Fig. 6-3*B*). In this case no one root is the largest. Many grasses, plantain, and other plants have fibrous roots. In some plants, like beets (Fig. 6-4*A*), radishes, and turnips, the roots become very large through the storage of food. Such roots are called **fleshy roots** and in most cases are taproots. In other plants, like the *Dahlia* and the sweet potato, several of the roots become fleshy. Such roots are sometimes called **clustered roots** (Fig. 6-4*B*).

Direction and extent of growth of roots Careful studies have shown that the extent of root development of common plants is often very great. In many cases the roots are actually much longer, more extensive, and greater in weight than the tops of the same plant (Fig. 6-5). A single corn-root system, for example, has been found to occupy thoroughly 230 cu ft of soil and the roots when placed end to end to have a total length of 1,320 ft. Many different factors influence the direction and extent of growth of roots. The species of plant in itself is important. All other influences being alike, one would not expect the root system of an oak tree to resemble that of a bean plant, or that of a dandelion to resemble that of corn. Many of our common Middle Western plants have root systems penetrating to a depth of 3 or 4 ft. Yet alfalfa, grow-

Fig. 6-4 A, fleshy taproot of the beet; B, clustered roots of Dahlia.

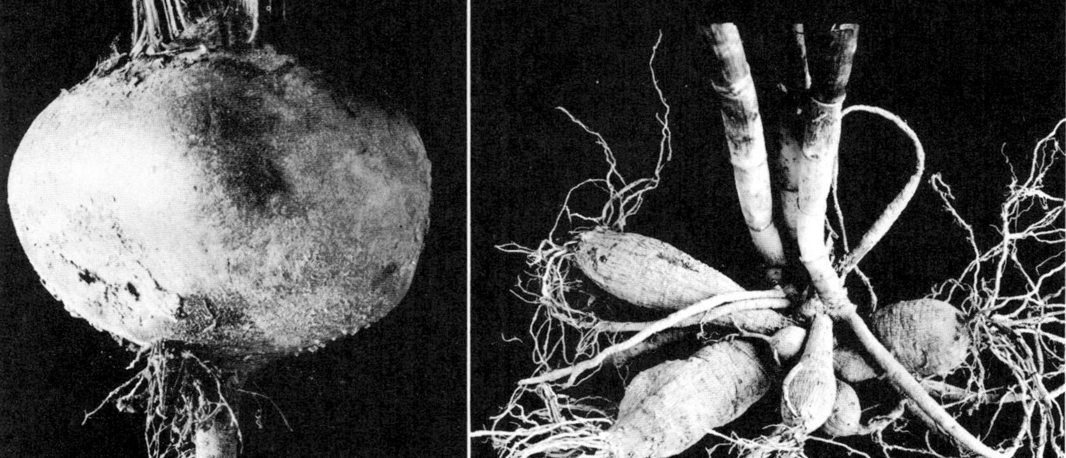

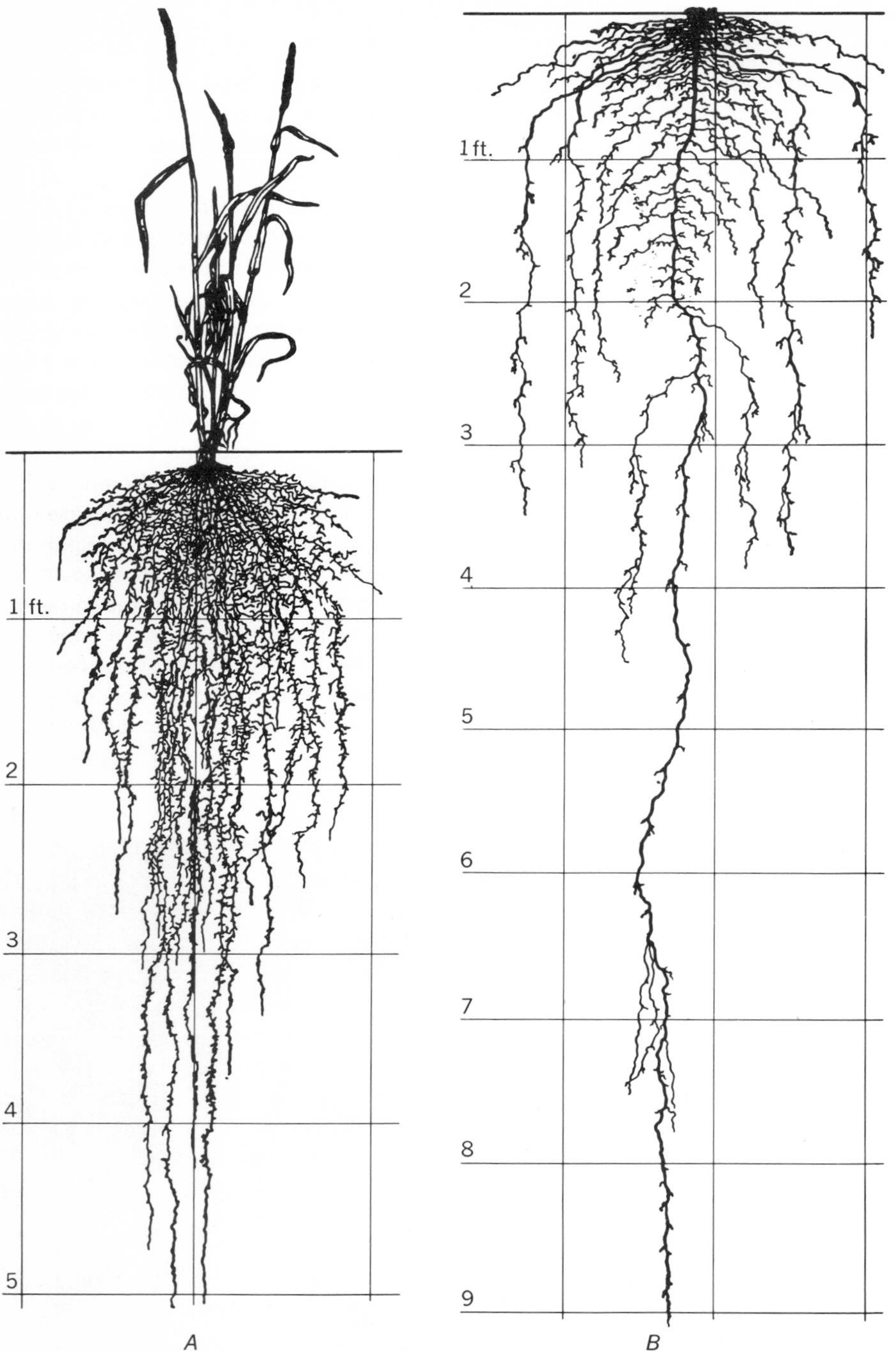

Fig. 6-5 A, *fibrous-root system of wheat;* B, *taproot system of red clover; numbers indicate depth of soil in feet. (From J. E. Weaver, "Root Development of Field Crops," McGraw-Hill Book Company, New York, 1926.)*

ing under similar conditions, has been found in some cases to have roots extending to a depth of 31 ft.

Numerous external or environmental factors are effective in determining the direction and extent of growth. These include gravity, light, temperature, soil texture, soil minerals and salts, oxygen supply, and moisture. Every root is subjected to all of these factors simultaneously and the actual growth made is the result of the combined action of all of them.

In response to gravity, roots in general tend to grow downward. This phenomenon is known as **positive geotropism** (Fig. 9-15). Some roots, if subjected to one-sided illumination, grow away from the light and exhibit what is known as **negative phototropism.** The majority, however, seem to be indifferent to light as far as direction of growth is concerned. Light indirectly affects the extent of root systems by supplying the tops with energy for photosynthesis. Thus food is manufactured which is transferred to the roots and utilized by them in making further growth. Roots will bend in the direction of the temperature most favorable to their growth, thus exhibiting **positive thermotropism.** Roots develop best in a loose soil and better in a fertilized soil than in a poor one. Nitrogen- and phosphorus-containing fertilizers seem to stimulate root development. Since we usually fertilize only the upper soil layers, we induce greater root development there; this may be disadvantageous in case of a long dry period.

Oxygen must be available for respiration in roots, which in turn is necessary in order that growth may occur. Experiments on numerous plants have shown that growth ceases when oxygen is removed by replacing it either with another gas or with water. The amount of oxygen necessary for growth varies with the species. Thus it has been found that the growth of cactus roots is checked when the supply of oxygen is reduced to 12 to 15 percent, but mesquite roots continue growth when the oxygen content of the air about them is only 2 percent. Roots at great depths or in waterlogged soils or in very compact soils are likely to suffer from lack of oxygen.

Roots do not seek water but will continue to grow in the direction of moisture supply. They are, therefore, **positively hydrotropic.** Moisture is an important factor in determining direction, depth of penetration, and lateral spread of roots. In general the less the rainfall, the less the penetration and the greater the lateral spread of roots. Conversely, where rainfall is great and where much of the rain penetrates into the soil and the water table is low, the longest and most deeply penetrating roots are found. In desert regions annual plants seldom penetrate to a depth of more than 8 in. with greatest development in the upper 2 or 3 in. In the semiarid regions of the Western Plains, Spanish-bayonet plants have been found to have roots only 18 in. deep but with a lateral spread of 30 ft and more. The depth of penetration of the average Middle Western plants has already been given as from 3 to 4 ft. The table shown on p. 112 gives further information concerning root systems of different kinds of plants.

Root hairs Several external features that distinguish roots from stems, such as the absence of leaves, regularly occurring buds, nodes, and internodes in the root, have already been pointed out. The external features of young roots can best be observed on seedlings. When seeds are germinated, the first structure to emerge is the root. Roots of this kind are usually white or colorless cylindrical organs more or less rounded off or pointed at the apex (Fig. 6-6). The apex itself is covered with a thimble-like tissue known as the **root cap.**

THE ROOT-HAIR ZONE Several millimeters behind the root cap will be found a multitude of fine white hairs radiating outward from all sides of the root (Fig. 6-6). These are called **root hairs.** The shortest root hairs are nearest the root cap. Farther back they increase in size until

*Extent of Root Systems**

plant	*maximum depth, feet*	*maximum spread, feet*	*type of root system*	*soil and locality*
Big bluestem grass	9	1	Fibrous	Prairies
Wild rye	2 to 3	1.5 to 2	Fibrous	Prairies
Rosinweed	9 to 14	3 to 4	Tap	Prairies
Spring wheat	4.8	1	Fibrous	Silt loam, Nebraska
Corn	8	4	Fibrous	Loess, Nebraska
Alfalfa	10 to 20	—	Tap	Silt loam, Nebraska
Potato	2 to 4.7	1 to 2	Fibrous	Loess, Nebraska
Cabbage	5	3.5	Fibrous	Nebraska
Strawberry	3	1	Fibrous	Silt loam, Nebraska
Pea	3 to 3.2	2	Fibrous	
Pumpkin	6	5 to 17.5	Tap and fibrous	Nebraska
Tomato	3	3 to 4	Fibrous	Ithaca, N.Y.

* Data from J. E. Weaver, "Root Development of Field Crops," McGraw-Hill Book Company, New York, 1926, and J. E. Weaver and W. E. Bruner, "Root Development of Vegetable Crops," McGraw-Hill Book Company, New York, 1927.

Fig. 6-6 Seedlings of mustard showing the root-hair zone a short distance back of the root tip of the primary root.

the maximum length is reached. The area covered by these hairs is known as the **root-hair zone.** Beyond this narrow zone no hairs are visible. This is explained by the fact that near the root apex new hairs are continually being produced, while in the older region of the root-hair zone the hairs are dying and disappearing. The root-hair zone is therefore constantly moving forward, keeping pace with the growth of the root apex. Since the root hairs are perhaps the most important structures concerned in absorption, it is necessary to consider them in further detail.

STRUCTURE OF ROOT HAIRS When seen under the microscope, the root hairs appear as very delicate, colorless outgrowths. Each hair is in reality an outward prolongation of a portion of an epidermal cell (Figs. 6-7, 7-4). The outer walls of the epidermal cells in this region are relatively thin and delicate. Consequently it is possible that the root hairs are formed by a bulging out of the epidermal cells resulting from turgor pressure from within. The outer, delicate walls of the root hair consist partly of pectic materials which are gelatinous in nature and enable the root hair to cling to soil particles and to absorb water and salts in solution. The protoplast of the cell furnishes the living content of the root hair, the nucleus of the epidermal cell usually migrating outward into the hair. The cytoplasm of the root hair usually lies next to the wall. There is also a plasma membrane. The central part of the root hair is occupied by a large vacuole or sometimes several vacuoles containing a solution of organic and inorganic salts. Such a structure is admirably suited to absorbing water by osmosis.

GROWTH AND DEVELOPMENT OF ROOT HAIRS Root hairs grow in a direction perpendicular to the surface of the cell from which they arise. In general they elongate until they come in contact with some solid substance such as a soil particle, against which they flatten out, thus presenting the greatest possible absorbing surface and coming into intimate contact with the film of moisture which surrounds the soil particle. As a result of this response to contact with soil particles, the root hairs assume very irregular forms and cling so tenaciously to the soil particles that it is almost impossible to remove a plant from the soil without leaving behind most of the root hairs. It is chiefly for this reason that plants pulled up out of the soil and transplanted usually wilt unless the transpiring surface is reduced by pruning to compensate for the reduced absorbing surface.

It is not unusual to find 200 to 300 root hairs per square millimeter of epidermis in the root-hair zone. In length they range between 0.1 and 10 mm and in diameter they average about 0.01 mm. The great numbers of the root hairs and their dimensions cause them to increase the absorbing surface of roots five to

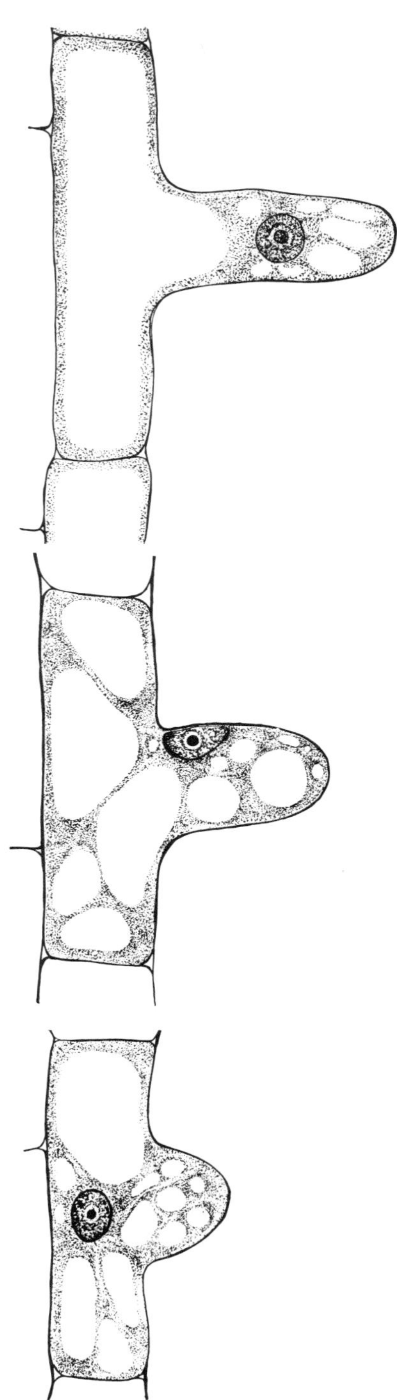

Fig. 6-7 Root epidermal cells showing development of root hairs. (Drawn by Helen D. Hill.)

eighteen times. The actual size and the intensity of production of root hairs depend upon the conditions under which they are formed. Of these conditions, the kind of plant, the temperature, the moisture, the oxygen supply, and the concentration of various minerals in the soil solution probably are the most important. A slow-growing root usually has a greater density of root hairs per unit area of epidermis than does a fast-growing root.

It has long been known that a saturated soil suppresses root-hair development. Some plants like *Elodea* and *Acorus* fail to develop root hairs at all when the roots are immersed in water; others, like corn, wheat, squash, and bean, will develop root hairs in water but not so well as when the roots are in a well-aerated soil. A few plants like duckweed and the yellow pond lily develop no root hairs under any conditions. A high concentration of the soil solution suppresses root-hair development. In general most plants develop root hairs best in a well-aerated, moist soil.

DURATION OF ROOT HAIRS Root hairs, as a rule, live for only a few days or at most a few weeks. Exceptions to this have been found in the root hairs of honey locust, Kentucky coffee tree, and redbud. The root hairs of these plants are very thick walled and persist as long as the root epidermis—several months. The root hairs of certain composites have been known to persist for three years.

The regions of the root In a lengthwise section of a root several different regions may be recognized (Figs. 6-8, 6-9). These are, in order from the apex toward the base of the root, the **root cap,** the **zone of cell division,** the **zone of cell enlargement,** the **zone of maturation,** the **zone of primary tissues,** and finally, in some roots, the **zone of secondary tissues.** The **root-hair zone** begins in the zone of enlargement and extends through the zone of primary tissues, reaching its full development in the latter zone. These various regions or zones of the root are not sharply differentiated but merge gradually into each other.

THE ROOT CAP The root cap (Fig. 6-9) is found on practically all roots except those of most aquatic plants, but never found on any other plant organ. It is a loose tissue or a thimble-shaped mass of cells like a cap which protects the growing point of the root. It may originate in one of several different ways. Whatever the method of origin of the root cap, it is clear that in all cases additions are made to it from certain special tissues of the root itself. These additions replace the outer cells of the root cap that are being continually worn away and sloughed off as the root tip is literally pushed through the soil. The walls of the outer cells of the root cap sometimes become gelatinous and furnish a slimy substance that may assist in overcoming friction.

ZONE OF CELL DIVISION Lying under and protected by the root cap is the active **promeristem,** or the **apical meristem,** of the root (Fig. 6-9). It is this region which constitutes the zone of cell division. The cells of the promeristem are all alike. They are relatively small parenchyma cells with rather dense cytoplasm and nuclei that are large in proportion to the size of the cells. The cells are in an active state of division, causing the root to elongate by the addition of new cells. As these new cells are added, they enlarge and the promeristem keeps moving forward, maintaining its position immediately behind the root cap.

ZONE OF CELL ENLARGEMENT Directly behind the zone of cell division is the zone of cell enlargement (Fig. 6-8). It is not sharply separated from the former, but in general three distinct regions or tissues can be differentiated, representing the first changes in form of the cells derived from the promeristem (Figs. 6-8*B*, 6-9*A*). These tissues are the **protoderm,** which later gives rise to the epidermis, the **ground meristem,** which gives rise to a more or less

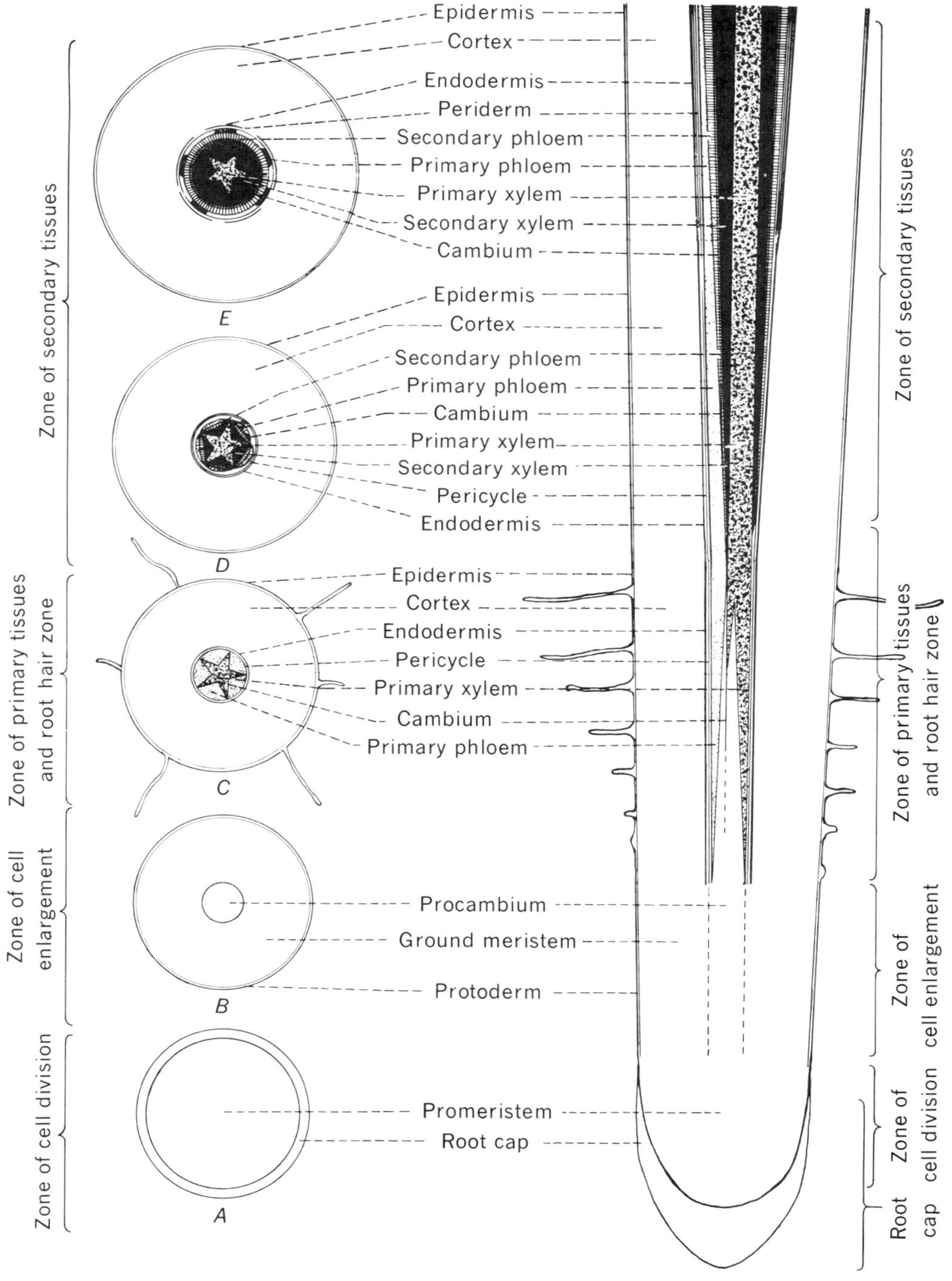

Fig. 6-8 Diagrammatic representation of the origin and arrangement of primary and secondary tissues in a young root of a dicotyledonous plant; longitudinal section of root at right, and transverse sections at different distances from the tip (A–E) at left. (Drawn by Florence Brown.)

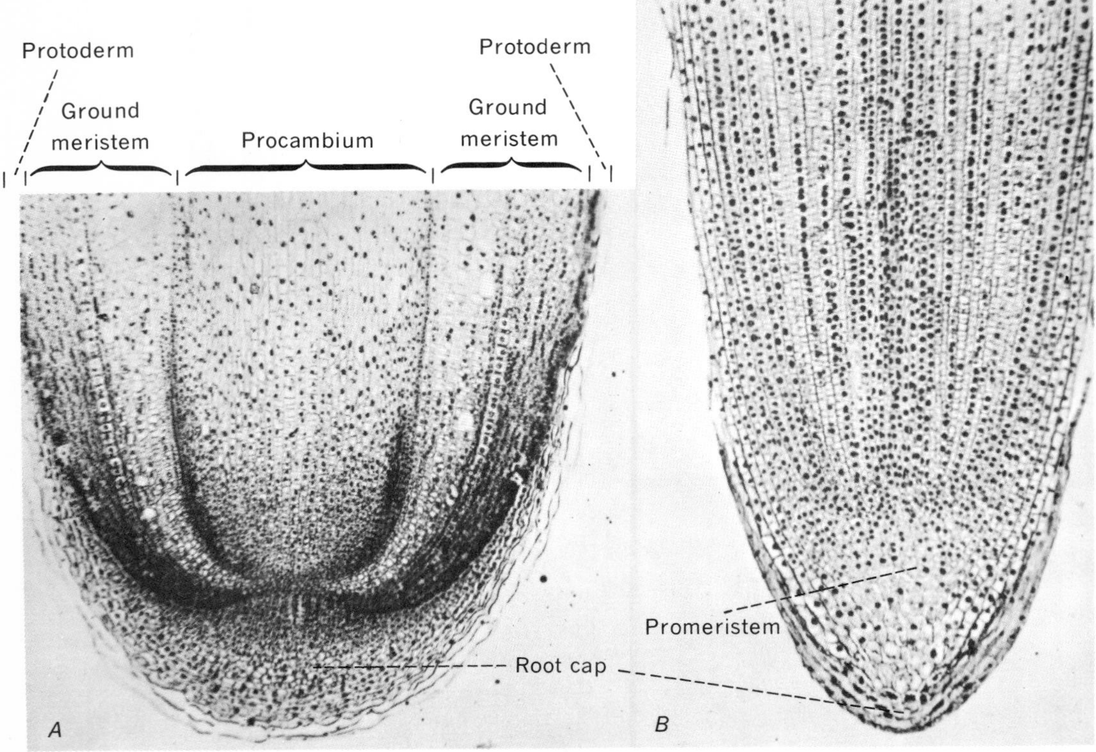

Fig. 6-9 Longitudinal sections of root tips. A, vanilla; B, onion.

extensive tissue lying under the epidermis and called the cortex, and the **procambium,** which gives rise to the central region of the root or vascular cylinder. Since these tissues are all continuing to change in form, they are to be considered as still meristematic. The zone of cell division, together with the zone of cell enlargement, comprises the actual growing region or zone of elongation of the root. It is worthy of note that this region in roots is short, only a few millimeters (1 to 10 mm) in length (Fig. 6-10), as compared with the elongating region of the stem, which is usually several centimeters in length.

ZONE OF MATURATION As the cells become older, they gradually begin to assume their mature characteristics and thus give rise to the zone of maturation. Here, for the first time, some of the cells are sufficiently differentiated to be recognizable as the mature tissues they will ultimately become. This is true of the epidermis and of the cortex. Root hairs are developing from the epidermis. In the vascular cylinder, both phloem and xylem are recognizable in part. The first xylem to be differentiated, called **protoxylem** (*proto,* "first" or "original"), appears in two or more points, on different radii, nearest the outer limits of the central cylinder. The first phloem, called **protophloem,** appears as groups of small cells alternating with the protoxylem cells. The central part of the root remains undifferentiated in the zone of maturation.

ZONE OF PRIMARY TISSUES The zone of maturation gradually merges into the zone of primary tissues (Fig. 6-8*C*). They are called primary tissues because they are developed directly from the apical meristem. It is in this zone of the root that the root hairs reach their full and final development. There is no further increase in length of the root in this or in older regions.

ZONE OF SECONDARY TISSUES In dicotyledonous plants and gymnosperms the entire older region of the root beyond the zone of primary tissues constitutes the zone of secondary tissues, so called because in this zone secondary tissues are added to the primary tissues. These secondary tissues are developed by the activity of cambium and cause the root to increase in diameter. They are described later. In the older parts of this zone, the root hairs have ceased to function because of a corky, waterproof secondary tissue which develops in the outer regions of the root. In monocotyledonous plants in general there is no zone of secondary tissues.

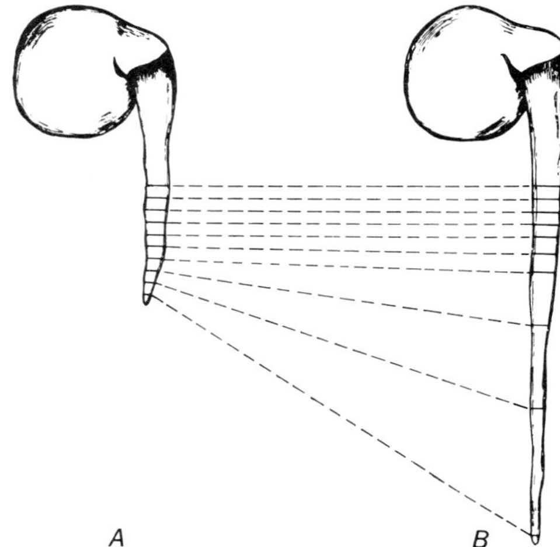

Fig. 6-10 Growing region of the root. A, a root marked into 1-mm divisions; B, the same root 24 hr later. (Drawn by Florence Brown.)

ANATOMY OF ROOTS

The primary tissues The general anatomical features of a young root can be studied best from the cross-sectional view (Figs. 6-8, 6-11 to 6-14). The primary tissues, derived directly from the apical meristems, consist of **epidermis, cortex, endodermis, pericycle, protophloem, metaphloem, cambium, protoxylem,** and **metaxylem.** The central tissues, bounded externally by the endodermis, constitute the **vascular cylinder.**

The **epidermis** is made up usually of a single layer of parenchyma cells constituting the outermost tissue of the root. Many of the cells of the epidermis develop root hairs which, as previously mentioned, are merely epidermal outgrowths. The **cortex** is a rather extensive fundamental tissue lying between the epidermis and the endodermis. It is made up of rather large, many-sided parenchyma cells sometimes nearly rounded and often with many intercellular spaces. The cortex at first functions in the transfer of water and minerals from the root hairs to the xylem but later is chiefly a food-storage tissue. Its cells are often filled with starch grains.

The **endodermis** is a rather conspicuous, well-developed tissue in roots. Usually it consists of a single layer of cells forming the inner boundary of the cortex. When fully differentiated, the cells are often much thicker walled than the cortical cells. In cross section they usually have an oval shape, the radial walls appearing thicker than the tangential walls. The radial and transverse walls of these cells are characterized by the presence of a strip of suberized or lignified material of varying thickness, often running all around the cell, called the **Casparian strip.** The cells of the endodermis opposite the protoxylem points are usually the last to be differentiated, thereby forming a direct passage through the thin tangential walls to the xylem for water and minerals absorbed by the root hairs and transferred through the cortex.

Next to the endodermis and external to the xylem and phloem, forming the outer boundary

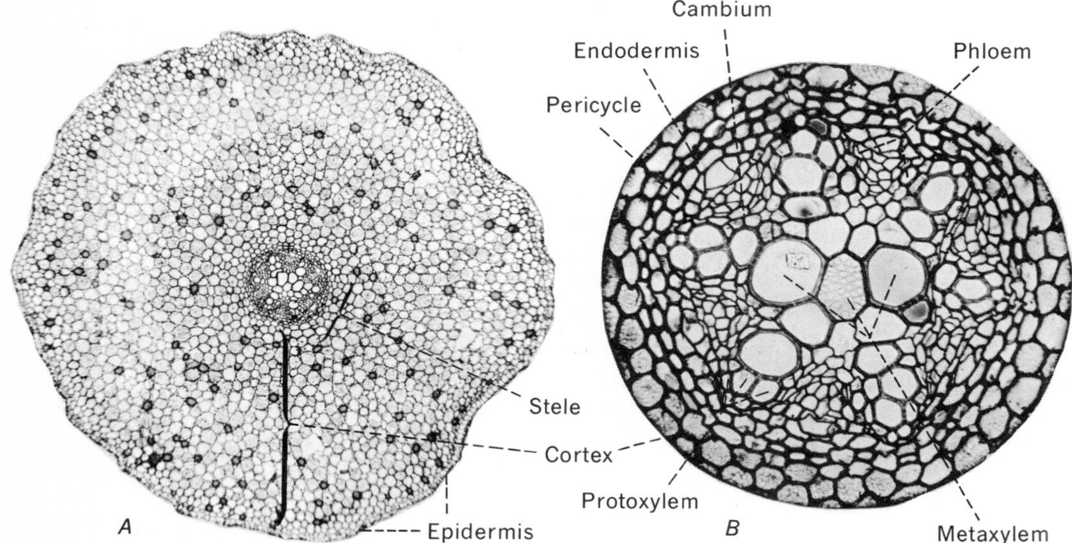

Fig. 6-11 Transverse sections of Ranunculus *root. A, entire section of root showing primary tissues; B, enlarged central portion of A.*

of the vascular cylinder, is the **pericycle** consisting of one or more layers of parenchyma cells. Often the pericycle of roots is a single layer of cells. Though not so conspicuous as the endodermis, it is a very important tissue in roots because of the fact that secondary roots originate in the cells of the pericycle and because it gives rise in most cases to the cork cambium, or phellogen.

The xylem occupies the center of the vascular cylinder and extends outward in two, three, four, five or more star-shaped rays to the pericycle. Depending upon the number of these rays, the condition of the xylem is referred to as **diarch, triarch, tetrarch, pentarch,** etc., (Fig. 6-14). The first xylem to be matured is the **protoxylem,** constituting the points of the rays. Later the more centrally placed xylem cells are matured, joining these rays and constituting the **metaxylem.** The protoxylem and metaxylem together are the **primary xylem.** In some roots in which the metaxylem does not differentiate to the very center, as in monocotyledonous plants and gymnosperms, there remains a central region of undifferentiated parenchyma cells (Fig. 6-12). This arrangement of the xylem, with the protoxylem external to the metaxylem, is called **exarch** and is characteristic of all roots. The kinds of cells that make up the xylem are described in the chapter on the cell. One of the principal functions of the xylem is the upward transport of water and inorganic substances absorbed from the soil by the roots. This upward transport takes place through tracheids and vessels, both of which have lost their original cell contents and are no longer living, but which carry on their function in an entirely passive and mechanical manner. The xylem also contains wood fibers that function in support and strengthening of the root.

In each of the angles formed by the star-shaped xylem, there is a group of somewhat smaller and thinner-walled cells comprising the **primary phloem** tissue. The outer part of this tissue, next to the pericycle, is the first to be fully differentiated and is called the **proto-**

phloem, while the inner part is the **metaphloem.** Protophloem and metaphloem are not so readily distinguishable as are protoxylem and metaxylem. The phloem functions mainly in the transport of manufactured food, which takes place through the sieve tubes already described in the chapter on the cell.

In dicotyledonous plants and gymnosperms the regions of phloem are separated from the regions of xylem by thin-walled, undifferentiated parenchyma cells, one layer of which, retaining its meristematic character, becomes the **vascular cambium.** In a cross-sectional view of the root this cambium seems to lie on and between the arms of the star-shaped xylem, separating it from the phloem. The cambium cells continue to divide and give rise to the secondary tissues of the vascular cylinder. In monocotyledonous plants there is no cambium.

The vascular skeleton of a plant is called a **stele.** The vascular cylinder of a root is known as a **protostele** with radial arrangement of tissues. This type of stele is characteristic of all roots.

Secondary tissues; increase in diameter of roots Increase in the diameter of roots occurs partly by the enlargement of the cells already present and partly by the addition of new cells. Most of the cells that enter into the structure of the older portions of the root have become more or less specialized and no longer divide to produce new cells. Of the primary tissues, only the cambium and the pericycle cells ordinarily continue to divide and it is from them that new tissues are developed. These new tissues are known as secondary tissues. In general, **primary tissues** are those derived directly from the apical meristems, while **secondary tissues** are those derived from the activities of cambium or from living tissues which were already fully differentiated. In general there are two principal types of secondary tissues developed in the roots of dicotyledonous plants and gymnosperms, viz., **secondary tissues in the vascular cylinder** and the **periderm.**

SECONDARY TISSUES IN THE VASCULAR CYLINDER Of all the cells of the primary tissues, the vascular cambium and the pericycle alone remain in the meristematic condition and thus continue to divide. The first lies between the phloem and the xylem, and by the division of its cells, new cells are added to both the phloem and the xylem. Consequently both of these tissues increase in size, but the latter grows much more rapidly than the former and eventually secondary xylem fills out the sections between the radiating arms of xylem originally occupied by the primary phloem (Figs. 6-8, 6-13). This pushes the phloem cells outward, until eventually the xylem region becomes cir-

Fig. 6-12 Stelar portion of a young root of Smilax, *a monocotyledonous plant, showing primary tissues.*

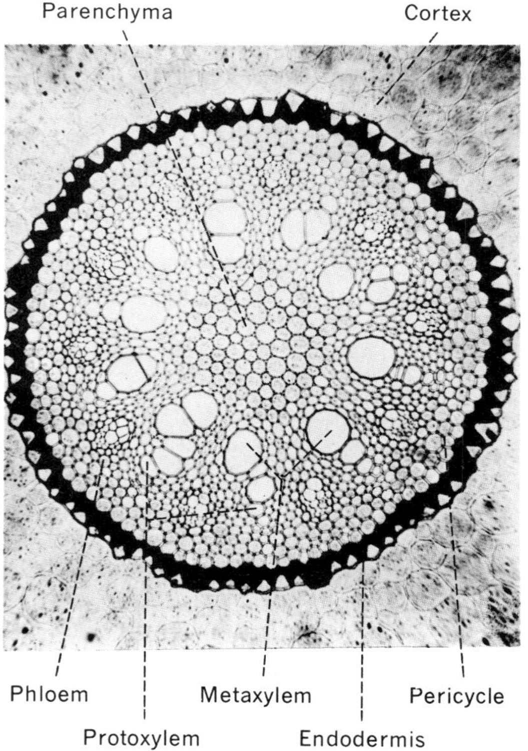

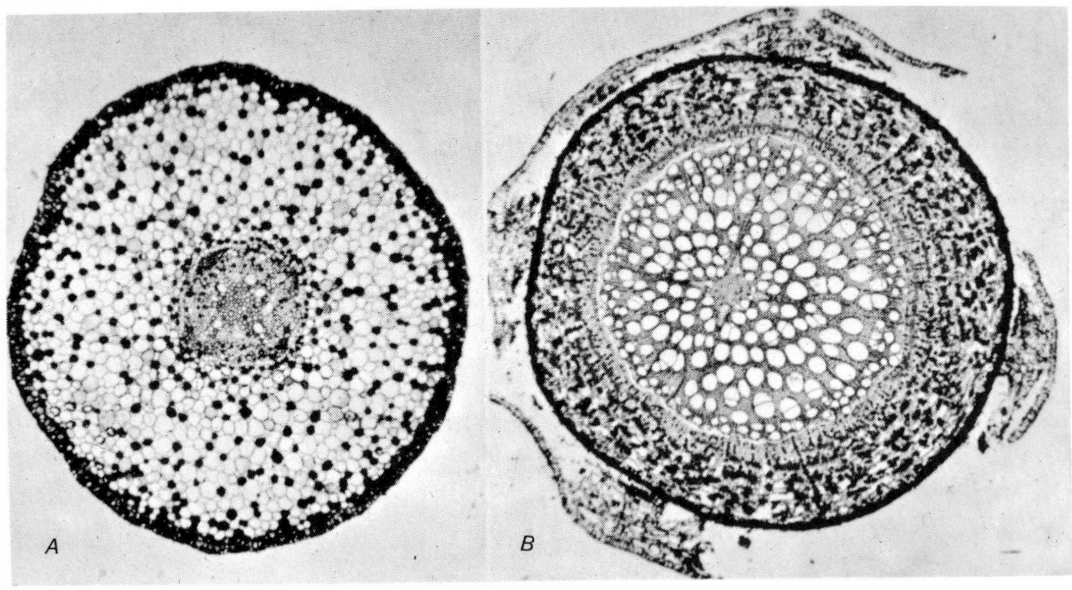

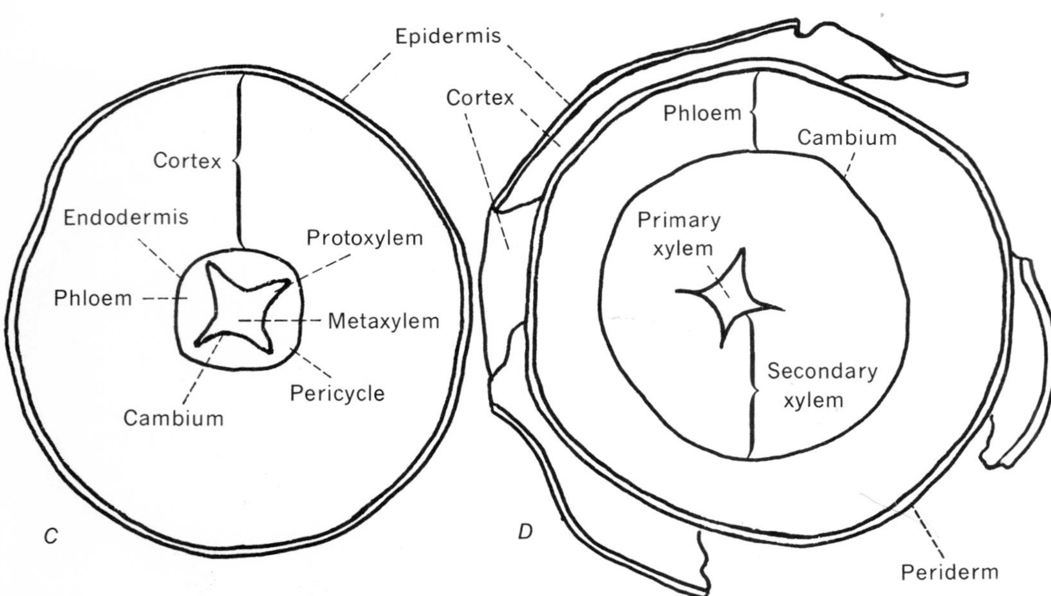

Fig. 6-13 Transverse sections of Salix *roots. A, young root, greatly magnified, showing only primary tissues; B, root one year old, less highly magnified than A, showing secondary tissues. Epidermis, cortex, and endodermis are torn as a result of internal expansion, the endodermis being no longer discernible. C, D, diagrams of A and B, respectively, indicating tissues. (Photomicrographs by Dr. D. A. Kribs.)*

cular in cross section. The cambium becomes completely continuous in this development, visible in the cross section of the root as a continuous layer of cells between xylem and phloem. Likewise the phloem, previously occurring only as groups of cells between the radiating arms of xylem, now also becomes a continuous cylinder of tissue external to the xylem. All the tissues external to the phloem are forced outward by this development. Since this increase occurs yearly in perennial roots, it follows that each year there is laid down a certain amount of new xylem or wood tissue, and the root comes to have in cross section the same evident annual rings characteristic of woody stems.

Monocotyledonous plants, lacking a cambium, have no true secondary growth of this type, and in general the root structure of monocotyledonous plants is quite different from that of dicotyledonous plants and gymnosperms. Most of the increase in diameter of roots of monocotyledonous plants takes place by the enlargement of cells of the primary tissues, but in a few kinds of more or less woody species there is a special type of secondary thickening.

THE PERIDERM In many roots, while the vascular cylinder is increasing in diameter by the addition of secondary tissues, a new secondary tissue arises, usually in the pericycle. The cells of the pericycle begin to divide, and when they do, the pericycle is transformed into a new lateral meristem known as the **phellogen,** or **cork cambium.** The phellogen usually remains as a single layer of cells, and new layers of cells are added both outwardly and inwardly by its divisions. Those developed outwardly, next to the endodermis and cortex, constitute the **phellem,** or **cork.** Those developed inwardly, next to the phloem, are called **phelloderm,** and the entire new tissue made up of phellem, phellogen, and phelloderm is called the **periderm** (Fig. 6-13*B*). The cells of the phelloderm are parenchyma and remain alive and active. The cells of the phellem become suberized, thus rendering them virtually waterproof, and at maturity they die, forming a rather impervious, protective layer around the outside of the root. Usually in roots there is relatively more phelloderm tissue developed than cork.

When the periderm is fully developed, the epidermis, with its root hairs, and the cortex die and are gradually lost, so that in old roots the periderm becomes the outermost tissue. Obviously the absorption of water and minerals by old roots of this type is reduced to a minimum. As the root keeps growing in diameter,

Fig. 6-14 Transverse sections of pine roots, showing secondary tissues. A, diarch; B, tetrarch.

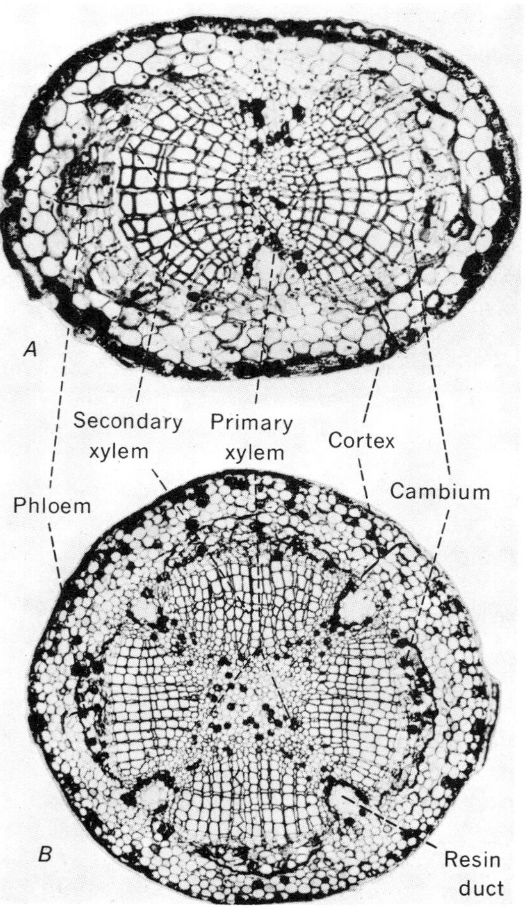

new periderm layers develop in the phloem tissue and the outer layers become broken up and furrowed as does the bark of the stems of trees. The outer dead tissues of the root, however, being underground, often decay.

FUNCTIONS OF ROOTS

The functions of **absorption, conduction, anchorage,** and **storage** are common to most roots. Many plants are also reproduced vegetatively from roots.

Absorption and conduction Perhaps the most important function of roots is that of absorption of water and inorganic substances. Since the higher plants grow almost exclusively on land, it is necessary for them to be in direct contact with the supply of water and inorganic substances that exists in the soil. While it has been shown that older, suberized roots can absorb water and inorganic salts under pressure and probably do function in this manner to some extent naturally, it is chiefly the young roots containing root hairs that take up the water and inorganic substances from the soil. The xylem of the root is continuous with that of the stem, and by this means water and inorganic substances are distributed throughout the plant.

Anchorage Roots serve to support the stem by anchoring it to the soil. Deep-rooted plants can serve this purpose better than shallow-rooted ones. Thus taprooted trees are not so readily wind thrown as are surface-rooted forms.

Storage In most plants part of the food manufactured in the parts aboveground is carried to the roots and stored. Such storage of food is particularly found in biennial and perennial plants in temperate regions. Biennials like the beet, mullein, and cabbage usually develop a rosette of leaves during the first year of their growth. During this year large amounts of food are made and stored in the roots. The following year these stored reserves are drawn on to develop an upright shoot on which flowers and seed are produced, after which the plant dies.

Very often roots become large and fleshy with stored food. This is true of sweet potatoes, beets, turnips, and many others. The internal structure of such roots becomes highly modified. Many of these plants are of considerable economic importance. The roots of many desert plants also store water, a fact which partly explains their ability to live under arid conditions.

Reproduction by means of roots Since many roots are capable of developing adventitious buds which give rise to leafy shoots, roots are sometimes a means of propagating plants. While the number of instances in which this feature occurs in nature may be limited, it is made use of by man in propagating many forms that otherwise might be difficult to propagate. Thus sweet potatoes are regularly grown from root cuttings much as the white potato is grown from tuber (stem) cuttings. Some plants, when cut down to the roots, develop adventitious buds which give rise to new plants. This is sometimes made use of in the propagation of roses and other plants. When the roots of some plants, for some reason or other, become exposed to light, adventitious buds often arise which give rise to new plants. This is true of poplars, apples, black locust, and others.

SPECIALIZED ROOTS

The functions of roots thus far considered are those that are more or less common to most roots. There are plants, however, in which at least some of the roots have become highly specialized in the performance of some one function; it may be one of those ordinarily performed by roots or it may be one entirely different. In either case such specialized roots may be greatly modified in form and appearance.

Storage roots An example of root modification has already been given in connection with food storage in roots. In the beet successive cambium layers or rings occur outside the original one. These rings give rise to xylem, phloem, and broad bands of parenchyma tissue which gradually become filled with carbohydrates as the root grows. The root thus becomes unusually large and fleshy.

The best examples of such fleshy storage roots are found among taprooted plants, of which the carrot and the turnip are common examples. Many wild plants also have fleshy storage roots of this type, which enable them to live over winter in a resting condition and to start renewed growth in the spring by utilizing the large stores of reserve food. Dandelion and burdock have roots of this type. In some plants several of the roots become fleshy. These have already been referred to as clustered roots. *Dahlia* (Fig. 6-4*B*), the rue anemone, and the sweet potato have clustered roots. In the sweet potato some of the adventitious roots developed at the nodes of stems lying on the ground become fleshy.

Aerial roots Roots are normally underground structures, but in certain species of plants they are found partly or entirely aboveground. These are spoken of as aerial roots. When the stems of corn have begun to grow rapidly, the first few nodes above the soil send down a cluster of aerial, stiltlike roots to the soil. These roots help to support the tall columnar stem but they also

Fig. 6-15 Specialized roots. A, a tropical epiphytic orchid with aerial roots; B, prop roots of the screw pine (Pandanus *sp.*).

grow into the ground and function in absorption of water and inorganic substances. Such roots are also found in the screw pine (*Pandanus*) and in other species. Roots of this type are sometimes called **prop roots** (Fig. 6-15*B*).

The mangrove tree, which grows along seashores in the tropics, sends down roots from its branches to the water beneath. These roots, formed in great numbers, catch sand and drift which cause the space to be filled up to form soil. For this reason the tree is often said to "march into the sea." The banyan tree of India also sends down roots from its branches. These roots send up shoots by means of which a single tree may soon develop into a grove of trees.

The best examples of aerial or air roots are found among some of the tropical orchids (Fig. 6-15*A*). These plants are **epiphytes;** i.e., they rest upon the trunks of trees or other supports and absorb water and minerals directly from rain as it falls on the aerial roots. These roots remain permanently in the air. They have a spongelike tissue near the exterior called the **velamen,** which enables them to absorb and hold water. In many of these plants the roots develop chlorophyll and carry on photosynthesis, a very unusual function for roots.

The **climbing roots** of such plants as the trumpet creeper, English ivy, poison ivy, and climbing hydrangea (Fig. 6-16) are also a form of aerial root. Large numbers of these adventitious roots are usually formed on the sides of stems growing along a wall or other support where they flatten out and hold the stem firmly to the surface. In some of the tropical climbers (*Ficus*) these roots become very long and sometimes grow around supports like tendrils.

Fig. 6-16 Climbing hydrangea with specialized roots used in climbing. (Drawing by Elsie M. McDougle.)

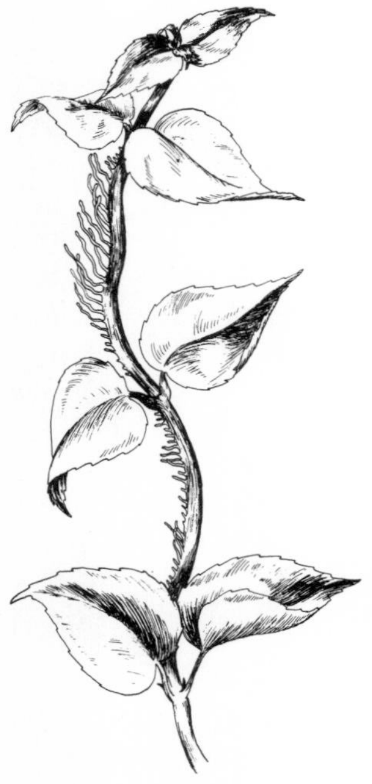

Roots of parasitic plants A highly specialized type of root is found in such parasites as mistletoe and the common dodder. The seeds of mistletoe germinate on the branches of trees and send their roots into the tissues from which water and minerals and possibly food are absorbed. The common dodder (Fig. 5-1*B*) is a parasitic twiner which sends out roots into the stem of the host plant wherever it comes in contact with it. These roots, called **haustoria,** penetrate to the vascular bundles and absorb nourishment from them. Dodder is sometimes a serious pest on clover. The leaves of mistletoe are green and carry on photosynthesis, but dodder depends almost entirely upon the host plant for its food.

Other specialized roots Other specialized roots occur in water plants. It is not unusual to find in the roots of such plants large air spaces which facilitate gaseous exchange and conduct air to the submerged parts. The internal anatomy of roots of this type is greatly altered.

The Spanish moss (*Tillandsia*) (Fig. 6-17), a

Fig. 6-17 Bald cypress (Taxodium distichum) *with the epiphyte Spanish moss* (Tillandsia usneoides). *(American Forests Magazine, Washington, D.C.)*

characteristic plant of the South, found hanging in masses from trees and often from telephone lines, has no roots at all. It is an epiphyte in which the ordinary functions of the root have been taken over by other parts of the plant.

QUESTIONS

1. When seeds are sown, the root is usually the first structure of the seedling to emerge from the seed. What is the advantage of this?

2. What is the advantage of the radial arrangement of tissues in the young root?

3. When a grown plant is pulled up out of the soil and the roots are placed in water, the plant will still often wilt. Why?

4. Root hairs ordinarily last for only a short time. Why does this not interfere with the absorption of water and inorganic substances?

5. Why do roots not develop well in a water-logged soil?

6. Epiphytes are much more common in the tropics than in northern regions. Why is this true?

7. Why are roots generally found more abundant in the upper layers of soil than at greater depths?

REFERENCES

ESAU, KATHERINE. 1960. Anatomy of Seed Plants. John Wiley & Sons, Inc., New York. Chapters 14 and 15 deal with roots.

ESAU, KATHERINE. 1965. Plant Anatomy, 2d ed. John Wiley & Sons, Inc., New York.

WEAVER, J. E. 1926. Root Development of Field Crops. McGraw-Hill Book Company, New York.

WEAVER, J. E., and W. E. BRUNER. 1927. Root Development of Vegetable Crops. McGraw-Hill Book Company, New York.

7

ABSORPTION OF WATER AND INORGANIC SUBSTANCES

Soil testing. (Courtesy of the Photographic Service, College of Agriculture, The Pennsylvania State University, University Park, Pa.)

This chapter considers the importance of water and inorganic substances and the mechanisms involved in their absorption; imbibition, osmosis, osmotic pressure, permeability, and the functions of the inorganic substances in the plant.

IMPORTANCE OF WATER AND INORGANIC SUBSTANCES

Of the various materials the plant gets from its environment, none is more important than water. The very composition of plants attests this, most herbaceous plants being made up of 70 to 85 percent water and even woody parts of plants consisting of as much as 50 percent water. Algae and other water plants frequently contain 95 to 98 percent water. Moreover plants growing in soil are constantly losing large quantities of water by transpiration. This water must be supplied through absorption by the roots if wilting of the plant is to be prevented. Water, being the most important solvent in nature, is the medium by which inorganic substances and foods are transported from one part of the plant to another. Without a constant supply of water the plant could not carry on any of its physiological activities such as photosynthesis, digestion, respiration, and growth.

Obviously a knowledge of the manner in which plants get water is important. It is no less important to know how inorganic substances are obtained since they are the raw materials out of which many of the constituents of plants are made. In the following paragraphs these matters are considered.

PRINCIPLES UNDERLYING ABSORPTION

Imbibition When a piece of hard, dry gelatin is dropped into a tumbler of water, it immediately absorbs water and thereby becomes soft and pliable. So long as the water is not heated, the gelatin will maintain its general shape although it will have swollen to many times its original size. This swelling of an apparently homogeneous, pore-free substance is called **imbibition.** The swelling is thought to be caused by the entering of water molecules between the particles of the gelatin itself, which will separate only to a definite limit determined by a balance between the attraction of the particles for each other and their attraction for water. The solid particles of gelatin in this case are thought to be groups of molecules or colloid particles, or the gelatin is said to be a **hydrophilic colloid** because of its affinity for water, the term hydrophilic meaning, simply, "water loving." The root hairs of plants likewise consist largely of hydrophilic colloids which enable them to soak up water much as gelatin does.

Although we are dealing here with absorption of water by roots, it should be mentioned that imbibition is important to the plant in many other ways also. Practically all parts of the plant, except the external waterproofed layers, contain hydrophilic colloids. Not only do such parts of the plant imbibe water, but they hold some of this water with considerable force. Water held in this manner is referred to as "bound water." The amount of bound water in a plant has much to do with the ability of the plant to withstand drying as well as freezing. Thus desert plants usually develop hydrophilic colloids which enable them to conserve water and live under conditions which would quickly kill other plants. The same may be said of winter-hardy plants. Hardy varieties are usually those that develop hydrophilic colloids which are able to bind some of the water and prevent freezing injury.

Imbibition also plays a part in the transfer and distribution of water in the plant, in the ascent of sap, in the general physical part of protoplasmic activity, in plant growth, and in many plant movements.

Diffusion of gases According to the kinetic theory, the ultimate particles of all substances are in a constant state of motion. When a gas, therefore, is liberated in a room, the molecules of the gas, by virtue of their own kinetic energy, spread abroad until they become evenly distributed throughout the room. This spreading abroad is called **diffusion.** Since there are more particles moving in a region of high concentration of a gas, it is obvious that diffusion will be more rapid from a region of high concentration toward a region of low concentration than in the reverse direction. When two or more gases are brought together, they will intermingle completely. In this case the direction of diffusion of each gas is not influenced by the presence of or direction of diffusion of the other gases, but each one will diffuse more rapidly from a region of high concentration of that particular gas, regardless of what the direction of diffusion of the other gases may be. For example, as a result of the liberation of oxygen in photosynthesis, the concentration of this gas tends to become higher inside the leaf than it is in the surrounding atmosphere. Consequently more oxygen moves out of the leaf than moves in. At the same time carbon dioxide is being consumed in photosynthesis. The concentration of carbon dioxide therefore tends to be greater outside the leaf than inside and therefore more of it moves into the leaf than moves out.

Here then we have oxygen gas moving more rapidly outward at the same time that carbon dioxide is moving more rapidly inward. At night, of course, the reverse is true. While the *direction* of diffusion is thus not influenced by other gases, the *rate* of diffusion *is* influenced. Any gas will diffuse more rapidly into a vacuum than it will into any other gas or mixture of gases. The greater the combined concentration of other gases, the slower will be the rate of diffusion of a particular gas into the other gases. Furthermore, not all substances diffuse at the same rate under the same conditions. For instance, hydrogen gas, having a density of 1, moves four times as rapidly as oxygen, the density of which is 16. In other words, the rate of diffusion of a particular gas is inversely proportional to the square root of its density. The rate of diffusion is also influenced by temperature, being directly proportional to the absolute temperature or increasing $\frac{1}{273}$ for every degree rise in temperature above 0°C. The higher temperature imparts greater energy to the diffusing particles and therefore increases the velocity.

Summarizing, we can say that the rate of diffusion of a gas depends upon the concentration of the gas, the temperature, the density, and the presence of other gases. The actual rate therefore will be the resultant of the combined action of all of these factors.

Diffusion of dissolved substances If a solid, liquid, or gas is introduced into a liquid in which it dissolves, it will behave like a gas diffusing in air; i.e., the molecules or ions of the substance will in time become equally distributed throughout the liquid. Thus a little sugar dropped to the bottom of a tumbler of water will soon dissolve and the molecules will diffuse through the water even though it is not stirred. The dissolved substance in this case is called the **solute,** the liquid in which it is dissolved, the **solvent,** and the whole system is called a **solution.**

A solution may contain many kinds of solutes just as air consists of several gases. Indeed in the soil and in the plant the solutions met with are always made up of many solutes. The rate of diffusion of a solute is determined by the same factors that govern the diffusion of gases, viz., concentration, temperature, density, and the presence of other solutes. In general the rate of diffusion of a solute is always greater from a region of high concentration of that particular solute to one of lower concentration, regardless of how many other solutes may be present. It cannot be overemphasized here that each solute in a complex solution moves independently of the movement of other solutes. When two or more solutions of different concentrations are brought together, diffusion of all solutes continues until the concentration of each solute becomes equalized throughout the liquid.

Diffusion through membranes Both gases and dissolved substances may diffuse through membranes in which the solvent is imbibed. The rate of diffusion in this case is determined by the nature of the membrane as well as by all the other factors that govern the rate of diffusion. Membranes differ greatly in the ease with which solutes and solvents can pass through them. A rubber membrane, for example, allows neither water nor any solutes dissolved in water to pass through it. Such a membrane is said to be **impermeable.** A filter paper, on the other hand, will allow both water and any substance dissolved in water to pass through it. In this case the membrane is said to be **permeable.** There is a third class of membranes which permit a solvent to pass but may prohibit the passage of many substances in solution. This type of membrane is said to be **semipermeable,** or **selectively permeable.** We may characterize the three types of membranes as follows:

1. Impermeable membranes allow neither solvent nor solute to pass.
2. Permeable membranes allow both solvent and solute to pass.
3. Semipermeable, or selectively permeable, membranes allow the solvent to pass but prohibit the passage of many solutes.

It should be said in connection with semipermeable membranes that no membrane is known to exist which bars the passage of all solutes. A membrane is said to be semipermeable, however, if it prevents the passage of many solutes or if it does not allow solutes to pass as readily as it allows the solvent to pass.

Cell membranes In plants all three types of membranes are found. In this connection, the only solvent we need to concern ourselves with is water. Suberized and heavily cutinized cell walls are practically impermeable as regards the passage of water and solutes. Ordinary cellulose cell walls, on the other hand, are readily permeable to both water and the substances dissolved in water. When cell membranes are spoken of, it is usually not the wall of the cell that is meant but the part lying next to the wall, called the plasma membrane. The plasma membrane of a living cell is selectively permeable. It is this membrane that is most significant in absorption of water and solutes by the cell, since, in the final analysis, it governs the exchange of all materials between the interior and the exterior of the cell. *Perhaps the most interesting feature about the plasma membrane is the fact that its permeability is not fixed and constant but fluctuates.* It may under certain conditions permit a certain substance to pass which under another set of conditions would be prohibited from passing. For instance, the cell sap of root hairs contains sugar which, under ordinary conditions, cannot escape from the root hair because the plasma membrane of the root hair is impermeable to sugar. If, however, the root is immersed in a single-salt solution of sodium or potassium, the sugar immediately passes out. In this case a change in the external environment of the root hair has changed the permeability of its plasma membrane.

Other factors or conditions bring about similar reversals and changes in the permeability of plant-cell membranes. When a membrane is permeable to a given substance, that substance, as far as is known at present, can pass equally readily through the membrane from either side. That is, if a plant cell permits the passage of a solute from the outside to the interior of the cell, it will usually also permit its passage from the interior to the exterior. The selective permeability of the plasma membranes is a function of the living cell only. Dead cells are permeable to solutes generally.

Osmosis and osmotic pressure When a *permeable* membrane is placed between a solution and a pure solvent, both the solvent and the solute will diffuse through the membrane until at equilibrium the concentration will be the same on both sides of the membrane. Thus in Fig. 7-1, if *a* is a *permeable* membrane and we place water in *A* and a water solution of sugar in *B,* water molecules will move from *A* to *B* as well as from *B* to *A.* Sugar molecules will also move through the membrane from *B* to *A.* After sugar molecules have passed into *A,* some of them will diffuse back again from *A* to *B.* At the start more sugar molecules will be moving from *B* to *A* while more water molecules will be moving from *A* to *B.* In other words, each diffusing substance will move faster from a region of high

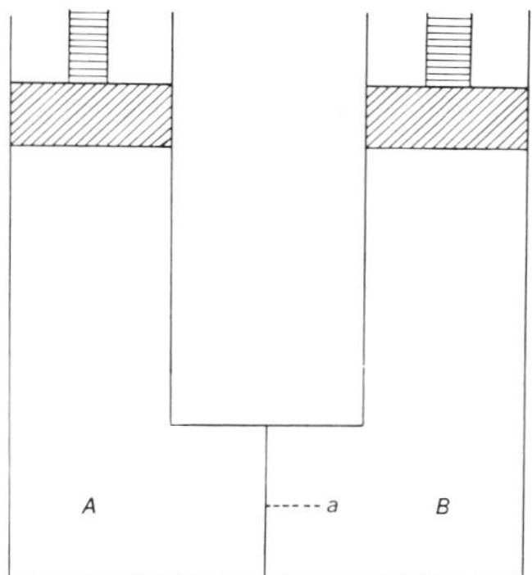

Fig. 7-1 See text for explanation.

concentration of that substance to a region of lower concentration than vice versa. When equilibrium is reached, there will be as much sugar in *A* as there is in *B*. Both sugar molecules and water molecules will still continue to diffuse but the level of the liquid on either side of the membrane will not change because at any given moment there will be as many molecules of water and of sugar passing through the membrane from one side as from the other. In other words, diffusion will take place in the same way as it would if no membrane at all were present.

Now suppose that we substitute a *selectively permeable* membrane at *a*, i.e., one that will allow water to pass but not sugar, and again place water in *A* and a sugar solution in *B*. Obviously water molecules will still continue to diffuse through the membrane from both sides, but since there are relatively more water molecules striking the membrane from side *A* than there are from side *B*, water will move more rapidly from *A* to *B* than it will from *B* to *A*, with the result that the liquid will rise in *B* and fall in *A*. The higher the concentration of the sugar solution, the higher will the liquid rise in *B*. Similarly, if two sugar solutions of different concentrations are separated by a membrane permeable only to water, water will move more rapidly toward the side of greater concentration of sugar and will set up a pressure on that side, which will raise the piston.

In the figure we could keep the level of the liquids equal by applying pressure to the piston on side *B*. This pressure would have the effect of speeding up the molecules in side *B* so that, even though there are fewer water molecules per unit volume on this side of the membrane, their increased speed would be such as to cause as many molecules of water to move from *B* to *A* as are moving from *A* to *B*. *The phenomenon, which results in a difference in rate of movement (or passage through the membrane) of solvent molecules from opposite sides of a selectively permeable membrane separating two solutions of different concentrations, is called* **osmosis.** The increased pressure resulting on the side of greater concentration is called **osmotic pressure.** The magnitude of osmotic pressure that can be developed by a given solution will depend upon the concentration of the solution as well as upon other factors.

It should be noted here that it is not the diffusion of the dissolved substance that is significant in osmosis but rather the movement of the water molecules. When solutes diffuse through a membrane, the phenomenon may be correctly called diffusion rather than osmosis. The term osmosis is thus restricted to conditions that result in osmotic pressure. Osmotic pressure will be set up only when the membrane is selectively permeable or when the membrane permits water to pass more freely than it permits solutes to pass. In the latter case the osmotic pressure set up will always be only temporary. *The term osmotic pressure is commonly used as a physical constant of a solution.* Thus we say that a liter of a solution containing a molecular weight in grams of any substance that does not ionize (i.e., a nonelectrolyte) has an osmotic pressure of 22.4 atm, meaning that a maximum osmotic pressure of 22.4 atm would be developed if this solution were separated from its pure solvent by a membrane permeable only to the solvent. *When the osmotic pressure of a solution is given, therefore, it always means the maximum osmotic pressure that this solution could develop when separated from its pure solvent by a membrane permeable only to the solvent.* Since a solution in a beaker does not create an osmotic pressure unless a selectively permeable membrane separates it from pure water, it would be more correct to refer to the **osmotic potential** of the solution rather than to its osmotic pressure. In other words, a liter of solution containing a molecular weight in grams of any nonelectrolyte has an osmotic potential of 22.4 atm.

THE GENERAL WATER RELATIONS OF THE CELL

The fact that plant cells are always completely enclosed structures introduces several additional forces that must be considered if we are to understand the mechanism of absorption by a cell. Mature living plant cells usually consist of a cell wall, which is permeable and elastic, a plasma membrane, which is selectively permeable, and a large central vacuole containing a solution of organic and inorganic solutes (Fig. 7-2).

By virtue of the osmotic potential of the cell sap, such a cell, when immersed in pure water or in a solution of lower concentration than that of the cell sap, is capable of absorbing water by osmosis and increasing in volume. As the cell enlarges, the wall stretches and, being elastic, tends to return to its original shape; that is, the wall resists extension and hence exerts an inward pressure which opposes the entrance of water. This force is called **wall pressure.** Further expansion of the cell will cease when the wall pressure is great enough to cause water to be forced out of the cell as fast as it is taken in by osmosis. When a cell is stretched in this condition, it is obvious that the liquid within the cell is pressing out against the cell wall. This actual hydrostatic pressure of the cell content against the wall is called **turgor pressure.** Turgor pressure obviously tends to force water out of the cell. When equilibrium is established, i.e., when the cell is neither increasing nor decreasing in volume, the turgor pressure just balances, i.e., is equal to the wall pressure.

It should be noted here that the turgor pressure is not the same as the osmotic pressure of the cell. The osmotic pressure is the maximum pressure which the cell could exert by virtue of its concentration of solutes if the cell were immersed in pure water. Turgor pressure can, under proper conditions, equal the osmotic pressure, but it can never exceed it. When a cell is not yet fully turgid, i.e., when the turgor pressure has not yet reached its fullest expression, it is capable of absorbing water from the exterior. The net force sending water into the cell in this case is equal to the difference between the osmotic pressures of the cell sap and the external solution minus the actual pressure (turgor pressure) already existing within the cell. This net force is sometimes called the **net suction force** of the cell. If we represent it by S and represent turgor pressure by T, internal osmotic pressure by P_i, and external osmotic pressure by P_e, the relation between these forces is given by the equation $S = (P_i - P_e) - T$. The term suction force was apparently derived from a literal translation of the German word *Saugkraft.* A better term for this force would probably be simply "water-absorb-

Fig. 7-2 Diagram of a mature living plant cell showing cell wall, cytoplasm, and a central vacuole containing a solution of organic and inorganic solutes. (Drawn by Florence Brown.)

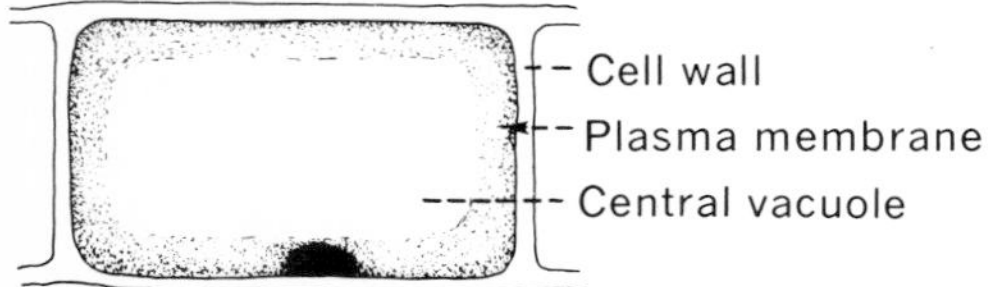

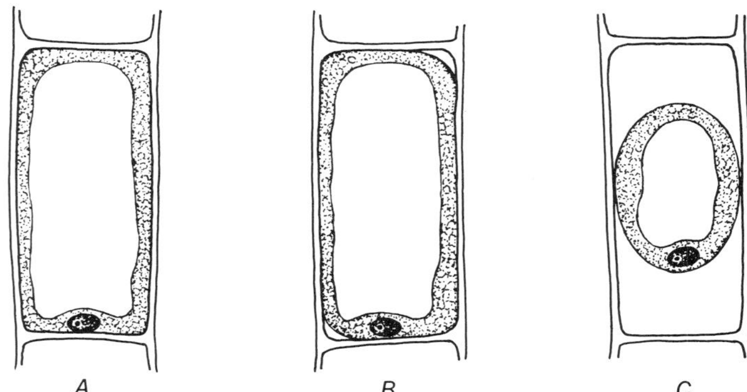

Fig. 7-3 Plasmolysis in corn root cells. A, turgid cell in water; B, cell in 2.3 percent KNO_3, showing slight plasmolysis; C, cell in 5 percent KNO_3, showing a high degree of plasmolysis. (Courtesy of F. D. Kern from "The Essentials of Plant Biology," Harper & Row, Publishers, Incorporated, New York, 1947.)

ing power." By some authors it is called "diffusion pressure deficit."

If a cell is immersed in a solution of greater concentration than that of the cell sap, obviously water will move out of the cell faster than it moves in and the cell will shrink in volume until the protoplasm actually shrinks inwardly, away from the cell wall. When this occurs, the cell is said to be **plasmolyzed** (Fig. 7-3). It is clear that the potential osmotic pressure of a plasmolyzed cell is very high, while the turgor pressure is zero.

ABSORPTION OF WATER AND INORGANIC SUBSTANCES BY LOWER PLANTS

Having considered the principal forces involved in absorption in general, we are now in a position to take up the manner in which these forces are involved in the movement of water and inorganic substances from the external medium into the plant. It should be emphasized at the start that water moves independently of the movement of dissolved substances. Furthermore each solute moves independently of all others.

In plants growing submersed in water, such as many of the algae, each cell usually receives water and minerals directly from the external medium. Water moves in by imbibition and by osmosis, and inorganic substances diffuse in through the plant-cell membranes. The metabolism of the cell is also involved in this absorption. The water relations of these plants are therefore primarily those already described for the cell.

ABSORPTION OF WATER BY VASCULAR PLANTS

In the higher vascular plants absorption becomes more involved because most of these plants grow on land. Whereas animals can move about from place to place to satisfy their needs, these plants remain fixed in one position and can obtain only materials that move to them. While the higher plants, unlike animals, synthesize all their own food, they are dependent upon the environment for water, inorganic substances, carbon dioxide, and oxygen. In these plants water and inorganic substances are absorbed by the roots and transferred to other parts of the plant. The root system alone remains in contact with water and minerals.

Condition of water in the soil Soil consists of weathered rock particles of various sizes together with more or less organic matter derived from the decay of plants and animals. Various living organisms, such as bacteria, protozoa, algae, fungi, insects, and earthworms are also found in soil and play a prominent part in the relation of soil to higher plants. Depending upon the size of the predominating particles present, soils may be classified as gravel, sand, silt, or clay. When there are approximately equal amounts of clay (or clay and silt) and sand present, the soil is called a loam. Loam soils are most favorable for the growth of plants. All soils contain many pores, or air spaces, in which oxygen, carbon dioxide, water vapor, and other gases are found. In clay soils these pores may be filled with water, resulting in poor aeration.

When water in the form of rain falls upon a soil, some of it runs off the surface, some sinks freely in response to gravity, while a considerable portion adheres, in the form of films, to the soil particles. The part that sinks in response to gravity is called **gravitational,** or **free, water.** After this water has drained away, the soil is said to be at its **field capacity.** Of the portion which remains, the part which is free to move by capillarity from one soil particle to another, as evaporation occurs at the surface or as water is removed from adjacent regions of the soil, is called **capillary water.** Even after a soil has become air dry, a certain amount of water will still remain in the form of very fine films on the soil particles. This is called **hygroscopic moisture.** The amount of hygroscopic moisture and the freedom of movement of the capillary water will depend upon the size of the soil particles, the chemical nature of the particles, the amount and nature of the organic matter present, and other factors.

In any case, when water is lost from a given region of the soil either by evaporation from the surface or by being absorbed by the roots of plants, the capillary water from adjacent regions tends to move toward that region because of ordinary capillary action. As this is a mass movement of water, it carries with it any substances in solution in the water. In other words, as the root absorbs water from the soil, some water may move toward the root from adjacent regions by capillarity, provided there is an abundance of water in the soil. Capillary movement of water, however, may be very slow and may be inadequate to supply the needs of a rapidly transpiring plant. If a soil dries out sufficiently, a condition may be reached when the soil particles themselves hold the water with sufficient force to prevent an adequate movement to the plant. Under such circumstances, unless water is added to the soil, the plant will wilt. Even though the plant wilts under such conditions, there is still a considerable amount of water left in the soil. The percentage of water left in the soil when plants permanently wilt in it under ordinary conditions is called the **wilting coefficient,** or the **permanent-wilting percentage,** of the soil. The wilting coefficient varies with different kinds of soil. It has been stated that the wilting coefficient is also influenced by the osmotic characteristics of the plant.

Water does not exist in a pure form in soil but is a solution containing various solutes, chiefly inorganic, commonly referred to as the **soil solution.** The osmotic pressure of the soil solution, while quite variable, averages around 0.2 to 1 atm in humid regions. Naturally, as a soil dries out, the soil solution becomes more concentrated and its osmotic pressure increases accordingly.

Absorption of water by root hairs In the higher vascular plants water is absorbed chiefly through the root hairs or through the very young root surfaces when root hairs are absent. The structure, position, and development of the root hairs have been mentioned on a previous page. There remains to be considered how they function in absorption.

The walls of the root hairs (Fig. 7-4) not only are thin but possess an external layer consisting largely of a hydrophilic colloid, calcium pectate, or other pectic material, which readily absorbs water by imbibition from the soil solution. This imbibed water furnishes a channel for the osmotic movement of water as well as for the movement of inorganic salts, since the wall itself is permeable. Just inside the wall of the root hair is a thin plasma membrane, colloidal in nature, which surrounds usually a single large central vacuole. While this membrane is permeable to most mineral salts, it is not permeable to many of the organic solutes inside the cell and hence behaves as a selectively permeable membrane. The large central vacuole contains cell sap which is a complex solution of inorganic and organic solutes among which may be mentioned sugar. The total concentration of all solutes in this cell sap is such as to give it an osmotic pressure of at least 4 to 10 atm. Since the osmotic pressure of the soil solution in humid regions is seldom over 1 atm and since the plasma membrane is selectively permeable with respect to many of the solutes inside the cell, it is obvious that there will be relatively more water molecules moving into the root hair at any given time than are moving out or, in other words, water will be taken up by osmosis.

The root hairs grow out into the pore space of the soil and, like the soil particles, become surrounded by films of water. As these films are removed by absorption by the root hairs, the films of adjacent soil particles are drawn on. These in turn draw on films of the particles adjacent to them, and thus water may move from considerable distances to the root hairs, provided there is an abundance of water in the soil. During active transpiration, or loss of water vapor from the plant, there is set up in the xylem a negative pressure or tension which probably promotes the entrance of water into the root hairs.

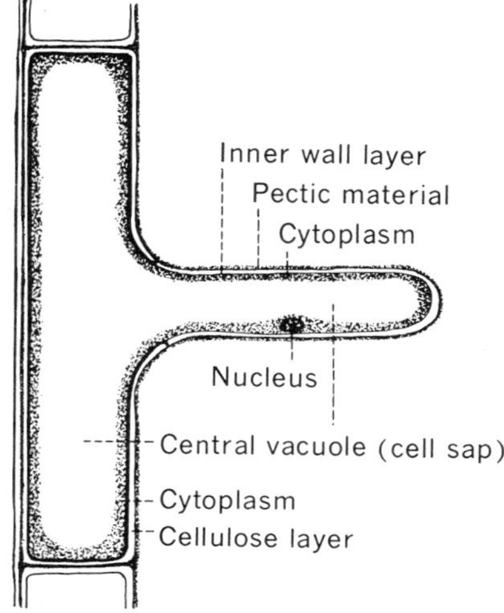

Fig. 7-4 Diagrammatic lengthwise section through a root epidermal cell, showing origin and structure of a root hair. Note that the outer wall layer of pectic material is continuous with that of the wall of the epidermal cell and the middle lamellae of adjoining cells. (Drawn by Florence Brown.)

Transfer of water to the xylem It will be recalled that root hairs are merely outgrowths or parts of epidermal cells. The epidermal cells are adjacent to the cortical cells and these extend to the endodermis. Inside the endodermis is the single layer of cells called the pericycle. At the protoxylem points of the root the pericycle lies next to the xylem (Fig. 6-11). This offers a direct channel for the passage of water and inorganic salts to the xylem at these points without the necessity of their passing through the phloem. As the water is absorbed by the root hairs, the root-hair cells tend to become more and more turgid. As the turgor pressure rises in them, water passes from the root-hair cell to the first cortical cell in contact with it. This cell in

Fig. 7-5 Effect of omission of essential elements on growth of buckwheat in sand cultures. The plants at the top, left, were grown in a complete solution containing all essential elements; those at the lowest extreme, right, were grown in distilled water with none of the essential elements. (Photograph by R. S. Beese.)

turn passes the water on to the next cortical cell and so on until it reaches the xylem. There is probably a gradient of suction force extending from the root hair to the xylem. As already mentioned in the preceding paragraph, the tension set up in the xylem by transpiration probably also promotes the transfer of water from root hairs to xylem. Through the xylem the water is carried to all parts of the plant.

ABSORPTION OF INORGANIC SUBSTANCES BY VASCULAR PLANTS

Inorganic substances required by plants From the soil vascular plants obtain, besides water, inorganic substances that are essential in the general metabolism of the plant. Knowledge of what minerals are required by green plants has been obtained by growing plants in solutions of known composition. Cultures in such solutions are commonly called **water cultures,** or **solution cultures.** The growing of plants in solution cultures is sometimes called **hydroponics.** It was very early discovered that, for the plant to be able to grow well, such solutions had to contain salts of phosphorus, potassium, nitrogen, sulfur, calcium, iron, and magnesium (Fig. 7-5). These elements, together with carbon, hydrogen, and oxygen, were looked upon as the 10 essential elements for plant growth. An essential element is one that is required for normal growth and development of a plant during its life cycle. Later work has demonstrated that there are many more than 10 essential elements. In addition to the 10 elements mentioned, there is now evidence that boron, copper, manganese, molybdenum, and zinc and possibly cobalt, silicon, sodium, chlorine, iodine, aluminum, and vanadium may be essential for some plants. The mineral elements are always taken up in combined form and never as elements. Some of them are required in much greater quantity than others. Those like boron, copper, zinc, and others that are required in very minute quantities are referred to as "microelements," or "trace elements." Such microelements as copper and zinc function as prosthetic groups of enzymes (see page 292) and are therefore essential for all plants.

Condition of inorganic substances in the soil The inorganic substances that are absorbed by plants are found dissolved in the soil solution. Most of them probably exist as ions, such as NO_3^-, $SO_4^=$, PO_4^{3-}, Cl^-, NH_4^+, K^+, Ca^{++}, Mg^{++}, Fe^{3+}, and others. Some of these ions are held on the surfaces of the soil particles by adsorption, a surface force common to such small particles. The presence of various colloids in the soil helps to increase the amount of adsorption and tends to prevent the leaching out of mineral substances that are needed by plants. Leaching out of inorganic substances is rather pronounced in a loose sandy soil containing little organic matter. Some ions, particularly $SO_4^=$, NO_3^-, and Cl^-, are held so weakly that they leach out readily and are lost in the drainage waters. Others, like PO_4^{3-}, K^+, and NH_4^+, are held more firmly.

Some of these minerals exist in forms that are not available to the plant. The chemical and physical forces operating in the soil, enhanced by activities of soil organisms, are important factors in rendering such substances available. This is of particular significance with regard to nitrogen. Nitrogen may exist in the soil in the form of gaseous nitrogen, ammonia, nitrates, nitrites, or complex organic forms such as occur in the soil humus. The nitrate form is the one most available to ordinary plants. While many plants probably absorb ammonium salts, gaseous nitrogen and organic nitrogen compounds are usually unavailable as such.

Nitrification Fortunately there exist in soils several groups of bacteria that are capable of converting the unavailable organic nitrogen into available nitrates. The process by which this is brought about takes place in several steps involving different kinds of bacteria and is called **nitrification.** In the first step in the process, the proteins of the soil organic matter are broken down to amino acids. This occurs during ordinary decay of organic matter and is brought about chiefly by bacteria and fungi. In the next step, by a process called **ammonification,** a group of bacteria known as **ammonifying bacteria,** together with certain fungi, converts the amino nitrogen to ammonia. The ammonia thus formed would escape from the soil in gaseous form were it not for the fact that other chemicals present in the soil react with it to form ammonium salts. Among these may be mentioned ordinary carbon dioxide and water, with which ammonia forms ammonium carbonate. In the third step of the process, the ammonia of the ammonium salts is oxidized to nitrite by bacteria known as the *Nitrosomonas* group. Finally the *Nitrobacter* or nitrate group of bacteria oxidizes the nitrite to nitrate. Thus nitrification finally results in the conversion of unavailable organic nitrogen to available nitrates. It should be observed here that nitrification does not increase the amount of nitrogen in the soil. It may actually result in a loss of nitrogen by leaching.

Nitrogen fixation Ordinary plants are unable to utilize atmospheric nitrogen. There exist in the soil, however, two types of bacteria that are able to assimilate the atmospheric nitrogen in soil air by converting it into proteins. These bacteria are known as **nitrogen-fixing** bacteria, and the process by which they assimilate free nitrogen is called **nitrogen fixation.** One type of these bacteria lives independently in the soil. Two prominent genera of this group are *Azotobacter* and *Clostridium.* The other type, consisting of species of *Rhizobium,* lives in enlargements called tubercles, or nodules, on the roots of leguminous plants such as peas, beans, and clover (Figs. 7-6, 7-7). Both of these groups tend to increase the amount of nitrogen in the soil. In the free-living forms the proteins formed from the atmospheric nitrogen are left in the

Fig. 7-6 Enlarged view of nodules on roots of a bean plant.

soil when the organisms die and by nitrification are made available to higher plants.

The nodule-forming bacteria, sometimes called symbiotic nitrogen fixers, enter the roots of legumes chiefly through root hairs and gradually work their way inward by means of an "infection thread" to the cortical cells. Here they cause abnormal growth which results in the well-known nodules, or tubercles. The bacteria utilize the carbohydrates and other food of the legume and undergo a change in form, considered a type of degeneration, resulting in V- and Y-shaped organisms. In this condition they utilize atmospheric nitrogen to synthesize proteins. Some of this nitrogen becomes available to the leguminous plant in which the bacteria are growing. When the plants are harvested, the roots containing bacteria are left in the soil, and when they decay, the organic nitrogen made by the bacteria becomes available nitrate through the process of nitrification. Each type of legume seems to require a particular strain of bacteria in its roots. If this strain is not present in the soil, it can be added artificially. Usually this is done by inoculating the seeds before they are sown.

From what has been brought out it is clear that a leguminous crop in a rotation possesses an advantage besides that of the value of the crop itself, since it increases the amount of nitrogen in the soil. This is one of the principal reasons why soybeans and sweet clover are used in a potato rotation or clover and alfalfa in an ordinary field-crop rotation.

Denitrification In addition to nitrifying and nitrogen-fixing organisms of the soil there are others that bring about a reverse process which results in the breaking up of nitrogenous sub-

Fig. 7-7 Nodules on roots of peanut plant; also shows peanut fruits developing below ground level. (Courtesy of R. M. Persell, U.S.D.A., Agricultural Research Service, Southern Utilization Research and Development Division, Beltsville, Md.)

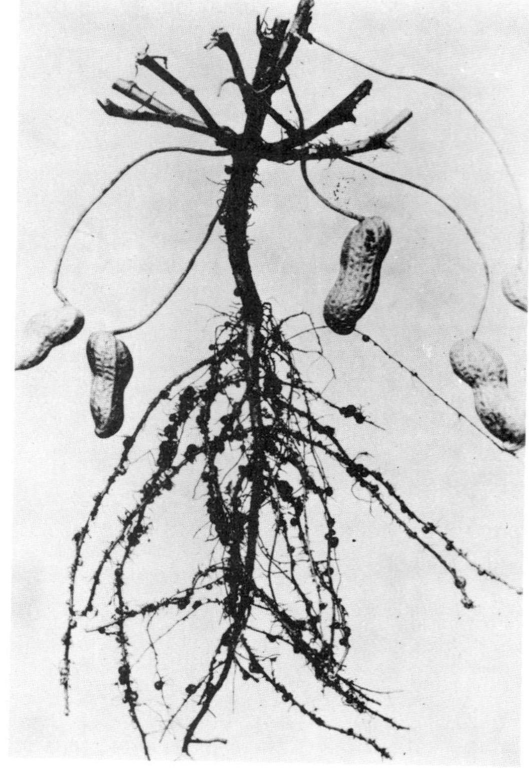

stances with the liberation of free nitrogen gas. These are called **denitrifying** organisms. Fortunately these organisms are chiefly anaerobes; i.e., they live in the absence of free oxygen and do not thrive when good aeration is maintained. Consequently, as long as a soil is kept in a good state of cultivation, the denitrifiers are suppressed and the nitrifying and nitrogen-fixing organisms are favored, leaving a balance of available nitrogen for higher plants.

Effect of roots on the soil Roots, themselves, growing in the soil have important effects on the chemical and physical nature of a soil which directly or indirectly affect absorption of inorganic substances by the plant. Thus it has already been stated that the decay of roots furnishes available nitrogen and other substances for later crops. By this means considerable amounts of organic matter are added to the soil each year. This organic matter influences both the physical and the chemical properties of the soil in addition to furnishing available solutes. Sometimes, however, complex organic compounds are formed from either living or decaying roots which prove to be toxic to later plants. These substances may be toxic only to the kind of plant which produced them. In such cases crop rotation overcomes the injurious effects of the toxic substances.

All living roots, in the process of respiration, give off fairly large quantities of carbon dioxide, CO_2. This carbon dioxide, uniting with water, forms carbonic acid, H_2CO_3, which acts as a weak solvent for some substances that would not readily dissolve in water alone. Thus it can be shown that after roots have been grown on a polished marble plate, the plate will be etched in all places where the roots came in contact with it. This etching is caused by the conversion of the relatively insoluble calcium carbonate, $CaCO_3$, or marble, to the more soluble bicarbonate, $CaH_2(CO_3)_2$, according to the following equation:

$$\mathbf{CaCO_3 + H_2CO_3 \rightarrow CaH_2(CO_3)_2}$$

This reaction takes place regularly in all limestone soils. The carbonate ion may also be exchanged by the root for NO_3^- or other ions needed by the plant.

The lower organisms in a soil also give off carbon dioxide in respiration. It is possible for the carbon dioxide content of the soil air to become great enough to be injurious to plants. This is caused partly through the relative decrease in oxygen content of soil air under these conditions. In a well-aerated soil the excess carbon dioxide escapes to the atmosphere, where it may be absorbed by the leaves and utilized in photosynthesis.

Factors affecting the absorption of inorganic substances by roots The processes involved in the absorption of inorganic substances by the root hairs are exceedingly complex and not yet well understood. They involve diffusion, but the metabolic activity of the absorbing organs is probably the controlling factor. The mere fact that a mineral is absorbed indicates that the plasma membrane of the root hair is permeable to it. This being true, it is obviously erroneous to speak of the entrance of minerals as taking place by osmosis. Minerals are not simply swept in with the water but diffuse in independently of the movement of water. In fact, it is possible for a mineral to be moving out of the root hair while water is moving in.

It has already been stated that some of the minerals of the soil exist dissolved in the soil solution. Only dissolved minerals are free to diffuse into the root hair. Yet not all dissolved minerals are equally able to enter. If the root-hair membrane is not permeable to them, they will not be able to enter at all, even though they are in solution. Furthermore the fact that the membrane is more permeable to some than it is to others will affect their rate of entrance. In general any mineral to which the root-hair membrane is permeable may enter, even though it may be injurious to the plant. Thus, if a sufficiently strong solution of copper sulfate, zinc sulfate, or other soluble toxic substance to which root-hair membranes are permeable is brought into contact with a root, it will be absorbed by the root hairs even though it kills the plant. The plant, in other words, has no power to select only such substances as are useful to it. Sometimes a plant will absorb a relatively useless ion more readily than a useful one. Thus sodium is absorbed by barley plants from some solutions much more readily than calcium is, and chlorine is more readily taken in than sulfate, even though sodium and chlorine may be less useful to the plant than calcium and sulfate. For this reason we may find substances in plants that probably are of no particular use to them but which existed in the soil solution and could not be excluded because the root-hair membrane was permeable to them.

One of the laws of diffusion states that a dissolved substance will always move more rapidly from a region of high concentration of that substance to one of lower concentration. Consequently, in order for a mineral to be absorbed, it must exist in higher concentration in the soil solution than in the cell sap of the root hair; otherwise the root hair would lose this mineral to the soil solution. It should be remembered here that each solute moves independently of the others and according to the relative concentration of that particular solute in the soil solution and in the root hair. For instance, if sodium nitrate is to be absorbed by the root hair, there must be a higher concentration of sodium nitrate in the soil solution than in the root hair, regardless of the concentrations of all other minerals.

If all the minerals that are absorbed by the root hair were to remain in the root-hair cell, a concentration would soon be reached which would cause these minerals to move out of the root hair as fast as they are moving in, and further net absorption would cease. This condition, however, probably does not occur because the minerals are free to diffuse inward to the adjacent cortical cells of the root. As soon as the concentration of a particular mineral becomes greater in the root-hair cell than it is in the adjacent cortical cell, this mineral will diffuse into the adjacent cortical cell. Similarly this latter cell will pass the mineral on to the cell adjacent to it and thus the minerals diffuse from cell to cell until they reach the xylem. There is thus established a decreasing concentration gradient of each mineral from the root hair to the xylem, permitting a continuing absorption of mineral salts from the soil solution.

In many of the lower plants growing in water, the absorbed minerals do remain in the cells in which they are absorbed. We should therefore expect to find that a concentration of a particular mineral would soon be reached which would equal the concentration of that mineral in the medium from which the mineral was obtained and that this would prohibit the entrance of any more of this mineral. Apparently, however, there are cases in which this is not true. For example, the common freshwater alga *Nitella* lives in a medium in which the chlorine content varies between 20 and 30 parts per million; yet the cell sap of the alga may have a concentration of chlorine of 3,500 parts per million. Moreover, under suitable conditions all the chlorine in the medium may be taken up by the alga, and yet none will pass out of the cell unless the cell is injured. Here we have an apparent contradiction of one of the laws of diffusion inasmuch as the chlorine seems to be taken up against a concentration gradient. A satisfactory explanation of this situation is difficult to find although it is possible that as soon as the chlorine enters the cell, it is changed to a form to which the cell membrane is not permeable. Whether this situation may occur also in root hairs is not definitely known. It is now definitely known that the metabolic activity of the absorbing cells influences the absorption of inorganic substances generally. This may explain the phenomenon of absorption against a concentration gradient.

Some of the inorganic salts occurring in the

soil solution are in an ionized condition. Thus sodium nitrate, $NaNO_3$, may ionize as Na^+ and NO_3^- ions. The two ions of a salt are not necessarily absorbed in equal proportions. Any large absorption of one ion in excess of another of opposite charge in equilibrium with it depends upon what in effect amounts to an exchange of ions between the soil solution and the root hair. If, for example, an excess of K^+ ions is absorbed from potassium sulfate, other cations like Na^+, Ca^{++}, or Mg^{++} may be displaced from the root hair. Similarly, NO_3^- ions or Cl^- ions may be absorbed and HCO_3^- ions from the plant exchanged for them. The production of carbon dioxide by roots may therefore play an important role in mineral absorption by the plant. It is also possible for a particular ion to retard or accelerate the absorption of another ion. Thus K^+ is absorbed by barley plants much more readily from the chloride (KCl) than from the sulfate (K_2SO_4). A relatively high concentration of Na^+ may depress the absorption of K^+ or Ca^{++}, and the presence of Ca^{++} may influence the entrance of Mg^{++}, K^+, or Na^+. This influence of one ion upon another is called **antagonism.** The acidity or alkalinity of the soil solution and many other factors probably also affect the entrance of inorganic solutes.

In general the relation between the root hair and the soil solution is a very complex one and the mechanism of absorption by higher plants difficult to understand. It has been demonstrated that the absorption of inorganic substances by the plant is not merely a physical process but probably a complex physiological process in which respiration plays a dominant role.

Fertilizers The composition of the soil solution varies greatly in different places and in the same locality at different times and under different conditions. Inorganic salts of such elements as iron, magnesium, sodium, chlorine, silicon, sulfur, manganese, and aluminum exist in available form in most soils. It is not at all uncommon, however, to find soils deficient in one or another of such salts as contain an available supply of potassium, phosphorus, and nitrogen. This is partly caused by the fact that plants probably absorb greater quantities of these minerals and partly because some of these salts are easily leached out of the soil. Sulfur may also be deficient in some soils, and in particular localities iron, magnesium, or manganese may be in such forms as to be unavailable to the plant. Minerals added to the soil to correct such deficiencies are called fertilizers. While fertilizers are added to soil for a variety of reasons, undoubtedly one of the principal functions is to furnish the plant with available inorganic salts. Fertilizers may be added in the form of farm manure, which contains most of the elements likely to be deficient in a soil; green manure, a green crop, preferably a legume, ploughed under; and commercial fertilizers which consist of salts of potassium, phosphorus, or nitrogen. A commercial fertilizer containing salts of all three of these elements is called a complete fertilizer. Commercial fertilizers are usually designated according to the percentages of nitrogen, phosphorus, and potassium they contain. Thus a 5–10–5 fertilizer contains 5 percent nitrogen, 10 percent P_2O_5, and 5 percent K_2O.

Nitrogen is added in such forms as sodium nitrate, ammonium sulfate, and many organic forms. Potassium is found in fertilizers in the form of potassium chloride or potassium sulfate, and phosphorus chiefly as phosphates. Sulfur is usually added in the form of gypsum. Lime is commonly added to soil to correct acidity and to improve the structure of the soil, but the calcium it contains is also very important in the life of the plant.

Utilization of inorganic substances by the plant Many of the inorganic substances the plant gets from the soil are used as raw materials out of which foods and other important plant constituents are made. The exact functions of each of the elements absorbed are not well known. Most of our knowledge of this matter has been obtained by the examination of plants grown in the absence of one or another of the minerals. It is customary to speak of the functions of the essential elements. It should be remembered that when this is done, it is the compounds containing these elements that are meant. None of the elements is absorbed as such.

Manganese is thought to be necessary for the proper function of respiratory enzymes in the plant. It has been found that a chlorotic condition of spinach, i.e., a yellowing of the plant caused by the failure of chlorophyll to develop, can be overcome by adding manganese salts to the soil.

Iron and *magnesium* are necessary for chlorophyll formation. Magnesium is a constituent of the chlorophyll molecule but iron is not. Just why iron is necessary for chlorophyll formation is not exactly known. Some soils, while containing an abundance of iron, have it in a form unavailable to the plant. Plants grown in such soils cannot develop chlorophyll and therefore do not thrive. The difficulty is usually overcome by spraying such plants with iron sulfate. Iron also functions in respiration as a part of the cytochrome oxidase system. High concentrations of magnesium are toxic to plants.

Phosphorus is a constituent of many organic compounds of the plant, such as phosphoproteins and phospholipids. Since the nuclei of cells contain phosphorus, lack of this element probably interferes with normal cell division and therefore checks growth. The combining of phosphates with sugars, adenosine, and other compounds plays a dominant role in the transformations of carbohydrates, fats, amino acids, and other organic substances in the general metabolism of the plant. It is particularly important in various energy transfers such as those occurring in photosynthesis and respiration. Phosphorus seems to increase root development and in many plants hastens maturity and ripening, particularly of grains.

Potassium seems to be necessary for the proper carbohydrate metabolism of the plant. Whether it directly affects the synthesis of carbohydrates or is chiefly of importance in facilitating the digestion and translocation of carbohydrates is not well known. At any rate, when potassium is deficient, storage organs such as roots, tubers, and seeds are small and shriveled. Potatoes, which are chiefly a carbohydrate crop, give increased yields with ample potash fertilization. It has been shown that in potash-starved plants the sieve tubes of the phloem, through which carbohydrates are carried to the different parts of the plant, degenerate to some extent. When plants have an ample supply of potassium, they have been reported to be more resistant to disease and insect injury.

Nitrogen is used chiefly in the building of proteins. Since all proteins contain nitrogen and since proteins are present in every living cell, the importance of an available supply of nitrogen to the plant is evident. It is absolutely essential to growth, affecting particularly the growth of the aboveground parts of plants. Nitrogen is also a constituent of chlorophyll. A deficiency of nitrogen soon manifests itself in the color of foliage, which becomes yellowish. Excess of nitrogen may cause excessive vegetative growth, which results sometimes in weak stems and tender, juicy foliage which is susceptible to both insect and fungus injury. In such crops as lettuce and cabbage, increased growth of this kind, if not carried to excess, may be an advantage, but in crops grown for seed, fruit, or storage organs, such as grain crops, tomatoes, and potatoes, it is a decided disadvantage since vegetative growth is enhanced and fruit and seed development and food storage are suppressed. Grain crops like wheat, oats, and barley have also a greater tendency to lodge when supplied with excessive nitrogen.

Calcium, usually added to soil in the form of lime, plays a very prominent part in absorption of other minerals from the soil. In the first place, it neutralizes acids which otherwise would prevent proper absorption of minerals. It also has an antitoxic effect on many poisonous substances in the soil, thereby preventing them from injuring the plant. Through its influence on the soil colloids it tends to loosen the soil in such a way as to provide better aeration, better drainage, and a more favorable temperature. In this way it facilitates root growth. In the actual process of absorption it is very important not only because it influences the entrance of other minerals but, through the formation of calcium soaps and calcium proteinates, it has much to do with maintaining the semipermeability of plant-cell membranes. In the absence of calcium the plant not only is unable to take in the needed minerals but allows many of the substances present in the root hair to escape.

It has already been mentioned that the root-hair cells contain calcium pectate, a hydrophilic colloid which enables them to imbibe water. This substance also forms the cementing material for holding cells together throughout the plant and is usually the first substance laid down during the formation of new cell walls. Consequently a supply of calcium is essential to good growth. In its absence the meristematic regions of the plant at the growing tips die. The general vigor of the plant is greatly influenced by calcium. In the cells of the plant it often unites with acids which result from ordinary metabolism, thereby neutralizing them and preventing any harmful action from them. Thus calcium oxalate crystals are found quite commonly in plants, having been formed by the union of calcium and oxalic acid. In these and in other ways calcium plays a dominant role in the life of the plant.

Sulfur is a constituent of at least three of the amino acids that occur in proteins. Since almost all the proteins naturally occurring in plants contain these amino acids, it is clear that sulfur is necessary for protein synthesis. The mustard oils, occurring particularly in the mustard family (Cruciferae), also contain sulfur. **Glutathione** is an autooxidizable substance which can take up oxygen from the atmosphere which may be yielded to other compounds in the cell and result in their oxidation. In this way glutathione may play a prominent part in respiration. These properties of glutathione rest upon the sulfur groups it contains. Sulfur is probably important to plants in other ways also. While most soils contain enough sulfur to supply the ordinary needs of plants, there are some localities in which sulfur is deficient in the soil. In parts of Washington and Oregon, greatly increased yields of alfalfa and other crops have been obtained by adding gypsum, $CaSO_4$, to the soil. These increased yields are attributed largely to the sulfur which gypsum contains.

In conclusion, it should be emphasized that if the plant is to do its best, it must be supplied with a proper balance of all the essential inorganic substances. If one of the inorganic substances is deficient, its lack will soon be manifested in the growth of the plant regardless of how much of all the other substances may be available.

QUESTIONS

1. Would a plant cell be able to take in water by osmosis if it were surrounded by a heavily cutinized cell wall?

2. Do the inorganic substances enter the root hairs by osmosis?

3. Of what advantage is the outer layer of pectic material on a root hair?

4. Why is it not advisable to apply a heavy application of commercial fertilizers during a prolonged dry period?

5. Why is it not advisable to apply commercial fertilizers to a wet lawn?

6. The atmosphere surrounding plants contains about 20 percent oxygen and only 0.03 percent CO_2; yet during the daytime CO_2 moves into the plant faster than it moves out and oxygen moves out faster than it moves in. Explain.

7. Spanish moss is a plant that has no roots.

How does this plant obtain water and inorganic substances?

8. What is the net suction force of a plasmolyzed cell having an osmotic potential of 10 atm when immersed in a solution of a substance to which the plasma membrane is not permeable and which has an osmotic potential of 2 atm?

9. What causes cells to become plasmolyzed?

10. How does the nitrogen gas of the air become available to plants?

REFERENCES

CORMACK, R. G. H. 1949. Development of Root Hairs in Angiosperms. *Botan. Rev.,* **15:**583–612.

CORMACK, R. G. H. 1962. Development of Root Hairs in Angiosperms. *Botan. Rev.,* **28:**446–464.

HAMBIDGE, G. (ed.). 1941. Hunger Signs in Crops. The American Society of Agronomy and The National Fertilizer Association, Washington, D.C. Although this book erroneously calls the inorganic substances the higher plants obtain from the soil "plant foods," it is an excellent symposium on mineral deficiencies in plants with unusually good illustrations, many in color.

KRAMER, PAUL J. 1949. Plant and Soil Water Relationships. McGraw-Hill Book Company, New York.

KRAMER, P. J. 1956. Roots as Absorbing Organs. *Encyc. Plant Physiol.,* **3:**188–214.

SLATYER, R. O. 1960. Absorption of Water by Plants. *Botan. Rev.,* **26:**331–392.

8

STEMS

Stems of cactus growing in Zion National Park, Utah.

The distinguishing features, external characteristics, and parts of stems; characteristics of annuals, biennials, and perennials; general anatomy of the stems of dicotyledons, monocotyledons, and gymnosperms, including origin of primary and secondary tissues, types of steles and vascular bundles; the physiology of stems, including conduction of water, minerals, and organic substances, root pressure, and the ascent of sap; and characteristics of some of the common specialized stems are considered in this chapter.

GENERAL FEATURES

Distinguishing features of stems The stem typically serves as a mechanical support for the leaves, flowers, and fruits and furnishes a path of conduction between these organs and the roots. While most stems are erect aerial structures, some remain underground, others creep along the surface of the ground, and still others are so short and inconspicuous that the plants bearing them are said to be stemless. Morphologically stems have certain external features by means of which they can be distinguished from other plant parts and particularly from roots, with which they are most likely to be confused, especially when underground. True stems arise from buds, they have nodes and internodes, they bear leaves and buds and sometimes roots at the nodes, and they have characteristic markings, such as leaf scars, bud-scale scars, and lenticels (Fig. 8-4). Internally stems are characterized by having a highly developed vascular system. Stemlike structures of lower plants lack the external markings and other features of true stems and have no vascular system.

Origin and growth of stems Stems develop from apical meristems, or buds. The first stem, or bud, is visible during the organization of the embryo and is commonly called the plumule. It may consist of several unelongated internodes and one or more leaf primordia. Growth in length of stems results from the activity of meristematic regions of terminal or lateral buds and consists not only of an increase in the number of cells but also in enlargement of cells already found in the bud. Consequently the stem tip may be divided into zones comparable with those of the root. There is, however, no structure like a root cap over the meristematic region of the stem. As a rule, the leaves of several nodes of the stem are folded over the stem tip in such a way as to protect this region even in actively expanding buds (Fig. 9-1). Whereas the region of elongation of the root averages around 2 to 10 mm in length, that of the growing stem is often 2 to 10 cm long and may involve several nodes and internodes (Fig. 8-1*A*). Below this growing region at the tip, the tissues become fully differentiated and then cease to elongate, although in dicotyledonous plants and gymnosperms, growth in diameter of the stem continues. In monocotyledonous plants elongation occurs at the bases of the internodes near the tip as well as at the tip of the stem (Fig. 8-1*B*). In many trees and shrubs the total growth in length of the stems during any one year occurs during a very short period in the spring. During the remainder of the growing season, new buds are differentiated at the stem tips and in the leaf axils.

Buds Buds are undeveloped and unelongated shoots, often in a dormant condition. Sometimes they are merely undifferentiated masses of meristematic cells. In other cases they are more elaborate structures, consisting of several nodes and short, unexpanded internodes, the whole enclosed by the closely packed, partly differentiated leaves. In perennial plants the buds are dormant during the unfavorable growing conditions of late autumn and winter. At such times the more prominent buds are protected by a series of overlapping scales which are usually specialized leaves. **Bud scales** (Fig. 8-4*B*) are often covered with hairs or wax, which increase their efficiency as protective organs. When buds are thus protected by scales, they are called **covered** or **protected buds** (Figs. 8-2, 8-3); those without scales are called **naked buds.** Herbaceous plants have naked buds. Covered buds are usually larger and more conspicuous than naked buds. Many buds are almost microscopic in size whereas others are very large. A head of cabbage, for example, is an unusually large bud.

The majority of the buds of a plant remain undeveloped for indefinite periods of time and for this reason are known as **dormant buds** in contrast to **active** or **developing buds.** They may

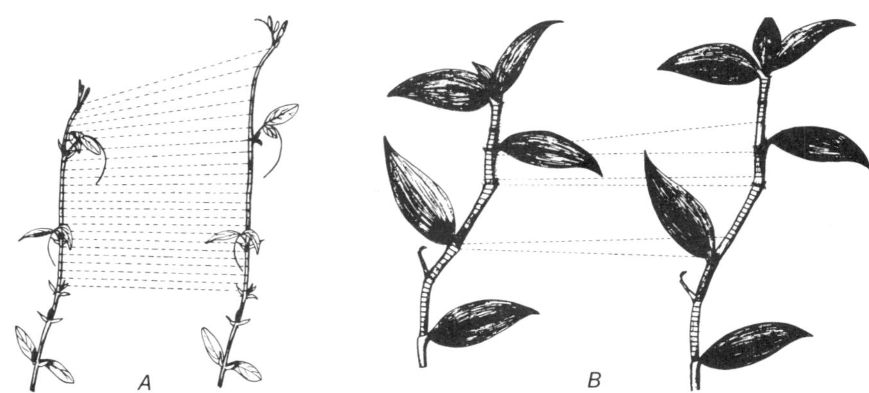

Fig. 8-1 A, region of elongation of the stem of a dicotyledonous plant, sweet pea (Lathyrus odoratus). *The stem tip, left, is shown as originally marked into 2-mm zones; right, the same stem is shown as it appeared two days later. Note that growth in length is restricted to the upper nodes and internodes, in this case involving about 24 mm of the stem tip. B, region of elongation in the stem of a monocotyledonous plant* (Tradescantia zebrina). *The plant, left, is shown marked into 2-mm zones; right, its appearance after three days. Note that the principal elongation, indicated by dotted lines, occurred at the base of each of the internodes near the tip. The growth at the apex is concealed by the leaves. Elongation at the base of the internodes results from the activity of intercalary meristems. (Drawn by Florence Brown.)*

never develop except under unusual conditions. If, for example, the end of a twig is broken off, buds which ordinarily would not have developed may renew growth.

Some buds give rise only to vegetative shoots consisting of stems and leaves. Such buds are often called **leaf buds,** although better names would be **stem buds** or **vegetative buds.** Others may develop flowers only. These are called **flower buds** or, since flowers may develop into fruits, **fruit buds.** Still others give rise to both vegetative shoots and flowers and are called **mixed buds.** The so-called fruit buds of the apple are in reality mixed buds. Most axillary buds are leaf buds, while buds occurring at the tips of stems are often mixed buds, as in many of the terminal buds of the horse chestnut. Peaches and some maples produce separate flower buds.

According to their position on the stem, buds may be classified as **terminal, axillary, accessory,** and **adventitious.** In a strict sense all buds of the stem except the terminal buds might be called lateral buds, but this term is most commonly used as a synonym for axillary buds.

Terminal buds, i.e., buds occurring at the tips of branches, are found on most plants and are particularly conspicuous on many trees and shrubs (Figs. 8-2, 8-3). They are often the largest buds on the plant and usually give rise to the principal growth in length of plants or branches bearing them.

Axillary, or **lateral, buds** develop in the axils of leaves (Figs. 4-1, 8-4). Axillary buds are often inconspicuous because of their small size. The majority remain dormant, but the removal or death of a terminal bud often brings them into activity. Since the axillary buds that do develop usually give rise to vegetative shoots, the type of branching of a plant is closely related to the position of these buds on the stem and hence

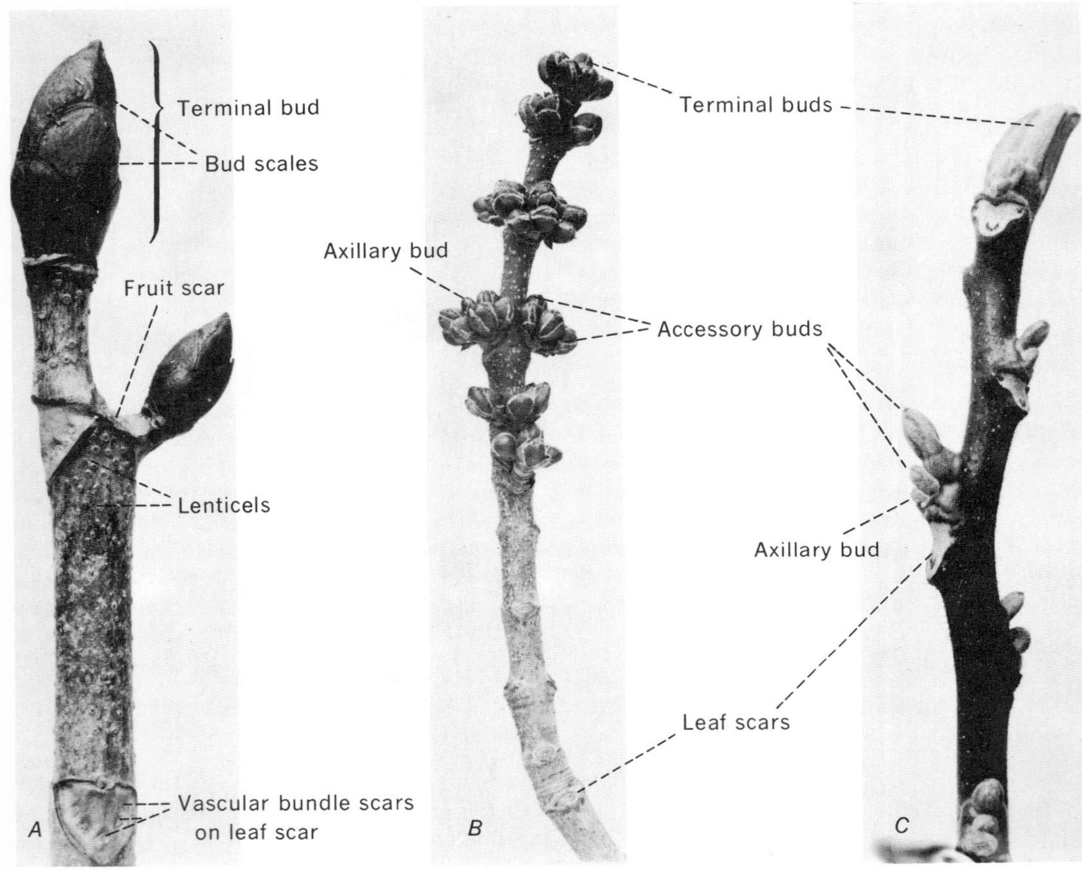

Fig. 8-2 Twigs showing buds and external markings when the trees are dormant. A, horse chestnut; B, silver maple; C, butternut.

to the phyllotaxy of the plant. Thus, since maples have opposite leaves with a bud in the axil of each leaf, the branches are also opposite.

Some species of plants regularly produce above or beside the axillary buds additional buds called **accessory** or, sometimes, **supernumerary buds.** In the silver maple and in the peach, for example, accessory buds are produced on both sides of the axillary bud (Fig. 8-2*B*). These accessory buds are the flower buds. In the butternut (Fig. 8-2*C*) and in the walnut, accessory buds occur above the axillary bud.

Adventitious buds arise irregularly on the plant not only on the stem but also on the root and, in some species, on the leaf. When branches of willow are cut back, many of these buds arise near the cut surface; when cuttings of sweet-potato roots are planted, adventitious buds give rise to the shoots which develop into the new plants. *Bryophyllum* leaves placed on moist soil also develop shoots from adventitious buds in the leaf notches (Fig. 4-26). Adventitious buds thus serve to propagate plants vegetatively. They also give rise to the common water sprouts of apple trees and other species.

Nodes and internodes **Nodes** are the places on the stem where the leaves arise. Axillary buds also occur at the nodes. The space between two succeeding nodes is called an **inter-**

Fig. 8-3 Opening buds of hickory, showing bud scales folding back and leaves emerging.

node (Figs. 4-1, 8-4). In most dicotyledonous plants there is no very great difference in appearance between nodes and internodes even in mature stems. In the monocotyledons, however, such as corn, the nodes, often called joints, are swollen somewhat and stand out in contrast to the thinner internodes. The entire internode of the grasses is often encircled by the sheathing leaf base, so that the leaf appears to be borne at the next upper node (Fig. 4-4, III). The number of leaves at the node is specific for each kind of plant. The nodes of plants with opposite or whorled phyllotaxy usually stand out more prominently than do those of plants with spiral phyllotaxy. The internal anatomy of a node is often different from that of the internode because of the connection of the vascular system of the leaf with that of the stem.

Scars and lenticels **Leaf scars** are the marks left on the stems of deciduous plants where fallen leaves were attached (Fig. 8-4). They mark the place where the abscission layer was formed at the base of the petiole. These scars, if examined carefully, will show **bundle scars** (Figs. 8-2*A*, 8-4*B*) left by the broken ends of the vascular bundles which formerly passed out of the stem into the leaf petiole. There are often as many of these bundle scars as there are main veins in the leaf. The shape of the leaf scar is specific for a given species of plant and may be used to some extent in identifying trees in winter.

Bud-scale scars occur in compact groups on the stem and mark the points of former attachment of the bud scales (Fig. 8-4*B*). The indi-

Fig. 8-4 A, twig of catalpa, showing sympodial growth; B, twig of horse chestnut, showing monopodial growth. Both twigs show characteristic stem markings. (Drawn by Elsie M. McDougle.)

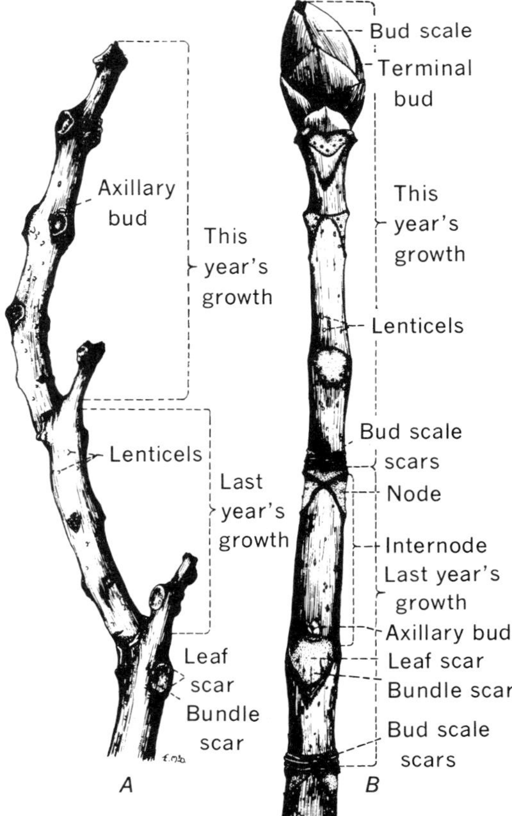

vidual scars are very small and narrow, but, being closely grouped, together they form a rather conspicuous ring around the stem. The majority of them are left by terminal bud scales, and since new terminal buds are formed each year, the age and yearly growth in length of the stem can be determined by observing the number of these rings and the distance between them.

Flower and **fruit scars** (Fig. 8-2*A*) mark the points of former attachment of flowers and fruits. They are sometimes quite conspicuous, as in the horse chestnut and in some fruit trees.

Lenticels appear on woody stems as small openings or pores, often slightly raised or ridged (Figs. 8-2*A*, 8-4). They facilitate gaseous exchange between the interior of the stem and the surrounding atmosphere. The size and shape of the lenticels are often characteristic for a given species. They are particularly conspicuous on young stems. On the bark of cherry and birch trees and on elderberry stems, they appear as prominent transverse markings. Lenticels are multicellular structures (Fig. 8-26) often arising on the stem in positions previously occupied by stomata.

Fig. 8-5 Trees, shrubs, and herbs in Pennsylvania hardwood forest. Trees include pine, maple, and oak; shrubs include laurel and huckleberry; herbs include ferns and bunch grass.

Size and form of stems Stems exhibit wide variations in size, form, and structure. In length they vary from less than an inch to several hundred feet, the latter dimensions being attained by the giant redwoods of the Pacific Coast. In thickness they vary from almost hairlike structures to trunks of trees 50 ft and more in diameter. Some stems are tender, fleshy, or watery, while others are hard and woody.

The classification of plants into **herbs, shrubs,** and **trees** (Fig. 8-5) depends upon the size and woodiness of stems. **Herbs** are plants with no persistent woody stem aboveground. Many are low-growing plants with succulent or fleshy stems. **Shrubs** and **trees** are perennials with woody stems, but shrubs are usually smaller than trees and consist of several main stems, while trees usually consist of a single main trunk or axis.

The large and conspicuous stems of trees assume various forms. The **columnar stem** is cylindrical, relatively unbranched, and usually bears at its summit numerous leaves. The palms, bamboo, and numerous other monocotyledons have this type of stem (Fig. 8-6). Branching stems are of two types, **excurrent** and **deliquescent.** The **excurrent** type characteristically consists of one principal vertical stem, called the trunk, which tapers from base to summit. From it smaller horizontal branches radiate outward. Typically the lowest branches are the longest and oldest and the uppermost ones the shortest and youngest, giving the whole plant a conical form (Fig. 8-7). Such evergreen trees as the spruce and the fir have the typical excurrent

Fig. 8-6 Columnar stems illustrated by three stems of bamboo.

type of stem. The **deliquescent** type consists of a vertical main stem or trunk which rises for some distance above the ground and then divides into several branches which in turn branch again and again, making the trunk seem to melt away or deliquesce (Fig. 8-8). This is typical of the stems of many deciduous trees, as, for example, the oak, the maple, and the elm. When a tree is growing in the open, its branching is likely to assume a different form from that found when trees are closely crowded together. In forests the lower branches, particularly, are often crowded or cut off from sufficient light and hence do not develop normally. Consequently the main trunk grows to a greater height. The best development of the deliquescent habit is usually seen in trees growing in the open.

Monopodial and sympodial branching As already indicated, the manner of branching of stems is closely related to the position of the vegetative buds. Since many of the buds are dormant and since those which do develop, for various reasons, do not grow to the same extent, it follows that branches are not so regular in arrangement as are buds and leaves. The manner of branching of cultivated plants is often considerably altered by severe pruning.

In some woody plants the terminal buds begin to differentiate very early in the growing season and by autumn consist of a definite number of nodes and internodes, together with all the leaves that will be produced by this branch the following spring. In the horse chestnut (Fig. 8-4*B*) the terminal buds may contain not only all the leaves produced by the branch but also the primordia of the bud scales of the next terminal bud. The actual elongation of such a stem takes place regularly, in a straight line, each year's growth being added as a direct continuation of the part of the stem already formed.

Fig. 8-7 Excurrent type of branching of hemlock (Tsuga canadensis). *(Photograph by C. F. Jenkins, from F. D. Kern, "The Essentials of Plant Biology," Harper & Row, Publishers, Incorporated, New York, 1947.)*

Fig. 8-8 Deliquescent type of branching of white oak (Quercus alba). *(Photograph courtesy of Pennsylvania Department of Forests and Waters.)*

Such growth is said to be **monopodial.** Lateral buds may give rise to branches in the same way. Sometimes a flower cluster is borne in the center of a terminal bud. When this happens, the development of the flowers and fruits brings the growth of the main branch to an end and the following year one or more of the lateral buds will develop into new side branches. Monopodial development occurs in the principal axis or trunk of evergreen trees with excurrent branching and in the branches of many deciduous trees with deliquescent or excurrent branching. Pine, spruce, fir, horse chestnut, and apple are examples, the first three having excurrent branching, the last two deliquescent branching.

In catalpa and other trees a different type of development takes place (Fig. 8-4*A*). None of the buds on stems of this type is very conspicuous, and the dormant buds are practically undifferentiated. The apex or tip of the stem consists of a mass of active meristematic cells which continue growth and the formation of nodes and internodes throughout the growing season. As a rule, the younger portions of the stem die at the end of the year, and the following season new growth proceeds from one or more of the lateral buds several nodes behind the apex. This manner of growth results in a series of short branches, each one attached to the side of the next preceding one, the whole forming an irregular or broken line. Often the dead portions of the branches persist for several years. This type of growth is called **sympodial.**

Annuals, biennials, and perennials Plants are classified as **annuals, biennials,** and **perennials,** depending upon the number of years they live. **Annual** plants are those which arise from seed, mature, and die in one growing season. Examples are bean, radish, and lettuce plants. Such plants generally have herbaceous stems. **Biennial** plants are those requiring two years to reach full development. During the first growing season the seeds germinate, and the young plants develop usually a rosette of leaves and store food in the roots. During the second growing season they reach maturity, produce flowers and seeds, and die. Examples of biennials are cabbage, beets, some foxgloves, and mullein. The stems of biennials are also of the herbaceous type. **Perennials** are plants which live for more than two years—in the case of some forest trees often for hundreds of years. In general, perennials are of two types, those with herbaceous stems and those with woody stems. The herbaceous type (sometimes called winter annuals) has aerial stems which die down to the ground each year and underground parts which live through the winter and the following growing season give rise to new shoots. Examples of herbaceous perennials are asparagus, rhubarb, and many grasses. The woody types have aerial stems which live for many years. Each year a new season's growth is added to that already made, and the stem increases in diameter, largely by the addition of new wood. Shrubs and trees are woody perennials.

ANATOMY OF STEMS

From what has been stated concerning the general diversity of form and manner of growth of stems, it is not surprising that there is also great diversity of internal structure. In this respect the stems of dicotyledons differ from those of monocotyledons as well as from those of gymnosperms and lower vascular plants. Even in the dicotyledons the internal structure of herbaceous plants is often quite different from that of woody plants, and the stems of vines are often unlike other stems. Within these larger groups there occur also many variations in arrangement and relative proportions of the tissues. In the discussion which follows, only a few representative forms are considered.

STEMS OF DICOTYLEDONS; PRIMARY TISSUES

Origin of the tissues If the extreme apex or tip of the stem is examined by cutting a thin section at right angles to the lengthwise axis and placing it under a microscope, it will be found to consist of cells that are all essentially alike (Fig. 8-10*A*). These cells constitute the **promeristem,** or **apical meristem.** The cells of the promeristem are parenchyma cells with relatively large nuclei and dense cytoplasm (Figs. 8-9*B*, 9-1*B*). If the stem is in an active, growing condition, many of these cells will be found dividing. Some of the daughter cells formed by these divisions enlarge and divide again, causing the whole growing tip to expand and move forward. The cells at the extreme tip retain their meristematic condition, but most of those somewhat behind the tip sooner or later cease to divide and develop into mature cells and tissues. If, therefore, a section is cut a short distance behind the tip, the cells will no longer be found to be all alike, but several well-defined regions may be discerned (Fig. 8-10*B*).

In the stems of dicotyledonous plants the outer layer or sometimes several outer layers of parenchymatous cells are called the **protoderm;** several groups or strands of smaller cells forming a circle as seen in cross section, the **procambium** strands; and a mass of parenchyma cells filling in all the remaining space, the **ground meristem.** These three regions constitute the **primary meristems** and they give rise, on further differentiation, to the **primary tissues.** They are thus a transitional stage between the promeristem and the primary tissues and may be considered simply regions of the apical meristem. All the cells of these regions, though showing some differentiation, are parenchyma cells and continue to divide, but the divisions occur less and less frequently, until they cease

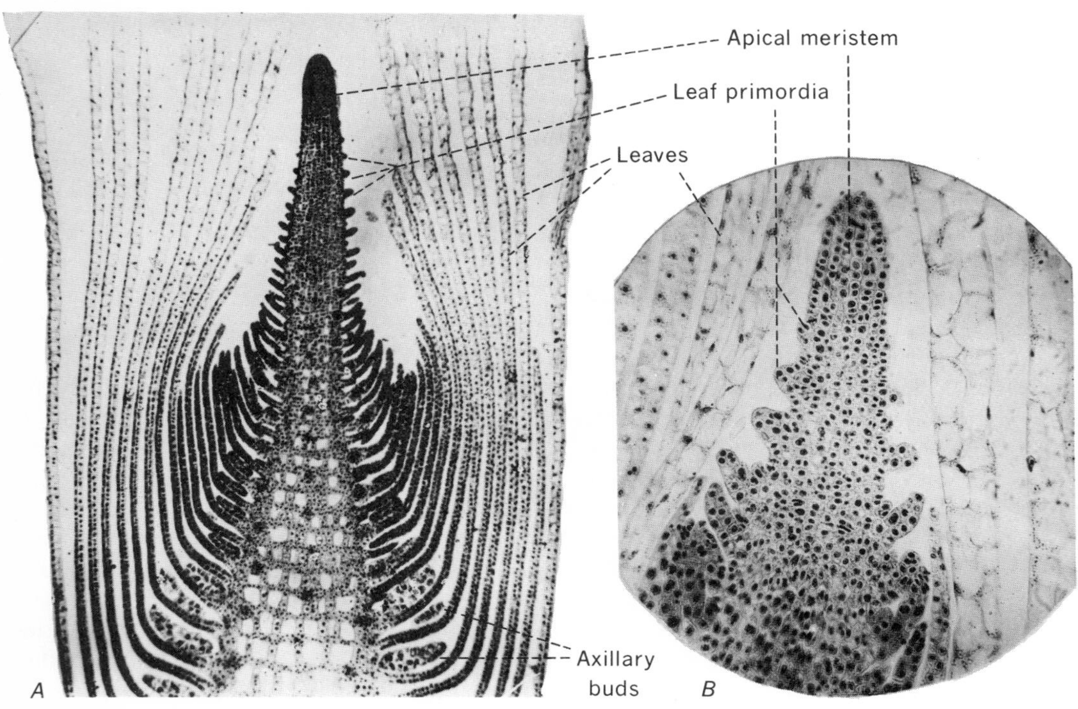

Fig. 8-9 A, Growing stem tip of Elodea canadensis. *B, enlarged view of terminal portion of A.*

altogether when the primary tissues are developed (Fig. 8-10*C*).

The protoderm continues to divide and gives rise to the **epidermis** and, in some cases, subepidermal layers as well. The procambial cells develop into the **vascular tissue,** consisting of xylem, cambium, and phloem. In some cases the vascular tissue may be arranged as **vascular bundles,** but in some stems it forms a complete primary cylinder. If a pericycle develops, it consists of parenchyma cells and lies immediately outside the primary phloem. Its presence in some stems is questioned, but in any event it would seem to be associated with the vascular tissue and would have its origin from procambium.

The ground meristem gives rise to the **cortex, pith rays,** and the **pith.** In some stems the innermost layer of the cortical tissue is recognized as the **endodermis,** but in some cases it is not clearly defined and certainly is not so easily recognized as it is in most roots.

The primary tissues of the stem are the epidermis, cortex, phloem, cambium, xylem, pith, and pith rays. If present, pericycle and endodermis are also primary tissues. They all have their origin from cells produced from apical-meristem activity and contribute essentially to the growth of the shoot. In most of the monocotyledons and in some of the lower vascular plants the primary tissues are the only tissues developed.

Arrangement of the primary tissues In general the primary tissues of the stems of dicotyledons have a concentric arrangement (Figs. 8-10*C*, 8-11). The pith, a mass of fundamental or parenchyma tissue, occupies the center of the stem.

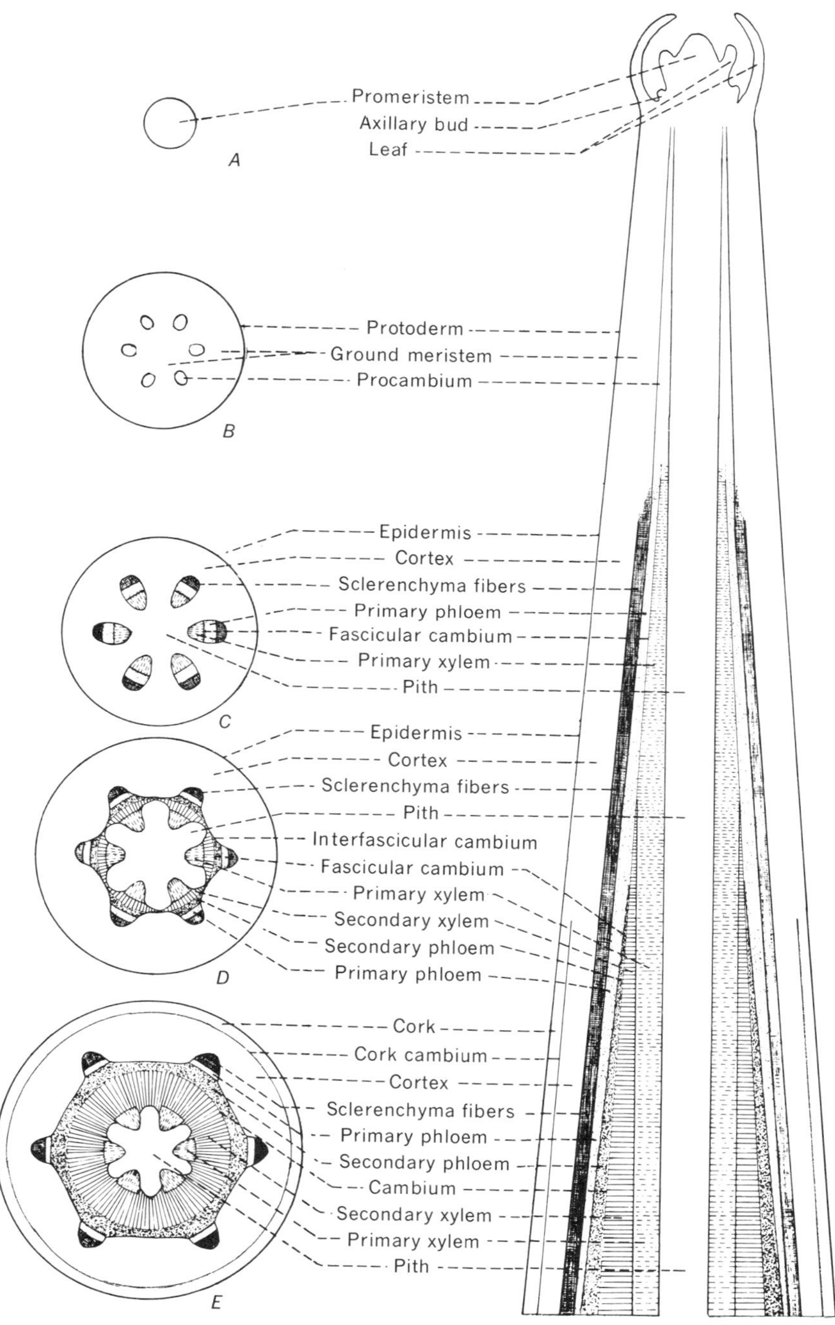

Fig. 8-10 Diagrammatic representation of the origin and arrangement of primary and secondary tissues in the stem of a dicotyledonous plant (modeled after a sunflower stem); longitudinal section of stem at right; transverse sections at different distances from apex (A–E) at left. (Drawn by Florence Brown.)

The vascular tissue is made up usually of separate vascular bundles, consisting of phloem, cambium, and xylem, and arranged in the form of a cylinder surrounding the pith and located between the pith and the cortex. The xylem of each bundle lies next to the pith and the phloem toward the cortex. The cambium, consisting usually of a single layer of cells that retain their meristematic condition and thus continue to divide, is located between the phloem and the xylem in each vascular bundle. The vascular bundles are separated from one another by radiating masses of parenchyma cells, called pith rays. The pith rays appear as extensions of the pith reaching out between the bundles toward the cortex. If present, the pericycle lies outside the vascular tissue, between the vascular tissue and the cortex. It may consist of several layers of parenchyma cells.

In some stems an endodermis occurs immediately outside the pericycle. While the endodermis is a prominent tissue in roots, it is seldom seen in stems and even when it does occur, usually does not have the well-defined structure it has in roots. The cortex lies external to the vascular bundles, forming a cylinder of fundamental tissue in the outer portion of the stem, in many respects resembling the pith. The epidermis consists usually of a single layer of cells surrounding all the other tissues and forming the outer layer of the young stem.

Structure of the individual tissues It should be remembered that there are great diversities in the structure of the tissues of stems of dicotyledonous plants. Woody plants, herbaceous plants, and vines differ from one another and, in addition, each of these groups shows variations. It is therefore not possible to speak of a typical stem of a dicotyledon. Certain structural features, however, are rather common. These are emphasized in the descriptions which follow.

THE EPIDERMIS In general the structure of the epidermis (Figs. 8-11, 8-12*A*) of the stem does not differ greatly from that of the epidermis of other parts of the plant. It consists usually of a single layer of parenchyma cells, somewhat elongated in the direction of the lengthwise axis of the stem. The inner walls, lying next to the cortex, remain thin, while the outer walls that are exposed to air are thicker and usually cutinized. Many stomata are present, which, in woody plants, are later replaced by lenticels.

THE CORTEX The cortex (Figs. 8-11, 8-12*A*) extends from immediately beneath the epidermis to the vascular tissue. The outer part of the cortex next to the epidermis is usually made up of collenchyma cells, the walls of which are thickened at the corners (Fig. 2-24*C, D*). The collenchyma cells aid in the support and strengthening of young stems. The majority of the cells of the cortex are parenchyma, little differentiated, except in size, as compared with the meristematic cells from which they originate. They are usually thin walled and more or less rounded in shape, but may become angular through pressure. Sometimes, however, there are sclerenchyma cells present consisting of stone cells or fibers. The outer layers of the cortical cells frequently contain chloroplasts and hence carry on photosynthesis. Food is stored in many of them.

THE ENDODERMIS An endodermis, although common in roots and in the stems of most lower vascular plants, is seldom found in the stems of seed plants. A few types of dicotyledons, like nasturtium (*Tropaeolum*), however, have a well-defined endodermis (Fig. 8-12*A*). When present, the endodermis forms the inner boundary of the cortex. It is composed of specialized parenchyma cells. The walls of the endodermal cells are frequently heavily thickened and the radial walls suberized. Characteristic thickenings on the walls occur in strips, called Casparian strips. The cells sometimes contain an abundance of starch.

THE PERICYCLE In the stems of many vascular plants the pericycle is difficult to determine

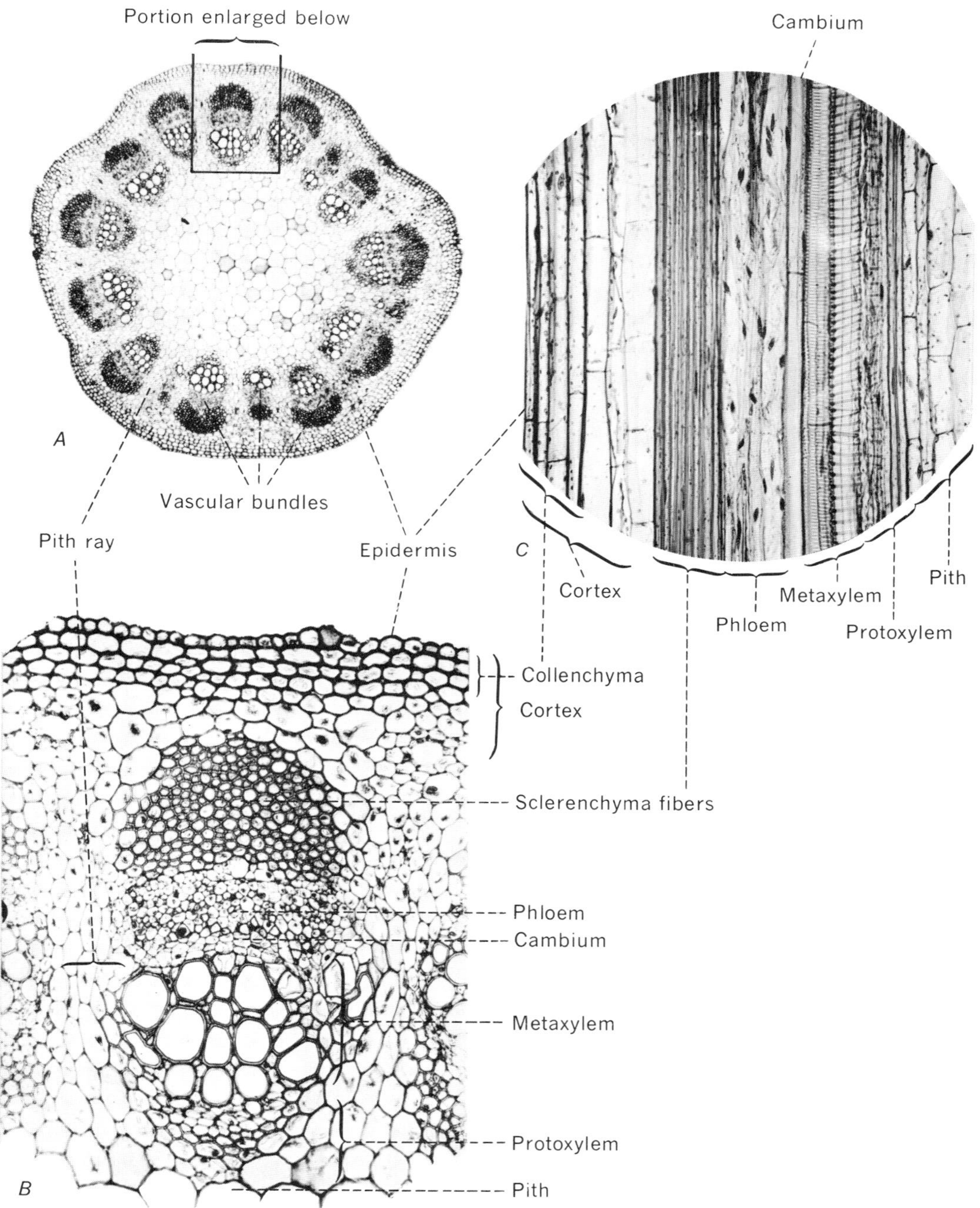

Fig. 8-11 The primary tissues of the stem of a dicotyledonous plant, sunflower (Helianthus annuus). *A, transverse section of entire stem; B, enlarged portion of A, showing structure of the individual tissues, including one vascular bundle; C, enlarged, median longitudinal section of A, from epidermis to pith and through the middle of a vascular bundle. (Photomicrographs by Dr. D. A. Kribs.)*

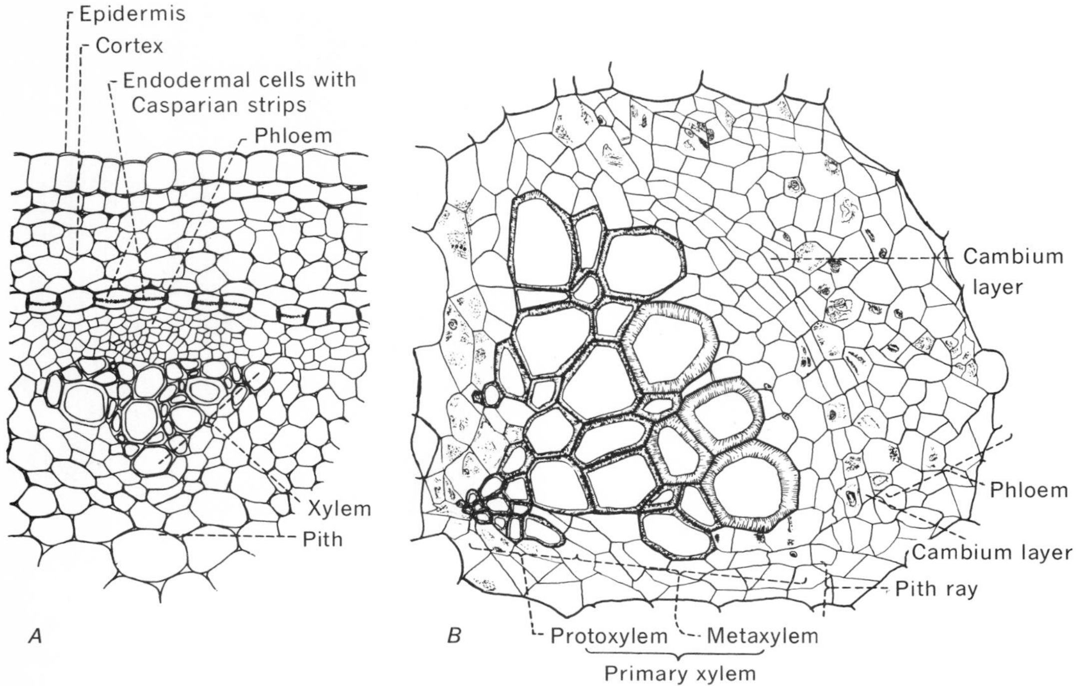

Fig. 8-12 A, portion of a transverse section through a young stem of nasturtium, showing a well-defined endodermis with Casparian strips and other primary tissues; B, transverse section of a single vascular bundle of castor bean (Ricinus communis) *showing the primary tissues. (A drawn by Helen D. Hill; B by F. J. Kilmer.)*

and by some anatomists is considered to be absent altogether. If present, it may consist of several irregular layers of parenchyma, or perhaps sclerenchymatous fibers.

SCLERENCHYMA Sclerenchyma occurs in the primary tissues of many stems. It may consist of fibers or sclereids (stone cells). In some stems with separate vascular bundles there may be a group of fibers just external to the phloem of each bundle, as in sunflower (Fig. 8-11). In other stems (Fig. 8-13) these fibers form a continuous or an irregular band around the vascular tissue. Sclerenchyma may be present in the cortex, in the phloem, or in the pericycle. The sclerenchyma fibers may be lignified, as in hemp, or consist mostly of cellulose, as in flax. They serve as strengthening elements in a young stem. As mentioned in Chap. 2, they are of considerable economic importance in such plants as hemp and flax, which are used in making thread, twine, and rope.

THE VASCULAR BUNDLES In most herbaceous stems and in many woody stems of dicotyledonous plants, the procambium strands very early differentiate into separate vascular bundles consisting of phloem, cambium, and xylem (Fig. 8-11). Separate bundles may not be formed; in this case concentric sections of these three tissues are developed. When separate vascular bundles occur, the primary xylem may lie next to the pith, with the primary phloem external to it. The cambium always lies between the phloem and the xylem. In some dicotyledonous plants, like members of the cucumber and nightshade families, another section of phloem is found in each vascular bundle next to the pith, i.e., between the xylem and the pith. Bundles of this type, in which phloem occurs on both sides of the xylem, are called **bicollateral bundles,** while those in which there is only one section of phloem external to the xylem are called **collateral bundles.** In the bicollateral bundle, cambium is ordinarily found only between the outer phloem and the xylem. The collateral bundle is the usual type in dicotyledonous plants.

The *phloem* of each bundle (Fig. 8-12*B*) consists of sieve-tube elements, companion cells, phloem parenchyma, and often fibers. A sieve-tube element and its companion cell are usually sister cells formed from the same procambial cell. Often the companion cell divides again, transversely, forming two or more companion cells adjacent to a sieve-tube element. Originally the sieve-tube cell has a nucleus and cytoplasm, but as the cell matures, only the cytoplasm remains. Cytoplasmic strands can often be seen extending from one sieve-tube cell to another through the sieve plates. The companion cells, even at maturity, have both nucleus and cytoplasm. The sieve tubes are the food-conducting channels of the stem. Phloem parenchyma often contains stored food.

The *cambium* (Fig. 8-12*B*) is usually a single layer of cells lying between the phloem and the xylem and retaining their meristematic condition. The cambial cells thus retain their ability to divide. They are thin-walled cells, rectangular as seen in cross section, but considerably elongated in the direction of the lengthwise axis of the stem. They usually have a prominent nucleus and dense cytoplasm. By the division of the cambial cells, new cells are added to the phloem and the xylem, giving rise to secondary tissues described later.

The primary *xylem* (Figs. 8-12*B*, 8-14) of the stems of dicotyledonous plants consists of tracheids, vessels, wood fibers of various kinds, and xylem parenchyma. It will be recalled that the tracheids and vessels are chiefly water-conducting cells, the tracheids also functioning in support. In very young primary xylem, the vessels are usually of the ringed or spiral type (Fig. 2-26*A–D*). Ringed or spiral vessels are usually formed during a period of rapid growth and elongation. Vessels formed after the stem has ceased rapid elongation are more likely to be of the scalariform or pitted types (Fig. 2-26*E, F*). In pitted vessels and other conducting cells there are thin areas in the walls, called pits. These pits sometimes have overhanging borders; in this case they are called bordered pits (Fig. 8-29*F*).

The wood fibers of the xylem are greatly elongated cells with small cavities or lumina and thick, hardened walls. The end walls taper and often overlap with the ends of the cells above and below. They are the principal strengthening elements of the xylem of dicotyledons, the

Fig. 8-13 Transverse section of stem of Aristolochia, *showing continuous ring (or section of cylinder) of sclerenchyma fibers outside the ring of vascular bundles.*

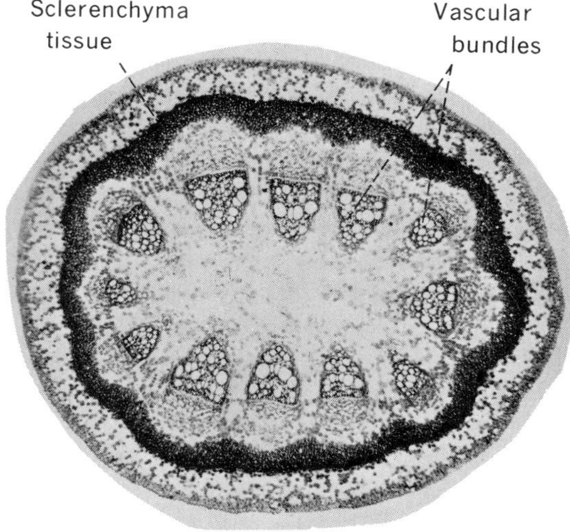

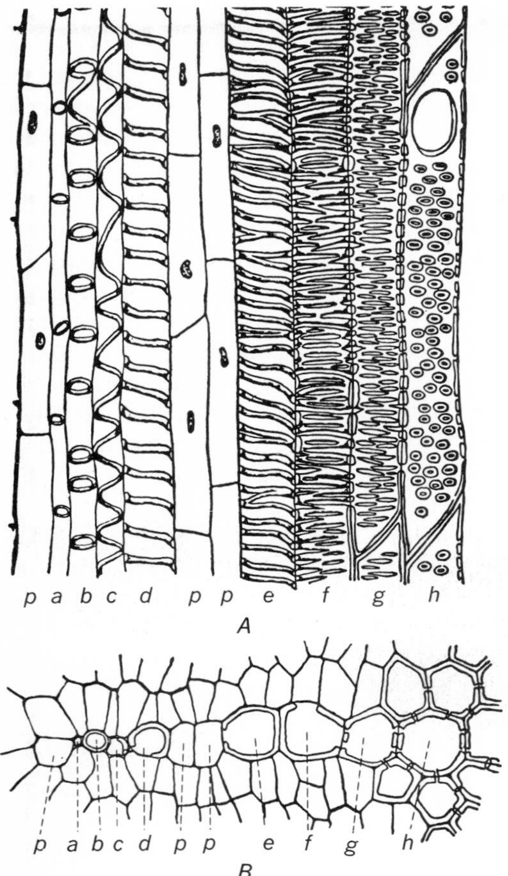

Fig. 8-14 Protoxylem and metaxylem in transverse, B, and longitudinal, A, section in Lobelia; *a, b, annular elements; c–e, spiral elements; f, scalariform element; g, scalariform-reticulate element; h, pitted vessel; p, parenchyma cell. (From A. J. Eames and L. H. MacDaniels, "An Introduction to Plant Anatomy," 2d ed., McGraw-Hill Book Company, New York, 1947.)*

overlapping of the cells assisting materially in this function. The thickening of the wood fibers, as well as that of tracheids and other xylem elements, consists of cellulose and lignin or lignocellulose. For this reason such cells are said to be lignified.

The xylem parenchyma consists of living cells that have not been differentiated into other xylem elements such as fibers and vessels. These cells can be recognized by the presence of square or horizontal ends, generally not tapering, and by their protoplasmic contents. The walls of most xylem parenchyma cells are somewhat thickened but not to the extent of those of other xylem elements. The xylem parenchyma is chiefly a food-storage tissue.

The first xylem to be differentiated from the procambial strands usually lies next to the pith. This xylem is the **protoxylem** (Figs. 8-12*B*, 8-14). Its cells are usually smaller in diameter than those of the xylem developed later. The balance of the xylem formed from the procambial strands gradually differentiates from the protoxylem toward the cambium and phloem. This xylem is the **metaxylem** (Figs. 8-12*B*, 8-14). The cells of the metaxylem are not only larger but also thicker walled than the protoxylem, being more heavily lignified. *The protoxylem and the metaxylem together constitute the primary xylem,* both of them having been derived directly from the procambial strands. It will be recalled that in the root the protoxylem is laid down next to the pericycle and that the metaxylem develops in a direction toward the center of the root, a condition called **exarch.** In the stem of dicotyledons the direction of development of the metaxylem is directly opposite, i.e., away from the center of the stem, and is spoken of as an **endarch** condition, the term endarch meaning "inner origin." The primary phloem may also be differentiated into **protophloem** and **metaphloem,** the metaphloem lying next to the cambium.

PITH AND PITH RAYS The pith (Figs. 8-11, 8-12*A*) is composed of fundamental parenchyma cells located in the center of the stem inside the vascular cylinder. In general the pith does not undergo great modification, although in a few instances the tissues are so greatly altered that their relation to fundamental tissue is difficult to recognize. This is true of the large central storage tissue of the potato tuber. The pith cells are usually rather large in comparison with the cells of the ground meristem from which they are derived and are frequently interspersed with numerous air spaces. They are often filled with starch grains and other foods. In older stems the pith often disintegrates and is finally lost.

In all stems of dicotyledons in which there are separate vascular bundles, masses of parenchyma cells occur between the bundles from the pith outward. These masses of cells appear as extensions of the pith, between the bundles, and they are known as pith rays (Fig. 8-11). In structure and function they are similar to the pith cells.

LACTIFEROUS DUCTS In a number of families of dicotyledons, as well as in a few families of monocotyledons, there is a much-branched system of tubes or ducts in which a milky sap called **latex** is found. This system usually runs all through the plant from roots to leaves and is even found in fruits. It is often especially abundant in the cortex and pith and is found in secondary tissues as well as in the primary tissues. Two types of lactiferous ducts are found, **latex vessels** and **latex cells.** The latex vessels originate in rows of meristematic cells, the end walls of which become dissolved and absorbed, in much the same way as xylem vessels are formed. The latex vessels, however, remain living and have many nuclei; i.e., they are coenocytic. These vessels are connected by many branches, forming an anastomosing system which usually extends all through the plant. This type of system is found in the poppy family (Papaveraceae), the chicory family (Cichoriaceae), the banana family (Musaceae), and a few others including the para rubber tree (*Hevea brasiliensis*), which is the chief commercial source of rubber (Fig. 8-15).

The latex cells are also long tubes, but in reality they are single cells which originate as minute structures in the embryo. As the plant grows, they elongate and branch, keeping pace with the growth of the plant until they form a branching system extending throughout the entire plant body. This type of duct system is found in members of the milkweed family (Asclepiadaceae) (Fig. 8-16), in the spurge family (Euphorbiaceae), and in several others.

Latex is a white, yellow, orange, or red viscous fluid, usually an emulsion, consisting of proteins, sugars, gums, alkaloids, enzymes, oils, salts, and other substances. The latex of a number of plants is collected and used in making commercial products, among the most important of which are rubber, chicle, opium, and papain.

The function of latex in the plant is not well

Fig. 8-15 Collecting latex from a rubber tree (Hevea brasiliensis). *The trees are tapped soon after sunrise. The latex stops flowing late in the morning and is then collected. (Photograph courtesy of F. D. Kern.)*

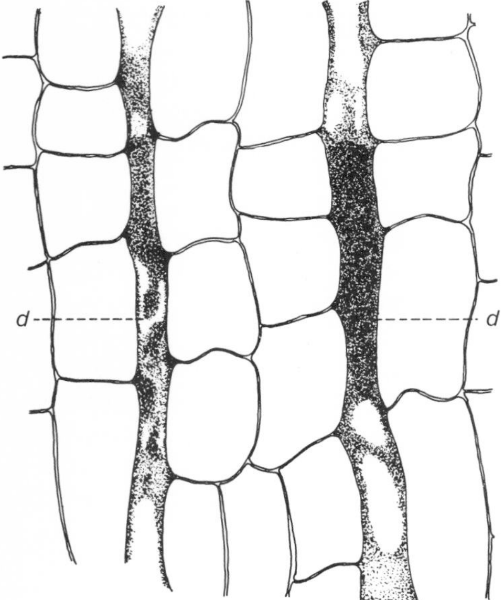

Fig. 8-16 Lactiferous ducts, d, of the latex-cell type, as seen in a longitudinal section of the cortex of milkweed (Asclepias *sp.*). *(Drawn by Florence Brown.)*

understood, but it is thought to be connected in some way with the nutrition of the plant and perhaps to be important in healing wounds and protecting wounded surfaces from attacks of parasites and from drying out.

Familiar examples of plants bearing latex are common lettuce, the milkweeds, the house rubber plant (*Ficus elastica*), and the wild bloodroot (*Sanguinaria canadensis*).

STEMS OF DICOTYLEDONS; SECONDARY TISSUES

The development of the primary tissues causes the stem to grow in length and to some extent in thickness. It gives rise to the primary plant body and to most of the branches of the main stem. In the majority of dicotyledons, and especially in woody plants, growth in diameter is brought about chiefly by the development of **secondary tissues.** Secondary tissues, it will be recalled, are those derived from the activity of cambium or from tissues already fully differentiated. There are two principal types of secondary tissues found in the stems of dicotyledonous plants, viz., **secondary vascular tissues** and **periderm.**

SECONDARY VASCULAR TISSUES

The cambial cylinder; fascicular and interfascicular cambium After the primary tissues have been formed and sometimes even before they are fully differentiated, the secondary tissues of the vascular cylinder begin to develop. In many stems of dicotyledons, especially in woody stems, a complete cambial cylinder is developed from the original procambial strands, although not all parts of this cylinder may be developed simultaneously. In stems in which there are separate vascular bundles, the complete cambial cylinder is formed from the cambium located within the vascular bundle **(fascicular cambium),** which developed from procambial tissue, and from additional cambial cells differentiated from the parenchyma of the pith rays **(interfascicular cambium)** (Fig. 8-17). When fully developed, this cambial cylinder consists of a layer of cells, appearing as a circle in cross section. The new cells formed by the division of the cambium differentiate into secondary xylem, which becomes continuous with the primary xylem, and secondary phloem, which becomes continuous with the primary phloem. These additions to the phloem and xylem cause the stem to increase in diameter (Figs. 8-10, 8-19).

The individual cambial cells are elongated in the direction of the lengthwise axis of the stem but appear rectangular in cross section, the tangential axis being the longer. Cambial cells usually divide along the tangential axis. When a cambial cell divides, one of the daughter cells becomes, on differentiation and perhaps on further division, secondary xylem, or phloem, while the other daughter cell retains

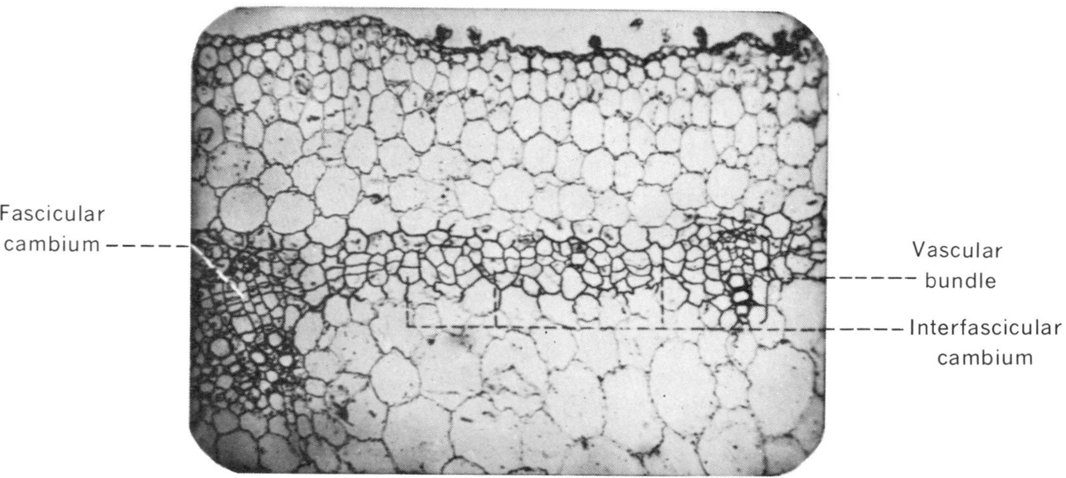

Fig. 8-17 Portion of mint stem, in transverse section, showing development of interfascicular cambium.

its meristematic condition. In this way the cambial cylinder perpetuates itself (Fig. 8-18). In some of the oldest trees a cambial cylinder has thus remained active for a thousand years or more, giving rise to new secondary tissues each year.

Fig. 8-18 Semidiagrammatic representation of the origin of xylem and phloem from the cambium. A–F, successive stages; B, division of the cambium, and C, enlargement of the daughter cells, the upper ones remaining cambium and the lower ones beginning to differentiate into xylem, x_1; D, the cambium has divided again, cutting off cells, p_1, which will become phloem; E, beginning of differentiation of the phloem, p_1; F, three further divisions of the cambium have taken place, resulting in the formation of the xylem cells, x_2 and x_3, and the phloem cells, p_2; in the meantime each of the cells, p_1, has divided again and cut off a companion cell, cc, the remaining daughter cell, s, becoming a sieve tube. (Drawn by Florence Brown.)

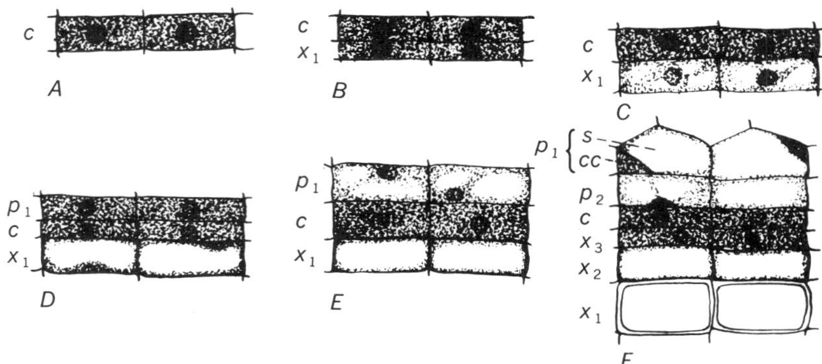

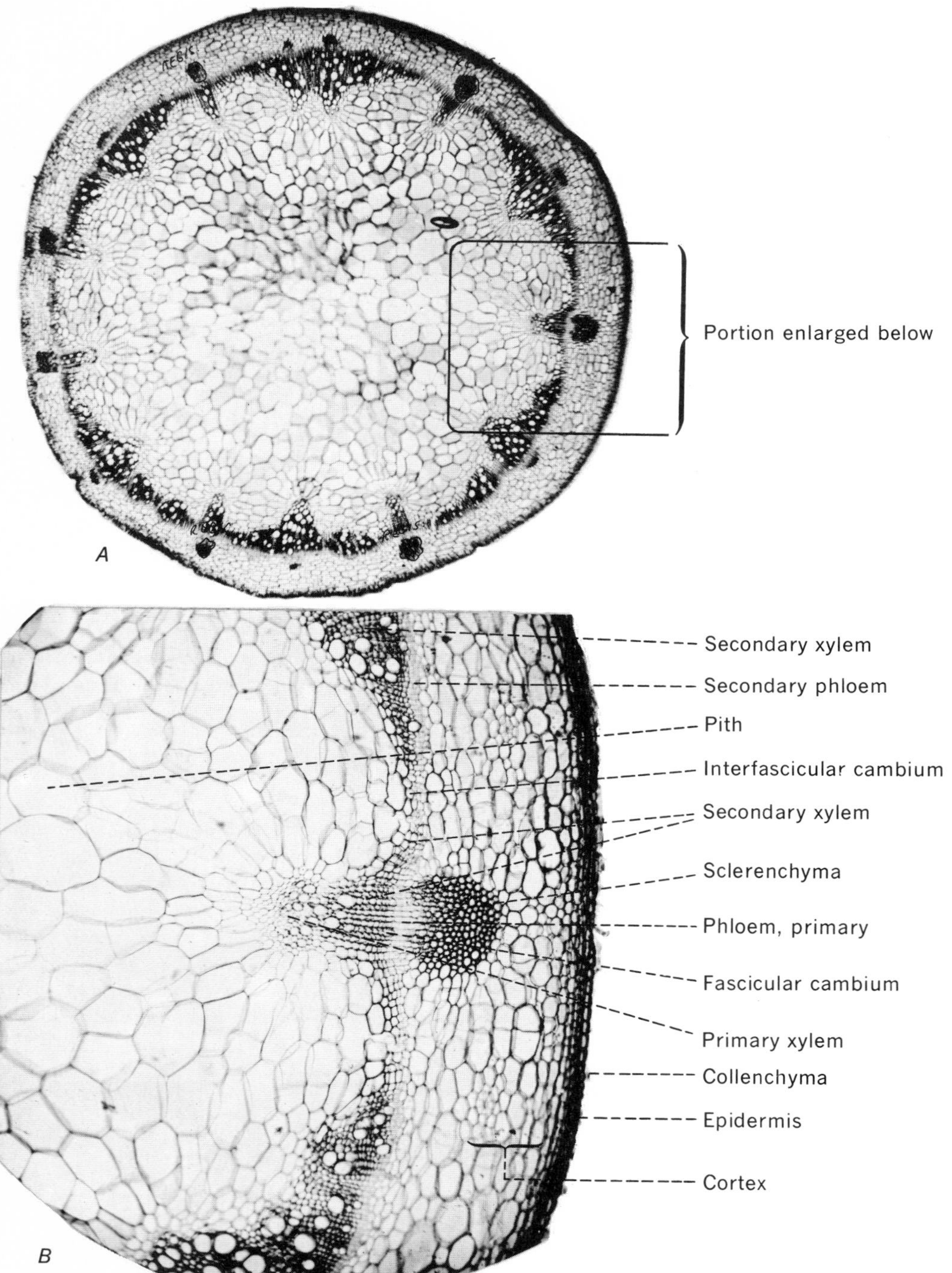

Fig. 8-19 Secondary tissues in the stem of sunflower (Helianthus annuus). *A, transverse section of the entire stem, showing beginning of development of secondary tissues as they are added to the primary tissues; B, enlarged view of a portion of A.*

Secondary phloem If the daughter cells of the cambium which lie next to the phloem become permanent cells, they develop into secondary phloem. Ultimately this secondary phloem, together with the primary phloem, forms a complete cylinder of phloem surrounding the cambial cylinder (Fig. 8-20). The secondary phloem, like the primary phloem, may consist of sieve tubes, companion cells, phloem parenchyma, and fibers. A sieve tube and companion cell may be formed from the same initial cell. Secondary phloem usually resembles the primary phloem to such an extent that it is not possible to determine where one begins and the other ends. Furthermore there is no marked difference, ordinarily, in the size of the cells or the relative proportions of the different cell types formed during different seasons of the year. Hence the yearly additions of phloem cannot be determined so readily as can the yearly additions of xylem. As the stem grows older, however, the primary phloem, as well as some of the older secondary phloem, is pushed outward and placed under greater and greater strain because of the increasing circumference of the stem resulting from cambial activity and the addition of new layers of wood, so that ultimately this phloem may become torn and functionless. The youngest phloem always lies nearest the cambium.

Secondary xylem By far the greater number of cells resulting from the division of cambium differentiate into secondary xylem. This means that the daughter cells of the cambium lying next to the xylem more often become the permanent tissues, while the other daughter cells remain cambium. These cells differentiate into secondary xylem, which becomes continuous with the primary xylem and with it forms a complete cylinder, as does the phloem. The secondary xylem soon comes to make up the bulk of the vascular tissue of woody stems.

The kinds of cells composing the secondary xylem are, in general, the same as those of the primary xylem, but the types of vessels formed and the relative proportions of vessels, tracheids, and wood fibers are often quite different. Also there are great variations existing among species in the relative proportions of the elements that are formed. The great differences in appearance of various kinds of wood are partly explained by these differences. In some species vessels predominate, while in others there are larger masses of thick-walled wood fibers. In general the vessels of the secondary xylem are chiefly of the scalariform or pitted type. The walls of the tracheids and wood fibers are usually also much thicker than those of these elements in primary xylem. The entire mass of secondary xylem usually has a much more regular organization than does the primary xylem.

Wood rays and phloem rays A prominent feature of the secondary xylem of most dicotyledons is the development of **wood rays** (Fig. 8-21). These consist of rows of living parenchyma cells usually containing stored foods and running radially through the xylem. Like the other xylem elements, they are formed by the division of cambial cells. In some cases they may extend from the cambium to the pith. In other cases they may begin with any year's development of xylem. They also extend into the phloem and are sometimes there referred to as **phloem rays** (Fig. 8-21). They originate in the cambium and, once started, usually continue to develop through all the secondary xylem and phloem later formed by additions of cells from the cambium. They are like thin ribbons of tissue extending radially from the center toward the periphery of the stem. While each one does not extend very far in the vertical direction, they are usually so thoroughly distributed through the xylem that any given cell is only a few cells at most away from a wood ray. Wood rays should not be confused with pith rays. Wood rays, originating in the cambium, are entirely secondary tissues.

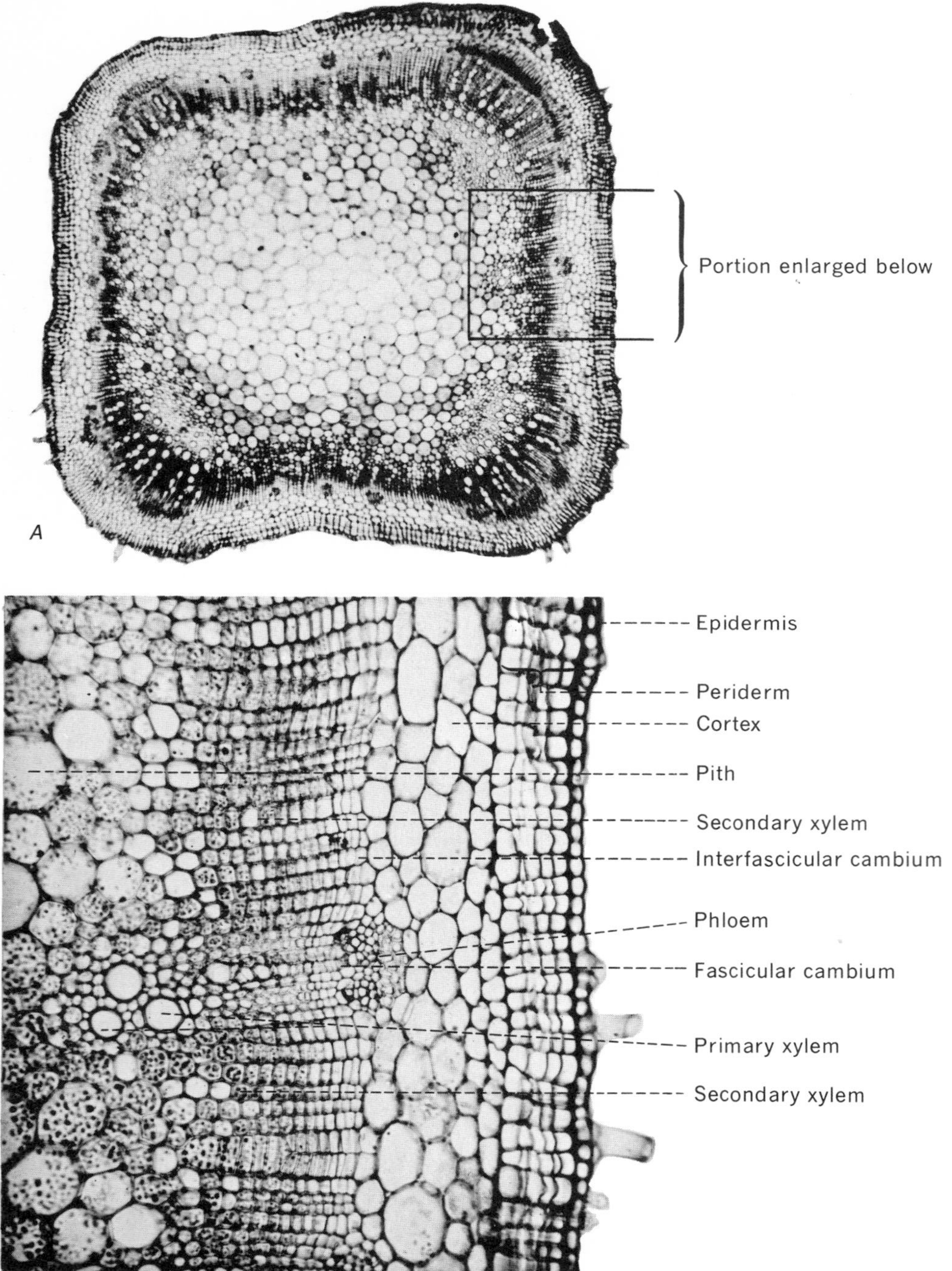

Fig. 8-20 Secondary tissues in the stem of a mint. A, transverse section of the entire stem, showing periderm and a complete cylinder of secondary vascular tissue; B, enlarged view of a portion of A.

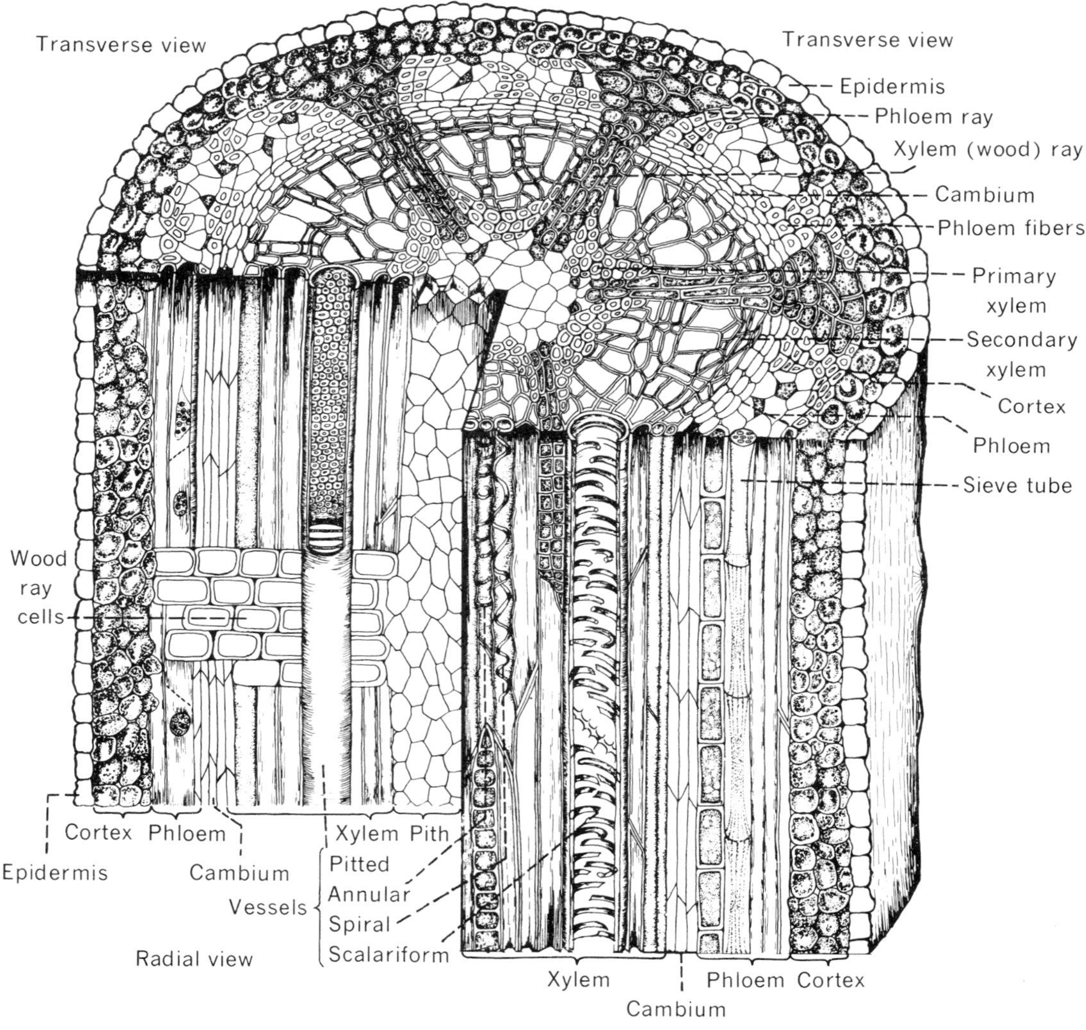

Fig. 8-21 Semidiagrammatic representation of a one-year-old stem of Liriodendron, *in transverse, radial, and tangential views. (Drawn by Edna S. Fox.)*

Annual rings One of the most conspicuous features of the secondary xylem, as viewed in cross section, is that it appears as a series of concentric layers (Fig. 8-22*A–C*). These layers are not always uniform in width around the stem but may be much wider in one place than in another. Each layer represents a year's addition to the xylem and is commonly called an **annual ring.** The width of an annual ring depends upon the amount of xylem made in the particular year. Each annual ring is made up of two parts, an inner layer of **springwood** and an outer, usually more compact layer of **summerwood.** The difference between these

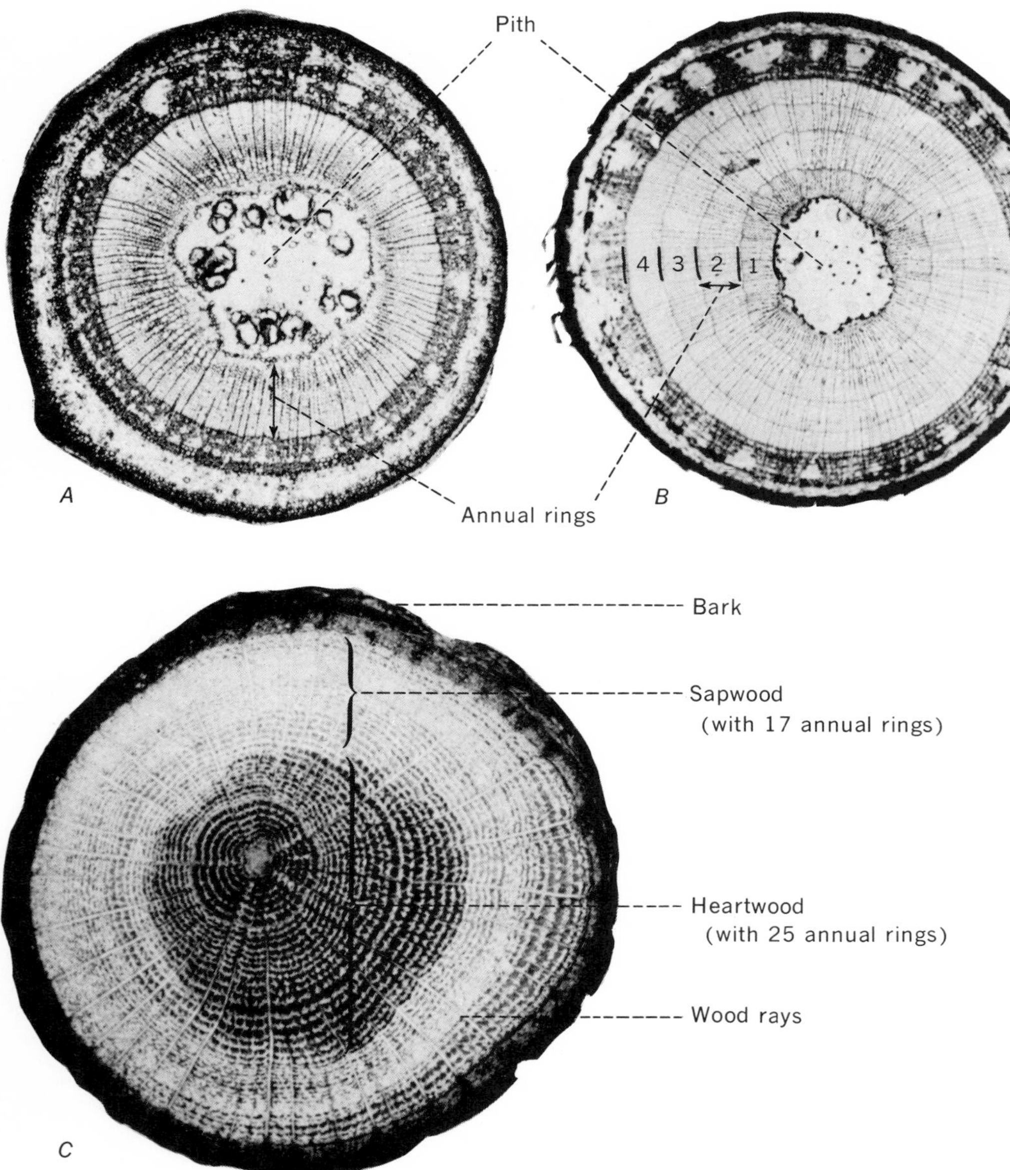

Fig. 8-22 Transverse sections of woody stems of dicotyledonous plants, showing annual rings of xylem. A, B, photomicrographs of Tilia *stems; A, enlarged more than B; A, one year old; B, four years old; C, transverse section of a branch of oak, showing annual rings, wood rays, bark and wood, heartwood and sapwood. (Photomicrographs A and B by Dr. D. A. Kribs; C, photograph by Homer Grove.)*

two parts is brought about by the fact that the tissues formed in the spring, when growing conditions are more favorable, are usually different in cell size, cell type, and cell arrangement from those formed later in the year. In many stems vessels predominate in the springwood and the cells generally are larger in diameter. In the summer and especially toward the end of the growing season, the cells do not become so large, are more compact, and are often thicker walled. There are often more wood fibers developed in the summerwood. Toward the end of the season the cambium ceases to divide and goes into a resting condition over winter. On the resumption of favorable growing conditions in the spring the cycle is repeated.

It is this alternation of active growth and rest and the difference between spring growth and summer growth which cause the yearly increments of xylem to stand out as annual rings. The age of a stem may be determined by counting these rings, although occasionally false annual rings may be formed by one cause or another, such as drought or defoliation by insects. While there is a similar increment of phloem each year, the annual increments do not stand out as rings because there is relatively less phloem formed each year and because there is not so marked a difference in the cells formed during the different periods. It should be mentioned that even during the first year of growth of a woody plant, much of the wood consists of secondary xylem. The first annual ring, therefore, is made up of all the primary xylem as well as the first year's secondary xylem.

Sapwood and heartwood In many stems the youngest or latest-formed xylem comprising only a few years' growth functions most actively in conduction and food storage. This wood is called **sapwood** (Fig. 8-22*C*). It often has a much lighter color. The older wood sometimes becomes stained through the deposition of oils, resins, and coloring matters; the cavities are often clogged with deposits of gummy materials and tyloses (Fig. 8-23), which interfere with conduction. Such xylem is called **heartwood.** In some species the heartwood dries out, while in others it remains wet. In still other

Fig. 8-23 Tyloses. A, transverse section; B, longitudinal section of secondary xylem of oak, showing wood fibers and vessels, the vessels containing tyloses.

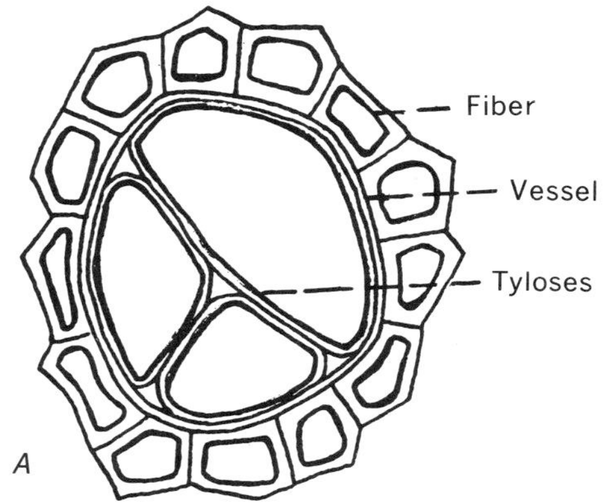

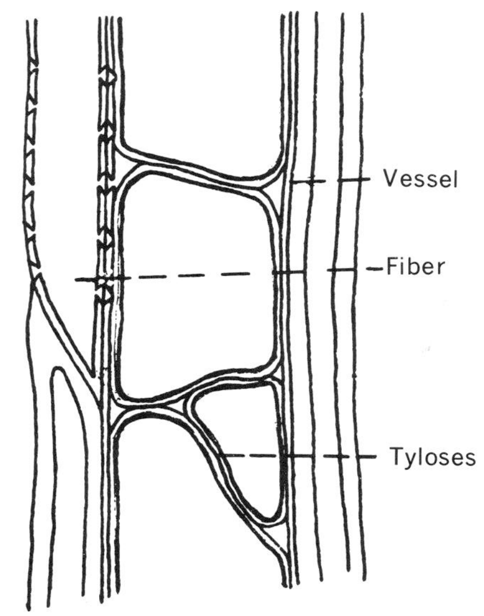

species the entire xylem continues to function. Heartwood is usually more durable as timber than sapwood and in general is commercially more valuable than sapwood. Both sapwood and heartwood are made up almost entirely of secondary xylem, the primary xylem having become altogether insignificant in amount in a tree old enough to show these regions.

Appearance of wood in different sections When the xylem or wood is viewed in **transverse** or **cross section** (Fig. 8-24*A, D*), the annual rings, as previously mentioned, stand out in concentric layers. The wood rays cross these layers, forming lines running radially from the center toward the circumference. The vessels often appear as large pores, while the wood fibers and tracheids form smooth, compact regions made up of smaller, thick-walled cells. Springwood and summerwood can usually be distinguished without difficulty.

In a lengthwise **radial section** (Fig. 8-24*B, E*), i.e., one cut along a radius of the stem, the annual rings appear as parallel bands which differ in different kinds of wood. The wood rays in this section are cut lengthwise and appear as broad bands running across the lengthwise axis. The individual cells of the wood rays appear brick-shaped in striking contrast to the ordinary xylem elements, most of which are greatly elongated in the direction of the lengthwise axis of the stem. Annual rings are readily discernible. Quarter-sawed lumber is cut radially. It is chiefly the position of the wood rays that gives such lumber its striking appearance.

In a lengthwise **tangential section** (Fig. 8-24*C, F*), i.e., one cut at right angles to a radius of the stem, the annual rings often appear as irregular figures. In this section the wood rays are cut in transverse section and hence appear as short chains of parenchyma cells running in the direction of the lengthwise axis or with the lengthwise axes of the other xylem elements. The individual vessels, tracheids, and wood fibers appear much as they do in the radial section. Plain-sawed lumber is cut tangentially.

THE PERIDERM

As the secondary tissues begin to form in the vascular cylinder, another type of secondary tissue starts to develop in the cortical region. In some herbaceous plants the epidermis and the outer cortical cells may continue to increase in size and number of cells for some time and thus keep pace with the increasing diameter. In most woody plants a new tissue, the **periderm,** develops. The periderm, as in roots, consists of three layers, viz., a meristematic layer called the **phellogen,** or **cork cambium,** the **phellem,** or **cork layer,** formed externally to the phellogen, and the **phelloderm,** formed next to the cortex (Figs. 8-20*B*, 8-25).

The phellogen The phellogen arises from relatively mature but living cells that have still retained their ability to divide. In the great majority of cases the phellogen arises from cortical cells immediately beneath the epidermis (Fig. 8-25*A*) (Leguminosae, *Pinus*), but in other instances the phellogen may be formed out of epidermis (apple) or from more deeply located cells (Fig. 8-25*B*), including the phloem tissue (Caryophyllaceae, *Berberis, Vitis*). It will be recalled that in the root the phellogen usually arises in the pericycle. Most of the divisions of the phellogen cells are along the tangential axis. One of the daughter cells remains phellogen, and the other develops into permanent tissue. If the latter happens to lie toward the periphery of the stem, it forms phellem, or cork, and if it lies toward the inside of the stem, it forms phelloderm. As a rule, in stems much more cork is formed than phelloderm.

The phellem The cells of the phellem or cork are very compact and, at maturity, nonliving. Their walls contain the waxy material, suberin, which tends to prevent the inner tissues of the stem from drying out. In a cross section of the

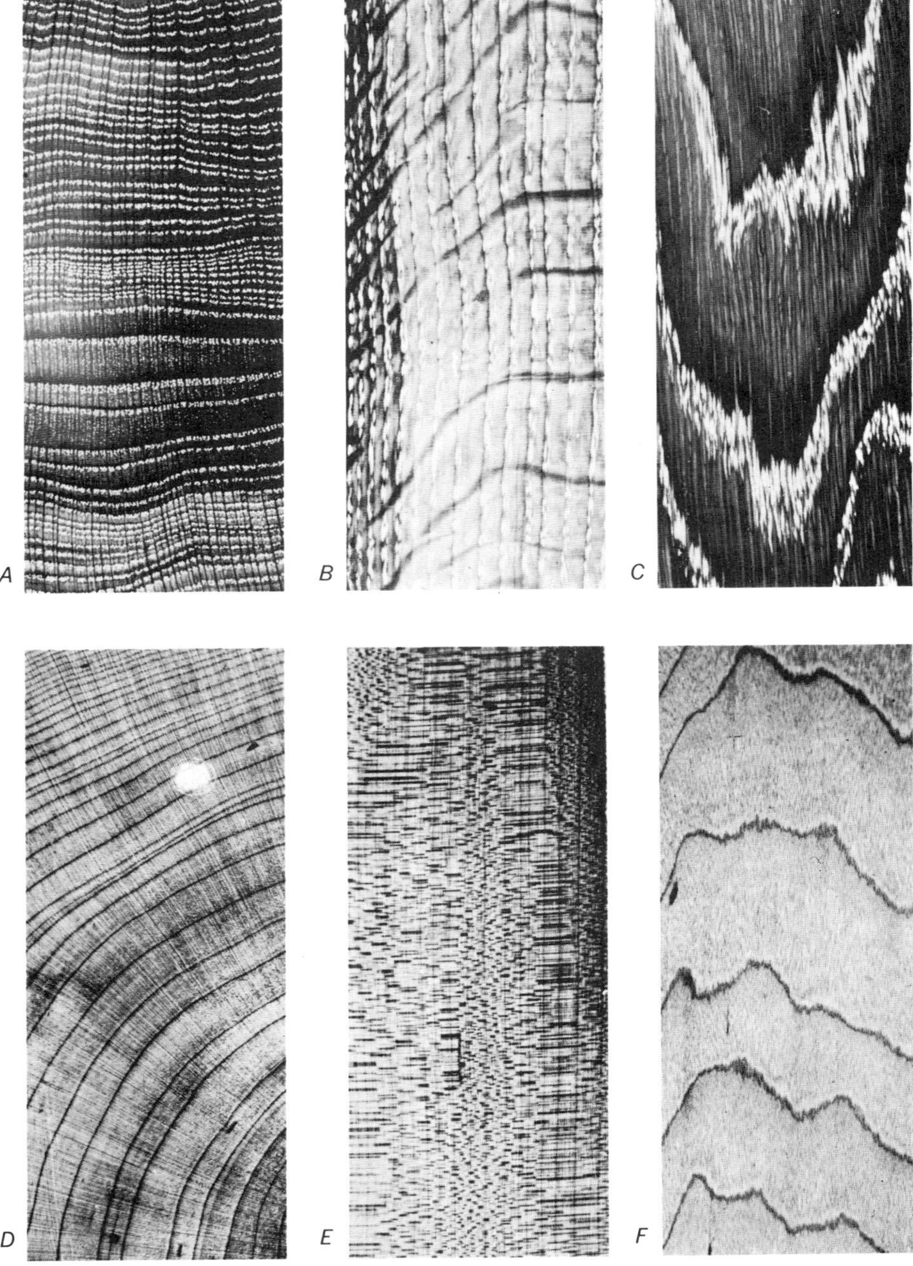

Fig. 8-24 Appearance of wood in different sections. A–C, yellow oak (Quercus muhlenbergii) *wood; A, transverse section; B, radial section; C, tangential section; D–F, silver maple* (Acer saccharinum) *wood; D, transverse section; E, radial section; F, tangential section.*

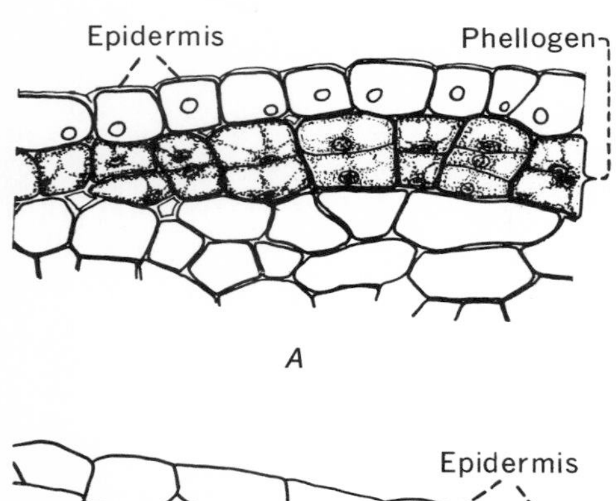

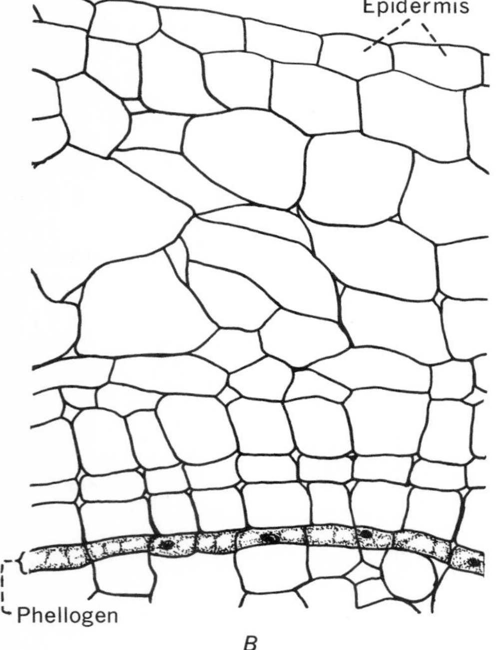

Fig. 8-25 Origin of phellogen and development of periderm. A, transverse section of outer portion of a stem of ash (Fraxinus *sp.*), *showing origin of phellogen in the layer of cells immediately beneath the epidermis; B,* Fuchsia *stem, showing origin of phellogen in inner cortex.*

stem, cork cells often appear roughly rectangular in shape, similar to the phellogen cells but somewhat larger. It is from the phellem of a species of oak (*Quercus suber*) that the cork of commerce is obtained.

The phelloderm The cells of the phelloderm remain living cells not unlike ordinary cortical cells. They sometimes function in photosynthesis and in food storage. In many stems relatively little phelloderm is formed.

Lenticels In older stems the periderm completely replaces the epidermis. The suberized cork cells of the phellem act as a moistureproof covering for the living tissues of the stem. Since this layer is also more or less impervious to gases, gaseous exchange between the exterior and the interior of the stem would be very difficult were it not for the development of lenticels. These structures are developed from the phellogen usually immediately beneath places in the epidermis where stomata occur (Fig. 8-26). The phellogen in such places develops, instead of cork, a mass of loose, thin-walled parenchyma cells with many air spaces. These often project slightly above the outer surface of the stem, appearing as small dots or ridges.

Duration of the periderm In some plants the first periderm persists for many years, adding new cork layers each year. It is obvious, however, that since the cork cells are dead, there is a limit to the extent to which they can stretch to keep pace with the internal expansion of the stem. In many trees the outer layers are soon broken and separated. Often new periderm layers develop deeper in the stem. These may be formed first in the inner cortex, then in the pericycle, then in the primary phloem, and finally in the secondary phloem. In old tree trunks the majority of the periderm layers are developed in the secondary phloem and all the tissues external to this become greatly broken and split up, making a very irregular, ridged outer surface such as that occurring on oak trees. The primary tissues external to the vascular cambial cylinder thus become entirely functionless and in many cases are sloughed off altogether and lost.

Bark and wood In all woody stems one or more years old, two well-defined regions can be discerned in cross section, the **bark** and the **wood** (Fig. 8-22*C*). The bark comprises all the tissues external to the vascular cambium and therefore includes, in young stems, the periderm, the cortex, the pericycle, the primary phloem, and the secondary phloem. In older stems the primary tissues external to the cambium are usually so disorganized as to be unrecognizable, the remaining tissues being periderm and secondary phloem. The wood, on the other hand, includes all the xylem, primary and secondary, the primary xylem being preserved near the center of the stem. In some stems pith may still be discernible, but in the case of most old woody stems, and especially in trees, the pith occupies so small a portion of the complete cross section as to be hardly visible, or it may be completely disintegrated.

STEMS OF GYMNOSPERMS

Much of the previous discussion concerning the origin and development of the primary tissues of dicotyledons is equally applicable to the stems of gymnosperms, such as the pines, firs, and spruces. Like the stems of dicotyledons, the stems of gymnosperms have a concentric arrangement of the primary tissues, with separate vascular bundles consisting of phloem, cambium, and xylem. The secondary tissues also arise in the same way as they do in dicotyledons. The chief difference between these kinds of stems is in the types of cells in the xylem and the phloem. The xylem of the gymnosperms (Figs. 8-27 to 8-29) is made up almost entirely of tracheids, with bordered pits on the radial walls. In a transverse section these tracheids appear rather thick walled and roughly square in shape. The xylem is traversed by many thin

Fig. 8-26 Lenticel of Prunus avium *in transverse section of stem. A number of successive layers of complementary and closing tissue have been formed, and the large amount of phelloderm dips inward into the cortex. (After Devaux, from A. J. Eames and L. H. MacDaniels, "An Introduction to Plant Anatomy," 2d ed., McGraw-Hill Book Company, New York, 1947.)*

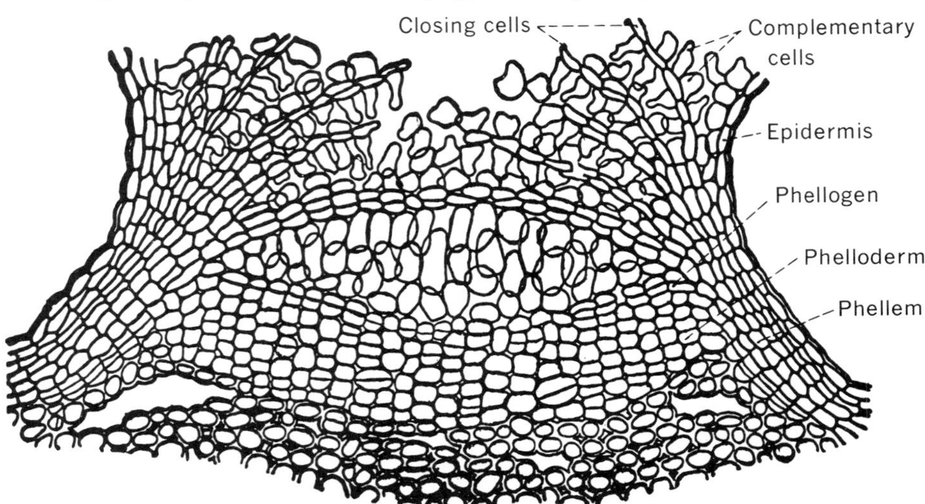

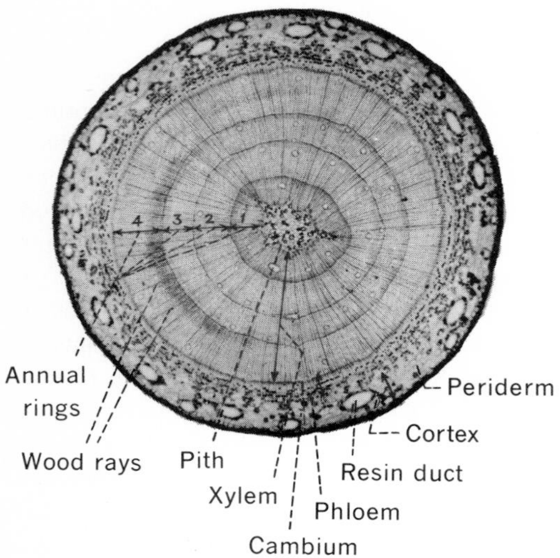

Fig. 8-27 Transverse section of the stem of pine, four years old, showing annual rings.

wood rays (Figs. 8-27, 8-28). In many species there are rather conspicuous resin ducts (Figs. 8-28, 8-29) which run both radially and longitudinally through the stem, but in other species resin ducts are entirely lacking. These resin ducts, or canals, are lined with parenchyma cells and are involved in the secretion and conduction of resin. Little or nothing is known of the function of this resin in the plant. The xylem contains neither vessels nor wood fibers, and in some species even the xylem parenchyma is lacking (Fig. 8-28). Annual rings are found (Fig. 8-27), caused by the difference in size of tracheids formed in the spring as compared with those formed in the summer.

Another conspicuous feature of gymnosperm stems is the complete absence of companion cells in the phloem. Sieve tubes with sieve plates on the side walls, however, are abundant (Fig. 8-29*B*). In general the gymnosperms have woody stems. Many of them attain great age and enormous size, like the giant redwoods and firs of the Pacific Coast and the cypress of the South. In these large trees the development of secondary wood is enormous in quantity. The conifers as a whole are among the most important forest trees used for timber.

STEMS OF MONOCOTYLEDONS

The stems of monocotyledons are, in general, made up entirely of primary tissues derived, as in dicotyledons, from apical meristems. In many monocotyledonous plants, however, and especially in the grasses, upright stems do not elongate much until the time of flowering, when a rapid development of the stem takes place, producing flowers at its summit. In this growth the development of the primary tissues of the nodes frequently lags behind that of the internodes, with the result that meristematic regions (of the nodes) alternate with regions in which permanent tissues have already been fully differentiated (in the internodes). These meristematic regions of the nodes are called **intercalary meristems.** Intercalary meristems are quite generally found in all growing regions of monocotyledons. Ultimately these meristems are completely transformed into permanent tissues. In the stems they occur usually only at the nodes (several to many) nearest the tip (Fig. 8-1*B*).

In general the stem of the monocotyledons consists of a mass of fundamental tissue very much like the pith of the dicotyledons, with isolated vascular bundles which pass vertically through this fundamental tissue (Fig. 8-30). An epidermal layer covers the outside of the stem. In many stems of monocotyledons a tissue consisting of several layers of hardened, thick-walled sclerenchyma forms a cylinder just beneath the epidermis.

The essential structural features of stems of monocotyledons are the arrangement of the bundles, the structure of the individual bundle, and, in general, the lack of cambium. In contrast to the concentric organization of tissues common in the dicotyledons, most of the monocotyledons have a scattered arrangement of

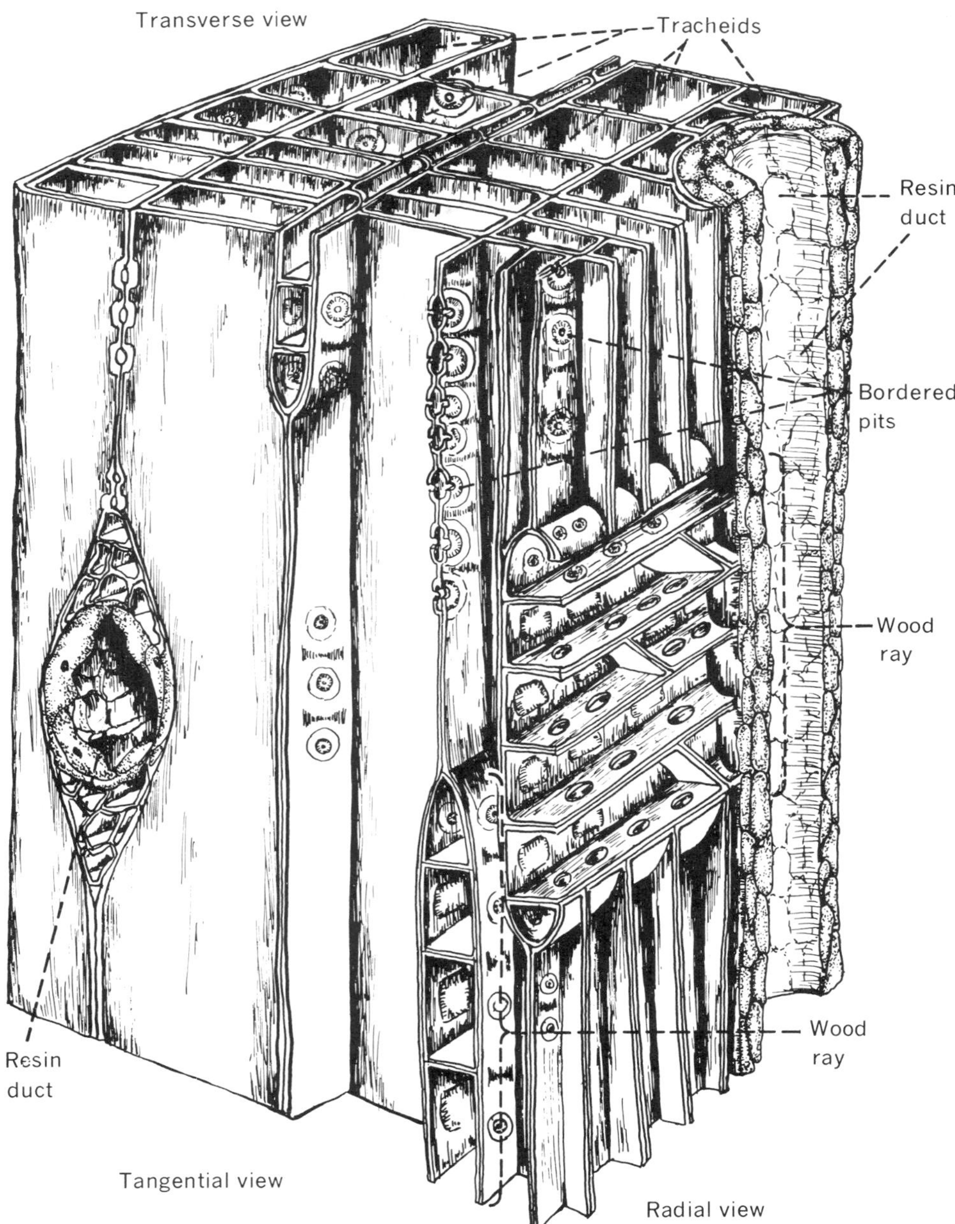

Fig. 8-28 Semidiagrammatic representation of a small block of white-pine wood as seen in transverse, tangential, and radial views. (Drawn by Marian A. Sutch.)

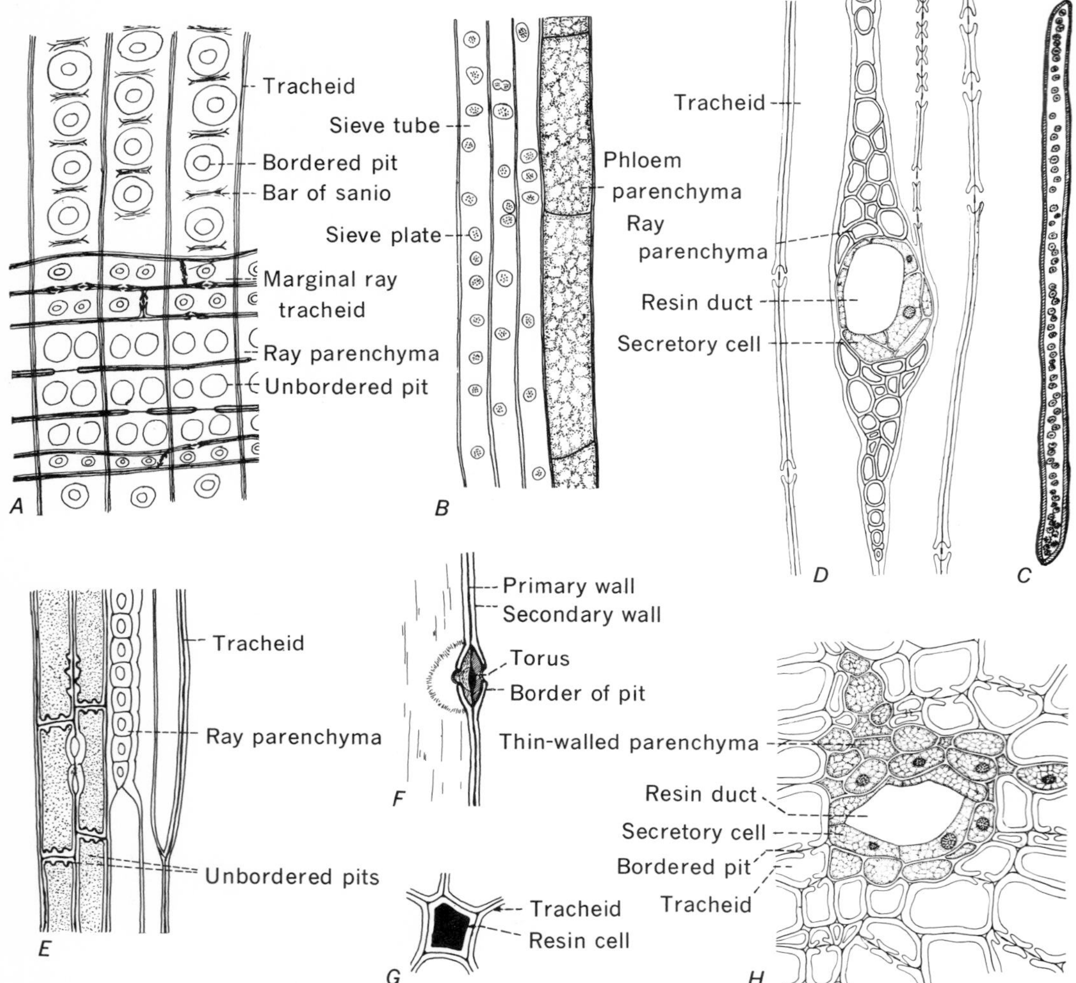

Fig. 8-29 Xylem and phloem elements in gymnosperm stems. A, radial section of pine wood showing tracheids running vertically and a wood ray running horizontally; B, phloem from a pine stem; C, radial view of a tracheid with bordered pits; D, tangential view of ray with horizontal or radial resin duct; E, tangential section of Taxodium *wood; F, diagram of a bordered pit; G, transverse view of a resin cell of* Sequoia *with adjacent tracheids; H, transverse view of a resin duct with surrounding cells. (A drawn by R. M. Gerfin; C by J. W. Holliday; D and H by A. J. Elser; E by M. A. Sutch; F and G by R. T. Quick.)*

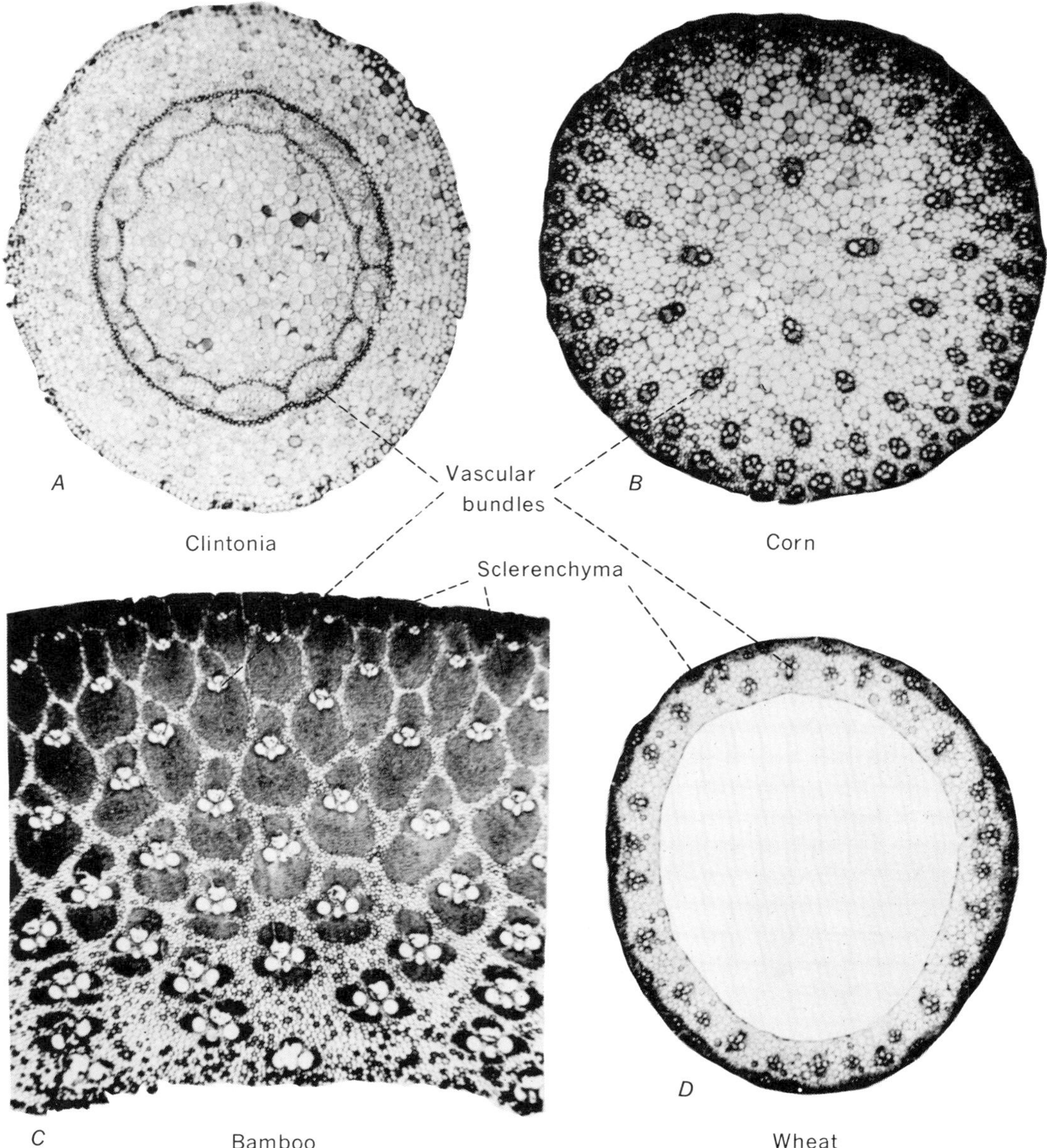

Fig. 8-30 Photomicrographs of transverse sections of stems of monocotyledonous plants, showing arrangement of vascular bundles. (Photomicrographs by Dr. D. A. Kribs.)

the vascular bundles, like that of corn (*Zea mays*) (Fig. 8-30*B*). In others, however, the bundles are grouped in the central portion of the stem, as they are in the rhizome of the calamus root or sweet flag (*Acorus calamus*). In this rhizome an endodermislike structure surrounds the central region. In *Clintonia* the bundles are close together and grouped in a definite cylinder or ring (Fig. 8-30*A*). The stems of certain monocotyledons, like wheat and some other grasses, at maturity have hollow centers (Fig. 8-30*D*).

The individual bundles of the stems of monocotyledons vary in structure with the several genera. The bundles vary also in structure in the different parts of the plant, and those of the internode frequently differ from those of the node of the stem. The bundle of the rootstock of *Acorus calamus* (Fig. 8-31) is a **concentric** bundle. In this bundle the phloem is the center core and a thin cylinder of xylem surrounds it. The phloem consists of sieve tubes, companion cells, and a considerable amount of undifferentiated parenchyma. The organization of phloem in the center, surrounded by xylem, constitutes the **amphivasal** type of concentric bundle (Fig. 8-31). It is frequently found in the underground stems of those types of monocotyledons possessing rhizomes. It is not characteristic of the aerial stems of monocotyledonous plants of the grass type, although it may occur in the internodal regions of these plants.

Fig. 8-31 Amphivasal bundle of sweet flag (Acorus calamus), *showing xylem surrounding the phloem.*

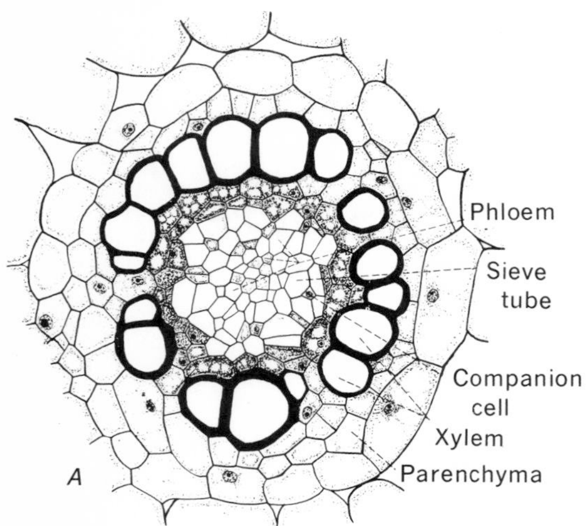

In the internodes of the stems of cereals, such as corn and wheat, and in other grasses the vascular bundles are collateral (Fig. 8-32). The phloem consists of sieve tubes and very distinct companion cells of smaller diameter, which usually are square in cross section. The xylem consists of ringed, spiral, and pitted vessels, some fibers, and parenchyma, and it is located toward the center of the stem. Frequently a large intercellular space, or lacuna, is formed in the area of the protoxylem cells. Usually there is a layer of sclerenchymatous tissue, several or more cells in thickness, which more or less completely surrounds the collateral bundle. In some monocotyledonous plants, viz., bamboo (Fig. 8-30*C*), this sheath is thick, strengthening the stem and providing considerable elasticity.

Vascular cambium is typically absent from the bundles and stems of monocotyledonous plants. No additional secondary tissues are produced in these closed bundles, and the plant axis is composed of primary tissues derived from apical meristems and intercalary meristems located at the nodes and near the leaf bases. Some monocotyledonous plants exhibit feeble cambial development in the fundamental tissue or vestigial cambium within the vascular bundles.

TYPES OF STELES FOUND IN PLANTS

The term **stele** is generally applied to the vascular skeleton and associated fundamental tissues as they appear in the primary axis of

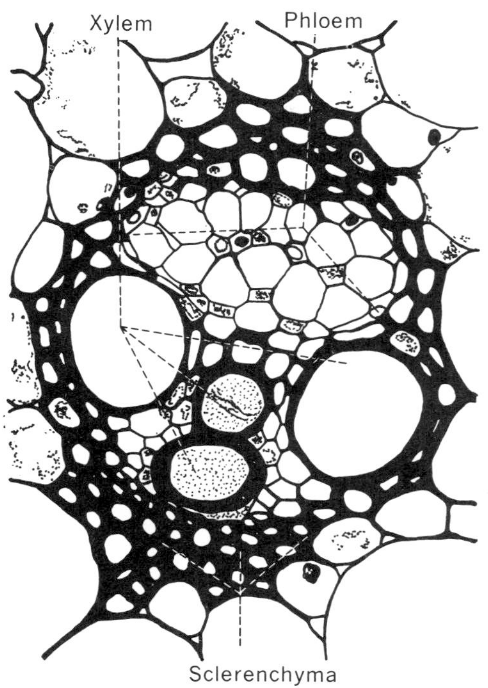

Fig. 8-32 Collateral bundle of corn (Zea mays).

the plant. In each portion of the plant axis there is a core of vascular tissue with varying amounts of intervascular (interfascicular) tissue, including the pith, if present. The stele is generally delimited by the pericycle, if present, and surrounded by the cortex. Different types of steles are recognized, depending upon the relative positions (and origin of these positions) of the different tissues and particularly the positions of the xylem and phloem with respect to each other. Among these are the protostele, the ectophloic siphonostele, the amphiphloic siphonostele, the dissected siphonostele, or dictyostele, the eustele, and the atactostele.

The **protostele** is the simplest and almost certainly the most primitive type. In its simplest condition it consists of a solid central cylinder of xylem, circular in transverse section and surrounded by a cylinder of phloem (Figs. 8-33*B*, 8-34*A*). There is no pith. It is found in some of the lower vascular plants of today and in some fossil genera. A modification of this simple form occurs generally in the roots of present-day vascular plants. In this type (Figs. 6-11, 8-33*A*), sometimes called an **actinostele,** or a protostele with radial arrangement of xylem and phloem, there is a solid central core of xylem with pointed, radiating arms extending radially toward the cortex, giving the xylem a star shape. The protoxylem forms the pointed arms of the xylem, and the metaxylem occupies the center of the stele. The phloem lies between the radiating arms, occupying radii that alternate with those of the protoxylem (Figs. 6-11, 8-33*A*). The number of radiating arms of xylem varies from two to many, the terms **diarch, triarch, tetrarch,** etc., being used to designate the number of such arms (Fig. 6-14).

The **siphonostele** is characterized by having the vascular tissues in the form of a hollow cylinder with a distinct pith in the center; i.e., the pith occupies the so-called hollow portion. The phloem and xylem form concentric cylinders. In the **ectophloic siphonostele** (Figs. 8-33*C*, 8-34*B*) the xylem cylinder lies next to the pith and is surrounded by the phloem cylinder. In the **amphiphloic siphonostele** (Figs. 8-33*D*, 8-34*C*) there are two cylinders of phloem, one lying next to the pith and surrounded by the xylem cylinder and the other externally to the xylem cylinder. The ectophloic siphonostele is found in the stems of some dicotyledonous plants and gymnosperms. The amphiphloic siphonostele is found in certain ferns and in such dicotyledonous plants as members of the cucumber family.

The siphonostele is often considerably broken up by the passage of the vascular elements from the stem to the leaf. Vascular strands passing into the leaf are termed **leaf traces.** When the leaf traces leave the vascular cylinder, they cause definite breaks, called **leaf gaps,** in the side of the cylinder from which they emerge. Naturally the leaf gaps occur above the leaf traces. The presence of leaf traces and leaf gaps

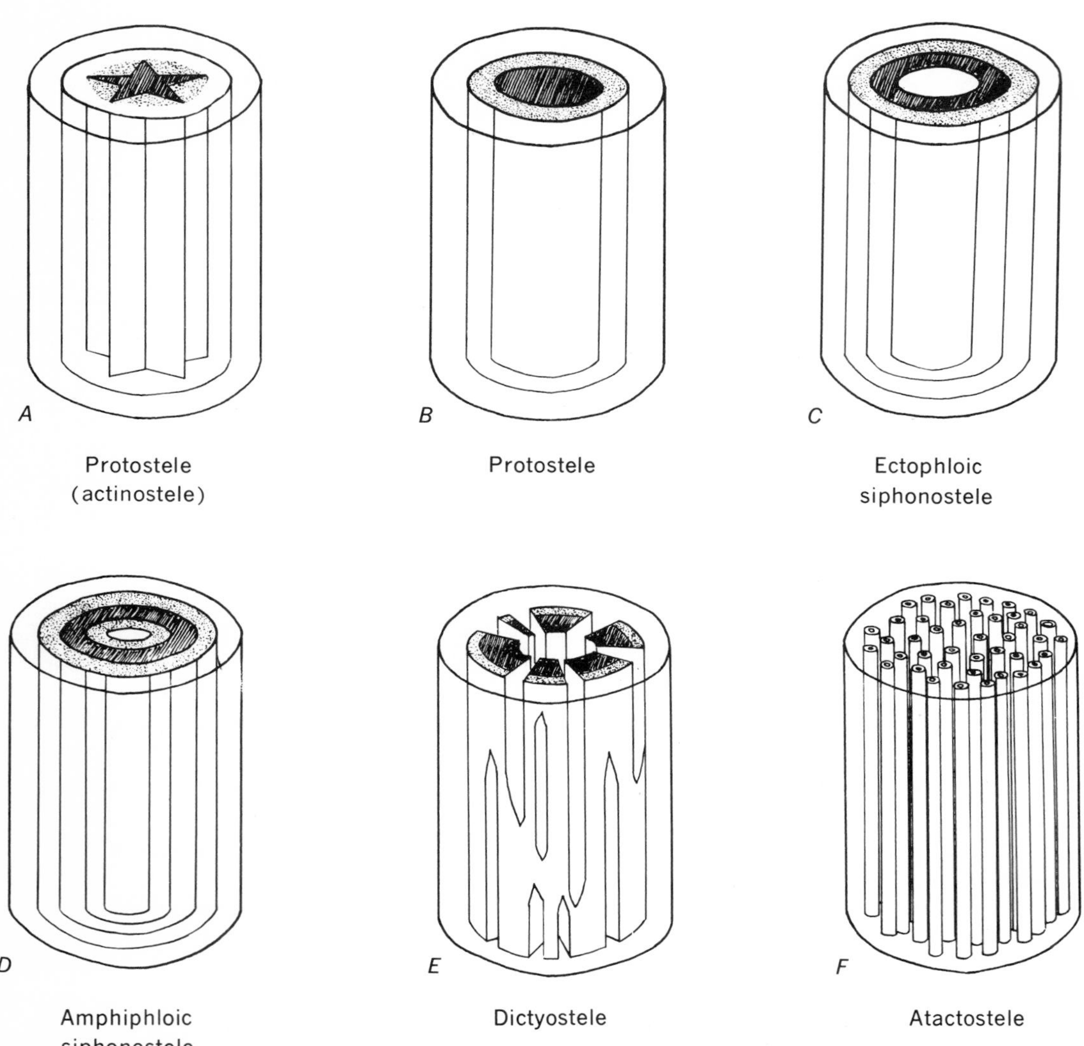

Fig. 8-33 Diagrammatic representation of types of steles. Phloem stippled and xylem shaded in each case. (Drawn by Christian Hildebrandt.)

changes the general topography of the stem in which they occur. In its simplest condition the siphonostele shows no dissection into smaller units. There are no leaf gaps or other interfascicular areas of parenchyma cells. In some vascular plants there may be leaf gaps, but these gaps do not extend far vertically and do not overlap. A cross section cut at an internode shows a continuous ring of vascular tissue (Figs. 8-33*C*, 8-34*B*). However, in some ferns and in other higher members of the vascular plants, leaf gaps are more numerous and extend vertically so that they do overlap. A cross section cut through an internode of such plants may show a network of separate vascular units. A siphonostele so broken up or dissected is called

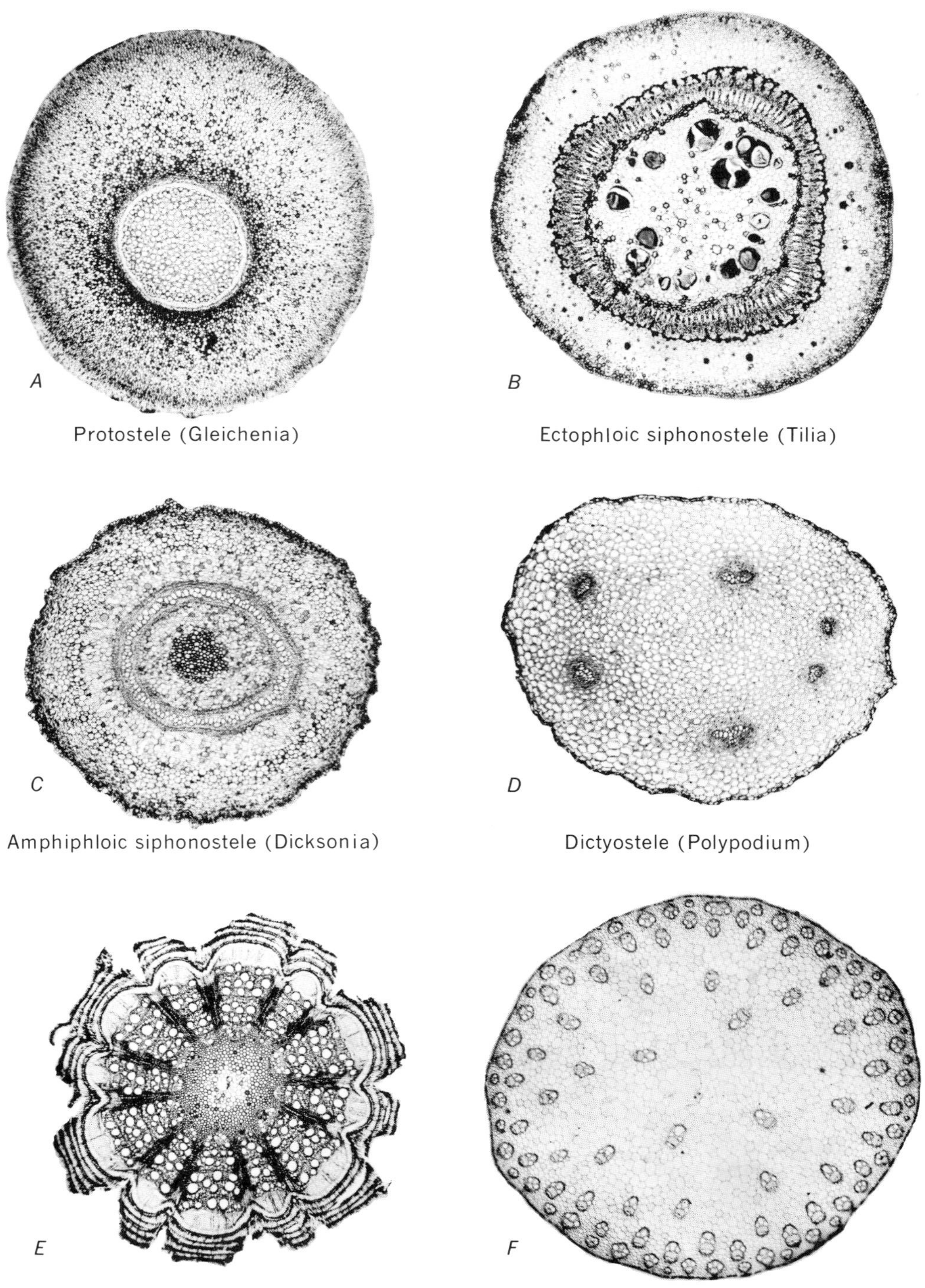

Fig. 8-34 Types of steles as seen in photomicrographs of transverse sections of stems. (Photomicrographs by Dr. D. A. Kribs.)

a **dictyostele** (Figs. 8-33*E*, 8-34*D*). The relative positions of vascular units and fundamental tissue as seen in transverse sections of the dictyostele vary at different levels in the stem.

The **eustele** (from the Greek, meaning "true stele") is somewhat similar in appearance to the dictyostele in transverse sections (Fig. 8-34*E*), but the separate vascular strands are the result of both leaf gaps and the presence of other interfascicular tissues which extend uniformly vertically so that a transverse section appears the same at all levels of the axis. In other words, there are separate bundles of vascular tissue alternating with fundamental tissue throughout the axis. The eustele is found in some dicotyledonous plants and in some gymnosperms.

In many monocotyledonous plants, like corn and other grasses, the stele is even more dissected, so that, in transverse section, the vascular strands or bundles lack any apparent order of arrangement but appear scattered through the fundamental tissue. Such a stele is called an **atactostele** (from the Greek *atactos*, "without order") (Figs. 8-33*F*, 8-34*F*).

TYPES OF VASCULAR BUNDLES

On the basis of the orientation of the xylem and phloem, the individual vascular bundles of the stele are classified as collateral, bicollateral, and concentric. If the bundles contain cambium, they are said to be open, but if no cambium develops, they are referred to as closed.

The collateral bundle (Fig. 8-35*B*) is characterized by having the protoxylem and the protophloem located on the same radius. The xylem tissue lies closer to the center of the axis, and the phloem toward the periphery of the stem. Collateral bundles are characteristic of the stems of dicotyledons and gymnosperms but are also found in some ferns and in monocotyledons like corn and other cereals.

The bicollateral bundle (Fig. 8-35*C*) has the same general features as the collateral bundle, but phloem is located both internally and externally to the xylem. The inner mass of the phloem, frequently of very limited extent, is next to the pith. The cambial layer occurs only between the xylem and the outer mass of phloem. All tissues lie on the same radius. Bicollateral bundles are found in the stems of certain ferns and in members of the cucumber family. The potato and the tomato have bicollateral bundles in which the internal phloem occurs in small scattered strands.

The concentric bundle is one in which either one of the vascular tissues surrounds the other. If the phloem surrounds the xylem, the bundle is **amphicribral** (Fig. 8-35*D*), but when the xylem surrounds the phloem, it is **amphivasal** (Fig. 8-35*E*). Amphicribral bundles are found in some of the ferns, and amphivasal bundles in the

Fig. 8-35 Arrangement of xylem and phloem in primary tissues; vascular bundles (diagrammatic). A, radial arrangement; B, collateral bundle; C, bicollateral bundle; D, amphicribral bundle; E, amphivasal bundle; phloem stippled, xylem crosshatched throughout. (Drawn by Florence Brown.)

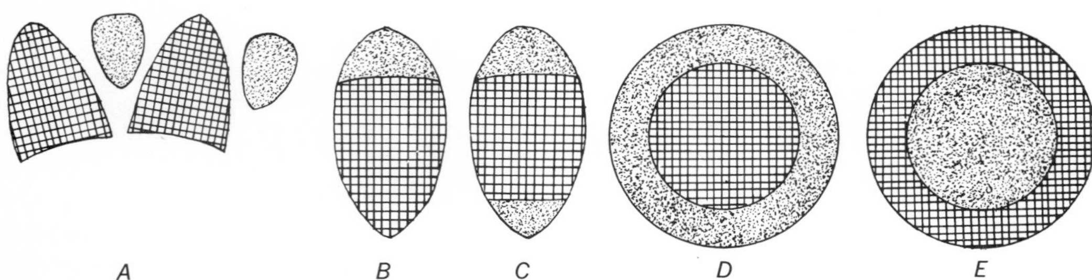

underground stems of some monocotyledons (Fig. 8-31).

Open and closed bundles An open vascular bundle is characterized by the presence of a cambium, formed out of procambial cells that have retained their meristematic abilities. New cell additions from the cambium account for enlargement of the primary bundles and may be responsible for new secondary bundles. Open bundles usually lack a bundle sheath. The collateral bundles of gymnosperms and dicotyledons and the bicollateral bundles of other dicotyledons are all open bundles.

The term closed bundle refers to the absence of cambium within the vascular bundle. Differentiation of cells is complete at the primary stage of development, and no additional cells are added to the bundle. Often, closed bundles have a definite external bundle sheath of heavy-walled sclerenchymatous cells, as in the bundle of corn. The vascular bundles of ferns and of monocotyledons are closed bundles.

The radial arrangement of xylem and phloem While the primary vascular tissue of the stems of plants most often is arranged in separate vascular bundles, the roots of plants show no such structure. Instead, the protoxylem and protophloem points are located on separate radii, and hence have most often been referred to as radial in arrangement. The primary xylem and phloem are often separated by parenchyma cells (Fig. 8-35*A*).

PHYSIOLOGY OF STEMS

General The stem provides mechanical support of the leaves, flowers, and fruits and the channel for the conduction of water, inorganic substances, and elaborated foods. It is chiefly through the growth and development of the stem that the leaves are brought into proper positions with respect to light while at the same time maintaining their connection with the water and mineral supply of the soil. Thus the leaves are enabled to carry on food synthesis. After the foods are made, the stem again provides the pathway through which these foods are removed from the leaves and carried to other regions of the plant. Similarly the stem supports the flowers, fruits, and seeds and provides the conducting channels through which these organs are supplied with necessary foods for development. In addition to this, stems are food-storage organs in the plant. Carbohydrates and other foods may be stored in the pith, the cortex, the phloem parenchyma, and the wood rays.

The stems of some species of plants are also used for water storage. This is true of most cacti and many other plants of dry regions. Young stems and especially those of herbaceous plants are green and carry on photosynthesis. Finally, the stem sometimes serves as a means of propagating the plant. While the number of species in which this occurs in nature is limited, it is made use of extensively in the commercial propagation of many fruits and greenhouse plants. Such plants as chrysanthemum, coleus, carnation, and rose are regularly grown from stem cuttings. Fruit trees are propagated by budding or grafting, both of which are types of stem propagation. The advantage to the grower of such asexual methods of propagation is that the new plants come true to the original type, whereas plants grown from seed often do not.

Tissues involved in water conduction It has already been mentioned that the xylem, or wood, is the water-conducting tissue of the plant. While, in many parts of the plant, water and dissolved inorganic substances pass through parenchymatous cells, it is chiefly through the tracheids and vessels that these substances are transported. The very structure of these elements indicates this, but it can also be demonstrated experimentally. If, for example, a cut stem is placed in a solution of thionin or any other nonpoisonous dye and allowed to stand for an hour or more, cut sections of the stem will reveal that only the tracheids and vessels are stained by the dye, proving that the liquid must have passed through these cells. That it did not move primarily in the walls of these cells can be proved by dipping the cut end of a leafy stem into gelatin, made liquid by warming it, and allowing the gelatin to be absorbed by the cut stem. If the gelatin in the stem is allowed to solidify by cooling in water, and a small piece of the end of the stem is then cut off so as to expose the cell walls, and the stem is again placed in water, the shoot will be found to wilt. This wilting results because the cavities of the tracheids and vessels have been clogged with gelatin. Similarly a leafy shoot that has been allowed to wilt will often fail to recover from the wilted condition when the stem is placed in water because the cavities of the cells are clogged with air. Still further evidence of the path of movement is furnished by girdling. Removing a ring of bark down to the cambium does not cause a plant to wilt, but if the stem is supported and the xylem carefully cut out through a complete section of the stem, without injuring the phloem and cortex, the plant will wilt. This proves that the xylem is the only tissue that can conduct water at a rate fast enough to prevent wilting.

In some trees only the sapwood functions in water conduction, the older wood becoming clogged with tyloses, which are bladderlike protrusions into the tracheids and vessels from the adjacent living parenchymatous cells. In other trees the entire xylem continues to function in water conduction. It is generally believed that most of the inorganic substances are also carried in the tracheids and vessels.

Exudation from stems; root pressure If the entire top of a small growing plant is cut off a few inches above the ground and a glass tube attached to the cut stump by means of rubber tubing, it will be found that sap gradually rises in the tube. If a manometer is attached to the cut surface, the actual pressure of the sap can be measured. In this case it will be found that this pressure seldom exceeds 2 atm and is often lower. The force causing the exudation of the sap from the cut stem is called **root pressure,** because it is thought to be caused by osmotic forces operating in the root. Sap also sometimes flows freely from twigs broken from plants during the spring. Such exudation may be particularly pronounced in pruned grapevines. It is sometimes called bleeding. Large volumes of liquid may be exuded over a considerable period of time. The exudate consists of small amounts of sugars and other organic and inorganic substances besides water.

Among the conditions necessary for such exudation may be mentioned an abundant supply of water, a favorable temperature, the presence of living cells in the roots, and, in general, conditions that would check transpiration. If the roots are killed, no exudation takes place. Exudation is greater at night than it is during the day in leafy plants. In woody plants it is far more pronounced in early spring before the leaves come out. Guttation, already described in the chapter on leaves (see page 80), and the flow of sap from some injured trees are also manifestations of root pressure.

The flow of sap from maple trees that are tapped for maple-sugar making is probably caused by stem pressures rather than by root pressure. It has been found to take place in tapped maple tree trunks completely severed from their roots and dipped into water. Maple sap flow is greatest on warm sunny days following freezing night temperatures. It ceases when night temperatures no longer fall below the freezing point, and it does not take place when temperatures are continuously either below or above the freezing point. Consequently maple tapping is done usually during the late winter months just before spring, when proper conditions prevail. A sugar maple may yield 5 to 8 liters of sap per day with a sugar content averaging 2 to 3 percent. This sugar is probably derived mainly from carbohydrates stored in the wood rays and parenchyma of the xylem.

Other examples of stem pressure may be seen

in the exudation of sap from palms and some other tropical plants. In the case of the palms the sap is exuded from the stem when the terminal inflorescence is cut off. Yields of from 6 to 10 liters of sap per day have been reported from a single palm. Palms are an important source of sugar in the tropics. The exuded sap of the century plant (*Agave americana*) is used extensively by the Mexicans in making a fermented liquor, pulque, and a distilled liquor, mescal.

The ascent of sap in stems In a previous chapter it was stated that an apple tree may lose as much as 96 gal of water per day through transpiration. This means that there must be in the stem some method of transporting large quantities of water to the leaves to replace this loss. The question as to how such trees as the giant redwoods of California or the eucalyptus of Australia, species that may attain a height of 300 to 400 ft, can lift the enormous quantities of water required by the leaves, at a rate fast enough to prevent wilting, has interested students of botany for many years. It is not the object of this discussion to present in detail all of the many theories that have been advanced to explain this phenomenon but to review some of the forces that may operate and to explain briefly the most likely method by which the sap rises to the tops of the tallest trees.

EXUDATION PRESSURE, OR ROOT PRESSURE From the fact that a stem may continue to give off water when the entire top is cut off, as mentioned in the preceding section, it might be thought that the sap is forced up from below by the osmotic forces operating to produce root pressure. Careful consideration of the matter, however, will show that root pressure is a wholly inadequate force in lifting sap to the tops of tall trees. In the first place, such pressure rarely exceeds 2 to 6 atm, whereas it has been estimated that about 20 atm are necessary to lift the sap to the tops of the tallest trees. Furthermore root pressure is lowest when transpiration is highest. In addition, negative pressures have frequently been registered in the stems of trees. Such negative pressures would not be found if the sap were forced up from below. That the roots are not necessary for the ascent of sap is indicated by the fact that the severed top of the plant does not wilt when placed in water. Accurate measurements show that such a severed top will lift the water with a force of several atmospheres.

ATMOSPHERIC PRESSURE If the outside of the stem were entirely sealed off from air, so that the xylem could act as a straight system of tubes from root to leaf, loss of water at the top of the column might be thought to create a vacuum, with the result that atmospheric pressure, acting on the lower end of the column, might force the water up in preventing the vacuum from being formed at the top. Unfortunately few of these conditions operate in the plant. There is no free surface at the bottom of the column upon which atmospheric pressure could operate, since the water is lifted from living cells of the roots. Furthermore the maximum height to which atmospheric pressure could lift the sap, even if all the necessary conditions were operating, would be less than 32 to 33 ft.

CAPILLARITY The fact that vessels and tracheids, through which the sap rises, are capillary tubes suggests immediately that some liquid could rise in these elements by capillarity. Undoubtedly capillarity does operate in the wood, but from the known diameters of the vessels and tracheids it is obvious that such rise could not exceed a few feet at the most. It is a well-known fact that the smaller the capillary tube, the higher will a liquid rise in it. We should, therefore, expect to find that the tallest trees would have the smallest vessels, but the reverse is often actually found to be true. Furthermore, in many trees the springwood, in which there is a greater movement of sap, is usually made up of larger vessels than is the late summerwood. For these as well as other reasons, capillarity may be dismissed as a force taking any active part in sap ascent.

OTHER POSSIBLE FORCES One of the older theories regarding sap ascent assumes that the water is vaporized at the bottom of the column and passes up as vapor to the top of the column where it is condensed back to a liquid. In addition to the difficulties involved in considering the effect of temperature on such vaporization and condensation, even if it did take place, such a method could not operate with sufficient speed to supply the demands of the plant during rapid transpiration and would account for the movement of pure water only and not the movement of inorganic substances dissolved in the sap of the vessels and tracheids.

The great German plant physiologist Julius von Sachs believed that the water moved in the walls of the xylem cells by imbibitional force. Water is undoubtedly found in the walls of the xylem cells; but from the fact that plants wilt when the cavities of the vessels are clogged, it is obvious that this method could not supply the water fast enough.

The plant physiologist Bose, of India, revived another of the older theories, viz., that the living cells all along the stem and, according to Bose, particularly those of the inner cortex provide osmotic forces which force the sap upward by a sort of system of relay pumps. This theory has little support anatomically or experimentally. It has been found that the sap still continues to rise after all the cells have been killed in long sections of the stem extending over many feet.

THE COHESION THEORY Up to the present time the most satisfactory explanation that has been offered to explain the ascent of sap is the cohesion theory. This theory assumes that the water is drawn up through the xylem by a pull applied at the top of the water column and that this pull is transmitted downward through the **cohering water** in the vessels and tracheids. The pull, at least during high transpiration, is caused by the evaporating power of water in the leaves but is limited in amount by the osmotic pressure of the leaf cells. The pull may also be created by a water deficit in the hydrophilic colloids of the cells at the top of the water column. This deficit tends to be relieved by creating a pull on the water in the vessels and tracheids. The water column in these vessels and tracheids does not break because of the tensile strength (cohesion) of the water column itself. It has been estimated that the total pull required to maintain the column at maximum transpiration does not exceed 20 atm. The osmotic pressure of the leaf cells has been found to average around 30 atm and hence is adequate to create this pull. Furthermore the imbibitional force of the colloids of the leaf and of the young twigs may reach a value of several hundred atmospheres.

The tensile strength of the sap has been measured and found to exceed 150 atm in some cases. This force also seems to be more than adequate. The theory assumes that there is an unbroken water column from roots to leaves. This apparently has also been found to be true, although many of the cells of the xylem may be filled with air without destroying the continuity of water. It might further be mentioned that the fact that negative pressures of 10 to 20 atm have been observed in stems supports a theory which assumes that the sap is pulled up from above rather than forced up from below. Furthermore the vessels, which are exclusively water-conducting cells, are usually strengthened by rings, spirals, and bars on the inside. This enables the vessels to withstand a pull from the top of the column rather than a pressure from below or within the cells.

The cohesion theory explains the ascent of sap exclusively on a physical basis. It excludes the living cells along the water column from taking any part in the lifting of the sap, although it requires the cells at the top of the column to be living cells. It should also be remembered that the water reaches the xylem at the bottom of the column through the living cells of the root, even though these cells may take no part in actually causing the sap to rise. Consequently sap ascent takes place only in a living plant.

At the present time there is more experimental evidence in support of the cohesion theory than of any other theory that attempts to explain the ascent of sap.

Conduction of food The xylem is involved chiefly with the conduction of water and inorganic substances. Since there are many storage cells in the wood rays lying next to the xylem vessels and tracheids, it is not unusual to find such foods as sugars in the sap of the xylem. It is generally believed, however, that the principal channel for the translocation of elaborated foods is the phloem. The structure of the sieve tubes is such as immediately to suggest conduction as their main function. Furthermore chemical analysis of the contents of the phloem cells shows them to consist largely of dissolved carbohydrates, proteins, and other nitrogen compounds. It is often found that sieve tubes are largest and the phloem better developed in stems in which there is obviously a rapid and large movement of food, as in the peduncles of some flowers and large fruits and in many of the elongated stems of cucurbits such as the pumpkin and the squash. That the phloem is the chief food-conducting tissue is also proved by girdling experiments. If a ring of bark is removed from the lower part of a cutting of willow or privet and the cutting placed in water, roots will develop only above the girdle, showing that the food necessary for their development can be obtained only from the upper part of the stem. If the xylem carried the food, roots should develop below the girdle as well as above it.

The fact that girdling of apple trees in early June sometimes increases the production of blossoms and fruit the following year also indirectly supports the contention that the phloem carries the food, since it is thought that the development of flower buds is conditioned by the carbohydrate supply in the stems at this time. When stems are left intact, much of the carbohydrate moves to the roots. Girdling prevents the movement to the roots and thus tends to increase the carbohydrate content of the stem above the girdle. If the girdling is done at any other season of the year than spring, it often kills the plant through starvation of the roots, again indicating that the phloem carries the food.

While there is considerable evidence that foods are transported in the phloem, little is accurately known concerning the mechanism of this movement or the forces that bring it about. There is some evidence that some of the inorganic substances may also be transported in the phloem. Apparently these minerals as well as the foods can move upward as well as downward in the phloem.

SPECIALIZED STEMS

Stems, like leaves and roots, may be specialized in form and function. Deviations from the types of stems thus far described are widespread and numerous. Some of the more prominent, important, or unusual ones are described in this section.

Underground stems There are four principal kinds of underground stems, viz., **rhizomes,** or **rootstocks, tubers, corms,** and **bulbs.**

RHIZOMES, OR ROOTSTOCKS The rhizome, or rootstock, is a horizontal stem growing beneath the surface of the soil. In some cases it is only partly covered. It may be compared with the prostrate or creeping aerial stem. Though often erroneously called roots, rhizomes are actually stems, as is evident from the fact that they consist of nodes and internodes. They bear scalelike leaves and axillary buds and often take root at the nodes.

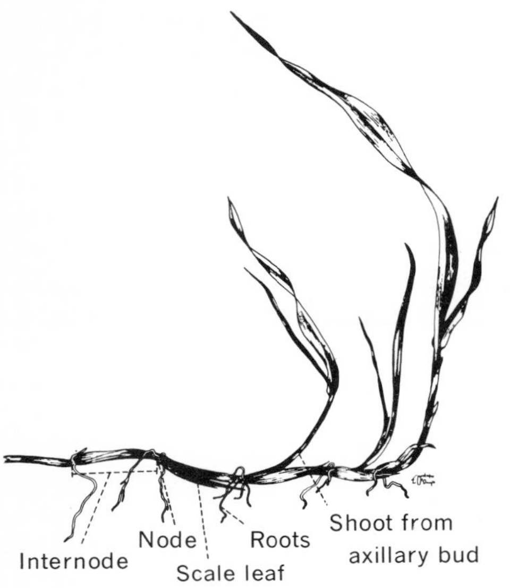

Fig. 8-36 *Quack grass* (Agropyron repens), *showing underground stem (rhizome) with nodes and internodes and bearing roots, leaves, and aerial shoots. (Drawn by Elsie M. McDougle.)*

Plants with rapidly elongating rhizomes, such as Canada thistle and quack grass (Fig. 8-36), spread quickly and widely. When cut up, as in plowing and cultivating, each piece produces a new plant. It is this feature which makes such plants obnoxious weeds, difficult to eradicate. Rhizomes are perennial structures that live over the winter. In the spring some of their buds develop into upright leafy shoots which eventually produce flowers and seeds, while others form new subterranean shoots. This is repeated again and again. In the meantime the older portions of the rhizomes often die, thus severing the connections between younger branches and giving rise to many separate plants. A rhizome may be cut into as many pieces as there are nodes and each piece will produce a new plant if there is sufficient food stored in it to enable the bud to develop into an aerial shoot. Furthermore pulling or cutting off the top of a plant with rhizomes does not kill it. New shoots are readily produced from the underground buds. Repeatedly cutting off the shoots as they develop from a rhizome will, however, ultimately cause the stored food reserves to be diminished and may kill the plant.

In some plants the rhizome elongates less rapidly, becomes short and stout, and contains a considerable amount of stored food, usually starch. The length of the living portion of such rootstocks varies from less than an inch to a foot or more. Iris, canna (Fig. 8-37), some mints, and Solomon's seal (Fig. 8-38) furnish examples of this type. Many forms of iris have rootstocks which are partly uncovered and bear true leaves which closely overlap each other because of the shortened internodes. Leaf scars, in the form of rings, mark the former points of attachment of the leaves. In the mints the rootstock is buried and only scale leaves are formed. In Solomon's seal the rootstock is also buried. Each year it sends up at the end of the rootstock a single shoot which bears the foliage and flowers and dies in the autumn, finally separating from the rootstock. A circular scar is left at the point where the shoot was attached. Because this scar looks something like the impression of a seal upon wax, the plant came to be called Solomon's seal.

Fig. 8-37 *Rhizome of* Canna indica.

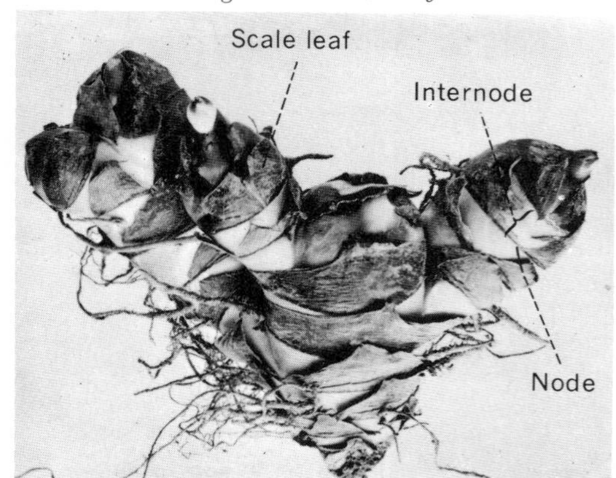

Since but one "seal" is formed each year, the limits of the years' growth of the rootstock can easily be determined.

TUBERS When rhizomes become enlarged at the growing tips by the accumulation of stored food, commonly starch, tubers are produced, like those of the Jerusalem artichoke or the white potato (Fig. 8-39*A*). The eyes of the potato are nodes at each of which several buds are produced in the axils of small scalelike leaves. Tubers are organs of food storage. In the autumn the rootstock dies except for the tubers which are left disconnected in the ground, each one capable of producing new plants in the spring. Potatoes are regularly propagated by cuttings of the tubers.

CORMS A corm, illustrated by the crocus, is a very short, thick rhizome, or rootstock, often much broader than long and usually growing upright rather than horizontally. Buds and roots are often produced at the nodes. The buds produce the new plants. Some of them, after a time, grow into new corms, and the old one dies. The scaly husk of the corm consists of the dead remains of leaf bases. These are quite prominent on gladiolus corms (Fig. 8-39*B*).

Fig. 8-38 Rhizome of Solomon's seal (Polygonatum *sp.) somewhat enlarged, showing circular scars (seals) left by abscissed aerial stems. Note also nodes and internodes and roots at nodes.*

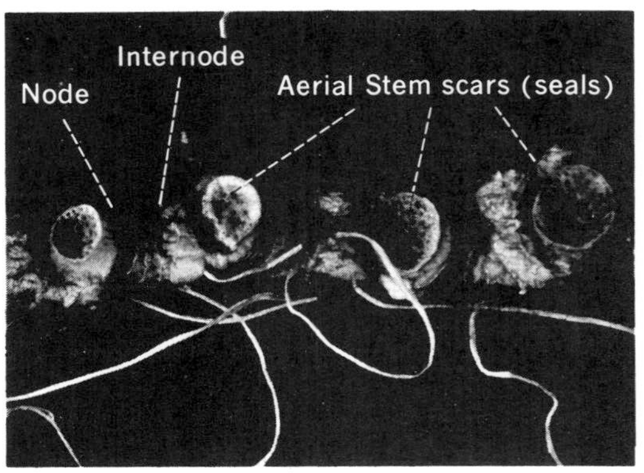

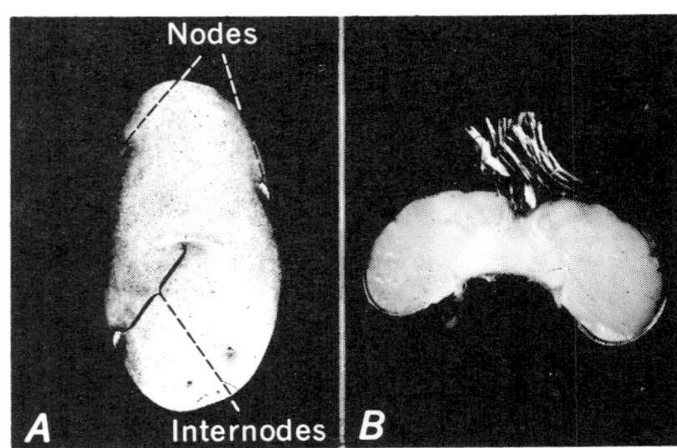

Fig. 8-39 A, potato tuber; note that the eyes are nodes with buds; B, Gladiolus *corm cut in half, showing central, solid stem with scalelike leaf bases.*

BULBS The bulb may be regarded as a short stem with fleshy leaf bases, commonly called scales. When the scales extend completely around the bulb, so as to appear in cross section as a closely compacted series of concentric rings, as in the onion, the bulb is said to be *tunicate,* or *coated.* The tulip, hyacinth (Fig. 8-40*A*), and leek are also of this type. When there are numerous narrow scales, not completely en-

Fig. 8-40 A, tunicate bulb of hyacinth, showing a partially differentiated flower cluster within the bulb; B, scaly bulb of Easter lily.

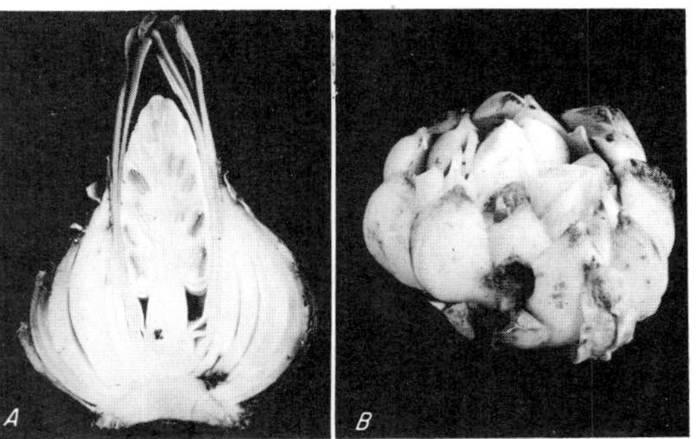

circling the stem, as in the lily, the bulb is said to be *scaly* (Fig. 8-40*B*).

Aerial stems of unusual form Aerial stems may be unusually long, as in plants with climbing or creeping stems. *Climbing stems,* or *vines,* usually rest upon or are attached to some support and often climb by means of special devices. Those of the *rambler* type simply rest on the tops of other plants, often on bushes and rapidly growing herbs. Many of these stems have epidermal outgrowths in the form of prickles or spines which enable them to stick to their supports. Certain roses, briers, and bittersweet are ramblers. *Root climbers,* such as climbing hydrangea (Fig. 6-16), English ivy, and poison ivy, climb by means of adventitious roots arising along the side of the stem in contact with its support. *Tendril climbers* climb by means of tendrils, which are sometimes specialized leaves or leaf parts, and sometimes specialized stems, as in the grape and Boston ivy (Figs. 8-41, 8-42). The tendrils of the Boston ivy are provided with adhesive disks. In the case of *twiners,* the entire stem winds about its support. Examples may be found in the morning glory, pole beans, and false bittersweet.

Creeping, or *prostrate, stems* trail along the surface of the ground and take root at the nodes. Trailing arbutus, ground pine, and partridgeberry have such stems. Those of the strawberry, called *runners,* eventually take root near the tips, form buds, and give rise to new plants. Stems like those of black raspberry, gooseberry, or dewberry bend over to the ground, take root, and send up a vigorous shoot which becomes an independent plant when the parent stem dies or is cut. Stems of this type as well as some

Fig. 8-41 Boston ivy (Ampelopsis tricuspidata), *showing branched tendrils with adhesive disks at their tips. The tendrils are specialized stems.*

Fig. 8-42 Growing tip of stem of grape (Vitis *sp.) with tendrils which are specialized stems.*

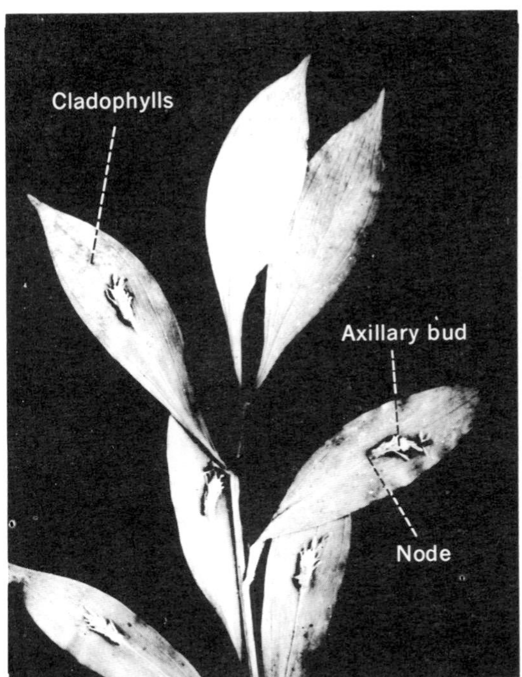

Fig. 8-43 Butcher's broom (Ruscus *sp.). The leaflike organs are specialized stems (cladophylls) with nodes bearing very small scalelike leaves and axillary buds.*

Fig. 8-44 Buckthorn (Discaria *sp.). The thorns are branches arising from axillary buds and bearing leaves at the nodes.*

creeping stems that regularly root at the nodes are sometimes called *stolons. Offsets* are shorter branches radiating outward along the ground in all directions from a main stem and forming clusters of leaves near their tips. These are formed in hen and chickens, some lilies, and houseleek.

Other types of unusual aerial stems include unusually short stems found in the so-called stemless plants, such as the dandelion and the common plantain, and *aerial bulblets,* which are similar to small underground bulbs and which are formed in the axils of tiger lily leaves and on the flowering shoots of onions. It is obvious that aerial as well as underground stems may be *organs of vegetative propagation.*

Stems may be the chief photosynthetic organs of the plant. In some cases represented by such plants as asparagus and butcher's broom (Fig. 8-43), these stems simulate leaves and hence are called *cladophylls.* Sometimes stems are stout and fleshy as in cacti (Fig. 8-45) and many other xerophytes. These stout, fleshy stems are frequently storage organs for food and water.

Finally, stems, as well as leaves, may be thornlike. *Thorns* may be unbranched as in the Osage orange and thorn apple or branched as in the honey locust and buckthorn (Fig. 8-44).

QUESTIONS

1. A popular writer stated in a book on Africa that deep, healed scratches made on the trunk of a tree by a lion sharpening his claws were found over 20 ft aboveground, that these scratches had been made years ago by the lion near the surface of the ground, and that the tree had grown in the meantime, carrying the marks upward. What is wrong with this statement?

Fig. 8-45 The giant cactus, saguaro (Carnegiea gigantea). *(U.S.D.I., National Park Service, photograph.)*

2. Does the arrangement of the branches of a tree follow exactly the phyllotaxy of the tree? Why?

3. During the growing season, how could you cause dormant buds on a tree to become active?

4. How would you prove that the xylem is the water-conducting tissue in a stem?

5. Why is the bark of an old oak tree furrowed and irregular?

6. Why are wood-ray cells not of much value in radial conduction of water and minerals?

7. Why do we not tap sugar-maple trees for sap in midsummer?

8. Why is heartwood preferred to sapwood for lumber?

9. What criteria do we use to determine whether an underground plant part is a root or a stem?

10. Enumerate some of the ways stems are of economic importance.

REFERENCES

CRAFTS, A. S. 1961. Translocation in Plants. Holt, Rinehart and Winston, Inc., New York. Takes up movement of inorganic, as well as organic, solutes.

DIXON, H. H. 1914. Transpiration and the Ascent of Sap in Plants. The Macmillan Company, New York. Gives a review of former theories and a detailed account of Dixon and Joly's theory of the ascent of sap.

ESAU, KATHERINE. 1960. Anatomy of Seed Plants. John Wiley & Sons, Inc., New York. Chapters 16 and 17 deal with stems.

ESAU, KATHERINE. 1965. Plant Anatomy, 2d ed. John Wiley & Sons, Inc., New York. Chapter 15 deals with stems. A more advanced reference book.

GREENIDGE, K. N. H. 1957. Ascent of Sap. *Ann. Rev. Plant Physiol.,* **8**:237–253. A critical review of recent theories regarding the ascent of sap.

ROMBERGER, J. A. 1963. Meristems, Growth, and Development in Woody Plants. U.S. Dept. Agr., Forest Serv., Tech. Bull. 1293. Review of anatomical, physiological, and morphogenic aspects of higher plants. Includes a list of correlative literature.

SACHS, ROY M. 1965. Stem Elongation. *Ann. Rev. Plant Physiol.,* **16**:73–96. Anatomical and physiological factors involved in stem elongation.

9

GROWTH AND MOVEMENT

Wheat, radishes, red clover, and alfalfa growing in a "Phytobiolab" growth chamber of the Kansas State University of Agriculture and Applied Science, Manhattan, Kansas. Plants photographed eight weeks from date of seeding. (Courtesy of Professor Charles V. Hall.)

The nature of plant growth and the internal and external factors affecting it are taken up in this chapter, with a short account of plant movements.

An exact definition of the term growth is scarcely possible. Fundamentally it is one of the attributes of living protoplasm. It is the sum of those activities of protoplasm whereby a living organism progresses from a less mature to a more mature condition. It can take place only where there is constructive metabolism, culminating in assimilation, which is the making of living substance out of nonliving. Growth usually involves an increase in mass and volume, i.e., an increase in weight and size; yet mere increase in weight or size is not necessarily growth. A seed, for example, may swell to many times its original size by the mere absorption of water before any growth takes place. Or, again, a seed developing on the plant may increase markedly in weight through the deposition of food without manifesting any growth.

Growth is more than a mere increase in weight or size. It involves a progressive change in form. This latter feature of growth is sometimes called **development.** It is manifested when an undifferentiated leaf primordium develops into a leaf or when a unicellular fertilized egg develops into a multicellular embryo. In both cases there is active growth. Furthermore the change in form is irreversible; i.e., the differentiated organ resulting from growth cannot revert back to the undifferentiated form from which it originated. Thus growth may be looked upon as an increase in mass or volume, accompanied by an irreversible change in form and structure, all resulting from the activities of protoplasm. True growth, therefore, is restricted to living organisms.

In the very young organism constructive metabolism exceeds destructive metabolism and growth is relatively rapid. As the organism matures, destructive metabolism approaches constructive metabolism in rate. When the two are equal, there is still growth even though no change in size occurs. In this case growth is manifested in tissue repair and maintenance. When destructive metabolism exceeds constructive metabolism, the organism gradually approaches senescence and death. In annual plants all this takes place within a single season.

In the measurement of growth, it is usually increase in weight or size that is measured. Increase in weight and size usually occurs even in such cases as the sprouting of potatoes in cellars or the germination of seeds in the dark. While the whole potato tuber or the whole seed in these instances is losing weight by respiration and transpiration, the actual growing organs are increasing in weight and size at the expense of the stored food reserves. In the growth of green plants in light, the increase in weight and size results from the absorption of materials from the exterior and from the synthesis of new substances within the plant.

The growth of organs of higher plants involves not only the enlargement of cells already present but also the formation of new cells. The formation of new cells by the division of those already present and their subsequent enlargement and maturation to their permanent forms provide the principal means of growth of any multicellular organ. Indeed, the entire plant body is developed from a single cell, the fertilized egg, which, by repeated division and differentiation, develops into the mature plant.

PHASES OF GROWTH

In the growth of a cell, an organ, or an entire organism, several stages or phases may be recognized, namely, a formative phase, an enlargement phase, and a maturation phase.

The formative phase of growth Cells that are in a formative condition are usually thin-walled parenchymatous cells with dense cytoplasm and small vacuoles. The nucleus occupies a relatively large part of the cell. Such cells are often in an active state of division. Cell division, which has already been discussed in Chap. 2, becomes a prominent part of the growth of all multicellular plants.

FORMATIVE REGIONS, OR MERISTEMS In many of the lower plants in which there is little or no tissue specialization, cell division may occur irregularly throughout the plant. Consequently such plants have no organs comparable with the roots, stems, and leaves of higher plants. In all multicellular plants in which there are specialized tissues, the formation of new cells by division is localized in rather definite regions, called **formative regions,** or **meristems.** In previous chapters attention has been called to the meristematic regions. It should be emphasized here that all the permanent tissues of plants are derived from one or another of these meristems. Consequently all the tissues of a plant may be classified as being either meristematic or permanent. The meristems of plants, in general, are the regions in which cell division takes place. They are the formative regions, therefore, where growth of a multicellular organ is initiated and from which all the tissues of the plant directly or indirectly originate.

According to their position, meristems may be classified as **apical, lateral,** and **intercalary. Apical meristems** (Figs. 6-9, 9-1) are those found at the tips of roots and stems, where they are commonly called **growing points.** They are responsible for the growth in length of the organs possessing them. In many of the lower vascular plants, like the ferns, the apical meristem consists of a single cell called the **apical cell.** In seed plants the apical meristem is made up of a group of cells. **Lateral meristems** are those situated along the sides of the stem and the root. The principal lateral meristems are the vascular cambium and the cork cambium, or phellogen. Lateral-meristem cells divide chiefly in one plane and thus cause the stems and roots in which they occur to increase in diameter. They are found chiefly in dicotyledonous plants and in gymnosperms. **Intercalary meristems** are parts of apical meristems which become separated from the apex by permanent tissues as the apical meristem moves forward. They are usually found at the bases of the internodes of the stems of many monocotyledonous plants (Fig. 8-1*B*) and in *Equisetum.* In many plants they also occur at the bases of the leaves. Like the apical meristems, they function in increasing the length of the organ in which they occur. Unlike most meristems, they ultimately wholly disappear as they are transformed into permanent tissues.

Fig. 9-1 Photomicrographs of lengthwise sections through the growing tip of Coleus, *showing apical meristematic region of the stem. B, enlarged view of the terminal part of A.*

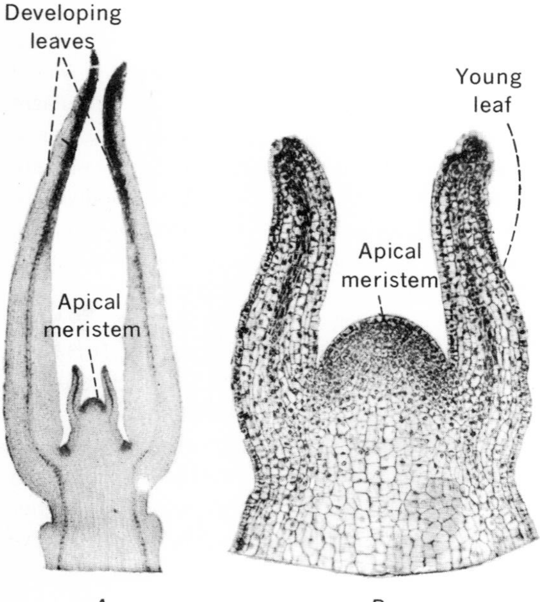

Meristems may also be classified as **primary** and **secondary.** The **primary meristems** are those which persist from the time of their original development in the embryo or young plant in the seed. The principal primary meristems are the apical meristems of stems and roots and the fascicular cambium of dicotyledons and gymnosperms. The intercalary meristems of the stems of monocotyledons are also primary meristems. Meristems that arise from permanent tissues are called **secondary meristems.** The most common secondary meristems are interfascicular cambium and the cork cambium, or phellogen. Both of these cambiums arise from cells that are already mature, the interfascicular cambium from parenchymatous cells lying between the original primary vascular bundles, and the cork cambium from the epidermis, cortex, phloem, or pericycle. Other secondary meristems may arise in any part of the plant as a result of wounding or from other causes. Any living parenchymatous cell is potentially able to develop into a secondary meristem. It is from such meristems that many abnormal growths, such as galls and tumors, develop. Such structures are often initiated by growth substances, which are described later in the text, and are often caused by insect injuries or by parasitic bacteria or fungi.

The primary meristems of seed plants, and particularly the apical meristems, originate in the embryo within the seed. In many seeds the entire embryo is in a meristematic condition. In others the apical meristems have already been developed at the tip of the rudimentary stem and the tip of the rudimentary root of the embryo. These apical meristems, as well as the principal lateral meristems, persist throughout the life of the plant, causing stems and roots to increase in length and in diameter. The meristems which give rise to leaves (leaf primordia) and the intercalary meristems of the stems of monocotyledons and a few other types of plants have a limited duration. As soon as the organs to which they give rise are formed, such meristems become completely transformed into permanent tissues and therefore cease to exist.

Enlargement When a cell divides, the two daughter cells together are at first only as large as the parent cell. Before growth can actually take place, the daughter cells must enlarge. Growth therefore includes not only cell division but also cell enlargement. In most of the meristematic regions of the plant, at least one of the daughter cells retains its meristematic condition and continues to divide. It is for this reason that such meristematic regions are perpetuated. The other daughter cell may also divide several times, but ultimately the cells arising from these divisions develop into permanent cells of one form or another. In regions where elongation is occurring, it is the daughter cells nearest the apex that remain meristematic. The elongation of the other daughter cells causes the apex to move forward. In the division of cambial cells, one of the daughter cells remains cambium and the other, either directly or after one or more divisions, develops into permanent cells.

The enlargement of the cells following division may take place in all dimensions or chiefly along one axis. The majority of the cells of such tissues as pith and cortex and the mature parenchymatous cells of other tissues result from approximately equal expansion in all dimensions. Fibers, tracheids, sieve tubes, and vessels are formed by the greater enlargement of the lengthwise axis of the cell. The shape and final form of all cells of permanent tissues are determined by the manner of enlargement of the original daughter cells resulting from division.

The enlargement itself results from the synthesis of new substances within the protoplasm and by the absorption of materials from adjacent cells or from the exterior. The first of these is sometimes called **growth by accretion.** The new materials synthesized often consist of hydrophilic colloids and osmotically active substances. These substances cause the cell to absorb water and therefore to stretch out. The expansion of the cell resulting from such absorption is called **growth by distension.** As the cell expands, the cell wall is thickened by the addition of new material, chiefly cellulose, made by the protoplasm. As the cell grows, vacuoles increase in size and many of them coalesce, until finally there may be one large central vacuole with the cytoplasm occupying a peripheral position in the cell.

Maturation As the cells enlarge, they gradually assume their permanent shapes and forms. This final phase of growth is usually called **maturation.** As maturation proceeds, the cells become fully differentiated. While all the cells are alike in the meristematic condition, their shapes, forms, and functions when they are fully differentiated are quite varied. The volume of the cell may have increased a hundred or a thousand times. The cells may remain living in the mature condition or may die. They may remain parenchymatous cells or become sclerenchyma, collenchyma, or highly differentiated xylem vessels, or sieve tubes. It is during the maturation phase of growth that all the different kinds of cells and tissues that have been described in the chapters on the leaf, the root, and the stem are developed.

FACTORS AFFECTING GROWTH

The rate at which a plant or an organ grows, as well as the shape or form it assumes, is determined by the combined operation of a multitude of complex internal and external factors. Internal factors are conditions existing within the plant, while external factors are conditions of the surroundings or environment of the plant. As stated by the German botanist Klebs, we may look upon a plant bud as a group of possibilities or potentialities, and what it becomes depends upon the factors brought to bear upon it. Of these factors, those which are external or environmental are more readily brought under control and hence have been more widely investigated. In the pages that follow, only a brief treatment can be given of the manner in which some of these factors operate.

INTERNAL FACTORS

Among the internal factors that affect growth may be mentioned **heredity,** the presence of **growth-regulating substances,** including **hormones, vitamins,** and other physiologically active substances occurring in minute quantities, the general **nutritional balance** of the plant, and the **correlation of plant parts.**

Heredity It is the heredity or inheritance of the plant that gives it its potentialities for developing into a certain form. Each plant has a group of hereditary factors which are capable of influencing the development of definite characters, provided the proper conditions are supplied to bring these characters out. For example, when the Peking variety of soybeans is grown in light from which the blue-violet end of the spectrum has been eliminated, the plants become twiners (Fig. 9-11*B*). Four-o'clocks, on the other hand, although they grow unusually tall, do not become twiners under the same conditions. The difference between these two species in their response to this environmental condition rests upon the fact that the soybeans have a hereditary factor for twining, while the four-o'clocks apparently do not. The Peking soybeans do not twine, however, under ordinary light. This shows the importance of environmental factors in bringing out hereditary characters. Many of the general hereditary characters of plants, however, appear so constant that even extreme variations in environmental factors fail to change them. A germinating bean seed, for example, cannot be made to develop into a pea plant or into any other kind of plant, no matter what environmental factors are brought to bear upon it. It is for this reason that species remain fairly constant under all conditions. Most of the structural and other distinguishing features of species are caused by hereditary factors that usually cannot be changed except by breeding. Heredity is therefore one of the most important factors affecting growth and development in the plant.

Nutritional balance The relative proportions of the foods—carbohydrates, fats, and proteins—in the plant body probably have much to do with the type of growth the plant makes. A "balanced ration" may be as necessary for the plant as it is for the animal. Since green plants synthesize all these foods themselves, the nutritional balance is conditioned by the supply of inorganic salts available to the plant as well as by the factors that influence photosynthesis. The importance of a proper balance between carbohydrates and nitrogen compounds in the plant has received the greatest attention of investigators. Thus it has been shown that when tomato plants are supplied with an excess of nitrate and an abundance of water, under ordinary conditions of light, they are likely to become excessively vegetative and unfruitful. In this case there is relatively too much soluble nitrogen in proportion to the amount of carbohydrate present to induce flowering and fruiting. On the other hand, if nitrogen is withheld from the plants, without preventing photosynthesis from taking place, i.e., by keeping the plants in good light, carbohydrates may accumulate in great quantities because they are not used in the synthesis of proteins and other nitrogenous substances, and the plants become very short and tough, nonvegetative and nonfruitful. They also become nonvegetative and nonfruitful, short and weak, when supplied with an abundance of nitrogen in the absence of sufficient light for active photosynthesis. In this case the carbohydrates are too low in quantity to permit the synthesis of higher organic compounds necessary for growth.

Between these extremes there exists a condition of balance between the amount of carbohydrate and the quantity of nitrogen compounds in the plant under which they become both vegetative and fruitful. This condition, obviously most desirable from the grower's viewpoint, is found when the plants are kept in good light and are supplied with a moderate amount of soluble nitrogen compounds.

The same correlation between a proper balance between carbohydrates and nitrogen compounds and growth and fruiting has been reported with other plants. Other relationships between foods and other compounds in the plant may also be effective in growth, but little is known about them.

Growth substances

HORMONES **Hormones** are substances which, though produced by the organism in extremely minute quantities, are capable of producing profound physiological effects. In the animal body they are sometimes referred to as "chemical messengers" because they are produced in one organ and are carried in the bloodstream to another organ upon which they have their effects. Well-known examples of hormones found in the animal body are thyroxine, epinephrine, and insulin. None of these has been found in plants. Plant hormones, or **phytohormones,** have been shown to play a prominent role in the metabolism and growth of plants, influencing root and stem growth, the elongation of cells, the production of flowers, movement of organs, the dominance of certain parts of plants over others, and the production of many abnormal growths, such as galls and tumors. In addition to the true hormones, many synthetic compounds, such as indolebutyric acid, α-naphthaleneacetic acid, α-naphthalene acetamide, and many others produce similar effects on plants. The general term **growth-regulating substances,** or simply **growth substances,** may be applied to all such substances as well as to the plant hormones and vitamins.

Prominent among plant growth substances are the **auxins,** which are considered to be essential for normal growth in length and to be effective physiologically in other ways. They have been found in the growing tips of stems, in pollen, in seeds, in leaves, and in other organs of the plant. They are found particularly in apical meristems and cambiums. Since they occur in extremely minute quantities, ordinary chemical methods cannot be used for their determination.

One of the earliest methods used for auxin determination is the so-called *Avena* test, in which young seedlings of oats (*Avena sativa*) are used. In the very young oat seedling, the growing shoot is surrounded by a sheath called the coleoptile. If 1 or 2 mm of the tip of the coleoptile is cut off, the coleoptile no longer elongates at a normal rate. Placing the tip back on the cut coleoptile restores the normal rate of elongation. If the tip is placed on only one side of the cut surface of the coleoptile, the side on which it is placed grows more rapidly than the other side, causing the coleoptile to bend away from the side to which the tip was applied. This indicates that something produced in the tip is transported downward into the coleoptile and affects its growth. If the tips of such coleoptiles are cut off and placed cut surface down on an agar gel for some time, some of the growth-promoting substance (auxin) diffuses out of the tips into the agar. If now a small cube of the agar is cut out and placed on one side of the cut surface of a coleoptile from which the tip has been removed, a bending of the coleoptile takes place away from the side to which the agar block was applied, indicating that the auxin has now diffused out of the agar into the side of the coleoptile to which it was applied, causing that side to grow more rapidly and thus causing a bending.

Similarly, if the growing tips of other plants are placed on agar in this manner, auxins will diffuse into the agar and tiny cubes of this agar will cause bending of decapitated coleoptiles. It has been found that, within limits, the degree of bending of the coleoptile is proportional to the amount of auxin present in the agar cubes. By measuring the degree of bending, it is possible to determine the amount of auxin present in the agar. This method furnishes a means of measuring quantitatively the amount of auxin which diffuses out of plant parts. In practice, it is necessary to control temperature and in other ways to standardize the method of procedure. Since the coleoptiles bend toward light, it is necessary to keep them in darkness during the procedure. It has been estimated that 1 part of a naturally occurring auxin from a stem tip in 110 million parts of water causes a bending of 10° in an oat coleoptile.

Auxins can also be extracted from plant tissues with chloroform or ether. In water extracts made at ordinary temperatures, the auxins are rapidly inactivated by oxidizing enzymes. One of the auxins, called heteroauxin, is indoleacetic acid. Since this can be obtained commercially in a pure form, auxin activity can be compared quantitatively with that of pure indoleacetic acid.

The regulatory effect which auxins exert on plant growth is considered to be brought about mainly through their influence on cell enlargement. For many years evidence has accumulated to indicate the existence of hormones that affect growth by promoting cell division. A number of investigators have found that extracts from coconuts, malt, yeast, wheat germ, and other plant parts have marked growth-promoting effects of this kind on excised embryos and other plant tissues. The term **kinins** (derived from *cytokinesis,* meaning "cell division") could be given to substances in general which promote cell division. One of these substances, **kinetin,** has been prepared in crystalline form from deoxyribonucleic acid obtained from herring sperm and from a yeast. Kinetin is effective in promoting cell division in plants in concentrations as low as one part per billion.

Growth substances (Figs. 9-2, 9-3) are con-

Fig. 9-2 Effects of growth substances. A, bending of stems and leaves of tomato resulting from treating one side of the stem with α-naphthaleneacetic acid in lanolin; 1, treated with lanolin only; 2–6, treated with 0.05, 0.1, 1.0, 5.0, and 10.0 mg of α-naphthaleneacetic acid per gram of lanolin, respectively; note epinasty (downward bending) of leaves; B, English holly (Ilex aquifolium), *showing roots induced by treatment with indolebutyric acid; untreated controls on left failed to root; C, fruit of Bonny Best tomatoes; left, control untreated, showing development of seeds following pollination; right, parthenocarpic (seedless) fruit developed after the flower cluster was treated with o-chlorophenoxypropionic acid, 50 mg per liter. (Photographs courtesy of P. W. Zimmerman, Boyce Thompson Institute for Plant Research, Inc., Yonkers, N.Y.)*

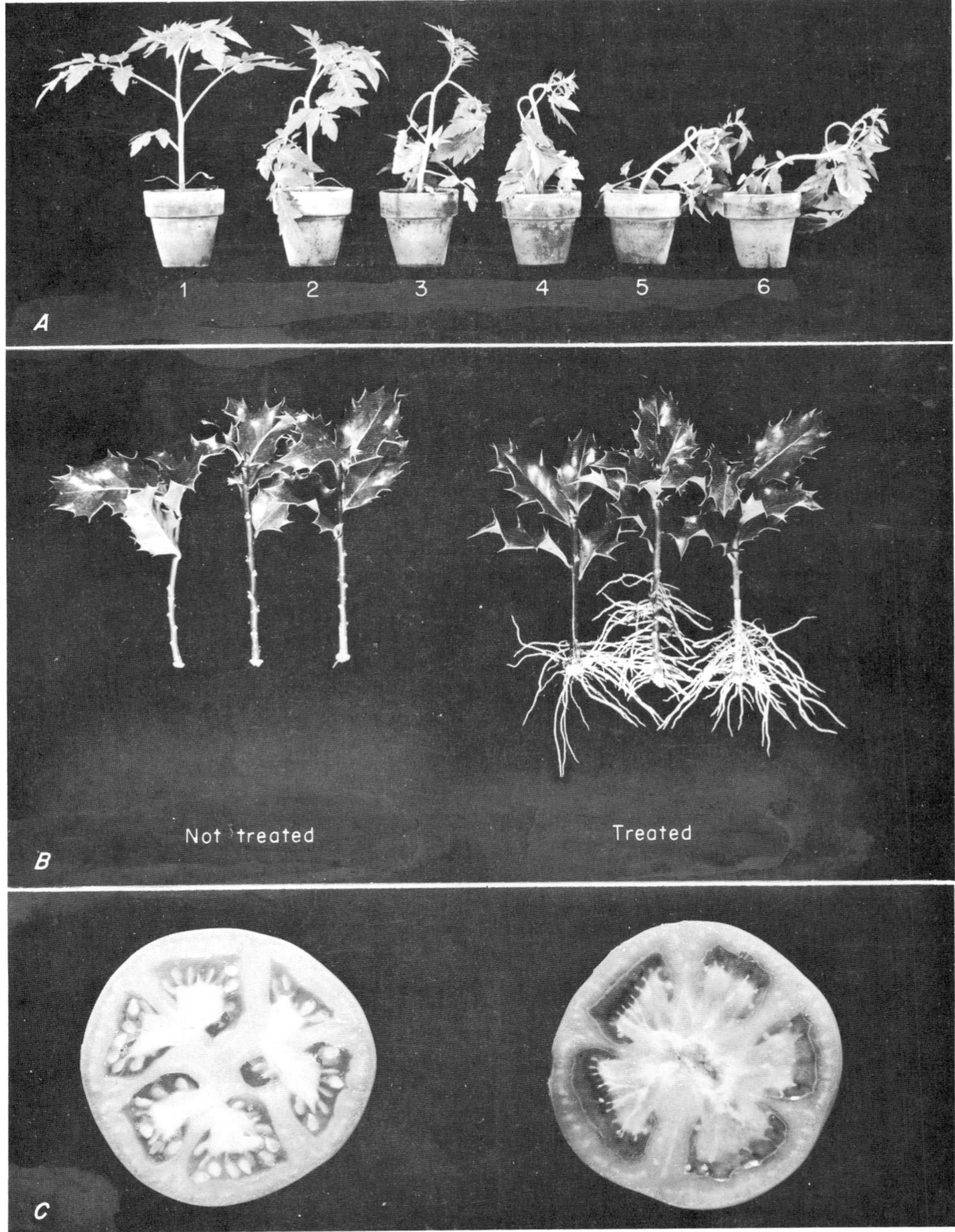
1
2
3
4
5
6
A
Not treated
Treated
B
C

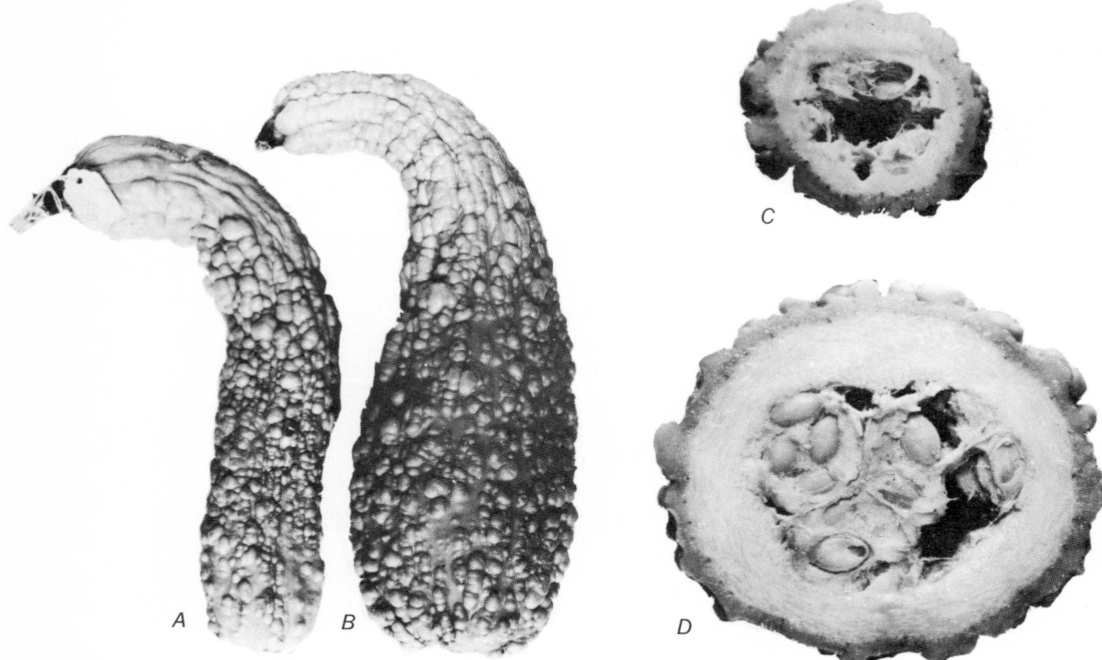

Fig. 9-3 Effect of growth substances on crookneck summer squash. A, parthenocarpic (seedless) fruit produced by application of indolebutyric acid to the style of an unpollinated flower; B, a normal fruit resulting from pollination; C, section of the fruit shown at A; D, section of the fruit shown at B. Note seeds in D and absence of seeds in C. (Photographs courtesy of Felix G. Gustafson, University of Michigan, Ann Arbor, Mich.)

sidered to be effective not only in stem elongation but also in the abscission of leaves and fruits, in the dominance of apical buds, in fruit development, in the initiation of roots on stems, and in tropisms, as is shown later in the text. They are also thought to be effective agents in causing abnormal growths such as the galls caused by insects, and tumors caused by fungi and bacteria. Such overgrowths have been produced experimentally by injecting growth substances into plant tissues. The root nodules of the legumes are also thought to be caused by growth substances. The use of growth substances in the rooting of cuttings has become an established practice. Often cuttings of hardwoods that do not root readily can be made to root if treated with growth substances. They are also used in a practical way to prevent the preharvest dropping of fruits such as apples. In fruits sprayed at the proper time with synthetic growth substances, the development of the abscission layer is delayed. The setting of fruit in the tomato has been enhanced, and the production of seedless fruits in a number of plants has been accomplished by the use of growth substances.

The growth substance **gibberellic acid** and other gibberellins obtained from the fungus *Gibberella fujikuroi* have been used extensively in experimental studies. Gibberellic acid, applied in very minute quantities to soil or solution cultures, or sprayed on leaves and stems of certain plants, causes excessive increase in height (Fig. 9-4). In many grass plants, like wheat and rice, it also causes increase in length of leaves. Stem internode length of dwarf varieties of garden peas has been reported to be increased by 200 to 400 percent. In some cases both fresh weight and dry weight are increased. It has no effect, however, on the growth of roots.

Some of the growth substances, depending partly upon their concentration, may inhibit growth. Thus auxin produced in terminal buds is thought to move downward in the stem through phloem and parenchymatous tissues to the lateral buds and prohibit their development. If the terminal bud is cut off, the lateral buds usually begin to grow. If, however, a block of agar into which auxin has been allowed to diffuse is placed on the cut surface, the lateral buds will not grow into branches, indicating that the auxin has diffused downward and prevented the lateral buds from developing. Whether this effect is produced by the auxin itself or by an inhibitor associated with the auxin, or by some other cause, remains to be proved. Auxin is also reported to prohibit root growth in concentrations that stimulate stem growth.

One of the best known of the growth inhibitors is **maleic hydrazide.** In proper concentration this substance has the interesting property of checking growth without necessarily injuring the plant. Because of its ability to prolong the dormancy of perennial plants, attempts have been made to use maleic hydrazide to delay growth of fruit-tree buds in spring to prevent them from opening and being injured by later freezing temperatures. Such attempts have not been very successful with fruit trees, but dormant raspberry plants have been delayed for several weeks in renewing growth. Maleic hydrazide has been used successfully as an herbicide against such plants as quack grass, wild onion, and garlic.

2,4-D (2,4-dichlorophenoxyacetic acid) is a growth substance that has come into wide use as an herbicide (Fig. 9-5). Its selective action on different species of plants makes it possible to use it to kill weeds in lawns without killing the grass. It is also used widely on field and other crops (Fig. 9-6). The related compound **2,4,5-T** (2,4,5-trichlorophenoxyacetic acid) is used especially to kill undesirable woody species like poison ivy.

The exact mechanism of the action of growth substances probably varies with the different kinds. There are hundreds of different chemical substances, many of them completely unrelated, that seem to have growth-promoting properties. There is some evidence that some of them affect respiration, enzyme activation, and other phys-

Fig. 9-4 Effect of gibberellic acid treatment on Vibrant chrysanthemum. Control plant on left, untreated; plant on right was sprayed at intervals from the fourth to the seventh week of short days with 100 ppm of gibberellic acid. (Photograph courtesy of Henry M. Cathey from a report by H. M. Cathey and N. W. Stuart, U.S.D.A., Agricultural Research Service, Beltsville, Md.)

Fig. 9-5 Weed control with 2,4-D. Left, oats sprayed with 2,4-D for control of wild mustard; right, unsprayed area. A few flowers of mustard still show in the sprayed area three days after spraying. These plants died later. (Photograph courtesy of G. F. Johnson, from G. H. Berggren and J. O. Dutt, "Chemical Weed Control," Penn. State Univ., Agr. Ext. Serv., Circ. 356, 1950.)

iological activity. For normal growth of a plant there must be an adequate supply of food, i.e., carbohydrates, fats, and proteins. These are the building materials and must be present in comparatively large quantities. The growth substances, which occur in minute quantities, might be thought of as exercising the control over the utilization of these building materials and perhaps effecting their actual incorporation into the structure of the plant body.

VITAMINS The **vitamins** are a group of organic substances that have been shown to have profound effects on the growth and health of animals. Since they were originally shown to be essential constituents in the diet of animals, although occurring in minute quantities, they were at first called accessory foods. Since they were obtained by the animal in foods, they were thought of as exogenous substances as opposed to hormones, which were endogenous, i.e., produced within the body of the organism. This distinction does not hold with plants, since most green plants apparently are able to synthesize vitamins within their tissues. Hence it is sometimes difficult to distinguish between vitamins and hormones in the plant.

Vitamins were defined by Willaman as "a class of substances, the individuals of which are necessary for the normal metabolism of certain living organisms but which do not contribute to the mineral, nitrogen, or energy factors of the nutrition of these organisms." By their not contributing to the energy factors of nutrition, Willaman undoubtedly meant that the vitamins are not oxidized by the organism for their contained energy as are the foods. Some of them, such as thiamine, nicotinic acid, and riboflavin, definitely take part in energy transfer which occurs in oxidative metabolism, i.e., in respiration. Although they are identified with the nutrition or metabolism of organisms, the vitamins, like the hormones, are effective in minute quantities and can be looked upon as growth substances. They are not foods. Unlike enzymes, the vitamins are used up by the organism. Many of them are definite chemical compounds that have now been synthesized artificially.

It has been known for a long time that plants are the principal source of most of the vitamins needed by animals. They are also essential to the life of plants. Among the important vitamins that are synthesized by plants and that probably take part in the metabolism and growth of plants are the following: vitamin A, a fat-soluble vitamin derived from carotene and essential for normal growth and vision of animals; B_1, or *thiamine,* the water-soluble antineuritic or antiberiberi vitamin; B_2, or *riboflavin,* a water-soluble growth vitamin; *nicotinic acid, pyridoxine,* or B_6, and *pantothenic acid,* all of which are members of the water-soluble B group and are the pellagra-preventive vitamins of humans, rats, and chicks, respectively; *biotin, inositol,* and *p-aminobenzoic acid,* which are also members of the water-soluble B group of vitamins; vitamin C, or *ascorbic acid,* the water-soluble antiscorbutic vitamin; vitamin D_2, or *calciferol,* the fat-soluble antirachitic vitamin; vitamin E, or *α-tocopherol,* the fat-soluble antisterility vitamin; vitamin K, a fat-soluble antihemorrhagic vitamin, and vitamin P, a flavone derivative concerned with permeability.

The effects of vitamin deficiency in animals are studied by withholding vitamins from the

Fig. 9-6 Chemical weed control in vegetables. One method of applying herbicides. (Photograph by R. S. Beese, courtesy of C. J. Noll, Pennsylvania Agricultural Experiment Station, University Park, Pa.)

diet. With many autotrophic plants, this method is not possible because such plants make their own foods and usually can also synthesize all the vitamins they need. Some of the fungi, however, are unable to synthesize certain vitamins. Roots and certain tissues of higher plants may depend upon other organs of the plant for their source of vitamins. Thus much of our knowledge of the importance of vitamins to plants has been obtained by studies with fungi, with isolated embryos or root tips, or with tissue cultures of higher plants on synthetic media.

From the culture of isolated root tips on synthetic media it has been demonstrated that vitamin B_1 or thiamine, is essential for the growth of roots *in vitro*. Thus, when a short root tip of a pea plant is cut off and transferred to a suitable medium containing the necessary inorganic substances and sugar, it will grow for a time at a comparatively fast rate. If now the tip of this root is removed and transferred to a fresh nutrient solution made up of inorganic substances and sugar, it will grow very little, and a third transfer may result in complete cessation of growth. If, however, a small amount of vitamin B_1 is added to the nutrient solution, the severed root tip will continue to grow at a fast rate. Obviously the original root tip contained enough vitamin B_1, which it received from the aboveground parts of the plant, to enable it to continue growth, but when this was used up, further growth was checked unless vitamin B_1 was supplied to the nutrient medium. This indicates that the root tip is unable to synthesize vitamin B_1 but must depend upon the aboveground parts of the plant for its supply.

Examination of root tips that have failed to grow because of lack of vitamin B_1 indicates that cell division has ceased. Thus it is shown that cell division in root tips depends upon a supply of B_1. Many other roots give the same response to vitamin B_1, but roots vary in the amount of this vitamin which is available to them while they are still attached to the plant. With most plants, an adequate amount is available, and therefore the supplying of additional vitamin B_1 to the roots of growing plants is often without effect. Some roots require for normal growth not only thiamine but also nicotinic acid and pyridoxine. There is apparently a diversity of vitamin requirements for roots of different kinds of plants.

Cultures of embryos detached from seeds, with cotyledons and endosperm removed, have also been made. Various workers have reported the need of such vitamins as thiamine, biotin, and ascorbic acid by embryos of different plants cultured in this manner. Studies of this type as well as other tissue cultures of higher plants have demonstrated the importance of vitamins to such plants.

Some of the earliest work on the relation of vitamins to plants was carried out with fungi, particularly with yeasts. Studies with fungi have been particularly enlightening. Some of the fungi apparently are able to synthesize all the vitamins they need for normal growth. Others, however, are distinctly limited in this respect and will not grow well unless they are furnished with certain vitamins in the culture medium. Thus *Phycomyces blakesleeanus*, a simple mold, cannot synthesize thiamine and will not grow unless this vitamin is supplied in the culture medium. Chemically the thiamine molecule is made up of two simpler substances, thiazole and pyrimidine. *Phycomyces* will grow if supplied with these two substances but not if only one or the other is supplied. Other fungi can synthesize thiamine if supplied with thiazole only, while still others can do so if supplied with pyrimidine only. Other vitamins that have been shown to be necessary for certain fungi include biotin, pantothenic acid, nicotinic acid, pyridoxine, and inositol. Wide differences apparently exist among the fungi in their requirements for different vitamins.

Some of the fungi have been found useful for vitamin bioassays, i.e., for the determination of the amount of certain vitamins present in foods or plant tissues. Such fungi have complete deficiencies for a given vitamin, and the amount of growth they make in a given time is proportional to the amount of this particular vitamin available to them. By measuring the amount of growth of such fungi in a culture medium to which is added a definite quantity of food or tissue, the vitamin content of which is to be ascertained, the vitamin content of the food or tissue can be determined. Thus certain races of yeast can be used for pyridoxine determination. *Phycomyces blakesleeanus* can similarly be used for vitamin B_1 assays.

The functions of many of the vitamins in plants have not yet been determined. Vitamin B_1 and possibly other members of the B group of vitamins operate through their relationship to vital enzyme reactions concerned in respiration. Some of them have been shown to be coenzymes. They are often very specific in their action and are sometimes effective in extremely minute amounts. Thiamine, for example, in some cases has been reported to be still active in a dilution of 1 part in 40 trillion. One gram of biotin dissolved in 25 million gal of water has been found to be sufficient for the normal growth requirement of yeast and of certain bacteria. The exact conditions under which they are synthesized in the plant are not well known, although such environmental factors as light and the supply of inorganic substances have been shown to affect the content of certain vitamins in plants.

Correlation of plant parts It has long been known that the removal of one part of the plant may affect the growth of other parts. Removal of the terminal parts of stems, for example, stimulates to growth lateral buds which normally would remain dormant. For this reason it is often possible to change radically the shape and form of a plant by pruning. A similar relationship exists between the development of vegetative shoots and flowers. Thus, if the flower buds of tomato plants are pinched off as they appear, the plant is stimulated to renewed vegetative growth. On the other hand, if the fruits are allowed to develop, vegetative growth is checked. In most annual plants the cessation of vegetative growth is correlated with the development of flowers and fruits on the plant.

Not only is the amount of growth an organ makes influenced by the removal of other organs but sometimes its whole form and structure as well. Thus the tendrils of pea leaves sometimes develop into broad leaflets if the foliage leaflets are cut away before the leaf has fully developed. Similarly the highly modified sporophylls (spore-bearing leaves) of some ferns become vegetative leaves if the ordinary leaves are removed before the sporophylls are fully differentiated. Many other examples could be given.

The effect on organs that remain when others are removed indicates that, in the normal growth of the plant, these organs must have a reciprocal influence on each other. The development of the vegetative shoot affects the development of roots; the growth of roots, in turn, affects that of the vegetative shoot; flower development affects vegetative-shoot development, and so on. Some of these effects have been attributed to the influence of hormones, but, in general, little is known as to how they are brought about. Undoubtedly competition among the different organs for the food supplies plays a part.

Periodicity of growth; the grand period The operation of internal factors is clearly manifested in the course of growth of a plant, as well as in the growth of any one of its parts, from the undifferentiated form to maturity. The growth of a cell, an organ, or a whole plant does not proceed at a uniform rate even when the conditions under which the plant is growing are kept uniform. Growth starts at a slow rate, gradually increases until a maximum rate is reached, and then falls off until it finally comes to an end. This can readily be shown by measuring, at regular intervals, the size of a growing organ or plant from the beginning of its growth until it is mature and by plotting the increase in size against time. When this is done, a characteristically shaped curve is obtained similar to the one shown in Fig. 9-7*A*. If the actual size of the organ or plant is plotted against time, an S-shaped curve is obtained as shown in Fig. 9-7*B*. Such curves are obtained no matter what measurement of size is used.

Sudden fluctuations in temperature or in other environmental conditions may cause irregularities in the curve, but if the whole period of enlargement is included, the general shape of the curve will remain the same. The total period of enlargement of an organ or a plant is called the **grand period of growth,** and the first type of curve mentioned above is referred to as a grand-period curve.

External conditions may shorten or prolong the grand period or they may change the ultimate size of the plant, but they have little or no effect on the course of growth as indicated. This fact has led to much speculation as to the underlying internal factors that bring it about. Most of the internal factors that have been mentioned, as well as others, have been individually championed by one investigator or another in explanation of it, but as yet no one of the explanations has met with general acceptance. The initiation of flowering, as mentioned in a previous section, has been definitely shown to be correlated with the falling off and final cessation of vegetative growth. Other correlations of plant parts and general food relationships also probably play a part.

In perennial plants periods of active growth are followed by periods of rest. In such plants each active period of growth follows the course of a grand period. The periodicity of perennial plants is most pronounced in temperate and arctic species but occurs also in many tropical species. The development of flowers and fruits and of other parts of the plant is also periodic in perennial plants. While periodicity may be influenced by external factors, it is probably governed largely by internal factors.

EXTERNAL FACTORS

General

The effects of external, or environmental, factors on growth have been much more thor-

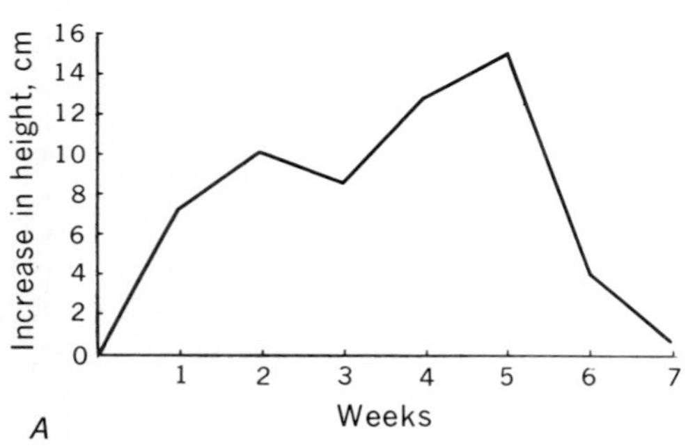

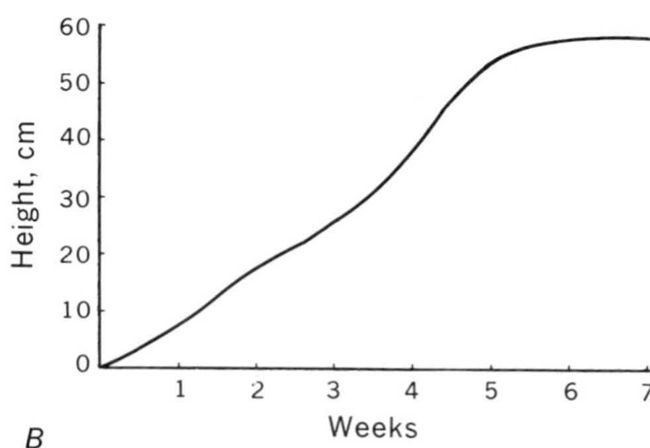

Fig. 9-7 Curves showing growth of Mandarin soybeans. A, increase in height plotted against time, giving a typical "grand period of growth" curve. The drop in the curve during the third week represents approximately the period at which the seedlings had exhausted the reserve food stored in the seeds and became independent plants. The drop during the sixth week is correlated with the initiation of flower production. B, the typical S-shaped curve obtained by plotting the actual height of these same plants against time. (Drawn by Florence Brown from data by H. W. Popp.)

oughly studied because many of these factors are more readily brought under control and their effects more readily observed. In the last analysis, however, the operation of external factors cannot be separated from that of internal factors. Any change in growth that results from the operation of an external factor can be explained only on the basis of a change in the internal conditions of the plant. In other words, the mechanism of the action of external factors rests on the influence such factors have on the internal mechanism of the plant. It is in this manner that external factors may affect not only the rate of growth and the ultimate size of a plant or an organ but also its general form. Changes in general structure and form are usually referred to as **formative** effects. Many of the external factors have a formative effect on the plant. It should be emphasized that the actual growth of the plant results from the simultaneous operation of all factors, internal as well as external. It is therefore not always possible to separate the influence of a single factor from the influence exerted by other factors.

Of the external factors influencing growth, **radiation, temperature,** and **moisture** have been most studied. The supply of **oxygen, inorganic salts,** and **carbon dioxide** is no less important. These have already been referred to in previous chapters. The oxygen supply influences the respiration of the plant and, through it, growth. It is considered in Chap. 12 and need not be taken up further here. The carbon dioxide supply affects photosynthesis, which in turn affects all organic synthesis within the plant, and hence growth. It has already been considered in Chap. 5. The effect of inorganic salts has been briefly taken up in Chap. 7. While it is definitely known that many of these salts are necessary for growth, the exact manner in which many of them influence growth has not been satisfactorily determined. A detailed discussion of their possible influences would be too involved in a work of this kind. **Electricity, gravity, mechanical agents, insect and fungus injuries,** and the presence of **toxic and stimulating substances** in the environment also affect growth. The present discussion is restricted more or less to radiation, temperature, and moisture.

Radiation

Importance of radiation Probably no other environmental factor in nature plays a more important role in the growth of higher plants than does radiation, and particularly visible radiation or light. It is the ultimate source of all energy stored by the plant in photosynthesis, upon which all forms of life are directly or indirectly dependent, and without which all growth would ultimately cease. Light affects the germination of some seeds. After the young plant has emerged from the seed, its future is again influenced by the kind and amount of radiation it receives. Not only is the rate of growth of the plant conditioned by the environmental radiation, but the size and form, the internal structure, the composition, the intensity of internal physiological processes, flowering, fruiting, and seed development of the plant are affected as well. The movement or orientation of plants and plant organs and the distribution of plants are likewise influenced by radiation. In short, radiation is a constant and important factor in the life of the plant from germination to maturity. It probably exerts a greater formative effect on plants than does any other external factor.

Variability of radiation The universal source of radiation of plants in nature is the sun. Much of the radiation from the sun is scattered by dust particles and moisture in the atmosphere. This diffused skylight is different in quality and intensity from direct sunlight. The combined radiation from the sun and the sky is commonly referred to as daylight. Both the quality and the intensity of daylight vary from hour to hour even on perfectly clear days. The intensity gradually increases from morning to noon and decreases from noon to night. The variation in quality is manifested by the difference in color of sunlight in early morning or late evening as compared with noon. Quality, intensity, and duration of daylight (length of day) also vary with the season of the year. On cloudy days quality and intensity may vary from minute to minute and over a wide range. These wide variations have made it difficult to study the effects of radiation on plant growth. Artificial light sources have been used for plant studies, but the radiation from these also varies widely. Furthermore no artificial source has yet been found which is capable of approximating the quality and intensity of noon daylight in summer.

Sunlight consists of infrared, visible, and ultraviolet radiation. Its spectrum extends from about 5,000 mμ in the infrared down to 291 mμ in the ultraviolet (1 mμ = 1/1,000,000 mm). The region of highest energy value in this spectrum is usually in the yellow to green region. The energy falls off gradually toward the infrared and rather sharply toward the ultraviolet. For the sake of comparison with other types of radiation, the approximate limits of different regions of the electromagnetic spectrum are given in the following table:

The Electromagnetic Spectrum

Radio waves	2,000,000 to 5,000 cm
Short electric waves	5,000 to 0.025 cm
Infrared (heat)	320,000 to 720 mμ
Visible region (light)	720 to 400 mμ
Red light	720 to 626 mμ
Orange light	626 to 595 mμ
Yellow light	595 to 574 mμ
Green light	574 to 490 mμ
Blue light	490 to 435 mμ
Violet light	435 to 400 mμ
Ultraviolet	400 to 13.6 mμ
X rays	1.32 to 0.007 mμ
Gamma rays of radium	0.137 to 0.002 mμ
Cosmic rays	Less than 0.002 mμ (?)
Sunlight spectrum	5,000 to 291 mμ

It should be emphasized that no one source of radiation consists of any more than a limited region of the whole electromagnetic spectrum. Furthermore the energy distribution in the spectrum of any two sources of the same range may be widely different. It is the spectral range and the energy distribution that are meant when quality of light is mentioned. Each region of the spectrum may affect plant growth. In the pages that follow, reference is made to effects of intensity, quality, and duration of radiation, with special emphasis on visible radiation or light.

Effect of darkness; etiolation Long before anything was accurately known concerning the effect of light on plants, the abnormal development of plants in darkness had been observed. The general appearance of plants grown in the total absence of light is familiar to everyone. In most dicotyledonous plants darkness causes excessive elongation of stems (Fig. 9-8), brought about chiefly by the abnormal lengthening of internodes. Petioles are also unusually long, but leaf blades small and undeveloped. The complete absence of chlorophyll gives the plants a pallid or yellow color. Rosette plants, which in light have unusually short stems, sometimes develop upright stems in the dark. In monocotyledons the leaves but not the stems become abnormally long in darkness (Fig. 9-8*C*, *D*), but the leaves are usually very narrow. Roots are usually poorly developed in all plants grown in the dark. Plants grown from seed fail to develop flowers in the dark; but when grown from buds in which flower primordia have been developed, such plants may produce flowers in the dark. The flowers in this case are usually paler in color. Many other abnormalities occur in specialized organs.

The tissues of plants grown in darkness are unusually soft, weak, and succulent. Cells are large, thin walled, and relatively undifferentiated. Strengthening tissues are poorly developed. Leaf mesophyll is of a homogeneous, loose structure with little differentiation of palisade and vascular tissue. As compared with plants grown in light, the dry weight of such plants is considerably lowered, and the per-

Fig. 9-8 A, B, effect of darkness upon a dicotyledonous plant, the common bean (Phaseolus vulgaris); *A, grown in weak, diffused light; B, grown in darkness; both planted at the same time. C, D, effect of darkness upon a monocotyledonous plant, wheat* (Triticum sativum); *C, grown in darkness; D, grown in weak, diffused light.*

centage of water increased. Sugars and soluble nitrogen compounds are usually higher in plants grown in the dark.

Plants having the characteristics enumerated above are said to be **etiolated.** The ability of etiolated plants to survive in the dark depends partly upon the quantity of stored foods available to them. Plants grown from storage roots, tubers, bulbs, or corms usually live longer in the dark than do seedlings. Unless light is provided, all higher plants sooner or later die. Death is hastened at high temperatures because of the rapid consumption of the food reserves brought about by increased respiration. It is interesting to note also that even a very small amount of light may profoundly influence the development of etiolated plants. Thus it has been shown that as little as 2 min of light per day is enough to cause etiolated seedlings of peas and broad beans (*Vicia Faba*) to start to expand the growing tip (plumule) and to unfold leaves which otherwise would not have developed in the dark.

The cause of etiolation is not yet fully known although it seems to be definitely tied up with the activity of growth hormones. That it is not caused by the failure of photosynthesis to take place is proved by the fact that plants grown in light but in the absence of carbon dioxide do not become etiolated. A type of etiolation is sometimes produced in plants grown at high temperatures or in the absence of the blue-violet end of the spectrum or in very weak light of any kind. Respiration and catabolic metabolism, in general, seem to proceed at a rapid rate in etiolated plants. The plants elongate rapidly as though they were reaching out for light. It is also interesting to note that leaves, which can function only in light, are usually not well developed in the dark. These features, though in no sense causal, are of some advantage to a seedling growing from seed lying deep in the soil. Etiolation caused by darkness occurs also in gymnosperms, lower vascular plants, mosses, algae, and fungi.

Effects of intensity of radiation The effect of low light intensity is similar to that of darkness in that it causes excessive elongation of stems, but it differs markedly in the effect it has on leaves and on chlorophyll development. Even in weak light, leaves completely unfold and develop an abundance of chlorophyll but remain thin and poorly differentiated internally. Leaves grown in dense shade have a loose palisade tissue consisting of a single layer of cells and a very loose spongy mesophyll, filled with air spaces. The epidermis is not heavily cutinized. The other tissues of the plant are also weak and tender. Roots are poorly developed when the tops of plants are in weak light.

As the light intensity is increased, the stems become shorter, there is a better development of roots, and all tissues are better differentiated. The plant takes on its ordinary appearance. Leaves reach a maximum size in moderate intensities. The final height attained by the plant is also often greatest in a medium intensity of light (Fig. 9-9). At very high intensities many plants remain very short and stocky. The leaves are likely to be smaller in area but much more compact and thicker. Often there are two or more palisade layers and very few air spaces even in the spongy mesophyll. High light intensity is conducive to a higher transpiration rate. All the structures enumerated tend to reduce this loss. The stems of plants grown under high light intensity are much thicker and the strengthening tissues are more fully developed. Within limits, the higher the light intensity, the greater is the dry weight of the plant. Roots and general storage organs are always better developed and flowering and fruiting are at a maximum in strong light. In very weak light, flowering and fruit development sometimes fail to occur altogether. The ratio of roots to tops of plants usually increases with increased light. The greater sturdiness of plants grown in full sunlight is probably caused in part by the increased photosynthesis under this condition. Shade plants are often low in carbohydrates and high in nitrogen compounds as compared with sun plants. The checking effect of light on stem elongation accounts for the fact that most plants grow more rapidly at night than they do during the day.

Plants do not all respond in the same manner to light intensity. Some require more light than others. Some species, like the sunflower, grow best only in full sunlight. Others may be able to live in as little as one hundredth of the intensity of full sunlight. Only plants that have a relatively low light requirement can develop normally in dense shade. Partly for this reason, the species that are found growing under the canopy of dense forests are limited. Similarly light intensity is one of the factors determining the succession of different associations of plants in a given area. When a beech, maple, and hemlock forest is cut down and the area is left unplanted, it is not replaced in the immediate future by a forest of the same type. An entirely new association of plants invades the area. Most of these new plants are able to grow in direct sunlight. This new association may be replaced by several others before the area, after many years, gets back to the original beech-maple-hemlock forest. The latter is able to maintain itself indefinitely, however, if not burned or cut down, partly because the seed-

Fig. 9-9 Effect of light intensity on the growth of Ito-San soybeans; plants six weeks old; approximate percentage of full daylight intensity received by each plant: 1, 66 percent; 2, 23 percent; 3, 8.6 percent; 4, 6 percent; 5, 3.8 percent. Note that the maximum height was attained by the plants under a medium light intensity (3).

lings of these species are able to develop in the shade of the mature trees. An association of this type that is able to maintain itself is called a **climax association.** It should be mentioned that many other factors besides light determine succession and the maintenance of climax associations.

Effects of quality of radiation The effects of different regions of the spectrum have been studied by growing plants under glasses that transmit limited regions. In all studies of this kind it is important that intensities be kept equal under the different kinds of radiation; otherwise it is impossible to separate quality effects from intensity effects. From studies of this kind it has been found that plants grown under the red-orange part of the spectrum become etiolated even at high intensities, although they become green. Those grown under the blue-violet end of the spectrum more nearly resemble plants grown in the full spectrum of daylight. Most striking results have been obtained when plants were grown in daylight from which the blue-violet end of the spectrum below 529 mμ was eliminated. Many plants grown from seed under this condition elongate more rapidly during the first two or three weeks of growth, as compared with plants grown in the full spectrum of daylight. Some plants, like soybeans, tomatoes, four-o'clocks, and coleus, attain much greater final height under the absence of blue-violet radiation, but others, like sunflowers, buckwheat, Sudan grass, and other sun plants do not. Soybeans become twiners under these conditions (Fig. 9-11*B*). All plants, regardless of their final height, have much thinner stems, poorly differentiated internal tissues, poor development of flowers, fruits, and general storage organs, much lower fresh and dry weights, and a higher percentage of moisture than do plants grown in full sunlight. In general they resemble plants grown in darkness (Figs. 9-10, 9-11). The same effects are produced when the region below 472 mμ is eliminated but not to so marked an extent.

On the other hand, plants grown in the absence of the red end of the spectrum, even at greatly reduced intensity, develop more nearly like normal plants. The blue-violet end of the spectrum seems to be much more efficient in dry-weight production than the red end of the spectrum, though not so efficient as the full spectrum. This is true in spite of the fact that the red end of the spectrum is more efficient in photosynthesis. The blue-violet end of the spectrum has a marked formative effect on the plant. It is chiefly this part of the spectrum that is responsible for the stunting effect of light on growth.

Plants grown in the absence of the ultraviolet portion of sunlight are little different from those grown in the full spectrum. This may be caused by the relatively low percentage of ultraviolet (usually less than 1 percent of the total radiation) present in sunlight at comparatively low altitudes. At higher altitudes it has been thought that ultraviolet radiation may exert a strong formative effect on the plant which may partly account for the reduced stature of plants in alpine regions. When a source of radiation rich in ultraviolet of shorter wavelengths than those found in sunlight is used, plants are seriously injured and often killed by it.

Infrared radiation of high intensity may also injure plants through the rise in temperature that it causes by being absorbed by the plant. The ordinary infrared of sunlight acts like darkness in its effect on growth. X rays in moderate doses have been reported to be injurious to many kinds of plants. X rays are much more penetrative than ultraviolet radiation and therefore are capable of more deep-seated effects on the plant. In general they check growth and cause abnormalities in the structure of the plant. In a few instances extremely light doses of X rays have been reported to stimulate the growth of some plants. X rays are not present in the ordinary radiation to which plants are exposed.

Effects of duration of radiation; photoperiodism In temperate regions plants growing in nature are subjected to alternations of light and darkness. Day lengths range from 15 hr in summer to about 9 hr in winter. It has been found that the growth of plants and particularly the development of flowers and fruits are markedly influenced by the length of day. This response of plants to length of the daily period of illumination has been called **photoperiodism.**

Normally the plant can attain the flowering stage only when the length of day falls within certain limits. These limits are reached only at certain seasons of the year. Some plants, like

Fig. 9-10 Effect of quality of radiation on plant growth. A, sunflower plants, 80 days old; B, carrots, 143 days old; 1, grown under ordinary greenhouse glass transmitting down to 312 mμ; 2, grown under a clear glass transmitting down to 296 mμ; 3, grown under a light-yellow glass which eliminated practically all ultraviolet radiation, transmitting down to 389 mμ; 4, grown under yellow glass which eliminated all violet and ultraviolet and part of the blue, transmitting down to 472 mμ; 5, grown under an orange glass which eliminated all ultraviolet, violet, blue, and half of the green, transmitting chiefly red, orange, and yellow—lower limit 529 mμ.

Fig. 9-11 Effect of quality of radiation on plant growth. A, four-o'clock plants, 65 days old; B, Peking soybean plants, 34 days old. Numbers refer to same conditions as in Fig. 9-10. Note that the elimination of only ultraviolet radiation (3) has little effect on the growth of the plants, while elimination of the visible blue and violet along with the ultraviolet (4 and 5) causes marked changes in growth and development, the soybeans becoming twiners.

radish and lettuce, bloom only during long days (12 hr or more) (Fig. 9-12*B*). Such plants are called **long-day plants** and bloom normally in midsummer. Others, called **short-day plants,** require a short day (less than 12 hr) for flowering and bloom normally either late in autumn or early in the spring. Examples of short-day plants are most of the spring flowers and such autumn-flowering plants as ragweeds, asters, cosmos, and scarlet sage (*Salvia splendens*) (Fig. 9-12*A*). Still other plants, like the tomato, respond to all day lengths and therefore may bloom all year round.

When short-day plants are kept under long-day conditions, they usually continue to grow vegetatively without flowering and may reach an unusual size. Many long-day plants when kept under a short day grow weakly vegetatively but fail to flower. Long-day herbaceous perennials like some of the sedums have been kept for eight to nine years under a short day without flowering, while controls receiving normal day lengths flowered every year during this time. One of these plants, after having been kept under a short day for eight years, was given the full day length of summer and bloomed normally at the same time as did the control plants. In all types the amount of vegetative growth is greatest under a long day. While many plants are on the border line between short-day and long-day plants, they probably have an optimum day length for flowering.

A day that is lengthened by means of artificial light, even of low intensity, seems to be as effective as the normal long day of midsummer in bringing long-day plants into flower. By this means it is possible to hasten the flowering of greenhouse long-day plants or to check the flowering of short-day plants. On the other hand, while the long days of midsummer may be shortened by not placing the plants in light so early in the morning or placing them in the dark in the afternoon, midday darkening does not produce the effect of a short day. Thus it has been reported that short-day plants placed in the dark between 10:00 AM and 3:00 PM in midsummer failed to flower or, in other words, responded as though they had been subjected to full summer light. The length of the dark period is as important as the day length in inducing flowering in short-day plants.

When plants receive alternations of light and darkness of 6 hr or less, they respond as though kept under a long day or under continuous illumination. When given 10 hr of light every other day or a full day (15 hr) of light every other day in midsummer, plants respond as though kept under short-day conditions.

Length of day has also been reported to have an influence on the formation of tubers and bulbs, the character and extent of branching, root growth, abscission and leaf fall, dormancy and rejuvenescence, and the habits of annuals, biennials, and perennials.

Plants have also been raised from seed to seed in continuous artificial light. In nature plants are not subjected to continuous illumination. Some plants are apparently able to grow normally in continuous light while others are not. Tomatoes are injured and finally killed in any day length over 19 hr. Many short-day plants fail to flower under continuous illumination.

Scientists of the Agricultural Research Service of the U.S. Department of Agriculture have discovered that photoperiodic responses of plants are controlled by a blue-green photoreceptive protein pigment they called **phytochrome.** It also regulates many other plant-growth processes, including stem elongation, germination, and pigmentation. Phytochrome exists in two forms —one with maximum action in the red region of the spectrum (660 mμ) and the other in the far red (730 mμ). Absorption of light by either form converts it to the other form. The factor that activates or inactivates phytochrome is a bile pigment, chromophore, which constitutes 1 percent of the phytochrome molecule. Phytochrome occurs in very minute quantities in the plant and probably acts as an enzyme.

Fig. 9-12 Effect of length of day on plants (photoperiodism). A, scarlet sage (Salvia splendens), *a short-day plant; B, lettuce, a long-day plant. The day lengths to which the plants were subjected are indicated on the pots. (Photograph courtesy of J. M. Arthur, Boyce Thompson Institute for Plant Research, Inc., Yonkers, N.Y.)*

Temperature

As previously stated, growth of the plant proceeds as a result of constructive metabolism. New materials are constantly being synthesized and others broken down in the building of protoplasm. This involves a series of complex chemical reactions, all of which are influenced by temperature. Temperature determines the rate at which these processes take place, and since the rate of growth is determined by the rate of these processes, temperature likewise affects the rate of growth. As will be seen later, it also exerts an important formative effect on the plant.

Temperature range and life duration The range of temperatures within which plants can live and grow varies with the species, with the different parts of the same plant, and with the general external and internal conditions to which the plant has been subjected either before or during the application of a change in temperature. It varies particularly with the condition of the protoplasm.

As a rule, plants do not thrive when moved much more than 100 miles north or south of the region to which they are indigenous, largely because of the difference in prevailing temperatures in different latitudes. Thus tropical plants usually fail to thrive in temperate regions and most plants of temperate regions cannot be made to grow in the tropics except at high altitudes. In high mountainous regions of the tropics, there is often the same range of plant associations extending from the lowland regions to the tops of the mountains as is found in going from the tropics toward the polar regions. Temperature is thus one of the most important factors determining the distribution of plants over the earth's surface.

In general, resting organs like seeds and spores can withstand a much wider range of temperatures than can growing plants. Cases have been reported of seeds that were able to germinate after having been subjected for a short time to a temperature as low as −180°C or as high as 103°C. Seeds taken by the Greely expedition in 1883 to Fort Conger, about 490 miles from the North Pole, were found in 1899 by the Peary party and sent to the United States, where they remained unplanted until the spring of 1905. At this time lettuce and radish seeds from the lot were planted. The lettuce failed to grow but about half of the radish seeds germinated and produced vigorous plants that produced seed. The ability of bacterial spores to withstand extreme temperatures is well known. Many spores can be boiled or frozen without injuring them. Resting organs like some tubers, bulbs, and the branches of deciduous trees in winter are also able to withstand wide ranges of temperature without injury. Succulent plants usually have a much narrower range within which they can live and grow.

The ability of growing plants to withstand low or high temperatures is governed by the conditions to which the plants are subjected. A sudden change in temperature is more likely to prove injurious than a gradual change. Many plants, when gradually subjected to lower and lower or higher and higher temperatures, are able to adapt themselves to it. Such adaptation involves fundamental changes in the condition of the protoplasm and particularly its degree of hydration. The ratio of bound water to free water is usually higher in resistant plant organs. Thus seeds that have imbibed water and swollen are much less resistant to extremes of temperature than are dry seeds in which much of the moisture present is bound water. More will be said later concerning bound water in connection with the discussion of subminimal temperatures. In addition to the water relations, many other factors, including age, the presence of osmotically active substances, and the general structure of the plant, play a part in determining the range of temperatures within which the plant can continue to live.

Cardinal points of temperature for growth The range of temperatures within which plants are able to grow may be designated by three cardinal points, viz., the **minimum,** or temperature below which no growth takes place, the **optimum,** which is usually considered the temperature which gives the highest rate of growth or is the best for growth, and the **maximum** temperature, above which no growth takes place. These cardinal points are not fixed during the life of the plant, nor are they always the same for all parts of the plant. The cardinal points for root growth may be different from those for stem growth. Those for flower development are frequently different from those for vegetative growth. This is clearly shown in many plants that flower very early in spring before the vegetative shoots have developed. Seeds will sometimes germinate at much lower temperatures than the temperature required by the more mature plants for growth. Although many variations occur among different species, the minimum temperature for growth of temperate plants is usually somewhat above the freezing point of water and the maximum around 45 to 50°C. The optimum usually averages around 30°C.

The optimum temperature for growth is conditioned by the length of time the plant is kept under the influence of this temperature. For example, plants may grow most rapidly for a short period of time at 35°C; but if this temperature is maintained, the rate may decrease below the rate obtained under a constant temperature that is lower. For this reason a fourth cardinal point is sometimes recognized which has been called the **maximum-rate** temperature and denotes the temperature at which the rate of growth reaches its highest intensity, though only for a short time. The true optimum temperature would then be the temperature at which growth proceeds most rapidly regardless of how long this temperature is maintained. In other words, the plant is able to maintain its most rapid growth indefinitely until maturity at the true optimum temperature.

In general the rate of growth of a plant is lowest near the minimum, gradually increases as the temperature rises toward the optimum, and falls off again toward the maximum. The optimum temperature is not always midway between the minimum and maximum but more often lies nearer the maximum than the minimum.

Supramaximal temperatures When plants are subjected to temperatures above the maximum, i.e., supramaximal temperatures, they are sooner or later killed, the time required depending upon how much above the maximum the temperature is. Ordinary plants are killed in 1 to 1½ hr when the temperature is 1 to 1½° above the maximum. When the temperature is very much above the maximum, death may be almost instantaneous. Every 10°C rise above the maximum shortens the time required to cause death from 10 to 150 times. From the fact that this relationship of temperature also holds in the coagulation of many proteins, it has been thought that death of plants under high temperatures may be caused by the coagulation or gelation of the protoplasm. A condition develops in the cells which is referred to as **heat rigor.** The protoplasm becomes set and a return to normal is impossible.

Subminimal temperatures Temperatures below the minimum also cause death if prolonged. In this case a **cold rigor** develops and the protoplasm becomes set as it does in heat rigor. This condition develops in thermophilic fungi and bacteria at ordinary temperatures, these plants having a high minimum temperature. Some tropical plants are likewise killed at temperatures above the freezing point, which indicates that freezing is not necessary to cause death. With many temperate plants, however, freezing is a common cause of injury and death. This is readily observed after the first heavy frost in autumn when many garden plants, previously green and vigorous, wilt, turn brown, and die. Some plants, however, and especially perennials are able to withstand severe freezing.

When death is caused at temperatures slightly above 0°C, it may result from wilting or from disturbed metabolism. When plants like tobacco, pumpkin, or beans have been growing under ordinary temperatures and are suddenly transferred to an air temperature of 2 to 4° C, they wilt. If the temperature of the soil in which these plants are growing is raised to 18°C without raising the air temperature above 2 to 4°C, they do not wilt. In this case it is evident that wilting results from a more rapid loss of water by transpiration than can be supplied by absorption at the low temperature, or in other words, at low temperatures absorption is reduced more than water loss, and the plant wilts. Many greenhouse plants like coleus, however, are killed by temperatures slightly above freezing before wilting occurs. In such plants the leaves usually become spotted when the temperature is too low, and these spots increase in size until finally the leaves turn brown and die. This happens in about four days at 3°C. In plants of this kind death is probably caused by disturbed metabolism.

Death resulting from actual freezing is much more common. There is a wide variation in plants in their ability to withstand freezing. Some can remain frozen all winter without being injured. In northern Siberia forest trees are not killed even though the temperature may fall to −62 to −64°C. In north temperate regions, wheat, planted in autumn, is able to survive severe winter weather. On the other hand, most annuals and many other plants are killed by freezing temperatures.

Since plants always contain salts in solution, the actual freezing point is usually several degrees below the freezing point of water. When freezing does occur, ice usually forms first in the intercellular spaces. If freezing is severe, ice may also form within the cells, and the whole tissue becomes frozen solid. The formation of ice in the intercellular spaces results in the removal of water from the cells. This removal of water is one of the most common causes of death, since it leads to coagulation and irreversible precipitation of the cell colloids. The increased concentration of salts under these conditions also contributes to this precipitation. Since water expands on freezing, there is also often a tearing or rupture of the cells resulting from it.

Hardening The ability of plants to withstand freezing depends upon their general structure as well as upon the condition of the protoplasm. Both of these in turn depend partly upon the conditions under which the plants have been grown. Plants with a thick epidermis and compact internal tissues are sometimes more resistant to freezing than are those loosely constructed. The resistance to freezing also increases with increased concentration of the cell sap and particularly with the ability of the tissues to bind water. These conditions can be developed in some plants by gradually subjecting them to lower and lower temperatures or, in other words, by **hardening** them. The gardener usually hardens his vegetables, before he transfers them to the open in spring, by first keeping them for a time in a cold frame. In nature, hardening naturally occurs in many plants during autumn and early winter. Some plants cannot be hardened in this manner. Such plants are easily killed by freezing temperatures. In species that are winter-hardy, it has been found that hydrophilic colloids develop during the process of hardening, which enable these plants to bind water so that it cannot readily freeze. Winter-hardy varieties of wheat, for example, are those which develop a relatively high percentage of bound water as the temperature is gradually reduced. Such plants also have a more compact structure and other features that enable them to withstand freezing. These conditions do not develop, however, even in winter-hardy plants unless the temperature has been gradually lowered. Winter-hardy plants as well as susceptible varieties, if first grown in a greenhouse and then suddenly subjected to freezing temperatures, are immediately killed.

Formative effects of temperature Temperature has a marked influence on the size, form, and general structure of the plant. These formative effects appear most striking when the temperature is near the minimum or near the maximum for growth. When plants are kept constantly under low temperatures, the length of the growing zone increases but the internodes remain short, resulting in a plant of comparatively low stature. On the other hand, when they are kept under high temperatures, the length of the growing zone decreases, the internodes become longer, and the plants become taller. Some plants develop a type of etiolation at high temperatures. This is true of potatoes. Potatoes develop tubers best at low to medium temperatures and hence can be grown more successfully in northern temperate regions like Maine. A European variety of potatoes has been reported to produce only tubers and no vegetative shoots when grown at a temperature of 6 to 7°C. Many plants develop excessive vegetative growth at high temperatures. Under these conditions there is probably a rapid rate of respiration and other metabolic changes that tend to deplete carbohydrates to such an extent that they do not accumulate and hence storage organs are poorly developed.

Temperature also affects flowering and fruiting. The temperature under which seeds are germinated may affect later growth and flowering of the plant. Different species of plants react differently to such temperature treatment. The seeds of some plants must be exposed to low temperatures if flowering is to occur when the plants mature. Such seeds, like wheat, are usually sown in the fall in temperate regions, being thus exposed to winter temperatures during early growth. Seeds are sometimes subjected to low temperatures during germination and then later planted. Such temperature treatment is called **vernalization.** Seeds of some tropical plants must be treated at high temperatures if flowering is to occur.

In many other ways temperature is important in the growth of the plant. It should be mentioned, however, that temperature always operates through the changes it induces in the internal conditions of the plant.

Moisture

The importance of water to the plant has been emphasized in Chap. 7. The loss of water by transpiration has also been considered in Chap. 4. These matters will not be considered further here. Growth takes place only in turgid cells, and this necessitates an ample supply of water. The amount of available water, like radiation and temperature, may influence to a marked degree the form, structure, and nature of growth of plants. The cardinal points of moisture have not been so accurately determined, but excessive moisture or excessive dryness are both likely to be unfavorable to the growth of most plants. Only plants with some degree of plasticity are able to adjust themselves to radical changes in moisture conditions. Those lacking such plasticity are often killed by drought or by a waterlogged soil. The conditions of plants that enable them to withstand drought have already been mentioned in connection with transpiration (Chap. 4).

In general, herbaceous plants growing under conditions of high moisture tend to have loose, succulent tissues with numerous air spaces. Strengthening tissues are often poorly developed, and cuticular and suberized membranes are reduced or entirely absent. Many plants fail to develop root hairs when the roots are in water. In very humid regions the stems of plants are likely to be more elongated and the leaves broader and thinner. Many of these features develop also under conditions of low light intensity. Since plants in shaded places are subjected both to lower light intensity and to higher humidity, it is not always possible to separate the effects of one from the other. Plants that have developed under conditions of this type are usually killed in a short time when subjected to a dry atmosphere. Plants that are capable of living both upon land and in water often assume an entirely different form on land from that which they have in water. Submersed leaves of such plants are often finely dissected. The internal structures are also decidedly different.

In contrast to plants of very humid habitats, those growing under very dry situations tend to be small of stature and compact in form. The stems are short and stout and the leaves small and thick. Often the leaves are mere scales or spines. Epidermal layers are heavily cutinized and sometimes covered with bloom or hairs. Stomata are few in number and in some species are sunken below the level of the epidermis. Such plants may also have a higher concentration of cell sap and therefore higher osmotic pressures within the cells. Hydrophilic colloids are often present in abundance and enable such plants to bind and hold such water as is available to them. Many of the features mentioned are also developed under high light intensity. Light and dryness probably both contribute toward causing such formative effects, since they are often both operative in the same locality.

A plant may have its roots in water and still develop as though it were grown in a dry situation. The stems and leaves of many swamp and bog plants resemble those of plants grown in arid regions. This may be explained by the fact that such plants may have difficulty in absorbing water because of the high osmotic pressure or the unfavorable acidity of the external medium or for some other reason. Situations of this kind are considered physiologically dry. Plants growing in salt marshes usually have high osmotic pressures. Some of the highest values of osmotic pressure that have been recorded in plants were found in plants growing along the edges of the Great Salt Lake. A situation which promotes excessive transpiration, even though there may be an abundance of water in the soil, may also be considered a physiologically dry one.

According to the abundance of available water in the habitats in which they naturally grow, plants may be classified as **hydrophytes,** those which grow directly in water or in very wet places; **mesophytes,** those found in ordinary humid regions with a moderate, though ample supply of water; and **xerophytes,** those growing in very dry or desert regions where conditions are conducive to excessive transpiration. A fourth group, **halophytes,** is sometimes added, which includes plants growing in salt marshes where the concentration of the soil solution is so high as to make absorption of water difficult. Halophytes are in reality xerophytes.

PLANT MOVEMENTS

Since the plants with which we are most familiar are rooted in the ground and hence do not move from place to place, the capacity of movement is often thought to be entirely lacking in plants. Among the simpler forms, however, there are entire plants which move from place to place, and the parts or organs of stationary plants may change their positions. Thus flower petals open and close and roots and stems may change their direction of growth and bend or curve under certain conditions. Ordinarily, however, the rate of movement is not rapid enough for observation except with the aid of some special device. Motion-picture films, requiring hours or days for exposure but projected in a period of a few minutes, give striking demonstrations of many plant movements, such as the opening of flowers and the nodding of stems.

Growth movements and turgor movements Many plant movements are growth movements. Increase in size in itself involves movement, since the expanding organ occupies more and more space. Growth movements may also be caused by a change in either the rate or direction of growth. Thus in a stem tip bending toward the light in response to one-sided illumination, the convex side is growing more rapidly than the concave side. In flower buds opening in response to an increase of temperature, the upper (inner) surface of the petals grows more rapidly than the lower surface, causing the petals to open outward. Such movements can occur only in the growing regions of plants and hence are termed **growth movements.**

Movements may also occur in mature organs and in such cases are usually not growth movements. Often these movements are the result of changes in the turgor pressure of certain cells, which in turn is thought to be caused by changes in the permeability of these cells. Turgor movements are usually more rapid than growth movements and may occur over and over again in the same organ. For this reason, they are sometimes called **movements of variation** or movements of alternation, in contrast to growth movements. The reaction of the sensitive plant (*Mimosa pudica*) (Fig. 9-13*A, B*) to the touch is a well-known example of a turgor movement. The opening and closing of stomata caused by changes in the turgor pressure of the guard cells are also movements of variation.

Autonomic movements and paratonic movements Growth such as the elongation of a stem or a root may be regarded as a slow movement performed by the growing organ. Such movements, occurring spontaneously because of activities going on within the plant and not directly related to any change in the environment, are called **autonomic movements.** Since the rate of elongation on all sides of a stem tip is not uniform, even under apparently constant environmental conditions, the movement of the stem tip as it elongates rarely occurs in the direction of a straight line. Instead, because of this unequal growth in different sections around the stem, the stem apex bends or nods from one side to another as it elongates, and the direction of growth is a loose spiral. This bending or nodding is an autonomic growth movement known as **nutation** or **circumnutation.**

Paratonic, or **stimulus, movements,** in contrast to autonomic movements, are caused by changes in the environment known as **stimuli.** Stimuli may be mechanical, chemical, or ethereal and include such agents as the force of gravity, contact, oxygen, water, acids, alkalies, salts, other chemicals, heat, light, and electricity. Plants move or respond to these stimuli because of one of the fundamental properties of protoplasm, termed **irritability.** Irritability may be defined as the sensitiveness of protoplasm or its capacity to respond to the influence of stimuli. The minimum length of time necessary for a stimulus to act upon a plant in order to bring about a later response is called the **presentation period,** and the period of time which elapses from the initiation of the stimulus until the visible beginning of the reaction is called the **reaction time.** The reaction or response of a plant to a stimulus may not begin until some time after the stimulus is applied and may continue after the stimulus is removed.

Types of paratonic, or stimulus, movements—nasties Paratonic movements of such a nature that the direction of movement is independent of the direction from which the stimulus is applied are called **nasties.** Nasties are responses of bilaterally symmetrical organs like leaves and flower petals, and the direction of movement is largely determined by the anatomy or struc-

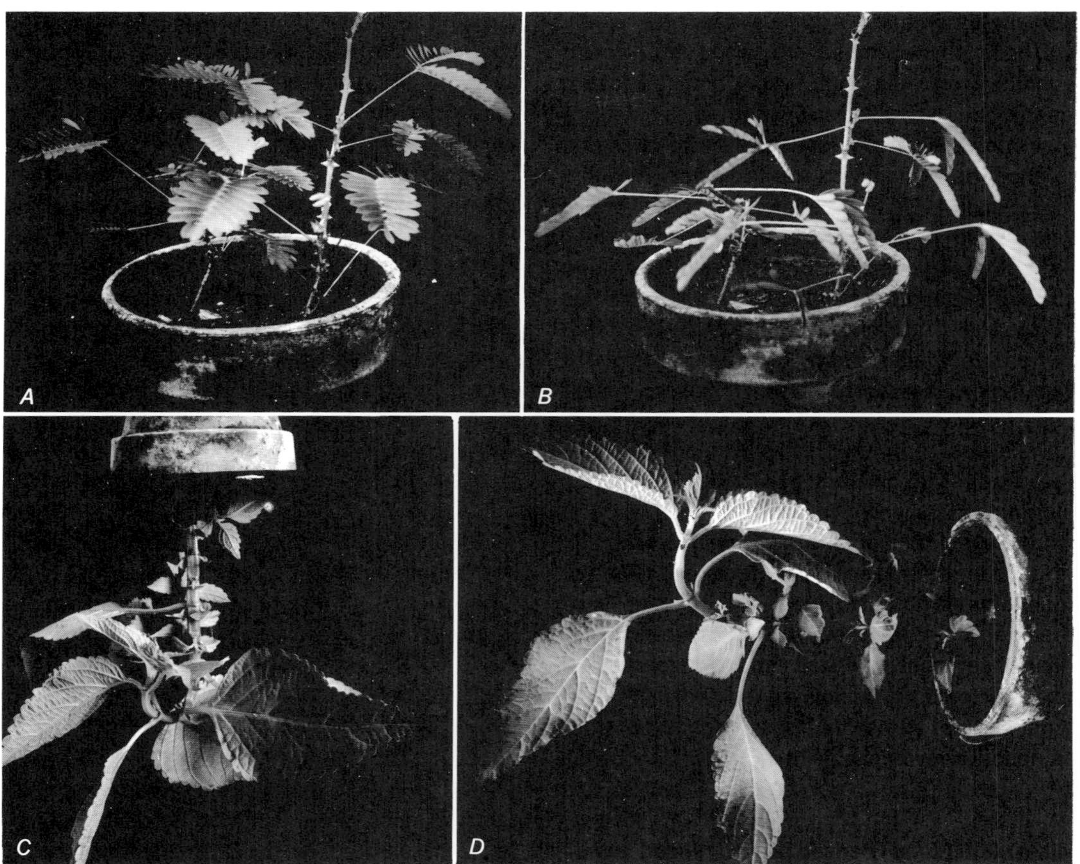

Fig. 9-13 A, B, sensitive plant (Mimosa pudica). *A, undisturbed plant; B, the same plant after the stem had been shaken; note that the leaves droop and the leaflets fold together as a result of the shock stimulus; C, D, geotropism as shown by stem of* Coleus; *C, curvature 24 hr after plant had been suspended vertically downward; D, curvature of stem 24 hr after plant had been placed in a horizontal position.*

ture of the organ. They occur more commonly in mature plant organs than in growing parts and hence in many cases are movements of variation or alternation rather than growth movements.

Nasties caused by changes in light intensity are called **photonasties.** For example, many flowers open under intense illumination and close in the dark or under weak illumination. The common dandelion opens on bright days but closes not only at night but on dull days or if artificially shaded. On the other hand, some flowers, like those of four-o'clocks or tobacco, open in light of low intensity and close under intense illumination.

Nasties caused by changes in temperature are called **thermonasties.** The rapid opening of certain flowers when brought into a warm room from a cold place is a thermonasty brought about by the increased rate of elongation of the upper side of the petals over that of the lower side.

Certain leaves as well as flowers may fold up at night. These so-called sleep movements of plants, brought about by the alternation of night and day, are the most common nasties and are termed **nyctinasties.** They are, of course, in no way related to the sleep of animals but are caused by changes in temperature or light intensity or both.

One of the most prominent and best known of the nastic turgor movements is furnished by the sensitive plant (*Mimosa pudica*), which rapidly droops its leaves when touched (Fig. 9-13*A, B*). In addition, the leaflets fold in pairs. A very strong irritation of one leaf may be transmitted throughout the plant, the leaflets folding and the leaflets drooping one after another. This type of movement is very unusual in plants.

Taxies and tropisms Paratonic, or stimulus, movements of such a nature that the direction of movement is determined by the direction of the stimulus, i.e., movements caused by unilateral stimulation, may be either **taxies** or **tropisms.** Certain unicellular algae, some bacteria, and certain reproductive cells of higher plants are provided with cilia or flagella and exhibit free locomotion of the entire body in response to and determined by the direction of stimuli such as light, temperature, or chemical constitution of the environment. These movements are called **taxies.**

Tropisms are typically responses of radially symmetrical organs, like roots and stems, and of the growing parts of these. Hence they are typically growth movements. Depending upon the nature of the stimulus, a tropism may be called **phototropism, geotropism, hydrotropism,** etc. The capacity of turning toward or away from light is known as **phototropism.** (When sunlight is the stimulus, the movement is sometimes called heliotropism.) Young growing stems usually bend toward the light (Fig. 9-14) and thus exhibit positive phototropism. It is chiefly the blue end of the spectrum that is responsible for the movement. Underground stems and roots are often indifferent to light, but some roots, such as those of mustard, bend away from the light and thus exhibit negative phototropism. Largely because of phototropic movements, the leaves of many plants form a so-

Fig. 9-14 Phototropism; effect of different colors of light. Turnip seedlings subjected to one-sided illumination from right; 1, to red light; 2, to green light; 3, to blue light; 4, to white light (sunlight).

called **leaf mosaic** (Fig. 4-9) in which the leaf blades are so situated that they rarely shade each other but fill in all spaces so that a solid pattern of leaf blades presents itself in a direction perpendicular to that of the light rays falling on them.

The capacity of roots and stems to orient themselves with regard to the force of gravity is called **geotropism.** Taproots, if directed in any way except vertically downward, generally turn downward and exhibit positive geotropism (Fig. 9-15). The greatest curvature is always in the region of greatest elongation. The fully grown parts do not bend. Similarly stems which normally grow vertically upward will, if moved into any other position, bend upward and exhibit negative geotropism (Fig. 9-13*C, D*). There are numerous other stimuli which may cause tropisms in growing organs.

The positive phototropism and negative geotropism of stems can be explained by differences in auxin activity on opposite sides of the stem under unilateral stimulation. Light, especially short-wavelength radiation, seems to check auxin activity. Thus, when a stem is exposed to one-sided illumination, the auxin activity is greater on the side away from the light, causing that side to grow more rapidly. This causes the stem to bend toward light. Similarly, when a stem is placed horizontally, auxin seems to accumulate on the lower side, causing that side to grow more rapidly. The stem therefore bends upward, or is negatively geotropic. Root growth is thought to be stimulated by much lower concentrations of auxin and to be inhibited by concentrations that stimulate stem growth. Thus, when roots are placed horizontally, auxin again accumulates on the lower side but this inhibits the growth on that side, causing the root to grow more on the opposite side. Roots therefore bend downward, or are positively geotropic. For the same reason many of them are negatively phototropic.

Hygroscopic movements, such as the twisting of awns of certain species of wild oats or the movement of the elaters of liverwort sporophytes, cannot properly be called stimulus movements since they are purely physical phenomena and not related to the irritability of protoplasm. They are usually caused by imbibition and loss of water.

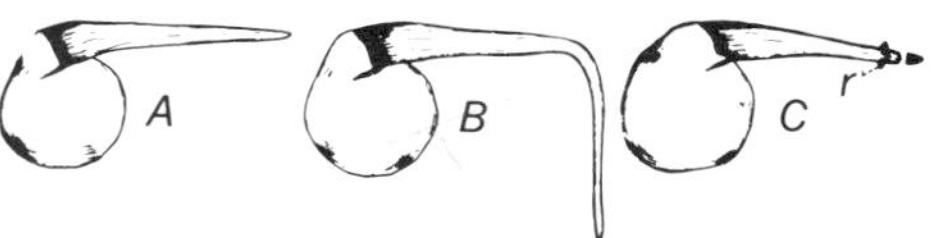

Fig. 9-15 Geotropism as shown by roots of peas. A, seedling of pea with root placed horizontally; B, the same root 24 hr later, the growing tip having bent down in response to gravity; C, a seedling with 4 mm of the tip removed and then placed horizontally; note that when the growing tip is removed, the root fails to respond to the stimulus; secondary roots have started to develop at r. Compare with Fig. 6-10, in which roots have been placed vertically. (Drawn by Florence Brown.)

QUESTIONS

1. Why does mowing a lawn not kill the grass? Could you mow beans in a vegetable garden in the same way without killing the plants?

2. Why are plants grown in the shade of trees often less vigorous than those grown in full sunlight?

3. In a dense stand of pines the lower branches of the trees often die. Why?

4. Why are forest trees on the lower side of a mountain usually taller than those near the summit?

5. Why do native plants fail to thrive when moved 100 miles north or south of the region in which they are indigenous?

6. Why is lettuce planted early in the spring rather than in midsummer?

7. If you were to plant scarlet-sage seeds at

two-week intervals beginning in early spring and continuing through the summer, would you expect to find a corresponding difference in the time of flowering of each planting?

8. Why do tobacco growers often "top" the plants (i.e., cut off the flowering portion) just before they begin to flower?

9. Most plants are more susceptible to freezing temperatures in the tender seedling stage, and yet winter wheat planted in late autumn is usually not killed by winter temperatures. Why?

10. Of what advantage would it be to spray greenhouse tomato plants with growth substances at flowering time?

11. Why is it not necessary to furnish plants with vitamins?

REFERENCES

FOGG, G. E. 1963. The Growth of Plants. Penguin Books, Inc., Baltimore. A comprehensive treatment of growth and factors affecting it.

LEVITT, J. 1951. Frost, Drought, and Heat Resistance. *Ann. Rev. Plant Physiol.,* **2**:245–268. A review of research carried out since 1941 on frost resistance, drought, and heat resistance of plants.

LOOMIS, WALTER E. (ed.). 1953. Growth and Differentiation in Plants. The Iowa State University Press, Ames, Iowa. A monograph on growth by the American Society of Plant Physiologists.

PARKER, JOHNSON. 1963. Cold Resistance in Woody Plants. *Botan. Rev.,* **29**:123–201. A review of all phases of low temperature resistance of woody plants.

ROMBERGER, J. A. 1963. Meristems, Growth, and Development in Woody Plants. U.S. Dept. Agr., Forest Serv., Tech. Bull. 1293.

SEARLE, NORMAN E. 1965. Physiology of Flowering. *Ann. Rev. Plant Physiol.,* **16**:97–118. "This review is restricted to developments that contribute to mechanistic concepts of the initiation of floral differentiation in short-day and long-day plants."

SIEGELMAN, H. W., and W. L. BUTLER. 1965. Properties of Phytochrome. *Ann. Rev. Plant Physiol.,* **16**:383–392. Summarizes recent work on this flowering pigment.

SINNOTT, EDMUND W. 1960. Plant Morphogenesis. McGraw-Hill Book Company, New York. Symmetry, polarity, correlations, etc., discussed on broad front.

WARDLAW, C. W. 1952. Morphogenesis in Plants. John Wiley & Sons, Inc., New York. Review of experiments in areas of embryology, shoot apex, leaf formation, gene action and morphogenesis, and others.

WENT, F. W. 1953. The Effect of Temperature on Plant Growth. *Ann. Rev. Plant Physiol.,* **4**:347–362. A brief review of some of the published papers on this subject.

WHITE, PHILIP R. 1963. The Cultivation of Animal and Plant Cells, 2d ed. The Ronald Press Company, New York. A concise, but inclusive, treatment of the methods used in cultivating cells outside the body.

(Many authors). 1964. Meristems and Differentiation. Brookhaven Symp. Biol. 16, Biology Department, Brookhaven National Laboratory. Available from Office of Technical Services, U.S. Department of Commerce, Washington, D.C. Includes all kinds of developmental studies, morphogenesis, growth and differentiation, regulated synthesis of RNA and protein.

(Many authors). 1961. Plant Growth Regulation: Fourth International Conference on Plant Growth Regulation. The Iowa State University Press, Ames, Iowa. Papers presented at the conference sponsored by the Boyce Thompson Institute for Plant Research, the New York Botanical Garden, and the Brooklyn Botanical Garden. Covers both natural and synthetic growth substances.

10
FLOWERS

Mountain Laurel.

The parts of a flower, their structure and arrangement, primitive and advanced floral structure, development of flowers on the stem, types of inflorescences, and reproduction by flowers, including pollination, development of male and female gametophytes, fertilization and the development of embryos, seeds, and fruits, are considered in this chapter.

Coloration in flowers.

Attention has been given in previous chapters to the vegetative parts of plants, i.e., to roots, stems, and leaves. Except as these organs function secondarily in vegetative propagation, the reproductive processes of plants have not been discussed. The flower is involved in the sexual reproductive processes of higher plants. Flowers lead to the formation of fruits and seeds.

GENERAL STRUCTURE OF FLOWERS

Parts of the flower Flowers may be composed of as many as four different sets of parts. These are, enumerated in order from outside to center of the flower, **sepals, petals, stamens,** and **pistils** (Figs. 10-1*A*, 10-2). The stem apex to which these are attached is known as the **receptacle.** The sepals collectively constitute the **calyx,** and the petals the **corolla.** The stamens collectively constitute the **androecium,** and the pistil or pistils collectively are called the **gynoecium.** These four sets of parts may be attached at different levels on a somewhat elongated receptacle, or they may be attached in more or less concentric rings or whorls on a flattened receptacle. In some flowers this arrangement is obscured, and the petals may seem to arise from the sepals, and the stamens from the corolla. Sepals and pistils can usually be seen attached to the receptacle.

When a flower has all four sets of parts, it is said to be **complete.** In many flowers one or more of the sets may be lacking. If any one of the four sets of parts is lacking, the flower is **incomplete.** Stamens and pistils are regarded as the essential parts of the flower. Both must be present, either in the same or in different

Fig. 10-1 Parts of the flower. A, flower of lily with one sepal and four stamens removed; B, front, or external, view of a stamen with versatile anther; C, view of stamen showing attachment of filament to anther; D, an immature anther cut in half showing the four pollen sacs and the connective, c; E, an anther after discharge of the pollen through the two lateral, longitudinal slits. (B, C, and E by Christian Hildebrandt.)

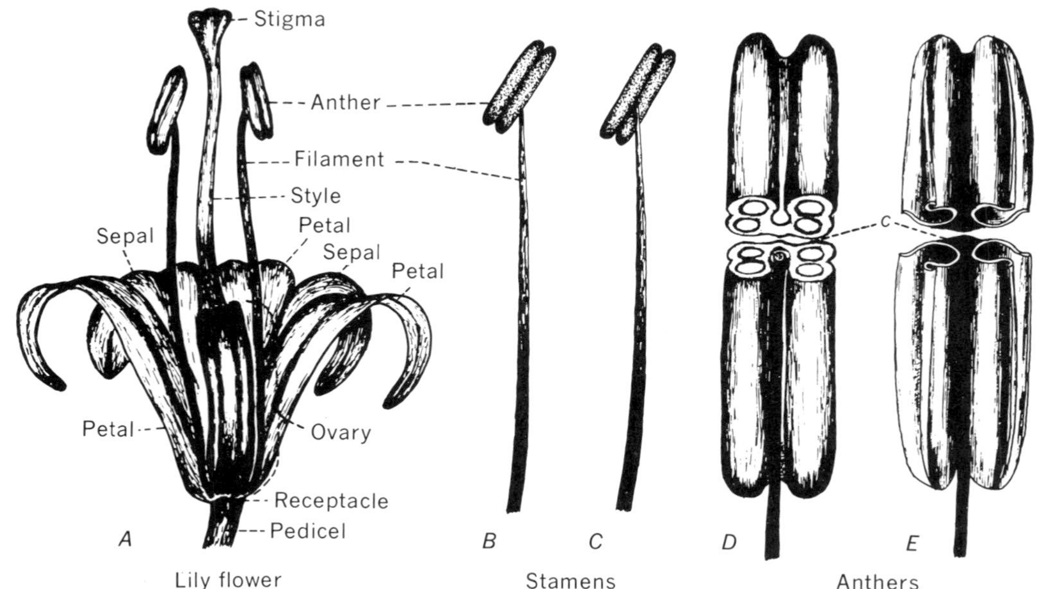

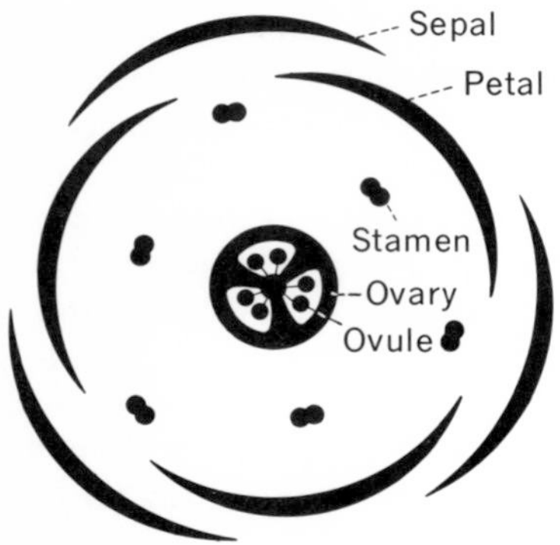

Fig. 10-2 Floral diagram of a lily flower, showing outer whorl of three sepals making the calyx, a whorl of three petals constituting the corolla, and two whorls of three stamens each and the three-part ovary which is part of the compound pistil and contains the ovules. (Drawn by Walter Westerfeld.)

flowers, in order to have sexual reproduction and the resultant formation of seed. If the flower lacks either stamens or pistils, it is said to be **imperfect.** When both stamens and pistils are present, the flower is said to be **perfect** regardless of whether or not sepals or petals are present. Therefore all complete flowers are perfect; incomplete flowers may be either perfect or imperfect, depending upon whether or not both stamens and pistils are present.

Imperfect flowers are of two types: those bearing pistils but not stamens **(pistillate flowers)** (Fig. 10-5*B*) and those bearing stamens but not pistils **(staminate flowers)** (Fig. 10-5*A*). When both these types of flowers are produced on the same plant, the species is said to be **monoecious,** a term meaning "one household." When only staminate flowers are produced on one individual plant and only pistillate flowers on another plant of the same species, the species is said to be **dioecious** (two households). Examples of monoecious species are corn (tassels bearing staminate flowers, the ear bearing pistillate flowers), cattails, alders, birches, walnuts, and hickories. Willows and cottonwoods are examples of dioecious species. Occasionally, as in the marginal flowers of the sunflower head and in the cultivated hydrangea and the snowball bush, neither stamens nor pistils are present. Such flowers are said to be **sterile.**

SEPALS AND PETALS Sepals and petals together constitute the **perianth.** The perianth parts are frequently spoken of as **accessory** flower parts, i.e., not essential to seed formation. Both may be lacking; but if only one set is lacking, the one present is always designated as the calyx. Typically sepals are green and leaflike; petals are often highly colored; occasionally both are colored and nearly equal in size, as in tulips and lilies (Fig. 10-1*A*). The number of sepals and petals is usually constant for a species. Flowers of monocotyledons usually have three sepals and three petals; those of dicotyledons usually have four or five of each.

The odors of flowers are caused by essential oils and other chemical substances that are formed in special secreting cells, usually of the petals. If nectaries, secreting a sugary liquid, are present, they are often situated at the bases of the petals on their inner surfaces, but they may occur at other places. Flower colors usually result from the presence of anthocyanin or carotenoid pigments. Colors and odors may be of importance in attracting insects that effect pollination, i.e., the transfer of the pollen from the anther of a stamen to the stigma of a pistil.

Both calyx and corolla, but especially the former, are protective structures prior to expansion of the flower bud.

STAMENS The stamens commonly occur in one or more whorls between the corolla and the pistil or pistils. The number of stamens per flower may or may not bear a relation to the number of sepals or petals. A flower may contain only a single stamen; more often the number is larger and definite, ranging from two in some species to many in others. Each stamen usually consists of a stalk or **filament** bearing at its apex an **anther** (Fig. 10-1*A–C*). The filament may be very short or may be entirely lacking, but in most cases it is elongated. The anther may be attached firmly at its base to the filament **(basal attachment)** or near its center, loosely permitting a rocking motion **(versatile anther)** (Fig. 10-1*B, C*), or in other ways.

Viewed externally, an anther usually consists of two elongated lobes united by a tissue extending from base to tip of the anther, called the **connective** (Fig. 10-1*D, E*). Examination of a cross section of a mature anther (Figs. 10-1*D*, 10-3*A*) reveals that each of these lobes consists

Fig. 10-3 A, photomicrograph of one half of a lily anther when pollen grains are about ready to be discharged. The two cavities containing the pollen grains have become one through the breaking down of the tissue between them. Liberation of the pollen occurs through the separation of the lip cells. B, a compound pistil of lily enlarged, showing parts. C, enlarged ovary of lily, almost mature, cut to show the three ovarian segments, each with a locule containing two rows of ovules. The ovules have been removed from one of the locules. (B and C by Christian Hildebrandt.)

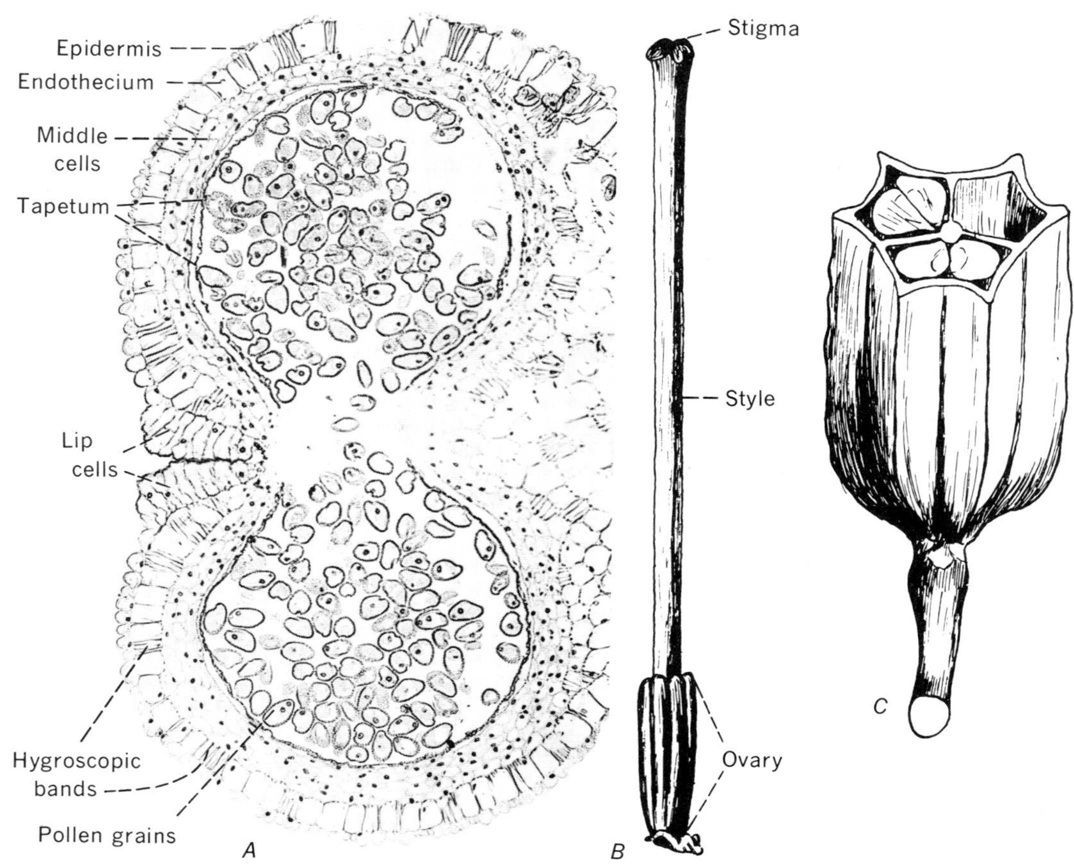

of two elongated cavities, or **pollen sacs,** containing **pollen grains.** The outer cell layer of the wall of each cavity is an epidermis. Just under the epidermis is a layer of larger cells traversed by irregular hygroscopic bands. This layer is called the **endothecium** and plays a prominent role in the final dehiscence (splitting open) of the anther. Next to it are usually several middle layers of cells which often flatten in the mature anther, and finally, next to the cavity itself, the remains of a nutritive tissue called the **tapetum.** Where the walls of the two adjacent cavities meet, externally, the cells are specialized as **lip cells.** At maturity the anther splits open at the lip cells. The tissue between the two cavities breaks down, the two cavities thus becoming one. The pollen is then released through a longitudinal slit on each side of the anther (Fig. 10-1*E*). In some cases, however, the pollen is shed through terminal slits or pores as in *Solanum;* by openings in tubular prolongations of the pollen sacs as in *Vaccinium;* or by hinged valves as in *Sassafras* and *Berberis.* In wind-pollinated flowers, like many of the grasses, the filaments elongate considerably at maturity, causing the anthers to become exserted beyond the surrounding floral tissue.

PISTILS The pistils are the central members of perfect and of pistillate flowers. In its ordinary form a pistil consists of three parts (Fig. 10-3*B*): the enlarged basal region is the **ovary,** the apex is the **stigma,** and the part between ovary and stigma is the **style.** The ovary contains one or more **ovules,** which later become seeds. The pistil in its simplest form is commonly considered to be a specialized leaf on which ovules are produced, the leaf having been inrolled and its margins or edges united in such a way as to enclose the ovules completely. A single ovule-bearing leaf of this kind is called a **carpel.** If the pistil consists of only one such leaf (carpel), it is a **simple pistil;** if it is made up of two or more carpels, it is called a **compound pistil.** The flower of the common buttercup (*Ranunculus*) has simple pistils; lilies have a flower with a compound pistil consisting of three carpels.

In a compound pistil the styles of the individual carpels may be separate throughout their length, or they may be united into a single **stylar column.** Even in the latter case there is likely to be an indication of the compound nature of the pistil in the number of lobes, or divisions of the stigma.

The stigma receives pollen grains carried to it by various agencies. It is frequently provided with a sticky stigmatic fluid and it may be fitted with grooves or depressions or with tiny, stiff hairs or with glands, all of which facilitate the reception of pollen. The style is sometimes much elongated, elevating the stigma above the other floral parts. Sometimes the style is lacking, as in the tulip flower; in this case the stigma is sessile on the ovary.

A cross section of an ovary reveals one or more carpellary cavities, or **locules** (often called cells), in which the ovule or ovules are contained (Fig. 10-3*C*). In general each carpel has its own locule and thus there are usually as many locules in a compound pistil as there are carpels, the contiguous walls of the individual carpels often forming partitions in the ovary. In some cases, however, these partitions may not be present and thus the ovary may have only one cavity or locule even though the pistil is compound.

The portion of the ovary to which the ovules are attached is known as the **placenta,** and the manner in which the ovules are distributed in the ovary is called **placentation.** The placentae usually occur on the inrolled, united margins of the carpel. In a simple pistil, therefore, the ovules may occur in a double vertical row along one side of the ovary, or they may arise singly from the top, sides, or bottom of the ovary. The positions of the placentae in compound pistils vary with the structure of the ovary. When the placentae occur on the central axis of an ovary which has several locules, as in lily (Figs. 10-3*C;* 10-4*A*, *E*), the placentation is **axile.** In com-

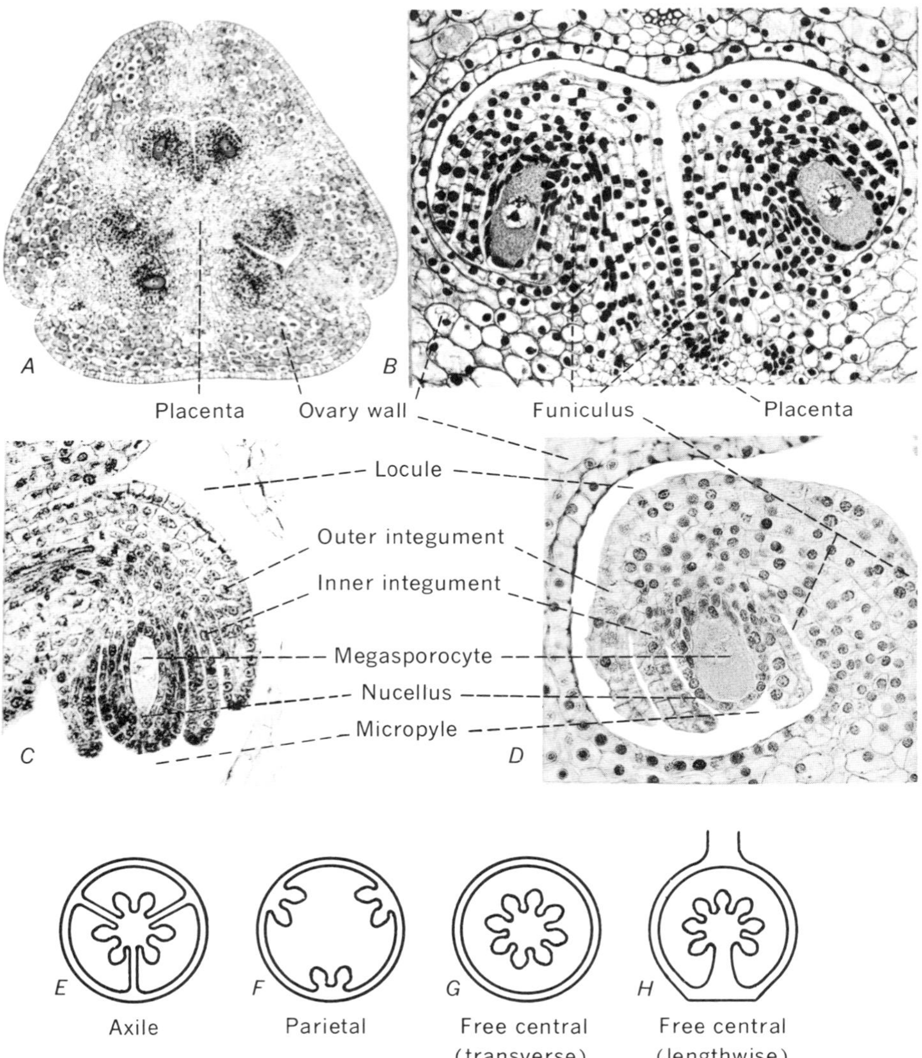

Fig. 10-4 Ovary and ovule structure; placentation. A, photomicrograph of a transverse section through a lily ovary, showing axile placentation with two rows of ovules in each of the three locules; B, an enlarged view of one locule of the ovary of A, showing two ovules, each with a megaspore mother cell or megasporocyte; C, single ovule of Fuchsia, *showing structure; D, single ovule of lily, showing structure; E–H, types of placentation, diagrammatically shown. (Photomicrographs by Dr. D. A. Kribs.)*

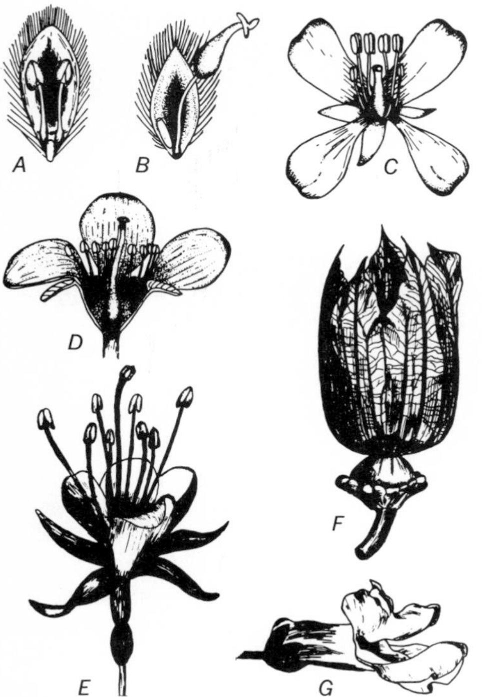

Fig. 10-5 Forms of flowers. A, a staminate flower of willow, consisting of two stamens borne in the axil of a scalelike, hairy bract; B, a pistillate flower of willow, consisting of a single-stalked pistil borne in the axil of a hairy bract; C, the hypogynous flower of Cardamine; *D, the perigynous flower of cherry; E, the epigynous flower of* Fuchsia; *F, the epigynous flower of the squash; G, the zygomorphic flower of snapdragon. (A–D redrawn by Christian Hildebrandt from drawings by Edna S. Fox; E–G, by Christian Hildebrandt.)*

pound pistils in which there is only one cavity, or locule, the placentae may occur on the wall of the ovary, as in some species of *Drosera* and of *Ribes;* in this case it is called **parietal** placentation (Fig. 10-4*F*), or there may be a central axis to which all ovules are attached. This is called **free central** placentation (Fig. 10-4*G, H*) and is found in pinks and in primroses.

The number of ovules contained in a single ovary varies from one, as in buckwheat flowers, to many hundreds, as in tobacco and poppy.

Form and arrangement of parts of the flower

FLORAL SYMMETRY In many flowers, like the lily (Fig. 10-1*A*), the members of each whorl of parts are all alike and are arranged radially around a central axis. Such a flower is said to be **regular** or **actinomorphic** or to have **radial symmetry.** When a flower of this type is cut vertically in any plane through the center, the resulting halves are somewhat like mirror images of each other. In other flowers, such as sweet peas, snapdragons (Fig. 10-5*G*), and violets, one or more of the petals, and frequently of the sepals and stamens also, are unlike the others in shape or form. Such a flower can be cut in only one vertical plane to obtain halves that are like mirror images of each other. The flower is then **irregular,** or **zygomorphic,** and has **bilateral symmetry.** Less often some of the sets of parts of a flower are spirally arranged. Thus in buttercups the stamens and pistils are spirally arranged but the sepals and petals are radially arranged. The whole flower in this case is still said to have radial symmetry. In flowers with spiral arrangement of parts there is usually a large and indefinite number of such parts.

UNION OF FLOWER PARTS In some flowers all the parts are attached directly to the receptacle and are entirely free from each other. Thus there may be separate sepals **(polysepaly),** separate petals **(polypetaly),** separate stamens, and separate pistils. In others the members of a whorl may be more or less united with each other, a condition referred to as **coalescence** of parts. Coalescence may involve sepals **(synsepaly** or **gamosepaly),** petals **(sympetaly** or **gamopetaly),** stamens, or pistils. In synsepalous flowers the bases of the sepals are united in such a way as to form a cup, or **calyx tube.** Similarly, in sympetalous flowers a **corolla tube** may be formed. The flowers of the common Jimsonweed have both a calyx tube and a corolla tube. The stamens may be united by their filaments into one **(monadelphous),** two **(diadelphous),** or several **(polyadelphous)** sets, as in members of the Malvaceae, Leguminosae, and Tiliaceae, respectively, or the anthers may be united **(syngenesious)**. as in the Compositae. Union of carpels **(syncarpy)** occurs in flowers with compound pistils.

In still other types of flowers, the members of one whorl of parts may be more or less united with members of another whorl, a condition called **adnation.** Thus the stamens may be attached to the petals as in flowers of the potato family and many other plants. In a few cases, as in orchids, the stamens are united with the carpels (Fig. 10-10*B*). More frequently sepals, petals, and stamens are united at their bases, forming a cup, or **floral tube (hypanthium).** In some cases the receptacle may form a part of the floral tube. The floral tube is often called a calyx tube, but the latter term should be restricted to cases in which only the calyx is involved.

HYPOGYNY, PERIGYNY, AND EPIGYNY Depending primarily upon the apparent position of the other parts of a flower with respect to the ovary or ovaries, flowers may be **hypogynous, perigynous,** or **epigynous. Hypogynous** flowers are those in which stamens, petals, and sepals are attached to the receptacle below and entirely free from the ovaries of the pistils (Figs. 10-1*A*, 10-5*C*). This arrangement results in **superior** ovaries. The term is applied both to flowers with many separate pistils, like the buttercup, and to flowers in which there is a single (simple or compound) pistil, as in peas and other legumes, which have a single, simple pistil, and in lilies (Fig. 10-1*A*), which have a compound pistil.

In **perigynous** flowers there is a floral tube, or hypanthium. Within this tube or cuplike structure, but free from it, the pistil or pistils are contained. The sepals, petals, and stamens appear to be borne on the margin or rim of the floral tube, but actually arise from the receptacle below the ovaries of the pistils. The ovaries therefore are superior as in hypogynous flowers. The cherry flower (Fig. 10-5*D*) is a good example of this type. In the perigynous flowers of saxifrage the lower part of the ovary is united with the floral tube.

In **epigynous** flowers there is also a floral tube, but in this case the floral tube is united with (adnate to) the wall of the ovary, so that the other floral parts appear to be attached to the top of the ovary (Figs. 10-5*E, F;* 11-6*A*). The ovary in this case is said to be **inferior.**

Primitive and advanced floral structure The earliest flowers are thought to have resembled a cone, or strobilus, which consisted of an elongated axis with a large and indefinite number of spirally arranged and separate stamens and pistils. Since the buttercups and magnolias (Fig. 10-6) have these characteristics, they are thought to represent this primitive type. Advances in floral evolution are indicated by the following characters.

1. A WHORLED OR CYCLIC ARRANGEMENT OF THE PARTS OF THE FLOWER The floral axis in flowers of this type is greatly shortened, and the parts appear in whorls or circles (Figs. 10-1*A*, 10-2). In the common buttercup and in the tulip poplar (Fig. 10-6) the sepals and petals are cyclic but the stamens and pistils have the more primitive spiral arrangement.

2. A REDUCED AND DEFINITE NUMBER OF FLORAL PARTS In the more primitive types of flowers there are usually large and indefinite numbers of flower parts (Fig. 10-6). More advanced types have fewer and definite numbers of floral parts (Figs. 10-1, 10-5*E*). In the higher members of the dicotyledons the number of each of the parts is often two or five or multiples of these, while in the monocotyle-

Fig. 10-6 The relatively primitive flower of the tulip poplar (Liriodendron tulipifera), *a member of the family Magnoliaceae, showing the numerous separate pistils spirally arranged on the elongated floral axis, and the numerous stamens of indefinite number. Pollen sacs are borne on the surface of the stamens. The three sepals and six petals have a cyclic arrangement.*

dons the number is usually three or multiples of three. The more advanced flowers also have fewer whorls or cycles of parts. Thus in many members of the heath family there are five whorls consisting of one whorl each of sepals, petals, and pistils and two whorls of stamens. In the phloxes and morning glories there are only four whorls of parts, a condition resulting from the presence of only one whorl of stamens. The willows (Fig. 10-5*A, B*) have greatly reduced floral parts.

A high degree of reduction of parts of the flower is seen in the grasses, in many of which there are imperfect flowers consisting of only stamens or only a pistil. In oats (Fig. 10-7) and other grasses the perianth consists of two minute scalelike structures called **lodicules** at the base of the ovary. There are usually three stamens and a single compound pistil consisting of three carpels, only one of which functions. There are only two (rarely three) feathery stigmas and two short styles. The ovary contains a single locule and bears a single ovule. The grasses are thought to represent the apex of an evolutionary line of monocotyledons.

3. DIFFERENTIATION OF THE PERIANTH The differentiation of the perianth into calyx and corolla probably occurred very early in the evolution of the flower. Today this condition of the perianth is found in what are considered primitive types of flowers (e.g., buttercups), as well as in many of the higher forms (most of the families of dicotyledons). In the lilies (Fig. 10-1) and in other monocotyledons there is often little difference between sepals and petals, both being showy and often highly colored. This condition is considered by some as more advanced than the one in which the calyx is green and the corolla more differentiated and showy (Fig. 10-5*C, D, G*). In some flowers the perianth is absent altogether, a condition thought to have been derived by reduction from a flower having a perianth.

4. COALESCENCE AND ADNATION OF FLORAL PARTS All the floral organs of primitive flowers are entirely separate from each other (Fig. 10-6). In more advanced forms, members of the same cycle are united or coalesced (Fig. 10-5*G*). On the basis of coalescence of

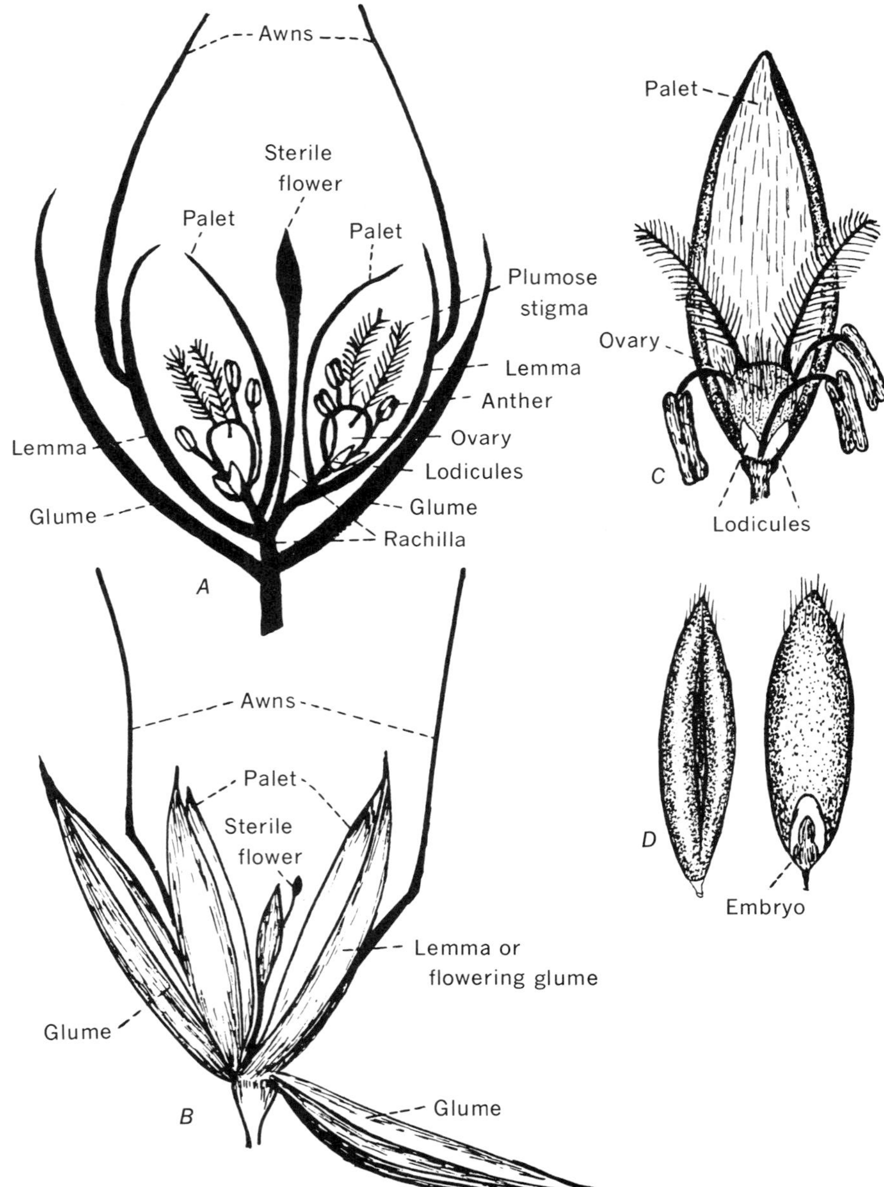

Fig. 10-7 Structure of a grass flower, oats (Avena fatua). *A, diagrammatic representation of a spikelet of oats with two fertile flowers and one sterile flower; B, enlargement of a single spikelet; C, enlargement of a single flower with palet behind; D, oat grain, or caryopsis, after removal of lemma and palet, the one to the right showing the embryo side.*

perianth parts, the dicotyledons are sometimes subdivided into two main groups, the **Archichlamydeae** or **Choripetalae,** in which the individual members of the calyx and corolla are entirely separate from each other or the perianth as a whole is poorly developed, and the **Metachlamydeae** or **Sympetalae,** in which the petals are united into a gamopetalous corolla.

Adnation, or the fusion of members of different cycles, is also thought to be an advanced character.

5. PERIGYNY OR EPIGYNY AS OPPOSED TO HYPOGYNY In general, hypogyny (Fig. 10-1*A*) is considered a more primitive condition than either perigyny (Fig. 10-5*D*) or epigyny (Fig. 10-5*E, F*), epigyny being the highest type. Hypogyny is found in buttercups, pinks, and mustards. Many members of the rose family have perigynous flowers. Epigyny occurs in the flowers of members of the orchid family, the carrot family, and the composites.

6. IRREGULARITY OR BILATERAL SYMMETRY AS OPPOSED TO REGULARITY OR RADIAL SYMMETRY Bilateral symmetry (Fig. 10-5*G*) is generally considered a more advanced condition in flowers than is radial symmetry (Fig. 10-1*A*) and is found in the higher members of both the monocotyledons and the dicotyledons. Bilateral symmetry in flowers is often correlated with a high degree of specialization in insect pollination.

The members of the sunflower family (Compositae) (Figs. 10-8, 10-9), which represents one of the highest levels of development found in the plant kingdom, have most of the advanced floral features mentioned. These include reduction in number of floral parts, coalescence and adnation of floral parts, epigyny, and bilateral symmetry.

In many of the families of angiosperms, the flowers may be quite advanced in some respects but primitive in others. Thus the mints have all the advanced features mentioned except that the flowers are hypogynous; the orchids (Fig. 10-10), usually considered among the most advanced monocotyledons, have the relatively primitive feature of separate petals. Many other combinations of advanced and primitive characteristics occur in other groups. In addition, some groups of angiosperms formerly thought to be primitive and ancient are now considered to have been reduced from types with many

Fig. 10-8 The dandelion, a composite. A, the open inflorescence (head); B–D, the head, enclosed by involucral bracts, representing successive closed periods of the head during the maturation of the flowers and fruits; E, head beginning to open as the fruits are maturing; F, fruits ready for dissemination.

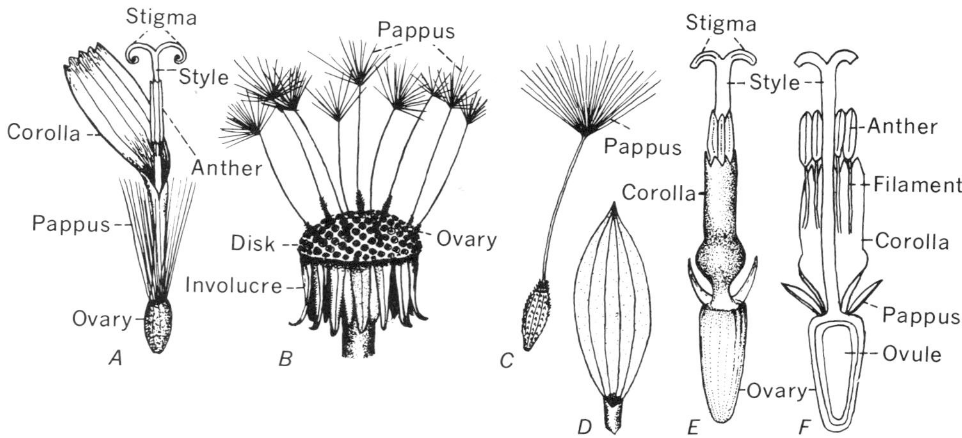

Fig. 10-9 Flower structure in Compositae. A–C, details of dandelion flower and fruit; A, single flower of dandelion removed from head; the pappus is believed to represent the calyx; B, head of dandelion after most of the fruits are disseminated; C, single fruit of dandelion; D–F, details of a single flower of sunflower removed from the head; D, a sterile marginal ray flower, consisting of only corolla; E, a disk flower; F, a disk flower with corolla split vertically to show stamen attachment and the ovary cut vertically to show the single ovule. (Drawn by Edna S. Fox.)

floral parts. Among these are the Amentiferae (ament, or catkin, bearers), including the beeches, oaks, chestnuts, hickories, walnuts, birches, willows, poplars, and alders. The cattails and grasses, in the monocotyledonous group, are also thought to have reduced flowers. A detailed discussion of the possible lines of development in floral evolution is beyond the scope of this book.

Development of flowers on the stem Flowers may be thought of as a group of specialized leaves borne on a stem tip. Floral development is similar to the development of a vegetative stem tip and, in early stages, is very difficult to separate from the usual vegetative growth. Flowers may arise at the apex of the main stem or at the apex of lateral branches. Flowers appear as buds, either as separate flower buds or as parts of a mixed bud.

The leaf primordium subtending the point of origin of a flower may or may not develop into a leaf. If it does and leaves are present on the flower cluster, they are usually of smaller size than the other leaves and are termed **bracts.** Occasionally these bracts are large, showy, and closely arranged so as to constitute the most conspicuous part of the flower cluster and are likely to be mistaken for petals, as in the Indian paintbrush (*Castilleja*), flowering dogwood, and *Poinsettia.* The chaff of grass is composed of bracts subtending flowers or flower clusters and the hooded portion (spathe) of the jack-in-the-pulpit is a single bract.

The various floral organs originate in a definite sequence at the apex of the floral stem. Commonly the origin of the floral parts follows an acropetal pattern: sepals, petals, stamens, pistils. In some cases this sequence is modified. For example, in some flowers, petal origin follows the formation of the other parts. Not only do the parts of a flower originate in some definite pattern, but their relative rate of development, after origin, is quite specific. Petals, for

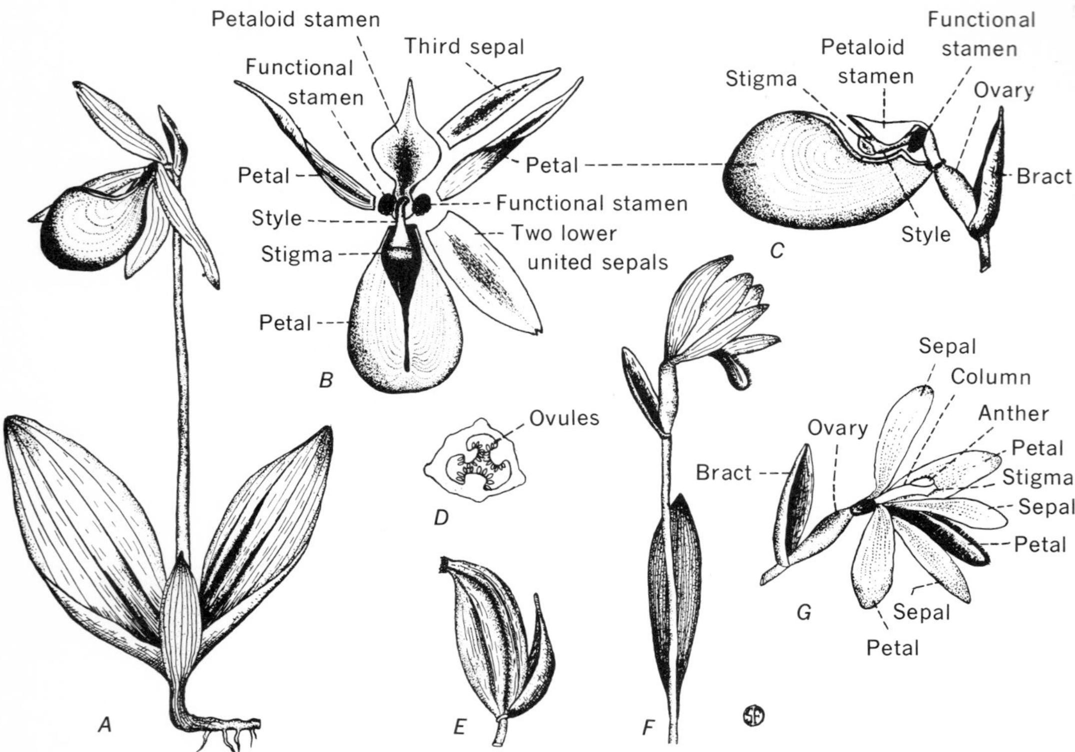

Fig. 10-10 Structure of orchid flowers. A–E, moccasin flower (Cypripedium acaule); *A, entire plant in flower; B, C, flower dissected to show parts, B, in front view, and C, in side view, with sepals and two petals removed; D, cross section of ovary; E, mature fruit; F–G, rose pogonia* (Pogonia ophioglossoides); *F, upper part of plant with terminal flower; G, the flower. (Drawn by Edna S. Fox.)*

instance, may originate before stamens and yet develop much more slowly.

At the outer edge of the receptacle, the sepal primordia are first seen as small protuberances of meristematic tissue at a number of points corresponding to the number of sepals that are to be developed (Fig. 10-11). These points of meristematic tissue are on a circle at the periphery of the receptacle. Elongation and differentiation in these several meristematic regions result in the development of the sepals. The members of the other floral sets originate in a similar manner and gradually assume their mature characteristics. The pistil or pistils occupy the central region, and therefore the apical growth of the central axis is usually stopped by their development. In this development of the flower, parts that are coalesced at maturity will be united from the beginning of their development. Similarly adnations are present from the start. The terms united, coalesced, fused, and the like, as applied to floral parts, do not imply that these parts were at one time separate in the flowers in which they occur, although it is possible that they might have been separate in the remote ancestors of these flowers.

Inflorescences Flowers are borne either singly, as in tulips, or in clusters, as in snapdragons. The flowering part of a plant, made up of one or more flowers, and especially the mode of arrangement of the flowers, is called an **inflorescence.** The central axis of an elongated inflorescence is known as a **rachis,** and the primary stalk which supports an inflorescence is termed a **peduncle.** If this stalk arises from the ground level and is nearly or quite leafless, it may also be called a **scape.** The stalks of the individual flowers of an inflorescence are called **pedicels.**

Some authors divide inflorescences into two classes, viz., **determinate** inflorescences, in which the flowers arise from terminal buds and thus terminate a stem or branch, and **indeterminate** inflorescences, in which the flowers occur in the axils of leaves, or bracts, thus arising from axillary buds. Some plants, like lilac and horse chestnut, have inflorescences which combine these two types and which are sometimes called **mixed** inflorescences. The solitary terminal flowers of some species and the cyme are considered determinate types, while many of the other types of inflorescences are considered indeterminate inflorescences. The different kinds of inflorescences are not always clearly distinguishable, nor can they always be identified as determinate or indeterminate. Thus some umbels are determinate while others are indeterminate. Some of the various types of inflorescences are illustrated diagrammatically in Fig. 10-12 and described as follows:

1. **Cyme.** Flowers arising from terminal buds, sometimes forming a flat-topped or convex cluster in which the central flowers bloom first,

Fig. 10-11 Origin of flower parts in Capsella. *A, sepals, s, first appear as small rounded swellings at the side of the stem apex; B, first appearance of stamens, st; C, sepal and stamen primordia enlarged; D, first appearance of carpels, ca; E, sepals, stamens, and carpels enlarging; F, diagram of longitudinal section through a greater portion of the stem apex, at this stage showing the first appearance of petal primordia, p; G, all parts enlarging.*

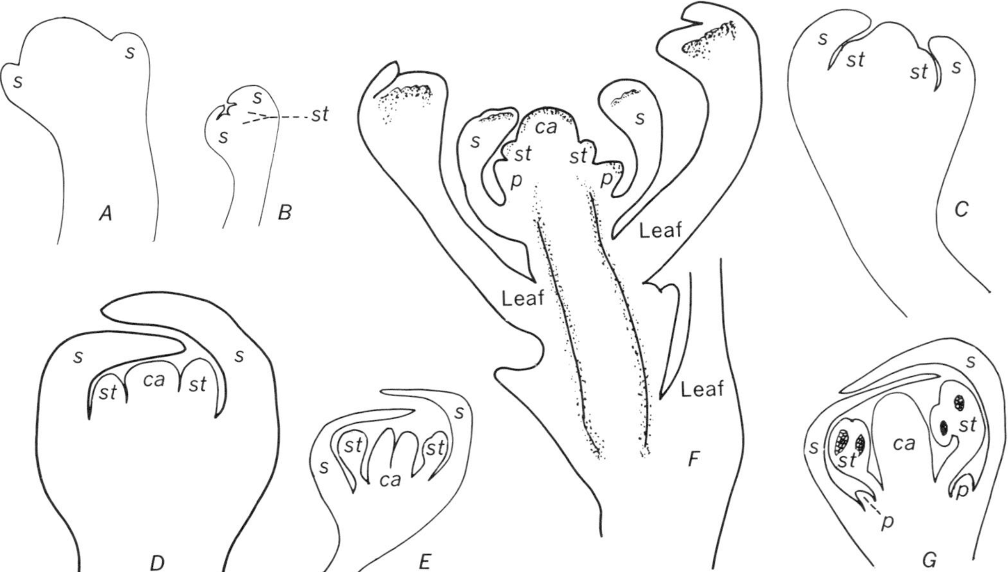

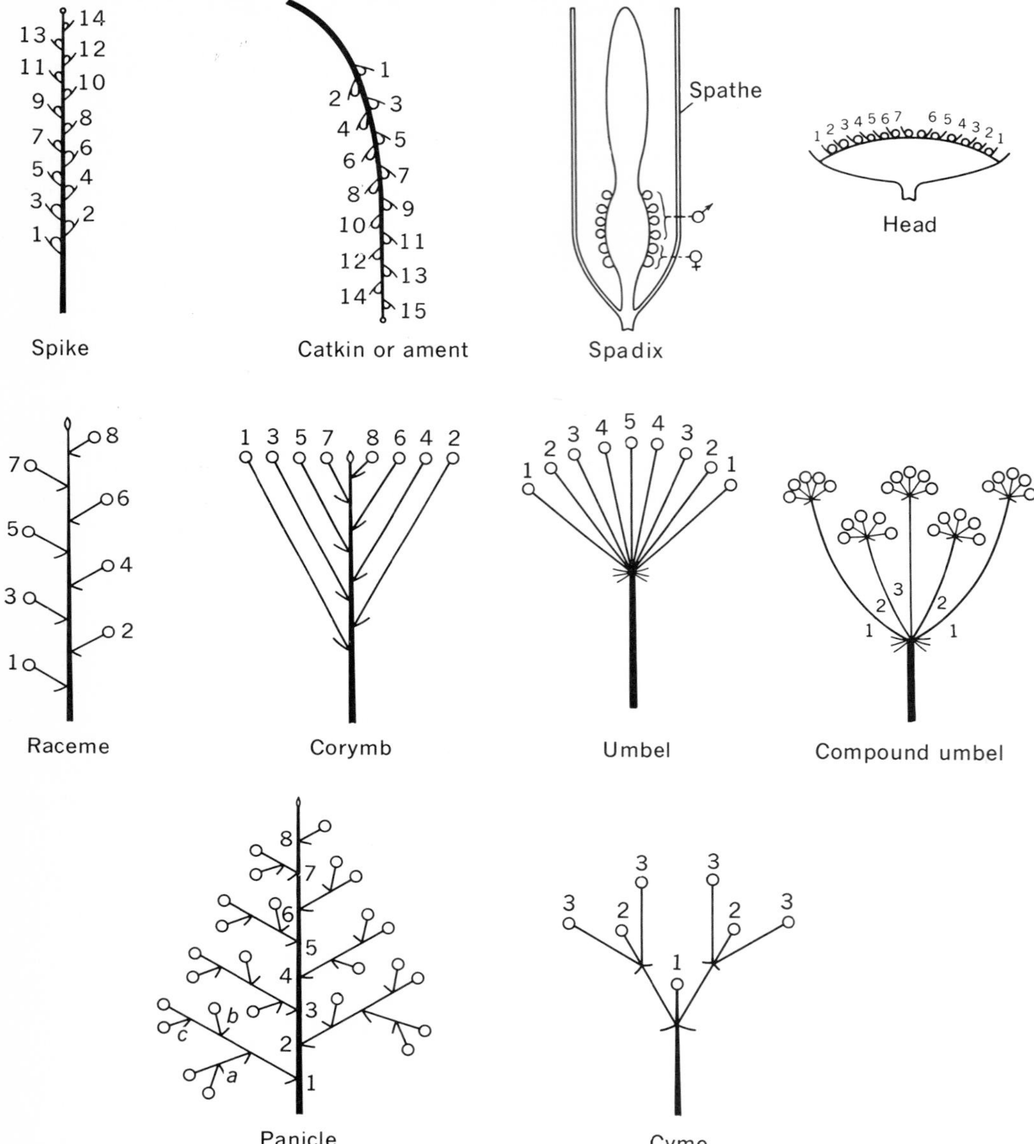

Fig. 10-12 Types of inflorescences, diagrammatically represented. Flowers are shown by small circles, and bracts by short slightly curved lines. Figures indicate the usual sequence of opening of flowers, number 1 opening first.

as in some members of the Saint-John's-wort and pink families.

2. **Scorpioid Cyme.** A cyme in which the terminal flowers have a single bractlet or none; when the bractlet is present, it occurs alternately to right and to left at the base of successive pedicels so that the structure bears flowers alternately to right and to left along one side. In some cases it coils (Boraginaceae) and in others it does not (Portulacaceae).

3. **Helicoid Cyme.** In this cyme, found in monocots, there is only one bractlet to a pedicel. Successive positions of the bractlets and pedicels occur in spiral arrangement.

4. **Spike.** Flowers sessile on a more or less elongated axis, as in the common plantains.

5. **Catkin** or **Ament.** A pendulous spike, sometimes scaly, consisting of either staminate or pistillate flowers, and characteristic of oaks, willows, poplars, birches, and related species.

6. **Spadix.** A fleshy spike or head with small, often imperfect flowers of one or both types, commonly surrounded by an enveloping sheath, called the **spathe,** as in calla lily and jack-in-the-pulpit.

7. **Head.** A dense cluster of sessile or nearly sessile flowers on a very short axis, as in dandelion, sunflower, other composites, and red clover.

8. **Raceme.** Flowers borne on an elongated axis on pedicels more or less equal in length, as in shepherd's purse and lily of the valley.

9. **Panicle.** An inflorescence somewhat like a raceme in which the pedicels have branched, the branching being somewhat irregular, as in *Yucca* and many grasses.

10. **Umbel.** An inflorescence in which the pedicels all arise from a common point, as in onions. Compound umbels occur in carrot and in other plants.

11. **Corymb.** Like a raceme, but the pedicels of the flowers become shorter from base toward the apex, resulting in a flat-topped or convex flower cluster, as in candytuft.

REPRODUCTION BY FLOWERS

Since the parts of the flower have been described in the previous section, there remains to be considered how the flower is involved in sexual reproduction. This involves a consideration of the development of pollen, pollination, the development of ovules, fertilization, the development of the embryo and endosperm, and finally the formation of fruit and seed.

Development of pollen grains The anther usually contains four elongated cavities, or pollen sacs, in which pollen grains are produced. An examination of these cavities while the anther is still young, usually while it is still in the flower bud, reveals the presence of many somewhat rounded cells called **microspore mother cells,** or **microsporocytes.** Each of the microspore mother cells, by two successive divisions, forms four cells, or **microspores** (Fig. 10-13*A–F*). The microspore mother cells have the same number of chromosomes as do all ordinary cells of the plant. This number is commonly called the **diploid,** or $2N$, number.

The process of division which takes place in the formation of microspores is called **meiosis** (see page 312) and results in the reduction of the chromosome number to the **haploid,** or $1N$, number, or half the diploid number. The microspores therefore are unicellular structures which have only half as many chromosomes as do the vegetative or somatic cells of the plant. The four microspores separate from each other, and each develops a characteristic shape or form which differs in different species of plants. The outer surface of the microspores may have spines, ridges, or furrows or may vary in other ways in different species. About the time the anther splits open (dehisces), the single nucleus of each microspore divides by mitosis, forming a **generative nucleus** and a **tube nucleus** (Fig. 10-13*G*), each of which has the haploid or $1N$ number of chromosomes. In this condition, the

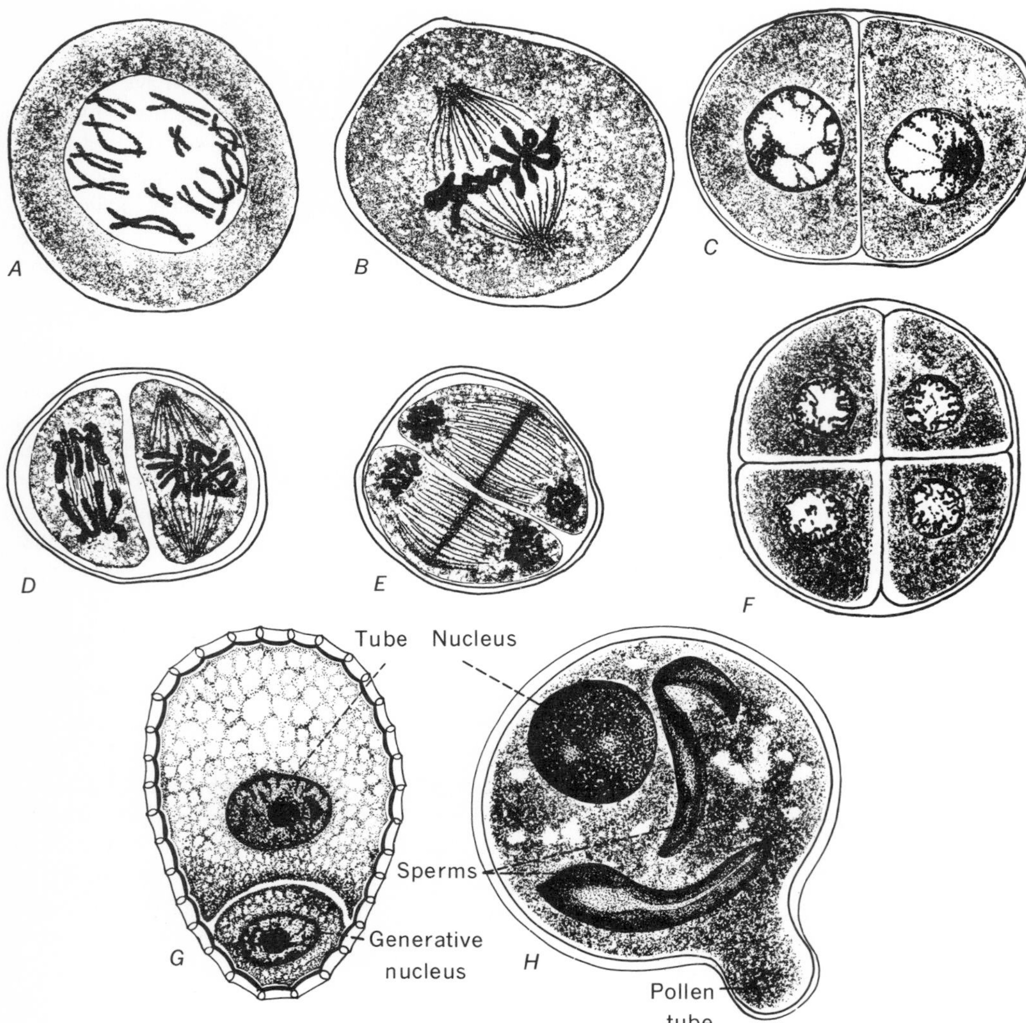

Fig. 10-13 Maturation of microspores and germination of pollen grains. A, prophase, and B, metaphase, of first meiotic division of the diploid (2N) microspore mother cell; C, diad; D, metaphase and anaphase of second division; E, telophase of second division; F, resulting tetrad of haploid (1N) microspores; G, pollen grain of lily; H, early stage in the germination of a pollen grain of chrysanthemum daisy. (G and H drawn by Helen D. Hill.)

microspores have become **pollen grains** and are shed from the anther. Hundreds of pollen grains are usually produced in each anther.

Pollination When the pollen grains are shed from the anther, they are disseminated by various agencies. Some of them may, by one means or another, finally reach the stigma of a pistil, either of the same or of another flower. *This transfer of pollen from anther to stigma is known as* **pollination.** Pollination ends when the pollen has reached the stigma. Plants are said to be **self-pollinated** when the pollen is transferred from an anther to a stigma of the same flower or to a stigma of another flower on the same plant; plants are **cross-pollinated** when the pollen from one plant reaches a stigma of a flower on another plant. Dioecious species are necessarily cross-pollinated. Monoecious species and plants with perfect flowers may be self-pollinated or cross-pollinated. Many, such as peas, tobacco, and many grasses, are commonly self-pollinated. Others, however, have devices that may induce cross-pollination where self-pollination might be possible. For example, the stigma may be receptive before the pollen in the same flower is mature; the anthers may mature and shed their pollen before the stigmas of the same flower are receptive; the flower may be so constructed that there is little chance that pollen can be deposited on a stigma of the same flower; the flower may be sterile to its own pollen because of an inhibition either of germination or of the growth of the pollen tube down to the ovules of the same flower.

In securing the transfer of pollen, various agencies may be brought into play. Wind, water, gravity, insects, and even birds may act in this transfer. Many grasses and many trees are wind pollinated. Flowers of these plants are often relatively inconspicuous, are often without nectaries, and produce great quantities of dry, powdery pollen. The stigmas are often feathery or much branched. The pollen grains of some wind-pollinated plants average around 0.025 mm in diameter and are sometimes, as in pines, provided with wings, which greatly facilitate their transport in air currents. Such pollen grains can be carried tremendous distances by the wind. How far they can be carried, however, is not so important a question as how long a period the pollen can remain viable when so carried. Under ordinary conditions pollen remains viable for only a comparatively short time ranging from a few days to several weeks. Some water plants, such as pond lilies, elevate their flowers above the water thereby facilitating wind pollination or insect pollination. Others (*Elodea*) develop flowers just at the surface of the water, in such a position that pollen is readily washed to them on the surface.

Seed plants as a whole are probably more dependent upon insects than upon any other one agency for pollination. Many irregular flowers are most efficiently pollinated by insects. Their construction usually involves such form of the flower and position of the nectaries as to result in the dusting of the insect with pollen that is carried to the stigma of the next flower visited. Many such flowers can be pollinated only by certain kinds of insects. Thus the bumblebee is the only highly effective pollinating agent for red clover. Such flowers as snapdragons, in which the throat of the corolla is nearly closed, can be sprung open only by an insect of considerable size alighting on the lower lip of the flower. The *Yucca* flower can be pollinated only by the *Pronuba* moth, which at the same time lays its eggs in the ovary among the ovules, some of which later are used as food by the emerging larvae. Butterflies and hummingbirds are active agents in cross-pollination in certain special cases. The common honeybee is one of the most general pollinators among insects.

Development of ovules An ovary may contain one or more ovules. Each ovule arises separately from the placenta as a minute, dome-shaped projection (Fig. 10-14*A*). The young ovule is

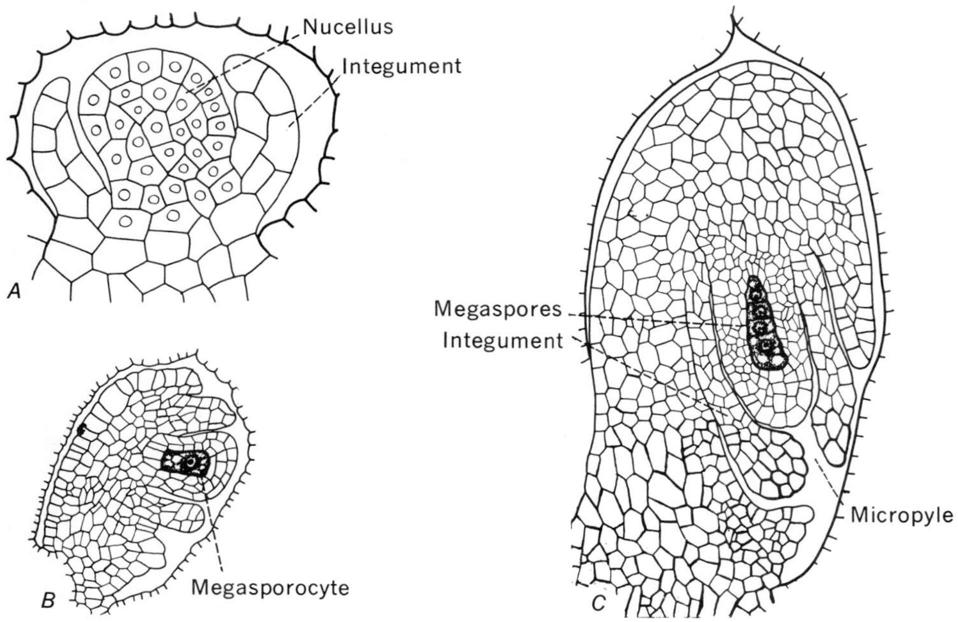

Fig. 10-14 Young ovules of Carex. *A, very young ovule with undifferentiated nucellus and integuments; B, ovule with integuments; the megaspore mother cell (megasporocyte) has been differentiated within the nucellus; C, ovule with integuments and micropyle. Four megaspores in linear arrangement have been developed as a result of the maturation divisions in the megasporocyte. (Drawn by Helen D. Hill, from slides prepared by Dr. H. A. Wahl.)*

attached to the placenta by a stalklike structure, called the **funiculus** (Fig. 10-4), through which a single vascular bundle passes and extends to the base of the nucellus. The **nucellus** (Fig. 10-14*A*) is a multicellular structure, and from its outer surface, only a little below its apex, one or more rings of tissue, called **integuments,** soon develop. The integuments continue to grow and completely cover the dome-shaped nucellus, leaving only a small opening, called the **micropyle** (Figs. 10-4, 10-14*C*). The end of the ovule, where the micropyle is located, is referred to as the micropylar portion or end. The opposite end of the ovule, where the funiculus, integuments, and nucellus merge, is called the funicular, or chalazal, end.

Before the integuments have enclosed the nucellus, there usually appears within the nucellus a single large cell called the **megaspore mother cell,** or **megasporocyte** (Fig. 10-4), which, by meiosis consisting of two successive divisions during which the chromosome number is reduced by one half, gives rise to four cells, usually arranged in a row, called **megaspores** (Figs. 10-14*C*, 10-15*A–F*). Each of the megaspores has the haploid ($1N$) number of chromosomes. The nucellus is the **megasporangium.** The development within the megasporangium, culminating with the production of four megaspores, is referred to as **megasporogenesis.** Subsequent development, in which one or more of the four megaspores may become involved, is called

megagametogenesis and ends with the formation of the female gametophyte or megagametophyte, with its megagamete, or egg.

The development of the **female gametophyte** (megagametophyte) from the megaspores may follow a number of patterns. Perhaps the commonest gametophytic development involves only one (monosporic) megaspore, but in other instances two (bisporic) or even four (tetrasporic) megaspores may contribute to the megagametophyte. Most often, in the monosporic type of development, the basal megaspore, the one farthest from the micropyle, immediately begins to enlarge and becomes the functioning megaspore. From this enlarging megaspore will develop the female gametophyte (megagameto-

Fig. 10-15 Development of the megaspores of Carex. *A, the megaspore mother cell as shown in the ovule in Fig. 10-14B; the nucleus is a diploid (2N) structure; B–F, meiotic division of the megaspore mother cell resulting in the formation of four haploid (1N) megaspores at F; G, H, enlargement of the functional megaspore and early disorganization of the remaining three nonfunctional megaspores toward the micropylar (lower) end of the ovule. (From slides prepared by Dr. H. A. Wahl.)*

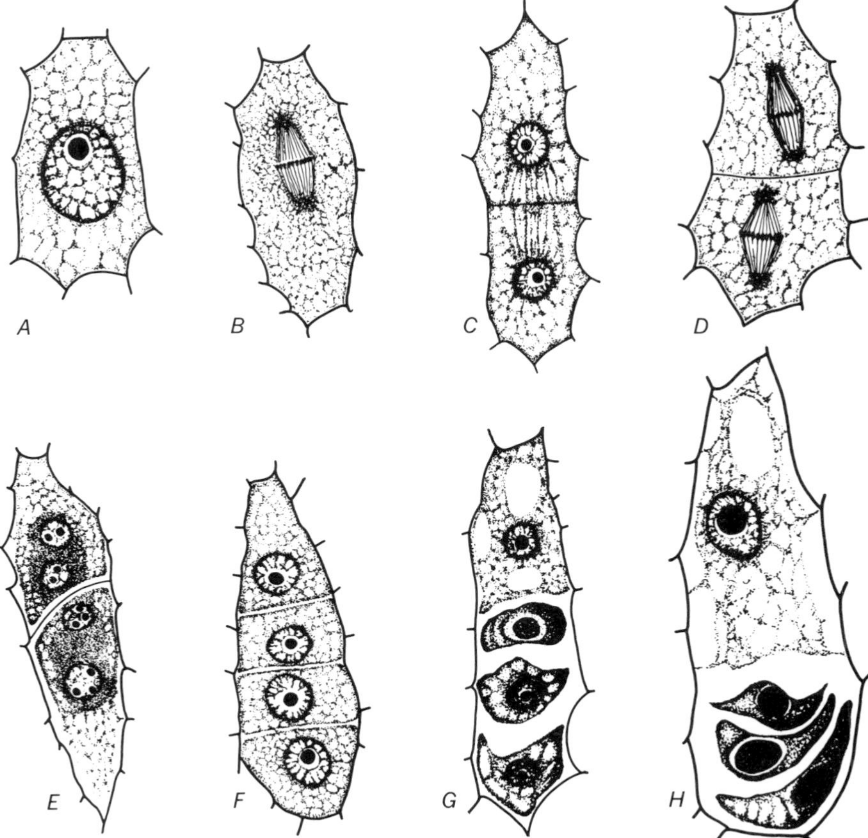

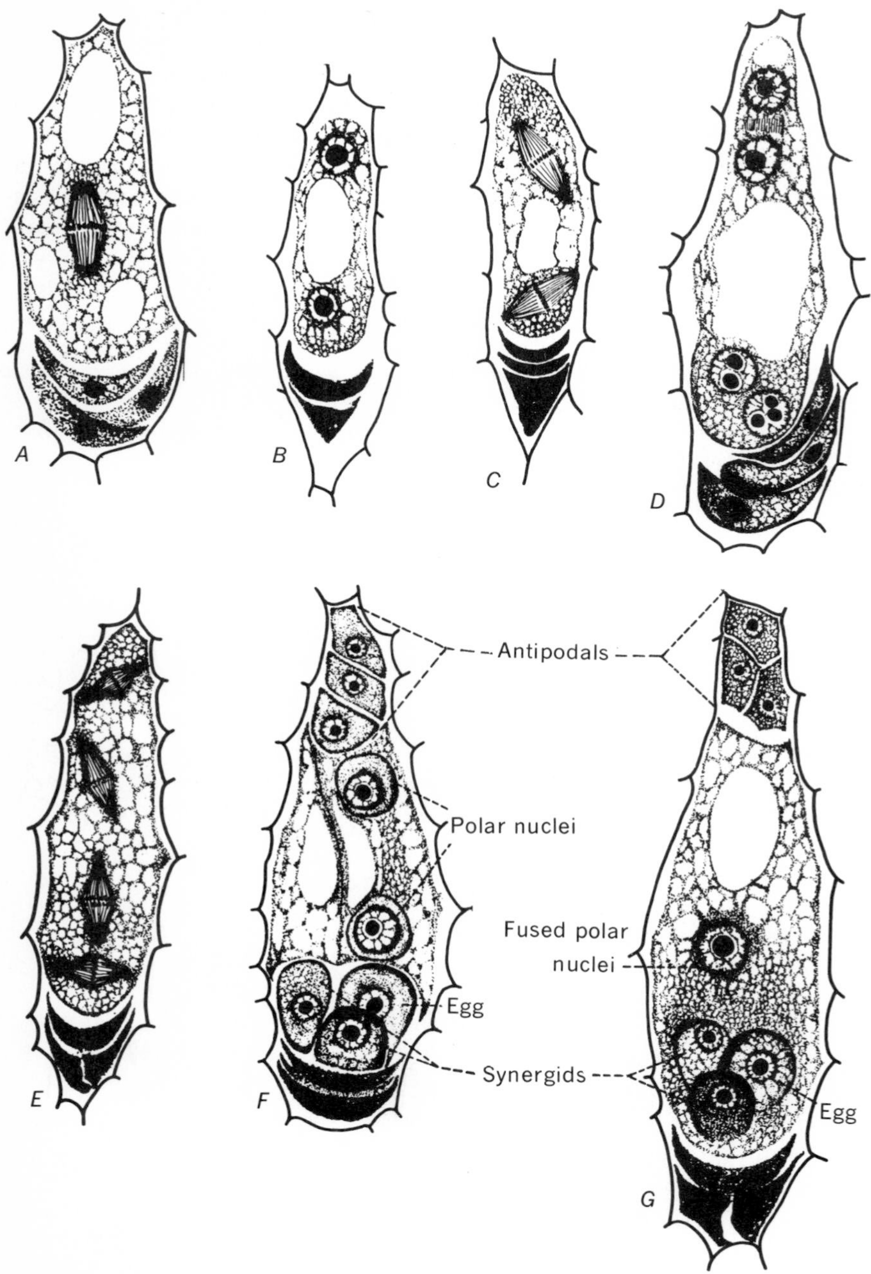

Fig. 10-16 Development of the female gametophyte of Carex. *Beginning with G of Fig. 10-15, three of the four megaspores have degenerated; A–E, mitotic division of the functional haploid megaspore into two, four, and finally, in F, eight haploid nuclei consisting of three antipodal nuclei, two polar nuclei, two synergids, and one egg nucleus; G, nuclei organized into the mature female gametophyte. (From slides prepared by Dr. H. A. Wahl.)*

phyte). The other three megaspores gradually disintegrate and lose their identity.

As the ovule grows, the functioning megaspore undergoes three nuclear divisions to produce eight nuclei. Of these, four are arranged toward the micropylar end of the ovule and four are closer to the funicular, or chalazal, end (Fig. 10-16). One nucleus from each group of four now moves toward the center of the ovule. These are the **polar nuclei.** The three nuclei remaining at the chalazal end of the ovule are called **antipodals** and probably will exhibit no further development. Of the three located at the micropylar end of the ovule, one serves as the **megagamete,** or **egg,** and the other two are called **synergids** (helpers). In most instances the synergids soon become disorganized. The whole structure consisting of three antipodals, two polar nuclei, an egg, and two synergids is the mature **female gametophyte** (megagametophyte). The female gametophyte is actually a greatly reduced female plant (Fig. 10-16*G*).

As the megagametophyte develops, parts of the ovule are replaced and probably resorbed. Most, or all, of the nucellus, including its epidermis, may be destroyed. In this case the female gametophyte is surrounded by integumentary tissue.

Events culminating in fertilization When a pollen grain reaches a receptive stigma, it germinates by developing a slender tube called the **pollen tube,** which breaks through a thin place in the wall of the pollen grain and penetrates the tissue of the stigma, ultimately growing down through the style (Fig. 10-17). The style may be hollow; but more often it consists of solid tissue, and in this case the pollen tube penetrates it by secreting enzymes which assist in dissolving away the stylar tissue as the pollen tube advances. Very soon after (and sometimes before) the pollen grain germinates, the generative nucleus divides to form two **male gametes,** or **sperms** (also called microgametes) (Fig. 10-13*H*). The germinated pollen grain with its tube nucleus and two male gametes, all of which are haploid ($1N$), is now the mature **male gametophyte,** or male gamete-producing plant. It is actually a greatly reduced male plant. The two sperms, together with the tube nucleus and most of the cytoplasm, keep moving down toward the tip of the pollen tube as it grows. Ultimately the pollen tube enters the ovule, usually at the micropylar end, penetrates the tissue of the nucellus, and reaches the female gametophyte (Fig. 10-17).

On reaching the female gametophyte, the pollen tube usually breaks at its tip. One of the sperms moves toward the egg and fuses with it. *This fusion of sperm and egg, and this alone, is* **fertilization,** or **syngamy.** The cell resulting from this fusion, called the fertilized egg, or **zygote,** ultimately develops into an **embryo,** or young plant. Since the sperm and the egg are each haploid ($1N$), the zygote resulting from their fusion is diploid ($2N$); i.e., it has two sets of chromosomes, one contributed by the male parent and one by the female parent. The other male gamete moves to the center of the female gametophyte and fuses with the two polar nuclei or with the nucleus resulting from their fusion if this has occurred. The resulting nucleus is triploid ($3N$) and is called the **primary endosperm nucleus** because it gives rise to a nutritive tissue called **endosperm.** The tube nucleus usually disintegrates.

Only ovules in which syngamy has occurred reach maturity and a pollen grain is required for each ovule. Thus it is not unusual to find many pollen tubes growing down through a single style. The time that elapses between pollination and fertilization varies from a few hours or days in some species to more than a year in others.

Variations in reproductive processes; the lily In the lily, which is widely used as illustrative laboratory material, meiosis results in the division of the megaspore mother cell into four haploid nuclei not separated by cell walls (Figs. 10-18 *A–E;* 10-19*A, B*). These four nuclei represent the four megaspores, and all four take part in the subsequent tetrasporic development of the

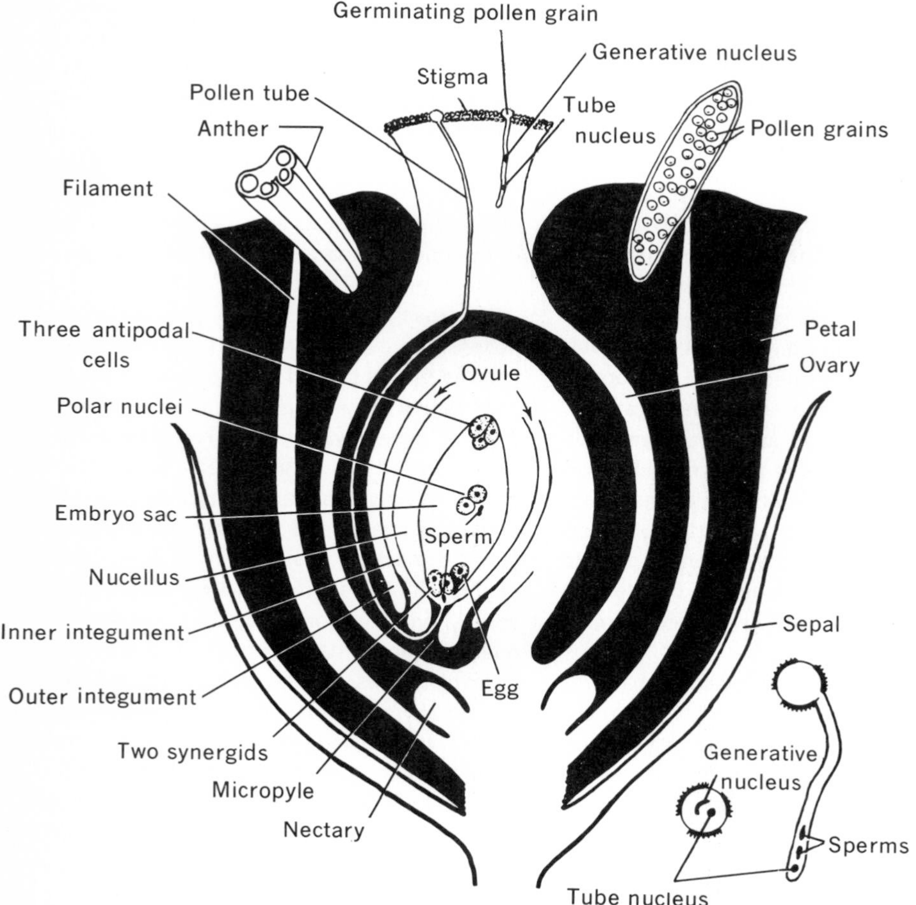

Fig. 10-17 Diagram of a sectional view of a flower to show germination of pollen grains (male gametophyte) and female gametophyte (embryo sac) at time of fertilization. (Courtesy of Dr. F. D. Kern from "The Essentials of Plant Biology," Harper & Row, Publishers, Incorporated, New York, 1947.)

female gametophyte. One of the four nuclei remains at the micropylar end of the developing female gametophyte, while the other three migrate to the funicular, or chalazal, end (Fig. 10-19*C*). This constitutes the so-called first four-nucleate stage. The fusion of the three megaspore nuclei at the chalazal end of the ovule is immediately followed by division. The chromosomes of the three nuclei are assembled on the same spindle and divide as a single nucleus (Fig. 10-19*D, E*). The two nuclei resulting from this mitotic division therefore have three times as many chromosomes ($3N$) as the two haploid ($1N$) nuclei resulting from the mitotic division of the micropylar nucleus. This now constitutes the second four-nucleate stage (Fig. 10-19*F*). It differs from the first four-nucleate stage in that two $3N$ nuclei of this stage are larger than the two $1N$ nuclei.

Each of the four nuclei of the second four-nucleate stage now divides mitotically (Fig. 10-20*A*) to form the eight-nucleate female gametophyte. In this stage there are four haploid ($1N$) nuclei at the micropylar end of the female gametophyte and four triploid ($3N$) nuclei at the chalazal end. One triploid ($3N$) nucleus from the chalazal end and one haploid ($1N$) nucleus from the micropylar end now migrate to the center of the gametophyte. These are the polar nuclei. The three haploid nuclei remaining at the micropylar end of the gametophytic sac form the so-called egg apparatus in which one serves as the female gamete or egg and the other two nuclei serve as synergids. The three antipodal nuclei remaining at the chalazal end of the gametophytic sac are functionless and soon become disorganized. This seven-celled, eight-nucleate structure is the mature female gametophyte (megagametophyte) (Fig. 10-20*B*).

Following the entry of the pollen tube into the female gametophyte, one of the liberated male gametes, or sperms, fuses with the female gamete, or egg. This fusion of gametes is fertilization, or syngamy. The second male gamete (microgamete) and the two polar nuclei fuse together in triple fusion to form the primary endosperm nucleus. Fusion of the male and female gametes, syngamy, results in a diploid ($2N$) zygote. The fusion of a $3N$ and a $1N$ polar nucleus with a haploid male gamete forms the

Fig. 10-18 The ovule and megaspore mother cell (megasporocyte) of Lilium. *A, longitudinal section of a young ovule; a large subepidermal cell differentiates as the megasporocyte; B, slightly older ovule with integuments developing; C, megaspore mother cell enlarging; D, E, first (meiotic) division of the diploid megaspore mother cell.*

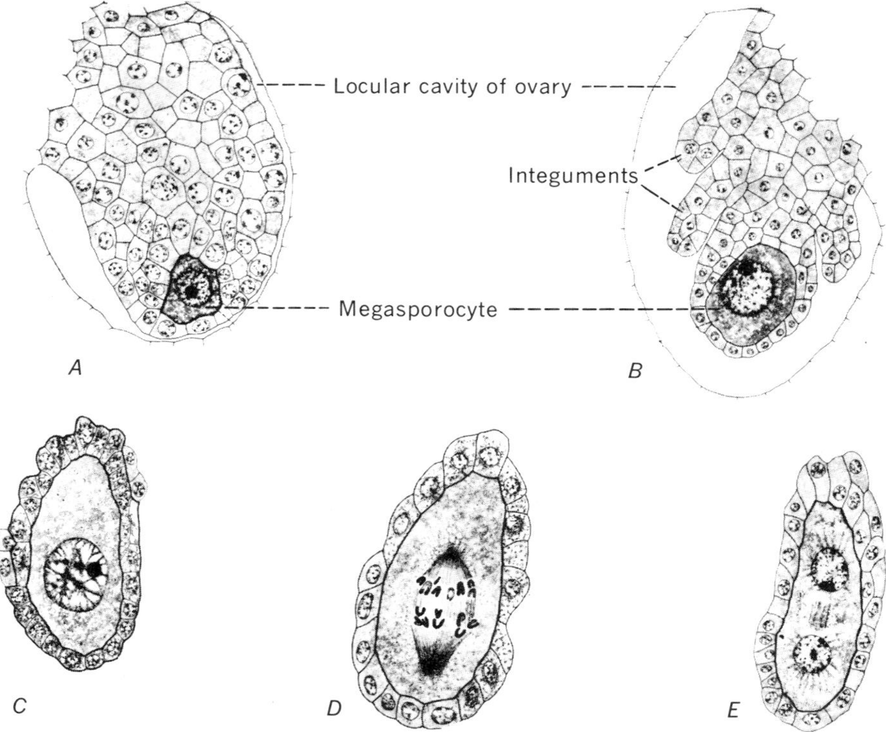

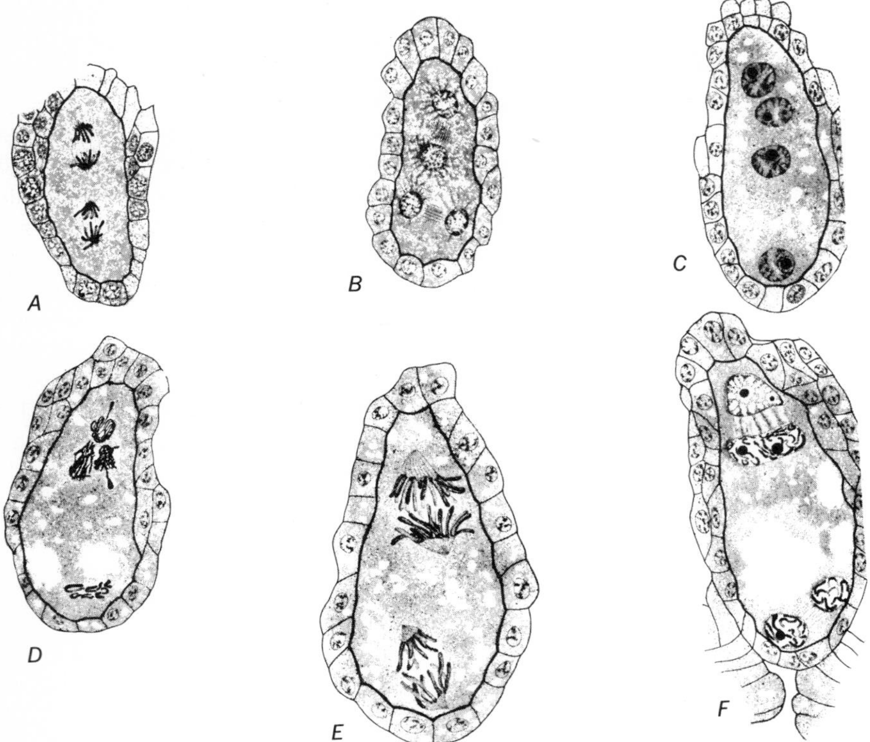

Fig. 10-19 Continued from Fig. 10-18 (Lilium). *A, B, second meiotic division resulting in four haploid megaspore nuclei. This is the first four-nucleate stage. C, three of the megaspore nuclei migrating toward the chalazal end of the developing megagametophyte, leaving one megaspore nucleus at the micropylar end. D, E, all four nuclei in mitotic division. In D the spindles of the three nuclei at the chalazal end are fusing to form a multipolar figure. F, the second four-nucleate stage. (C, D, and F drawn from slides furnished by G. H. Conant.)*

5N primary endosperm nucleus. These two fusions are called double fertilization (Fig. 10-20C).

Development of the embryo Following syngamy, the zygote undergoes a series of divisions which lead to the development of the **embryo.** The female gametophyte now becomes the **embryo sac,** or structure in which the embryo is contained. With rare exceptions, the first division of the zygote forms two cells, the basal cell, which lies closest to the micropyle, and the terminal, or apical, cell. In general, subsequent development is such that the **suspensor** of the embryo, which attaches the embryo in the embryo sac, is formed from the basal cell. The terminal, or apical, cell divides to produce the embryo proper (Fig. 10-20E). Often the basal cell divides transversely and the terminal cell longitudinally to form a four-celled embryo. Further divisions of the basal cell add additional cells to the suspensor, and divisions of the embryo-proper cell will produce quadrant, octant, and other stages of the embryo itself. The exact method of the origin of the parts of the embryo varies in different groups of seed plants, but certain early stages are essentially the same.

In *Capsella* (Fig. 10-21), which is commonly used to illustrate embryo development in dicotyledonous plants, the first division of the zygote produces the basal and terminal cells. A second transverse division of the basal cell and a longitudinal division of the terminal cell result in the formation of a four-celled embryo. The two cells produced by the division of the terminal cell (embryo proper) divide twice more to make the so-called octant stage (eight cells), and the division of each of these cells results in the presence of sixteen cells, eight of which

Fig. 10-20 Continued from Fig. 10-19 (Lilium). *A, B, mitotic division of the four nuclei, of the second four-nucleate stage, to form eight nuclei of the female gametophyte. The three antipodal nuclei and one of the polar nuclei are 3N. The three micropylar nuclei and one of the polar nuclei are 1N. C, immediately prior to double fertilization. One sperm nucleus lies next to the two polar nuclei and one sperm nucleus is contiguous with the egg nucleus. D, the primary endosperm nucleus (5N) in first mitotic division. E, development of the embryo and the endosperm; the zygote has divided several times to form a young embryo. Synergid nuclei and antipodal nuclei have disintegrated. Nuclei shown are endosperm nuclei.*

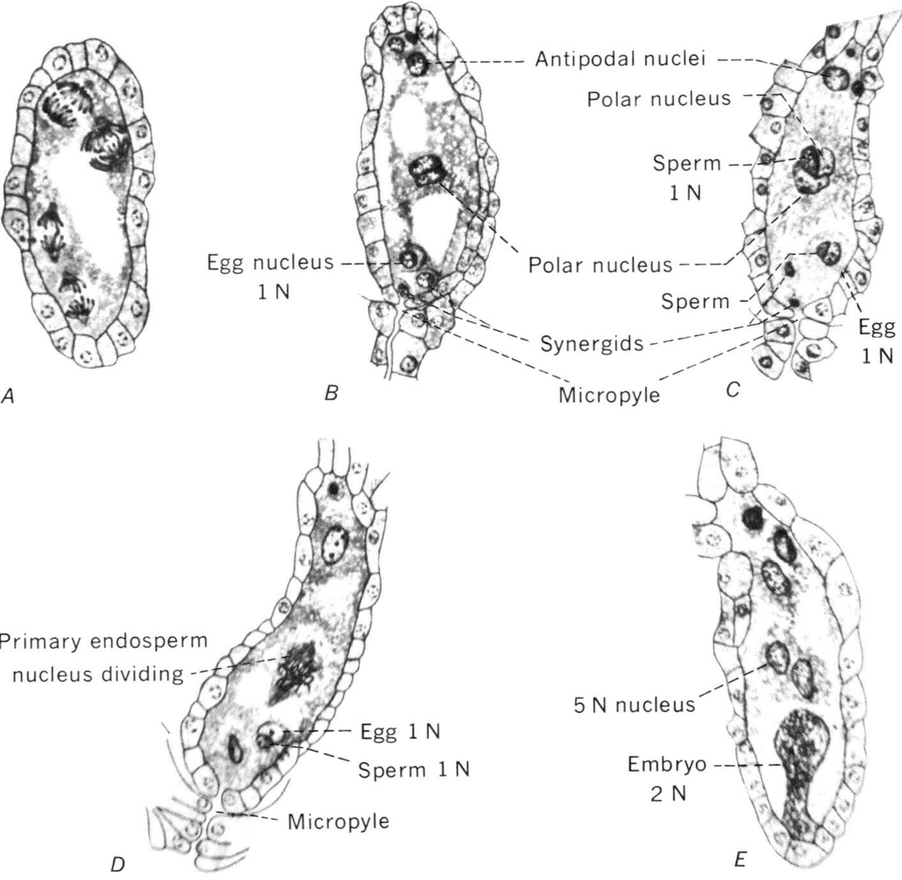

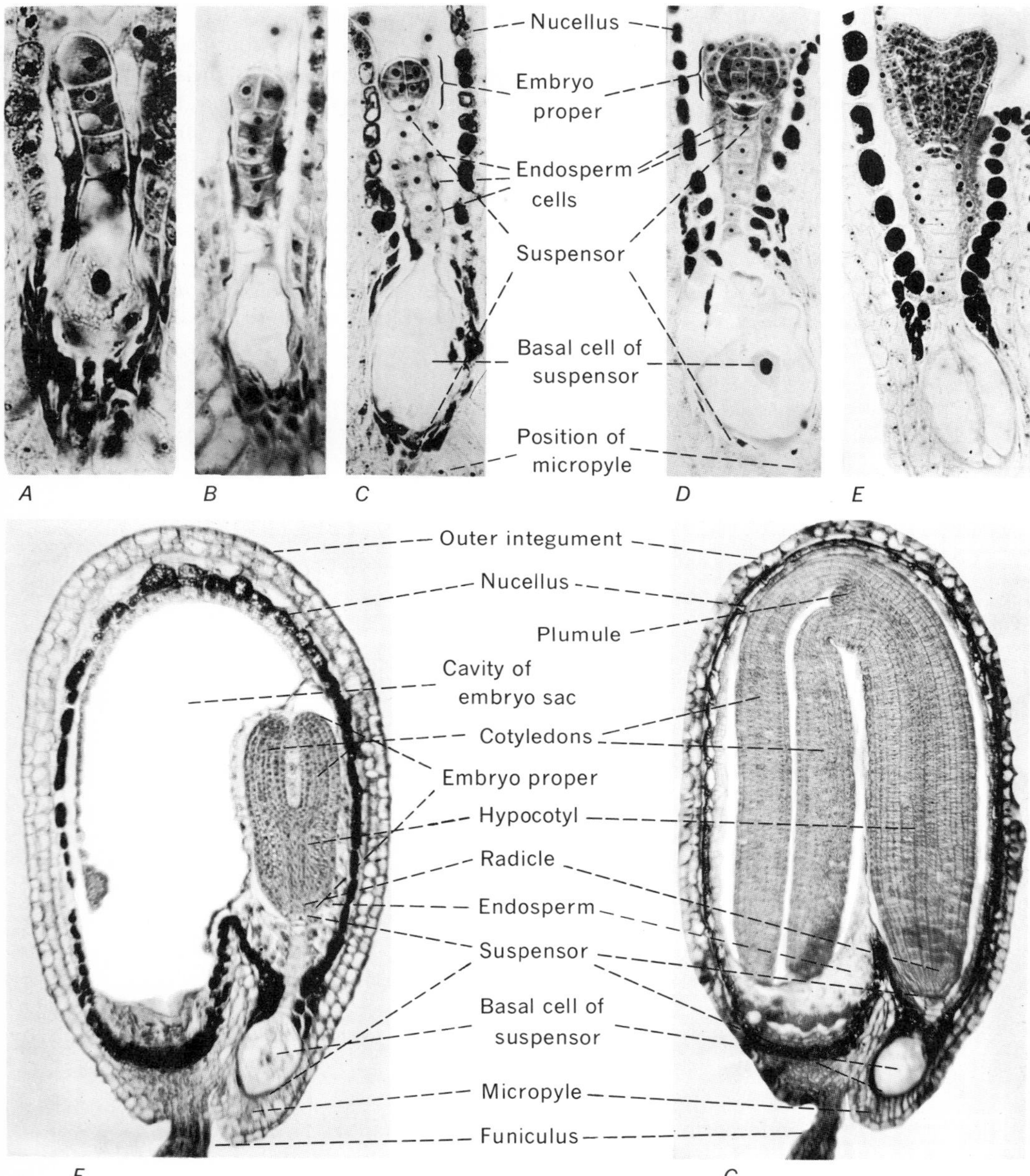

Fig. 10-21 Photomicrographs showing development of the embryo in Capsella; *A, proembryo; B–G, successive stages under decreasing magnification; F, G, lengthwise sections of entire ovules, showing position of embryo within the ovule. (Photomicrographs A–E by Dr. D. A. Kribs, from slides prepared by Dr. H. A. Wahl.)*

lie toward the center of the more or less sphere-shaped embryo proper and eight of which lie outside these, essentially surrounding them. In essence, the outer group of eight cells, during future development, gives rise to the epidermis, and the inner group of eight cells provides the fundamental and procambial tissues of the developing plant.

While this part of the embryo is developing, the upper cells of the suspensor divide and several of them contribute to the formation of the root-tip end of the embryo. The end or basal cell of the suspensor, next to the micropyle, usually fails to divide but enlarges to form a conspicuous vesicular cell (Fig. 10-21).

The degree to which the embryo develops before the seed ripens varies greatly in different plants. In some cases, it develops only a little beyond an early, or proembryo, stage. In others all the organs of the embryo are formed, and the embryo may occupy the entire embryo sac.

Formation of endosperm Immediately following fertilization and often before the embryo starts to develop, the primary endosperm nucleus (Fig. 10-20*D*) undergoes a series of mitotic divisions that result in the development of many free nuclei. Later, cell walls develop between the nuclei. In this way a storage tissue called the **endosperm** is developed, in which reserve food, available for the growing embryo, is stored. In most plants the endosperm in its growth replaces all the nucellar tissue, but in a few cases, like that of certain water lilies, the nucellus persists and forms a considerable part of the mature ovule. The endosperm itself may be entirely consumed by the developing embryo or may remain as a food-storage tissue. In peas, beans, and other legumes the endosperm disappears entirely before the ovule ripens. In the morning glory, the castor bean, and all grasses the endosperm persists in the fully ripened ovule.

Formation of fruit and seed During growth of the embryo and endosperm, there is also a rapid growth of the other tissues of the ovule and an increased translocation of foods to the embryo sac. The ovule usually increases markedly in size, and many physical and chemical changes take place in its tissues. The integuments expand and then frequently harden and dry out. Soluble foods carried into the ovule are converted into insoluble storage forms. Thus sugars may be converted into starch, amino acids into proteins, and fatty acids and glycerol into fats and oils. The relative amount of water in the tissues decreases, and the ovule gradually changes from a relatively soft, succulent structure to a hard, relatively dry body. As this development progresses, the physiological activity within the ovule gradually decreases and the embryo becomes dormant. The ovule thus ripens into a **seed** in which the integuments become protective seed coats, and the interior is occupied by a resting embryo together with stored food. This food may still remain in the endosperm or it may be stored entirely in the embryo itself.

Fertilization stimulates a rapid growth of the whole ovary as well as the ovules. Such flower parts as the stamens, petals, and sepals, together with the stigma and style of the pistil, usually wither and fall off after pollination and fertilization take place, although in some plants the bases of the stamens, petals, and sepals or the receptacle may be stimulated to renewed growth and development. In some cases the style of the pistil may likewise be retained and enlarged. The ovary usually increases greatly in size, and its tissues may become highly differentiated in the production of parts involved in the protection and the dissemination of the seeds. The mature ovary containing the seeds, together with any accessory structures developed from the receptacle or other parts, is now called the **fruit.** The different kinds of fruits are described in the following chapter.

When fertilization fails to occur, the entire flower usually dies, or as is commonly stated the flower fails to "set fruit." Thus there are seldom as many fruits on a plant as there were flowers. There are some plants, however, that produce fruits regularly even though fertilization has not taken place. This condition is known as **parthenocarpy,** and the fruits so produced are said to be **parthenocarpic.** Among the plants which normally develop parthenocarpic fruits may be mentioned the common banana, the navel orange, the seedless raisin grape, and the pineapple. In recent years parthenocarpic fruits have been produced artificially by spraying, or applying in other ways, certain chemicals (growth substances) to the pistils of the flowers or to the ovaries directly. In this way seedless tomatoes, cucumbers, peppers, melons, and other fruits have been produced (Figs. 9-2, 9-3).

The development of seeds and fruits is an exhaustive process which usually checks the growth of the vegetative organs of the plant. Annual plants usually die soon after the seeds and fruits mature. In such plants the seeds alone remain living and serve to perpetuate the species.

QUESTIONS

1. How does a flower bud compare with a vegetative bud?

2. What is the difference between a perfect flower and a complete flower?

3. Draw a floral diagram of a regular flower consisting of five sepals, five petals, ten stamens, and a compound pistil consisting of five carpels.

4. What characteristics of flowers are associated with wind pollination?

5. Are perfect flowers necessarily self-pollinated?

6. What advantages accrue from cross-pollination?

7. Why is the common buttercup considered to be a relatively primitive type of flower?

8. Which of the following are haploid and which are diploid: *a.* tube nucleus; *b.* cells of the endothecium; *c.* cells of the nucellus; *d.* synergids; *e.* embryo?

9. How does weather at blossoming time affect the set of fruit in fruit trees?

10. Some varieties of fruit trees do not bear well unless there are other varieties near them or unless another variety is grafted on them. Why?

REFERENCES

EAMES, A. J. 1961. Morphology of the Angiosperms. McGraw-Hill Book Company, New York. Extensive treatment of the morphology and reproductive process in the flowering plants.

ESAU, KATHERINE. 1965. Plant Anatomy, 2d ed. John Wiley & Sons, Inc., New York. Chapter 18 deals with the structure of flowers.

JOHANSEN, D. A. 1950. Plant Embryology. Chronica Botanica Company, Waltham, Mass. A detailed presentation of types of embryos and their development.

MAHESHWARI, P. 1950. An Introduction to the Embryology of the Angiosperms. McGraw-Hill Book Company, New York. Sound, comprehensive treatment of the embryos of flowering plants.

RICKETT, H. W. 1944. The Classification of Inflorescences. *Botan. Rev.,* **10:**187–231.

WARDLAW, C. W. 1955. Embryogenesis in Plants. John Wiley & Sons, Inc., New York.

WILSON, C. L., and T. JUST. 1939. The Morphology of the Flower. *Botan. Rev.,* **5:**97–131.

WODEHOUSE, R. P. 1935. Pollen Grains. McGraw-Hill Book Company, New York.

11

FRUITS, SEEDS, AND SEEDLINGS

Fruits and seeds of milkweed. (Photograph by Robert S. Beese.)

The kinds of fruits, their structure; the structure, dormancy, germination, and economic importance of seeds; and the development of seedlings are taken up in this chapter.

FRUITS

Definition Botanically a fruit is a matured ovary with or without seeds. In many fruits, however, other structures derived from other flower parts or from the axis (receptacle) may become a part of the fruit. Thus in fruits derived from inferior ovaries, the enlarged hypanthium, or floral tube, is usually still present in the fruit. In some cases these accessory structures may become a prominent part of the fruit, as in apples and pears. Furthermore a fruit may consist of several matured ovaries remaining together as a unit and may even include the matured ovaries of an entire inflorescence. A **fruit** *may therefore be defined as a structure made up of one or more matured ovaries together with any accessory structures closely associated with them.*

From the definition of a fruit it is clear that the popular use of the term does not always coincide with the botanical. Thus tomatoes, cucumbers, snap beans, pea pods, peppers, squashes, and many kinds of nuts that appear in our markets are botanically fruits. Furthermore many single-seeded fruits, like those of lettuce, buckwheat, sunflower, and grains like corn and wheat are commonly sold in the markets as seeds. The fact that they are all matured ovaries clearly proves that all these examples are fruits.

Structure of the pericarp The matured wall of the ovary in the fruit is called the **pericarp.** The structure of the pericarp varies greatly in different kinds of fruits. In some, three distinct layers may be differentiated, viz., an outer layer called the **exocarp,** a middle layer called the **mesocarp,** and an inner layer called the **endocarp.** The **exocarp,** though sometimes more complex, often appears as a single layer of epidermal cells, sometimes heavily cutinized and sometimes hairy. The **mesocarp,** or middle layer, may be very thin or it may be a well-developed tissue several centimeters thick. It usually contains vascular bundles, and in some fruits it is fleshy. The **endocarp** also varies greatly in structure in different fruits; sometimes it consists of a single layer of cells and sometimes of many layers. In some fruits the endocarp becomes very tough or hard. In others it is fleshy. These three layers are most clearly differentiated in such fruits as peaches, plums, and cherries in which the fleshy part of the fruit is the mesocarp.

In some fruits, and particularly those arising from inferior ovaries, the floral tube may persist in the fruit, forming the outermost layers of the fruit. In such cases the pericarp may be difficult to differentiate.

KINDS OF FRUITS

Fruits are of many kinds. The structure of a fruit can be understood only by a knowledge of the structure of the flower from which the fruit arises. The presence of accessory structures in a fruit may obscure the true nature of the fruit in its mature condition and may render the fruit difficult to classify unless its development is followed step by step from the flower. Even then, interpretation of structures by different investigators may lead to difficulties in classification.

In general all fruits may be classified into three groups: (1) **simple fruits,** which are developed from a single (simple or compound) pistil and thus consist of a *single matured ovary* together with any accessory structures closely associated with the ovary; (2) **aggregate fruits,** which consist of a *number of matured ovaries* aggregated as a unit on a common receptacle, together with any accessory structures, all developed from a *single flower* with many separate pistils, as in raspberries and blackberries; (3) **multiple fruits,** which consist of all the matured ovaries of *several flowers* grouped into a single mass, together with any accessory structures, the whole being developed from an entire inflorescence as in pineapples, mulberries, and figs.

The term **accessory fruit** is sometimes applied to fruits in which a major part of the matured fruit has not developed from the ovary. This type of fruit does not represent a group distinct from the three just mentioned but is represented in all three groups. Thus the simple fruit of the apple, the aggregate fruit of the strawberry, and the multiple fruit of mulberry are all accessory fruits. The major portion of the fleshy part of an apple fruit consists of the enlarged bases of the sepals, petals, and stamens; the fleshy part of the strawberry is the receptacle; and in the mulberry the fleshy sepals form a considerable part of the fruit.

Of the three groups mentioned, simple fruits constitute by far the largest group and the most diversified. They may be subdivided into **dry fruits,** in which the pericarp and any accessory structures become more or less dry when mature, and **fleshy fruits,** in which a part or all of the pericarp and any accessory structures become fleshy at maturity. The dry fruits are further subdivided into **dehiscent fruits,** which split open at maturity, and **indehiscent fruits,** which do not split open at maturity. The line (or lines) along which a dehiscent fruit opens at maturity is called a **suture,** and the resulting segments of the pericarp are known as **valves.**

Simple, dry, dehiscent fruits

LEGUME The **legume** fruit is characteristic of the members of the legume or pea family. In legume flowers, such as those of the pea (Fig. 11-1*B*), the gynoecium consists of a single, simple pistil with a superior ovary containing a single cavity, or locule. By the maturation of this pistil, the fruit, or pod, is formed (Fig. 11-1*A, C*). The remains of stigma and style can often be seen at the free end of the pod. The seeds are developed along the ventral suture, which is the one along which the margins of the carpel are united. The dorsal suture corresponds to the midrib of the carpel. At maturity the fruit splits along both dorsal and ventral sutures into two valves. Some leguminous plants have fruits that show considerable variation from this type. In tick trefoil (*Desmodium*) the fruits are constricted between the seeds and finally break up crosswise into distinct parts. A legume of this type is called a **loment.** Some legume fruits are single seeded. Others have many seeds. The legume is distinguished from other simple, dry dehiscent fruits by the fact that it develops from a simple pistil and splits open at maturity along both dorsal and ventral sutures.

FOLLICLE The **follicle** resembles the legume in that it develops from a single, simple pistil but differs from the legume in that it splits open at maturity along only one suture. Examples of follicles are the fruits of larkspur (Fig. 11-1*F, G*), columbine (Fig. 11-1*E, I*), and milkweed (Fig. 11-1*H*).

CAPSULE The **capsule** differs from the legume and the follicle in that it is developed from a *compound pistil.* Any simple, dry dehiscent fruit that develops from a compound pistil may be called a capsule (Fig. 11-2). Capsules therefore always consist of two or more ovarian segments, which are the ovaries of the individual carpels. At maturity a capsule may open by means of pores, as in poppy (Fig. 11-2*E*) **(poricidal dehiscence),** or by a regular transverse circular line of division, as in purslane (*Portulaca*) (Fig. 11-2*A, B*) **(circumscissile dehiscence).** The latter type of capsule is called a **pyxis.** More commonly, capsules dehisce by lengthwise division into valves. If the lengthwise splitting occurs down the back, or dorsal, suture of each ovarian segment directly into the loculi, as in iris and lily, the dehiscence is said to be **loculicidal** (Fig. 11-2*G, g*); if the splitting occurs through the partitions and between the loculi, thus dividing the capsule into its component ovarian segments, as in Saint-John's-wort and some species of *Yucca,* the dehiscence is **septicidal** (Fig. 11-2*H, h*). When the valves of either of the last two types break away from the partitions, these remaining attached to the axis of the fruit, as in morning glory, the dehiscence is **septifragal** (Fig. 11-2*I, i*).

The fruits of members of the mustard family (Cruciferae) are capsules consisting of two ovarian segments which separate at maturity

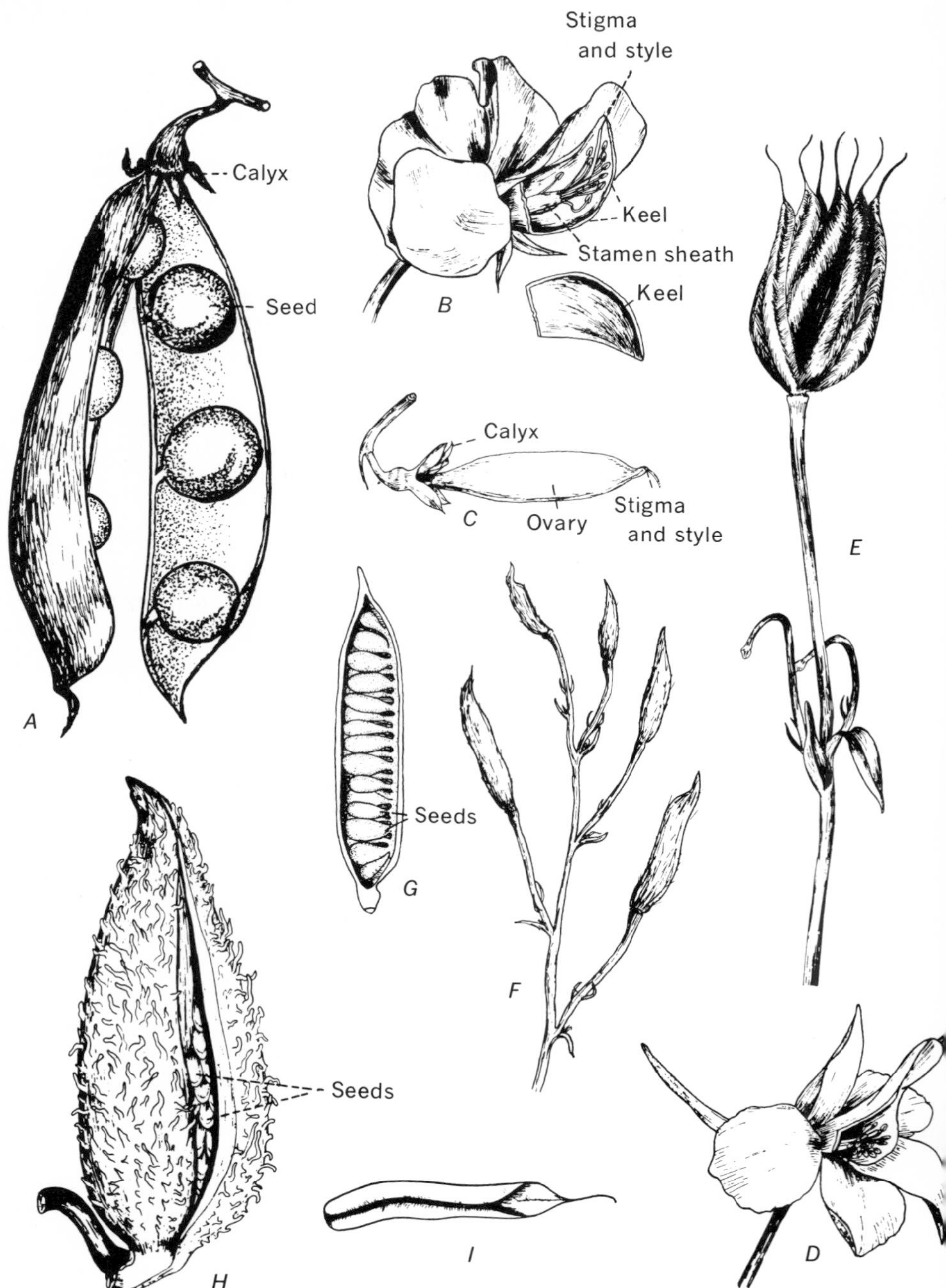

Fig. 11-1 Legumes and follicles. A, mature legume fruit (pod of pea); B, flower of pea with portion of keel cut away to expose the essential organs; C, young pod of pea; D–I, follicles; D, flower of larkspur with one side of a petal removed to show several stamens and pistils; E, mature follicles of columbine, all from one flower; F, follicles of larkspur; G, follicle of larkspur in longitudinal section; H, follicle of milkweed beginning to open; I, old follicle of columbine that has opened and discharged the seed. (A, B, E, H drawn by Elsie M. McDougle; C, D, F, G, by Christian Hildebrandt.)

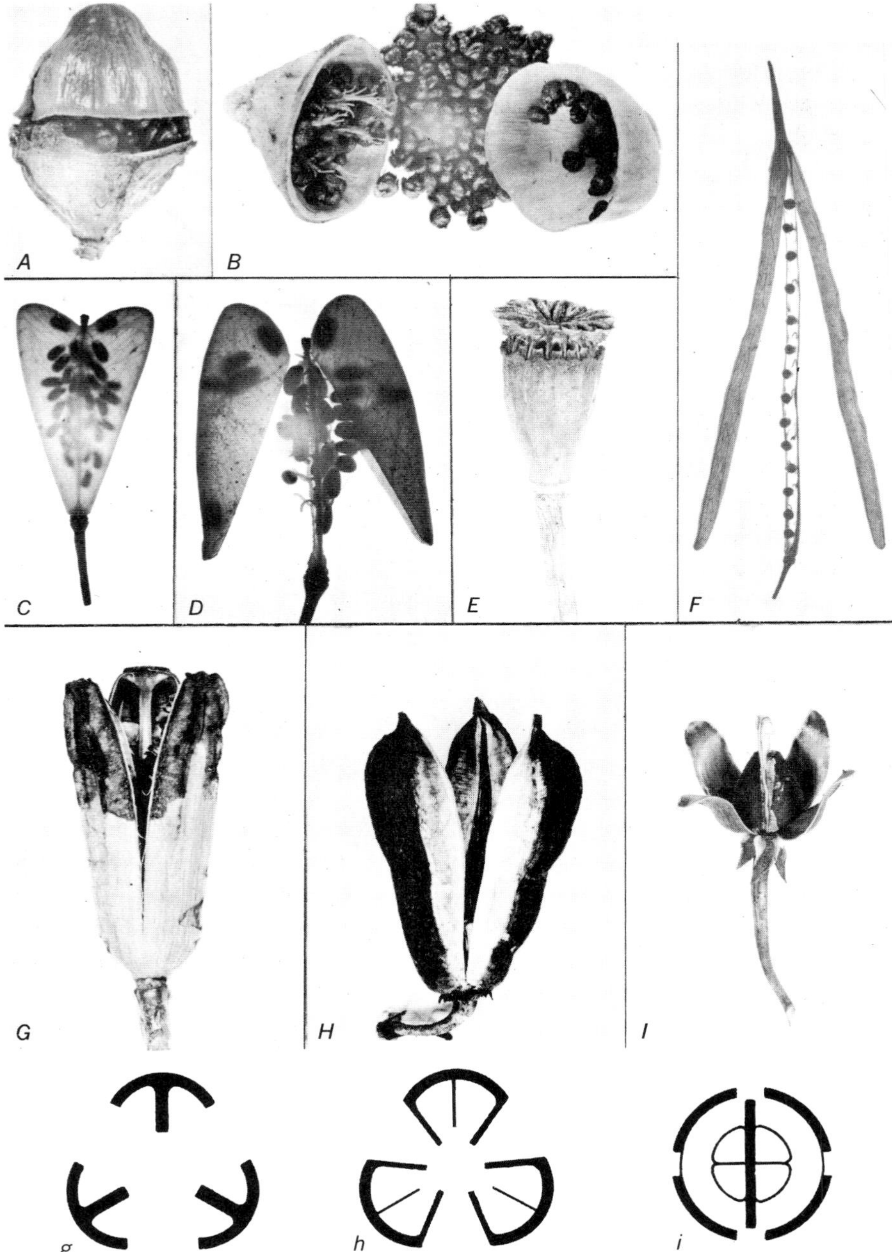

Fig. 11-2 Capsules. A, pyxis of Portulaca *showing circumscissile dehiscence; B, the same opened, showing discharged seeds (both A and B greatly enlarged); C and D, silicle of shepherd's purse; C, before opening; D, showing dehiscence from below upward, exposing the seeds; E, capsule of poppy, showing poxicidal dehiscence; F, silique of rutabaga; G, capsule of lily showing loculicidal dehiscence; H, capsule of a species of* Yucca, *showing septicidal dehiscence; I, capsule of morning glory opened to show septifragal dehiscence; g–i, diagrams of transverse sections of G, H, and I, respectively, with seeds removed in g and h. The four central structures in i are seeds. (Photographs E–I by Homer Grove.)*

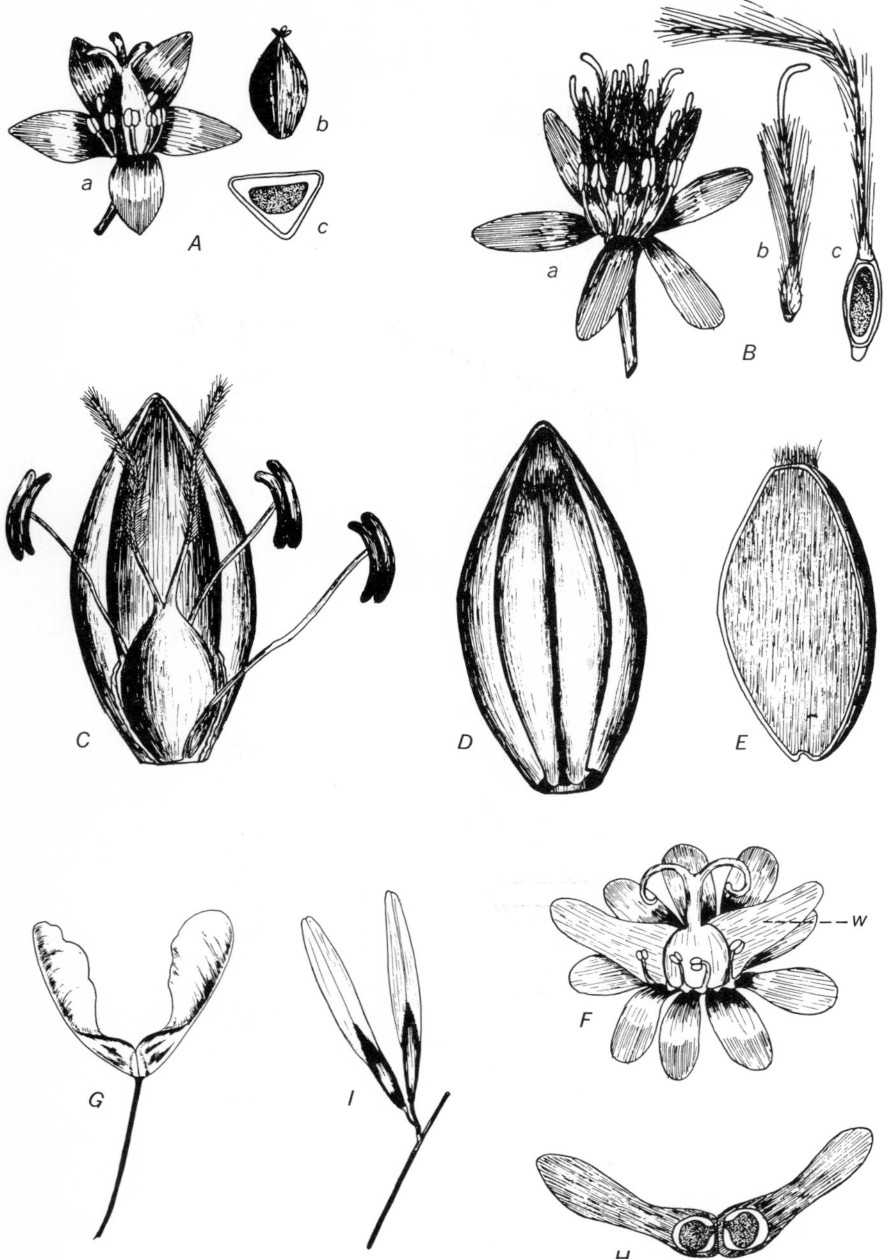

Fig. 11-3 Achenes, grains, and samaras. A, flower and achene type of fruit of buckwheat; a, flower of buckwheat, showing sepals, stamens, and the pistil with three styles and stigmas; b, the mature three-angled fruit with remnants of styles at apex; c, cross section of the fruit with the single seed; B, flower and achene fruit of Clematis; *a, flower, showing six petal-like sepals, a circle of several stamens, and numerous simple, separate pistils clothed with long hairs below the protruding stigmas; b, a single pistil; c, a pistil with the ovary cut vertically to show the single locule and the single seed;*

into two valves opening from below upward, exposing a thin partition with the two placentae to which the seeds are attached. This type of capsule is called a **silique** if it is much longer than broad, as in cabbage, rutabaga, and *Cardamine* (Fig. 11-2*F*), and a **silicle** if it is short and broad, as in shepherd's purse (Fig. 11-2*C, D*).

Simple, dry, indehiscent fruits

ACHENE **Achenes** (Fig. 11-3*A, B*) are small, dry, indehiscent, one-seeded fruits in which the seed is attached to the pericarp at only one point. They are derived from pistils in which only one ovule develops. This type of fruit is characteristic of the buckwheat and sunflower families and occurs also in the buttercup and rose families. Some achenes, like those of buckwheat (Fig. 11-3*A, b, c*), are derived wholly from ovary tissue, while in others, such as members of the sunflower family, the floral tube is fused to the pericarp. Since the seed is attached at only one point, the pericarp is readily separable from the seed coat. Achenes are commonly mistaken for seeds.

GRAIN, OR CARYOPSIS **Grains** (Fig. 11-3*C–E*), like achenes, are simple, one-seeded, dry fruits derived from single pistils, but differ from achenes in that in grains the seed coat is permanently fused to the pericarp over its entire surface, and thus cannot be easily separated from the seed. Grains are characteristic of all grasses, such as corn, wheat, oats, barley, rice, and rye.

SAMARA, OR KEY FRUIT The **samara** (Fig. 11-3*F–I*) is essentially a winged achene, although the term is sometimes used for any indehiscent dry fruit furnished with a wing. It is the characteristic fruit of maples and ashes and occurs also in elms, birches, and in the tree of heaven (*Ailanthus*). The wing is usually an outgrowth of the ovary wall. In some cases, as in elms and birches, the wing extends all around the pericarp. In ashes and maples it develops on one side only. In maples two carpels develop side by side, each forming a samara. For this reason the maple fruit is sometimes called a double-key fruit. The two parts may separate at maturity. Samaras usually have only one seed but occasionally have two.

SCHIZOCARP The **schizocarp** (Fig. 11-4) is the characteristic fruit of the carrot family. It consists of two carpels, each of which has a single locule usually containing a single seed. The carpels of the fruit separate, each being

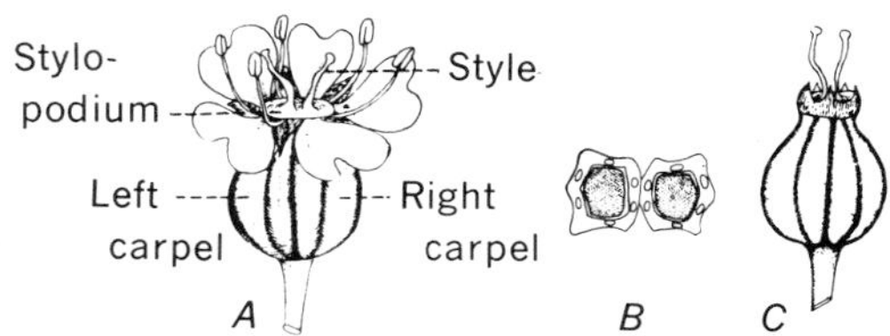

Fig. 11-4 The schizocarp fruit of the carrot family. A, flower of Cicuta; *B, cross section of fruit, showing the two ovarian segments insecurely attached to each other and the single seed in each; C, ripening schizocarp of* Cicuta; *the free part of the calyx shows as a crown of five small lobes at top of the inferior ovary. (Copied by Christian Hildebrandt from drawing by Edna S. Fox.)*

Fig. 11-3 Continued
the persistent styles become plumose in the fruit; C, flower of wheat, showing the three stamens and the single pistil with two plumose styles partially enclosed by the chaffy bract; D, the maturing wheat grain enclosed in the bract; stamens and stigmas have disappeared; E, longitudinal section of the grain; the outermost layer is the pericarp; F, flower of maple showing sepals, petals, stamens, and the pistil with two styles and stigmas, the wall of the ovary having grown out on either side to form the wing, w, of the samara fruit; G, mature samaras of maple; H, mature ovaries of maple samaras cut lengthwise to show the single seed in each; I, mature samaras of ash. (A–F and H drawn by Christian Hildebrandt; G, I by Elsie M. McDougle.)

called a **mericarp,** which is indehiscent and resembles an achene. The fruits of mallows are of similar structure but involve a number of carpels, each containing a single seed.

NUT **Nuts** (Fig. 11-5) are dry, indehiscent fruits in which the pericarp is hard or crustaceous throughout. At maturity they commonly have but one locule and one seed but are usually developed from a compound pistil, only one carpel of which develops. Examples are acorns, beechnuts, hazelnuts, and chestnuts. All these examples are developed from inferior ovaries and hence the wall of the fruit consists of the floral tube as well as the pericarp, both being hard. In addition, all have at the base an **involucre,** which is an accessory structure not developed from the ovary. The cup of the acorn is the involucre. It consists of numerous coalesced bracts (Fig. 11-5*F–I*). In the beech two separate one-seeded fruits, each coming from a separate flower, are usually enclosed within a single involucre (Fig. 11-5*A–E*). The structure of the chestnut is similar, but each bur (involucre) usually encloses three nuts. In the hazelnut (Fig. 11-5*J, K*) the involucre is made up of leafy bracts. Popularly, the term nut has been applied to many structures that botanically are not this type of fruit. Thus peanut pods are legumes, Brazil nuts are seeds, and almonds are parts of drupes.

Fig. 11-5 Fruits of the nut type. A, two pistillate flowers of beech surrounded by bristly involucral bracts, with only stigmas and styles protruding; B, beech flowers cut vertically, showing the ovary and the three styles of each of the two flowers; C, single flower of beech taken out of the involucral bracts; the presence of a calyx is plainly indicated by the three small calyx lobes below the bases of the three styles; D, bur of beechnut at time of ripening of the two nuts; E, cross section of the nut showing the single seed; F, oak pistillate flower composed of a single pistil with three styles and surrounded by the closely set bracts that form the cup; G, more mature pistil in cup; H, mature acorn fruit in cup; I, vertical section through fruit and cup of acorn showing the single large seed; J, hazelnut fruit entirely enclosed in the prominent involucral bracts; K, vertical section through hazelnut with its single seed. (A–G by Christian Hildebrandt; H–K by Elsie M. McDougle.)

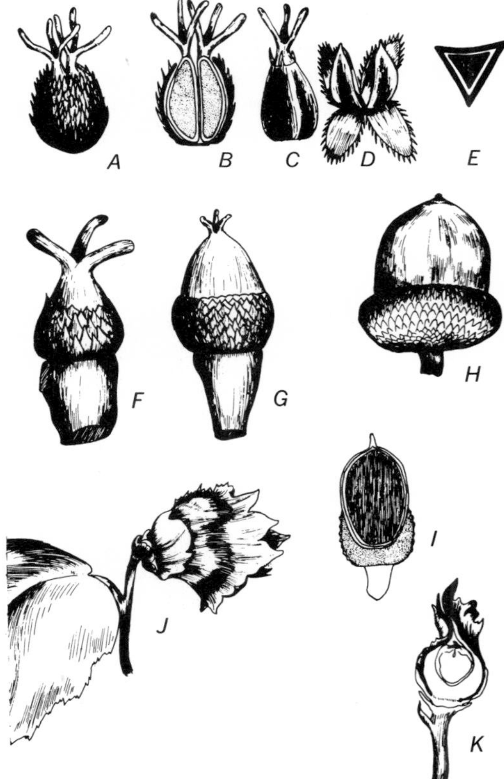

Simple, fleshy fruits

POME **Pome** fruits (Fig. 11-6) are characteristic of that portion of the rose family to which the apple, pear, and quince belong. The fruit is developed from a compound pistil consisting of two or more carpels and an inferior ovary. The fleshy outer portion of the fruit develops from the bases of the calyx, corolla, and stamens (hypanthium) which surround the ovary. The outer part of the pericarp also becomes fleshy, while the endocarp becomes more or less cartilaginous, forming the core of the fruit, containing several to many seeds.

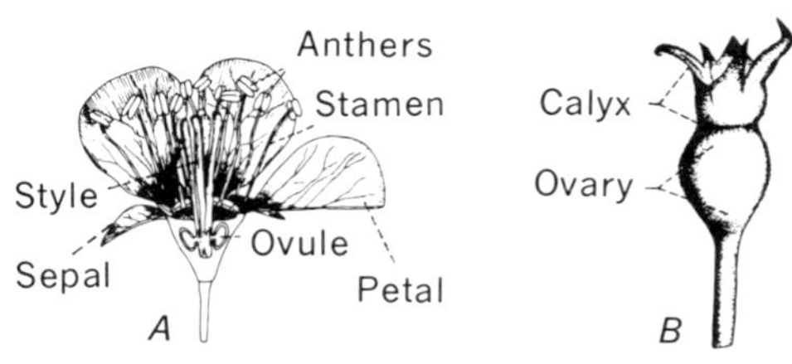

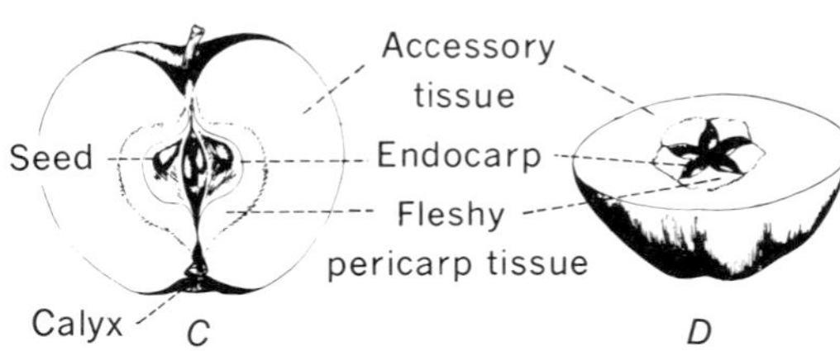

Fig. 11-6 The pome fruit (apple). A, apple flower cut vertically; B, young apple fruit; C, apple fruit cut vertically; D, apple fruit cut transversely. (A, B copied by Christian Hildebrandt from drawings by Edna S. Fox; C, D by Elsie M. McDougle.)

Since a major part of the fruit is not developed from the ovary, the pome is an accessory fruit.

DRUPE The **drupe** (Fig. 11-7*A–C*), commonly called a stone fruit, is a fleshy fruit in which the pericarp consists of a thin exocarp, usually skinlike, a fleshy mesocarp, and a hard, stony endocarp, as in cherries, plums, and peaches. In the examples mentioned, the fruits develop from single, simple pistils of perigynous flowers (Fig. 10-5*D*). Since these all have superior ovaries, no accessory structures are present in the fruit. Usually only one seed develops within the ovary. Some drupaceous fruits, however, develop from inferior ovaries and several carpels may originally be involved, as in *Viburnum*. Usually only one carpel matures. The almond, the olive, and the apricot are also drupes.

BERRY The **berry** is a simple fleshy fruit in which all parts of the pericarp are fleshy or pulpy except the exocarp, which is often skinlike. Berries originate from flowers with simple or compound pistils, the ovaries of which may have one or many ovules. More often the fruits are many seeded. Grapes, tomatoes (Fig. 11-7*D*, *E*), gooseberries, and cranberries are true berries. In the grape and in the tomato the fruit is developed from a superior ovary, while in the cranberry and in the gooseberry it develops from an inferior ovary. In the latter case the floral tube invests the matured ovary externally.

The fruit of squashes, pumpkins, cantaloupes, watermelons, and cucumbers is a berry invested with a rind that is not readily separable from the pericarp. The rind is made up mostly of floral-tube tissue. This type of fruit is called a **pepo.** Similarly the fruit of oranges, lemons, and grapefruit is surrounded by a leathery rind (in this case part of the pericarp since the ovary is superior) which is separable. This fruit is called a **hesperidium** (Fig. 11-7*F*).

From the description just given, it is clear that the popular use of the term berry does not always coincide with the botanical. Mulberries, strawberries, blackberries, and raspberries are not berries in the botanical sense.

Aggregate fruits Aggregate fruits are derived from single flowers with many separate pistils (Fig. 11-8*A*, *B*) and thus consist of a number of similar small fruits all on a common receptacle and maturing together as a single unit. Most aggregate fruits are fleshy. Good examples are strawberries, raspberries, and blackberries.

In the strawberry (Fig. 11-8*A–D*) the receptacle enlarges greatly, forming the edible part of the fruit. The true fruits of the strawberry are achenes, which are popularly thought of as seeds. In the fruit they appear scattered over the surface of the receptacle. The strawberry is also an accessory fruit.

In the raspberry and the blackberry (Fig. 11-8*E*) the individual fruits are drupelets which adhere together. In the blackberry these drupelets also adhere to the receptacle and cannot be

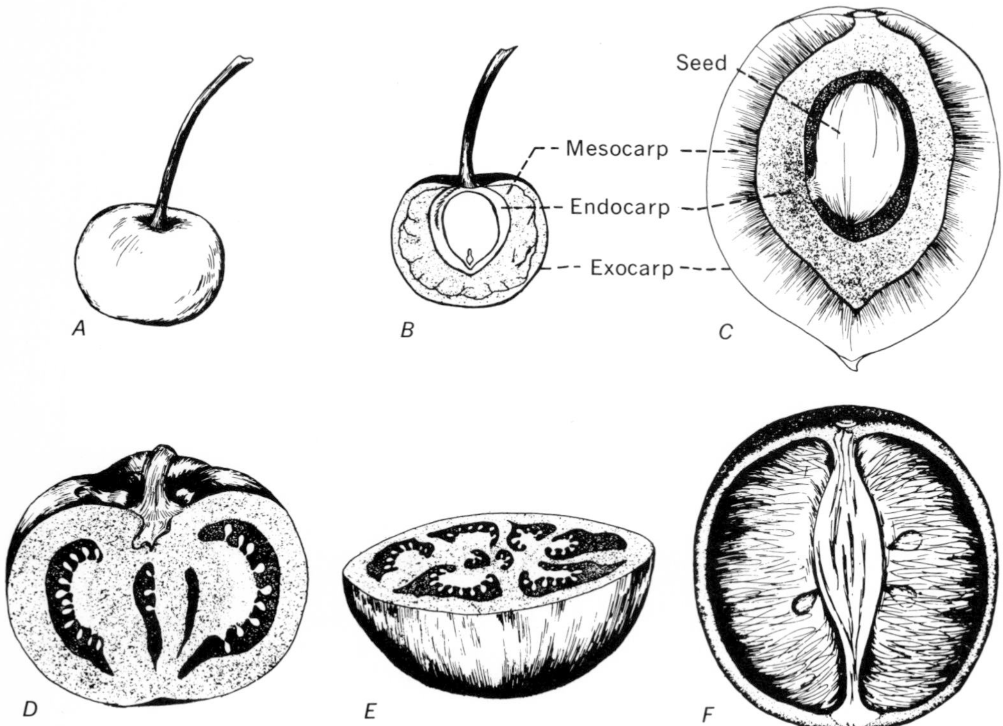

Fig. 11-7 Drupes and berries. A, external view of drupe fruit of cherry; B, cherry cut vertically; C, drupe fruit of peach cut vertically; D, E, the berry type of fruit; D, vertical section of a tomato; ovary fleshy throughout; E, transverse section of the tomato; F, the orange, a hesperidium, or berry, with a leathery, separable rind, cut vertically. (Drawings by Elsie M. McDougle.)

Fig. 11-8 Aggregate and multiple fruits. A–E, aggregate fruits; F–H, multiple fruits. A, flowers of strawberry, showing a large number of centrally located, separate pistils surrounded by a varying number of stamens; petals and sepals are both present; B, vertical section through the strawberry flower; C, the mature strawberry fruit; D, the strawberry fruit cut vertically, showing the minute achenes on the surface of the receptacle tissue; E, blackberry fruit cut vertically to show structure; F, drawing of a longitudinal section through the mulberry fruit; G, a single pistillate flower of mulberry, showing the single pistil with two styles and surrounded by the fleshy calyx; H, photograph of mulberry fruit. (B by Edna S. Fox; C–E by Elsie M. McDougle; F, G by Christian Hildebrandt.)

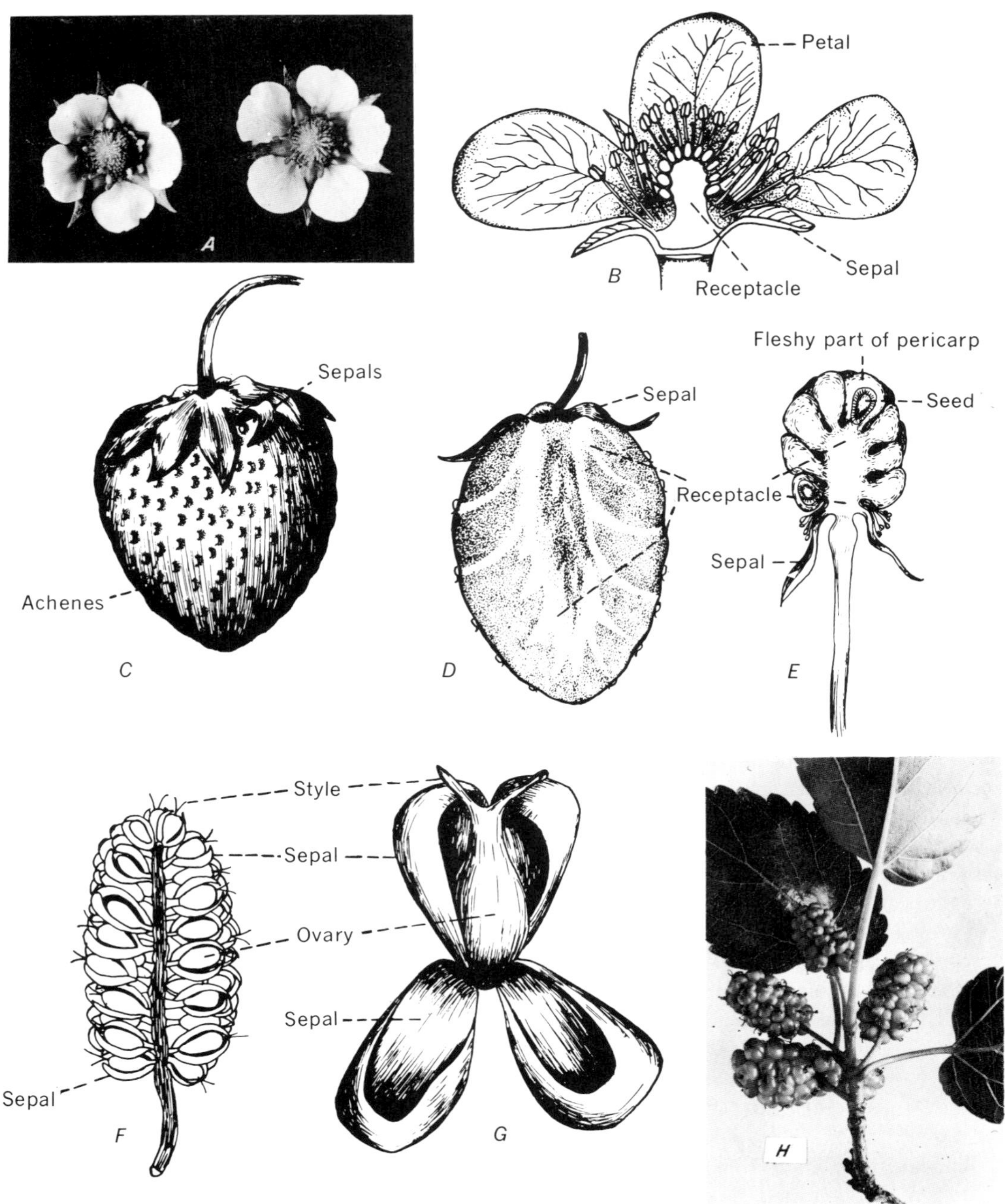
A
Petal
B
Receptacle
Sepal
Sepals
Achenes
C
Sepal
Receptacle
D
Fleshy part of pericarp
Seed
Receptacle
Sepal
E
Style
Sepal
Ovary
Sepal
Sepal
F
G
H

readily separated from it. In the raspberry the adhering drupelets can be lifted readily as a unit from the small domelike receptacle.

In the rose fruit, commonly called a **hip,** the floral tube surrounds the matured ovaries (achenes) and becomes fleshy, the whole forming an aggregate fruit which is also an accessory fruit.

Multiple fruits Multiple fruits develop from entire flower clusters and thus differ from simple fruits and aggregate fruits which develop from individual flowers. In the multiple fruit the matured ovaries of the entire flower cluster remain aggregated in a single mass. Most multiple fruits are fleshy and are also accessory fruits. Pineapples, Osage oranges, mulberries, and figs are good examples. In pineapples the flowers are sessile on a central axis which is leafy at the apex. At maturity the ovaries as well as axial structures are enlarged and fleshy, forming the fruit.

In the mulberry staminate and pistillate flowers are borne in separate inflorescences, the pistillate flowers in short, dense spikelike clusters (Fig. 11-8*F, H*). Each pistillate flower (Fig. 11-8*G*) consists of four sepals (no petals) and a pistil of two carpels, only one of which develops. The two stigmas of the pistil usually persist in the fruit. At maturity the sepals as well as parts of the one-seeded ovaries become fleshy, the entire cluster ripening as a unit.

In the fig the upper part of the peduncle becomes fleshy and completely envelops the fruitlets, the whole structure ripening into a single multiple fruit called a **syconium.**

RÉSUMÉ OF FRUIT CLASSIFICATION

I. **Simple Fruits.** Those which are developed from a single (simple or compound) pistil, and which consist of a single matured ovary together with any accessory structures closely associated with the ovary.

- *A. Dry Fruits.* Those in which the pericarp and accessory structures become more or less dry when mature.
  - 1. *Dehiscent Fruits.* Those which split open at maturity.
    - *a. Legume.* Developed from a simple pistil; splitting along two sutures into two valves (peas, beans, and black locust).
      - (1) *Loment.* A segmented legume (tick trefoil).
    - *b. Follicle.* Developed from a simple pistil; splitting along one suture (milkweed, columbine, and larkspur).
    - *c. Capsule.* Developed from a compound pistil (poppy, purslane, iris, Saint-John's-wort, and morning glory).
      - (1) *Silique.* The elongated two-loculed capsule of the mustard family (cabbage, cardamine).
      - (2) *Silicle.* A short, broad silique (shepherd's purse, peppergrass).
      - (3) *Pyxis.* A capsule with circumscissile dehiscence (*Portulaca*).
  - 2. *Indehiscent Fruits.* Those which do not split open at maturity.
    - *a. Achene.* Small, one-seeded; seed attached to pericarp at one point only; pericarp readily separable from seed coat (sunflower, lettuce, and buckwheat).
    - *b. Grain,* or *Caryopsis.* Small, one-seeded; seed coat fused to pericarp over its entire surface (corn, wheat, oats, rye, and barley).
    - *c. Samara.* A winged achene (ash, maple, elm, and birch).
    - *d. Schizocarp.* Two (occasionally one) carpels invested by the floral tube; the carpels separate at maturity into two indehiscent *mericarps* (members of carrot family); in mallows

there are several carpels, each becoming a mericarp.

e. Nut. Pericarp hard or crustaceous throughout; usually from a compound pistil only one carpel of which develops; mostly one-seeded, usually with an involucre (chestnut, hazelnut, acorn, and beechnut).

B. Fleshy Fruits. Those in which a part or all of the pericarp and any accessory structures become fleshy at maturity.

1. *Pome.* Developed from a compound pistil with two or more carpels and an inferior ovary; floral tube forming major fleshy part of fruit; outer part of pericarp fleshy, endocarp cartilaginous (apple, pear, and quince).
2. *Drupe,* or *Stone Fruit.* Mostly one-seeded fruits in which the exocarp is usually thin and skinlike, the mesocarp fleshy, and the endocarp stony (peach, plum, cherry, and olive).
3. *Berry.* All parts of the pericarp fleshy or pulpy except the exocarp which is often skinlike (grape, tomato, gooseberry, and cranberry).

a. Pepo. A berry with a thick, inseparable rind (pumpkin, squash, cucumber, and melons).

b. Hesperidium. A berry with a leathery, separable rind (orange, lemon, and grapefruit).

II. **Aggregate Fruits.** Those consisting of a number of similar small fruits (fruitlets), all of which are developed from a single flower with many separate pistils and which mature together as a single unit on a common receptacle (together with any accessory structures), mostly fleshy (strawberry, raspberry, and blackberry).

A. Hip. The fruit of the rose, the fleshy floral tube surrounding the matured ovaries, which are achenes.

III. **Multiple Fruits.** Those consisting of the matured ovaries of an entire flower cluster (together with any accessory structures), all adhering together in a single mass. Mostly fleshy (pineapple, mulberry, Osage orange, and fig).

A. Syconium. The multiple fruit of the fig, in which the upper part of the peduncle becomes fleshy and completely envelops the fruits.

SEEDS

Economic importance of seeds Seeds are so generally found separated from the plants that produce them and they differ so markedly in appearance from vegetative organs that we often fail to associate them with the plants producing them. In many cases the seed or the fruit is the only part of the plant commonly known. When such plants as the pea, the bean, corn, wheat, or rice are mentioned, it is usually the seed or fruit rather than the whole plant that is visualized. This situation has resulted from the fact that since the dawn of history, seeds have ministered greatly to the needs of man (Fig. 11-9). Indeed, the use of the cereals by man antedates all written historical records. In the earliest known records of the civilization of the Tigris and Euphrates Valleys, man was already cultivating wheat for bread. Similarly the cultivation of rice was practiced by the ancient Chinese civilization of the valleys of the Hwang Ho and the Yangtse Kiang and the cultivation of corn by the Mayan civilization of the New World.

It is significant that each of these three prominent civilizations of antiquity should have originated in a region in which one of these three cereals seems to have had its native home or to have been cultivated since prehistoric times. Barley, oats, and rye are also cereals that date back to ancient or prehistoric times. The name cereal itself implies antiquity, since it was given to the grains by the Romans, who derived it from the name of their goddess Ceres, the giver

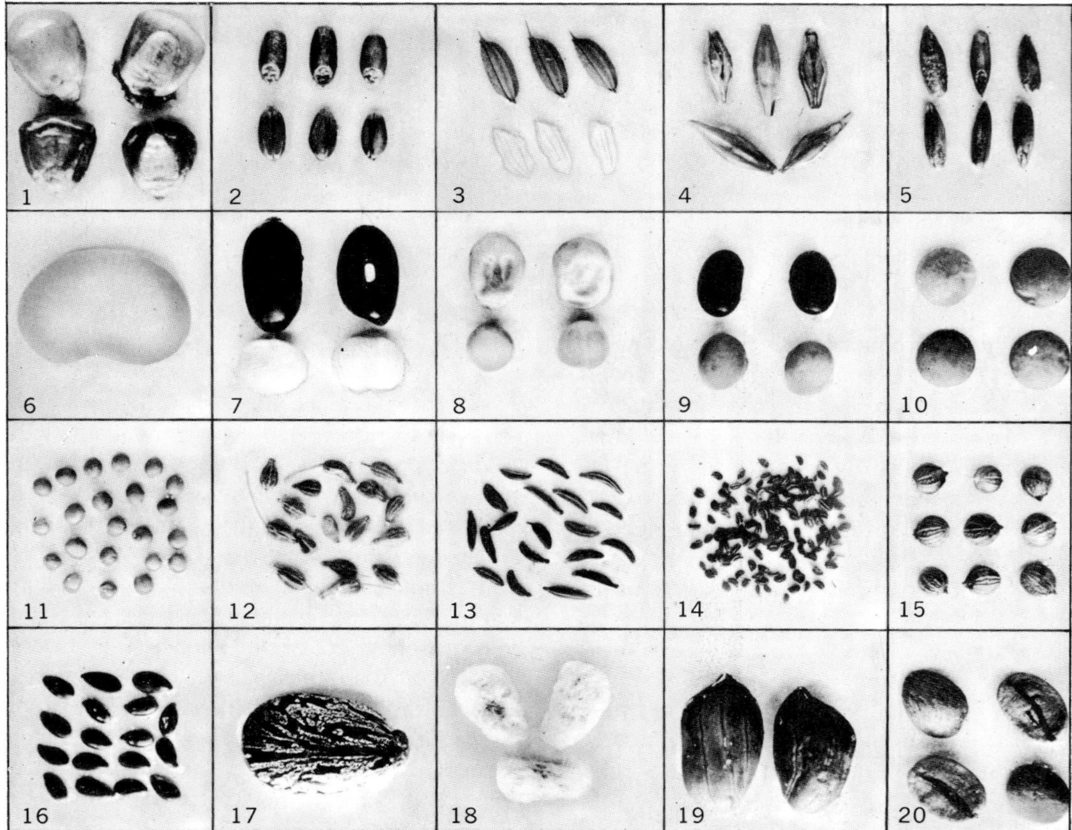

Fig. 11-9 Economically important seeds and seedlike fruits.

Top row, Cereals

1, Corn (Zea mays); *upper two, field corn; lower two, sweet corn; 2, wheat* (Triticum sativum); *3, rice* (Oryza sativa); *4, barley* (Hordeum sativum); *5, rye* (Secale cereale). *(All are seedlike fruits used by man as food since prehistoric times.)*

Second row, Legumes

6, Lima bean (Phaseolus lunatus); *7, beans* (Phaseolus vulgaris); *upper two, string beans; lower two, field beans; 8, peas* (Pisum sativum); *upper two, garden peas; lower two, field peas; 9, soybeans* (Glycine soja); *upper two, variety Peking; lower two, variety Ito-San; 10, lentils* (Lens esculenta).

Third row, Spices

11, Mustard (Brassica alba); *12, anise* (Pimpinella anisum); *13, caraway* (Carum carvi); *14, celery* (Apium graveolens); *15, coriander* (Coriandrum sativum). *(12–15 are seedlike fruits, all belonging to the parsley family, Umbelliferae.)*

Fourth row, Coffee and Seeds That Are Important Sources of Vegetable Oils

16, Flax (Linum usitatissimum); *17, castor bean* (Ricinus communis); *18, cotton* (Gossypium herbaceum); *19, peanuts* (Arachis hypogaea); *20, coffee* (Coffea arabica).

of grain. Even today the cereals occupy one of the most prominent positions in the diet of man.

In addition to the cereals, coconuts, peanuts, almonds, walnuts, filberts, chestnuts, peas, beans, and lentils have been used by man as food since ancient times. While many of these are botanically fruits, it is actually the seed that serves as food.

At the present time not only are seeds the most important sources of food but they are used as drugs and in the making of beverages, medicines, paints, varnishes, clothing, ornaments, and many other commercial products. Oils are extracted from such seeds as the coconut, cottonseed, corn, flax, peanut, castor bean, soybean, and almond. These oils are used in various ways, some as food products, others in making soaps, varnishes, paints, and linoleum. Some seeds, like mustard, caraway, anise, coriander, and celery, are used as condiments. Buttons have been made from vegetable ivory obtained from the seeds of a palm. Cotton, with its many uses, is the fibrous material surrounding the seeds of the cotton plant. In these and in other ways seeds are of great economic importance. Indeed, seeds are used by man more than any other part of the plant.

General features of seeds The seeds of plants differ so greatly from each other in size, form, color, and other general features that the characteristics of the seed may be taken as one of the distinguishing features of a species. Even the varieties within a species can often be differentiated on the basis of their seed characteristics. In size they vary from the dustlike seeds of some of the orchids to the enormous seeds of the coconut. They may be spherical, ovoid, elliptic, elongated, disklike, or very irregular in shape. The outer walls may be smooth or rough and may be sculptured or covered with outgrowths in the form of spines, hooks, or fibers. All parts of the spectrum are represented in their colors. They may be of uniform color or mottled. Although no two species of seeds are exactly alike in all these features, they all resemble each other in that they are ripened ovules consisting of the same general kinds of parts and in that they all serve the functions of dissemination, protection, and reproduction of the species.

GENERAL STRUCTURE OF SEEDS

A mature seed consists of an embryo, surrounded by one or more seed coats and stored food. The food may be stored either in the endosperm tissue or entirely within the embryo.

The seed coats The seed coats are developed from the integuments of the ovule. There are usually two of these coats, corresponding to the two integuments of the ovule, but in some ovules there is only one integument and hence only one seed coat. The outer seed coat is called the **testa.** When two seed coats are present, the inner one is often much thinner than the outer. The seed coats sometimes become very dry and hard, thereby effectively protecting the more delicate embryo within. At the place where the seed breaks off from the funiculus (now often called the seedstalk) there is a scar, called the **hilum,** and below the hilum a small pore, the **micropyle** (Fig. 11-10). It will be recalled that the micropyle is the former point of entrance of the pollen tube into the ovule. In seeds in which the ovule is bent over into a position parallel with that of the seedstalk, there is usually a ridge visible on the seed where the tissues of the funiculus continued on into the ovule. This ridge, now directly above the hilum, is called the **raphe.** The upper end of the seed, where the vascular tissues spread out into the ovule, at the junction of the nucellus, integuments, and funiculus, is called the **chalaza.**

The embryo Of the internal parts of the seed, the embryo is the most representative as well as the most important. It is the young plant developed from the fertilized egg and, in the mature resting seed, is always in a dormant condition. The degree to which the embryo is developed in the seed varies in different plants. Sometimes the seeds are shed from the plant before the embryo has been differentiated into well-defined parts. In such cases further development can occur only after the seeds start to germinate. In its most advanced stage in a seed, the embryo consists of four distinct parts, viz., the **plumule,** the **cotyledons,** the **hypocotyl,** and the **radicle** (Fig. 11-10).

It is convenient to consider that the embryo, in its simplest condition, consists of an axis bearing one or more **cotyledons** (seed leaves). The tip of the axis (both root tip and shoot tip) is a meristem with considerable potential for future growth. A developed embryo shows some elaboration. That part of the axis bearing cotyledons is the **hypocotyl-root axis.** The **hypocotyl** is the part of the axis below the attachment of the cotyledons. The embryonic root at the basal tip of the hypocotyl is referred to as the **radicle.** The embryonic shoot (epicotyl), above the point of cotyledon attachment, is made up of unelongated internodes and one or more leaf primordia; it is called the **plumule.**

The cotyledons are leaves and bear the same relationship to the embryonic shoot (plumule) as other vegetative leaves bear to vegetative stem tips. Many cotyledons are specialized in shape and form and differ markedly from the vegetative leaves of the plant. In some cases they are food-storage organs, in other cases they are absorbing (haustorial) organs, and in still other instances they may serve as photosynthetic structures following seed germination.

Dicotyledonous plants are characterized by the presence of two cotyledons, and monocotyledonous plants by the occurrence of a single cotyledon. In spite of the fact that in some monocotyledons the large single cotyledon appears to be terminal on the axis, it is in fact lateral. The embryonic shoot meristem (epicotyl) represents the morphological tip of the axis and is set in a depression of the cotyledon. Gymnosperms have a variable number of cotyledons.

The stored food The stored food of the seed may be in the form of **endosperm** or may be entirely in the embryo, chiefly in the cotyledons. If, at the time of embryo development, some nucellar tissue remains intact and provides food for the germinating embryo, it is referred to as **perisperm.** When endosperm is lacking in seeds, it indicates that the embryo developed to such a stage as to absorb the endosperm completely before the seeds ripened. This condition is found in practically all legumes. In all the cereals and other grasses the food is stored largely in the endosperm tissue and is not used by the embryo until germination begins.

The foods stored in seeds consist of *carbohydrates, fats,* and *proteins,* the percentage of each varying greatly in different species.

The *carbohydrates* may occur as starch, hemicelluloses, or sugars, with starch predominating. Starch is always found in definite plastids, the starch grains, which have characteristic forms peculiar to a given species. In fact, the source of any particular commercial starch can be determined by the shape and form of the starch grains. Starch occurs in the cotyledons of legume seeds and in the endosperm of cereals. The cereals are especially rich in starch. Corn starch and wheat starch are familiar examples. Hemicelluloses are found in the walls of endosperm cells of such seeds as dates, coffee, and onions, making these walls very tough and hard. While hemicelluloses are used like true cellulose for strengthening cell walls, unlike cellulose, they may be digested to simpler compounds and hence may be used as reserve foods. Sugars are stored in the seeds of sweet corn and peas and in many nuts. The most common sugar found in seeds is sucrose.

Fats are found more abundantly in seeds than in any other part of the plant. Vegetable fats are obtained almost exclusively from seeds rich in stored fat. The fat usually occurs in the form of oil globules. As mentioned in a previous chapter, fats are more efficient storage foods than are carbohydrates, yielding 2¼ times as much energy per unit weight as do the carbohydrates. In the cereals most of the fat occurs in the embryo. In other plants it may be found also in the endosperm. The seeds of sunflower, flax, peanuts, soybeans, and the castor-oil plant are very rich in fats.

Proteins occur in all seeds, since they are an essential constituent of all living protoplasm, but some seeds, like the legumes, are richer in proteins than others. In the cereals the storage proteins occur in small granules called *aleurone grains,* which are found in a single layer of cells (the aleurone layer) comprising the outermost portion of the endosperm. Since proteins are important foods in the building of protoplasm, they are essential for the embryo during the germination of the seed.

SPECIFIC STRUCTURE OF REPRESENTATIVE TYPES OF SEEDS

Common bean The common bean (Fig. 11-10) may be taken as a typical dicotyledonous seed lacking endosperm. Such seeds, like those of other legumes, are formed within the pod, which is the ripened ovary. Each one is attached along the ventral suture of the pod by the funiculus or seedstalk. When the seeds are shed, the funiculus breaks off, leaving a prominent scar, the hilum. Just below the hilum can be seen the micropyle, and above the hilum is the ridge formed by the raphe. The seed coats have characteristic colors which vary with different varieties of beans but are commonly variations of brown, black, and white.

When the seeds are soaked in water, they swell considerably, and the seed coats become soft. In this condition the seed coats are easily removed. The entire interior of the seed is occupied by the embryo and chiefly by the two fleshy cotyledons or seed leaves, which may easily be separated. On the side of the seed opposite the raphe is found the radicle, with its tip directed toward the micropyle, and continuous with it is the hypocotyl. The plumule has differentiated two well-defined leaves which fold over the growing tip. These become the first true leaves of the bean plant on germination.

In this seed and in all seeds of this type, there is no endosperm, this tissue having already been consumed by the developing embryo. Most of the food of the seed is stored in the two large cotyledons, which in this case never function as true leaves.

Castor bean Castor beans (Fig. 11-11) (really not beans in the ordinary sense) are produced in a capsule consisting of three carpels, each one of which bears a single seed attached by a slender stalk to the center of the base of the

Fig. 11-10 Seed structure of common bean. (Drawn by Elsie M. McDougle.)

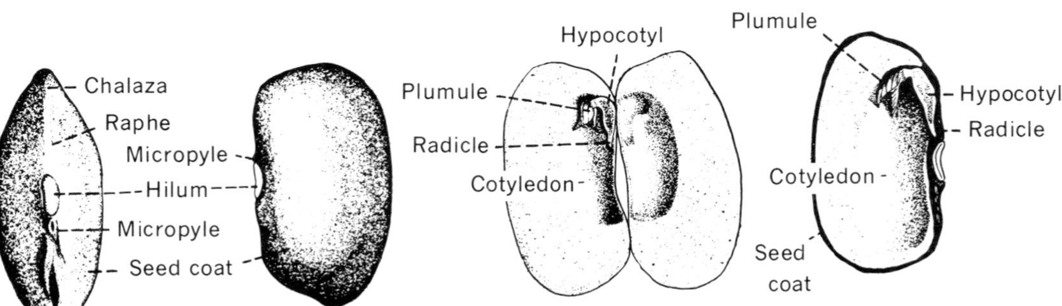

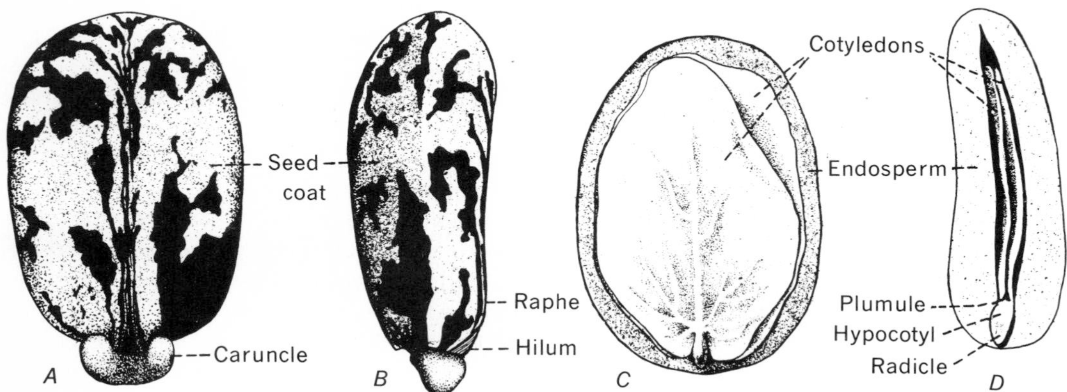

Fig. 11-11 Structure of castor-bean seed. A, view of flat surface; B, edgewise view; C, the seed split vertically after removal of seed coats, showing the two thin cotyledons lying against the surrounding endosperm; D, the interior of the seed cut longitudinally along its narrow axis. The micropyle is concealed by the caruncle. (Drawn by Elsie M. McDougle.)

capsule. At maturity the capsule bursts open, forcibly ejecting the seeds. At the base of the seed is a more or less soft, spongy structure, called the **caruncle,** developed from the outer integument and not commonly found on seeds. The caruncle absorbs water readily and may therefore be of some use in promoting germination. It usually obscures the micropyle. The hilum and the raphe are easily discernible. The seed coats are very hard and brittle and are often brightly colored and mottled, giving the seed some resemblance to a beetle, from which the plant received the name of *Ricinus,* which is a Latin term for a kind of insect or mite.

When the seed coats are broken away and the seed sectioned lengthwise, the interior is found to consist chiefly of a white, oily endosperm with the embryo in the center (Fig. 11-11*C, D*). The embryo consists of two thin leaflike cotyledons, an entirely undifferentiated plumule, and a very short hypocotyl and radicle. This structure is typical of dicotyledonous seeds having endosperm. The endosperm of the castor bean may contain from 40 to 50 percent of oil and 15 to 20 percent of protein in the form of aleurone grains. It also contains a poisonous substance, ricin. The castor oil of commerce is extracted from these seeds.

Corn The familiar corn grain (Fig. 11-12) is actually a fruit, since it consists of an entire ripened ovary containing a single seed. The

Fig. 11-12 The corn grain. A, view of the embryo side of grain; B, longitudinal section cut at right angles to the face of the grain. (Drawn by Elsie M. McDougle.)

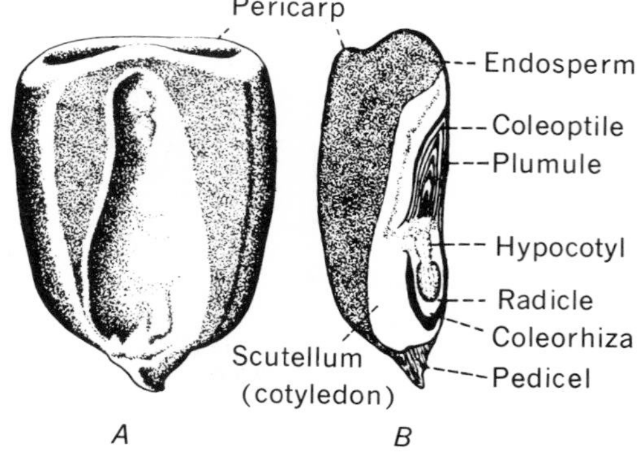

outer hull is the pericarp, or ovary wall, fused to the seed coat. Originally there are two seed coats; but as the grain matures, the outer one disappears and the inner integument becomes flattened through pressure from within the seed. At maturity the pericarp is so firmly fused to the remaining inner seed coat that the whole appears as a single tissue. Only by examining sections under a microscope can the actual structure be ascertained. Part of the nucellar tissue usually remains immediately inside the seed coat. The stalklike structure at the bottom of the grain is the pedicel of the spikelet and not the funiculus.

The grain is usually concave on the side of the embryo. A lengthwise section of the grain through the center and at right angles to the broad axis (Fig. 11-12*B*) reveals the embryo embedded in a large mass of endosperm. The outermost part of the endosperm lying next to the nucellus consists of a single layer of cells, called the **aleurone layer,** filled with aleurone grains, which are mostly protein. The remaining endosperm is often made up of two well-defined regions, an outer **horny endosperm** and an inner **starchy endosperm.** The relative positions and the amounts of these two types of endosperm vary in different varieties of corn. The horny endosperm is of a tougher consistency and contains more protein than does the starchy endosperm.

The embryo consists of a single cotyledon, called in grains the **scutellum,** a well-developed plumule, a very short hypocotyl, and a radicle. The scutellum or cotyledon is a broad, flat absorbing organ lying against the endosperm. It never emerges from the seed, but absorbs food from the endosperm and transfers it to the growing parts of the embryo during germination. The plumule, consisting of the growing stem tip and one or more leaf primordia, is completely covered over by a sheath called the **coleoptile.** Similarly the radicle is enclosed by a sheath called the **coleorhiza.** These sheaths are characteristic of all the cereals and other grasses. The hypocotyl is extremely short and does not elongate even during germination. The stem of the seedling develops from the plumule. In general the structure of the corn grain is typical of monocotyledonous seeds containing endosperm.

DISSEMINATION OF SEEDS AND FRUITS

The higher plants are not able to move about but remain fixed in the place in which they grow. The fruits and seeds of these plants furnish almost the only means by which they can be spread from one place to another. The fruit aids in the dissemination of the seed, although the seed itself may, at maturity, possess structures that facilitate dispersal. The actual dispersal of the seeds is brought about in a great variety of ways but chiefly through the agencies of wind, water, and animals.

The wind is probably the most important agency in seed dissemination in nature. Many species of trees, like the maple, elm, birch, ash, and ailanthus, have winged fruits (Fig. 11-13 *D–F*); others, like the trumpet creeper, catalpa, and many conifers, have winged seeds (Fig. 11-13*G, H, O*) that are carried considerable distances by the wind. The hop hornbeam, the bladdernut (*Staphylea*), and other plants produce their seeds in bladderlike fruits (Fig. 11-13*P*), while the black locust (Fig. 11-13*A*) and honey locust have thin, dry elongated pods. Both of these types of fruits are easily blown about by the wind. Sometimes the seeds themselves, like those of many orchids, are so tiny and light as to be blown through the air like dust. In the dandelion, goldenrod, wild lettuce, aster, and other plants, tufts of hairs are found on the small, single-seeded fruits (Fig. 11-13*M, U*) which act like parachutes that permit the wind to carry them great distances.

It is interesting to note that some of the worst weeds and most widely spread plants have seeds or fruits that are distributed by this

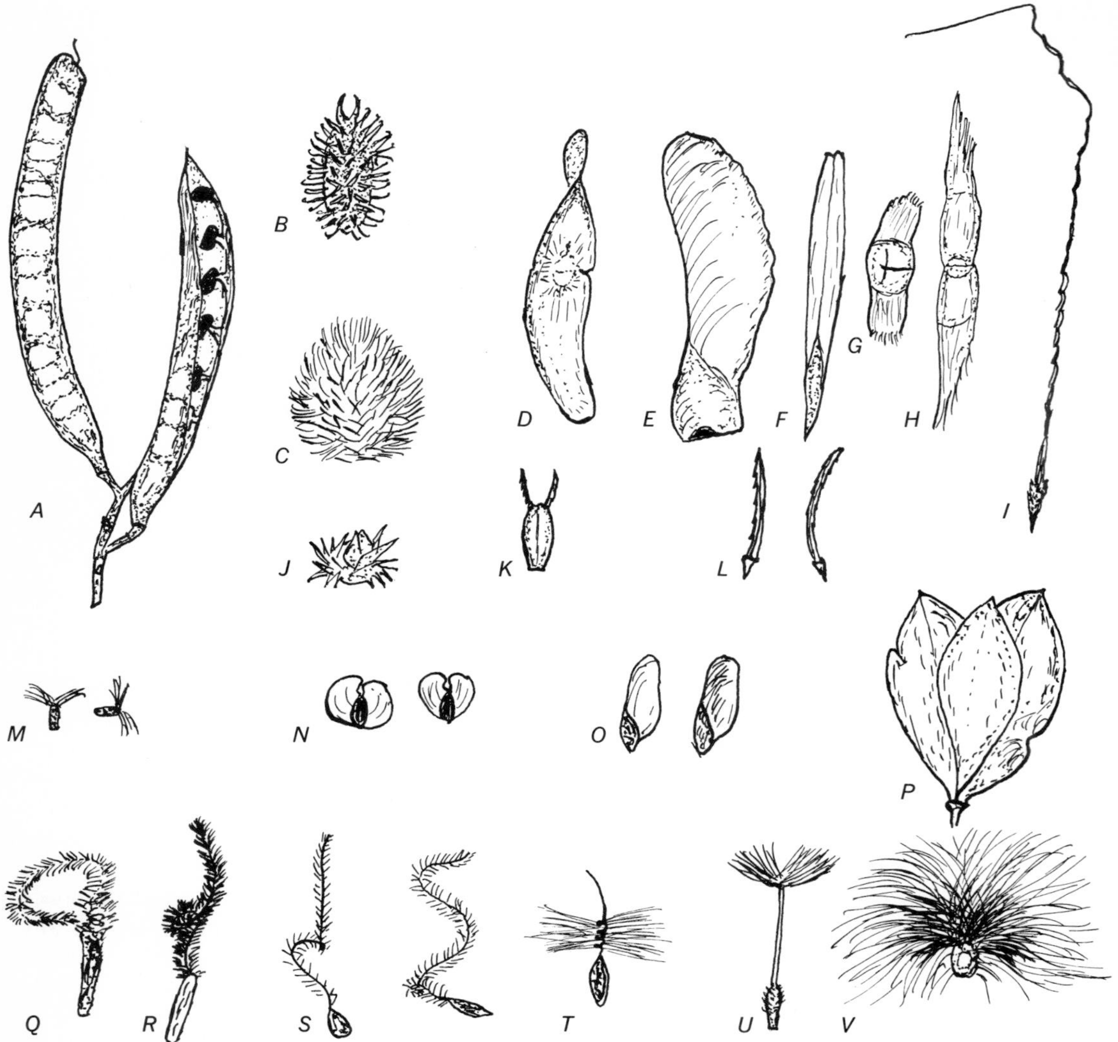

Fig. 11-13 Seed and fruit dispersal. A, pods of black locust; B, C, spiny fruits of cocklebur and burdock, respectively; D–F, winged fruits of tree of heaven, maple, and ash, respectively; G, H, winged seeds of trumpet creeper and catalpa; I, fruit of the grass, Stipa, *with twisted awn; J, prickly fruit of the sand bur; K, fruit of beggar-ticks with sharp barbs; L, fruits of sweet cicely with minute barbs; M, parachutelike fruit of goldenrod; N, winged fruit of birch; O, winged seeds of hemlock; P, bladderlike fruit of bladdernut; Q, R, plumose fruits of* Erodium; *S, T, plumed fruits of* Clematis *and* Geranium, *respectively; U, parachutelike fruit of dandelion; V, plumed seed of milkweed.*

method. In *Clematis, Geranium,* and *Erodium* (Fig. 11-13*Q–T*) the style of the flower persists on the fruit and becomes plumelike, functioning like the tufts of hairs on the fruits just mentioned. In the milkweed (Fig. 11-13*V*) and in cotton the tufts of hairs are outgrowths from the seed itself. In the so-called tumbleweeds, like some of the pigweeds, some grasses, and false indigo, the whole plant breaks off at the surface of the soil and, forced by the wind, rolls over the ground, distributing the seeds as it goes. The poppy, evening primrose, toadflax, and many other plants produce tiny seeds in capsules that open at the apex. When these capsules are shaken by the wind, the seeds are scattered in all directions.

Many fruits and seeds that are carried by the wind fall into streams in which they may be transported farther. This is especially true of seeds and fruits of plants growing on the banks of streams or near them. The light fruits of ragweed and the fruits of many sedges are probably carried this way. The seeds of some water lilies have buoyant coverings that enable them to float. The distribution of the coconut along the shores of tropical seas has been thought to be at least partly attributable to the buoyant, saltwater-resisting outer husk of the fruit.

Another less common method by which plants disperse seeds is by means of explosive fruits. In some of the vetches and other legumes, as the pod ripens, unequal forces and strains are set up in the tissues which finally cause the pod to burst open forcibly and scatter the seeds in all directions. After this has happened, the two valves of the pod remain curled up or twisted. The capsules of witch hazel and of the castor-oil plant forcibly eject their seeds in a somewhat similar fashion. In the squirting cucumber and in touch-me-not the seeds are ejected while the fruit is still soft and succulent. In this case osmotic or turgor forces probably cause the explosion of the fruit.

Animals are instrumental in disseminating seeds in a variety of ways. In some cases, as in many fleshy fruits, the seeds may be passed through the digestive tract uninjured when the fruit is eaten. Birds distribute the seeds of many berries and other fleshy fruits in this manner. Sometimes only the fleshy part of the fruit is eaten and the seeds discarded in places where they give rise to new plants. Squirrels and other animals are instrumental in disseminating such species as dogwood, hickory nuts, walnuts, and oaks by collecting the fruits and hiding them in various places. Many species of plants produce fruits or seeds that adhere to the fur or hair of animals or the clothing of man and are distributed in this way. In the so-called beggar's lice, in Spanish needles, in cocklebur, and in burdock, this is brought about by the hooks or barbs of the fruit (Fig. 11-13*B, C, J–L*). The grains of many of the grasses have long awns (Fig. 11-13*I*) and other structures that cause them to adhere to the wool of sheep and to other animals. The seeds of mistletoe are sticky and adhere to the feet of birds, by which they are transported from one tree to another.

Many other examples of this type of dispersal might be given. It should perhaps be mentioned that the widest dissemination of useful species of plants has been brought about by man himself through his agricultural and industrial operations. At the same time he has also been instrumental in introducing many bad weeds that were present as impurities in crop seed and in other agricultural products.

VIABILITY, LONGEVITY, AND DORMANCY OF SEEDS

Viability and longevity of seeds As previously stated, the seed contains a dormant embryo. Under proper conditions, this embryo is able to germinate, i.e., to continue growth and develop into a new plant. So long as these conditions are not met, the seed may remain dormant or, if the conditions are adverse, the embryo may die. By the **vitality,** or **viability,** of the seed is meant its capacity to renew growth or germinate. By **longevity** is meant the length of time the seed can remain dormant and still be viable. Both of these are variable factors in any seed, since they are governed not only by the inherent characteristics of the plant but also by the conditions under which the seeds were developed on the plant, as well as by the conditions to which the seeds are subjected after they are shed from the parent plant.

Immature seeds, as well as those produced on weak, spindling plants, are often deficient in stored food reserves and in other ways may be weakened in such a way as to reduce their vitality. When planted, such seeds usually give rise to weak plants. The temperature and other conditions under which the seeds developed on the plant also affect their vitality since they affect the physiological conditions that accompany seed development within the ovary.

Many seeds retain their viability best under relatively dry storage conditions and under a medium to low temperature. The amount of water in the seed is closely correlated with the ability of the seed to withstand extreme temperatures. When seeds are very dry, they can be subjected to extremely low or to relatively high temperatures without destroying their viability. In fact, the range of temperatures that dry seeds will withstand without injury is greater than that of any other part of the plant. Dry sugar-beet seeds, for example, have been known to give 96 percent germination after having been subjected for 30 min to a temperature of −180°C. At the other extreme of temperature, lotus seeds have been reported to still germinate after having been subjected for 16 hr to a temperature of +103°C. Most seeds, when dry, are not injured by temperatures as low as −10 to −20°C. It is partly for this reason that they are able to withstand unfavorable conditions that would probably kill vegetative organs.

It has been found that crop seeds maintain their viability best when stored under uniform conditions and preferably at a comparatively low temperature and low humidity. On the other hand, the seeds of willows, poplars, and some maples are killed if allowed to dry out. The seeds of many weeds are able to lie in the ground and be subjected to all sorts of adverse conditions, sometimes for years, without losing their viability.

Aside from the effect of external conditions, great variation exists among different species of plants as regards the longevity of their seeds. In some cases the seeds remain viable only a short time, no matter what the external conditions may be. In other cases seeds retain their viability for many years. All lose their viability in time. The longest recorded case of longevity of seeds is that of Indian lotus seeds which were able to germinate after having been buried in a peat bed for probably more than 200 years. The longevity of most seeds is decidedly less than this, although some legumes have been known to retain their viability for 50 years or more. Seeds of black mustard, the common pigweed, shepherd's purse, chickweed, and other weeds have been found to germinate after being buried more than 30 years in the soil. Onion seed, on the other hand, loses its viability rapidly after a single year and many seeds of garden vegetables remain viable for only 3 to 5 years even under the best conditions of storage. Willow and poplar seeds remain viable only a short time after they are formed on the plant. Unless they germinate soon after falling to the ground, they die.

The cause of the ultimate death of seeds has not yet been adequately determined, although in some cases it is thought to be the gradual denaturing of the proteins and possibly other constituents of the protoplasm of the embryo. Other chemical changes may also contribute. That it is not the gradual consumption of the stored food reserves, resulting from the continuous though feeble respiration of the seed, is proved by the fact that large food reserves often remain after the seeds have lost their viability.

Dormancy of seeds The seeds of some plants are able to germinate as soon as they are matured on the plant. It is not uncommon, in a very rainy season, to find sweet corn germinating in an ear that has fallen to the ground or peas and. beans sprouting in the pods in which they are formed. Wheat germinates readily in the shock in rainy weather. In the majority of species, however, and especially in many wild plants, the seeds require a distinct rest period during which they fail to germinate even when supplied with the best of ordinary germinative conditions. This period of dormancy may last only until the following spring or may extend over a period of several years. The delayed germination of such seeds may be caused by the nature of the seed coats, by the condition of the embryo, or by a combination of both of these factors. We may refer to the first of these types as **seed-coat dormancy** and to the second as **embryo dormancy.**

SEED-COAT DORMANCY When the delayed germination is caused by the nature of the seed coats, it may be brought about by (1) **the impermeability of the seed coat to water,** (2) **the impermeability of the seed coat to oxygen,** or (3) **the mechanical resistance of the seed coat to the expansion of the embryo and seed contents.** In all of these cases germination results when the seed coats are filed, rubbed over abrasive materials, soaked in strong sulfuric acid, broken in any other manner that will not injure the embryo, or removed altogether.

The seeds of many of the legumes, like the Kentucky coffee tree, black locust, honey locust, Judas tree, and hard seeds of clovers and alfalfa belong to the first category. It has been found that when seeds of clovers and alfalfa are hulled by hand, 60 to 70 percent of the seeds are hard (i.e., dormant) whereas less than 20 percent are hard when they are hulled by machinery. The machine in this case injures the seed coats sufficiently to make them permeable to water. The seeds of canna and of many other plant species are also impermeable to water.

The best-known example of a seed the coats of which are impermeable to oxygen is furnished by the common cocklebur (*Xanthium*). The spiny fruit of this plant contains two seeds, each of which has a minimum oxygen requirement for germination, the upper seed requiring more oxygen than the lower. If the seed coats are removed or if the seeds are subjected to high temperatures or to pure oxygen, they usually germinate readily. In either case sufficient oxygen is able to reach the embryos, whereas when the seed coats are intact it is not. In nature the lower seed, having a lower oxygen requirement, usually germinates the year after it is formed, and the upper seed the following year, germination being facilitated by frost, heat, or other conditions that gradually modify the seed coats. This type of dormancy is found also in other composites and in some grasses. It is possible that the failure of carbon dioxide to escape from the seed may also contribute to the dormancy of such seeds.

The seed coats of the seeds of water plantain (*Alisma plantago*), of pigweed (*Amaranthus retroflexus*), and probably of many other plants permit both oxygen and water to enter the seed, but the coats are so strong that the expanding contents are often unable to break through them and hence the seeds fail to germinate. In this case germination takes place readily when the seed coats are broken or weakened in some manner that will not injure the embryo.

EMBRYO DORMANCY Embryo dormancy may be caused by a very rudimentary or undeveloped embryo or by the failure of the embryo to awaken or emerge from its resting condition.

In the chapter on the flower it was mentioned that some seeds ripen and are shed from the plant before the embryo has developed much further than the proembryo stage. In some cases there is little development beyond the fertilized egg. When this is true, the seeds are always delayed in their germination because of the time involved during which the embryo is developing after germinative conditions have been provided. It may take several weeks or even months before the embryo is able to emerge from the seed. This type of dormancy is rather widespread through the various groups of seed plants, being found in gymnosperms (*Ginkgo* and *Gnetum*), monocotyledons (dogtooth violet, *Erythronium denscanis;* and other Liliaceae), and dicotyledons (winter aconite, *Eranthis hyemalis;* buttercup, *Ranunculus Ficaria;* and others).

Some of the most interesting cases of dormancy, as well as the most difficult to overcome, are those in which the embryos are persistently dormant. Such seeds fail to germinate even when the seed coats are removed and the seeds placed under the best ordinary germinative conditions. Germination takes place only after a series of changes, usually called afterripening, have taken place in the embryo. To this class belong many trees, shrubs, and other wild plants as well as many cultivated species. Among the forest species may be mentioned linden, tulip poplar, ash, pines, hemlock, and other conifers, dogwood, hawthorn, and viburnums. The seeds of such fruits as apples, pears, peaches, plums, and cherries and of flowering plants like holly, iris, lily of the valley, Solomon's seal, and hundreds of other species also have embryos that must be afterripened before they will germinate. In nature afterripening sometimes occurs during the winter, allowing the seeds to germinate in spring. Often the process continues over a period of years, some germination taking place each year. This is a very efficient method of perpetuating a species and tiding it over unfavorable growth conditions.

The changes that take place during the afterripening of seeds have not been adequately worked out with many species. In the hawthorn the acidity of the cell sap of the cotyledons gradually increases as the seeds become capable of germination. This increased acidity apparently hastens the physiological processes of the embryo which are necessary for further growth.

It has been found that many seeds of this type can be made to germinate if stratified at low temperatures. A temperature range of 0 to 10°C has been found the most effective for the majority of the species studied. By stratification is meant the placing of the seeds between layers of sand, sawdust, peat, or other material and keeping them moist and at a low temperature. Peat has been found to be much more effective for this purpose than any other material. Afterripening also proceeds better when the seeds are scattered through or mixed with the peat. The period of time required for afterripening of seeds during stratification varies between one and six months, depending upon the species. This method of overcoming the dormancy of embryo-dormant seeds is of considerable practical importance to nurserymen in the propagation of species with seeds of this type.

SEED GERMINATION

Conditions necessary for germination The resumption of the growth of the embryo after it has been dormant in the seed is called seed germination. Even with seeds that are not dormant in the sense described in the foregoing section, certain conditions must prevail before this renewed growth will take place. Most prominent among these are a **supply of water,** a **supply of oxygen,** and a **favorable temperature. Light** also is necessary or at least influential in hastening the germination of some seeds while darkness is necessary for others.

WATER SUPPLY Mature seeds usually contain 15 percent or less water. This low water content is one of the most influential factors keeping the seeds dormant. The protoplasm and its constituents as well as the stored food are in so concentrated a form in the resting seed as to reduce markedly the intensity of physiological activity. Before growth of the embryo can proceed, therefore, water must be absorbed. It is not necessary, however, for the seeds to be immersed in water. In fact, it is far better for them to be rather in a moist atmosphere, since covering them with water limits the oxygen supply to the seeds. Some seeds have a remarkable ability to absorb water. Often they can obtain almost as much water from an apparently dry soil as they can from pure water, an air-dry soil furnishing enough water for germination. It is this moisture-absorbing capacity of seeds that sometimes causes them to start to germinate when they are stored in a humid atmosphere. In many cases the seed coats consist of hydrophilic colloids. The absorption of water by such coats causes them to swell and weaken, thus permitting the expanding embryo to break through them. Even in seeds with coats that do not absorb water readily or swell, the expansion of the interior of the seed through water absorption is usually great enough to cause the seed coats to burst.

The stored food reserves are most often in an insoluble form. Before they can be used by the embryo they must be rendered soluble and diffusible. The process by which this is brought about is **digestion.** This process is described in the following chapter, but it may be said here that it can proceed only when there is an abundant supply of water present. The digestion of the foods is also necessary to provide the materials for the increased **respiration** of the seed which always occurs during germination. Thus a supply of water is of first importance to the seed in germination. In fact, the swelling of the seed resulting from water absorption is the first indication that germination is taking place.

OXYGEN SUPPLY Even in its most quiescent state, a seed carries on respiration. The respiration in the dormant seed, however, is usually so feeble as to be hardly detectable. Respiration is the oxidation of food as an immediate source of energy to maintain life. When the embryo resumes its growth, much more energy is needed and hence respiration increases. For this increased respiration a supply of oxygen is necessary. In many seeds the seed coats must imbibe water before they permit oxygen to diffuse through them readily, but as previously mentioned this does not mean that the seeds must be immersed in water. A few kinds of seeds are known to be able to germinate in the total absence of oxygen, but few seeds are able to complete the germination process under these conditions.

FAVORABLE TEMPERATURE The temperature that is required by seeds for germination varies with different kinds of seed. Each kind has a range of temperatures within which the seeds will germinate. The lowest temperature of this range is called the minimum, the best temperature the optimum, and the highest temperature the maximum. These three temperatures are called the cardinal points of temperature. The cardinal points of temperature for germination are not absolutely fixed for a species but fluctuate with different samples of the same species because of the great differences resulting from conditions under which the seeds were formed on the plant or stored. Most seeds fail to germinate at a temperature as low as 0 to 5°C or as high as 45 to 48°C. The optimum for many seeds is between 25 and 30°C. There is so much difference in different kinds of seeds, however, that these figures have a very limited application. The best temperature for germination must be determined for each kind of seed. The effect of temperature is primarily on the physiological processes going on within the seed during germination.

LIGHT Light is probably of greater importance in seed germination than has heretofore been supposed. The German investigator Kinzel, who studied for many years the influence of light on seed germination, found that, of 965 species of plants studied, the seeds of 672 were favored by light, 258 were inhibited, and 35 were indifferent. A wide range of plants was included in this study. It is interesting to note that a large majority of the seeds (70 percent) was favored by light. Among these are many grasses, evening primrose (*Oenothera biennis*), yellow dock (*Rumex crispus*), common mullein (*Verbascum Thapsus*), loosestrife (*Lythrum salicaria*), and the Chinese lantern plant (*Physalis Franchetti*). The mistletoe (*Viscum album*) and a few others will not germinate at all in the absence of light.

Light interferes with the germination of seeds of several species of *Phacelia*, chive garlic (*Allium schoenoprasum*) and other Liliaceae, fennel flower (*Nigella sativa*), several members of the pigweed family, and the Jimsonweed (*Datura Stramonium*). The small grains, corn, and many legumes such as beans and clover germinate as well in light as in darkness.

Fig. 11-14 Stages in germination of bean seed and establishment of seedling. (Drawn by Elsie M. McDougle.)

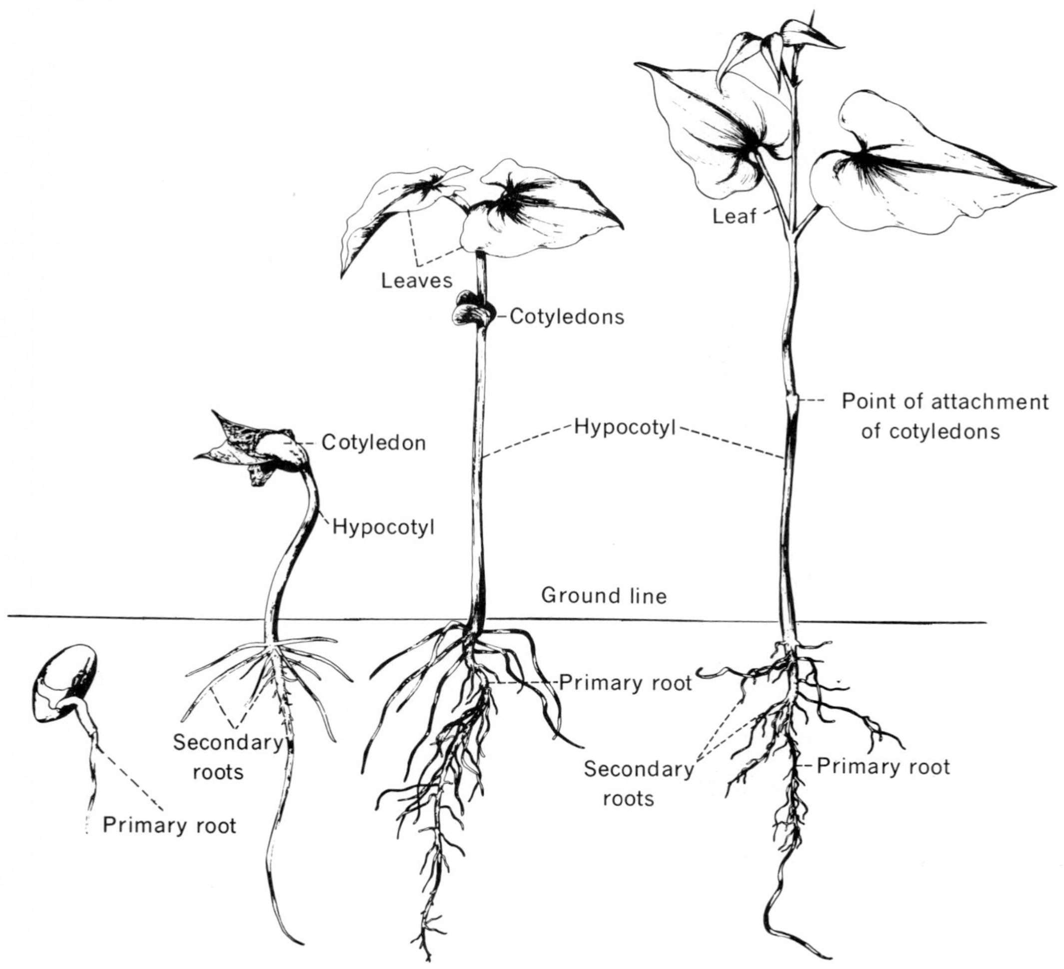

It has been found that certain conditions may partly or entirely displace the effect of light in light-sensitive seeds. These vary with different species, but among them may be mentioned removal or injury of the seed coats, keeping the seeds in an atmosphere of pure oxygen, treatment of the seeds with acids and other chemicals, especially nitrogen compounds, intermittent temperatures or high constant temperatures, and afterripening in dry storage. Some of these are effective with one type of seed and some with another.

Activities accompanying germination When all of the necessary conditions for germination have been supplied, the seed may start to sprout. To supply these conditions, we usually place the seeds in moist soil or in a suitable germinator. The first visible indication of germination is usually the swelling of the seeds and often the softening of the seed coats through absorption of water. In seeds with hard seed coats, the size of the seed may not change until the swelling of the interior of the seed bursts the seed coats.

After the seeds have imbibed water, greatly increased physiological activity develops within them. By means of enzymes, the stored foods are digested to soluble and diffusible substances and thus put into a form that can be assimilated by the embryo. Respiration increases markedly, as can be shown by the heat generated and by the elimination of large quantities of carbon dioxide by the seed. The embryo itself increases greatly in size through absorption of water, and the protoplasm of its living cells becomes more dilute and therefore capable of renewed physiological activity. Through the transfer of the digested foods to the growing points of the radicle and the plumule, these organs begin to grow. The radicle usually starts its growth first and is the first part of the embryo to emerge from the seed. In some seeds, like the bean, the hypocotyl as a whole greatly elongates, carrying the cotyledons and the plumule with it out of the seed, after the radicle has emerged (Fig. 11-14). In others, like the pea, the hypocotyl does not elongate but the plumule itself elongates, leaving the cotyledons, still surrounded by the seed coats, in the soil (Fig. 11-15). In seeds containing endosperm, the stored foods are digested and transferred to the cotyledons of the embryo and from them to the radicle and the plumule. In the grasses the single cotyledon remains attached to the endosperm as an absorbing organ, and only the plumule and radicle elongate. In this instance the cotyledon remains permanently beneath the soil.

Much of the increase in size of the embryo results from expansion of cells already present. The whole embryo of the seed consists of undifferentiated or only partially differentiated cells. Complete differentiation proceeds with germination. At the growing tips of the radicle and the plumule, cell division takes place. The growth of the embryo therefore results from both cell division and cell enlargement. As a result of this activity, the radicle becomes a functioning root, the plumule a functioning vegetative shoot producing stem and leaves, and the whole embryo develops into a seedling. A seed is said to have germinated when the radicle has emerged from the seed coat, but germination is usually considered complete only when the seedling has become an independent plant. The time required for the completion of germination varies greatly with different species. Seeds like mustard or radish will often begin to germinate within a day and complete the process in a week or less. Other seeds may require a much longer period. In the case of seeds with very rudimentary embryos many weeks may elapse before the embryo emerges from the seed coats.

SEEDLINGS

As previously stated, the seedling is the young plant that has just emerged from the seed during germination, having developed from the renewed growth of the embryo. The seedling stage of the plant lasts from the time the embryo emerges from the seed until it becomes independent of the stored food reserves of the seed. Seedlings are of several types, depending upon the manner in which the cotyledons function during germination. Figures 11-14 to 11-17 illustrate the different types.

In the common bean (Fig. 11-14) the food is stored largely in the cotyledons. After digestion it is translocated to the hypocotyl, the radicle, and the plumule. The radicle emerges first, after which the hypocotyl elongates, carrying upward with it the two cotyledons and the plumule. Under the stimulus of gravity, the radicle bends downward into the soil, while the hypocotyl elongates in the opposite direction. Many root hairs are formed on the radicle, and secondary roots begin to develop almost immediately. Together they tend to hold the seedling in position. At first there is a distinct bend in the hypocotyl, but as germination proceeds, it straightens out and the two cotyledons spread apart so as to occupy opposite positions on the hypocotyl. During this time the plumule develops rapidly and differentiates the first two functioning foliage leaves. The cotyledons often develop chlorophyll and thus carry on photosynthesis for a while, but they never assume the shape and form of true leaves. The food that is stored in them is gradually digested and transferred to the growing parts of the young plant where it is consumed. Finally they shrivel up and fall off. By this time the plumule has become an actively growing vegetative shoot, the hypocotyl a functioning stem, and the radicle a root with many branches. In this stage the seedling is an independent plant.

Fig. 11-15 Stages in germination of pea seed and establishment of seedling. (Drawn by Elsie M. McDougle.)

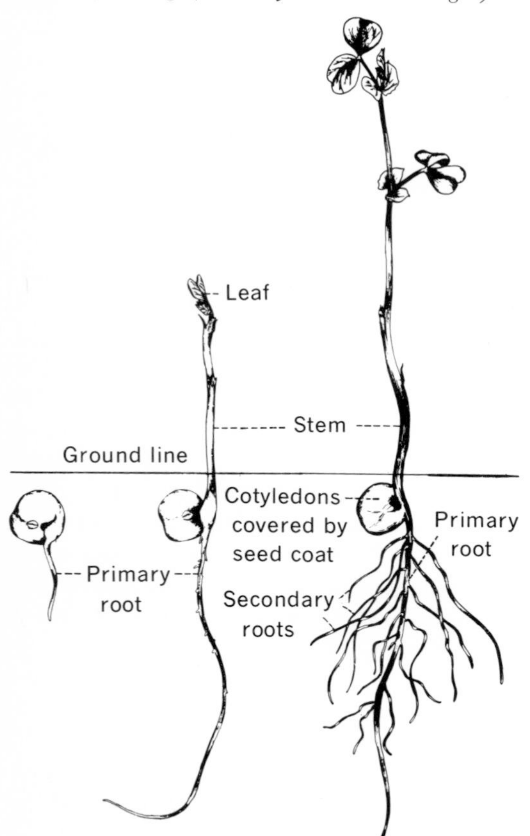

In the bean, as in practically all other dicotyledonous plants, the first two foliage leaves formed are entirely different in shape and form from those later formed. As in the case of other plants having compound leaves when mature, the first two leaves of the bean are simple.

In the pea (Fig. 11-15) the hypocotyl never elongates and therefore the cotyledons remain underground. The radicle emerges first, as in the bean, but the plumule develops much more rapidly, giving rise to the first true stem of the plant as well as to the leaves. The cotyledons finally disintegrate as the food in them is consumed by the developing seedling.

The germination of the castor bean (Fig. 11-16) is somewhat similar to that of the bean except that the cotyledons are relatively thin structures that function first as absorbing organs and later as true leaves. The endosperm of

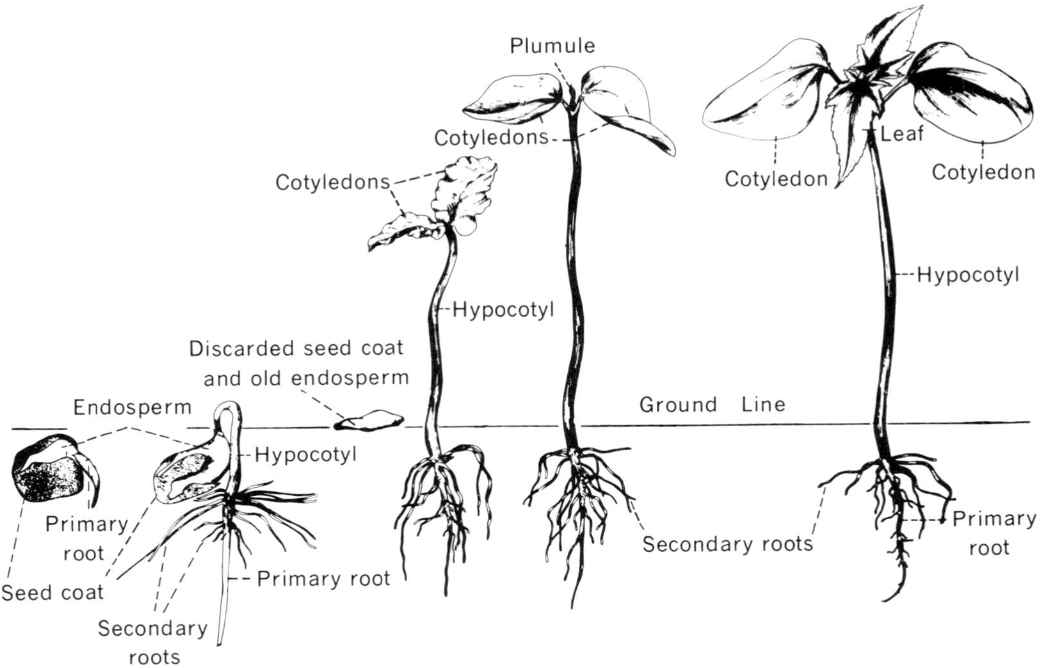

Fig. 11-16 Stages in germination of castor-bean seed and establishment of seedling. Cotyledons become the first green leaves. (Drawn by Elsie M. McDougle.)

this seed is carried up with the cotyledons as the hypocotyl elongates. The cotyledons do not spread apart entirely until the endosperm has been almost entirely consumed. When they finally do separate, the remaining endosperm dries up and falls off. Often remains of the endosperm and of the broken seed coats adhere for a while to the cotyledons. The plumule of the castor bean requires considerable time to get under way, since it is hardly more than a mass of undifferentiated cells in the seed. The cotyledons become green and function as foliage leaves for a long time. This method of germination is characteristic of dicotyledonous seeds having endosperm.

The germination of corn (Fig. 11-17) may be taken as typical of all cereals and other grasses. In this case there is only one cotyledon, the scutellum, and it functions exclusively in absorbing the digested food from the endosperm and transferring it to the growing parts. All growth of the seedling results from the elongation of the radicle and of the plumule. The hypocotyl, as in the pea, does not elongate. The radicle emerges first but is followed almost immediately by the plumule. The radicle gradually breaks through the coleorhiza and forms the primary root. Adventitious roots begin to develop at once from the point of origin of the radicle on the hypocotyl. These adventitious roots develop very rapidly, forming a rather extensive fibrous root system. The primary root is usually of short duration. Later, adventitious roots develop also at the first nodes of the stem. The coleoptile keeps pace with the growth of the plumule for some time but is ultimately broken through as the first true leaves emerge. When germination is complete and the seedling has

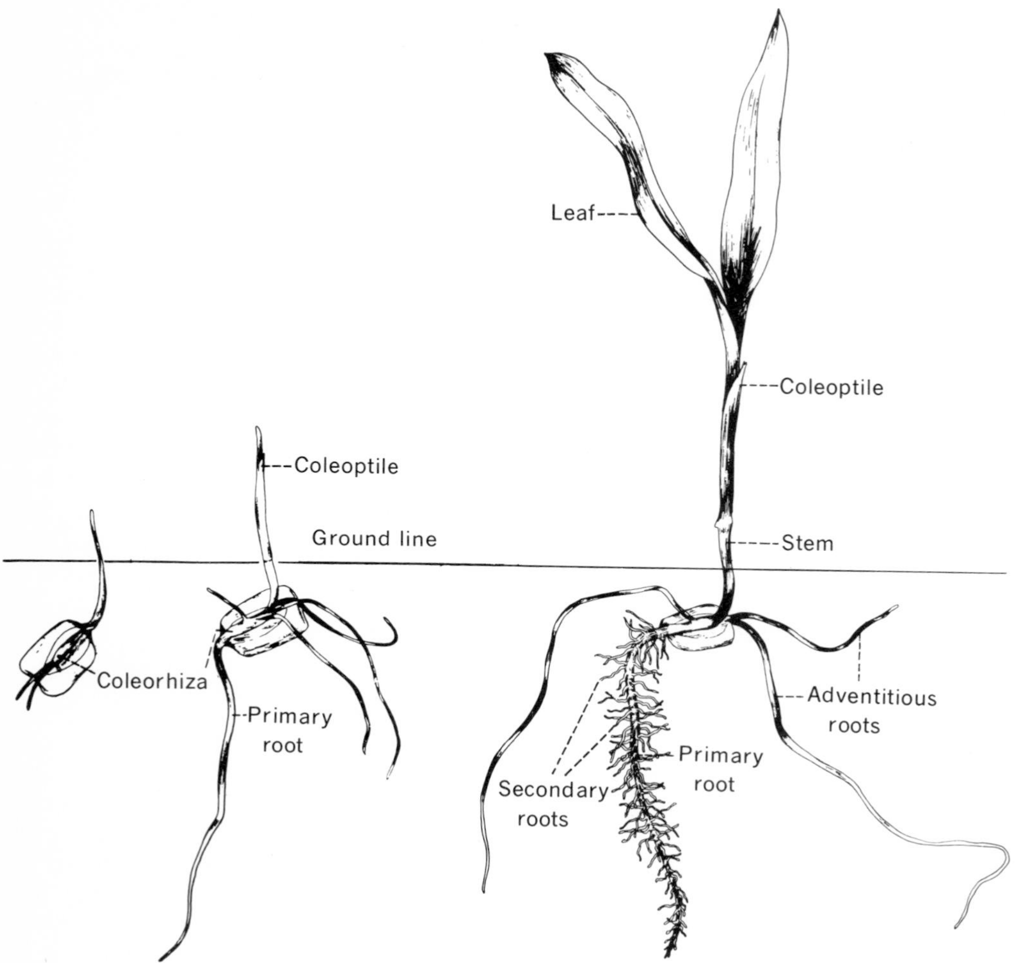

Fig. 11-17 Stages in germination of corn grain and establishment of seedling. (Drawn by Elsie M. McDougle.)

become independent of the endosperm, what is left of the seed gradually decays in the soil.

QUESTIONS

1. How do we determine whether a plant body is a fruit?

2. Of what advantage to the strawberry plant is the fleshy part of the strawberry fruit?

3. Why are dandelions and daisies more commonly found in our fields than are other wild plants?

4. Of what advantage to the plant are seeds with coats impermeable to water?

5. Why is the corn grain a fruit rather than a seed?

6. On burned-over or heavily logged forest

areas in Pennsylvania there is often a rapid growth of aspen. Since these trees are not planted by man, how do you account for them?

7. In the so-called Barrens of central Pennsylvania large areas, once covered with pines and other tall forest trees which were cut years ago by the lumbermen, are now overgrown with scrub oak. Acorns are not carried far by wind, nor do they remain viable long if they dry out. There are no streams in these areas. Could you suggest how this scrub oak might have started on these areas?

8. Would you expect wheat found in the ancient Egyptian tombs still to be viable?

9. What type of storage conditions would you recommend for crop seeds?

10. When seeds lose their viability, is the loss caused by a depletion of their food reserves?

REFERENCES

CROCKER, WILLIAM. 1948. Growth of Plants. Reinhold Publishing Corporation, New York. Contains the results of this investigator's many years of work on seed longevity and dormancy as well as that of others at the Boyce Thompson Institute for Plant Research.

ESAU, KATHERINE. 1965. Plant Anatomy, 2d ed. John Wiley & Sons, Inc., New York. Chapters 19 and 20 deal with the structure of fruits and seeds.

FOREST SERVICE, U.S. DEPARTMENT OF AGRICULTURE. 1948. Woody-plant Seed Manual. U.S. Dept. Agr. Misc. Publ. 654. Contains general information on seeds and seed germination, dispersal and dormancy, and specific information on some 444 species and varieties of woody plants as to their seed and seed-germination characteristics.

MAYER, A. M., and A. POLJAKOFF-MAYBER. 1963. The Germination of Seeds. The Macmillan Company, New York. Deals with the structure of seeds, dormancy, metabolism, ecology, and factors affecting the germination of seeds.

Warburg apparatus, used in gas-exchange studies, measures the amount of gas evolved or absorbed in cell respiration. (Courtesy of Bronwill Scientific Division, Will Scientific, Inc., Rochester, N. Y.)

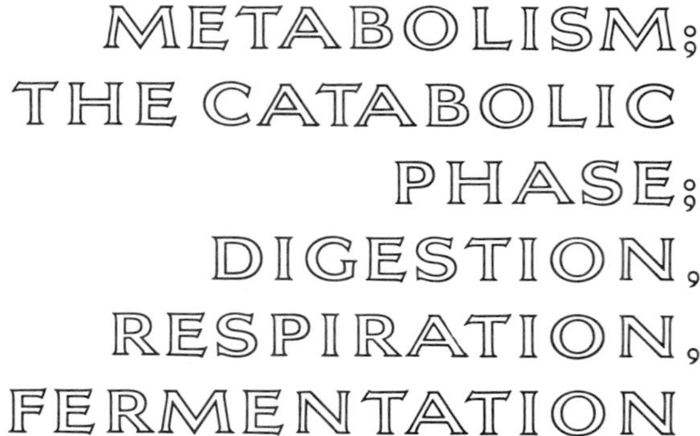

12 METABOLISM; THE CATABOLIC PHASE; DIGESTION, RESPIRATION, FERMENTATION

The catabolic phase of metabolism; the nature of enzymes; coenzymes; digestion of carbohydrates, fats and proteins; anaerobic and oxygen respiration are discussed in this chapter.

The anabolic phase of metabolism was taken up in Chap. 5. It will be recalled that the anabolic phase of metabolism consists of building-up, or synthetic, processes. The catabolic phase, which is the subject of the present chapter, consists of tearing-down processes. The principal catabolic processes are digestion, respiration, and fermentation. Both phases are essential to the proper physiological balance of the plant, which culminates in assimilation and growth and which keeps the plant in a healthy, living condition. It should be kept in mind that the ability to carry on metabolism as a whole is one of the fundamental properties of protoplasm, and hence metabolism takes place in every living cell. Metabolic processes, however, reach their highest intensity in actively growing regions of the plant. Since the embryos of germinating seeds are in an active state of growth, they furnish excellent material for studying such processes as digestion and respiration. It is partly for this reason that the discussion of the catabolic phase of metabolism has been postponed until germination has been considered.

ENZYMES

General nature of enzymes Because all metabolic processes involve the activity of enzymes, a knowledge of the general properties of enzymes is essential to an understanding of metabolism. Enzymes have been defined as catalysts produced by living organisms. A **catalyst** is a substance which is capable of accelerating the rate of a chemical reaction by its presence. Catalysts usually are found unchanged after the reaction is completed, and hence are able to bring about almost unlimited chemical change without themselves being used up, although their action is usually checked by the accumulation of the end products of the reaction. As an example of the action of a simple inorganic catalyst may be mentioned the preparation of oxygen gas from potassium chlorate. When potassium chlorate is heated, it very slowly gives off oxygen; but if a little manganese dioxide is mixed with it, the gas is evolved very rapidly. The manganese dioxide, while greatly accelerating the reaction, can be recovered unchanged when the reaction is completed. In other words, the manganese dioxide acts as a catalyst.

Enzymes likewise accelerate chemical reactions without appearing chemically combined with any of the end products when the reaction is completed. Enzymes, however, differ in many ways from inorganic catalysts. To begin with, they are very sensitive to temperature. Low temperatures check their action but do not destroy them. A temperature of 100°C is high enough to destroy most enzymes completely, and many of them are entirely inactivated at a temperature as low as 70°C. Enzymes, furthermore, are colloidal substances that ordinarily do not pass through dialysis membranes which permit passage of small organic molecules or ions. They are sometimes prepared by precipitation from water extracts of plant or animal tissues in which they occur. In recent years many enzymes have been obtained in a pure, crystalline state (Fig. 12-1). Among these are *urease, amylase,* starch *phosphorylase, pepsin, trypsin, papain, tyrosinase, catalase, peroxidase,* and *ascorbic acid oxidase.* All enzymes thus far isolated have been found to be proteins. Many of the properties of enzymes can be attributed to their protein nature.

Specificity of enzymes One of the most interesting features of enzymes is that they can be very specific in their action. The enzyme *urease,* for instance, acts on urea and on nothing else. *Catalase,* an enzyme of wide distribution in plants and animals, accelerates the decomposition of hydrogen peroxide to free oxygen and water but has no effect on other compounds. Similarly, most enzymes accelerate only a particular chemical reaction and have no effect on other reactions. On the other hand, enzymes are capable of causing a reaction to take place in either direction. For example, *lipase,* under certain conditions, causes the hydrolysis of fats to fatty acids and glycerol, but under other conditions may bring about the condensation of fatty acids and glycerol to form fats.

How enzymes are named Enzymes are named either according to the type of substance upon which they act or according to the type of reaction they accelerate. Thus enzymes that cata-

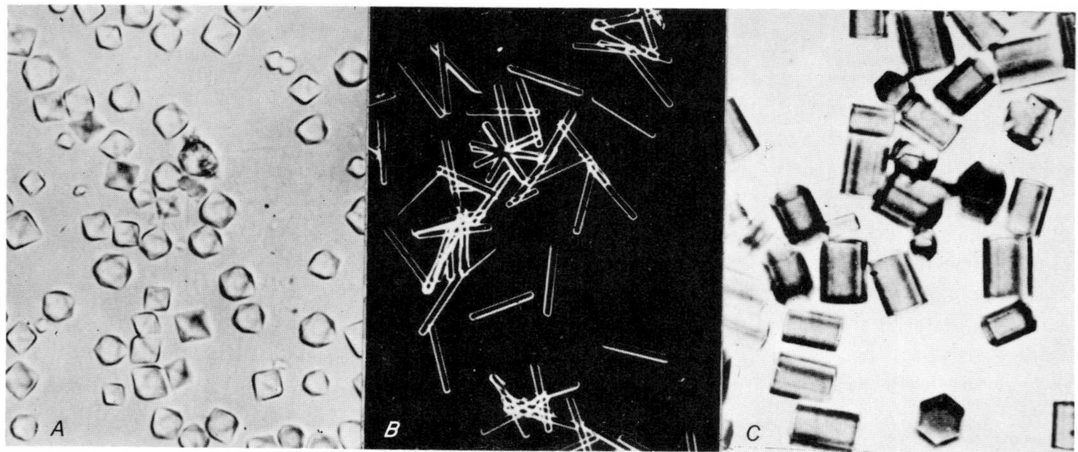

Fig. 12-1 Crystallized enzymes. A, urease, the first enzyme ever to be prepared in crystalline form; B, trypsin; C, catalase (from beef erythrocytes). (Photomicrographs courtesy of Dr. James B. Sumner, Laboratory of Enzyme Chemistry, Cornell University, Ithaca, N.Y. Dr. Sumner was the first to prepare any enzyme in pure crystalline form.)

lyze the splitting of carbohydrates are called *carbohydrases.* Similarly the ending *-ase* is substituted for the ordinary ending of any other substrate upon which a particular enzyme acts, and this name is given to the enzyme. The enzymes *sucrase, cellulase,* and *amylase* were named in this way. Enzymes which cause hydrolysis are called *hydrolases;* those which cause oxidation are called *oxidases,* or *dehydrogenases* if they are involved in removing hydrogen; and so on. Many of the enzymes were known before this system of naming enzymes was adopted. The names of such enzymes were not changed under the new system. In this category belong *pepsin, rennin, trypsin,* and *bromelin.*

Factors affecting enzyme activity Among the factors which affect the rate of activity of enzymes may be mentioned temperature; concentration of the substrate; concentration of the enzyme itself; accumulation of end products of the reactions; the acidity or alkalinity of the medium; light; and the presence of certain substances which may act as accelerators or inhibitors of enzyme action. Salts of the heavy metals such as silver nitrate, mercuric chloride, and copper chloride retard the action of many enzymes. Most enzymes are sensitive to the hydrogen-ion concentration of the medium in which they occur, some requiring a slight acidity and others a slight alkalinity for optimum activity. Light, and especially ultraviolet radiation, may destroy some enzymes or modify their activity.

PROSTHETIC GROUPS; COENZYMES The catalytic activities of many enzymes depend not only upon the basic chemical structure of the enzyme but also upon the presence of certain substances which combine specifically with this basic structure. These **prosthetic groups** may be simple inorganic substances like copper or zinc. Thus copper acts as the prosthetic group of *ascorbic acid oxidase,* an enzyme which catalyzes the oxidation of ascorbic acid (vitamin C). Magnesium forms the prosthetic group for the action of *phosphatases* or *phosphorylases* in liberating or transferring phosphate groups. Others of the prosthetic groups are complex organic substances. These are the ones usually referred to as **coenzymes.** Among them are a number of the vitamins of the B complex. Thus *cocarboxylase,* which is the pyrophosphate ester of thiamine (vitamin B_1), is the coenzyme of *carboxylase,* an enzyme capable of splitting off CO_2 from the carboxyl group (COOH group) of α-keto organic acids like pyruvic acid ($CH_3 \cdot CO \cdot COOH$).

Prosthetic groups in general have been found associated chiefly with oxidizing enzymes, and not with hydrolyzing enzymes such as the majority concerned in digestion. With enzymes requiring coenzymes for their action, the part of the enzyme without the coenzyme, which, in most cases, is the protein part of the enzyme, is sometimes called the **apoenzyme;** the intact enzyme, consisting of apoenzyme and coenzyme, is referred to as the **holoenzyme.** Apoenzymes are completely inactive in the absence of the proper coenzymes.

Some of the most important coenzymes involved in metabolism are mentioned in the following paragraphs and in the discussion on respiration. Among them, and of great importance in metabolism, are two phosphopyridine nucleotides, namely, **nicotinamide adenine dinucleotide (NAD)** and **nicotinamide adenine dinucleotide phosphate (NADP).** They were originally referred to as diphosphonucleotide (DPN) and triphosphonucleotide (TPN), respectively. Both are complex organic derivatives of the vitamin nicotinic acid (niacin), and function in several dehydrogenation reactions brought about by *dehydrogenases.* Under proper conditions they are easily oxidized and reduced. Since oxidation involves a loss of electrons or a loss of hydrogen atoms and reduction involves a gain of electrons or a gain of hydrogen atoms, these coenzymes function as electron- or hydrogen-transferring agents. Thus they are alternately oxidized and reduced as they react with specific products of metabolism such as certain carbohydrates, amino acids, and other organic acids, these products being, in turn, reduced or oxidized when they react with the coenzymes.

The flavin coenzymes, such as **flavin adenine dinucleotide (FAD),** are derivatives of the vitamin riboflavin (vitamin B_2). They are also coenzymes of the pyridine nucleotide type and act as electron or hydrogen carriers. They are thus able to function in the oxidation and reduction of NAD and NADP. They are important in the final stages of aerobic respiration.

The **cytochromes** are metalloprotein respiratory pigments in which the effective group is an iron-containing hematin. They are usually designated as cytochromes a, b, c and are related chemically to chlorophyll and to the hematin of blood hemoglobin. They are part of the cytochrome system and act as hydrogen or electron carriers in biological oxidation-reduction reactions. They are especially important in the final stages of aerobic respiration, in which they give up hydrogen to oxygen to form water.

Another important coenzyme, called **coenzyme A,** which is a derivative of the B-complex vitamin pantothenic acid, reacts with the acetyl group (the CH_3CO— part of the acetic acid molecule CH_3COOH) to form acetyl coenzyme A. One of the functions of this coenzyme is to accept or release the acetyl group; that is, it transfers acetyl groups from one compound (e.g., a fatty acid) to another compound (e.g., another fatty acid). In this manner it plays a part in the synthesis and breakdown of fatty acids and fats. It is also important in respiration and in other metabolic processes.

Prominent in many metabolic reactions are the adenosine[1] phosphates. There are three of these, namely, **adenosine monophosphate (AMP),** containing one phosphate group, **adenosine diphosphate (ADP),** with two phosphate groups, and **adenosine triphosphate (ATP),** with three phosphate groups. These substances, while not actually coenzymes, act as agents of phosphate transfer between metabolic products. Thus ATP, acting with *hexokinase,* can transfer some of its phosphate to the sugar, glucose, yielding glucose-6-phosphate, and thereby becoming ADP. On the other hand, ADP can accept phosphate from an-

[1] Adenosine is a nucleoside consisting of the pentose ribose and the pyrimidine adenine.

other phosphate-containing compound, and thereby be transformed into ATP. In the reactions in which they are involved, ADP and ATP serve as agents of energy transfer in the cell. In this manner energy-yielding reactions of metabolism are coupled with the production of ATP from ADP and phosphate.

Thus ADP and ATP serve to transfer energy between energy-requiring and energy-yielding metabolic reactions. An example of this is seen in the conversion of light energy into chemical energy of ATP by the process of photosynthetic phosphorylation. Still further examples will be seen in the reactions involved in respiration, which furnishes the energy to drive many other metabolic processes, such as the synthesis of proteins from amino acids and the synthesis of complex fatty acids from simple fatty acids.

Most of our knowledge of enzymes and coenzymes has been obtained from *in vitro* studies. When we consider the complex chemical environment of enzymes occurring within living cells of the plant and the hundreds of coupled enzymatic reactions that are occurring simultaneously within the cell, it is not remarkable that our knowledge of the precise mechanism of any metabolic process as it occurs within the plant is far from complete and that many of the conclusions which are now considered to be correct will probably be modified in the future.

DIGESTION

Introduction In Chap. 5 an attempt was made to show how the plant synthesizes the foods needed for its existence. These foods are always made within living cells. Not all of the living cells of the plant are capable of synthesizing all the foods they need to maintain life. Carbohydrates, for example, can be synthesized out of carbon dioxide and water only in cells containing chloroplasts. In order that cells lacking chloroplasts, such as the cells of the root, may be supplied with carbohydrates, it is necessary for carbohydrates to be transported to them from the chlorophyll-containing cells. But in order that this may be accomplished, it is necessary that these foods pass through living protoplasmic membranes. As has already been mentioned, in most of the dicotyledons carbohydrates are temporarily stored in the cells of the leaf mesophyll in the form of starch. Starch, being insoluble in water, is incapable of passing through a protoplasmic membrane. Hence it must first be changed to a soluble compound before it can be moved out of the mesophyll cells.

Similarly, when starch is stored in roots, in stems, or in any other organ of the plant, it must first be changed before it can be utilized by the plant. The same may be said of fats and proteins. Not only must these foods be rendered transportable but they must be put into such form that they can be oxidized for their stored energy (respiration) or utilized in the building up of other organic substances and protoplasm. *The processes involved in rendering foods soluble and diffusible so that they may be transported or utilized in the general metabolism of the plant are collectively called* **digestion.**

Seat of digestion in plants In the higher animals, including man, primary digestion of food takes place in a definite set of organs known as the digestive tract from which the digested products are passed into an elaborate circulatory system extending to all parts of the body. There is nothing comparable with this in plants. No specific set of organs is set aside for digestion. The process occurs chiefly in the parts of the plant where foods are stored, although it may occur in any living cell. Digestion of stored food occurs very rapidly in germinating seeds and in tubers and roots renewing growth in spring. At such times large amounts of food are necessary for the construction of new cells and for the energy required in the rapid metabolism occurring in the growing regions.

Digestion in plants is sometimes classified as **intracellular** and **extracellular. Intracellular** digestion occurs inside the cell. It is found to take place especially in cells containing food reserves, such as those which occur in storage roots and tubers and in seeds. Intracellular digestion takes place also in the mesophyll cells of the leaf. When the foods are digested outside the cell and the digested products absorbed by the cell, the digestion is said to be **extracellular.** This type of digestion occurs in insectivorous plants, such as the pitcher plants and sundew. The insects that are trapped by these plants are usually digested on the outer surface of the leaf and the digested product is absorbed. Extracellular digestion is the common form used by the fungi.

General nature of digestion Chemically all digestion is **hydrolysis;**[2] i.e., it involves a change or splitting of compounds into simpler compounds by the chemical addition of water. This change may be illustrated by the digestion of cane sugar (sucrose) as follows:

$$\underset{\text{sucrose}}{C_{12}H_{22}O_{11}} + H_2O \rightarrow \underset{\text{glucose}}{C_6H_{12}O_6} + \underset{\text{fructose}}{C_6H_{12}O_6}$$

Hydrolysis does not result in a complete decomposition of foods with release of all their energy but simply transforms them into more soluble or available forms. The combined end products still retain almost as much energy as the original substance hydrolyzed. Thus the heat of combustion of a gram-molecular weight[3] of sucrose is 1,349,600 cal, while the heat of combustion of the glucose and fructose resulting from the hydrolysis of the sucrose is 1,348,600 cal.

Hydrolysis of this type implies an abundant supply of water, but if the foods, in a pure state, were simply mixed with water, the process would go on at an extremely slow rate. The rate is tremendously increased in both plants and animals by the action of enzymes.

Digestion of carbohydrates Of the carbohydrates found in plants, glucose, fructose, sucrose, starch, cellulose, and hemicelluloses occur in greatest abundance. Of these, glucose and fructose are the simplest, being monosaccharides. These two sugars are soluble, diffusible, and in a form capable of being used directly in the metabolism of the plant. Hence they require no further digestion. In fact, they are the most common end products of the digestion of other carbohydrates. Sucrose, starch, cellulose, and hemicelluloses must first be digested before they can be utilized.

[2] Not to be confused with **hydration,** which involves union with water without chemical decomposition.

[3] A gram-molecular weight equals a molecular weight in grams, and is called a **mole** or **mol.**

SUCROSE DIGESTION As mentioned in a previous chapter, sucrose, $C_{12}H_{22}O_{11}$, or cane sugar, is the most widely distributed disaccharide found in plants. It is a common storage carbohydrate in many monocotyledonous plants like sugarcane and occurs also in the stems, roots, leaves, and fruits of many of the dicotyledons. Being readily soluble in water, it is ordinarily diffusible, but some plant-cell membranes are impermeable to it. Sucrose as such is nonfermentable. The advantage to the plant in having it digested may rest on the probability that it cannot always be used directly in respiration or assimilation or in the synthesis of other organic substances.

The digestion of sucrose yields glucose and fructose according to the following equation:

$$\underset{\text{sucrose}}{C_{12}H_{22}O_{11}} + H_2O \rightarrow \underset{\text{glucose}}{C_6H_{12}O_6} + \underset{\text{fructose}}{C_6H_{12}O_6}$$

This reaction is accelerated by the enzyme *saccharase* (also called *invertase, sucrase,* or *β-d-fructosidase*). This enzyme has been found to occur in green leaves, fruits, grains, stems, potato tubers, some roots, pollen, and such lower plants as fungi and bacteria. It is especially abundant in yeast. Salts of the heavy metals (silver, copper, mercury) inhibit its action. Maximum activity is obtained with low concentrations of sucrose (5 to 10 percent). *Saccharase* also hydrolyzes the sugars raffinose, gentianose, and stachyose and, to some extent, inulin.

STARCH DIGESTION Starch is the most common storage form of carbohydrate found in plants. Being insoluble in water, it must always be changed before it can be used or transported. It is always found in the plant in the form of grains, the starch grains, which vary in form in different species (Fig. 12-2). Starch grains usually contain, in addition to starch, less than 1 percent of adsorbed other substances, including fats, proteins, tannins, phosphates and other minerals, and hemicellulose. The starch exists in the grain in colloidal condition, probably as colloidal aggregates of starch molecules in some way bound up with phosphates. Chemically starch consists entirely of glucose units. Most natural starches are made up of at least two chemically different components, viz., **amylose,** consisting of a spiral, unbranched chain of glucose units, and **amylopectin,** which is a branched chain of glucose units. The proportions of these two components apparently differ widely in starches of different plants. Some starches contain 10 to 20 percent or more amylose and 80 to 90 percent or less amylopectin.

Fig. 12-2 Starch grains. A, potato; B, bean; C, corn; D, oats. (Drawings courtesy of F. D. Kern from "The Essentials of Plant Biology," Harper & Row, Publishers, Incorporated, New York, 1947.)

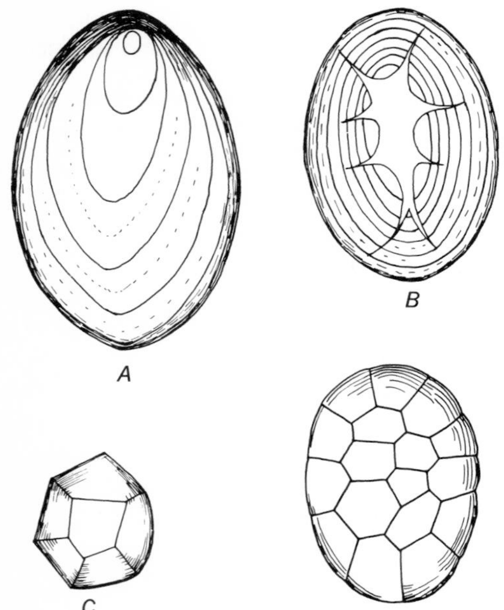

The enzyme complex concerned in the digestion of starch is commonly called *diastase. Diastase* is found widely distributed in plants. It is found in the majority of seeds and leaves, in the juices of roots and tubers, in fruits, in wood and bark of trees, in germinating pollen, in the latex of rubber trees and poppies, and in some of the fungi. The seeds of legumes and cereals are especially rich in *diastase.* The amount of *diastase* increases greatly during the germination of starchy seeds, where the digestion of starch proceeds rapidly. A young seedling may have a *diastase* content one thousand times greater than that of the seed from which it grew. The maximum *diastase* content is usually reached at the time of the formation of the first leaves of the seedling. In barley this occurs about 11 or 12 days after the beginning of germination.

Malt diastase is one of the most common forms used in experiments on starch digestion. It is prepared from germinating barley grains in various ways, usually by extracting with water and precipitating several times with alcohol. *Takadiastase,* a form commonly used in medicine, is prepared by growing the fungus *Aspergillus oryzae* on steamed bran or rice for 40 to 48 hr and then extracting the enzyme with water and precipitating with alcohol.

Diastase is not a single enzyme but consists of several separate enzymes with different properties. How many of these separate enzymes may be present in a *diastase* preparation has not yet been determined, but prominent among them are the *amylases* (so named from the Latin *amylum,* which means starch), including the dextrinogenic (dextrin-forming) *α-amylase* and the saccharogenic (sugar-forming) *β-amylase.* There may also be present *phosphatase* such as the starch-liquefying *amylophosphatase,* which has been reported to be able to break down the colloidal structure of the starch, and *phosphorylases,* which have been shown to function in the synthesis of starch out of sugars. *Maltase,* an enzyme which hydrolyzes maltose to glucose, is also usually found in tissues containing *diastase*.

In general, starch is ultimately broken down, in digestion in the plant, to glucose, with the intermediate production of, possibly, soluble starch, various dextrins, and maltose. A number of different enzymes of the *diastase* complex are involved. Starch may be converted first into soluble starch by the action of *amylophosphatase*, although the presence of this enzyme has not been fully established.

The action of *β-amylase* consists of the progressive removal of maltose units from the ends of chains of glucose molecules. The unbranched component of starch, amylose, is completely hydrolyzed by *β-amylase* to maltose. The branched component, amylopectin, on the other hand, is converted to maltose to the extent of only about 50 percent, the residue being a relatively long-chain dextrin, called dextrin A or α-amylodextrin. The incompleteness of the hydrolysis of amylopectin by *β-amylase* is ascribed to the presence of the cross-linkages in the amylopectin. These cross-linkages, or points of branching of the chains of glucose molecules, in amylopectin are thought to interfere with the further action of the enzyme. When *β-amylase* acts directly on starch, which contains both amylose and amylopectin, about 60 percent of the starch is converted to maltose, the remaining residue being α-amylodextrin, which still gives a blue-violet color with iodine.

The action of *α-amylase* is somewhat more complex than that of *β-amylase*. In the early stages of its reaction it produces only short-chain dextrins but no maltose. The short-chain dextrins are called reducing dextrins because they reduce Fehling solution. They give no iodine reaction. In later stages of reaction of *α-amylase*, the dextrins themselves may be hydrolyzed with the production of maltose and perhaps some glucose. As a result of the action of the *amylases*, maltose tends to accumulate, but if *maltase* is present, it converts the maltose to glucose, maltose being made up of two units of glucose. In the plant the glucose resulting from starch digestion may immediately be converted into fructose or sucrose. Maltose and dextrins are found rarely and in small amounts in plants.

The digestion of starch can best be studied by using a starch paste made up by mixing a gram of starch with a little water and stirring it into 200 cc of boiling water and allowing it to cool. When starch grains are used, the action of *amylases* is very slow, often causing localized corrosion of the grains and resulting in an empty hull after digestion is complete, owing to an outer layer of the grains, which may be a less hydrated form of starch. When a starch paste is made, action is more rapid. If the digestion is studied by means of iodine tests, color changes occur from blue through violet, red, and red-brown to colorless. These changes are associated with the production of definite fragments of the starch molecule, which are dextrins and maltose. Fragments containing 6 units or less of glucose give no color reaction with iodine; those containing 8 to 12 units give a reddish coloration, and those containing 12 units or more give a violet to blue reaction. The different kinds of dextrins are simply fragments of the basic starch molecule, consisting of 3, 4, 8, 12, or possibly 16 to 17 glucose units. At the end of the reaction, the presence of reducing substances, mostly maltose, can be detected with Fehling solution.

DIGESTION OF CELLULOSE AND HEMICELLULOSE Celluloses and hemicelluloses are both found in the walls of the cells of plants. Cellulose is chiefly a structural material, but the hemicelluloses may sometimes be used as reserve food.

The enzymes concerned in the digestion of cellulose have been isolated chiefly from bacteria and fungi that cause decay. It is these organisms which break down the relatively insoluble celluloses in plant remains in the soil and in decaying timber.

Cellulose is hydrolyzed by the enzyme *cellulase* to a disaccharide called cellobiose. The enzyme *cellobiase* then converts the cellobiose into two molecules of glucose. This may be represented as follows:

(1) $(C_6H_{10}O_5)_n + nH_2O \rightarrow n(C_{12}H_{22}O_{11})$
cellulose (action of *cellulase*) cellobiose

(2) $C_{12}H_{22}O_{11} + H_2O \rightarrow 2C_6H_{12}O_6$
cellobiose (action of *cellobiase*) glucose

The exact mechanism of the digestion of cellulose has not been as thoroughly determined as has the digestion of other carbohydrates. It is likely that it is much more complex than is indicated by the two equations above and that other enzymes are involved.

The digestion of the hemicelluloses is brought about by enzymes called *cytases* and results in such end products as glucose, mannose, galactose, and pentoses.

Digestion of fats and oils The digestion of fats and oils can best be observed in the germination of fatty seeds in which relatively high percentages of fat occur. It has been mentioned in a previous chapter that fats are esters of glycerol and fatty acids. Before a fat can be utilized by the plant or translocated to some other region, it must first be rendered soluble and diffusible. This is brought about in the plant by the enzyme *lipase*.

Plant *lipase* has been isolated chiefly from germinating seeds, such as castor beans and the seeds of other members of the spurge family (Euphorbiaceae), soybeans, seeds of cucurbits, flax, hemp, rape, poppy, and corn. *Lipase,* it will be recalled, is the same enzyme that causes the synthesis of fats from glycerol and fatty acids. Whether it causes synthesis or hydrolysis is at least partly governed by the relative water content of the tissue in which it occurs. When the water content is diminishing, as when seeds are being formed on the plant, synthesis of fats occurs. When the water content is increasing, as when seeds are placed under germinating conditions, *lipase* hydrolyzes the fats that are present. It is for this reason that the acid content of fatty seeds increases during germination. The hydrolysis proceeds very slowly at first, but as the percentage of acid formed increases, the rate increases. In other words, a little acid accelerates the action of *lipase*. The digestion of the fat may be represented as follows:

$$\begin{array}{c} H \\ | \\ H{-}C{-}OOC\cdot R \\ | \\ H{-}C{-}OOC\cdot R \\ | \\ H{-}C{-}OOC\cdot R \\ | \\ H \end{array} + 3H_2O \rightarrow \begin{array}{c} H \\ | \\ H{-}C{-}OH \\ | \\ H{-}C{-}OH \\ | \\ H{-}C{-}OH \\ | \\ H \end{array} + 3R\cdot COOH$$

any fat (action of *lipase*) glycerol fatty acid

If palmitic acid were the fatty acid involved in the above reaction, R would be $C_{15}H_{31}$ in each case. Since a fat is usually made up of several different fatty acids, mixtures of these fatty acids appear as end products.

Many of the fatty acids appearing as end products of the hydrolysis of fats by *lipase* are just as incapable of passing the plasma membranes of the cells in which they are formed as were the original fats. Consequently it is likely that the fatty acids are still further broken down to carbohydrates before they can be utilized. Such experimental evidence as is available supports this assumption, but relatively little is known about the mechanism of the process. Glycerol is diffusible and hence can probably be translocated as such, but it is likely that it, too, is changed before it is assimilated.

Digestion of proteins While proteins occur in every living cell, they are seldom found in large quantities even in storage regions of plants and hence a study of their digestion within the plant is attended with difficulty. It will be recalled that basically the proteins consist of chains of amino acids. In many of the proteins other substances are chemically combined with the amino acids, forming side chains which may exert a considerable effect upon the specificity of proteolytic enzymes. When the naturally occurring proteins are broken down by enzyme action, they may yield, at first, **proteoses** and **peptones.** Both proteoses and peptones are soluble in water and are noncoagulable by heat, but the proteoses can be precipitated by saturating their solutions with ammonium sulfate, while the peptones cannot. Both the proteoses and the peptones are still chains of amino acids but are shorter than those of the original protein. Both are still rather complex compounds. Compounds consisting of two or more amino acids, the carboxyl group of one being united with the amino group of another (peptide linkage), are called **peptides.** Those consisting of two such combined amino acids are called **dipeptides,** and those consisting of many amino acids are called **polypeptides.** The peptides are less complex than the proteoses, peptones, or original proteins.

The hydrolytic enzymes that digest proteins and their derivatives have been classified in several ways. In general, enzymes that attack the original native proteins are called *proteinases.* Examples of these are *papain, pepsin,* and *trypsin.* Those which hydrolyze peptides are called *peptidases.* If they hydrolyze dipeptides, they are called *dipeptidases,* and if they hydrolyze polypeptides they are called *polypeptidases. Carboxypeptidases* act only on the free carboxyl group of a polypeptide chain, and *aminopeptidases* act only on the free amino group of a polypeptide chain. The proteolytic enzymes are sometimes separated into two broad groups, the *exopeptidases* and the *endopeptidases.* The *exopeptidases* require a free α-amino or α-carboxyl group adjacent to the sensitive peptide linkage in the main part of the molecule of the substrate and are therefore restricted in their action to terminal peptide bonds. The *endopeptidases* do not require free terminal amino or carboxyl groups and are therefore capable of splitting central peptide bonds of proteins and suitably substituted peptides. Most of the *proteinases* are *endopeptidases,* while most of the *peptidases* are *exopeptidases.*

The *amidases* are a group of enzymes that usually split off ammonia from *amido* or *amino* compounds. In some cases urea appears as an end product of their action. *Amidases* are found in many plants, including the fungi as well as higher plants. Among them are *arginase,* which splits the amino acid arginine into ornithine and urea; *asparaginase,* which forms aspartic acid and ammonia from asparagine; *aspartase,* which converts aspartic acid into ammonia and fumaric acid; and *urease,* which hydrolyzes urea to ammonia and carbon dioxide. *Urease* was the first enzyme to be prepared in a pure crystalline form.

The general term *protease* is commonly used for all enzymes that catalyze the hydrolysis of proteins and of protein hydrolytic products such as proteoses, peptones, and polypeptides. Many *proteases* have been found in plants. Some of the best-known are *pepsin,* reported to be present in insectivorous plants, *papain,* obtained from the latex of *Carica papaya* (Fig. 12-3); *ficin,* from the milky sap of several species of fig trees; *bromelin* from ripe pineapples; and *solanain* from the horse nettle (*Solanum eleagnifolium*).

In the plant the end products of protein digestion are the amino acids, with the intermediate production of proteoses, peptones, and peptides. The amino acids are carried to growing parts of the plant and are utilized in the resynthesis of proteins, enzymes, and probably other compounds.

Fate of the products of digestion As has been seen in the preceding paragraphs, digestion goes on particularly in regions of the plant where foods are stored. As soon as these foods are digested, they are in a form which enables them to be transported to other parts of the plant, especially to regions where growth is proceeding. Here they are mostly used in building up new protoplasm, new cells, and new tissues; but a large portion of the digested food is also used in these regions to supply the energy needed for growth and for normal physiological functioning of the plant. In this case the foods are oxidized

Fig. 12-3 Papaya, or papaw (Carica papaya) *from the latex of which papain is obtained. The fruits are also used as salads or as melonlike breakfast fruits. (Photograph courtesy of F. D. Kern and the Department of Agriculture and Commerce, Puerto Rico.)*

or respired. Other foods are digested in one part of the plant, transported to other parts, and converted back again to the original compounds in new storage regions. Thus the starch which accumulates in the leaves of many dicotyledonous plants during the day is digested at night and carried to various storage regions, such as roots, stems, and tubers, and is there reconverted into starch.

In addition, all the various chemical compounds that are synthesized in the plant are made out of the digested foods.

We may summarize the uses made by the plant of the digested foods as follows:

1. They may be carried to other parts of the plant (translocation).
2. They may be oxidized for their contained energy (respiration).
3. They may be converted back to the original compounds in new storage regions (food storage).
4. They may be used in the synthesis of new organic compounds (organic synthesis).

5. They may be used to build new protoplasm (assimilation and growth).

RESPIRATION AND FERMENTATION

Introduction As long as an organism is living, it must be supplied with energy to carry on the ordinary life processes. In a previous chapter reference was made to the fact that plants containing chlorophyll store up radiant energy in the making of carbohydrates by photosynthesis. This stored energy in carbohydrates and other compounds is made available by the process of **respiration.**

Respiration is essentially an oxidation process. In its ordinary form oxidation implies a chemical reaction in which oxygen combines with some of the constituent elements of the substance being oxidized, thereby forming oxides and liberating energy. Such oxidation occurs during the burning or combustion of organic substances and results in the liberation of considerable energy in the form of heat. If, for example, we burn sugar, the oxygen of the air combines with the sugar to form carbon dioxide and water, and much heat is liberated. The same thing takes place in the plant but at a much slower rate and at ordinary temperatures. Such oxidation at lower temperatures is made possible by the action of enzymes, or in other words, respiration in the plant is enzymatic in nature. Furthermore much of the energy liberated does not appear as heat, but is used in carrying on the work of the living cells of the plant. *The term respiration is used in a comprehensive way to include all oxidation, or decomposition, of materials resulting in the liberation of energy, i.e., all catabolic changes involving substantial energy release, and any gaseous exchange accompanying this.* It should be emphasized that the most important feature about respiration is that energy is released by it, and not that oxygen is taken in and carbon dioxide liberated.

Oxidation may take place in the absence of free oxygen, involving loss of hydrogen or loss of electrons. There are two types of respiration, depending upon whether free oxygen is involved and to what degree the oxidation proceeds, viz., aerobic, or oxygen, respiration and anaerobic respiration, or fermentation. **Oxygen respiration** involves the utilization of free oxygen at least in the final stages of the process and results in complete oxidation to carbon dioxide and water. **Anaerobic respiration,** or **fermentation,** takes place in the absence of free oxygen and results in incomplete oxidation, thus yielding compounds capable of being still further oxidized, such as alcohols and organic acids.

Seat of respiration in plants In the higher animals and man, the lungs, tracheae, and nostrils are often looked upon as organs of respiration, and the inhalation of air and exhalation of carbon dioxide as respiration. In a strict sense, however, this mechanical exchange of gases is not respiration but breathing. In the plant there are no organs comparable with lungs, but the stomata, lenticels, and intercellular spaces function in gaseous exchange. True respiration takes place in every living cell of plant or animal at all times, day and night. So much is this true that we may look upon a cell as dead if respiration no longer goes on within it. There are, in other words, no special organs of respiration in the plant, but all parts of the plant that consist of living cells carry on the process. In the cell the mitochondria play a dominant role in respiration. The mitochondria contain the respiratory enzymes and coenzymes and are involved in terminal aerobic respiration.

Gaseous exchange; the respiratory quotient We may represent the complete oxidation of glucose by the following equation:

$$C_6H_{12}O_6 + 6O_2 \rightarrow 6CO_2 + 6H_2O + 673{,}000 \text{ cal per mole}$$

It will be seen from this equation that oxygen is consumed and CO_2 liberated and that the volume of oxygen used is equal to the volume of CO_2 liberated. It is customary to refer to the relation of CO_2/O_2 as the respiratory quotient. Obviously, in the foregoing equation the respiratory quotient would equal 1. In determining the respiratory quotient, it is necessary to measure the oxygen consumed by the plant as well as the CO_2 liberated. When this is done, the values obtained for the respiratory quotient are sometimes less than 1 and sometimes greater.

This may be brought about by one or more of the following causes: (1) The carbohydrates may be incompletely oxidized to organic acids or other compounds in which the production of CO_2 does not take place. This would make the quotient less than 1. A typical example of this is found in some of the cacti. In these plants the respiratory quotient is always less than 1 at night because carbohydrates are oxidized to organic acids. The following day these organic acids, together with unchanged carbohydrates, are completely oxidized to CO_2 and H_2O and, since less oxygen is required to oxidize the acids, the quotient becomes greater than 1. A similar condition has been found to take place in developing green apples. (2) Respiration may take place in the total absence of oxygen, but with the liberation of CO_2, as in fermentation, making the quotient greater than 1. (3) Other substances besides carbohydrates may be oxidized, yielding different quotients, greater or less than 1. For instance, when fats are oxidized, as frequently happens, the quotient will be less than unity because of a greater amount of oxygen needed to oxidize fats and sometimes because of the fixation of some of the oxygen by unsaturated fatty acids. The respiratory quotient of germinating fatty seeds like sunflower or flax is often as low as 0.3. On the other hand, the oxidation of some of the organic acids yields quotients greater than unity. Thus the complete oxidation of oxalic acid, which rarely occurs in plants, would give a quotient of 4 as seen in the following equation:

$$2\begin{array}{c}\mathbf{COOH}\\|\\\mathbf{COOH}\end{array} + \mathbf{O_2} = 4\mathbf{CO_2} + 2\mathbf{H_2O}$$

oxalic acid

In general the respiratory quotient is equal to unity only when carbohydrates are completely oxidized.

Substances used in respiration From the preceding statements of the fluctuations in the value of the respiratory quotient, it is clear that plants are able to use different substances as a source of energy. Probably the most common material oxidized by the higher plants is either glucose or fructose. Other sugars are also used. Fats, amino acids, other organic acids, and even proteins are oxidized by some plants. In cases of starvation, protoplasm itself may be broken down, and some of the products used in respiration.

Among the bacteria there exist specialized groups that utilize inorganic materials as a source of energy. Thus the hydrogen bacteria use hydrogen, the sulfur bacteria sulfur, the methane bacteria methane, and the hydrogen sulfide bacteria hydrogen sulfide. These substances are all oxidized as a source of energy.

If the whole plant kingdom is included, we could say that the substances used in respiration range all the way from the simplest inorganic materials to the most complex organic material. In general, however, we may look upon the sugars as the most common respiratory material in the higher green plants.

Factors affecting the rate of respiration; internal factors The rate at which respiration goes on in the plant is governed by a number of internal as well as external factors. Among the internal factors may be mentioned the amount of respirable material, especially carbohydrate, the amount of cell matter actually respiring, the activity of respiratory enzymes, the acidity of the cell sap, the activity of the plant, and the age of the plant. The internal factors have not all been thoroughly investigated, although some knowledge has been obtained. It has been found, for example, that leaves containing small amounts of sugar, as a result of being kept in the dark, respire very feebly, but when supplied with sugar (in the dark), the rate of respiration immediately goes up. High concentrations of sugars, however, decrease the rate again, probably through osmotic effects, since high concentrations of mineral salts behave similarly. It is natural to assume that the continuance of respiration will depend upon an adequate supply of food and hence the rate of respiration falls when the supply becomes inadequate. In general, young plants respire at a more rapid rate per unit of dry weight than do older plants, partly because a greater percentage of their tissues is active physiologically and hence requires a greater supply of energy. Similarly, dormant organs like seeds and buds respire very feebly, but when growth is renewed, the rate of respiration immediately goes up.

Among the most important internal factors affecting respiration are the activities of **respiratory enzymes** already mentioned. While respiration is oxidation, many of the substances used as a source of energy must first be hydrolyzed before they can be oxidized in respiration. This involves many of the hydrolyzing enzymes already mentioned. Some of the respiratory enzymes catalyze oxidation involving the addition of oxygen, while others catalyze oxidation involving loss of hydrogen or loss of electrons. There are two principal classes of oxidizing enzymes, the *oxidases* and the *dehydrogenases.* These different types of enzymes operate together in bringing about oxidation of chemical compounds which are ordinarily considered to be stable. Examples of oxidizing enzymes of the *oxidase* type are the *iron oxidases,* such as *cytochrome oxidase* and the various *peroxidases,* and the *copper oxidases,* such as *tyrosinase, laccase,* and *ascorbic acid oxidase.* Examples of *dehydrogenases* are *alcohol dehydrogenase, glucose dehydrogenase,* and *lactic dehydrogenase. Catalase,* although not an oxidizing enzyme, is usually classified with *oxidases* because its action is closely connected with physiological oxidation. It decomposes hydrogen peroxide into water and gaseous oxygen.

External factors Of the external factors that affect the rate of respiration, the most effective are temperature, light, oxygen supply, water supply, carbon dioxide concentration, toxic and stimulating substances, and agents of disease and injury.

TEMPERATURE In general the rate of respiration increases as the temperature increases, approximately doubling in rate for every 10° rise in temperature. Plants may still continue to respire at very low temperatures, although the rate is very low. This is especially true of dry, dormant structures like seeds. Since some seeds are able to withstand temperatures of −50°C and lower, and, since respiration continues as long as life continues, some respiration must go on even at such low temperatures, even though it is too feeble to be detected by ordinary means. As the temperature is increased, the rate of respiration increases until a maximum point is reached. Beyond this the rate remains the same until a temperature is reached that kills the plant. While there is some evidence that at higher temperatures the rate of respiration may fall off in some plants, in general it has been found that there is no optimum for respiration. That is, the rate continues to rise with increasing temperature to the maximum rate and stays at that point. In other words, there is no one temperature at which the highest rate of respiration is obtained.

LIGHT Respiration increases in chlorophyll-containing plants in light, but the effect of light is an indirect one in that it probably operates through supplying respiratory material in the form of carbohydrates. Plants lacking chlorophyll do not respond in this manner. That light is not necessary for respiration is seen from the fact that it proceeds at night as well as in the daytime. It is possible that part of the effect on rate is caused by the rise in temperature resulting from the absorption of radiant energy by the cell constituents. Spoehr explained part of the effect of light by the fact that the sun's rays cause ionization of atmospheric oxygen and this causes autooxidation in the protoplasm. As a result of the effect of light, shade plants often respire at a lower rate than do sun plants.

OXYGEN SUPPLY While it is possible for respiration to continue for a time in higher plants even in the total absence of oxygen, normally a supply of oxygen must be available for the higher plants to continue to live. Whether or not oxygen is present greatly affects the type of respiration and the end products produced. As a rule, there is much more oxygen present in the atmosphere than is needed by plants. Concentrations as low as 1 percent have been found to be sufficient in some cases. It is probable that many of the interior tissues of plants normally do not obtain very high concentrations of oxygen and hence could not respire normally if large quantities of oxygen were needed. There are some bacteria and fungi that normally live in the total absence of oxygen. Such forms carry on only anaerobic respiration.

WATER SUPPLY The effect of water content on respiration is well known in its effect on seeds. Dry seeds respire at a very slow rate, but as water is absorbed, the rate increases up to a maximum which varies with different seeds. A further increase in water may then diminish the rate. It is this increased respiration that causes stored grain sometimes to heat up when it becomes moist. The effect of water in this instance is partly explained by the fact that the water may increase digestive processes which provide respiratory materials. The respiration of the growing plant is also affected by the water content of its tissues. If water is lost by transpiration to such an extent as to render the cells flaccid, growth is checked and with it the respiratory activity.

CARBON DIOXIDE CONCENTRATION The accumulation of carbon dioxide gas resulting from respiration may check further respiration. This is shown by the fact that germination of seeds is entirely checked if the carbon dioxide is allowed to accumulate around them. When this carbon dioxide is removed, however, germination continues. Green plants in light can stand rather high concentrations of carbon dioxide without injury. The checking effect of carbon dioxide on respiration may partly explain the fact that protoplasmic streaming is stopped by carbon dioxide.

TOXIC AND STIMULATING SUBSTANCES The toxic effect of certain chemicals upon plants as well as animals often results from the effect of these substances on respiration. While in strong doses such substances may completely check respiration and thereby cause death, it is interesting to note that many of them in very weak doses speed up the rate of respiration for a while. For example, it has been found that a dose of 0.1 cc of chloroform in 970 cc of air causes an increase in the output of carbon dioxide from leaves. After a time this increase falls back to the normal rate. A dose as great as 1 cc of chloroform in 970 cc of air also stimulates carbon dioxide production for a while, but proves to be toxic because there is no recovery of the leaf from it. If the dose is raised to 10 cc in 970 cc of air, there is no stimulation at all, and respiration ceases immediately. Ether, acetone, formaldehyde, caffeine, paraldehyde, cocaine, morphine, quinine, solanine, and other substances have been found by various workers to produce similar effects; i.e., a very small dose acts as a stimulant, a large dose as a narcotic. There have been attempts to explain this action of poisons, but none of them is well established.

Warburg believed that boundary-surface phenomena play an important role in respiration and that the action of poisons is explained by the fact that they are strongly surface active and hence interfere strongly with respiration. Some toxic materials probably operate through their effects on respiratory enzymes.

AGENTS OF DISEASE AND INJURY Disease and injury to the plant often also stimulate respiration. The degree of stimulation depends upon the extent of the injury and the nature of the tissue injured. In some instances injury induces renewed growth, which results in healing of the wound. This increased growth implies increased respiration in that more energy is being used. In some cases, as in the potato, it has been shown that injury causes an increase in sugars in the injured portion, probably caused by a stimulation of *diastase* activity. These sugars furnish respiratory material and may contribute to the increased rate of respiration. If the injury or disease is sufficiently widespread in the plant, it may become quite detrimental through the loss of materials resulting from increased respiration, although many other factors contribute to the weakening of the plant.

Anaerobic respiration; fermentation Many plants, when deprived of free oxygen, still continue to give off carbon dioxide and often produce alcohol and other compounds. This has been found to be true of seeds, fleshy fruits, leaves, flowers, woody parts of plants, and many fungi and bacteria. Respiration of this type is called anaerobic respiration, or fermentation. Well-known examples of such fermentation are alcoholic and lactic acid fermentations of sugar. These may be represented by the following equations:

$$\underset{\text{glucose}}{C_6H_{12}O_6} \rightarrow \underset{\text{ethyl alcohol}}{2C_2H_5OH} + 2CO_2 + 28{,}000 \text{ cal per mole}$$

$$\underset{\text{glucose}}{C_6H_{12}O_6} \rightarrow \underset{\text{lactic acid}}{2CH_3CHOHCOOH} + 18{,}000 \text{ cal per mole}$$

It will be observed that in each case no free oxygen is involved and that much less energy is liberated than when the sugar is completely oxidized. Both processes are enzymatic. The first is utilized in the commercial production of alcohol by yeast. The second occurs when milk sours in the presence of lactic acid bacteria.

Fermentation processes are used widely in commerce. The best-known of these processes is alcoholic fermentation brought about by the enzyme complex *zymase,* which consists of a whole series of enzymes and coenzymes. The fermentation of sugar to alcohol has been practiced by man for centuries. While yeast is commonly employed for this purpose, many other plants are also capable of fermentation activity, particularly other fungi and bacteria, as well as higher plants. Pea seeds, after being soaked in water and placed in an inverted test tube of mercury, where they have no access to oxygen, will, in a short time, produce much carbon dioxide and alcohol. Some plants will die if kept too long under anaerobic conditions. The products of the incomplete oxidation which takes place probably contribute to the cause of death, many of these products being toxic. Other plants, especially some of the fungi and bacteria, are not injured until considerable end products accumulate.

The foregoing equations used as examples of fermentation give merely the initial substrates and the end products of the reactions. In each case a complex series of intermediate steps occur before the end products are reached. These steps are indicated in the accompanying diagram (Fig. 12-4). In fermentation as well as in aerobic respiration these steps may be divided into two phases: (1) **glycolysis,** in which the carbohydrates are oxidized to pyruvic acid,

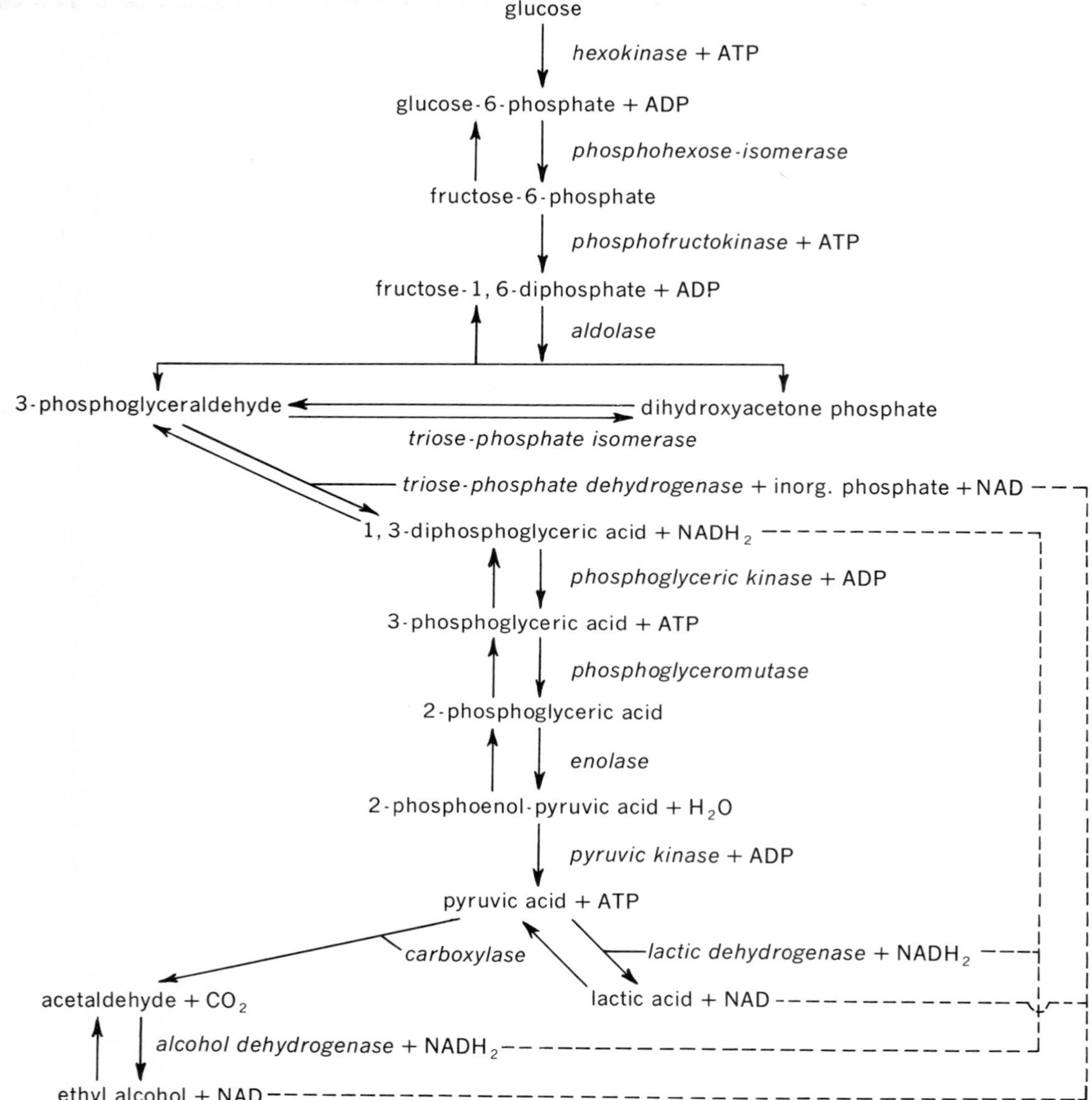

Fig. 12-4 Steps in glycolysis of glucose and two pathways of anaerobic metabolism of pyruvic acid.

and (2) the subsequent **metabolism of the pyruvic acid.** Glycolysis involves no free oxygen and follows the same pattern in both aerobic and anaerobic respiration. The metabolism of pyruvic acid, on the other hand, although it may follow a number of patterns, follows a different pattern in aerobic respiration from the one it follows in anaerobic respiration. This is indicated in a com-

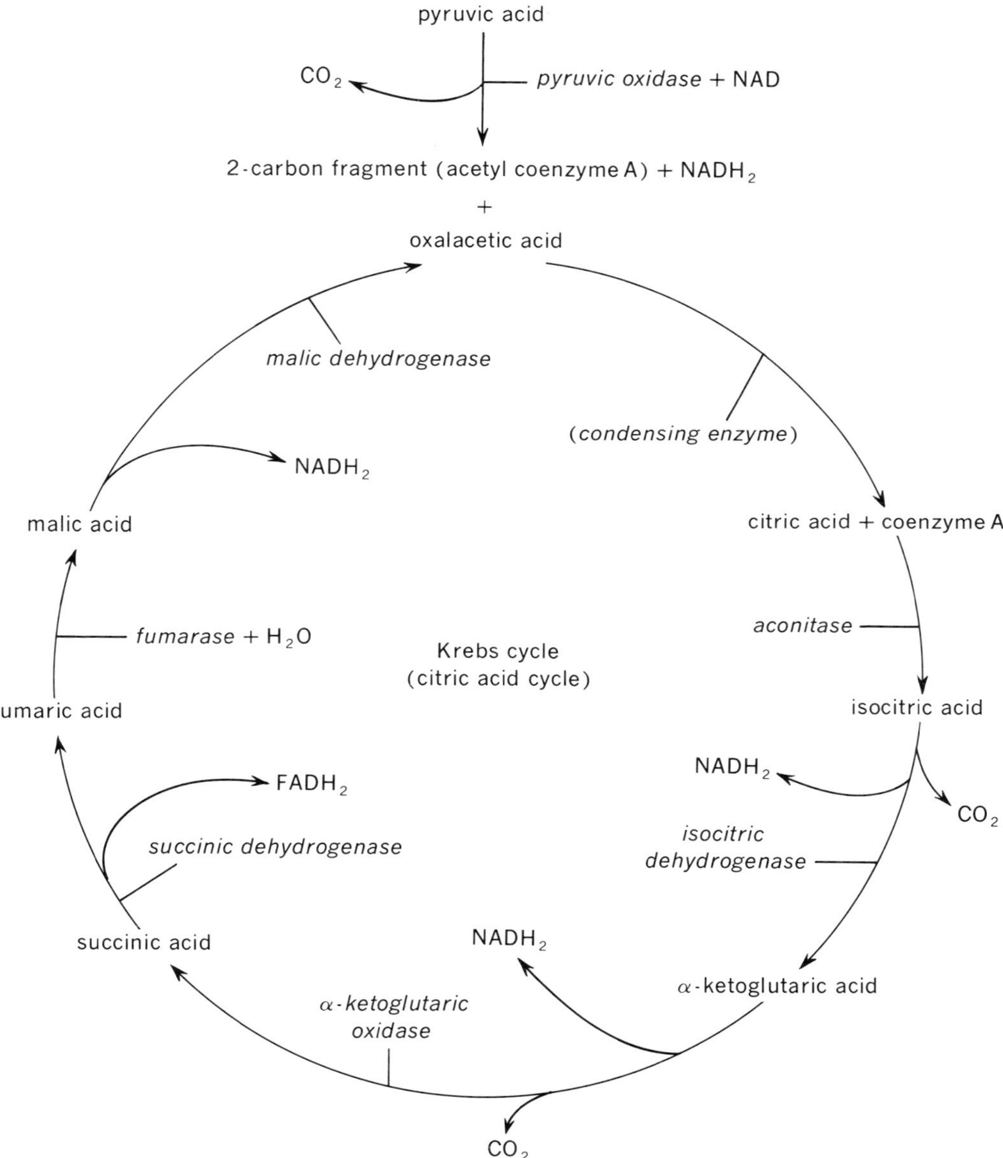

Fig. 12-5 Aerobic oxidation of pyruvic acid.

parison of Figs. 12-4 and 12-5. In all types of fermentation the metabolism of pyruvic acid takes place in the absence of oxygen.

During the initial steps of glycolysis the sugars are phosphorylated and ultimately converted into fructose-1,6-diphosphate. These changes are brought about by the transfer of phosphate from adenosine triphosphate (ATP) to the sugar, the ATP thus being converted to adenosine diphosphate (ADP). Throughout all the following steps of glycolysis up to the formation of pyruvic acid, the derivatives of the original carbohydrate remain in the phosphorylated state.

By the action of the enzyme *aldolase,* the fructose-1,6-diphosphate, a 6-carbon compound, is split into two 3-carbon compounds, or trioses, namely, dihydroxyacetone phosphate and 3-phosphoglyceraldehyde. Some of the dihydroxyacetone phosphate may be ultimately transformed into glycerol and used in the synthesis of fats, but some of it may also be converted into 3-phosphoglyceraldehyde as indicated in Fig. 12-4. By the loss of hydrogen, the latter is then oxidized to 1,3-diphosphoglyceric acid. In this step, nicotinamide adenine dinucleotide (NAD) acts as the hydrogen acceptor, thereby becoming $NADH_2$. Finally, by a series of reactions involving several enzymes, as indicated in Fig. 12-4, pyruvic acid is formed. In the final step of glycolysis the phosphate is taken up by ADP, which becomes ATP. The ATP is then free to give up its phosphate in further reactions resulting in the release of more usable energy.

The metabolism of the pyruvic acid, as previously stated, may follow a number of different patterns. In alcoholic fermentation, by the action of the enzyme *carboxylase,* it is first converted into acetaldehyde (CH_3CHO) through the removal of CO_2 from its carboxyl group (COOH group). The CO_2 escapes as a gas, and the acetaldehyde, by the action of *alcohol dehydrogenase,* is reduced to ethyl alcohol. In the latter reaction $NADH_2$ gives up its hydrogen, thereby becoming oxidized to NAD. The $NADH_2$ may be the same that was formed during glycolysis. In lactic acid fermentation, *lactic dehydrogenase* converts the pyruvic acid into lactic acid.

It should be emphasized regarding anaerobic respiration (fermentation) that it is a much less efficient method for the plant to obtain energy than is aerobic respiration. Whereas the complete oxidation of glucose, as previously mentioned, yields 673,000 cal of energy per mole, the fermentation of glucose to ethyl alcohol yields only 21,000 to 28,000 cal per mole. It is likely, therefore, that when a growing plant is placed under anaerobic conditions, its death may be due partly to the fact that it is unable to obtain sufficient energy to continue its normal physiological activity. Nonetheless, there are some plants that make use of this method exclusively to obtain energy. Among these are some of the bacteria. The lactic acid bacteria are a well-known example. Such plants cannot live in an atmosphere containing free oxygen.

Aerobic, or oxygen, respiration In contrast to anaerobic respiration (fermentation), aerobic respiration takes place only in the presence of free oxygen and results in complete oxidation of sugars to CO_2 and H_2O. As in anaerobic respiration, there is a glycolytic phase and a pyruvic metabolism phase; the latter, however, proceeds by an entirely different pathway, a much more complicated one. The glycolytic phase may proceed in the manner already described under anaerobic respiration (Fig. 12-4), although there are alternative pathways for glucose breakdown such as the pentose-phosphate pathway, which need not be discussed here. The probable steps in the oxidation of pyruvic acid are indicated in Fig. 12-5. Through the action of the enzyme system *pyruvic oxidase,* the pyruvic acid is oxidized to a 2-carbon fragment with the liberation of CO_2. During this reaction NAD is reduced to $NADH_2$. The 2-carbon fragment combines with oxalacetic acid and enters a cycle of reactions known as the citric acid cycle, or the **Krebs cycle**

(so named for the man who first described it). The Krebs cycle involves the action of many enzymes and the formation, step by step, of a series of organic acids, ultimately resulting in the production of more oxalacetic acid, which is able to react with more of the 2-carbon fragments of pyruvic acid oxidation and repeat the cycle. During this process CO_2 is liberated in each of three of the reactions.

The hydrogen removed from the various intermediates of the cycle, together with some developed in the glycolytic phase, combines with NAD to form reduced NAD, which is $NADH_2$. Finally, the hydrogen from $NADH_2$, through the action of the cytochrome enzyme system, combines with oxygen to form H_2O, and by this action the $NADH_2$ is oxidized to NAD. Associated with these reactions and forming a very important part of the process, oxidative phosphorylation takes place in which adenosine diphosphate (ADP) combines with inorganic phosphate to form adenosine triphosphate (ATP). Because of the high-energy compound ATP, the number of ATP molecules formed is of great importance, since this energy can be used when ATP reacts with other compounds to drive metabolic reactions. It will be observed (Fig. 12-4) that some ATP is left over in the glycolytic phase of respiration. Much more is developed in the aerobic oxidation of pyruvic acid. (This is not shown in Fig. 12-5.) Thus aerobic respiration yields much more energy than does anaerobic respiration.

For every molecule of pyruvic acid oxidized in the above reactions, 2½ molecules of oxygen are consumed and 3 molecules of CO_2 and 2 molecules of water are liberated. The overall reaction may be summarized as follows:

$$\underset{\text{pyruvic acid}}{\mathbf{CH_3 \cdot CO \cdot COOH}} + 2\tfrac{1}{2}\,\mathbf{O_2} \rightarrow 3\mathbf{CO_2} + 2\mathbf{H_2O} + \text{energy}$$

In summary, aerobic respiration, like anaerobic respiration, involves a glycolytic phase in which sugars are phosphorylated and then broken down to pyruvic acid. In anaerobic respiration the pyruvic acid is incompletely broken down to ethyl alcohol or to organic acids. The entire process of anaerobic respiration yields only a small amount of energy. In aerobic respiration, the pyruvic acid, through a complex series of reactions involving the Krebs cycle and the cytochrome enzyme system, is ultimately oxidized to CO_2 and H_2O. The end products of aerobic respiration are thus CO_2 and H_2O, and most of the energy originally present in the sugar is liberated for use in metabolism and growth.

Energy relations in respiration It has already been stated that the most important feature about respiration is that it provides energy for the immediate use of the plant. The production of this energy is brought about by reactions involving adenosine diphosphate (ADP) and adenosine triphosphate (ATP). The general characteristics of these substances have already been mentioned. When ATP is converted to ADP and phosphate, approximately 8000 cal of energy per mole are released. On the other hand, when ADP is converted to ATP by the addition of phosphate, approximately 8000 cal per mole are stored in the ATP. During glycolysis of glucose, already discussed, certain of the degradation products are able to react with ADP to produce ATP. Thus the energy of glucose is transferred, by coupled reactions at certain sites in units of about 8000 cal, to ATP. During the incomplete oxidation of glucose in anaerobic respiration, only a small fraction of the energy available in glucose is recovered in ATP, whereas in aerobic respiration a very much larger portion is recovered. This occurs specifically by oxidative phosphorylation during the oxidation of $NADH_2$ by oxygen in the presence of the cytochrome system.

ATP is the most directly utilizable source of energy of the living cell. Once produced, it is able to drive energy-requiring reactions by being reconverted to ADP and phosphate and thus giving up its phosphate bonds. Innumerable energy-requiring reactions utilize the energy of

ATP produced in respiration. Among such energy-requiring reactions in the plant cell which are directly demonstrable to be dependent upon a supply of ATP are the synthesis of proteins from amino acids and the synthesis of complex fatty acids from simple fatty acids. It has also been shown that contraction of muscle fibers, the bioluminescence of fireflies and some fungi, and the production of electricity by certain eels are processes which are directly dependent upon a supply of ATP. The maintenance of body temperature in warm-blooded animals and the energy required for many physiological processes in plants are probably also dependent upon a supply of ATP. Thus ATP is the energy-producing unit of the cell and is utilized in the cell wherever biochemical work must be performed.

The total amount of energy released in respiration depends upon the kind of material oxidized and upon the completeness of the oxidation. Not all substances yield the same amount of energy. Thus a gram of carbohydrate yields, on complete oxidation, 4,100 cal, a gram of alcohol 7,100 cal, a gram of fat 9,100 cal, and a gram of protein 5,800 cal. If the fats are compared with the carbohydrates, it is obvious that a given amount of fat, on complete oxidation, will yield 2¼ times as much energy as the same amount of carbohydrate. This is because the fats contain relatively less oxygen.

In general, compounds rich in hydrogen or low in oxygen yield more energy than compounds relatively low in hydrogen and high in oxygen. The degree of oxidation, as already indicated, is also important. Thus, if glucose is completely oxidized, as in aerobic respiration, it yields 673,000 cal per mole, while if it is oxidized to ethyl alcohol, as in fermentation, it yields only 21,000 to 28,000 cal per mole. In other words, about twenty-five times as much sugar would be used in fermentation to obtain the same amount of energy as is yielded in aerobic respiration. This may help to explain why most plants are unable to survive under anaerobic conditions.

SUMMARY OF THE CATABOLIC PHASE OF METABOLISM

In contrast to the anabolic phase of metabolism, which consists of building-up, or synthetic, processes, the catabolic phase involves only tearing-down processes. The principal catabolic processes are digestion, aerobic respiration, and anaerobic respiration (fermentation). Digestion includes all processes involved in rendering foods soluble and diffusible, so that they may be transported or utilized in the general metabolism of the plant. Such processes involve the hydrolysis of carbohydrates, fats, and proteins by enzymes, which are organic catalysts. Insoluble carbohydrates are hydrolyzed to soluble sugars. Fats are converted into glycerol and fatty acids, the latter being often still further transformed into sugars. Proteins are ultimately broken down into amino acids. Very little energy release is involved in any of these changes. The digested products may be translocated to other parts of the plant and used in respiration, in the synthesis of other organic substances, or in the building of new protoplasm (assimilation). Some of them may be converted back to insoluble forms and stored in various parts of the plant. Digestion is a function of every living cell but takes place particularly where foods are stored, as in germinating seeds.

Respiration begins where digestion leaves off and involves a much more drastic breakdown of the foods with the release of energy usable by the plant. Chemically it is oxidation and includes all catabolic changes involving substantial energy release and any gaseous exchange accompanying this. There are two types of respiration, namely, aerobic, or oxygen, respiration and anaerobic respiration, or fermentation. The former, which is the common type of all higher plants and animals as well as of most of the lower plants, requires a supply of free oxygen and results in the complete oxidation of sugars and other substrates to CO_2 and water, with the release of most of the contained energy of the substance oxidized. The latter takes place in the absence of free oxygen and results in incomplete oxidation to CO_2 and alcohol or to various organic acids, with the release of much less energy.

Both types of respiration involve a glycolytic phase and a pyruvic metabolism phase. The glycolytic phase proceeds by the same pathway in both types of respiration and ends with the production of pyruvic acid. In anaerobic respiration, by the action of various enzymes, the pyruvic acid is ultimately broken down to CO_2 and alcohol or to lactic acid or other organic acids. In aerobic respiration the pyruvic acid is completely oxidized to CO_2 and water after going through a complex series of steps involving the Krebs cycle and many different enzymes. The final stage of oxidation is brought about by the cytochrome (electron transport) system. In this step hydrogen is united with oxygen to form water. CO_2 is liberated during several steps of the complete aerobic process. Oxidative phosphorylation also takes place, by which adenosine diphosphate (ADP) combines with inorganic phosphate to form adenosine triphosphate (ATP). ATP is the most directly utilizable source of energy of the living cell. It is through ATP that the original energy of the respired foods is made available to drive all other metabolic processes involving a supply of energy.

It should be emphasized that the anabolic and catabolic phases of metabolism are completely interdependent. The general metabolism of the cell proceeds by the simultaneous operation of both phases. Only for the convenience of study do we separate them. The ultimate outcome of all metabolism is assimilation and the maintenance of life and growth.

QUESTIONS

1. What anabolic and catabolic processes go on simultaneously in a leaf in the daytime?

2. How do enzymes differ from inorganic catalysts?

3. Of what advantage is it to the plant to store its carbohydrate as starch rather than as sugar?

4. Why is the starch in a mature seed not converted into sugar while the seed is still on the plant?

5. Where does digestion occur in a plant?

6. What advantage has a fat or oil over a carbohydrate as a storage food?

7. Why can the higher plants not survive in an atmosphere lacking oxygen?

8. Why is it not necessary for the higher plants to have an elaborate digestive system like that of the higher animals?

9. What is the principal function of respiration?

10. Would you expect to find large amounts of pyruvic acid in plants?

REFERENCES

BEEVERS, H. 1961. Respiratory Metabolism in Plants. Harper & Row, Publishers, Incorporated, New York. Considers the biochemical events important in respiration.

STILES, WALTER. 1960. Respiration, III. *Botan. Rev.*, **26**(2):209–260. A comprehensive review of all phases of respiration in plants.

WHITE, A., P. HANDLER, and E. SMITH. 1964. Principles of Biochemistry, 3d ed. McGraw-Hill Book Company, New York.

YEMM, E. W., and B. F. FOLKES. 1958. The Metabolism of Amino Acids and Proteins in Plants. *Ann. Rev. Plant Physiol.*, **9**:245–280. Includes breakdown and synthesis of proteins as well as general nitrogen metabolism.

13 GENETICS

Assyrian bas-relief. Two figures of Ashur-nasir-pal, attended by priests in ceremonial attire, assisting at the pollination rite before a conventionalized pistillate date tree (center). The right hand of each of the priests carries the staminate inflorescence of the date, the left hand holds the pollination tray or basket. (Courtesy of the Trustees of the British Museum.)

Fundamental principles of heredity are briefly presented, including Mendel's laws, the physical and chemical basis of heredity, mutation, monohybrid and dihybrid ratios, incomplete dominance, interaction of genes, hybridization, linkage, polyploidy, and population genetics.

Introduction; meiosis In Chap. 10 meiosis was mentioned in connection with the reproductive processes in flowers. Since a knowledge of meiosis is essential to an understanding of heredity, a more detailed discussion of it follows here. The numerous species of plants and animals have a characteristic number of chromosomes in the nuclei of the cells composing their bodies. The gametes of species of plants and animals are likewise characterized by a definite number of chromosomes, which is normally exactly half of that in the nuclei of the body cells. The individual chromosomes differ in size, shape, and many minor structural features. These chromosomes constitute a "set," or more technically, a **genome** in each gamete. For each individual chromosome in the male gamete there is a corresponding chromosome in the female gamete. These two matching individual chromosomes are very much alike. They are similar not only in their external features but also in their generally equivalent hereditary units, or genes.

Union of the gametes brings these types of chromosomes together in the zygote as pairs. Thus each parent, male and female, furnishes one member of every pair of chromosomes in an organism. Because of their origin and similarity in size, form, and general structural features, the two chromosome members are said to be **homologues,** which together constitute a pair of **homologous chromosomes** (Fig. 13-1). For well-known organisms these pairs of homologous chromosomes have been numbered. For example, in cultivated Indian corn, or maize, there are 10 chromosomes as the haploid, or $1N$ number, in each gamete, and 10 pairs, or 20 individual chromosomes, as the diploid number in the zygote and other cells of the sporophyte. These chromosome pairs have been numbered according to their size, with the largest designated as 1 and the smallest as 10.

An understanding of the process of meiosis is essential to the study of reproduction and heredity in plants and animals. Every organism arising by sexual, or gametic, reproduction develops from a $2N$ zygote, or fertilized egg, which is the product of the union of a $1N$ male gamete and a $1N$ female gamete originating in the male and female parents. Fertilization, or union of the gametes, is thus that critical point in the life cycle at which the number of chromosomes is increased from the $1N$, or haploid, to the $2N$, or diploid, number. During *mitosis* the number of chromosomes remains constant from one cell generation to another throughout the growing body of an organism. The nucleus of each cell in the embryo and in the adult developing from it thus has the $2N$ number of chromosomes.

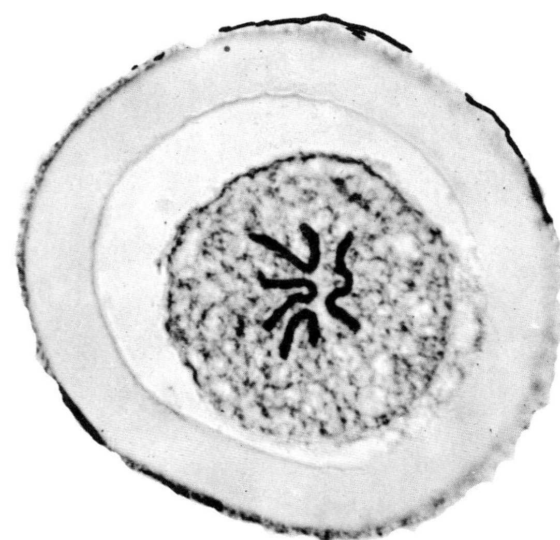

Fig. 13-1 Photograph showing two pairs of homologous chromosomes in Ascaris, *a worm. Note two large chromosomes, upper and right, and two smaller ones, lower and left. (Courtesy of Dr. Mervin Reines.)*

When, however, such diploid plants produce spores or animals produce gametes, the chromosome number is reduced to the haploid. This reduction in chromosome number is accomplished by two successive nuclear divisions, called the first and second meiotic divisions (Fig. 13-2). Together, the two nuclear divisions accompanied by only one division of the chromosomes constitute the process of meiosis. The term meiosis, from Greek *meion,* "lessening" or "reduction," and *osis,* "condition," thus literally a reduction condition, refers to the reduction in the number of chromosomes from the $2N$ to the $1N$ number. The process is often referred to as the reduction divisions. In each of the two meiotic divisions there is a prophase, metaphase, anaphase, and telophase. In these features the meiotic divisions resemble the mitotic divisions occurring in growing tissue. In the mitotic divisions, however, the number of chromosomes remains the same (see also Chap. 2).

In meiosis all the pairs of homologous chromosomes are involved. In the following discussion details of the behavior of only one pair of homologues are given, but all pairs of homologous chromosomes are simultaneously undergoing the same process. Reference to exchange of pieces between chromosomes and chiasmata and the relationship of these to crossing-over is purposely omitted in this discussion.

THE FIRST MEIOTIC DIVISION AND THE SEPARATION OF HOMOLOGOUS CHROMOSOMES Cells that undergo meiotic divisions are generally designated as **meiocytes** (Greek *meion,* "reducing," and *cyte,* "cell"). The meiocytes of plants, most often called spore mother cells (sporocytes), are formed in the sporangia. Haploid, or $1N$, meiospores in plants are produced as a result of meiosis in the meiocytes (spore mother cells).

One fundamental feature of chromosome behavior during prophase of the first division in meiosis, which differs from that in mitosis, is the pairing of the homologous chromosomes. During synapsis, as the process of pairing is called, the two homologues, or synaptic mates, of each pair come into intimate contact throughout their length (Fig. 13-2*B*).

While in contact, each member of the chromosome pair is replicated, forming two chromatids. The replication involves all of the original chromosome except the centromere, which remains intact throughout this stage. The chromatids thus held together in the region of the centromere form chromatid pairs, or diads. Because the two members of the pair of homologous chromosomes are in contact at the time of their replication, there are actually two pairs of chromatids forming a four-partite structure, the **chromosome tetrad,** or perhaps more properly, the **chromatid tetrad** (Fig. 13-2*C*). Although each chromosome is now composed of two chromatids, as long as they remain attached in the region of the centromere, this bipartite structure may still be considered as a single chromosome. As prophase of the first meiotic division draws to a close, the chromosome tetrads move toward the equator of the division spindle. During metaphase of the first division, the chromosome tetrads are located at the equatorial regions of the spindle, forming the equatorial plate.

ANAPHASE AND REDUCTION IN CHROMOSOME NUMBERS During anaphase of the first meiotic division, the original chromosome members of each pair of homologous chromosomes are separated and, with their two chromatids still held together at the centromere, move to opposite poles of the division spindle and eventually into different nuclei (Fig. 13-2*D*). In this way the four chromatids of the chromosome tetrad are separated into pairs, which may now be called **chromosome diads,** or **chromatid diads.** Since the two chromatid members of each diad pair actually constitute an original chromosome, their separation during anaphase of the first division results in the separation, or disjunction, of the two members of the original pair of homologous chromosomes. Each secondary meiocyte, formed as a result of the first meiotic division, receives one chromosome diad, representing one of the members of each original pair of homologous chromosomes (Fig. 13-2*E, F*). Thus the $2N$ number of chromosomes found in the primary meiocyte is reduced to the $1N$ number in each of the secondary meiocytes.

During telophase of the first meiotic division, the chromosome pairs are rounded up into the nuclei of the secondary meiocytes. Accompanying the division of the nucleus, cytokinesis generally, but not always, occurs and two secondary meiocytes are formed from the original primary meiocyte. Secondary meiocytes in plants may or may not be separated by cell walls.

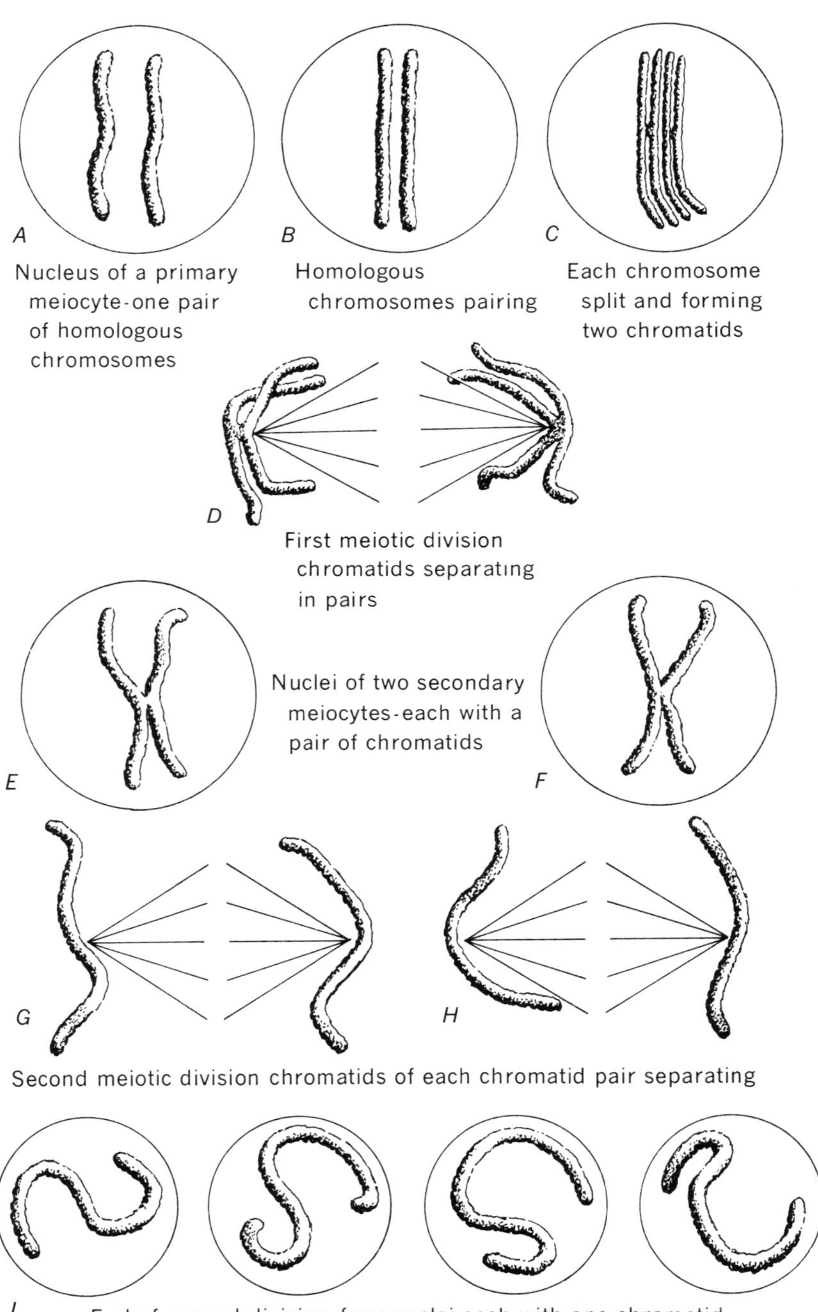

Fig. 13-2 Meiosis. Diagrammatic scheme illustrating behavior of chromosomes during the first and second meiotic divisions. Result is the reduction in number of chromosomes to one half of the original number. (From J. Ben Hill and Helen D. Hill, "Genetics and Human Heredity," McGraw-Hill Book Company, New York, 1955.)

THE SECOND MEIOTIC DIVISION AND THE SEPARATION OF CHROMATIDS In plants a second meiotic division with prophase, metaphase, anaphase, and telophase occurs in each of the two secondary meiocytes, or secondary sporocytes, as they are often called. An important feature distinguishes the second meiotic division from the first. Since only one member of each original pair of homologous chromosomes is present in the secondary meiocytes, there is no pairing of chromosomes. During the second division, however, the centromere holding the chromatids of each chromosome diad is divided. The chromatids of each pair thus released are separated, each passing intact to one of the poles of the division spindle and eventually into a different nucleus (Fig. 13-2*G*, *H*).

The four chromatid members of the original chromosome tetrad are thus separated and each one is distributed to one of the four nuclei that result from meiosis (Fig. 13-2*I*). Consequently the number of chromosomes is reduced to the 1*N*, or haploid, number in each megaspore or microspore. The meiotic divisions are important not only in reducing the number of chromosomes from the diploid to the haploid number but also as a mechanism for the segregation and assortment of the contained hereditary units, or genes. An understanding of meiosis is thus fundamental in studies of heredity.

During telophase of the second meiotic division, each of the four chromatid groups rounds up to become the haploid chromosome set in the nucleus of each of the four meiospores formed from an original meiocyte. The four meiospores are frequently referred to as a quartet (or tetrad) of spores. The appearance of the quartet of spores varies in different kinds of plants. In some cases the spores are in a linear arrangement (Fig. 10-15*F*); in others they are tetrahedral and arranged in a spherical mass often surrounded by the wall of the original meiocyte (Fig. 10-13*F*). When spores formed as a result of meiosis germinate, they grow into gametophytes of varying degrees of complexity.

COMPARISON OF FERTILIZATION AND MEIOSIS Fertilization and meiosis are compensating processes in reproduction that keep the chromosome numbers in a state of equilibrium from generation to generation. Union of haploid gametes brings the number of chromosomes in the zygote up to the diploid number. Then, at the time in the life cycle when meiosis occurs, the chromosomes are reduced to the haploid number. Relationship of fertilization and meiosis in point of time also varies in different organisms. In some of the lower plants such as some of the green algae, fertilization forming the zygote is shortly followed by meiosis in this zygotic cell. This type of meiosis is called zygotic, or initial, meiosis. In many other plants meiosis occurs in spore mother cells formed within sporangia. Because in these cases meiosis occurs later in the life cycle, at some time after fertilization, and the meiotic products are 1*N* meiospores which may grow and develop into haploid gametophytes, meiosis in these plants is designated as sporic, or intermediate.

COMPARISON OF MITOSIS AND MEIOSIS The processes of mitosis and meiosis are similar in that each has the characteristic prophase, metaphase, and anaphase, with nuclear division spindles, and a final stage, or telophase, during which new nuclei are organized. Mitosis and meiosis differ, however, in several details. Mitosis occurs in all meristematic (or growing) tissues, both somatic and reproductive, whereas meiosis is normally restricted to the final two divisions of the nuclei in the sporangia of plants. Further differences between the two processes are found in the behavior of the members of the pairs of homologous chromosomes. During mitosis, not only in haploid, or 1*N*, cells, but also in diploid, or 2*N*, cells where both members of each pair of homologous chromosomes are present, the individual chromosomes act independently. Each chromosome and the genes it contains are replicated, and one chromosome of each pair passes into a new nucleus during the nuclear division. No pairing of homologous chromosomes occurs in mitosis. Successive nuclear divisions are independent. In mitosis, characteristically, a single nuclear division follows each replication (Figs. 2-18, 2-19). This type of nuclear division may continue through many cell generations without change in chromosome number.

During normal meiosis, which usually occurs in $2N$ tissue, the members of the pairs of homologous chromosomes act together, pairing previous to the nuclear divisions. Pairing is accompanied by a replication of each individual chromosome and results in the formation of a group of four chromatids, the chromosome tetrad (Fig. 13-2*C*). This event is followed by two complete nuclear divisions, the first and second meiotic divisions, in which the chromosome behavior in the second division is definitely dependent on the characteristics of the first. This relationship of a single replication of chromosomes followed by two complete nuclear divisions makes possible the reduction in chromosome numbers (Fig. 13-2).

Mendel's contribution The modern study of heredity begins with the publication in 1866 of a paper by Gregor Mendel, an Austrian monk. In this paper the author reported the results of some very important experiments in heredity. Somehow the full significance of these experiments was not comprehended by scientists until about 1900, when they were rediscovered and confirmed independently by De Vries, Correns, and Tschermak, European biologists; it was Bateson, a British zoologist, who, through his publications, was largely instrumental in bringing Mendel's conclusions to the general attention of biologists. In his investigation of the inheritance of the somatic characters of the common garden pea, Mendel established certain principles which have since been shown to be applicable to a great diversity of organisms, both plant and animal, including man. In fact, since 1900 the results of investigations of inheritance in a large number of plants and animals have so uniformly supported the fundamental principles of heredity which Mendel established that they have come to be regarded as some of the great generalizations of biological science.

The facts of heredity In his study Mendel selected plants showing contrasting characters, such as tall plants and dwarf plants, plants bearing red flowers and plants bearing white flowers, plants producing smooth seed and plants producing wrinkled seed. His method of experimentation was to mate plants (hybridize) which differed in one pair only of these contrasting characters, e.g., tall plants mated with dwarf plants. Later in his experiments he mated plants which differed in regard to two or more pairs of contrasting characters, e.g., tall plants bearing smooth seed mated with dwarf plants bearing wrinkled seed.

Artificial mating, or crossing, in plants is accomplished by removing the young stamens or anthers from the flower to be used as the female, or seed, parent. This prevents self-pollination, or selfing. The flower is then covered with a small bag to prevent natural, or open, pollination (Fig. 13-3). Later, when the stigma is mature, or receptive, pollen from the selected male parent is placed on the stigma of the female parent. Seed produced by the female plant may produce hybrid plants. This technique is called hybridization, or making a cross.

The offspring resulting from seeds produced by the mating of parent plants with contrasting characters, $\mathbf{P_1}$, are referred to as the **$\mathbf{F_1}$ generation,** or first filial generation of hybrids. In such hybrids it was found that often only one of the contrasting characters appeared in the F_1 generation; that is, in a cross between tall pea plants and dwarf pea plants all the F_1 hybrids were tall plants (Fig. 13-4). Mendel called the parental trait which thus appeared in the F_1 generation the **dominant** trait or character. The parental trait which did not appear in the F_1 hybrids he called the **recessive** trait. Tallness is therefore dominant over dwarfness in peas.

When the seeds produced by the selfing of tall dominant F_1 hybrids were planted, the sec-

Fig. 13-3 Bagging of plants to prevent natural cross-pollination. (Courtesy of U.S. Regional Pasture Research Laboratory, Agricultural Research Service, U.S.D.A., University Park, Pa.)

ond filial generation, or the **F_2 generation,** of hybrids was obtained. An interesting observation was made by Mendel when he discovered that both of the original parental traits were found in plants of the F_2 generation, in this case some tall pea plants and some dwarf pea plants (Figs. 13-4 to 13-6). The tall plants were apparently exactly like the original tall plants of the parental strain and the tall plants of the F_1 hybrid generation. The dwarf pea plants were apparently exactly like the original dwarf plants of the parental strain. These tall and dwarf plants were, however, not equally numerous. In every F_2 generation studied, there were about three times as many dominant tall plants as there were recessive dwarfs. Mendel found F_2 totals of 787 tall and 277 dwarf plants, i.e., 2.84 tall to 1 dwarf or approximately a 3:1 ratio from all crosses.

Mendel studied the third filial generation of hybrids grown from seeds produced on the selfed F_2 hybrid plants. Seeds from the dwarf F_2 pea plants produced all dwarf plants; i.e., they were pure for the recessive dwarf trait. One third of the selfed tall F_2 pea plants (or one fourth of the total F_2 population) bore seeds which produced tall F_3 plants; i.e., they were pure for the dominant tall trait. Two thirds of the selfed tall F_2 pea plants (or one half of the total F_2 population) showed by this breeding test that they were not pure for either the tall or the dwarf trait. The seeds from this group of plants produced tall plants and dwarf plants in the ratio of three tall plants to one dwarf plant. This behavior indicated that they were actually like the original F_1 hybrids as regards their breeding qualities, for the original F_1 hybrid produced a 3:1 ratio of tall to dwarf plants.

Segregation In each cell of the pea plant there is the $2N$ number of 14 chromosomes. When maturation occurs in the microspore mother cells of the anthers and in the megaspore mother cells in the ovules of a pea plant, meiotic divisions reduce the number of chromosomes from the $2N$, or diploid number of 14, to the $1N$, or haploid number of 7.

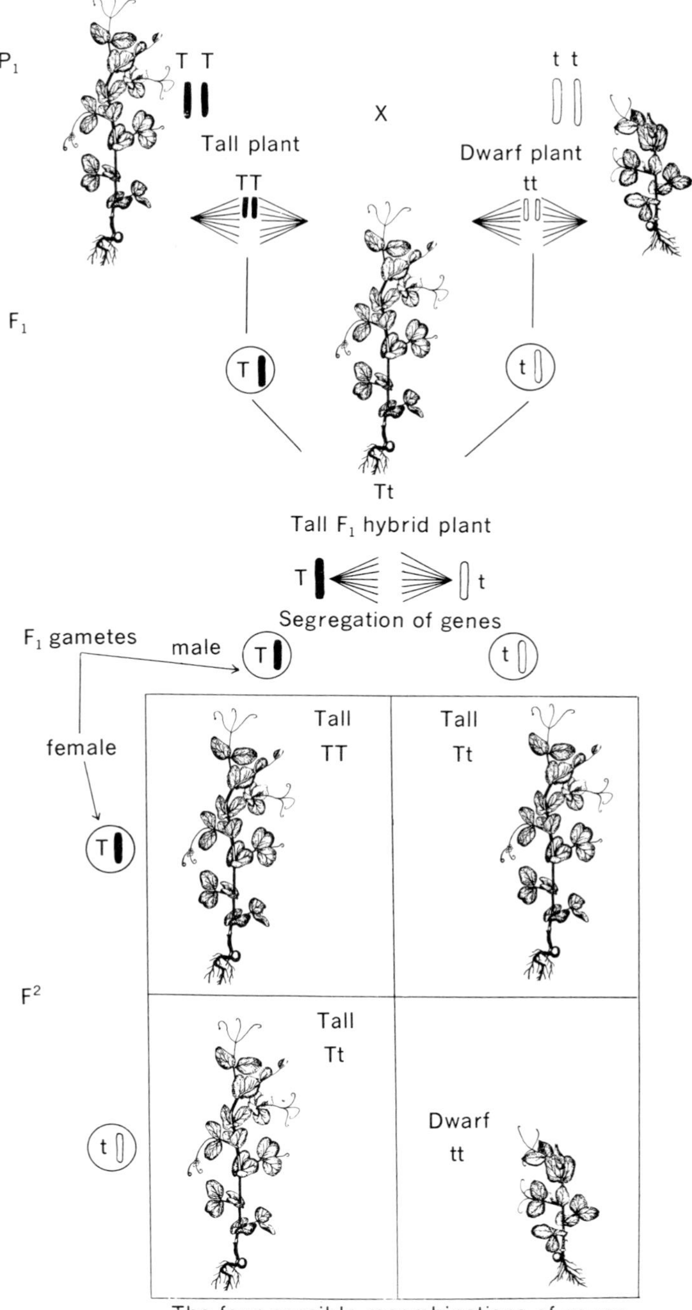

Fig. 13-4 Inheritance in tall and dwarf peas; above, the tall and dwarf parents, P_1; center, the tall first-generation hybrid, F_1; below, the two classes of the second generation, F_2, consisting of tall and dwarf plants in the ratio of three tall to one dwarf. (From J. Ben Hill and Helen D. Hill, "Genetics and Human Heredity," McGraw-Hill Book Company, New York, 1955.)

Fig. 13-5 F_2 *generation, tall and dwarf segregates from cross of pea plants differing in height. Plants supported in upright position by dead branches of trees. This "brush" treatment permits observation of differences in size. (Photograph from cultures of Dr. E. M. East, courtesy of Dr. Orland E. White.)*

Fig. 13-6 Diagram showing relation of genes to chromosomes and their behavior in inheritance; above, gametes of parents, P_1*, uniting to form fertilized egg which will develop the first hybrid generation,* F_1*; center, the two meiotic divisions in* F_1 *hybrid during which the homologous chromosomes with genes* T *and* t *are separated (segregation) and eventually spores are produced; the spores develop the gametophytes and eventually the gametes; below, recombination of genes as a result of self-fertilization in* F_1 *hybrid; checkerboard shows the four classes of* F_2 *progeny;* T*, the gene conditioning tall, and* t*, dwarf.*

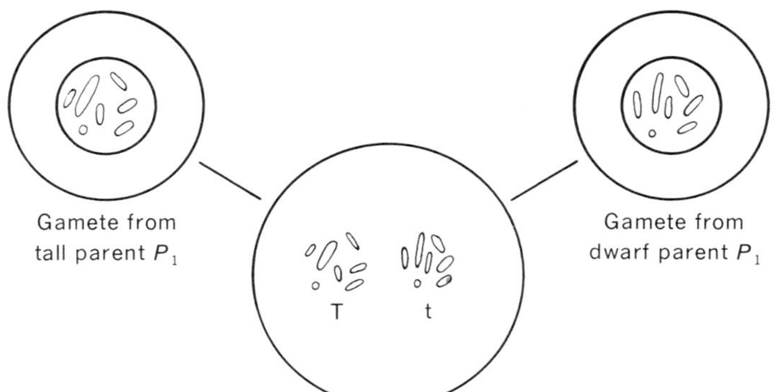

Fertilized egg which may develop into the F_1 hybrid plant

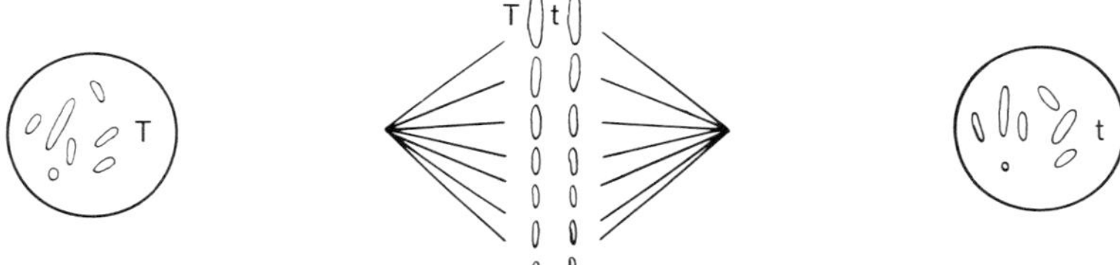

Diagram of reduction division in the F_1 plant

Equation division in F_1 plant

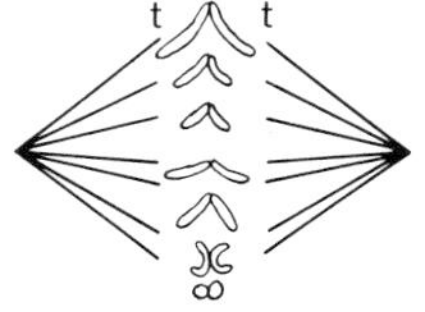

Equation division in F_1 plant

1 N spores formed by the F_1 plant

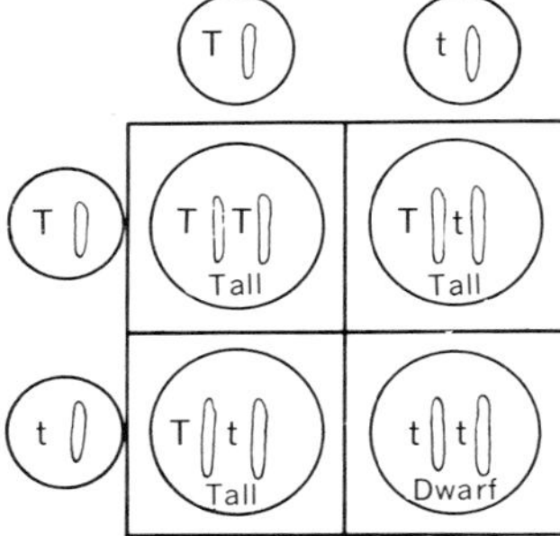

Diagram showing the union of F_1 gametes to form the F_2 plants

F_2 ratio 3 tall to 1 dwarf

Not only is the number of chromosomes reduced to the half number during meiosis, but in all normal meiotic divisions the two members of each of the seven pairs of homologous chromosomes, or synaptic mates, are separated from each other, passing to opposite poles of the division spindle and ultimately into different megaspores and microspores, from which they pass on into different gametes. Since the two genes *T* and *t*, for tallness and dwarfness, respectively, are carried in different members of a pair of homologous chromosomes, they too are carried apart at meiosis. The separation of the genes *T* and *t*, called the **segregation of allelic genes,** is the basis of **Mendel's first law of heredity.** This separation is of the utmost importance, for it ensures that these genes, derived from different parents and associated together in the F_1 hybrid, are finally completely separated and pass into different gametes.

After Mendel's time further investigations into size and other characteristics of pea plants showed that there are very tall varieties, real dwarf kinds, and intermediate ones called half-dwarfs, whose heights may actually differ considerably among the varieties. It has been demonstrated that height in peas depends largely upon two inherited structural features: (1) the number of internodes and (2) the lengths, long and short, of the internodes. The very tall varieties have numerous long internodes. The dwarfs have relatively few short internodes, and the half-dwarfs an intermediate but variable number of medium-long internodes, with the lengths varying from variety to variety.

Dr. Orland E. White studied inheritance in approximately 250 varieties of peas showing about 75 distinct characteristics. He thought that at least five pairs of allelic genes, acting within the environment, control development of height in peas. He believed that the investigators of height in peas, including Mendel, studied the length of internodes. Plants with short internodes were generally regarded as dwarfs, those with long internodes as tall in most cases, regardless of the relative number of internodes involved. Crosses between these half-dwarf types may yield 3:1 F_2 ratios (Fig. 13-5). White thought that probably Mendel's cross was between a long-internode half-dwarf, regarded as tall, and a short-internode half-dwarf, regarded as dwarf. This cross yielded an F_2 generation with the ratio of three half-dwarfs with long internodes to one half-dwarf with short internodes. Mendel interpreted this as a 3:1 difference in height.

Monohybrid ratios in other characters Pea plants bearing round seeds in contrast with wrinkled ones have distinct hereditary units, or genes, that determine the dominant round trait (Fig. 13-7). These genes may be designated *RR*, where *R* indicates the round trait. The allelic genes determining the contrasting recessive trait wrinkled may be indicated as *rr*. The seeds of peas differ in color as well as shape. There are dominant yellow-colored seeds and recessive green seeds. The allelic genes determining these contrasting traits are *Y-y*, where *Y* determines the yellow and *y* the green color.

Following hybridization of two pea plants, one bearing dominant round seeds *RR* and the other recessive wrinkled seeds *rr*, the P_1 female plant produces pods containing all round seeds with heterozygous F_1 embryos *Rr* (Fig. 13-8). Self-fertilization of F_1 plants grown from these seeds produces pods with both round and wrinkled seeds, frequently with both types in one pod. The embryos in these seeds are *RR*, *Rr*, and *rr* in the ratio of 1:2:1. Because of full dominance of *R* over *r* there will be three round seeds—one *RR* and two *Rr*—to one wrinkled, *rr*. Furthermore, because the characteristics are traits of the embryos, or seed coats, hybridization involving dominant yellow, *YY*, and recessive green, *yy*, in peas yields three yellow to one green seed in the pods produced by selfed F_1 *Yy* plants.

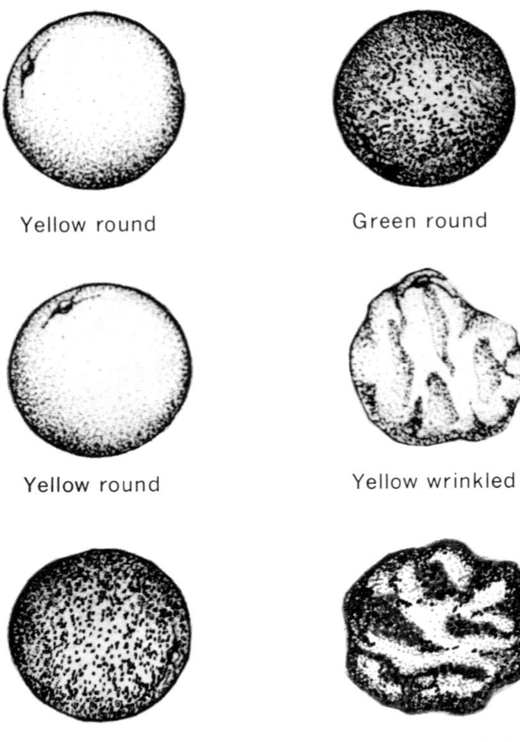

Fig. 13-7 Some characters of peas. Studies of inheritance have included experiments with these and other characters of peas.

Dihybrid ratios Combinations of two or more independently inherited traits may occur in one organism, as, for example, tallness or dwarfness of plant with shape and color of seeds. Tall pea plants may bear either round or wrinkled seeds, which may be either yellow or green. Likewise, the seeds produced by dwarf pea plants may be round yellow, round green, wrinkled yellow, or wrinkled green, depending on the hereditary factors, or genes.

Mendel, interested in learning whether two pairs of contrasting traits were independently inherited, crossed a pea plant bearing dominant round yellow seeds, *RRYY,* with one producing recessive wrinkled green seeds *rryy*. Following this hybridization, seeds produced all showed the dominant round yellow characteristics. The embryos within these seeds were heterozygous, *RrYy* for both pairs of genes, and were therefore **dihybrids.**

When these seeds were planted, their F_1 embryos developed into adult F_1 plants. The seeds borne by the F_1 plants contained F_2 embryos and represented the F_2 generation with various recombinations of shapes and colors of seeds. Mendel found that the F_1 plants "yielded seeds of four sorts, which frequently presented

Fig. 13-8 Inheritance of round and wrinkled characters of peas; above, round and wrinkled parents, P_1; center, the round first-generation hybrid, F_1; below, the two classes of the second generation, F_2, consisting of three round seeds to one wrinkled seed.

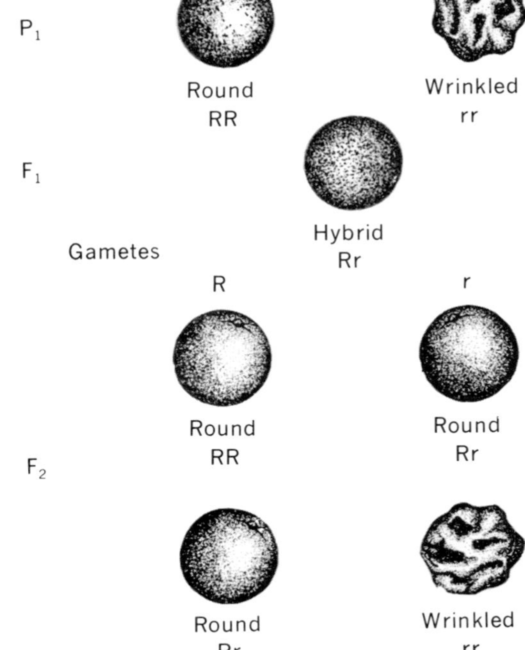

themselves in one pod." In a total of 556 seeds, there were 315 round yellow, 101 wrinkled yellow, 108 round green, and 32 wrinkled green seeds. This represents an approximate ratio of 9:3:3:1.

In seeking an explanation of the 9:3:3:1 F_2 ratio of traits found when parents that differ in two pairs of characteristics are crossed, it is essential to consider the relationship of the genes to the chromosomes. The relationship of genes to chromosomes is such that the gene R is located on a certain chromosome and the gene r is located on the other member of this pair of synaptic mates, or homologous chromosomes. Likewise, Y and y are located on the two members of another pair of synaptic mates. During meiosis the two members of a pair of synaptic mates separate; that is, the chromosome bearing the gene R separates from the chromosome bearing the gene r. At the same time, separation of the synaptic mates is occurring in all six pairs of the remaining chromosomes including the pair bearing the genes Y and y. The chromosome bearing the gene Y is separated from the chromosome bearing the gene y. This chromosome behavior results in the simultaneous segregation of the two pairs of genes R-r and Y-y. Since there is no relationship between the two pairs of synaptic mates, or pairs of homologous chromosomes, concerned, the accompanying segregation is called the independent segregation of two pairs (or more) of genes. This independent segregation of synaptic mates leads to the **chance,** or **free, assortment** of the genes into the resulting microspores and megaspores and finally, of course, into the gametes produced by the F_1 hybrid plant.

Simultaneous inheritance of two characters has been studied in a variety of characters and organisms. Among these may be mentioned summer squash, which has fruits of disk and spherical shapes which may be white or yellow. In the F_2 generation of crosses between plants with white disks and yellow spheres, progenies of four phenotypic classes were produced (Fig. 13-9), in the dihybrid 9:3:3:1 ratio. There were nine plants with white disk fruits, three with white spheres, three with yellow disks, and one with yellow spheres.

The physical basis of heredity Mendel believed that there must be some substance in the germ plasm or gametes, which was related to the development of each somatic trait he studied. Mendel's reputation as one of the great biologists of modern times rests upon the fact that he formulated a theory of the behavior of the hereditary factors. This theory is in accord not only with observed phenomena of breeding but with the facts which have been more recently learned about the behavior of the chromosomes in the reproductive cells.

The presentation in 1900 of Mendel's original work stimulated great interest in the study of heredity and led directly to the establishment of the experimental method in this field.

The scientific study of the problems of heredity by experimental methods, combined with a careful study of the behavior of the reproductive cells, has made possible the expansion of Mendel's original theory into one of the most important generalizations of biological science.

The essential feature of this generalization is the conception that the factors determining hereditary characters are associated with the chromosomes of every cell in the body. Since these hereditary factors are inseparably associated with the chromosomes, they move with them through all the cell divisions of the maturation processes which lead in plants to spore production (and in animals to the production of gametes). The union of the gametes at fertilization brings the chromosomes from each parent, with their associated hereditary factors, into the fertilized egg (zygote) which develops into a new individual of a new generation. In this way the hereditary factors pass from generation to generation. The behavior of the chromosomes during meiosis and fertilization forms the physical basis for the transmission of the factors which

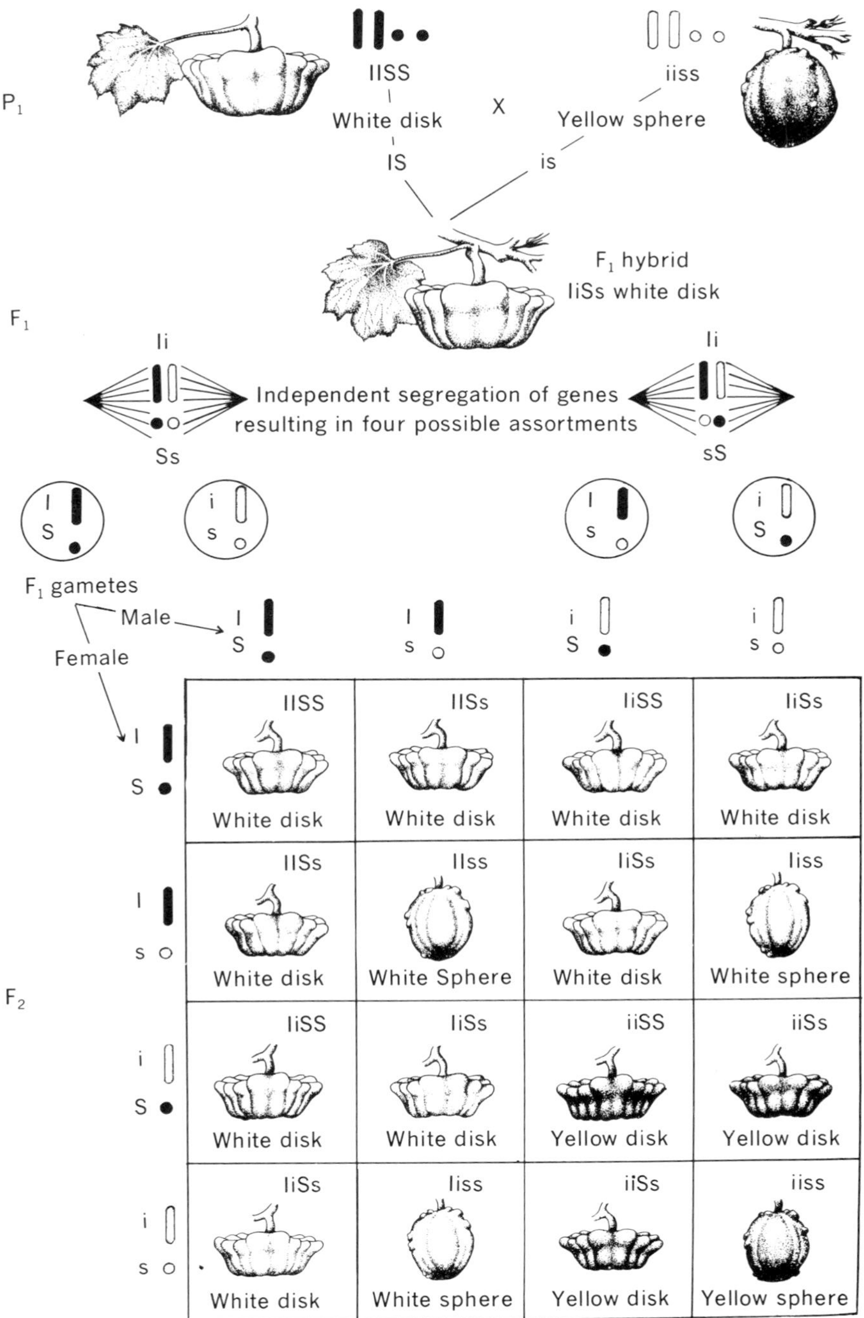

Fig. 13-9 Simultaneous inheritance of two pairs of contrasting characters in summer squash. (From J. Ben Hill and Helen D. Hill, "Genetics and Human Heredity," McGraw-Hill Book Company, New York, 1955.)

determine or influence the development of hereditary characters (Fig. 13-6). This conception of the relations of hereditary factors to the chromosome is known as the **chromosome theory of heredity** and is considered as offering the most plausible explanation of the observed facts of heredity.

In this theory the hereditary unit, or material substance most closely related to the development of a trait, is assumed to be located in the chromosomes. This material is called a determiner, a hereditary factor, or a **gene.** Genes occur in pairs on homologous chromosomes, one allele on each. The term **allele** emphasizes the paired relationship of the genes. Each gene is the allele of the other, and the two genes constitute a pair of alleles, or allelic genes. Since the genes are located in the chromsomes, their passage from one generation to another is dependent upon passage of the chromosomes by the gametes from one generation to the following.

Chromosome mapping Location of the genes on the chromosomes, or **chromosome mapping,** develops into a specialized science in which the study of chromosomes and inheritance of characteristics are combined. Chromosome mapping depends on information about the linkage (see later) of genes, which is dependent on the physical behavior of the chromosomes in relation to the reappearance of characters in generations subsequent to the mating of known parents. The fruit fly, *Drosophila,* and the corn plant, *Zea mays,* are among the organisms with best-mapped chromosomes.

Hereditary variation Genetics deals not only with the stability of genes and chromosomes but also with the evidences of their change, or mutation. Stability in inheritance accounts for the repeatability of specific characters of succeeding generations of plants and animals. Mutability, on the other hand, is involved in organic changes which provide the physical basis for variations in the genic material which is heritable. Not only does hereditary variation result from the chance recombination of chromosomes (genes) following independent segregation or assortment during meiosis, but it also results from mutations or changes in genes or chromosomes. **Gene mutations** result from changes in the chemical structure of a gene and **chromosomal mutations** from changes in either the structure or number of chromosomes.

Gene mutations, or changes in chemical structure resulting from molecular rearrangements, may occur at any time in any cell of the organism. Few of the changes are seen and only those changes that appear in the reproductive cells are transmitted from parent to offspring through genetic reproduction.

Many gene mutations are not seen because, frequently, only one gene of an allelic pair mutates and very often the mutation is recessive. Some mutations are **lethal** and may result in such a severe alteration of a metabolic process that survival of the organism and, consequently, of the mutation is impossible. If a dominant mutation occurs in certain vegetative cells, such as those of a flower bud, the subsequently developed flower or fruit may exhibit characteristics considered to be desirable by man. Such **bud mutations** have on occasion given rise to new kinds of fruits, such as nectarines; new or more desirable flowers, as a more showy sweet pea; or a desirable tree, as in the case of the Lombardy poplar.

Genes are remarkably stable but they can and do change, or mutate. When a change occurs under natural conditions, it is referred to as **spontaneous mutation.** Changes can also be induced by the artificial application of agents, such as ordinary or radioactive chemicals, and certain kinds of radiation, such as X rays and ultraviolet radiation. Agents that induce mutation are called **mutagens;** however, the differences, if they exist, between spontaneous and induced mutations are not clear.

In evolution the importance of variation within a population is most important and unless variation is present, it is impossible for natural selection to operate. Evolution has been defined as a change of the frequency of genes within a population. It is obvious that mechanisms that contribute to heritable genetic variability have been and are prerequisite to organic change.

The nature of genes The presence of a structural entity related to the reappearance of characters in successive generations was postulated by Mendel. This postulate is a significant part of Mendelian inheritance and is conceived of as particulate in nature. Thomas Hunt Morgan and his co-workers laid the groundwork for the rather general acceptance of genes as these structural entities in the chromosomes. The existence of genes has been assumed, but their existence has not been demonstrated with certainty. We know that genes are a part of the large DNA molecule and that there is more than one kind of gene. A gene might be conceived of as a molecule of DNA (Fig. 2-5), but it is possible that a single molecule of DNA serves or functions as more than one gene. Biochemical studies have offered explanations of how genes function (see also Chap. 2). These explanations assume that genes carry "codes" which relay particulate information that dictates what chemical ingredients, along with their arrangements, contribute to the formation of proteins (Fig. 2-17). Different codes are possible because the nucleotides, adenine, guanine, thymine, and cytosine (Fig. 2-16) may be linked together in variable arrangements. Since a molecule of DNA may have as many as 20,000 nucleotides, all of which may have a variety of arrangements of the code units (A-T, T-A, C-G, G-C), the number of possible combinations is tremendous. Thus a biochemical explanation has been offered for the possible structure and behavior of a gene.

Other actions of genes in the development of traits Besides the allelic relationship of dominance and recessiveness, which may be either full or incomplete, genes may show other actions. Although generally recognized by its determination of some rather prominent trait, a single gene may influence the development of more than one character. As an example, certain genes may determine the development of color in the flower or in the seed and also influence the production of color in other parts of the plant, such as the stem or the leaves. In other cases certain genes may determine the development of some visible characteristic like color or form and in addition interfere with some *vital function*. Genes that act in this way cause the death of the organisms that receive them. They are called **lethal genes.** Genes that act in this way usually yield a 2:1 F_2 ratio instead of the expected 3:1. This is because one of the organisms in each four dies as a result of the lethal action of the genes. It should be emphasized, however, that genes which influence more than one trait are not necessarily lethal in their effects.

The interaction of genes in the development of characteristics It is now recognized that many heritable traits are *dependent upon more than one gene.* Possibly every gene in the body exerts an influence in the development of every trait, with some one, two, or a few *major* genes having a predominant effect. Many cases are known in which a given characteristic is dependent upon two, three, or several distinct pairs of major genes. In these cases the genes act to *complete, supplement,* or even *inhibit* the action of other major genes of the group. A group of genes acting together in this way is called a **gene complex,** and the **interaction** often results in the numerical modification of the expected Mendelian ratios.

Some modifications of the ordinary F_2 dihybrid ratio of 9:3:3:1 are as follows:

9:7	13:3
9:3:4	12:3:1
9:6:1	15:1

Although the numerical relationships of visible traits are modified in the above ratios, *they constitute no exceptions to Mendelian principles. Segregation, assortment, and recombination of genes occur regularly; the modifications of the ratio are the result of the action and interaction of more than one pair of genes* in the development of the visible characteristics.

Incomplete dominance of the genes Since Mendel's time other investigators have found that genes do not always show complete dominance and recessiveness as exhibited in the case of tallness and dwarfness in peas. The genes determining or conditioning many characters are only partly, or incompletely, dominant. This is true in the case of many bright-colored flowers such as red-flowered four-o'clocks. When plants such as the four-o'clock, bearing red-colored flowers, are crossed with plants bearing white flowers, the plants of the F_1 hybrid generation bear intermediate pink-colored flowers. In the F_2 generation the ratio is 1:2:1, i.e., one plant bearing red-colored flowers, two plants bearing pink flowers, and one plant bearing white flowers. The red- and white-flowered plants continue to breed true in succeeding generations, but the pink-flowered plants, being actually hybrids, in succeeding generations continue to segregate into the red-, pink-, and white-flowered plants in the ratio of 1:2:1.

These instances of **incomplete dominance,** now known to be very numerous, are due to the action of the genes. Neither the gene *w*, conditioning white-colored flowers, nor the gene *W*, conditioning red-colored flowers, is completely dominant or completely recessive when in hybrid combination as *Ww* in the F_1 generation. Both genes seem to react together, making pink-colored flowers. Segregation of the genes during meiosis takes place as expected, *W* separating from *w*. Recombination at the time of fertilization results in the formation of one *WW* zygote which upon maturity produces red flowers, two *Ww* zygotes producing ultimately pink flowers, and one *ww* zygote producing white flowers.

Linkage During recent years the investigation of the inheritance of the numerous characters of many different organisms, both plants and animals, has expanded and extended the conception of heredity as proposed by Mendel in 1866. The fundamental idea of heredity, viz., segregation, has remained substantially as Mendel proposed it. With the increased knowledge of the chromosomes and their behavior, certain new ideas of heredity have come to be held generally among students of genetics. It has been found that occasionally two characters, or traits, tend to be inherited together, not segregating as normally expected. When characters tend to be inherited together, they are said to be linked and the condition is called **linkage.** The physical, or material, basis of linkage is to be sought in the relation of the genes to the chromosomes, as described in the following paragraph.

Since the number of chromosomes in any species of organism is limited and the characters, or traits, of complex organisms are very numerous—from a practical working basis almost unlimited—it follows that there are a great many more genes conditioning these numerous traits than there are chromosomes. In the pea, for instance, the $2N$ number is 14, in corn, 20, in the lily, 24, in wheat, 42. It has been found for all organisms thoroughly studied that each chromosome carries a definite and an extensive group of genes (Fig. 13-10). Two or more genes carried on a certain chromosome are said to be linked. All the genes on a pair of homologous chromosomes collectively form a **linkage group.** In general the genes on a chromosome are inherited exactly as a single gene. This follows because the behavior of a single gene or a group of linked genes is determined by their association with the chromosome on which they occur. When reference is made to the inheritance and behavior of a single gene on a certain chromosome, it should be kept in mind that actually there is a whole group of genes in the chromosome but that only one gene is being considered at the time.

Linkage of genes leads to numerical modifications of the dihybrid ratios. These modifi-

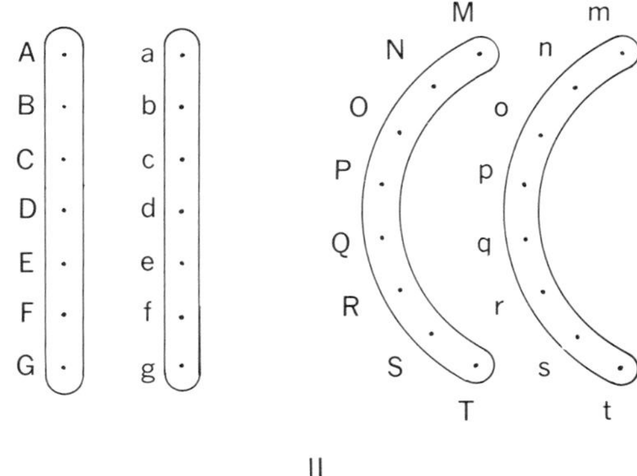

Fig. 13-10 Diagram of the relationship of genes and chromosomes forming linkage groups. I, a pair of homologous chromosomes, each with a group of genes A and a to G and g. These genes constitute a linkage group; II, a second pair of homologous chromosomes of different shape with genes M and m to T and t forming a second linkage group.

cations constitute an exception to Mendel's second law of heredity, the assortment of genes. This exception is based upon (1) the relationship of the linked genes to the chromosomes and (2) the behavior of chromosomes during meiosis, which influences the segregation of genes.

Chromosomes and genes in sex determination

Except for hermaphrodites, or bisexual individuals, animals are regularly of two sexes, sperm-producing males and egg-producing females. In the gamete-bearing, or gametophytic, phase of lower plants, as in some algae and bryophytes, there are also distinct sexes, sperm-producing male and egg-producing female plants. Among species of the angiosperms, such as common garden asparagus, cultivated hemp, campion (*Lychnis*), and a few others, two kinds are found and popularly called male and female. These are sporophytic plants, or spore-producing and not gamete-producing. The spores produced differ in size, and the plants are correctly termed microsporous and megasporous. In both plants and animals the sexual condition is generally determined on a chromosome basis and is therefore inherited.

In several groups of animals and in some plants, one of the several pairs of chromosomes has been definitely related to sex determination and they are therefore called the sex chromosomes. The female generally has a pair of homologous sex chromosomes, designated as X because originally their function was unknown. The male usually has a pair of sex chromosomes composed of an X chromosome and another, which may be larger, smaller, or sometimes odd shaped, called the Y chromosome. As regards the sex chromosomes, a female is therefore XX and a male XY. In some cases the male may lack the Y chromosome and thus has one chromosome less than the number normal for the female. This condition is usually designated as XO. Actually, the characteristics maleness and femaleness are conditioned predominantly by genes, or hereditary factors, carried in the X member of the sex chromosomes. The Y chromosome is, perhaps, very nearly a blank.

When an XX female produces gametes, each egg cell normally contains one X chromosome. As regards the sex chromosome, the female gametes are all alike. When a male with XY sex chromosomes forms gametes, however, the sperms are of two kinds, equally numerous, one with an X and the other with a Y chromosome. During reproduction an X egg has equal chances of being fertilized either by an X or a Y sperm. For this reason, in most groups of sexually reproducing organisms, the sex ratio is 50:50, that is, one female to one male. Sometimes, because of unequal viability, the sexes may not be equally numerous at birth or shortly thereafter.

Among such social insects as the bee, the sexes differ conspicuously in the number of chromosomes. The males, which develop parthenogenetically from eggs that have not been fertilized by a sperm, have only one half (haploid) the diploid number of chromosomes characteristic of the females that result from fertilized eggs. Thus, in the social insects, femaleness appears to accompany a normal diploid complement of chromosomes and maleness a deficiency or abnormality in the chromosome complement.

SEX LINKAGE The X and Y chromosomes carry many of the genes primarily concerned in sex determination and are therefore known as the sex chromosomes. Besides the sex-determining genes, the sex chromosomes may also carry other genes, which collectively are called **sex-linked genes,** and the traits they condition are known as **sex-linked characters.** Most of the sex-linked genes are in the X chromosomes and are more specifically designated as X-linked genes. The Y chromosome generally carries comparatively few genes, which are called the Y-linked genes. Thus sex-linked genes are mostly associated with the X chromosome and pass through the generations with it.

Sex linkage differs from ordinary, or autosomal, linkage. In sex linkage one or many genes conditioning certain body, or somatic, traits are also carried in the X chromosomes and are thus associated, or linked, with the sex-determining genes as well as with one another. In ordinary, or autosomal, linkage, two or more genes determining somatic characters are associated together in the autosomal chromosomes, or autosomes. Sex-linked inheritance was discovered when some traits, especially recessive characters, manifested themselves in only one of the sexes, usually the male, often missing the female in that generation.

The explanation was found in the chromosomal constitution. The female has the XX constitution, and the male the XY, or in some cases XO, that is, only an X with the Y entirely lacking. Since the female usually has two X chromosomes, she may carry either two dominant genes and be homozygous dominant, one dominant and one recessive gene and be heterozygous, or two recessive genes and be homozygous recessive for any one or several of a considerable number of sex-linked characteristics. The male with only one X chromosome and an odd or Y chromosome may have one dominant or one recessive gene carried in the X chromosome but never both, because the Y is largely devoid of genes. Normally he cannot be heterozygous.

Sex determination is best known in animals.

Fig. 13-11 Photomicrographs illustrating polyploidy in certain species of Bromus, *a genus of the grasses. (Courtesy of Dr. Frank L. Barnett, U.S. Regional Pasture Research Laboratory, Agricultural Research Service, U.S.D.A., University Park, Pa.)*

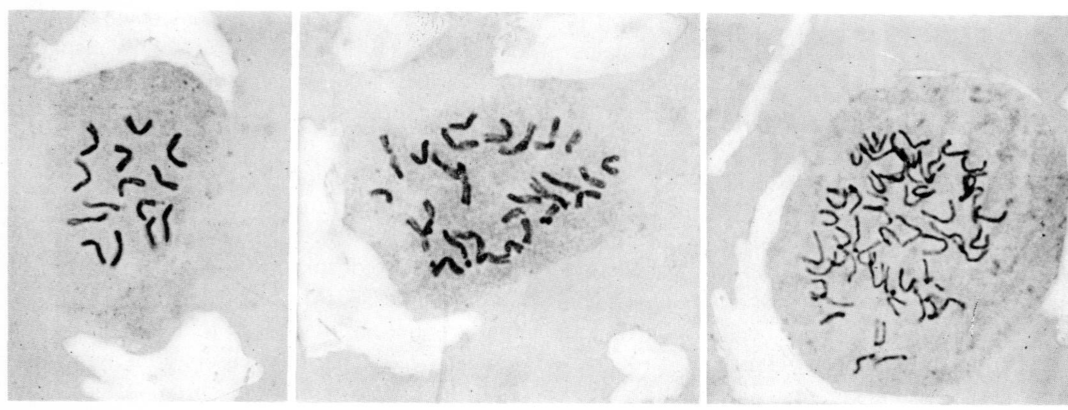

B. anomalus $2N = 14$ | *B. severtzonii* $2N = 28$ | *B. inermis* $2N = 56$

Fig. 13-12 Size differences related to chromosome numbers in plants. From left to right, tetraploid, diploid, and haploid tomato plants, Lycopersicon esculentum. *(Courtesy of E. W. Lindstrom, from J. Ben Hill and Helen D. Hill, "Genetics and Human Heredity," McGraw-Hill Book Company, New York, 1955.)*

A few cases have also been reported in plants. The discovery of sex linkage and its relation to chromosomal constitution was one of the cornerstones in the building of the chromosome theory of heredity.

Polyploidy Studies of chromosome numbers in plants are of importance in several aspects. These numbers vary from the 1 set, or haploid, through the 2 set, or diploid, the 3 set, or triploid, the 4 set, or tetraploid, to the many set, or polyploid (Fig. 13-11). *Ploid* comes from the Greek meaning "fold" or "number." The chromosome sets are usually referred to as $1x$, $2x$, $3x$, and $4x$ and higher ploids. The chromosome sets are associated with many characters, size, fertility, and productivity, since the chromosomes carry the genes that condition these traits. Size of the organisms carrying the different sets ($1x$, $2x$, $3x$, etc.) may range from smaller haploids to larger diploids, usually regarded as the normal and still larger triploids, tetraploids, and higher polyploids (Fig. 13-12). There are, however, exceptions to the generalization that size increases with additional chromosomes, and some of the higher-ploid types are of small size. When the chromosome sets are of such a nature that pairing in the meiotic division is irregular, especially in those with uneven numbers as $3x$ and $5x$, fertility may be reduced or lacking, and seed and fruit production are impaired.

Fig. 13-13 Colchicum autumnale. *Drawing of the meadow saffron from an eighteenth-century engraving showing roots, bulb, and flowers of the plant. This plant is the source of an alkaloid, colchicine. (Courtesy of John Wyeth and Brothers, Inc., Philadelphia, from J. Ben Hill and Helen D. Hill, "Genetics and Human Heredity," McGraw-Hill Book Company, New York, 1955.)*

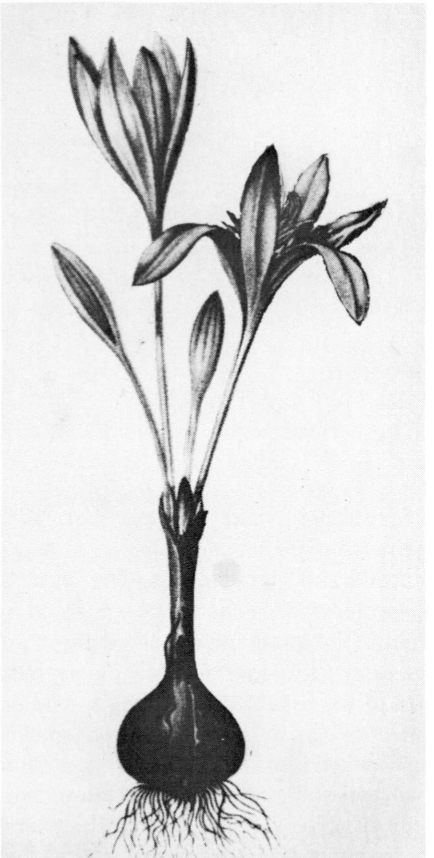

Colchicine, which is an alkaloid obtained from the plant *Colchicum autumnale* (Fig. 13-13), induces polyploidy by preventing spindle formation, and it has been used in attempts to obtain large plants or plants with large flowers or fruits. Perhaps, more often, polyploidy has been induced as an aid in the hybridization of plants with different chromosome numbers.

Cytoplasmic inheritance Studies involving maternal inheritance of some plastid characteristics (chloroplasts notably) have been interpreted as demonstrating extranuclear or extragenic inheritance (Fig. 13-14). Studies in which cytoplasmic inheritance has been reported usually involve the reappearance of characters associated with the maternal parent. It should be

Fig. 13-14 Chlorophyll deficiency shown by alternate green and nongreen stripes in leaves of maize. (Courtesy of E. G. Anderson, from J. Ben Hill and Helen D. Hill, "Genetics and Human Heredity," McGraw-Hill Book Company, New York, 1955.)

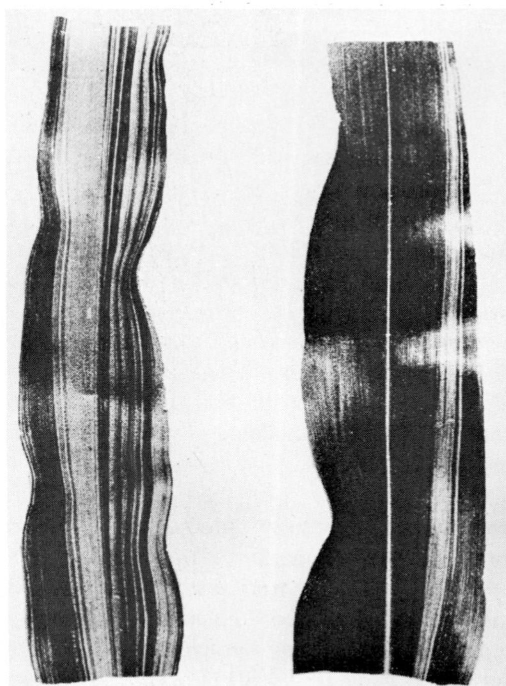

Fig. 13-15 Albino and green seedlings of Zea mays.

emphasized, however, that some chlorophyll deficiencies are controlled by genes on the chromosomes and are not cytoplasmic in nature. Some genetically controlled characters are so extreme that seedlings are albinos and die as a result of their inability to carry on photosynthesis (Fig. 13-15).

Population genetics More recently our knowledge of genetics has been applied to the study of populations of interbreeding individuals. The investigation of such communities of plants and animals is as significant as the study of a more closely related group, limited to a few individuals or, perhaps, a single individual. The field of population genetics is making an effective approach to an understanding of the changes that take place in populations and thus in evolution. A better understanding of the concept of a species, the degree of variability within a population of interbreeding individuals, the distribution of recessive and dominant characters within a population, as well as other significant biological problems, are the concern of this area of genetics.

Hybridization and its applications Hybridization, as illustrated in Mendelian heredity, has been of importance to man for many centuries. Some of its practical values were known long before the scientific studies of hereditary traits in plants and animals were begun. Hybridization is useful in the development of new varieties of plants and breeds of animals. Many sorts of apples, peaches, pears, grapes, strawberries, potatoes, and other crop plants were produced by hybridization in the original species to combine different characteristics, with later selection of the most desirable individuals. Through vegetative propagation these new varieties remain constant except for an occasional bud mutation, or variation. Among such economic field

Fig. 13-16 Mature corn plants illustrating increased hybrid vigor or heterosis of F_1 plants in center which were developed from the cross of two inbred lines at the sides. (Courtesy of D. F. Jones, from J. Ben Hill and Helen D. Hill, "Genetics and Human Heredity," McGraw-Hill Book Company, New York, 1955.)

crops also as corn, wheat, oats, and barley, better and more valuable kinds have been originated through hybridization.

Hybridization of corn has added materially to the value of this important farm crop both in quantity and quality. In the production of hybrid corn, strains are inbred for several generations and then these inbreds are hybridized. The resulting hybrids may include some that are more vigorous and more productive than the pure varieties from which they originated (Fig. 13-16). This condition is referred to as **hybrid vigor,** and it is one of the manifestations of **heterosis** (*hetero,* "different," *osis,* "condition"; literally, a condition different from that of the parents).

QUESTIONS

1. How is knowledge of heredity related to the production of hybrid corn? Name other crops besides corn in which the techniques of hybridization have resulted in improvement.

2. What does pollination have to do with plant breeding? What techniques would need to be used to keep self-pollinated plants from being selfed?

3. How could you establish whether plant characters are inherited together or independently?

4. In which parts of plants can mitosis be studied? Meiosis?

5. Mendel was not the first person to study inheritance. Why are his contributions of historic importance?

6. What might hopefully be the eventual outcome of chemical explanations for the structure and behavior of genes?

7. How are genes assumed to affect the production of characteristics of organisms?

8. If you were to plant a seed taken from a McIntosh apple, would the seedling develop into a McIntosh apple tree?

9. How might mutation influence organic change?

10. Do polyploids appear in nature or are they only the products of genetic experimentation?

REFERENCES

BEADLE, GEORGE W. 1949. The Genes of Men and Molds. *Sci. Am.,* **179**(3):30–39.

DOBZHANSKY, T. 1950. The Genetic Basis of Evolution. *Sci. Am.,* **182**(1):32–41.

HILL, J. BEN, and HELEN D. HILL. 1955. Genetics and Human Heredity. McGraw-Hill Book Company, New York.

JUKES, T. H. 1965. The Genetic Code, II. *Am. Scientist,* **53:**477–487.

SONNEBORN, T. M. 1950. Partner of the Genes. *Sci. Am.,* **183**(5):30–39. Cytoplasmic inheritance.

SRB, ADRIAN M., RAY D. OWEN, and ROBERT S. EDGAR. 1965. General Genetics, 2d ed. W. H. Freeman and Company, San Francisco.

SWANSON, CARL P. 1964. The Cell, 2d ed. Foundations of Modern Biology Series. Prentice-Hall, Inc., Englewood Cliffs, N.J.

TAYLOR, J. HERBERT. 1958. The Duplication of Chromosomes. *Sci. Am.,* **198**(6):36–42.

14

PLANT TAXONOMY; DIVISIONS OF THE PLANT KINGDOM

CAROLI LINNÆI
S:Æ R:GIÆ M:TIS SVECIÆ ARCHIATRI; MEDIC. & BOTAN. PROFESS. UPSAL; EQUITIS AUR. DE STELLA POLARI; nec non ACAD. IMPER. MONSPEL. BEROL. TOLOS. UPSAL. STOCKH. SOC. & PARIS. CORESP.

SPECIES PLANTARUM,
EXHIBENTES
PLANTAS RITE COGNITAS,
AD
GENERA RELATAS,
CUM
DIFFERENTIIS SPECIFICIS,
NOMINIBUS TRIVIALIBUS,
SYNONYMIS SELECTIS,
LOCIS NATALIBUS,
SECUNDUM
SYSTEMA SEXUALE
DIGESTAS.
TOMUS I.

Cum Privilegio S. R. M:tis Sueciæ & S. R. M:tis Polonicæ ac Electoris Saxon.

HOLMIÆ,
IMPENSIS LAURENTII SALVII.
1753.

Title page of Linnaeus's "Species Plantarum."

The importance of taxonomy to botanical science is considered through a brief review of historical developments. Kinds of systems of classification are presented with examples, and a relatively detailed summary of the system followed in this textbook is included. A brief consideration of plant identification and nomenclature is presented with examples. Alternation of generations is discussed.

PLANT CLASSIFICATION

The need of a system of classification Probably 375,000 species of plants are known to science, and many more remain to be discovered, named, and classified. These species present great variation as to size, structure, methods of reproduction, mode of life, and ecological and geological distribution over the earth. Obviously it is necessary to refer the plants of such a large and diverse group to a system of classification. It is easier to convey to another a clear idea of a kind of tree or a chemical compound by referring to it as a particular oak or as a carbohydrate compound, respectively, than by trying to differentiate it in succession from all other kinds of trees or from all other chemical compounds. It is necessary in the interest of brevity to have some way of identifying an organism and of indicating its relationship to other living organisms. An important reason for a system of classification of living organisms is convenience.

Systems of classification and nomenclature Taxonomy was one of the first disciplines within the field of botany. Classification systems, based on form or habit, were introduced by the Greeks and for approximately one thousand years these formed the basis of systems introduced by Albertus Magnus in the thirteenth century, Otto Brunfels in the fifteenth century, Andrea Cesalpino in the sixteenth century, and others to the time of Carolus Linnaeus (1707–1778).

The classification by Aristotle and Theophrastus of plants into herbs, shrubs, and trees was such a type, but it is now known that many trees are more closely related to certain herbaceous plants than they are to other trees. The arrangement of flowers on the basis of color, which is common in popular flower guides, is another example, but the red rose certainly is more closely related to the white rose than it is to a red poppy. Plants may be classified on the basis of their relation to water. The ecological groups, i.e., hydrophytes (plants that live in, or partly in, water), mesophytes (plants that live in moist terrestrial locations), and xerophytes (plants that live in dry or desert situations), represent systems based on form and habit. In the same way, some plants are referred to as parasites or saprophytes, depending on the source of their organic food. The parasitic Indian pipe and the likewise parasitic bacterium causing diphtheria are, at best, only distantly related.

The publication of "Genera Plantarum" by Linnaeus in 1737 introduced a second era of plant classification, characterized by the introduction of artificial systems based on numerical classification. Linnaeus's rather inaptly termed sexual system of classification was an artificial system based essentially on the number of stamens and styles in each flower. It was relatively simple and convenient and made it possible to determine the name of a plant.

The period of numerical classification was followed by an era in which systems of classification were based on the form of plants. It was brought to a close by the findings of Darwin and the publication, in 1859, of his book "On the Origin of Species by Means of Natural Selection."

In 1875 August Wilhelm Eichler proposed a system of classification based essentially on genetical relationships and evolutionary concepts. By about 1900 a system which was based on the work of Eichler and introduced by Adolf Engler was generally accepted and remains the most important system used today in the herbaria of this country.

In general two types of plant classification may be recognized, i.e., the **artificial** and the **natural** systems. The term natural is essentially synonymous with the term **phyletic.** A series of plants may be so arranged within a system of classification as to make evident their relationship to each other. It is a fundamental fact of biological science that these relationships are present in varying degrees in all living organisms. A system that attempts to show what is thought to be a natural, or phyletic, relationship is known as a natural system.

About 1880 three names, all of which had been introduced into the literature at some earlier date, were used to designate the divisions of the cryptogams (plants in which reproductive organs were thought to be hidden). These terms —**Thallophyta, Bryophyta,** and **Pteridophyta**— were widely adopted and are still used by some botanists.

Although the term Bryophyta is generally acceptable and is used, considerable doubt exists relative to the proper use of the terms Thallophyta and Pteridophyta. The belief by many botanists that the fungi have not evolved from the algae has led more recently to a separation of the algae into seven divisions and the fungi into three divisions. For many years the ferns, lycopods, and horsetails were thought to be sufficiently closely related (the ferns and their allies) to be recognized as a single division, the Pteridophyta. Evidence has accumulated that proves that, although these plants are characterized by the presence of vascular tissue, they are, in fact, divergent lines from the psilophytes. Some botanists have established all four lines: psilophytes, lycopods, horsetails, and ferns as separate divisions. Other botanists have designated the division as **Tracheophyta** (all vascular plants), with the subdivisions **Psilopsida, Lycopsida, Sphenopsida,** and **Pteropsida.** This system is followed in this textbook.

Included within the subdivision Pteropsida are the classes **Gymnospermae** and **Angiospermae** in addition to the **Filicinae,** the ferns. The discovery that certain Paleozoic fernlike plants, the pteridosperms, reproduced by means of seeds and contained leaf gaps in the stele seemed, in a broad sense, reason enough to refer all ferns, as well as the seed plants represented by the Gymnospermae and the Angiospermae, to the Pteropsida.

Plant classification is very old, and, in its development over hundreds of years, there have been many changes. Plant classification is not static. As botanists learn more about plants and discover important evidences of natural relationships, there will be many significant changes in the future.

AN OLDER CLASSIFICATION OF THE PLANT KINGDOM

DIVISION THALLOPHYTA
- Subdivision Algae
  - Class Euglenineae
    - *Euglena*
  - Class Cyanophyceae. Blue-green algae
    - *Gleocapsa, Oscillatoria, Rivularia,* etc.
  - Class Chlorophyceae. Green algae
    - With several orders and genera as *Volvox, Ulothrix, Oedogonium, Spirogyra,* etc.
  - Class Chrysophyceae. Yellow-green algae
  - Class Bacillariophyceae. Diatoms
  - Class Phaeophyceae. Brown algae
    - *Ectocarpus, Laminaria, Fucus,* etc.
  - Class Rhodophyceae. Red algae
    - *Nemalion, Batrachospermum, Polysiphonia,* etc.
- Subdivision Fungi
  - Class Schizomycetes. Bacteria
  - Class Myxomycetes. Slime molds
  - Class Phycomycetes. Algalike fungi
    - *Saprolegnia, Rhizopus,* etc.
  - Class Ascomycetes. Sac fungi
    - *Peziza,* powdery mildews, etc.
  - Class Basidiomycetes. Club fungi, rust fungi, mushrooms, coral fungi, etc.
  - Class Fungi Imperfecti. Imperfect fungi

DIVISION BRYOPHYTA
- Class Hepaticae. Liverworts
  - *Marchantia, Riccia, Pellia,* etc.
- Class Musci. Mosses
  - *Sphagnum, Mnium, Polytrichum,* etc.

DIVISION PTERIDOPHYTA
- Class Psilophytinae
  - Order Psilophytales
    - Several fossil genera
  - Order Psilotales
    - *Psilotum* and *Tmesipteris*

Class Lycopodinae

Order Lycopodiales. Club mosses

Lycopodium and *Phylloglossum*

Order Selaginellales. Little club mosses

Selaginella

Order Lepidodendrales. Giant club mosses

Fossil forms, *Lepidodendron, Sigillaria,* etc.

Order Isoetales. Quillworts

Isoetes

Class Equisetinae

Order Equisetales. Horsetails

Equisetum

Order Sphenophyllales

Fossil forms, *Sphenophyllum,* etc.

Order Calamitales

Fossil forms, *Calamites,* etc.

Class Filicinae

Order Ophioglossales. Adder's-tongue ferns and grape ferns

Ophioglossum, Botrychium, etc.

Order Marattiales. Marattiaceous ferns

Marattia, Danaea, etc.

Order Filicales. True ferns

Polypodium, Pteris, Marsilea, etc.

DIVISION SPERMATOPHYTA

Class Gymnospermae

Order Cycadofilicales. Pteridosperms or seed ferns

Fossil forms, *Lyginodendron,* etc.

Order Bennettitales

Fossil forms, *Williamsonia,* etc.

Order Cycadales. Cycads

Zamia, Cycas, Dioön, etc.

Order Cordaitales

Fossil forms, *Cordaites,* etc.

Order Ginkgoales

Fossil forms and the living genus *Ginkgo*

Order Coniferales. Conifers

Pinus, Tsuga, Taxus, etc.

Order Gnetales

Gnetum, Ephedra, Welwitschia

Class Angiospermae. Flowering plants

Subclass Dicotyledoneae. Dicots

Orders Ranales, Magnoliales, Rosales, etc.

Subclass Monocotyledoneae. Monocots

Orders Graminales, Liliales, Orchidales, etc.

CLASSIFICATION OF THE PLANT KINGDOM FOLLOWED IN THIS TEXTBOOK

DIVISION CHLOROPHYTA. The green algae. Plants with cells containing definite nuclei and plastids colored by chlorophyll and carotenoid pigments; photosynthetic products and storage materials, starch and oil; asexual reproduction by fragmentation and spore formation; spores often motile; sexual reproduction through the fusion of motile and nonmotile gametes; many orders and genera, such as *Volvox, Ulothrix, Oedogonium, Spirogyra, Chara,* etc.

DIVISION EUGLENOPHYTA. The euglenoid algae.

DIVISION PYRROPHYTA. Greenish-tan to golden-brown algae.

DIVISION CHRYSOPHYTA. The yellow-green to golden-brown algae. Starch is never formed, and food reserve is commonly leucosin, a complex carbohydrate. Asexual reproduction by motile or nonmotile spores, and sexual reproduction commonly isogamous with motile or nonmotile gametes. Autogamy (the fusion of two sister nuclei in a cell) occurs, and a number of genera are coenocytic. Cell wall generally formed by two overlapping halves. Represented by such forms as *Vaucheria* and the diatoms. Recently it has been suggested that the xanthophytes of this division be placed into a new division, the **Xanthophyta.** Differences in pigmentation and flagellation suggest a closer relationship of the so-called yellow-green algae to the Chlorophyta.

DIVISION PHAEOPHYTA. The brown algae. Plants with cells containing definite nuclei and plastids, containing chlorophyll and a carotenoid pigment, fucoxanthin, of yellowish-brown color masking the green; photosynthetic products are sugars. Foods are stored as simple and complex carbohydrates, such as laminarin, and complex alcohols, such as mannitol. Genera *Ectocarpus, Laminaria, Fucus,* etc.

DIVISION CYANOPHYTA. The blue-green algae. Cells without definite nuclei and without plastids; blue-green and red pigments (phycobilins) and yellow pigments present; photosynthetic products and storage materials, glycogen and glycogen products; no sexual reproduction; multiplication by fragmentation and through nonmotile resting cells; no motile cells. Genera *Gleocapsa, Oscillatoria, Nostoc, Rivularia,* etc.

DIVISION RHODOPHYTA. The red algae. Plants with cells containing definite plastids, with chlorophyll, carotenoids, and red and blue pigments (phycobilins), which mask the others; products of photosynthesis stored in form of a complex carbohydrate, floridean starch. Genera *Batrachospermum, Nemalion, Polysiphonia,* etc.

DIVISION SCHIZOMYCOPHYTA (Schizomycota). The bacteria. Unicellular plants; cells without definite nuclei.

DIVISION MYXOMYCOPHYTA (Myxomycota). The slime molds. Plants in which the vegetative body is a naked amoeboid plasmodium; reproduction by a large number of spores produced in a sporangium.

DIVISION EUMYCOPHYTA (Eumycota). The true fungi. Definite cell walls formed in all stages of vegetative development. Plant body generally a branching thallus, multicellular or unicellular and coenocytic. Asexual reproduction by many kinds of spores, some of which are formed within definite sporangia. Certain spores (ascospores and basidiospores) are formed immediately following meiosis.

Lower fungi. Formerly the class Phycomycetes but now divided into six classes as follows: Chytridiomycetes, Hyphochytridiomycetes, Oömycetes, Plasmodiophoromycetes, Zygomycetes, Trichomycetes. A group of true fungi; plant bodies filamentous, nonseptate structures with numerous small definite nuclei; asexual reproduction by a large indefinite number of small spores produced in a sporangium; sexual reproduction general. Genera *Rhizopus, Saprolegnia, Albugo,* etc.

Higher fungi.

Class Ascomycetes. The sac fungi. A group of true fungi; plant bodies filamentous, septate; cells with definite nuclei; asexual reproduction frequently by single-celled spores borne terminally on stalks; sexual reproduction results in the development usually of eight spores, produced in a sac, or ascus. Genera *Peziza, Penicillium, Aspergillus,* etc.

Class Basidiomycetes. The basidium fungi. A group of the true fungi; plant bodies threadlike or composed of filaments; sexual reproduction typically results in the development of four spores produced on a stalk, the basidium. Genera *Amanita, Agaricus, Puccinia,* etc.

Class Deuteromycetes. The imperfect fungi. A large artificial group of fungi in which sexual reproduction has not been determined. A form class.

The lichens. Association of a fungus (Ascomycetes or Basidiomycetes) and an alga (Chlorophyta or Cyanophyta). Relationship of fungus and alga controversial; may be one of parasitism or a consortium (symbiotic in nature). Lichens of considerable ecological importance as pioneers.

DIVISION BRYOPHYTA. The mosslike plants. Plants with thalloid bodies; asexual multiplication by means of multicellular buds, branching, and fragmentation; sexual reproduction by means of gametes produced in multicellular sex organs, antheridia and archegonia; the zygote regularly forms an embryo with the juvenile stage retained within the archegonium.

Class Hepaticae. The thallose and leafy liverworts. Gametophytes generally with prostrate dorsiventral thalli; spores released by longitudinal splitting of the capsule.

Order Jungermanniales. Most of the known liverworts. Gametophytes generally dorsiventrally flattened with two rows of lateral leaves and usually a row of ventral leaves, termed amphigastria. Genera *Calypogeia, Scapania, Bazzania, Porella,* etc.

Order Metzgeriales. Most abundant in the tropics. Gametophyte usually thallose. In leafy forms there are two rows of lateral leaves but amphigastria are absent. Genera *Riccardia, Fossombronia, Pellia, Mylia,* etc.

Order Marchantiales. The thallus liverworts. Genera *Marchantia, Riccia, Ricciocarpus, Asterella, Conocephalum, Plagiochasma,* etc.

Class Anthocerotae. The horned liverworts.

Order Anthocerotales. Genus *Anthoceros.*

Class Musci. The mosses.

Subclass Sphagnidae. Genus *Sphagnum.*

Subclass Bryidae. Contains most of the mosses.

Subclass Polytrichidae. Genus *Polytrichum.*

DIVISION TRACHEOPHYTA. The vascular plants. Plants with well-developed vascular system present in the sporophytic plant body characteristic of the group.

SUBDIVISION PSILOPSIDA. The primitive vascular plants. The sporophytes generally have no roots and lack true leaves; no leaf gaps present in the vascular cylinder; spores all of one type produced in sporangia borne terminally on stems.

Class Psilophytinae.

Order Psilophytales. Fossil forms. A group of fossil forms including the oldest and most primitive vascular plants known. Genera *Rhynia, Psilophyton, Asteroxylon,* etc.

Order Psilotales. Living forms. Small, frequently epiphytic, tropical and subtropical plants; gametophytes cylindrical tuberous structures, subterranean or aerial, with a mycorrhizal fungus growing in the tissues. Genera *Psilotum* and *Tmesipteris* the only living representatives of the order.

SUBDIVISION LYCOPSIDA. The microphyllous plants. The sporophyte has roots, stems, and small true leaves; the vascular cylinder is without leaf gaps; sporangia borne on the upper or adaxial surface of leaves or sporophylls that are generally aggregated into cones.

Class Lycopodinae.

Order Lepidodendrales. Fossil forms. Giant tree forms with roots, stems, and generally small leaves, preserved as fossils from the Carboniferous period; plants heterosporous with large and small spores; sporophylls aggregated in cones. Genera *Lepidodendron, Sigillaria,* etc.

Order Lycopodiales. Living forms. Small herbaceous plants with roots, stems, small leaves, and sporophylls bearing sporangia that contain a single type of spore; sporophylls, generally, in cones. Genera *Lycopodium* and *Phylloglossum.*

Order Selaginellales. Living forms. Small herbaceous, generally tropical plants, frequently epiphytic; plants with roots, stems, leaves, and sporophylls; heterosporous, with two kinds of spores, large and small; sporangia produced on sporophylls aggregated in cones. Genus *Selaginella.*

Order Isoetales. Living forms. Small herbaceous plants growing partly submerged; plants with roots, stems, and relatively large, slender leaves, each bearing a sporangium; heterosporous, with large and small spores. Genus *Isoetes.*

SUBDIVISION SPHENOPSIDA. The wedge-leaved plants. The sporophytes have roots, rhizomes, aerial stems without leaf gaps, and generally small, often wedge-shaped leaves developed in whorls at the nodes of aerial stems and branches; aerial stems generally articulate or jointed; with few exceptions plants are homosporous, with spores all of one type borne in sporangia on specialized structures; the sporangiophores either loosely aggregated or in compact cones.

Class Equisetinae. The horsetails.

Order Sphenophyllales. Fossil forms. A group of small fossil forms with small wedge-shaped leaves. Genus *Sphenophyllum.*

Order Calamitales. Fossil forms. A group of large treelike forms preserved as fossils. Genus *Calamites.*

Order Equisetales. Living horsetails or scouring rushes. Small herbaceous plants. Genus *Equisetum* the single living representative of the subdivision Sphenopsida.

SUBDIVISION PTEROPSIDA. The macrophyllous plants. The sporophyte has roots, stems, and, generally, large leaves; vascular stele has leaf gaps; plants homosporous and heterosporous with spores produced in sporangia often borne on the lower or abaxial surface of sporophylls; this subdivision of the vascular plants includes all types of ferns and the seed plants.

Class Filicinae. The ferns. This class includes plants with erect aerial stems, radial in form, and horizontal underground rhizomes; leaves generally large; numerous sporangia are produced on the underside of leaves; although a few forms are heterosporous, most are homosporous; gametophytes are generally macroscopic; fertilization by swimming sperms.

Order Ophioglossales. The adder's-tongue and grape ferns. Plants with macroscopic gametophytes and sporophytes; gametophytes tuberous, mycorrhizal, and subterranean; sporophytes consist of roots, short upright stems, and leaves; plants homosporous with spores produced in eusporangiate structures in a specialized spike as a fertile part of a leaf. Genera *Ophioglossum, Botrychium.*

Order Marattiales. The marattiaceous ferns. Plants with macroscopic gametophytes and sporophytes; gametophytes, thick expanded thalli, resembling liverworts; sporophytes with roots, radial stems, and large leaves; homosporous with the single type of spores produced in large numbers in synangia. Genera *Marattia, Danaea,* etc.

Order Filicales. The true ferns. Plants generally with macroscopic, physiologically independent gametophytes and sporophytes; leaves usually large, generally homosporous but exceptionally heterosporous. Genera *Polypodium, Pteris, Marsilea,* etc.

Class Gymnospermae. Plants with microscopic gametophytes and large macroscopic sporophytes; ovules, developing into seeds, produced on the upper side of an open scale; pollen airborne.

The Cycadophyte Line.

Order Cycadofilicales, or *Pteridosperms.* Fossil forms representing the oldest seed plants. Generally small plants with large fernlike leaves bearing well-developed ovules at their tips; pollen deposited on nucellus of ovule; fertilization probably by swimming sperms. Genera *Lyginopteris, Heterangium,* etc.

Order Bennettitales. Fossil forms. Plants with short, thick stems, large leaves; megasporophylls and microsporophylls often with spiral arrangement on the same plant; fertilization probably by swimming sperms. Genera *Williamsonia, Cycadeoidea, Bennettites,* etc.

Order Cycadales. The living cycads. Dioecious plants with distinct staminate and ovulate plants bearing large fernlike leaves generally on short, thick stems; sporophylls aggregated into cones; ovules ripen into seeds covered by an outer fleshy layer; fertilization by swimming sperms. Genera *Cycas, Dioön, Zamia,* etc.

The Coniferophyte Line.

Order Cordaitales. Fossil forms. Tall, slender trees forming extensive Paleozoic forests; leaves slender, leathery, often small structures; plants monoecious and dioecious with sporophylls borne in male and female cones; fertilization probably by swimming sperms. Genera *Cordaites, Mesoxylon,* etc.

Order Ginkgoales. One living representative. Tall trees, distinctly staminate and ovulate; microsporangia produced several in flexuous cones; ovules borne in pairs on a short, slender stalk, ripen into seeds covered by an outer fleshy layer; fertilization by swimming sperms. *Ginkgo,* the single living genus.

Order Coniferales. The living cone-bearing trees and shrubs—trees and shrubs generally with small, often needle-shaped leaves; monoecious plants with sporophylls aggregated into cones; fertilization by nonmotile sperms carried to the archegonium by pollen tubes; ovules mature into seeds that are generally dry. Genera *Pinus, Larix, Tsuga, Picea,* etc.

Order Gnetales. Living forms. A small group of staminate and ovulate plants with characteristics of both gymnosperms and angiosperms; both tracheids and vessels found in the vascular tissues. Genera *Ephedra, Gnetum,* and *Welwitschia.*

Class Angiospermae. The flowering plants. The plants with microscopic gametophytes and large macroscopic sporophytes; leaves generally relatively large; plants monoecious or dioecious; pollen mostly airborne or carried by insects and deposited on the stigma or upper part of the carpel; fertilization by nonmotile sperms carried to female gametophyte by a pollen tube; ovules developing into seeds enclosed by carpels.

Subclass Dicotyledoneae. The dicots. Flowering plants characterized by many features among which are: two cotyledons in the embryo; flower parts mostly in fours and fives; leaves with netted veins; starch generally the photosynthetic storage product; many orders, families, and genera, such as:

Order Ranales. Genus *Ranunculus,* or buttercup, etc.

Order Rosales. The roses. Genus *Rosa,* the rose, etc.

Order Geraniales. Genus *Geranium,* the common geranium, etc.

Subclass Monocotyledoneae. The monocots. Flowering plants characterized by many features among which are: one cotyledon in the embryo; flower parts mostly in threes, and parallel veins in the leaves; sugar often a photosynthetic storage product; many orders, families, and genera, such as:

Order Graminales. The grasses. Genera *Zea* (corn), *Triticum* (wheat), *Avena* (oats), *Oryza* (rice), *Poa* (bluegrass), etc.

Order Liliales. The lilies. Genus *Lilium,* the lily.

Two systems of classification, arranged for comparison, are given in this chapter. One is an older system of classification of the plant kingdom into four divisions and the other is the one used in this textbook.

The characters that result in a plant being designated as a plant simultaneously place it in the plant kingdom. These characters are broad, common, inclusive characters which serve to identify all plants and to separate them from animals. In instances where broad, inclusive characters used to identify plants overlap with characters used to identify animals, questions exist as to whether an organism is one or the other. Bacteria, some fungi, and certain algae are representative of kinds of organisms the identity of which as a plant or an animal is often debated. The uniflagellates, especially, include dubious transitional forms, and it has been suggested by some biologists that three or even four kingdoms should exist. Notable exponents of a four-kingdom scheme of classification and the times of their proposals include

H. F. Copeland (1938, 1956), F. A. Barkley (1939), W. Rothmaler (1948, 1951), and R. H. Whittaker (1959).[1] Included among the suggested names for kingdoms in these plans are Protista, Protobionta, Protoctista, Monera, and Mychota.

Characters which emphasize the similarity of plants result in a greater number of kinds of plants being referred to a taxon. Characters which have a smaller range of application, usually characters that separate and distinguish plants from each other, result in an increase in the number of taxonomic groups to which they are referred and reflect a closer relationship of the plants within such a taxon. In practice the attempt to group plants which are naturally related into the same taxon has determined the system; the system does not determine or establish the natural relationship between plants.

Categories of classification (each of which is a taxon) are as follows: division, subdivision, class, subclass, order, family, genus, and species. The names of species and genera have diverse endings, among which *–a, –ea, –ia, –i, –is, –um,* and *–us* are common.

The term taxon was introduced by Dr. H. J. Lam in 1950. By some it is used to designate only minor categories in classification; by others it is used to designate any taxonomic entity and is a substitute for the more cumbersome term taxonomic group.

In 1753 there appeared a monumental work entitled "Species Plantarum," by the great Swedish naturalist Linnaeus (Carl von Linné), in which there was emphasized the scheme of the present binominal system (binary nomenclature) and in which specific names were uniformly assigned to plants. The idea underlying this scheme is to designate each individual kind of plant with a binominal, i.e., a name consisting of two parts. The first part of the name is the genus, or generic part of the name, and the second part of the name is the species, or specific part of the name.

Moreover the name (or an abbreviation) of the person first publishing the description of the plant is often written following the binomial. Thus the abbreviation L. following the name *Rosa alba* signifies that the botanist Linnaeus originally gave the name *Rosa alba* to this particular kind of rose. The rules regulating botanical nomenclature provide that once a binary combination has been used, the same name can never legitimately be applied to any other plant; i.e., there can be but one kind of plant named *Rosa alba,* and this plant to which Linnaeus gave the name in 1753 is still so known today. Not all kinds of plants are designated by a binomial. Some are recognized by names that include an additional descriptive term designating the cultivar, subspecies, form, or clone. In general the generic name is capitalized but the specific name is not. When a specific name is derived from a person's name, it is often capitalized.

The following illustrates this usage:

Division Tracheophyta
  Subdivision Pteropsida
    Class Angiospermae
      Subclass Dicotyledoneae
        Order Chenopodiales
          Family Chenopodiaceae
            Genus *Beta*
              Species *vulgaris* L.
                Cultivar Cicla

Beta vulgaris L. is the scientific name, or binomial, of common beet. The cultivar of beet, called Cicla, is Swiss chard.

Binary nomenclature—the binomial Prior to the middle of the eighteenth century the plants known to botanists were relatively few in number. Usually it was sufficient to refer to each plant with a short descriptive phrase, as a farmer with a half dozen cows might refer to each animal by some descriptive phrase, such as the long-horned red cow. But when his herd of cows increased to a larger number, he might have

[1] See review by A. Cronquist in references at the end of this chapter.

more than one long-horned red cow and he would be more inclined to designate each cow with a short individual name, not necessarily descriptive in character. Thus we find that earlier botanists designated a species of mint with the following Latin phrase: *Prunella magna flore albo* (large prunella with white flower). Such a method was cumbersome, and as new kinds of plants were recognized, the inadequacy of the method was apparent.

Scientific names are not derived from the English language but come mostly from either the Latin or the Greek. At the present time Latin words or latinized words from other languages are almost universally used in naming new species of plants.

Minor variations often occur between a number of individuals of the same species. Such variations may be too slight to consider the individuals to be new species, and yet they are of sufficient importance to be recognized. In this instance an additional category designated as the variety, subspecies, cultivar, or clone is added to the binomial and the whole name becomes in fact a name with three parts rather than two. Thus *Pisum sativum* telephone designates a cultivar of the common garden pea.

How names are chosen The names that are chosen (either scientific or common) are usually significant when their meaning is known. They are generally chosen to represent one of the following situations: (1) The name may refer to one of the chief characteristics of the plant, as the white rose (*Rosa alba*), the lobed spiraea (*Spiraea lobata*), the showy orchid (*Orchis spectabilis*), or the large rhododendron (*Rhododendron maximum*). (2) It may be chosen from the character of the habitat, i.e., the kind of substratum on which the plant grows; for example, the water buttercup (*Ranunculus aquaticus*). (3) It may be chosen to designate the geographical location, as in the New England aster (*Aster novae-angliae*) or the Missouri aster (*Aster missouriensis*). Such epithets may not have the same meaning today as when they were first applied. Not all plants named *missouriensis* grow in Missouri. (4) It may be taken from the name of the discoverer of the plant, as Drummond's aster (*Aster Drummondii*) or Harper's carex (*Carex Harperi*). Other origins are sometimes met with, but these are the common ones.

Common versus scientific names Two names have often been used for the same plant. One, the common name, has been used more often by the amateur botanist or the layman, and the scientific name has been used largely by the more technically trained student. The names aster, citrus, catalpa, chrysanthemum, rhododendron, and many others are used interchangeably as part of the common and scientific names. Neither common nor scientific names are always so simple as are these examples. Common names may consist of a noun and any number of qualifying adjectives, such as the purple-flowered milkweed and the small southern yellow orchid. A scientific name often consists of only two parts, the binomial. When a word is added to the binomial to designate form, subspecies, or variety, this becomes a part of the scientific name. The scientific name of a plant refers to one kind of plant (taxon) and is worldwide in application. The same is not true of the common names. The same common name may be used for several different plants, or different common names may be used for the same plant. The definiteness of scientific names is one of their great advantages over common names.

The terms algae, fungi, bacteria, mosses, ferns, ground pines, flowering plants, and seed plants are also common, or vernacular, names. Some confusion in the use of the words algae and fungi exists because they were used previously as formal taxonomic terms.

PLANT IDENTIFICATION

Although plant classification and nomenclature are parts of the science of taxonomy, there is a third and equally important area to be considered, i.e., that of plant identification. A common procedure is to use manuals, or keys, and to compare the unknown plant with identified plants in a herbarium. Keys are devices used in helping to identify an unknown plant. At the present time the most widely accepted type is the **dichotomous key,** in which each member of the dichotomy is a **lead.**

1. Sepals and petals of flower present
   a. Perianth colored
   a. Perianth not colored
2. Sepals and petals of flower not present
   b. Flower perfect
   b. Flower imperfect

The interest and research of taxonomists may deal with many kinds of problems. Some studies, for example, deal with the species of a genus or the genera within a family. The roses of the world (a monograph) may be the basis of investigation, or the flora of a county, mountain, or valley (a floristic study) may demand the time of a taxonomist.

ALTERNATION OF GENERATIONS

Sporophytic and gametophytic generations The life history of a plant is an account of events over a period of time. An account of the complete life history of a plant includes all its behavior from the occurrence of a given event in the life cycle through all the various stages to the repetition of this event. One of the most notable features of the life histories of plants is the almost universal occurrence of two phases, or alternating generations, that differ in many features. They are contrasted in size, in appearance, and in cell features. Gametes are produced by one of these generations and meiospores are produced by the other generation. The gamete-producing plant is the gametophyte and the meiospore-producing plant is the sporophyte (Fig. 14-1).

Fig. 14-1 Diagrammatic life cycle showing alternating generations.

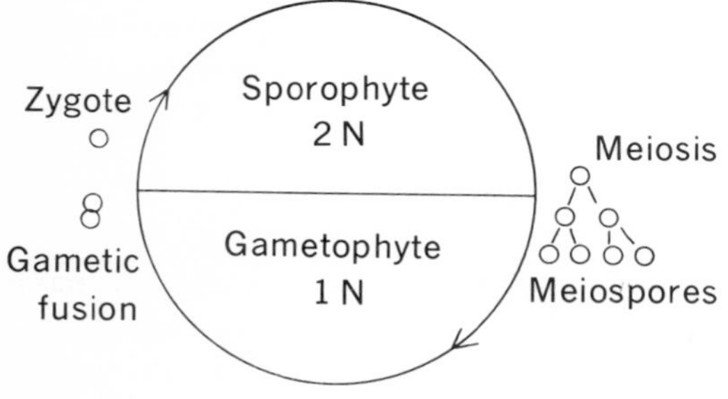

Relative extent of sporophyte and gametophyte Plants of the various divisions of the plant kingdom differ greatly in the proportion of the total time of their life histories that is represented by each of the two different phases. Each generation in turn varies in its relative size and differentiation. No single statement can be made that is accurate or reflects all of the exceptions.

In general many plants of the lower groups have a conspicuous gametophyte. This is especially true of some of the algae, fungi, and the bryophytes. In these plants the sporophytic structures usually are relatively small, inconspicuous, and of short duration. In many of the plants of the lower groups, however, the two alternating generations are independent plants that are often evenly balanced as to size and conspicuousness. In the bryophytes the generations are more evenly balanced.

In seed plants the entire plant represents the sporophyte, and the gametophytic structures are reduced to microscopic proportions. The gametophytes of the seed plants are represented by the germinated pollen and the embryo sacs.

Relative independence of sporophyte and gametophyte In the algae and the bryophytes the gametophytic structures are photosynthetic and manufacture their own food and are therefore capable of an independent existence. They are also independent as regards the absorption of inorganic substances. The sporophytic structures in these groups are usually small and in most cases more or less dependent for all food upon the gametophytic thallus. In the common ferns both gametophyte and sporophyte lead an entirely independent existence in their adult condition. But in heterosporous forms the male gametophytes, at least, are entirely dependent upon the stored food in the spore. The female gametophytes are small and, in most cases, probably do not synthesize much carbohydrate food, although they are chlorophyllose. In seed plants the sporophyte is complex and well developed and leads an entirely independent existence. But the gametophyte is only a few cells in extent and consists of structures which are, in all cases, entirely dependent upon the sporophyte.

Nuclear differences in sporophyte and gametophyte As early as 1894 Strasburger made the generalization that, for plants showing alternation of generations, the generations differ in nuclear constitution. This difference lies in the fact that the nuclei of all cells of the sporophytic generation contain twice the number of chromosomes present in the nuclei of the cells of the gametophytic generation. The sporophytic generation is initiated by fertilization, or syngamy, which is the fusion of two $1N$ gametes, resulting in the formation of the zygote, which is $2N$. Sooner or later meiosis takes place whereby the chromosome number is reduced by half in the formation of meiospores. This marks the end of the sporophytic generation and the beginning of the gametophytic generation. The meiospore is thus the first cell of the gametophyte. The gametophyte produces gametes, which fuse to form a new sporophyte. The two critical points in the life cycle of plants are, therefore, syngamy, or fertilization, and meiosis.

Alternation of generations is a fact, but the origin of the alternating generations is still a very live question. One interpretation suggests that the sporophytic generation is a new generation, interpolated into the life history between fusion of gametes and meiosis. Presumably it arises as a result of a delay in meiosis which allows time for the development of the sporophyte by vegetative growth. This interpretation is called the **antithetic theory of alternation.**

In contrast, many botanists disagree with the suggestion that the sporophytic generation is new or more recently interpolated into the life history of a plant. The **homologous theory of alternation** postulates that plants have evolved from ancestral algal stock in which both gametophytic and sporophytic phases were already differentiated. The continued development of the gametophytic and sporophytic phases, the so-called correlative phases, has resulted in the degree of development exhibited by plants today.

A critical examination of the above discussion reveals that little is determined concerning the origin of the alternation of generations. If we assume the presence of the gametophytic generation, it is then reasonable to suggest that the origin of the sporophytic generation results from the fusion of gametes. We have, however, already assumed the presence of one of the two generations. Advocates of the homologous postulate assume the presence not only of the gametophytic but also of the sporophytic generation. If the existence of only one generation is assumed, then it becomes necessary to establish the origin of the alternate; if the existence of both generations is assumed, there is no need to establish the origin of either generation. As a result, a discussion of the antithetic theory of the alternation of generations is often concerned with the origin of the sporophyte from the existing gametophyte. Advocates of the homologous theory of the alternation of generations usually discuss the development and further elaboration of both generations since they have no need to consider the origin of either generation.

QUESTIONS

1. What impact did the publishing of "The Origin of Species" by Darwin have on the development of schemes of classification?

2. What are the advantages of scientific and common names?

3. What confusion exists in the use of the terms algae and fungi?

4. Construct a key, using the dichotomous system with leads, to classify the following: tree, bluegrass, pinecone, leaf, red flower, chair, paper, clothing, xylem, and potato.

5. What is the significance of syngamy and meiosis in the sexual life history of a plant?

6. Why is the term taxon a useful one?

7. Do you think it is more useful to design a system of classification to reflect the common characters of plants or those characters which indicate diversity among plants?

8. Why are botanists concerned with alternation of generations?

REFERENCES

BAILEY, L. H. 1949. Manual of Cultivated Plants. The Macmillan Company, New York. A manual of commonly cultivated plants; very useful.

BENSON, LYMAN. 1962. Plant Taxonomy. The Ronald Press Company, New York. An excellent text dealing with the principles of taxonomy.

CRONQUIST, ARTHUR. 1960. The Divisions and Classes of Plants. *Botan. Rev.,* **26**(4):425–482. An excellent review article citing many additional references.

GLEASON, H. A. 1963. Illustrated Flora of the Northeastern United States and Adjacent Canada. The New Britton and Brown, 3 vols. Published for the New York Botanical Garden by Hafner Publishing Company, Inc., New York. A manual valuable for plant identification.

GRAY, ASA. 1950. Gray's Manual of Botany, 8th ed. Largely rewritten and expanded by Merritt L. Fernald. American Book Company, New York. A standard manual for the identification of plants.

LAWRENCE, GEORGE H. M. 1951. Taxonomy of Vascular Plants. The Macmillan Company, New York. Covers the principles of taxonomy.

SIMPSON, GEORGE GAYLORD. 1961. Principles of Animal Taxonomy. Columbia University Press, New York. A modern, sound treatment of taxonomy; sophisticated but presented in a logical, understandable manner.

WAHL, HERBERT A. 1965. Alternation of Generations—Again. *Turtox News,* **43**(8, 10):206–209, 248–250. A stimulating discussion of the alternation of plant generations.

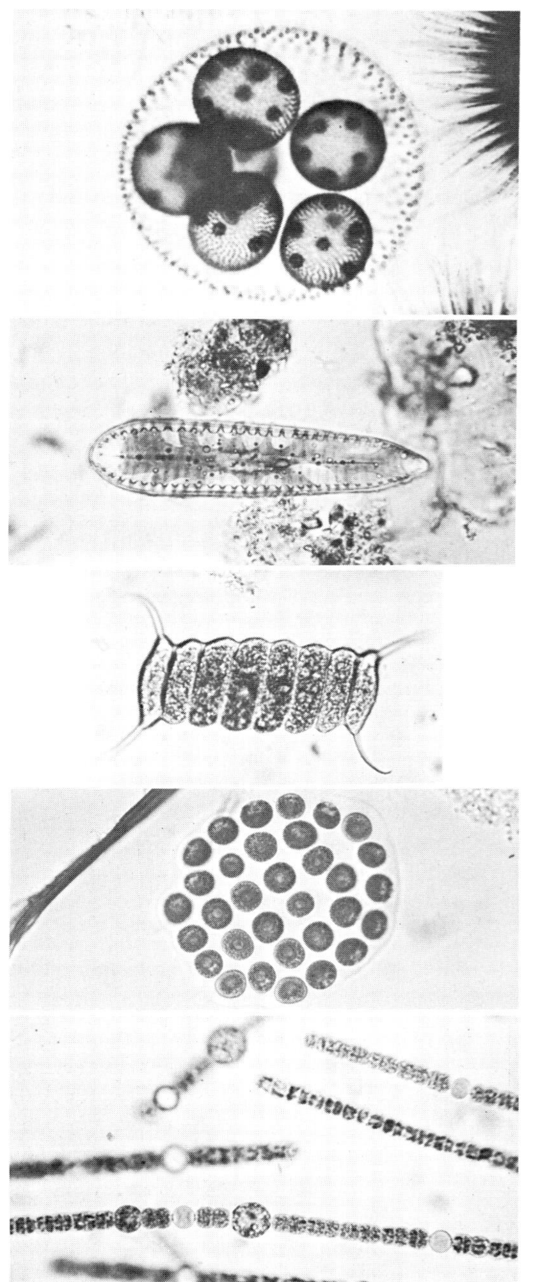

Types of Algae. *(Courtesy of Dr. J. D. Dodd.)*

15

ALGAE

The kinds of algae, their distribution, life histories, and economic importance are discussed. Some attention is devoted to the products of metabolism, storage products, cellular organization, including evidence from electron-microscope studies, and fundamental considerations of the kinds of gametic reproduction.

General features Within the algae are found some of the oldest and most primitive types of plant life. The algae vary in size from microscopic forms to large seaweeds which, in length of main axes, equal or surpass the height of the tallest tree. In mass and dry weight, they are only a fraction of these large land plants.

Typically the algae grow in water or on damp soil and sometimes as endophytes in the tissues of other plants and animals. Algae are common in lakes and ponds, as well as in most streams, and are residents of even semipermanent pools of water. The small free-floating algae, called **phytoplankton,** contribute to the total suspended plant and animal cells of both fresh and salt water. Plankton is very important as a source of food and serves as an early step in the food chain of larger animals, especially fish. Other aquatic algae grow attached to a substratum. Many biologists have suggested that algae might be raised by man as a food, and in some parts of the world algae form a part of the local diet.

Algae are of economic concern if they develop in reservoirs in large numbers, subsequently clogging drains and filter beds. They can be of considerable value in helping to aerate water or in fixing elemental nitrogen in their bodies which, upon liberation, improves the growth of crop plants. The development of algae in sewage disposal plants is encouraged because they utilize nitrates and phosphates in their growth. Oxygen liberated as a by-product of photosynthesis helps to facilitate aerobic bacterial decomposition of raw sewage. It is this principle that has demanded considerable time of scientists in an effort to construct a biological system to be used during extended flights in space. In this system it would be beneficial to use algae to assist in the decomposition of human wastes, to liberate oxygen as a by-product of photosynthesis, and to produce, through growth of the algae, an occasional harvest of food. Some algae are valuable as living material for basic biological research, and some are used as the living systems for the production of biochemical compounds. The relatively small space required for their culture and their somewhat less demanding environmental requirements make them adaptable to research. The large cells of some algae, such as *Nitella,* are especially valuable in cellular research. Because of their ability to utilize radioactive carbon in their metabolic processes, algae are used commercially to produce radioactive (identifiable) lipids that, through purification in the laboratory, serve as standards in such procedures of investigation as chromatography.

Algae are plants of simple structure and organization; many consist of a single cell. Cellular organization, except in the Cyanophyta, is essentially the same as that of other plants, and electron-microscope studies show that such components as the cytoplasm, nucleus, membranes, chloroplasts, and other organelles are basically the same as those studied elsewhere in the plant kingdom. Some algae are loose aggregations of cells essentially alike or usually exhibiting only slight differentiation. Certain species of large brown algae, however, show diversified organization of the plant body. They have strong holdfasts attaching them to rocks, long flexuous stipes, and an expanded blade portion. Some of these plants have considerable differentiation of tissues. But they are without well-developed tissue systems and especially lack the xylem and phloem of the vascular tissue. Because of the relatively simple tissue structures and the absence of a well-developed vascular system, the plant body is regarded as a thallus.

The algae exhibit a variety of organizational levels; i.e., there are single-celled algae; loosely associated colonies of cells, which are rather easily separated from each other; colonies in which cells are fastened to each other by cytoplasmic connections; more highly specialized forms, such as the brown algae. Simultaneously the question may be raised as to whether the total of cellular activity represents the activity of the organism, whatever its degree of complexity, or whether there is an organismal level of activity and control of which the cell is a part. The multinucleate condition of some plants, as a result of the failure of cellular cross walls to develop, suggests that the importance of organization and behavior at the organismal level is greater than the sum of the activity of the individual cells found within an organism. This opinion is further confirmed by the fact that electron micrographs show more cytoplasmic connections between cells than were seen earlier with the light microscope. If the concept is accepted that cytoplasmic connections reflect the interdependence of cells, we might then conclude that organismal activity is something more than the simple sum of the activity of its cells.

Reproduction The gametes are produced within unicellular gametangia. Even when a number of gametangial cells are organized into a more elaborate structure than that represented by a single cell, there is no sterile jacket or covering surrounding them. Fusion of gametes is called syngamy or fertilization.

In some algae the gametes seem to be similar in size and structure and are called **isogametes.** Their fusion is usually called **isogamy.** In other algae the gametes are obviously dissimilar, i.e., in size, motility, behavior, and structure, and are called **heterogametes.** Their union (in sexual reproduction) is most often referred to as fertilization and one specific type is called **oögamy.** Commonly, in oögamy, the smaller and usually motile gamete is called the microgamete, or sperm, and the larger nonmotile gamete is referred to as the macrogamete, or egg. Motility of the single cells of algae is made possible by the presence of one or more flagella. The flagellum may be smooth, or it may be covered with many small extensions; in this case it is a tinsel flagellum.

Whatever the nature of the gametes that have fused, the product of their fusion is a zygote, and its chromosome number is twice that of either of the gametes. The diploid zygote may grow immediately into a new plant, may become a resting cell, or may divide meiotically. It is common practice to refer to the resting zygote formed from the fusion of isogametes as a **zygospore** and to the resting zygote formed in oögamy as an **oöspore.**

Whereas fusion of gametes, along with the doubling of the chromosome number (during syngamy), is one feature of sexual reproduction, the other is meiosis, with the reduction of the chromosome number. If the zygote undergoes meiosis, with the production of meiospores, the division is referred to as **zygotic meiosis.** If meiosis is delayed until some later stage of development in the life history of the plant (alga), the division is called **intermediate meiosis.**

Associated with zygotic meiosis is the subsequent growth of one or more of the meiospores to produce haploid plants, whereas germination and growth of an unreduced zygote result in the formation of a diploid plant.

Asexual, or *vegetative, reproduction* is common in algae. Neither fusion of cells nor reduction of chromosome number is involved. In unicellular algae a mitotic division of the nucleus, accompanied by cellular division, produces two cells. In some algae, especially the long filamentous ones, **fragmentation,** or breaking of the filament, takes place, with each new fragment produced behaving the same as the longer filament from which it was derived. A somewhat more elaborate method of asexual reproduction is the formation of asexual spores of various types. These spores differ from meiospores in that they are not the product of meiosis but contain exactly the same number and same complement of chromosomes as the parent cell from which they were derived. The product of their growth is an exact duplicate of the plant by which they were produced. In many algae motile asexual spores are produced, called **zoospores,** whereas others may be nonmotile **(aplanospores).**

Classification The separation of algae into divisions is largely determined by the physiological characters of the vegetative cells and the behavior of the reproductive cells and their flagellation. Cellular characters of importance include the pigments present, the nature of the food reserve, and the chemical composition of the cell wall. In the present discussion of the algae, five divisions are considered briefly: Chlorophyta,

Chrysophyta, Phaeophyta, Cyanophyta, and Rhodophyta.

CHLOROPHYTA

General features The Chlorophyta, or grass-green algae, have their photosynthetic pigments localized in chloroplasts, and the reserve food usually is stored as starch. Gametic reproduction is common within the division, and motile cells, both zoospores and gametes, usually have two or four anterior whiplash flagella of equal length. Most of the grass-green algae are freshwater forms, but some orders, such as the Ulvales and Siphonales, are predominantly marine.

Green algae may form scum on the surface of quiet, stagnant water or grow firmly attached to submerged rocks, pieces of wood, and similar matter in swift-flowing streams or on lake shores. Although most are aquatic, some grow in moist terrestrial habitats, such as damp soil and the shaded sides of rocks and trees.

The green algae show a considerable range of variation in the form and structure of the plant body. Many of the lower members of the division are single-celled plants (Figs. 15–1, 15–14).

The Chlorophyta are characterized by a well-defined nucleus. Chloroplasts are present and in many algae are comparatively large and of various characteristic shapes. The process of carbohydrate manufacture occurs in green algae, as in higher plants. Frequently starch accumulates around the characteristically rounded **pyrenoids,** and these proteinaceous bodies are thought to be intimately associated with the elaboration of starch. Vacuoles are present in the cytoplasm of most kinds of green algae, some of which at least are contractile. Some primitive forms also contain an eyespot (Fig. 15-1).

Asexual reproduction In some filamentous species of the green algae there is a tendency for the longer filaments to fragment into individual cells or into short sections of a few cells each. Such fragments may grow into long filaments. In other species accidental breaking of the colony may take place as a result of injury or water currents. The formation of various kinds of spores (zoospores, aplanospores) or of gametes may cause rupturing of cells and fragmentation of the filament, or colony.

The most common method of asexual reproduction is by formation of zoospores. Zoospores are usually formed in vegetative cells, morphologically the same as other cells of the filament, but they are referred to as zoosporangia. One or more zoospores may be formed within a zoosporangium. However, if more than one is produced, the number is usually some multiple of two (4, 8, 16).

Frequently nonmotile spores are formed, called aplanospores, and are considered by some to be abortive zoospores. Aplanospores with greatly thickened walls are often called hypnospores. Vegetative cells may thicken their walls and become sporelike resting stages, but they are distinguished from the spores mentioned above by the fact that the newly formed wall around the protoplast is fused with the parent-cell wall. These are called **akinetes,** and unlike the aplanospore the akinete is not considered to be a modified zoospore but rather a direct specialization of a vegetative cell.

Sexual reproduction Gametic union is widespread among the Chlorophyta. In the most primitive case there is fusion of motile and similar gametes. In algae exhibiting isogamy, it is difficult to make morphological distinctions between the gametes. In anisogamy the gametes involved in syngamy are similar in form but differ in size and are distinguishable. In the case of oögamy, the gametes involved in syngamy differ in size and often in motility. The oögamete is usually larger and nonmotile and the microgamete is usually smaller and may be motile or not. Isogamy, anisogamy, and oögamy are sometimes arranged in this order to represent the fusion of more highly differentiated gametes and are interpreted by many botanists to represent increasing complexity in gametic reproduction. If the fusing gametes are from the same parent plant (thallus), the species is **homothallic;** if the gametes are derived from two separate filaments of opposite mating types, the species is **heterothallic.**

The zygotes—products of gametic fusion, or syngamy—may be thin walled or thick walled. The thin-walled zygotes germinate within a short time, but those with thick walls may remain dormant for a considerable length of time before germination. Most often the thick-walled zygotes will undergo meiosis on germination (zygotic meiosis). The thin-walled zygotes, on germination, often will form new cells by mitosis, with accompanying cellular division and no reduction in the chromosome number.

Origins and evolution The biological interest in the green algae lies in the primitive nature of the group. They exhibit many primitive structural features and methods of reproduction. Although the present genera and species are probably not identical with the ancestors of the higher green algae, nonetheless the simple plant bodies of the lower forms of the present day may be very similar to the kinds that appeared in early geological periods. It is supposed that there was a common primitive stock from which all plant life has been evolved. It is possible that the complex and varied array of modern plants may have originated from relatively simple plants similar to some of the lower green algae. The motile single-celled forms and unspecialized colonial types are the ones generally regarded as points of origin of the higher green algae. Evolution is generally traced from the lower types into the filamentous forms.

Classification of the green algae Within the Chlorophyta various evolutionary lines of ordinal rank have been distinguished. Nine of these are considered here.

Volvocales

General characteristics The order Volvocales is of interest in that its members are simple forms of aquatic plant life. All are motile unicellular or colonial plants, with the colonies often of definite shape but never filamentous (Figs. 15-1, 15-4, 15-5). Asexual reproduction is by cellular division, and sexual reproduction may be isogamous, anisogamous, or oögamous.

The unicellular members of the group are of particular interest because they represent the most primitive of living green plants. It is believed that they are very similar to the forms from which green plants arose early in the earth's history. The vegetative forms of members of this order are diverse, varying from single cells to colonies consisting of thousands of cells.

The unicellular Volvocales *Chlamydomonas* is the best-known unicellular member of the Volvocales. It is widely distributed in fresh water and in damp soil. The vegetative cell is rounded, ellipsoidal, or pear shaped in outline. Motility is achieved by means of two flagella of equal length, which are longer than the length of the cell and are attached to the anterior end of the body (Figs. 15-1, 15-2). Internally there are present a red eyespot in the cup-shaped chloroplast, a nucleus, cytoplasm, one or more contractile vacuoles, and other organelles (Fig. 15-3).

Asexual reproduction is by longitudinal division of

Fig. 15-1 Chlamydomonas. *A, in vegetative condition; B, cell dividing; C, cell in resting condition. (Drawing by Christian Hildebrandt.)*

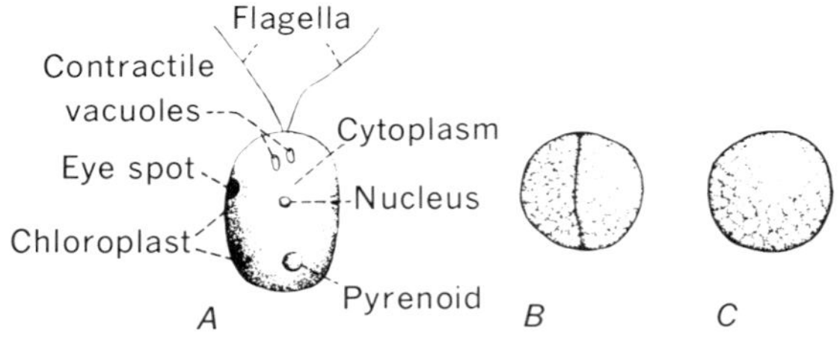

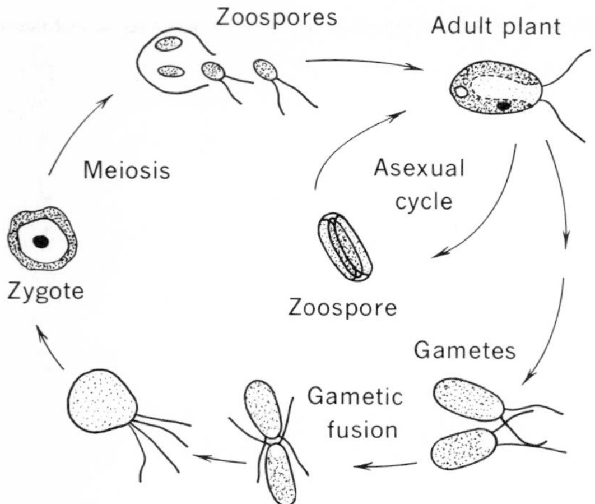

Fig. 15-2 Chlamydomonas. *Life-history diagram.*

the contents of the parent cell into two, four, or eight daughter protoplasts. Each newly formed protoplast develops a wall and a neuromotor apparatus of its own before being liberated. Failure of the daughter cells to escape (especially in forms growing on soil) may result in the establishment of a **palmella** stage, in which continued divisions might result in the production of hundreds of cells.

Sexual reproduction may involve the fusion of gametes from one parent cell (homothallism) or from two parent cells (heterothallism). The gametes may be similar or not, resulting in isogamous, anisogamous, or oögamous types of gametic reproduction. The wall of the zygote may thicken and the zygote may undergo a period of dormancy. Division of the zygote, at the time of germination, is meiotic, and usually four meiospores are produced, which escape and behave as adult haploid plants (Fig. 15-2).

The colonial Volvocales The cells of the colonial Volvocales are arranged either in the form of a disk or as a hollow sphere. The vegetative cells of the colony are similar to the individual cells of *Chlamydomonas,* and all are biflagellate. Three genera—*Pandorina* (Fig. 15-4), *Eudorina,* and *Volvox*—may represent an evolutionary line of three methods of development: (1) an increase in the number of cells of the colony, (2) the specialization of certain cells for reproduction, and (3) advancement from isogamy to oögamy. *Volvox* is the most elaborate of the colonial forms, showing considerable advancement from the more simple and primitive Volvocales.

Volvox

GENERAL CHARACTERISTICS Colonies of *Volvox* are spherical and are about the size of a small pinhead. The colony may consist of as many as 50,000 cells, and these cells are arranged in a single layer just within the gelatinous matrix that gives the colony substance. Some species have the cells joined together by connecting cytoplasmic strands (Fig. 15-5*B*). The cells are differentiated, with only certain ones taking part in reproduction.

ASEXUAL REPRODUCTION Of the many cells of a *Volvox* colony, only a few differentiate as asexual reproductive cells; these are called **gonidia.** A gonidium divides to produce a number of daughter cells, all of which are held together as a new colony (Fig. 15-5*F*), but since a number of gonidia divide at the same time, several new colonies are formed simultaneously. After division ceases, the young colony turns inside out by inversion through a small pore, or opening, the **phialopore.** It soon develops flagella and later escapes from the parent colony.

SEXUAL REPRODUCTION Sexual reproduction in the genus *Volvox* is oögamous. The gametes are sharply differentiated, the female gametes, or eggs, being nonmotile and very much larger than the male gametes, or sperms, which are motile. Some species are homothallic, with compatible gametes being produced within one colony, whereas others are heterothallic.

The gametes are produced within specialized cells of the colony called gametangia (Fig. 15-5*G*). The content of the male gametangium

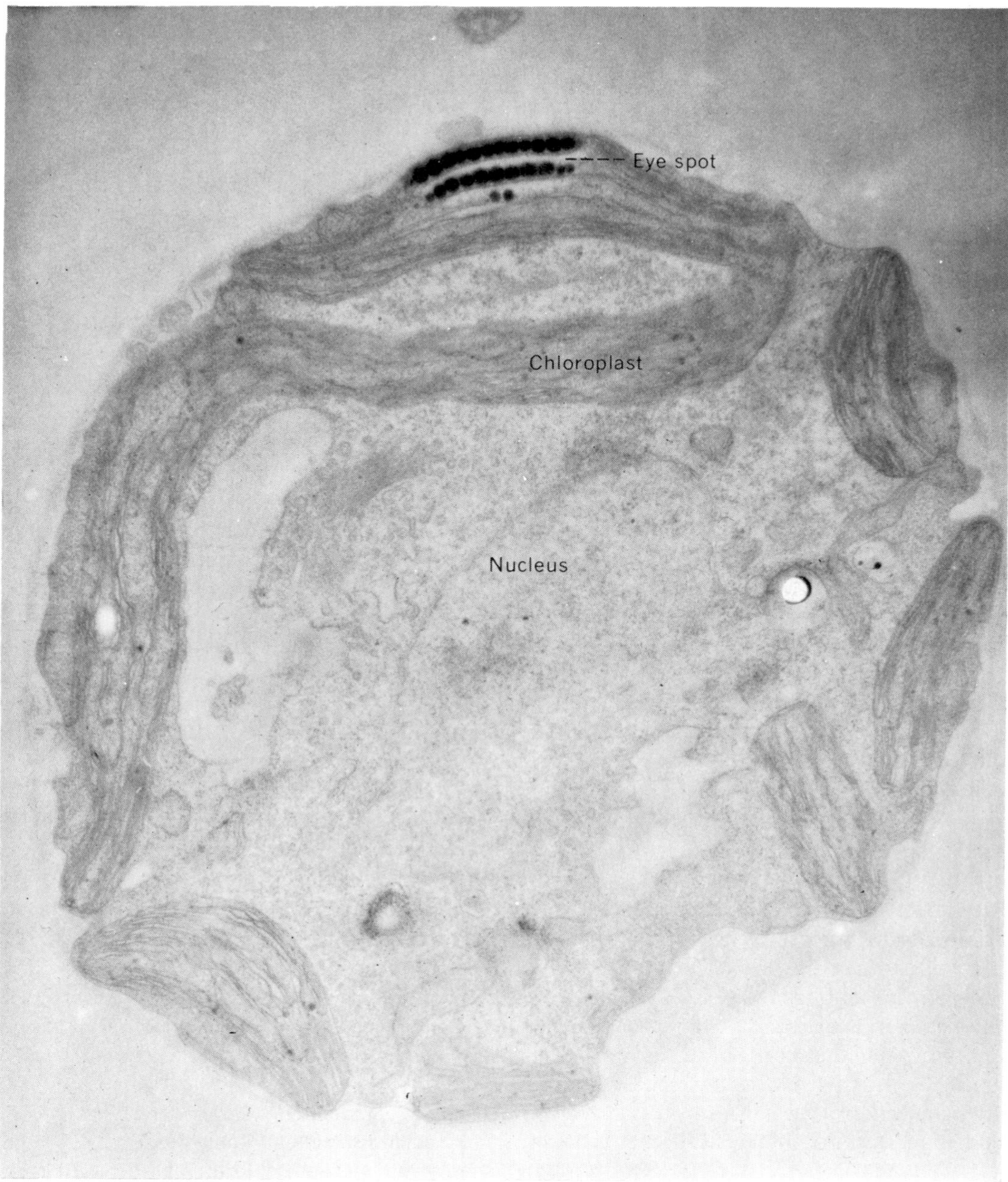

Fig. 15-3 Electron micrograph of Chlamydomonas. *(Courtesy of Dr. Paul Grun.)*

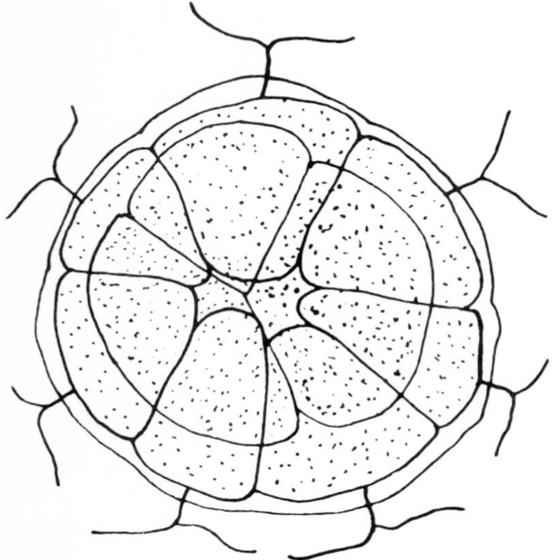

Fig. 15-4 Pandorina. *Diagram of colony.*

divides many times to produce at maturity the small biflagellated sperms. The content of the female gametangium fails to undergo division but enlarges to become an egg. Fertilization of the egg by a single motile sperm results in the production of a zygote which develops a thick, often stellate wall and undergoes a resting period (Fig. 15-5*H*). On germination the first divisions are meiotic and a new colony of haploid cells is produced.

SIGNIFICANCE OF THE VOLVOCALES There is some interest in algae because they are a primitive group and an understanding of their development and behavior may form a part of the knowledge necessary to an understanding of the evolution of all green plants. Forms like *Chlamydomonas* have provided excellent material for significant biological investigation in cytology and ultrastructure, metabolism, and behavior.

Tetrasporales

General characteristics The algae belonging to this order are, for the most part, made up of cells united into nonfilamentous colonies that may have definite shape or may be more or less amorphous. A few genera have solitary cells. The vegetative cells are nonmotile, although at times these metamorphose into a flagellated motile stage.

Some kinds, such as *Tetraspora* and *Palmella* (Fig. 15-6), occur in extensive gelatinous colonies growing in quiet fresh water. The vegetative cells of *Tetraspora* bear two long rigid cytoplasmic processes called pseudocilia, which may not extend beyond the watery gelatinous matrix of the colony. *Palmella,* on the other hand, has no such cytoplasmic extensions.

The typical vegetative condition in these genera resembles the temporary palmella stage occasionally found in the unicellular members of the Volvocales. For this reason the two orders are thought to be closely related.

The genus *Tetraspora,* as its name indicates, is characterized by a tendency of its vegetative cells to occur in groups of four within the matrix of the colony. However, in many colonies the cells are irregularly scattered. In addition to the loose, gelatinous, palmella-type colony, some genera form branched colonies; still others grow as single-celled organisms attached as epiphytes on larger aquatic plants.

Asexual reproduction Asexual reproduction is by means of zoospores or aplanospores. At almost any time in the life of a colony, its vegetative cells may become motile as biflagellate zoospores. These zoospores escape from the colonial matrix and swim around for a short time before they become nonmotile and secrete a new gelatinous matrix. Growth of old and new colonies continues with cellular division.

Sexual reproduction Sexual reproduction is by biflagellate gametes. The protoplast of a vegetative cell divides to form four or eight pear-shaped gametes. Colonies may be dissimilar as to sex, some producing male and others female gametes. Although there is no difference in their size, there are, however, physiological differences. The gametes escape from the matrix and

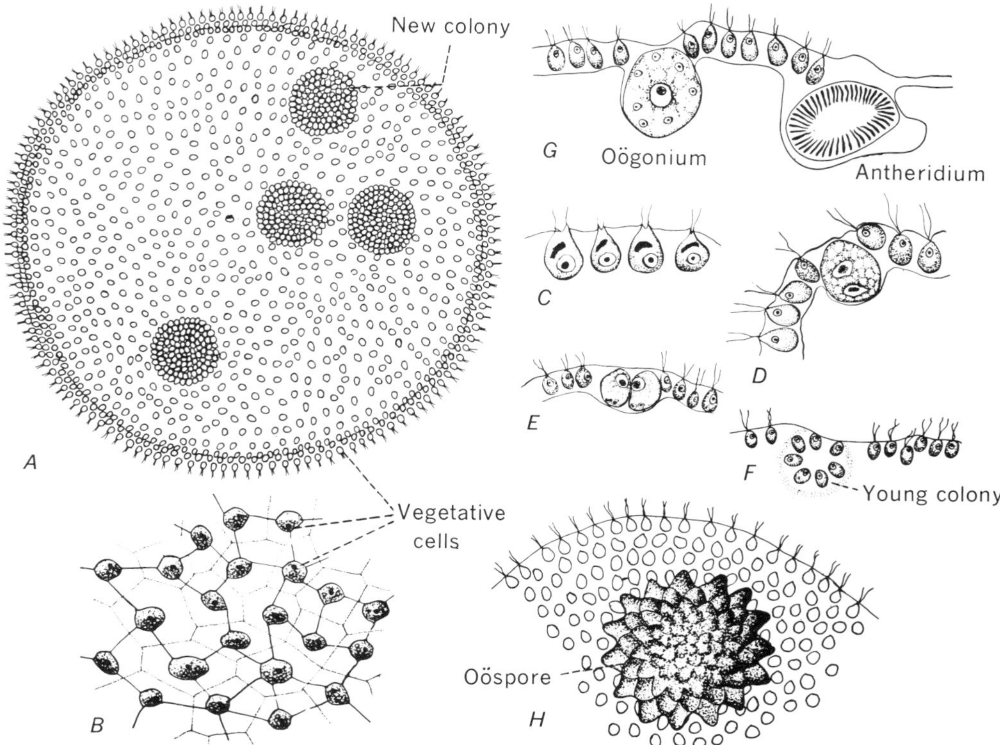

Fig. 15-5 Volvox. *A, vegetative colony with four younger colonies; B, a portion enlarged, showing coalescent sheaths and cytoplasmic connections from cell to cell; C–H, sections of portions of colonies; C, vegetative cells; D–F, beginning of asexual reproduction by formation of new colonies; D, vegetative cell enlarged; E, cell similar to D after the first division; F, further divisions to form young colony within parent colony; G, showing antheridium containing sperms and oögonium containing egg; H, the warted oöspore resulting from fertilization of the egg by a sperm. (A, B, and H by Helen D. Hill.)*

fuse to form zygotes, which usually retain their motility for a short time before coming to rest. They increase in size and secrete a protective wall. At the time of germination of the zygote, four or eight nonmotile cells (aplanospores) are formed which, by additional divisions, produce a new colony.

Ulotrichales

General characteristics Most of the members of approximately 80 genera of the Ulotrichales are freshwater algae. However, some genera

Fig. 15-6 A, Tetraspora; *B,* Palmella; *both showing cells embedded in a gelatinous matrix, with* Tetraspora *showing typical grouping of four cells. (Drawing by Helen D. Hill.)*

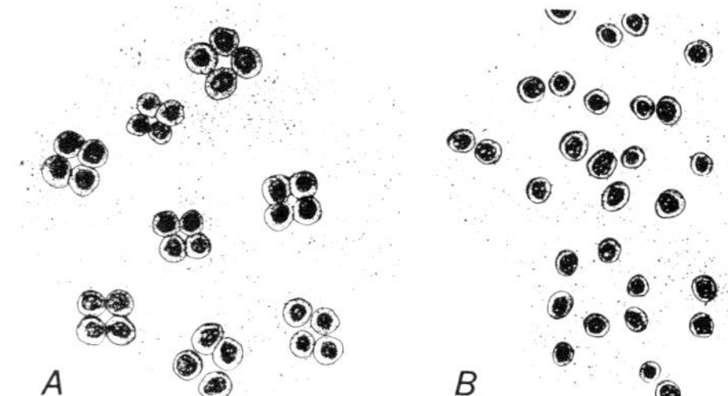

include both freshwater and saltwater forms and several genera are exclusively marine in their distribution. Some occur in cold running water, some in quiet water. Others are epiphytic upon larger water plants, and a few grow within the tissue of other plants. Most species are attached to some sort of substratum, and in many genera there is specialization of the basal cell to form a **holdfast.**

The Ulotrichales have uninucleate cells usually with a single, laminate, parietal chloroplast. They show considerable diversity of form, ranging from unbranched filaments to predominantly branched filaments (Fig. 15-8). In a few species the plant body is a discoid thallus (Fig. 15-10), and in one, *Protococcus* (Fig. 15-9), the plant body is unicellular.

Branching represents a more advanced development than that exhibited by the unbranched filamentous forms. Often branching forms will have their plant body divided into a prostrate and an upright, or erect, portion, a condition said to be **heterotrichous.** They reproduce asexually by means of quadriflagellate zoospores, and some genera, as *Ulothrix,* form both macrozoospores and microzoospores. Aplanospores and akinetes are not uncommon. Sexual reproduction ranges from isogamy to oögamy. Some genera, including *Ulothrix* (Fig. 15-7), have a life cycle exhibiting the alternation of a multicellular

Fig. 15-7 Ulothrix. *A, portion of a filament showing the parietal chloroplast with numerous pyrenoids; B, sexual reproduction; formation of gametes and their escape; C, asexual reproduction. (Drawn by Christian Hildebrandt.)*

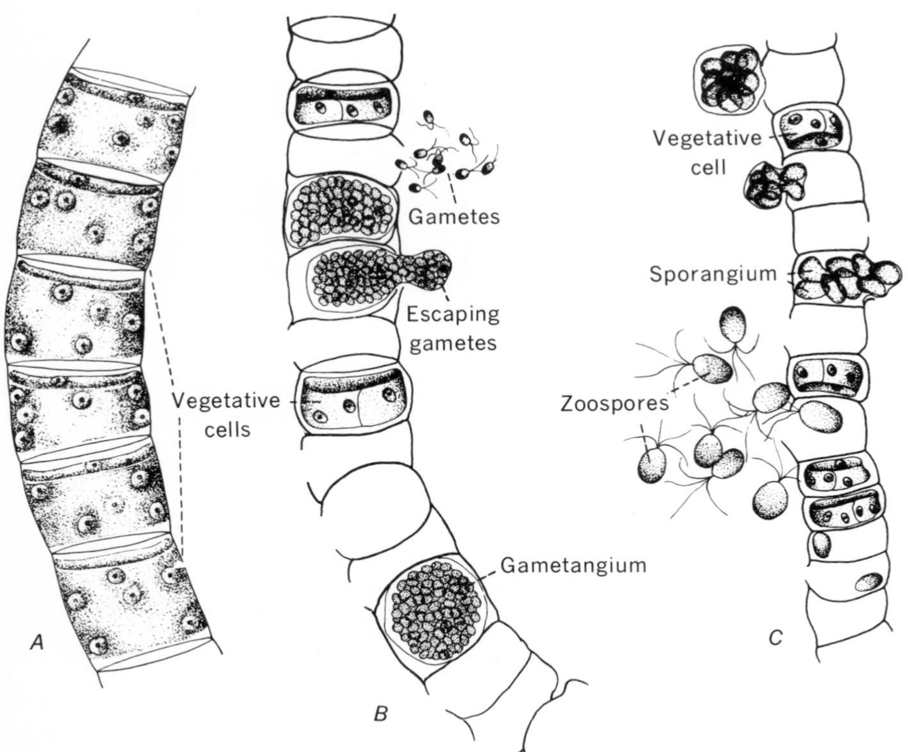

haploid plant with a single-celled diploid plant.

Ulothrix

CELL STRUCTURE *Ulothrix* is a well-known genus of the unbranched filamentous type. The filament is composed of cylindrical cells placed end to end. Each cell consists of a wall and protoplast with a single nucleus and a band-shaped or platelike parietal chloroplast, curved to fit the rounded contour of the cell (Fig. 15-7*A*). Usually one or more pyrenoids are found in the chloroplast.

ASEXUAL REPRODUCTION Asexual reproduction, as in all members of the order, is by formation of zoospores. The larger zoospores are called **macrozoospores** and the smaller ones **microzoospores.** They are formed in separate sporangia, with 1, 2, 4, or 8 macrozoospores per sporangium and 16 to 32 microzoospores per sporangium. The zoospores are formed by repeated division of the protoplast of the vegetative cell (sporangium) (Fig. 15-7*C*). Each zoospore is a single cell with a protoplast surrounded by a thin membrane and provided with four flagella. The protoplast consists of a nucleus, cytoplasm, and a chloroplast containing a red eyespot. When they are mature, the zoospores escape from the sporangium through a rupture of the wall. After emerging, each zoospore swims rapidly for a short period, then comes to rest, loses its flagella, becomes attached to some convenient substratum, and grows into a new gametophytic plant.

In some species, when zoospores are not liberated from the parent cell, they may secrete a wall and become aplanospores. Ordinarily these aplanospores germinate later within the cell. Under other circumstances, usually conditions unfavorable for growth, vegetative cells form thick-walled aplanospores. Upon the return of conditions favorable to growth, these spores germinate by forming daughter protoplasts, which, in some instances, function as zoospores and eventually reproduce the vegetative filaments.

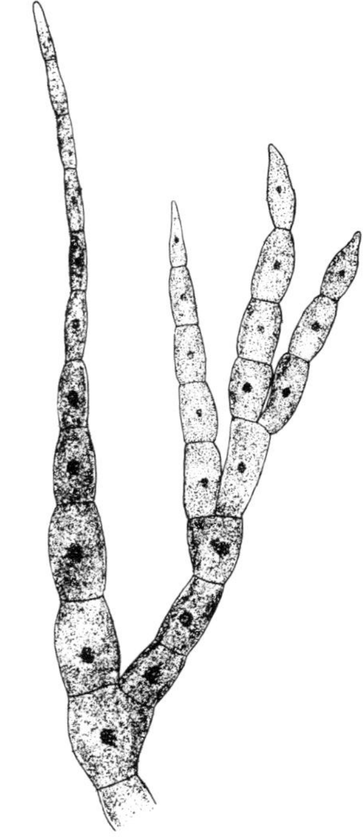

Fig. 15-8 Branched filament of Chaetophora.

SEXUAL REPRODUCTION Sexual reproduction is isogamous and involves the formation of small biflagellate isogametes. These, like the zoospores, are formed by repeated division of the protoplast of a vegetative cell, which thus becomes a gametangium producing 32 to 64 gametes (Fig. 15-7*B*). After emerging from the gametangium, two gametes from different filaments fuse to produce a zygote. The zygote remains motile for a short time after its formation but soon becomes a resting cell and later germinates (first divisions are meiotic) to form several motile cells, each of which produces a new haploid plant.

In this and in certain other genera with isogamous reproduction of this type, the similarity of gametes to zoospores suggests that sexual reproduction may have originated by fusion of zoospores in response to unfavorable environ-

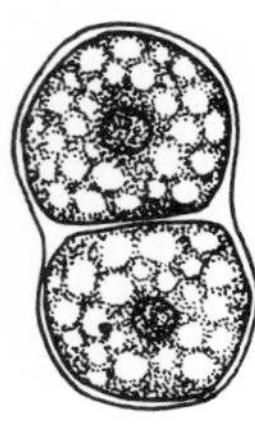
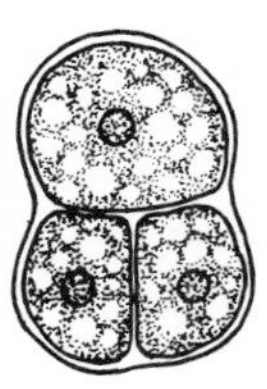
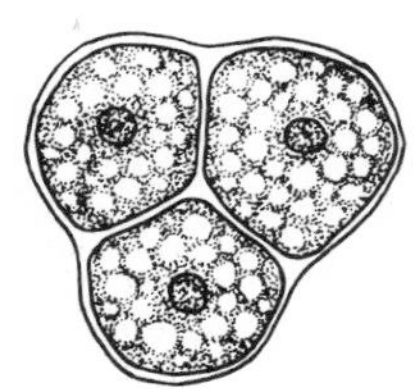
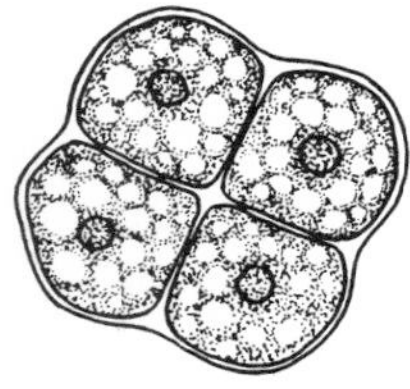

Fig. 15-9 Protococcus. *Single cell and cells united in colonies of two, three, and four, each cell showing a nucleus and vacuolate cytoplasm.*

mental conditions. It is not impossible that sexuality arose after this manner in more than one order of green algae.

Protococcus This alga grows in moist locations on trees, old fences, stone walls, and in similar situations. Usually the plants are single spherical cells, occurring either as isolated individuals or in small compact masses or groups consisting of three or more cells (Fig. 15-9). Occasionally, when the plants are submerged in water, cell divisions may continue until many cells form a colony and irregular branches may develop. Largely as a result of this branching under certain environmental conditions, *Protococcus* is interpreted to be a reduced, simplified form. It is thought to be derived from a filamentous ancestor in which both vegetative structures and reproductive functions retrogressed. *Protococcus* apparently has lost its capacity for all types of reproduction except cell division.

Fig. 15-10 Vegetative plant body of Coleochaete *consisting of a flat plate of cells.*

Ulvales

General characteristics The cell structure of some genera of the Ulvales is so similar to that of the Ulotrichales that many botanists feel that these forms should be placed in a single order. However, the majority of the species of this order are marine, with only a few found in brackish or fresh water. The uninucleate cells of representatives of the Ulvales divide in two or three planes and produce thalli that are platelike, hollow tubes or solid cylinders.

Ulva, the sea lettuce, has a number of species which grow in salt water or in brackish water. It exhibits an alternation of isomorphic generations, in which the gametophytic plant produces biflagellate gametes and in which the meiospores are produced following meiosis in some cells of the sporophytic plant. Species of *Ulva* are heterothallic and gametic reproduction is isogamous or anisogamous. Fusion of gametes results in the formation of a motile cell (zygote) which after a short time comes to rest and develops into a sporophytic plant. Meiospores are quadriflagellate and motile.

Cladophorales

General characteristics The Cladophorales are filamentous green algae, which may be either simple or branched filaments composed of cylindrical cells placed end to end. The cells of the filament are coenocytic with numerous small nuclei and generally a reticulate chloroplast that encircles the protoplast. Many pyrenoids are present. Some species are marine, whereas others are found exclusively in fresh water. Asexual reproduction is by quadriflagellate zoospores, aplanospores, or akinetes, and sexual reproduction is isogamous or anisogamous.

Cladophora Some species of *Cladophora* grow abundantly in the shallow water of lakes and streams; several are marine. The plants grow attached to rocks and structural timbers of various sorts, such as the piles supporting piers, submerged logs, and other substrata, and they may attain a length of several inches. The plant body is a coarse, regularly branching, filamentous structure with cross walls separating the coenocytic segments (Fig. 15-11).

Asexual reproduction in *Cladophora* is accomplished by the production of numerous small, uninucleate, quadriflagellate zoospores formed in the terminal segments of the filaments (Fig. 15-11*B*). When they are liberated, these motile cells soon come to rest and grow into a new plant. In *Cladophora* and some related genera, asexual reproduction is also accomplished through the formation of akinetes.

Cladophora is isogamous and heterothallic with the production of many small biflagellate gametes that fuse in pairs to form diploid zygotes. The zygote germinates soon after its formation, and the division of the zygote nucleus is mitotic, forming a $2N$ plant directly.

Cytological investigations indicate that some species of *Cladophora* have alternating plant forms with definite haploid and diploid chromosome numbers. The diploid plant develops from the diploid zygote and produces haploid meiospores as a result of meiotic divisions. Haploid plants developing from the $1N$ zoospores (meiospores) produce gametangia containing $1N$ isogametes. Gametes from different thalli fuse to form the $2N$ zygote, completing the life cycle.

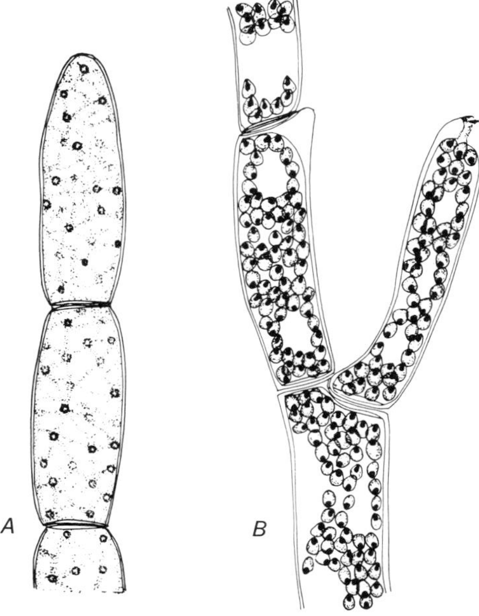

Fig. 15-11 Cladophora. *A, terminal cells of a vegetative filament with numerous pyrenoids and nuclei; B, portion of filament showing numerous zoospores in each cell. (Drawings by Helen D. Hill.)*

Oedogoniales

General characteristics There are only three genera within this order, and *Oedogonium* is the only one with unbranched filaments (Fig. 15-12). The plants grow in quiet water, usually attached to other plants, as upon the leaves and petioles of water lilies and similar aquatics. The outstanding characteristic feature in this order is a peculiar flagellation of the motile bodies. The numerous flagella of all the motile reproductive bodies, i.e., the zoospores, the androspores, and the male gametes, are attached in the form of a ring, or "crown," at the anterior end of the cell.

Asexual reproduction is by zoospores, and sexual reproduction is always oögamous, with certain species producing their male gametes in the terminal cell of special dwarf filaments.

Form, structure, and cell differentiation Considerable cell differentiation is found in both the branched and unbranched types. Usually the basal cell serves as a holdfast (Fig. 15-12*A*). The gametangia in all genera arise by the specialization of vegetative cells. The vegetative cells are characterized by a netlike parietal chloroplast, containing pyrenoids in varying numbers (Fig. 15-12). Cell division may take place in any cell of the filament except the basal cell. However, in some species at least, it appears that cell division is much more common in terminal or near-terminal cells of the filament. Often an individual cell undergoes several successive divisions, each leaving a characteristic hemicellulosic apical cap. These caps, because of the

Fig. 15-12 Oedogonium. *A, two young vegetative plants, one of a single cell, the other two cells, showing a single large nucleus and numerous pyrenoids on the netlike chloroplast; the basal cells in each case are holdfast cells; B, portion of macrandrous filament showing cell content of vegetative cell and the formation of antheridia; C, portion of a filament showing oögonium, egg, vegetative cell, and two dwarf male plants, attached below the oögonium; the terminal cell of a dwarf male plant becomes an antheridium. (A drawn by Helen D. Hill, B and C by Christian Hildebrandt.)*

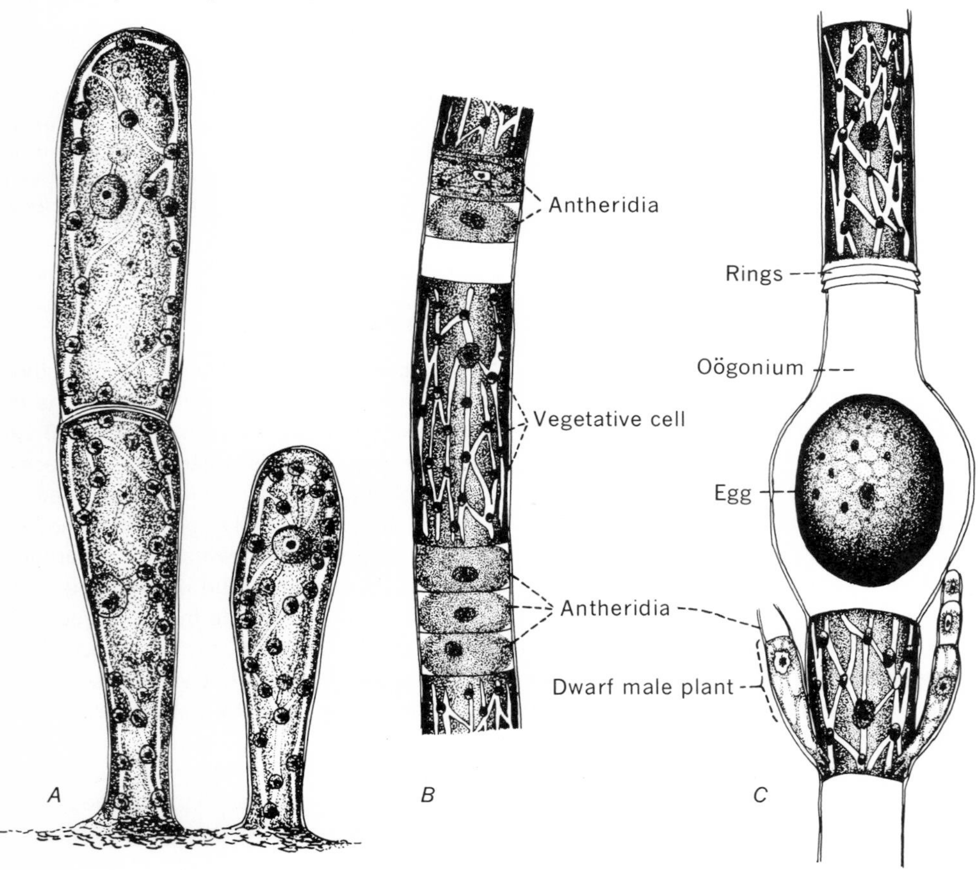

repeated divisions of the protoplast, form a series of rings at the end of the cell (Fig. 15-12*C*).

Asexual reproduction Asexual reproduction in the genera of this group is by formation of zoospores. As represented by *Oedogonium,* any vegetative cell may become a zoosporangium. In this event the protoplast rounds up into a single zoospore which attains motility by means of the characteristic crown of flagella at the anterior end. The zoospore escapes from the zoosporangium and swims away, retaining motility for only a short time. Soon it comes to rest, loses its flagella, and grows into a new filament. The basal cell forms the organ of attachment, or holdfast, of the new plant.

Sexual reproduction Sexual reproduction is oögamous, with the production of large nonmotile eggs and small motile sperms. Sexual reproduction may be **macrandrous,** with antheridia produced on regular-sized filaments, or **nannandrous,** with antheridia produced on special dwarf filaments. The sexual reproduction may be further complicated by the fact that macrandrous species may be homothallic or heterothallic. All three conditions occur in the genus *Oedogonium* (Fig. 15-12*B, C*).

In macrandrous species each antheridium is a short cylindrical segment of the filament, and although the content of an antheridium might produce a single sperm, more often it divides to produce two motile sperms. In nannandrous species dwarf male plants develop from an **androspore,** which is smaller than the usual zoospore and apparently definitely associated with sexual reproduction. Single androspores are produced in androsporangia, which are special sporangia formed by division from the ordinary vegetative cells. They are smaller than the regular zoosporangia and larger than the antheridia of the homothallic species, and, in size, the androspores are between ordinary zoospores and gametes. They are provided with the characteristic crown of flagella and are motile. When androspores are liberated, they swim to the female filaments, where they attach themselves either to the oögonia or in their vicinity. The dwarf male plant develops from the androspore, forming one or more antheridial cells at its tip and vegetative cells below these. Each antheridium produces two small motile male gametes (antherozoids), which bear the characteristic crown of apical flagella.

The oögonia are highly differentiated structures, each one originating from a vegetative cell which by division forms two cells, the basal one becoming the supporting cell of the oögonium and the upper one the oögonium proper (Fig. 15-12*C*). The protoplast of the oögonium becomes the female gamete. The egg is nonmotile and remains within the oögonium, which breaks at maturity near the region of the so-called receptive spot. The motile sperm approaches the oögonium and is received in a bit of protoplasm that protrudes from the receptive spot. The male gamete penetrates the egg, and fertilization is effected when the nucleus of the sperm fuses with the nucleus of the egg. The zygote develops a heavy wall and becomes a resting cell.

Germination of the zygote Through decay or fragmentation of the vegetative filament, the resting zygote, or oöspore, is set free, but the thick-walled cell may not germinate for a year or two. Meiosis precedes or accompanies germination, and four haploid meiospores are formed, each of which may develop into a mature filamentous adult.

Zygnematales

The members of this order are strictly freshwater plants of fairly wide distribution. *Spirogyra,* although in many respects not typical of the green algae, is almost universally studied in elementary laboratory courses in botany. The outstanding features of the order are the absence of flagellated cells and isogamous reproduction by means of amoeboid gametes.

The chloroplasts are distinctive and consist of three general types. Some chloroplasts are spirally twisted and occupy a peripheral position in the cell (Fig. 15-13*A*). Others are in the form of an axial plate that extends the length of the cell, and in some there are two stellate chloroplasts axial to each other (Fig. 15-13*D*).

No asexual reproductive bodies are formed by members of the Zygnematales, and all sexual reproduction is by nonflagellated isogametes that exhibit amoeboid qualities. The gametes are most often referred to as aplanogametes. Gametic union may take place through tubular connections between cells, or the gametes may escape from the gametangia before fusion to form the zygote. The zygote becomes a thick-walled zygospore. Meiotic division of the protoplast of the zygote before germination results in the production of one, two, or four haploid plants at the time of renewed growth.

The order is often divided into two families, one containing the filamentous forms and the other the unicellular forms.

Filamentous forms (Zygnemataceae)

GENERAL FEATURES The members of this family are usually unbranched filamentous plants. Instances of branching have been reported but the branches are never well developed. The filament is composed of cylindrical cells growing end to end. Most species grow unattached in quiet water in thick free-floating masses, but a few species have holdfasts attaching the filaments to stones and other objects in the shallow water where they thrive. *Spirogyra* is probably the best-known genus.

CELL STRUCTURE The cylindrical cells composing the filaments are in general alike in all the genera, with slight differences in the disposition of the parts of the cell due to the nature of the chloroplasts. The cells of the several genera and species vary in size. Each cell consists of a cell wall and a protoplast, with its nucleus, cytoplasm, and one or more chloroplasts. In *Spirogyra* the center of the cell is occupied by one or more vacuoles. The chloroplasts of all species are large and conspicuous and are beautiful objects when viewed with a microscope. Each chloroplast has from one to several prominent pyrenoids.

Spirogyra, Zygnema, and *Mugeotia,* the principal genera of the family, illustrate the forms of chloroplasts found in the group (Fig. 15-13). The chloroplasts of *Spirogyra* are in the form of spiral bands, the edges of which are often scalloped. Pyrenoids are present in each chloroplast. The nucleus is generally located in the center of the cell and is embedded in a mass of cytoplasm, from which radiating strands pass to the chloroplasts in the region of the pyrenoids. The chloroplasts of the cells of *Zygnema* are radiate, or star shaped, two in each cell, centrally placed, with the nucleus embedded in a mass of cytoplasm suspended between them. The chloroplast of *Mugeotia* is a single flat band in axile position in the cell.

GROWTH, CELL DIVISION, AND ASEXUAL MULTIPLICATION Growth and cell division are not restricted to any particular part of the filament. Cell division is preceded by division of the nucleus. Division of the chloroplasts normally follows that of the nucleus. The new cell wall is then formed as a ringlike growth which gradually approaches the center of the cell and finally separates the chloroplast and the cytoplasm into two equal parts (Fig. 15-13).

Cells of *Zygnema* may form a thick-walled resting cell with markings similar to those of the zygospore. The protoplast of a cell rounds up and contracts before secretion of the wall to produce a **parthenospore.** Often, when gametes fail to fuse, they form parthenospores and, in

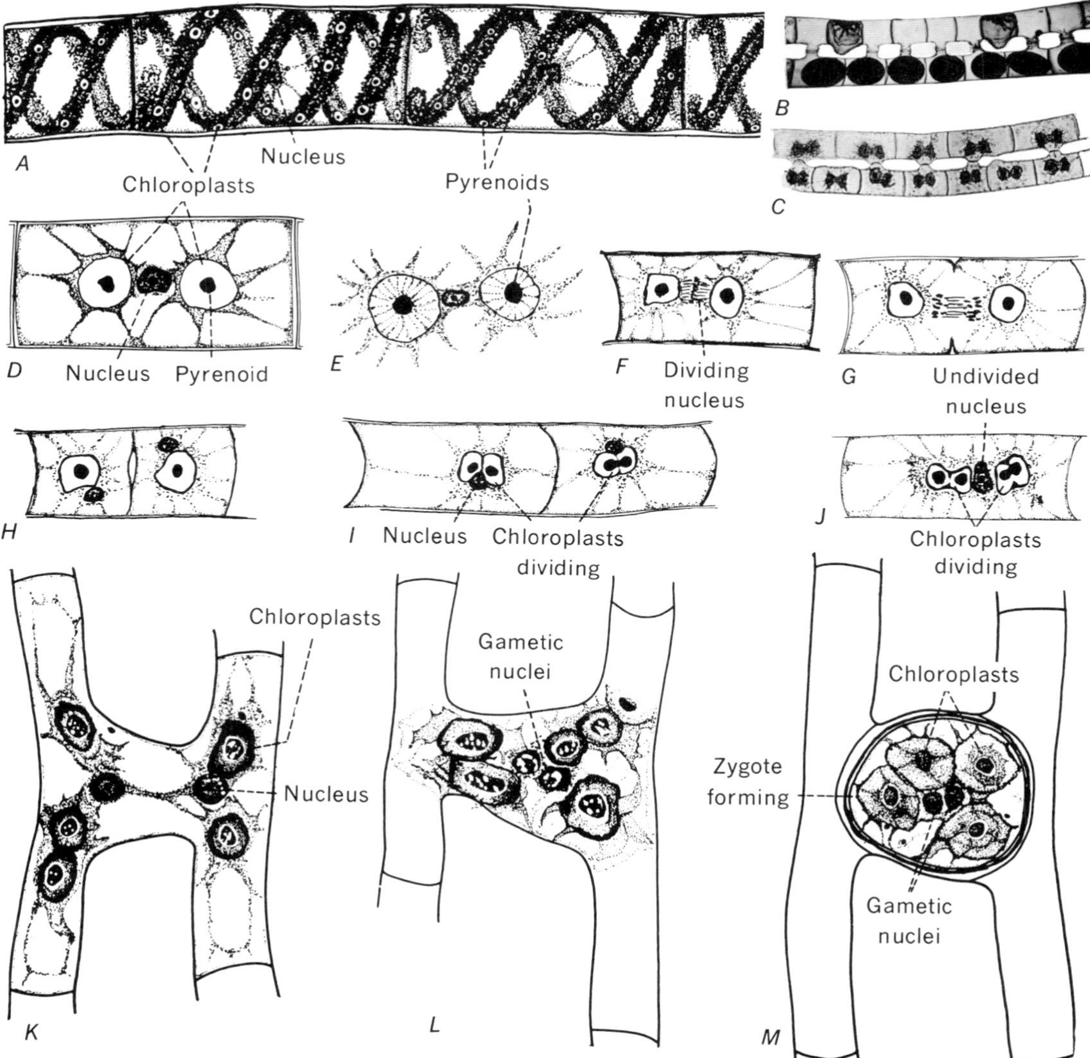

Fig. 15-13 Spirogyra *and* Zygnema. *A, portion of a vegetative filament of* Spirogyra; *B,* Spirogyra *conjugating, showing dark, ellipsoid zygotes; C,* Zygnema, *early stage of conjugation; D–J,* Zygnema, *details of cell structure and cell division; D, vegetative cell; E, structure of the plastids; F–I, normal cell division, plastid division following nuclear division; J, abnormal division, plastid division preceding nuclear division; K–M,* Zygnema *species, zygote forming in the conjugation tube.*

some species of *Zygnema,* no zygotes are produced. Multiplication of filaments by means of fragmentation occurs within the genus.

SEXUAL REPRODUCTION Sexual reproduction in the genera of the Zygnemataceae is accomplished by **conjugation,** an isogamous method, which is characteristic of the order. In the process two vegetative cells combine their contents to form a zygote. In *Spirogyra* (Fig. 15-13*B*) this fusion of cell contents takes place within one of the two conjugating cells; in some forms it occurs midway between the two participating cells. In either case the gametes are formed from the undivided protoplasts of vegetative cells. Conjugation may occur between cells of two adjacent filaments or, much more rarely and then only in certain species, between two adjacent cells of the same filament. In the former case two parallel filaments lie adjacent to each other, the cells touching at various points. Short tubelike projections develop from one or both of the cells at these points of contact. By further growth of these tubes the filaments are gradually pushed apart, but the ends of the tubes remain attached. Soon the end walls of these coalesced tubes are resorbed and a continuous passageway is left between the two cells or gametangia. Thus the conjugation tube is formed.

In *Spirogyra* and sometimes in *Zygnema* the contents of one cell pass through the conjugation tube and fuse with the contents of the other cell (Fig. 15-13). The conjugation of adjacent cells of the same filament is called lateral conjugation, but generally conjugation is between the cells of two filaments (scalariform conjugation).

After the gametes come together in one of the gametangia or in the conjugation tube, they fuse, forming a zygote. The fusion of the nuclei is frequently delayed. The chloroplasts contributed by one gametangium are said to disintegrate, but those from the other persist in the zygote. The zygotes (Fig. 15-13), ellipsoidal or spherical in shape, develop a heavy wall, and they rest for some time, possibly until the spring following their formation. Meiosis usually occurs before germination of the zygospore, and plants produced at the time of germination are haploid.

Unicellular forms (Desmidiaceae)

GENERAL FEATURES The members of the Desmidiaceae are collectively known as **desmids.** They are single-celled, nonmotile plants of very diverse form and rare beauty but are particularly characterized by the fact that the cells are usually divided into two symmetrical halves by a median constriction (Fig. 15-14). The two halves are referred to as **semicells,** the constriction is called the **sinus,** and the narrow portion connecting the two halves is known as the **isthmus.** The cell wall consists of two layers, an inner one of cellulose and an outer one of cellulose impregnated with other materials, frequently iron. Numerous small pores occur in the cell wall and through these an enveloping mucous or mucilaginous material is extruded. This gelatinous sheath is sometimes referred to as a third layer of the cell wall.

The margins of the cell wall are lobed, indented, or have other diverse patterns. When observed from the end view, the cells are often triangular in outline and frequently show radiating processes. The diversity of the forms of the desmids is so great that it is impossible in a brief discussion to describe the various genera adequately.

CELL STRUCTURE The cell structure is very diverse. In general the cell consists of the sculptured wall, the protoplast with its nucleus, cytoplasm, one or two chloroplasts, and one or more vacuoles (Fig. 15-15). Most of the desmids have a single nucleus, located centrally, and two relatively large chloroplasts, one located in each semicell. The nucleus is embedded in a small mass of cytoplasm and in the case of constricted forms is located in the isthmus. The remainder

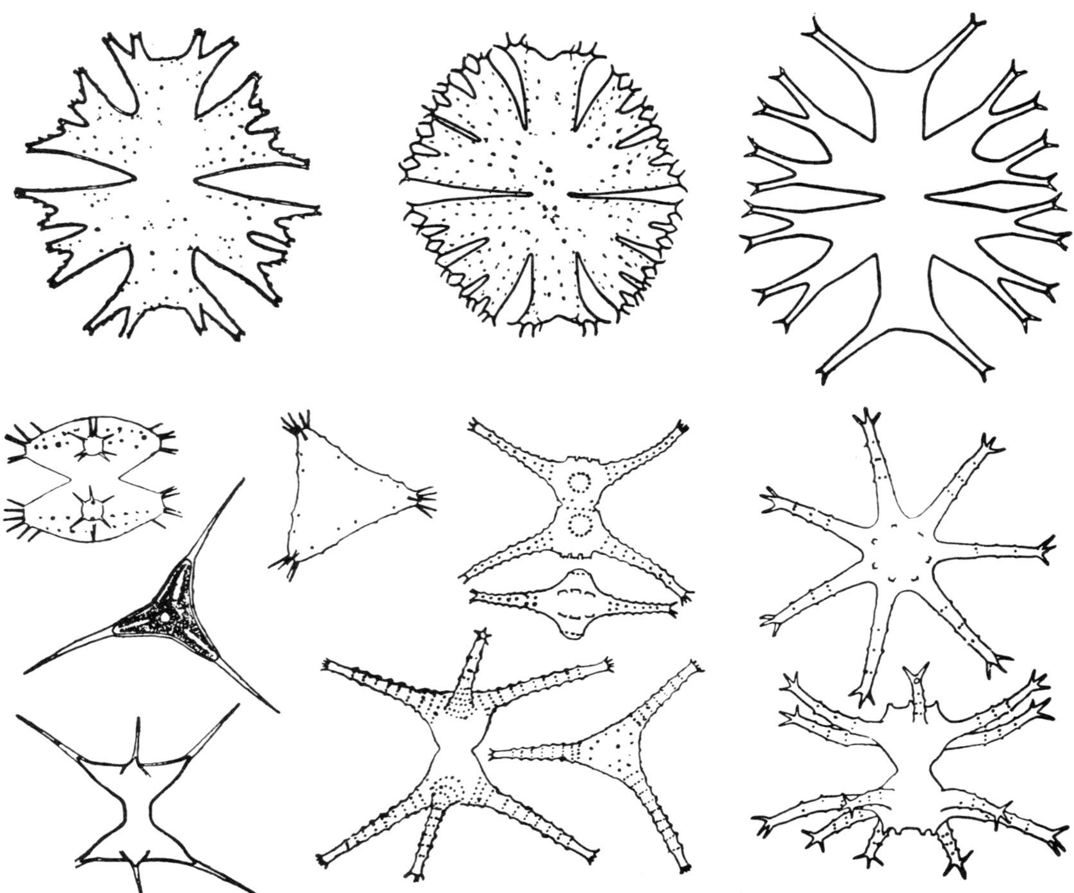

Fig. 15-14 Various forms of desmids. Top three figures, species of the genus Micrasterias; *all others species of* Staurastrum. *(From Gilbert M. Smith, "The Fresh-Water Algae of the United States," 2d ed., figs. 237 and 239, McGraw-Hill Book Company, New York, 1950.)*

of the cytoplasm is variously disposed in the cell. The chloroplasts are provided with pyrenoids which are frequently numerous.

GROWTH, CELL DIVISION, AND ASEXUAL REPRODUCTION Of the two semicells that constitute the desmid plant, one is always younger than the other. The younger semicell is at first very small, but it gradually attains the same size as the other semicell. This feature is due to the manner of cell division. When the desmid divides, the nucleus undergoes division first. The isthmus where the nucleus is located elongates and a constriction forms on the isthmus wall between the two new nuclei. At this constriction a wall is laid down across the isthmus. Each half of the isthmus then begins to enlarge and grows into a new semicell the size

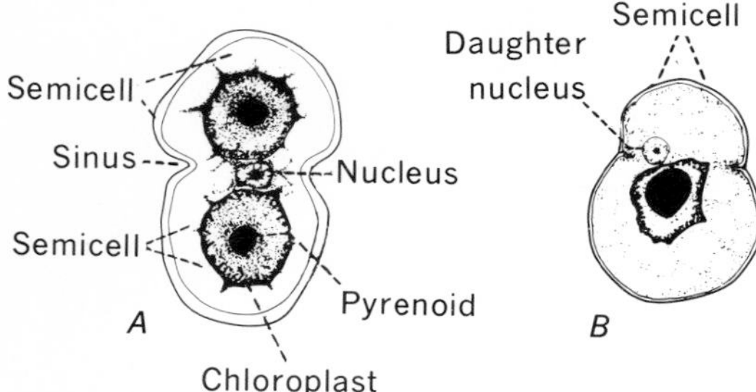

Fig. 15-15 Cosmarium. *A, cell before division, showing the two semicells, the sinus, the centrally placed nucleus, and the large chloroplasts in semicells, each chloroplast containing a single pyrenoid; B, one semicell of the original cell and a new semicell growing from the region of the sinus; the single nucleus of the parent cell has divided, leaving a daughter nucleus for the new cell; other parent semicell not shown.*

of the old one. A new chloroplast is formed in each of the new semicells which soon separate at the point of contact, and asexual multiplication is accomplished (Fig. 15-15). Parthenospores have been reported for a few species.

SEXUAL REPRODUCTION Sexual reproduction in desmids is by **conjugation,** a process of isogamous reproduction in which two entire cell contents behave as gametes. The cells which will conjugate come together and generally become embedded in a common mucilaginous mass. The cells open at the isthmus, the semicells separating and allowing the protoplasts to emerge. Each protoplast acts as a gamete. They come together and fuse, forming a zygote which surrounds itself with a heavy wall, frequently covered with spines, warts, or other outgrowths (Fig. 15-16). A few species develop a conjugation tube and, generally, the zygote is formed within it. The walls of the four empty semicells, being held together by a gelatinous envelope, frequently remain near the newly formed zygote. Fusion of the nuclei of the two protoplasts, although it is delayed, takes place eventually and fertilization is accomplished. After a period of dormancy the zygote or zygospore undergoes meiotic divisions and germinates with the production of two or more young desmid plants. Sexual reproduction in the desmids is more rarely observed than in the filamentous Zygnematales.

Chlorococcales

General characteristics The outstanding feature of members of this order is the total absence of vegetative cellular division. The only cell divisions in protoplasts are those preceding the formation of such reproductive cells as gametes and zoospores. In some instances there is nuclear division, unaccompanied by cellular division, and, consequently, there is a tendency for some types to become multinucleate. Cells may be solitary or arranged in nonfilamentous

Fig. 15-16 Diagram to show conjugation and the germination of the zygote of a species of Cosmarium. *(From Gilbert M. Smith, "The Fresh-Water Algae of the United States," 2d ed., fig. 225, McGraw-Hill Book Company, New York, 1950.)*

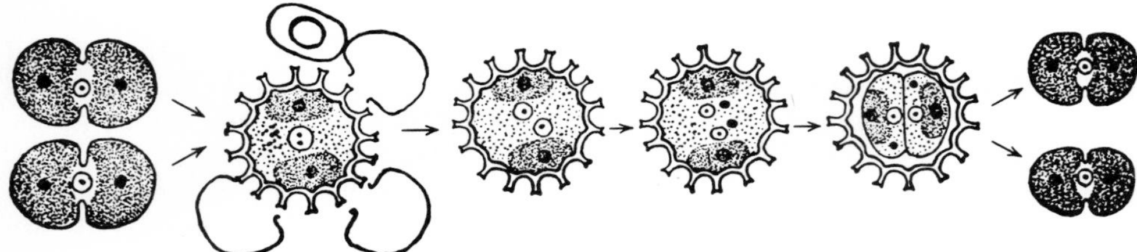

colonies with definite or indefinite numbers of cells. Most species are freshwater forms and are common in the phytoplankton of ponds and lakes. Some of the smaller types grow attached to other algae and on larger aquatic plants, and some often exist as endophytes within the bodies of such animals as *Paramecium* and freshwater sponges.

Reproduction Asexual reproduction is by means of zoospores or aplanospores that frequently are **autospores,** i.e., having the same distinctive shape as the parent cell. Sexual reproduction, if present, may be isogamous, anisogamous, or oögamous.

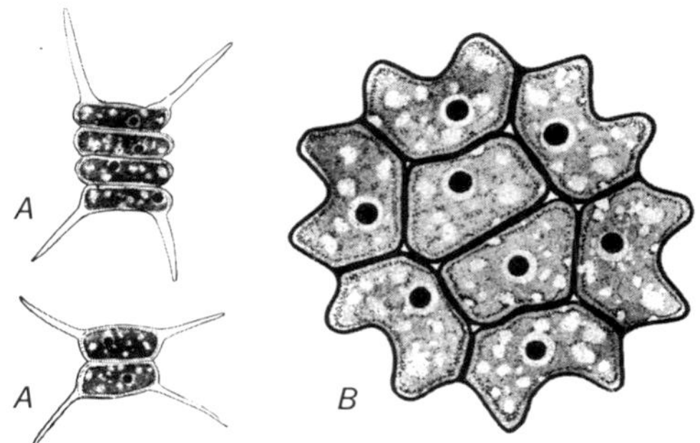

Fig. 15-17 A, Scenedesmus, *two- and four-celled colonies, with stout spinelike appendages on the terminal cells; B,* Pediastrum, *eight-celled colony.*

Chlorella *Chlorella* is a single-celled alga that has been widely used in the study of respiration and photosynthesis. It has also gained some notoriety as a possible source of food for man, especially in hostile environments. The cells are small and generally spherical with a single parietal chloroplast. Autospores are produced and represent the only method of reproduction. Two to sixteen such spores may be produced and they are liberated by rupture of the parent cell.

Scenedesmus Most freshwater collections will contain *Scenedesmus.* The usually flat plate of elongated cells, arranged with their long axes parallel, forms the **coenobium.** The number of cells present is always a multiple of two, and the only method of reproduction is the division of a protoplast to form a daughter coenobium, which may or may not have the same number of cells as the parent coenobium (Fig. 15-17*A*).

Pediastrum Colonies of *Pediastrum* (Fig. 15-17*B*) vary from as few as 2 to as many as 128 cells. They are flat, one cell thick, and free floating. The outer cells of the colony often differ in shape from the internal ones, exhibiting radiating processes. Young cells have a single parietal chloroplast and are uninucleate, but older cells may be multinucleate. All cells of the coenobium may give rise to biflagellate zoospores, but only rarely is division of the protoplast of the parent cells simultaneous. After division of the protoplast, the zoospores, enclosed in a thin retaining membrane, escape from the mother cell. The zoospores swim (swarm) around inside the saclike structure for a short time, then arrange themselves in a position like that of the cells of the parent colony, becoming motionless. The cells then grow to adult size.

Sexual reproduction of *Pediastrum* is isogamous. Spindle-shaped biflagellate gametes are produced and, following fusion, the zygote greatly increases in size. Division of the zygote protoplast produces a number of motile cells which are liberated and swim in all directions. Later the content of each of the first-formed motile cells redivides to form additional cells which are never liberated as individuals but are retained within a vesicle to form a new coenobium.

Hydrodictyon *Hydrodictyon* is a large alga. The word *dictyon* means "net"; thus, as the name implies, the vegetative plant body grows as a netlike structure usually floating on the surface of quiet, fresh water (Fig. 15-18*A*). Young cells contain a single nucleus and a single pyrenoid, but older cells become multinucleate with a large

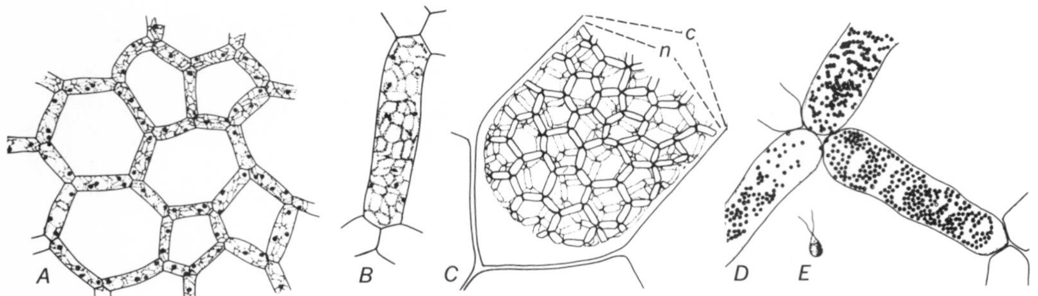

Fig. 15-18 Hydrodictyon. *A, small portion of the saclike net characteristic of the water-net alga, the cells all in vegetative condition; B, single cell from a vegetative net enlarged, showing pyrenoids and chloroplasts; C, portion of a cell, c, from a net showing a young net, n, within, formed by the conjoining of zoospores; D, three cells from a net showing isogametes within the cells; E, a single gamete. (Drawings by Helen D. Hill.)*

reticulate chloroplast containing many pyrenoids (Fig. 15-18*B*). Reproduction is essentially the same as in *Pediastrum* (Fig. 15-18*D, E*).

Other green algae

Many of the green algae are almost exclusively marine, and genera belonging to these groups show great diversity in the development of their plant bodies. Of the nearly 400 species of the Siphonales, most are marine, and these largely are distributed in tropical or subtropical oceans. *Codium* (Fig. 15-19*A*) has a plant body composed of siphonaceous filaments interwoven to form a structure resembling a small rope of considerable length. Gametic reproduction is isogamous. *Valonia* (Fig. 15-20) represents the order Siphonocladales, which contains about 150 species, all of which are marine and are found in tropical and subtropical oceans. *Valonia* consists of an irregular series of inflated, lobed segments. These forms are of very irregular size

Fig. 15-19 Species of marine algae. A, Codium; *B,* Acetabularia.

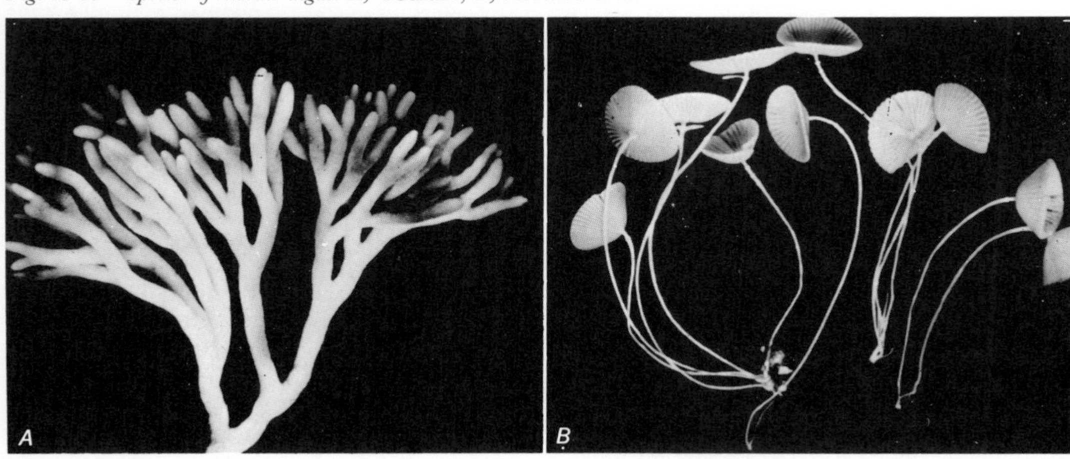

Fig. 15-20 Bottle cells of Valonia.

and shape, resembling small bottles, and are often referred to as "sea bottles." *Acetabularia,* a genus of the order Dasycladales, consists of a stalked portion with an expanded caplike top (Fig. 15-19*B*). All representatives of the order Dasycladales are marine and are found in warm seas.

Charales

General characteristics All members of the class Charophyceae are placed in the single order Charales. The family Characeae includes all the living genera, whereas the other families contain only fossil representatives. The Charales, represented by the well-known genera *Chara* and *Nitella,* constitute a group of green plants that have uncertain affinities with other plant groups. They grow submerged, attached by rhizoids to the mud, forming thick masses at the bottom of small pools and slowly flowing streams. The plant body is slender and flexuous, with individual plants attaining lengths of 6 to 12 in. The multicellular plant has a whorl of short branches at each node (Fig. 15-21*A*).

Internodes consist of a single elongated cell, but several short cells enter into the structure of the nodes. The short branches also have nodes and internodes similar to those of the main axis. In some genera, but not all, the internodal cells become overgrown by sheath cells arising from basal nodes of the whorl of

Fig. 15-21 Chara. *A, vegetative plant; B, male (globule) and female (nucule) fructifications. (From Gilbert M. Smith, "Cryptogamic Botany," 2d ed., vol. 1, figs. 66A and 69A, McGraw-Hill Book Company, New York, 1955.)*

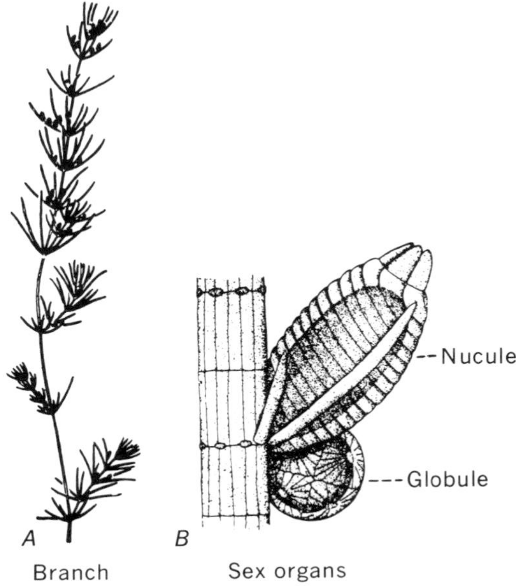

branches. In cross section an ensheathed internode shows the central internodal cell as the main axis, surrounded by smaller sheath cells.

Incrustrations of calcium carbonate typically found in some genera such as *Chara* are reflected in their common name stonewort. Other genera like *Nitella* are free of such deposits. Cells of the Charales are multinucleate with numerous small chloroplasts embedded in a layer of peripheral cytoplasm. An inner layer exhibits very conspicuously a rotary streaming movement.

Reproduction The only known method of asexual reproduction in the Charales involves detachment of various vegetative outgrowths which develop into new plants. Sexual reproduction is oögamous. The male fructification is called a **globule** and the female fructification a **nucule.** Both consist of a gamete-producing part surrounded, or enveloped, by a multicellular sheath derived from cells below the sex organs. The use of the terms antheridium and oögonium is misleading, particularly if applied to the whole so-called multicellular fructification. Most species are homothallic. In some genera, like *Nitella,* the globule is regularly borne above the nucule, but in others, like *Chara,* it is the reverse (Fig. 15-21*B*).

Fusion of the nonmotile egg and motile sperm forms the zygote, which secretes material to form a thick wall. Later the thick-walled zygote, along with other thick-walled parts of the nucule, drops to the bottom of the pond or stream. Before germination the zygote nucleus divides into four nuclei. At the time of germination the four nuclei present in the zygote contribute to the initial stage of development of the multicellular plant, but the three nuclei in the basal cell soon disintegrate and the new plant is formed as a result of continued divisions of the uninucleate lenticular cell.

CHRYSOPHYTA

General characteristics The Chrysophyta vary in color from yellow-green to golden brown as a result of the predominance of carotenes and

Fig. 15-22 Vaucheria sessilis. *A, branching coenocytic filament; B–D, asexual reproduction; B, C, zoospore escaping from sporangium; D, zoospores germinating to form vegetative filament; E, sexual reproduction with antheridium and oögonium. (Drawn by Christian Hildebrandt.)*

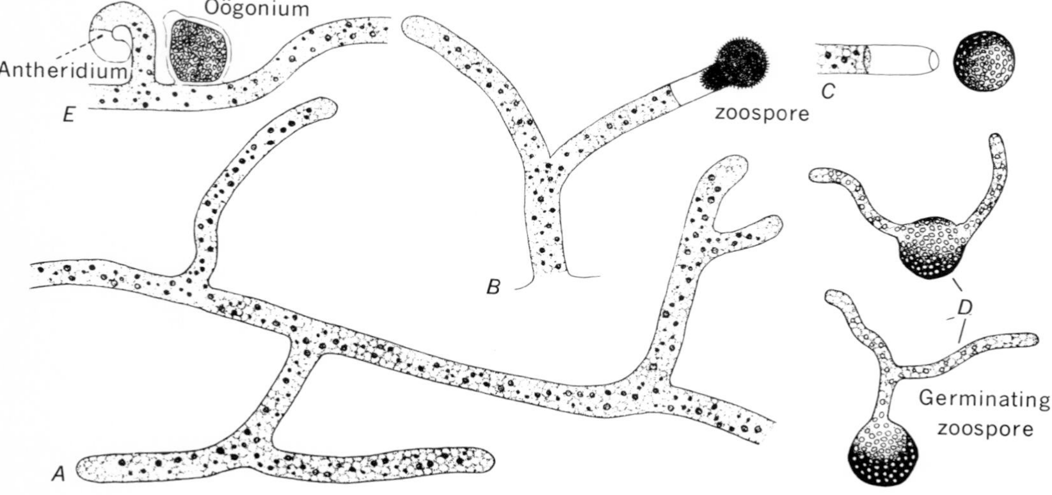

xanthophylls. The food reserves include leucosin, a complicated carbohydrate, and oils; starch is not formed. In some representatives the cell wall consists of two overlapping halves and is sometimes impregnated with silica. Considerable variation exists in differentiation, resulting in both motile and nonmotile cells, as well as unicellular and colonial types of either definite or amorphous shapes.

Reproduction Asexual reproduction is associated with the production of both motile and nonmotile spores, and gametic reproduction is usually isogamous, but it may be anisogamous or oögamous. Gametes may be flagellated or nonmotile.

Classification Most species are freshwater forms, but about one fourth of the representatives of the class is marine. Some botanists question the organization of the division to represent a natural grouping. But similarities in the nature of the food reserves formed, the formation by many representatives of a unique spore, the statospore, and the overlapping of halves of vegetative cells seem reason enough to place the representatives into three classes within the division. The three classes are Chrysophyceae, or golden brown algae, the Xanthophyceae, or yellow-green algae, and Bacillariophyceae, or the diatoms. Only the last two classes are considered here.

Heterosiphonales

General characteristics The order Heterosiphonales includes all the siphonaceous and multinucleate unicellular Xanthophyceae. The genus *Vaucheria* is one of the best-known representatives. A few species are marine, but most of them are distributed in fresh water or are terrestrial. Often terrestrial species grow in recently plowed fields, where they form rather extensive dark-green feltlike mats.

The plant body of *Vaucheria* is a sparingly branched, tubular structure with no septations, or, in other words, a single **coenocyte** (Fig. 15-22*A*). The central part of the cell is occupied by a large vacuole, and many nuclei and small disk-shaped or elliptical chloroplasts are found within the parietal cytoplasm. Many oil droplets are found within the cytoplasm. Cross walls occur only in the event of injury or in connection with the formation of reproductive structures.

The most common asexual method of reproduction is by means of large motile zoospores which are formed singly in zoosporangia, cut off at the ends of the tubular thallus by a cross wall (Fig. 15-22*B*, *C*). The contents of this terminal zoosporangium consist of an oval mass of cytoplasm with numerous nuclei and chloroplasts. Two flagella are developed opposite each nucleus of the structure, and the whole body, when it is mature, escapes through an opening at the end of the sporangium. After a very short period of motility, followed by a subsequent loss of the flagella, it forms a new plant by germination. In germination the zoospore forms one to several tubular outgrowths that may continue to grow indefinitely to produce an adult plant (Fig. 15-22*D*).

In terrestrial species, apparently, the ecological conditions bear some relationship to the formation of the asexual reproductive cells. Frequently the entire contents of a sporangium develop into a thin-walled aplanospore instead of a motile zoospore. In time the protoplast of the sporangium divides to form many small aplanospores, and, in other instances, fragmentation of the plant body occurs with formation of short segments, each of which becomes thick walled.

Sexual reproduction occurs in all species of *Vaucheria* but less frequently in those species growing in flowing water. All species are oögamous, but freshwater and terrestrial species are homothallic, whereas several marine species are heterothallic.

The female gametangia, oögonia, and the male gametangia, antheridia, are generally produced on short branches close together on the filament. In some species, as in *Vaucheria geminata*, the gametangia are produced together on a short branch, or stalk, of the filament (Fig. 15-22*E*). In most species the oögonium is formed terminally on a very short branch, or stalk, which is separated from the rest of the plant by a septum. In shape the oögonium is spherical or ovoid, with a short, rounded beak which opens to receive the sperm at the time of fertilization. At maturity the contents of the oögonium consist of cytoplasm, a single rather large nucleus, numerous chloroplasts, and stored food in the form of oil. The contents of the oögonium form the large spherical female gamete, or egg (also called the oösphere). The antheridium of *Vaucheria* is a slender, curled tubular structure separated by a septum from the short, slender branch upon which it is produced (Fig. 15-22*E*). The contents of the antheridium consist of cytoplasm, numerous nuclei, and chloroplasts. At maturity the numerous small male gametes, or sperms, are formed. Each sperm consists of a nucleus and a small amount of cytoplasm surrounded by a membrane. The sperms are biflagellate. The antheridia develop at the same time as the oögonia. When the oögonium is mature, the end of the beak opens slightly and it has been reported by some botanists that a single sperm enters. Actual fertilization, which may be delayed, consists of the fusion of the nuclei of the sperm and egg. Following fertilization, the zygote develops a heavy wall and becomes a dark resting cell. After several weeks the zygote germinates and directly forms a new *Vaucheria* plant. Cytological behavior at the reproductive stage has been insufficiently investigated. It is likely, however, that meiosis is zygotic. In such cases the vegetative filaments are haploid.

Diatoms

The Bacillariophyceae, commonly known as the diatoms, are single-celled microscopic algae of diverse forms. They are variously colored but generally are yellow or brown. Some of them exhibit slight motility. They occur in both fresh and salt water and in other damp places, including such aerial habitats as walls, rocky cliffs, the bark of trees, mosses, or damp soil. Some grow attached to various substrata in the intertidal zone, and others are epiphytic on red and brown algae. The diatoms form an important part of the plankton, the basic food of aquatic fauna.

The preservation of the remains of diatoms as fossils in vast deposits of diatomaceous earth is an indication of their abundance in and persistence through past geological ages. Such deposits have been formed by both freshwater and marine diatoms of the past, but those of the oceans have contributed to much thicker deposits. The Lompoc deposits in California are over 700 ft in thickness and are mined by open-pit methods. Oil drilling in the Santa Maria field in California has revealed much deeper deposits, at least 3,000 ft in thickness.

Diatomaceous earth is used in such materials as toothpaste, plastics, paints, and insulating material, and in the filtration of liquids, especially in sugar refineries.

It has been estimated that there are about 170 genera and 5,500 species of diatoms, both fossil and living. Two large, and probably artificial, groups are recognized, the orders Centrales and Pennales. The members of the Centrales occur as round, generally nonmotile forms (Fig. 15-23). In this order the markings on the walls present a radiate appearance. In the Pennales the elongated organisms are boat shaped or needle shaped with markings appearing pinnately (Fig. 15-23). Motility is common among the Pennales.

The nature of the cell wall An outstanding feature of diatoms is the nature and structure of the cell wall, which consists of two overlapping parts or valves, one slightly larger than the other, which are described as "fitting together like the two parts of a pillbox." This structural characteristic is the basis for the name diatom, which

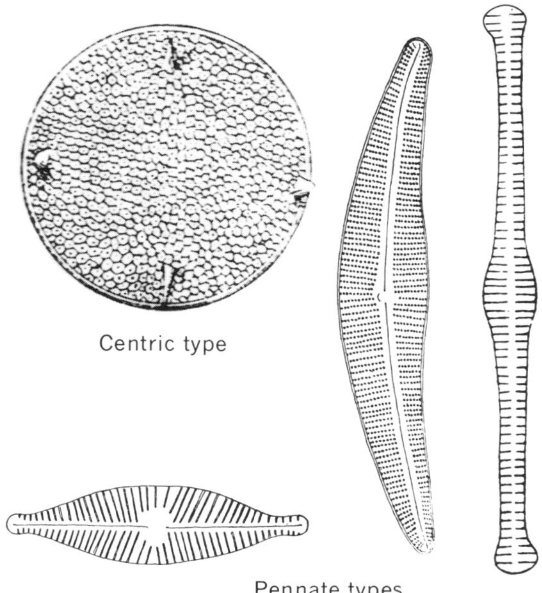

Fig. 15-23 Diatoms. (From Gilbert M. Smith, "Cryptogamic Botany," 2d ed., vol. 1, figs. 113B and 114B–F, McGraw-Hill Book Company, New York, 1955.)

means "cut in two." Because of the structure of the wall, diatoms present two views to the observer, from the top or bottom the valve view and from the sides where the parts of the boxlike wall overlap, the girdle view.

Chemically the wall of a diatom is composed of the organic substance pectin and the inorganic compound silica. Opinions differ as to the disposition of these two substances. Some investigators think of the wall as consisting of an outer silicated structure and an inner layer of pectin. Others regard the wall as being composed of pectin impregnated with silica. If viewed under the proper magnification, conspicuous striations on the walls of diatoms appear to be composed of dots (groups of pores). Among the Centrales the dots are usually large and arranged radiately. In the Pennales, where the dots may be extremely small, they can be seen only with high magnification. Arrangement in this group is pinnate, the dots appearing in regular rows near the edges of the valves and extending inward to near the middle.

On the back of each valve of most of the elongated types, there is an additional marking called the **raphe.** The word means "seam," or "stitch." Structurally the raphe is a V-shaped slit or cleft in the wall which extends from end to end of the valve and which is V shaped in cross section. **Nodules,** which are also structural modifications of the cell wall, terminate the raphe and interrupt it near the midpoint. Some of the distinguishing features of the orders, families, and genera include the presence or absence of raphes, the nature of the nodules, and the size and distribution of the dots constituting the striations on the walls of the valves.

The structure of the protoplast The protoplast of a diatom consists of cytoplasm closely appressed to the cell wall and one or several large vacuoles, often crossed by cytoplasmic strands. In species with elongated cells a central bridge of cytoplasm contains the nucleus and associated organelles. The nucleus is often irregular in shape, is bounded by a "double envelope" with pores, and contains a nucleolus. Depending upon the species, the cell may contain a single large chromatophore, a few medium-sized chromatophores, or several small ones. These chromatophores usually occupy a parietal position. Pyrenoids occur in the chromoplasts of some diatoms. Food reserves are stored as fats, oils, and chrysolaminarin. The chlorophyll which they contain is often masked by the presence of several xanthophylls, among which fucoxanthin is most common.

Motility of diatoms Members of the order Centrales apparently are nonmotile, motility being restricted to the Pennales. Many but not all of the pennate types are motile, with movements slow and jerky and confined to forward and backward directions. The best explanation of their movements appears to be one associated with the flow of currents of water set up by the circulation of cytoplasm and involving the raphe in some manner.

Methods of multiplication Diatoms multiply by a peculiar method of cell division. As the cell increases slightly in diameter, the valves separate. The nucleus undergoes mitosis with the division spindle parallel to the short axis of the cell in a manner that ensures longitudinal division of the protoplast. If there is a single plastid or several large plastids, each one of them undergoes division, a half passing eventually into each daughter protoplast. When there are numerous small plastids, some of them pass into each of the daughter protoplasts and there divide to restore the original number.

Division of the nucleus is followed by division of the protoplast along the median line of the longitudinal axis. Thus two protoplasts are formed from the original one. When division has been completed, each cell has one siliceous valve, but the side along the plane of cell division is bounded only by a protoplasmic membrane. The peculiarity of the process is found in the manner of new wall or valve formation. Each daughter protoplast grows a new valve to match the one remaining from the original cell wall. In each case this new valve fits inside the original with the result that one new diatom is the same size as the parent and one is slightly smaller. Cell division is reported to take place at night, and it is said to be fairly rapid. Because of frequent multiplication and the tendency for some of the cells to become reduced in size, part of the population of diatoms becomes more diminutive.

Auxospore formation and rejuvenescence When the small diatom cells reach a size incompatible with physiological processes, auxospores are formed and restoration of normal size and physiological rejuvenescence follow. The prefix *auxo-* in the term auxospore comes from Greek and means "grow." An auxospore then is a growth spore, or an enlarged spore, as compared with the diminutive gametes that fuse and give rise to it. The sexual process leading to the formation of the auxospore may be isogamous or anisogamous, and an oögamous union of gametes has been demonstrated for several centric-type diatoms.

The vegetative diatom is a diploid structure. Variation in cytological details occurs in the different genera of diatoms, but it may be assumed that meiosis with a reduction of chromosome numbers takes place with the formation of gametes. The haploid cells thus formed escape from the valves and fuse. The resulting cell grows into an enlarged protoplast that develops into the auxospore, or renewal spore. Normal vegetative cells are formed from the auxospore.

PHAEOPHYTA

General characteristics The Phaeophyta, or brown algae, are a widely distributed group of marine forms that attain their greatest development in the colder waters of continental coasts. They are the predominant flora in part of the arctic and antarctic seas, and there are only three rare freshwater forms known. Their color, varying from dark brown to olive green, results from a brown carotenoid pigment, fucoxanthol. This pigment is located in plastids, where it tends to obscure the green chlorophyll. The brown algae show great diversity in form and structure. *Ectocarpus* is a small filamentous plant, whereas *Laminaria, Macrocystis,* and *Nereocystis* are massive seaweeds called the giant kelps. Although some species of kelps are relatively small plants, only 3 or 4 ft in length, many are giant forms reaching lengths of 300 ft or more. Besides a holdfast, the kelps have a long, slender, flexuous stipe and an expanded blade that may be entire or dissected (Fig. 15-29). The blade portion grows from the activity of a group of meristematic cells located at its base. This growing region may also renew blades frayed by wave action.

Another group of brown algae, known as the rockweeds, is commonly found along the rocky coasts of the temperate zones. The species of *Fucus* (Fig. 15-31*A*) are the best known of this group of the brown algae. The genus *Ascophyllum,* closely related to *Fucus,* grows in similar habitats. These plants grow attached to rocks at the shoreline, where the plants are often alternately exposed and submerged, owing to the ebb and flow of the tides (Fig. 15-30). *Sargassum,* the gulfweed (Fig. 15-31*B*), either grows attached to rocks or is free floating. *Sargassum* makes up a large part of the conspicuous vegetation of the Sargasso Sea, a great ocean eddy in the southern part of the North Atlantic Ocean. The plants in this group, represented by *Fucus, Ascophyllum,* and *Sargassum,* are of medium size, ranging from a few inches to 1 or 2 ft in length. In general form the plant body is a ribbonlike thallus, with holdfasts and air bladders.

Generally brown algae grow attached by holdfasts to rocks near the ocean shoreline, where at least at high tide they may be submerged. Some kinds, like *Fucus* and *Sargassum,* may live for some time as free-floating plants. Many of the smaller filamentous ones, like *Ectocarpus,* grow as epiphytes on the larger seaweeds.

Cell differentiation is more pronounced in the thalli of the kelps than in the thalli of any other algae. In the stipe, differentiation of cells makes an epidermal layer (meristoderm), a cortical region, and a central region (medulla) recognizable. Besides these, connecting filaments that often extend in a radial direction and an elaborate system of mucilage canals are developed in the stipes. Some conducting cells have the structure and appearance of the sieve tubes of the higher flowering plants, with protoplasmic strands passing through the openings of the sieve plates. These tissue differentiations are continued into the blades of the kelps, though less markedly (Fig. 15-27).

Many brown algae grow as a result of meristematic activity at the bases of the filaments. Other Phaeophyta grow as a result of divisions of a single apical cell or of a transverse row of such cells. In the kelps growth is not apical but is located at the juncture of the stipe and blades or sometimes at the base of the stipe.

Cells of the brown algae have nuclei, cytoplasm, and chloroplasts. The chloroplasts contain, in addition to the green chlorophyll pigments, yellow xanthophyll that is probably in excess of the green. A brown carotenoid pigment, fucoxanthol, which is also present in the chloroplasts, masks the green chlorophyll and produces the characteristic brown color of the Phaeophyta. The products of photosynthesis in the brown algae are sugars. Food reserves are stored as simple sugars, the alcohol manitol, and complex polysaccharides, such as laminarin. This is in contrast with the green algae, where the carbohydrate food reserve is generally starch.

The kelps and rockweeds are of limited economic importance. Iodine and potash, once extracted from some brown algae, can now be obtained more economically from accessible mineral deposits. Brown algae do serve as the only important source of algin or alginic acid. The salts of alginic acid, the alginates, are colloidal gels that are used extensively in the baking industry and as a stabilizer in the manufacture of ice cream. They are used in various commercial processes and products, including the fireproofing of clothes, in soaps and shampoos, in pharmaceutical preparations, in the manufacture of some plastics, and as a substitute for laundry starch. Brown algae are used as a fertilizer for gardens and for agricultural land along the coastal areas of certain European and Asiatic countries. In the Orient a food product from the kelps, called kombu, is eaten as a vegetable or mixed with fish and meat.

Reproduction All brown algae, except members of the order Fucales, may have in their life history both a free-living sporophytic plant and a free-living gametophytic plant. When reproductive cells are formed within **plurilocular sporangia** on sporophytic plants, the motile zoospores formed are diploid and grow to form new sporophytes. These spores have been called **neutral spores** because they produce new plants of the same generation and not gametophytic plants of the alternate generation. In some genera of brown algae no plurilocular sporangia are formed.

In isogamous brown algae, gametes that fail to fuse with other gametes usually disintegrate but they may develop into new plants of the same generation. In anisogamous brown algae, the larger gamete may, with the failure of syngamy, develop into a new gametophyte. In sexual reproduction the gametes are formed in gametangia located on gametophytic plants. Gametes are formed in **plurilocular gametangia** (Fig. 15-24*B*). Syngamy may be isogamous or anisogamous. Gametic fusion may also be oögamous, with a small motile (flagellated) male gamete fusing with a larger nonmotile female gamete.

The zygote, formed as the product of syngamy, may grow into an independent diploid, or sporophytic, plant. Meiosis, which commonly takes place in unilocular sporangia (Fig. 15-24*C*), results in the formation of haploid meiospores, which with development form gametophytes. In the genus *Fucus* (see following discussion), the relationship between sporophyte, meiosis, gametophyte, and gamete is not so easily seen as in *Ectocarpus*.

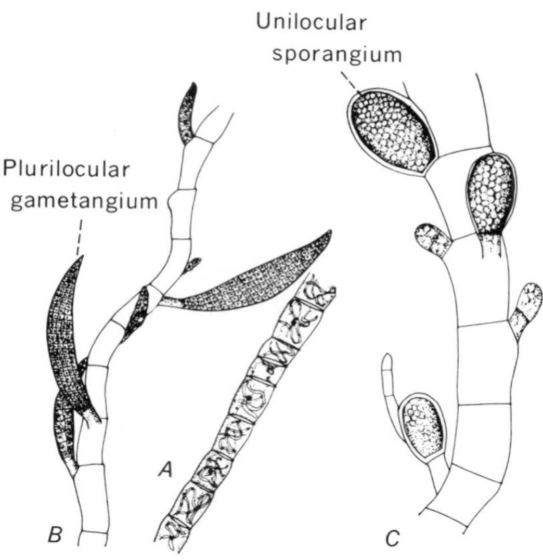

Fig. 15-24 Ectocarpus. *A, a small portion of a vegetative filament; B, plurilocular gametangia on a gametophytic thallus; C, unilocular sporangia on a sporophytic thallus. (Drawings by Helen D. Hill.)*

Ectocarpales

Ectocarpus *Ectocarpus* is one of the best-known brown algae. The plant body is composed of an irregular prostrate portion and small branched filaments which grow upright in tufts. These filaments may be single strands of cells, or they may consist of several rows of cells joined together. Each of the cells of the filament contains a protoplast consisting of a nucleus, cytoplasm, and chloroplasts. The chloroplasts are platelike or band shaped (Fig. 15-24*A*).

Ectocarpus has an alternation of isomorphic generations. The gametophytes and sporophytes are essentially alike in appearance, but, in the case of the rather thoroughly investigated *Ectocarpus siliculosis,* environmental conditions, probably temperature, commonly inhibit the development of one of the generations.

Biflagellate zoospores (meiospores) are produced within unilocular sporangia borne on diploid plants (Fig. 15-24*C*). Meiotic divisions result in the production of four meiospores, each one of which undergoes several additional divisions to form a total of 32 or 64 zoospores, which, under suitable environmental conditions, may grow into the alternate, or gametophytic, plant of *Ectocarpus*. Diploid plants also produce plurilocular sporangia, and the biflagellate zoospores germinate to form diploid plants, similar in every way to the parent plant.

The plurilocular gametangia borne on haploid or gametophytic plants develop similarly to the plurilocular structures borne on diploid plants. Both develop from the terminal cell of a lateral branchlet, with repeated transverse and vertical divisions, resulting in several hundred small cubical cells arranged in from 20 to 40 transverse tiers. The plurilocular reproductive structure is elongated and is circular in cross section (Fig. 15-24*B*).

Each compartment of the plurilocular gametangium gives rise to a single motile gamete. Although the fusing gametes are the same size and are morphologically similar, gametes from the same plant never fuse with each other. Gametes that fail to fuse may develop parthenogenetically into new gametophytic plants. Zygotes develop into sporophytic plants.

Not only is there a regular alternation of diploid with haploid plants by means of gametic fusion and the production of meiospores, but there may also be a reduplication of each generation. Diploid zoospores germinate to form sporophytic plants, and gametes may develop parthenogenetically to form gametophytic plants (Fig. 15-25).

Cutleriales and Dictyotales

These orders are placed in the same class as Ectocarpales. Representative genera include *Cutleria, Zanardinia,* and *Dictyota. Cutleria* and *Zanardinia* are closely related genera, each exhibiting an alternation of generations. In the case of *Cutleria* the alternating plants are somewhat dissimilar, but in *Zanardinia* they are identical. The sporophytic plants of both genera produce only unilocular sporangia. The gametophytic generation is heterothallic, and gametic reproduction is anisogamous (Fig. 15-26*A, B*).

Dictyota exhibits an alternation of isomorphic generations, with the growth of erect thalli initiated by a single apical cell (Fig. 15-26*C*). A vertical division of the apical cell or a branch produces two daughter cells that give rise to dichotomous branching by continued independ-

Fig. 15-25 Ectocarpus. *Life-history diagram.*

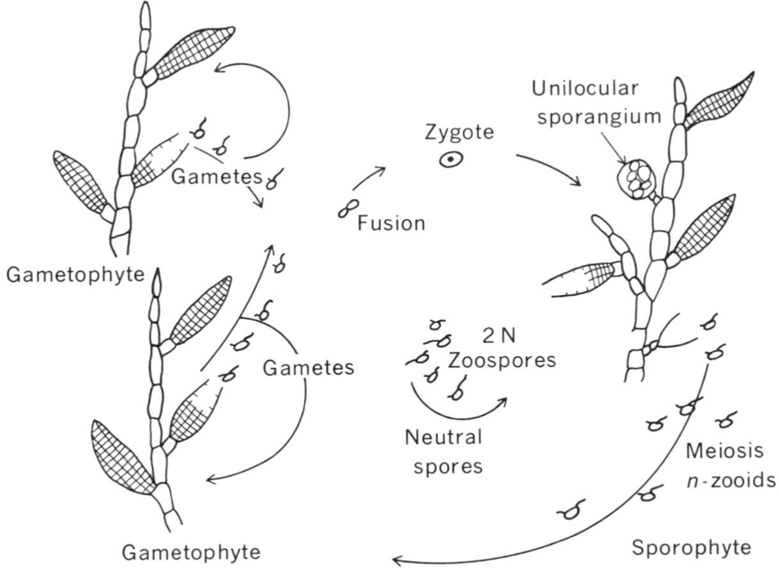

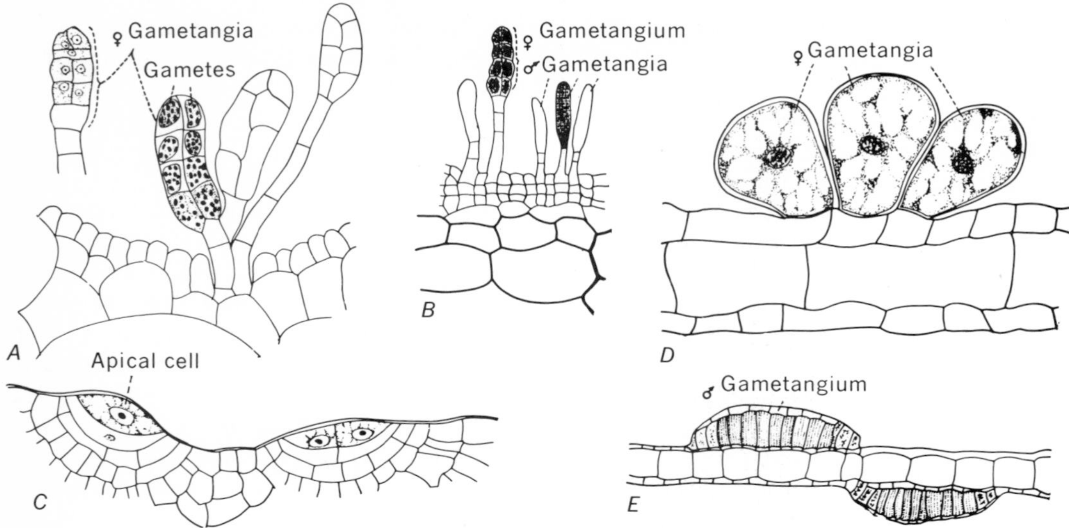

Fig. 15-26 Details of sexual reproduction in some genera of the brown algae, as seen in sectional views. A, Cutleria, *showing the superficial female gametangia containing gametes; B,* Zanardinia, *showing the superficial male gametangia and female gametangia; C,* Dictyota, *apical region, showing origin of dichotomous branching; D, E, sexual reproduction in* Dictyota; *D, female gametangia, and E, male gametangia on both sides of the thallus.*

ent division. Four meiospores are produced in the unilocular sporangium of the sporophyte as a result of meiotic divisions in which there is a genotypic determination of sex. At the time of germination of these spores, two produce male and two produce female gametophytes. Gametic reproduction is oögamous (Fig. 15-26*D, E*), and the resulting zygote germinates to form a fertile diploid plant.

Laminariales

The **kelps** include about 30 genera and are common in the colder oceans of the world. Many of them occur along the Pacific Coast, of which *Macrocystis* and *Nereocystis* (Figs. 15-27, 15-28) are well known. Fewer forms are found along the Atlantic Coast. *Laminaria* (Fig. 15-29), from which the order name is derived, occurs along both coasts of the United States.

Most members of the order have a sporophytic plant body differentiated into a holdfast, stipe, and blade. Growth occurs at intercalary meristems, usually located between the blade and stipe. Only unilocular sporangia are produced by the sporophytic plants, and the gametophytes are usually filamentous and microscopic in size. Gametic reproduction is oögamous.

Fucales

Fucales is the only order of the class Cyclosporeae and is represented by the relatively well-known genera *Fucus, Ascophyllum,* and *Sargassum* (Figs. 15-30 to 15-32). Life histories of the related genera are essentially the same. The class is characterized by the absence of a free-living, multicellular, gametophytic generation. Most Fucales grow attached to rocks of the intertidal zone, but some occur at deeper levels. In certain species of *Sargassum* the thallus is free floating.

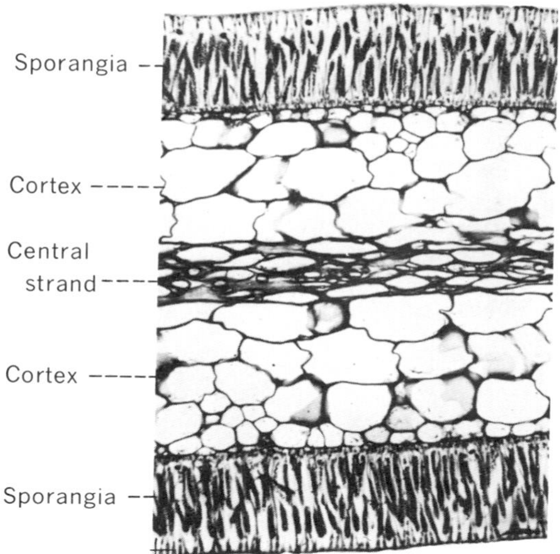

Fig. 15-27 Nereocystis. *Transverse section of a small portion of thallus showing sporangia. Tissue differentiation with central strand surrounded by a cortical region.*

Fig. 15-28 Nereocystis. *Enlarged view of portion of thallus, showing zoosporangia with zoospores and paraphyses. (Drawing by Helen D. Hill.)*

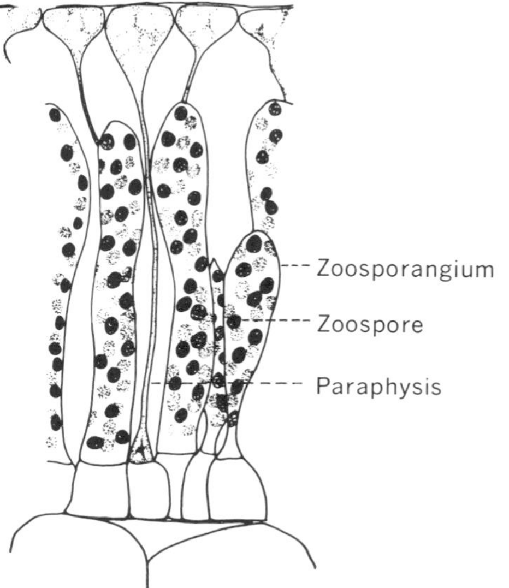

The plant body of *Fucus* consists of a flat, dichotomously branched thallus of an olive-green or brownish color. The macroscopic plants normally grow firmly attached by an expanded holdfast to completely or partly submerged rocks. When covered, the thallus is buoyed up in the water by bladderlike structures, or floats.

The only known method of **asexual reproduction** in *Fucus* is by fragmentation. When this occurs, broken pieces may continue to live as floating plants. This is found especially in the related genus *Sargassum*. **Sexual reproduction** is oögamous. Small actively motile microgametes, or sperms, fertilize very much larger nonmotile macrogametes, or eggs.

Before the significance of the chromosome cycle was recognized, the reproductive structures of *Fucus* were thought to be gametangia, antheridia, and oögonia. Investigations show, however, that meiosis occurs and meiospores

Fig. 15-29 A species of Laminaria, *one of the kelps, showing the well-developed, branched holdfast, the short stem, and the well-developed blade.*

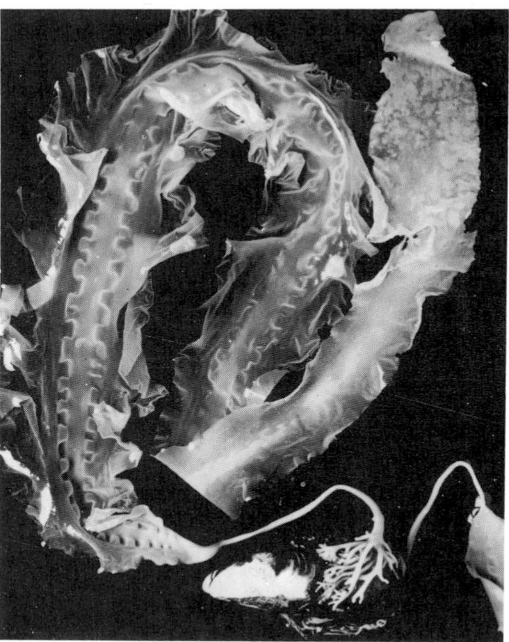

Fig. 15-30 Rockweeds. Fucus *and* Ascophyllum *in the intertidal zone. (Courtesy of Dr. C. J. Hillson.)*

are produced in what are properly called sporangia. The *Fucus* plants are sporophytes. Since the spores and their final products are of two different sizes, the terms microsporangia and megasporangia may be properly applied to them. These sporangia together with sterile filaments, called **paraphyses,** are produced in small cavities known as **conceptacles,** which are aggregated at the tips of the branches in swollen conical structures, the **receptacles** (Figs. 15-31*A,* 15-33). The numerous microsporangia are produced terminally on very small branched filaments arising from inner walls of the conceptacle.

Meiosis, with the production of four meiospores, occurs in the nucleus of the young microsporangium, reducing the number of chromosomes from the diploid number to the haploid number. The meiospores are retained within the microsporangium, where each undergoes a series of four mitotic divisions with the formation of 16 haploid cells. This group of cells may be regarded as a very much reduced gametophyte. Since each of the four nuclei behaves in the same way, there are four of these 16-celled gametophytes making a total of 64 haploid cells, each of which eventually becomes a microgamete, or sperm, with the reduced number of chromosomes. The sperms are minute, slightly elongated structures that owe their motility to two laterally placed flagella. At maturity the sperms escape from the microsporangium and swim to the eggs. Meiosis also occurs in the nucleus of the young megasporangium with the production of four meiospore nuclei. Again, as in the case of the meiospores in the microsporangium, the spores are not shed but are retained in the megasporangium. Through a single mitotic division which immediately follows meiosis, each meiospore forms a small two-nucleate gametophyte. Since in this instance there is only one cell generation in the gametophytic tissue, there is a total of eight nuclei. Cleavage walls later divide the cytoplasm into eight blocks, each with one of the haploid nuclei. When these cell masses round up, they become the nonmotile macrogametes or eggs. At maturity they are set free from the sporangium and float in the water.

When the gametes are expelled from the conceptacles, the large nonmotile eggs floating free in the water are soon surrounded by hundreds of the minute sperms which are attracted to them. The sperms are described as attaching one flagellum to the surface of the egg while the second remains free, lashing the water. The sperms, attached in this way, by reason of their motility set the nonmotile eggs in rotation. This may continue until a sperm penetrates the egg and fertilizes it. The union of the sperm and egg ends the remarkably short haploid, or gametophytic, phase of *Fucus*. The zygote immediately begins to grow and to develop the sporophytic plant that becomes attached to a suitable support, grows into a mature *Fucus* plant, and repeats the life cycle (Fig. 15-32*B*).

The unique feature of reproduction in *Fucus* is the retention of meiospores, their growth forming reduced gametophytes with the eventual formation of gametes within both microsporangia and megasporangia. The sporangia thus serve both as spore cases and as gametangia.

CYANOPHYTA

General characteristics Approximately 1,500 species are placed in the single class Cyanophyceae, and both marine and freshwater forms

Fig. 15-31 A, a species of Fucus, *showing the fruiting tips or receptacles, which contain the conceptacles that open to the exterior through the white raised pores. Note the dichotomous branching of the thallus. B, a species of* Sargassum, *showing the stemlike and leaflike parts, as well as the rounded swollen bladders, or floats.*

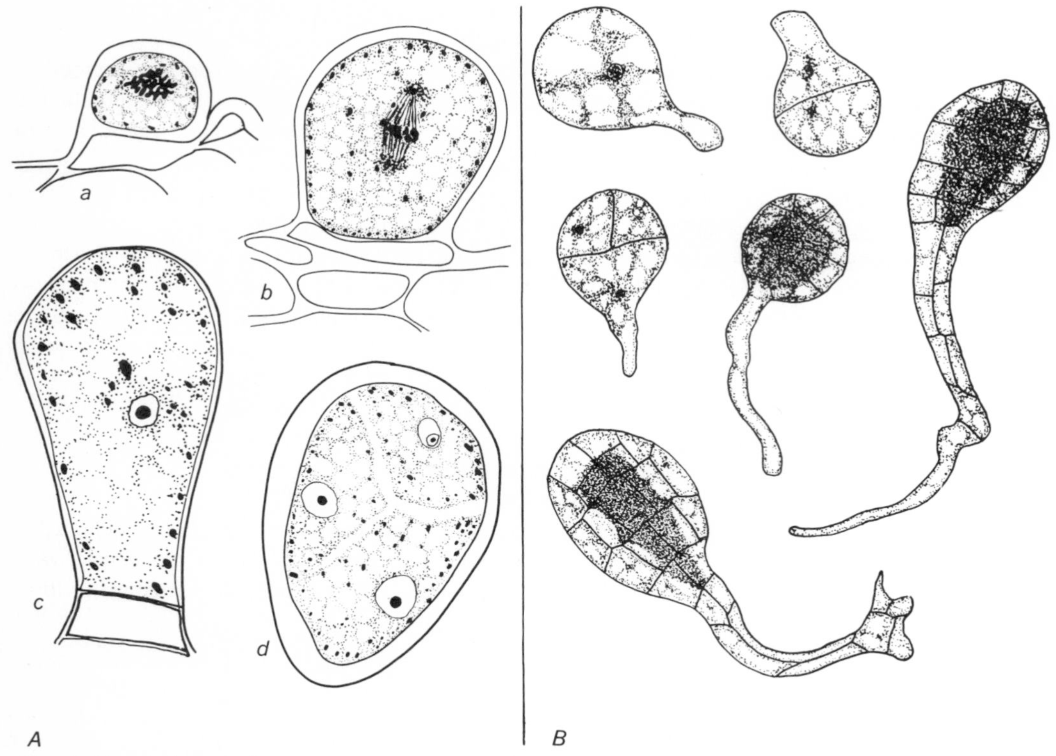

Fig. 15-32 *A,* Ascophyllum; *a, young megasporangium; b, nuclear division in oögonium reducing the number of chromosomes from 2N to 1N; c, young megasporangium, showing development of basal cell; d, megasporangium, containing four cells, each with 1N number of chromosomes; B,* Fucus *sporophytes, showing progressive development from one-celled stage to many-celled plant with small holdfast.* (*B drawn by Helen D. Hill.*)

are represented. It has been suggested that the blue-green algae are specialized and evolutionarily advanced, but the consensus is that they are relatively simple, so-called primitive, at about the same evolutionary level as bacteria. A number of forms have been reported from the Precambrian (Fig. 15-34), representing the oldest of the fossil algae. The blue-green algae are widely distributed on many substrata, such as rocks, flowerpots, the bark of trees, and the soil. Many occur in fresh water, and free-floating (planktonic) forms may be present in large quantities. They inhabit environments having a wide variation in temperature, growing on snow and in hot springs. They are especially abundant in stagnant waters abounding in decaying organic matter. Some blue-green algae live intimately with other organisms, such as bryophytes, ferns, and gymnosperms, occurring within colorless cells as **symbionts.** Others are relatively common algal symbionts of some lichens.

The blue-green algae are probably most important as primary food producers within the environments where they occur most abundantly. If the growth is excessive, they may prove to be a nuisance in water-supply systems, and, in extreme conditions, the water may become toxic and unfit for humans or livestock and may result in the killing of aquatic organisms. The blue-green algae are important as fixers of elemental nitrogen in the soil, where they may occur to a depth of several feet, and in semi-aquatic environments, such as rice paddies.

Flagellated cells are absent. However, in some filamentous forms lacking an encapsulating sheath, a peculiar oscillating motion is characteristic. No single satisfactory explanation of this motion has been agreed upon, but suggestions offered by different investigators include: the streaming of cytoplasm, secretion of mucilaginous materials, contraction and expansion rhythms, and oscillations of submicroscopic particles.

Chloroplasts are absent but pigments are present in numerous flattened chromatophores. In addition to chlorophyll *a*, a number of water-

Fig. 15-33 Fucus, *sexual reproduction. A, a transverse section of a receptacle, showing several conceptacles; B, a conceptacle enlarged, showing numerous microsporangia and sterile hairs, or paraphyses; C, transverse section of a receptacle, showing four conceptacles; D, one conceptacle enlarged, showing numerous megasporangia and sterile hairs, or paraphyses.*

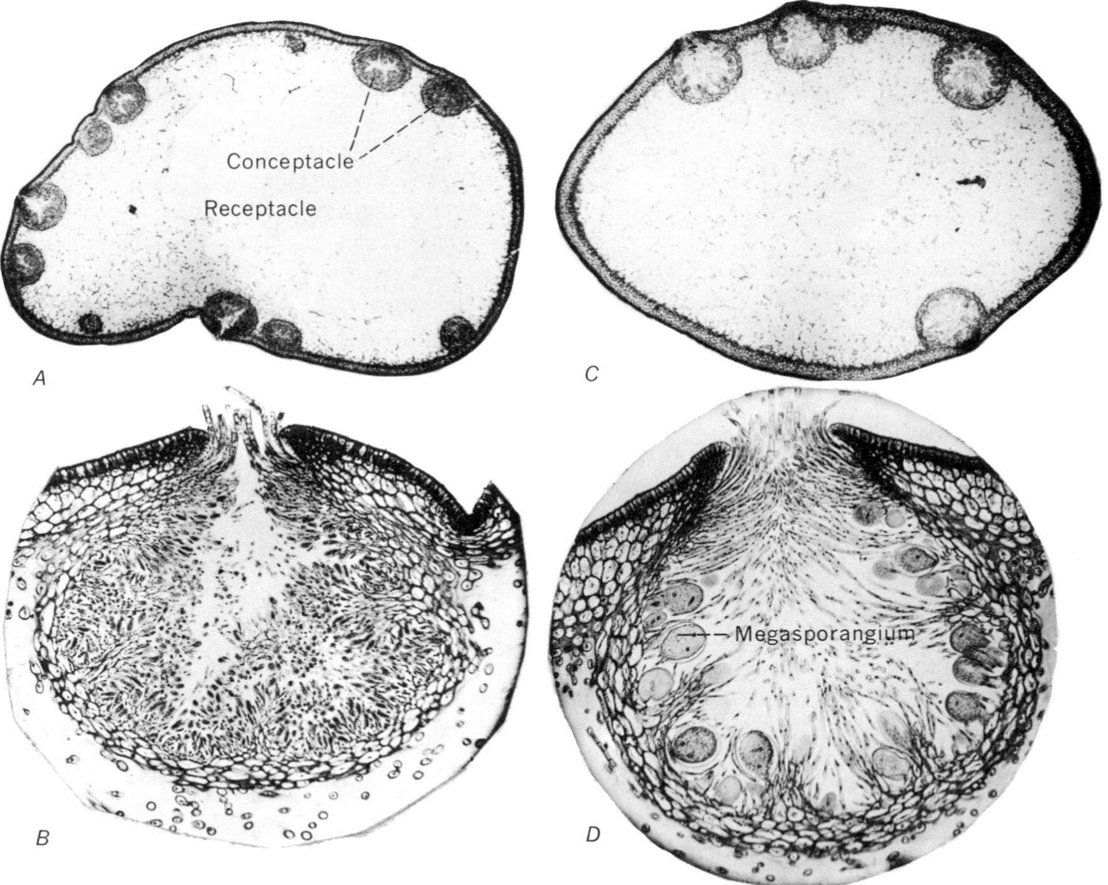

Fig. 15-34 Precambrian rock containing fossil blue-green algae. Glacier National Park.

soluble pigments, called **phycobilins** (biliproteins), including phycocyanin and phycoerythrin, are present. The luxuriant growth of the algae, a bloom, will often result in the greenish blue or red color of water, as in the case of the Red Sea.

Structural development and form The single isolated cell is the simplest condition, although by cell division, these individuals may become aggregated into loose masses. Each cell consists of a single protoplast surrounded by a gelatinous sheath (Fig. 15-35). From such isolated individuals or indefinite colonies, we may regard the development of colony formation as proceeding along three lines: (1) an aggregation of cells, culminating in a rather definitely spherical colony of either a few or a large number of cells, as in *Gloeocapsa* and other forms; (2) an aggregation of cells all in one plane, forming a flat, square colony of a single layer of cells, as in *Merismopedia;* (3) a single line of cells placed end to end, forming a long chain or filament, as in *Oscillatoria* and other genera (Fig. 15-35).

The development of these types of colonies in relation to cell division may be thought of as caused by the progressive restriction of the planes and directions in which cell division takes place.

Differentiation of cells in the colony has proceeded further in some of the filamentous colonies than in any of the other types. For example, in *Gloeotrichia* (Fig. 15-35*E*) and in *Rivularia,* the **heterocyst,** a specialized cell, always occupies a basal position and resting cells are always located next to it. The vegetative cells gradually taper in size to a whiplike extremity at the other end.

Cell structure The investigation of the cell structure of the blue-green algae has been an interesting, though difficult, problem. The interest centers around the fact that these plants are recognized as constituting a primitive group, and the investigation of the cell structure should, therefore, shed some light on the question of the origin and development of the highly organized protoplast of the higher plants. These investigations have often led to contradictory results or to conflicting interpretations; consequently the problems that have been considered are by no means all solved.

Compared with the cells of most other plants, there is little cellular differentiation. With the light microscope the cell seems to consist of a centrally located, clear area surrounded by a denser portion. It has been known for many years that the central region, the **centroplasm,** contains chromatin and that the pigments are concentrated in the peripheral region, or the **chromoplasm.** Recent studies with the electron microscope have provided more detail concerning the fine structure of the cytoplasm and the nature of the cell wall, but the research is still inconclusive and interpretation is difficult in some instances.

From studies of a limited number of genera, we might conclude that the cytoplasmic matrix is distributed throughout the cell and that no membrane-bound organelles exist of such categories as the chloroplast, nucleus, or mitochondria. The central region, centroplasm, shows crystalline granules and areas of low electron density, resembling the nucleoid areas of bacteria, and it contains DNA. Chromosomes have not been identified. Photosynthetic lamellae and various granules are found in the chromoplasm. Ribosomelike granules are scattered throughout the cytoplasmic matrix but are more numerous in the area surrounding the centroplasm. Intercellular cytoplasmic connections that involve the plasma membrane have been reported by some investigators, whereas others report tubelike connections between cells but doubt that actual cytoplasmic connections exist. Other inclusions, such as vacuoles, structured granules, and polyhedral bodies, have been seen in electron micrographs but their function is in doubt and is still under investigation.

The **cell envelope** surrounding the cytoplasmic matrix is composed of two layers: the cell sheath and the inner investment. The inner investment is considered to have three regions: the outer, middle, and inner. The cell sheath may be composed of at least two regions, the

Fig. 15-35 Blue-green algae. A, Spirulina, *vegetative form; B,* Anabaena, *vegetative filament with heterocyst, h; C,* Cylindrospermum, *filament mainly composed of vegetative cells, but with basal heterocyst, h, and resting cell, r; D,* Merismopedia, *with cells in early stages of division at right and later stages at left, all embedded in a gelatinous matrix, g; E, entire filament of* Gloeotrichia, *showing gelatinous sheath, g, and basal heterocyst, h; F, cells of* Gloeothece, *each central protoplast surrounded by one or more gelatinous sheaths, g.*

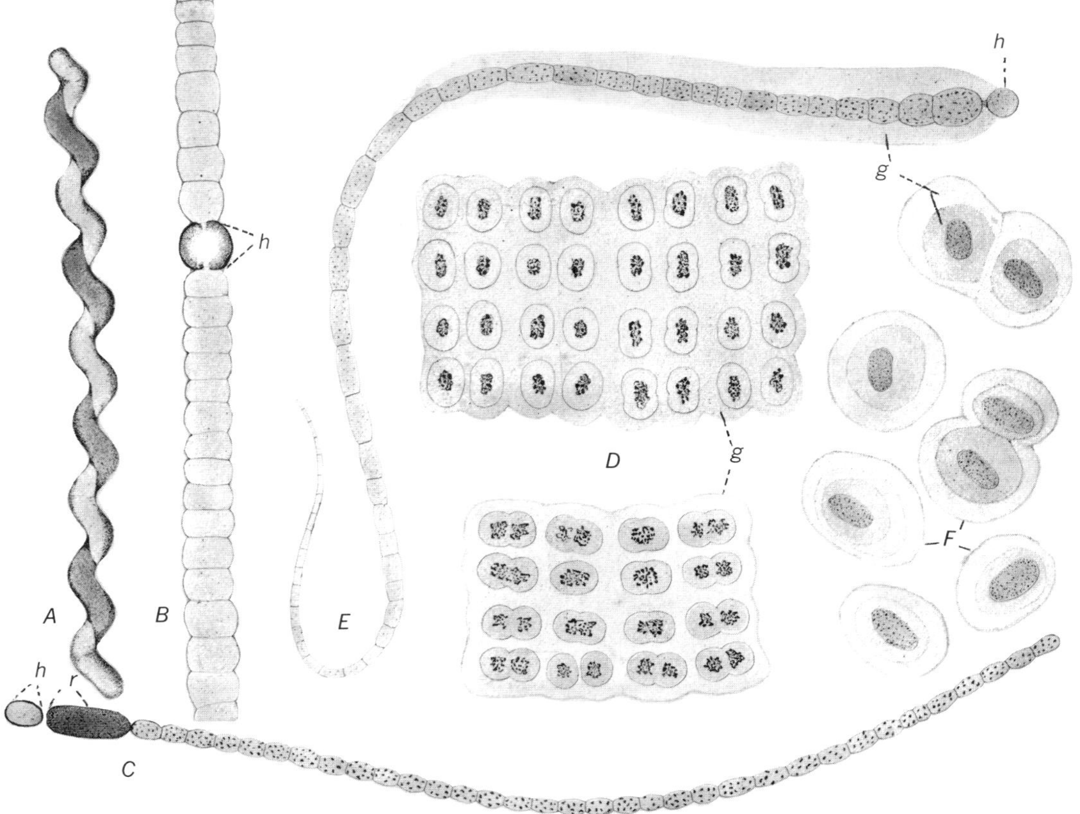

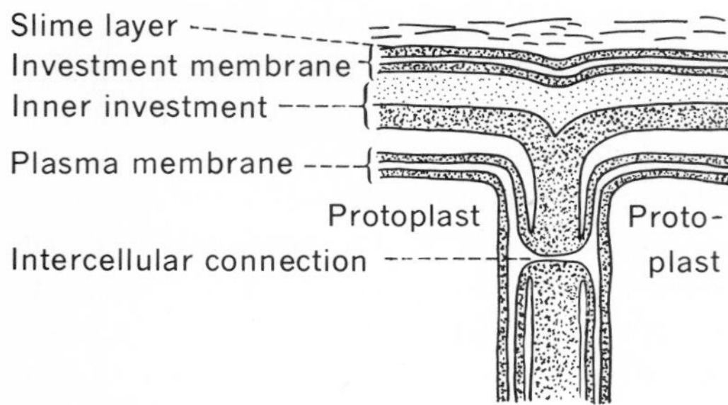

Fig. 15-36 Diagram of the cellular envelope of a blue-green alga. (Redrawn from D. C. Wildon and F. V. Mercer, Australian J. Biol. Sci., pp. 585–596, 1963.)

outer of which is less structured (Fig. 15-36).

Cell differentiation Some, though not extensive, differentiation of the cells is found in the colonial forms of the blue-green algae. In some of the filamentous forms, as in the genera *Nostoc, Cylindrospermum,* and *Anabaena,* certain cells, called **heterocysts,** occur at irregular intervals in the filament (Fig. 15-35*B*). These heterocysts are somewhat larger than the ordinary vegetative cells of the filament and are filled with a homogeneous material. In the early stages of their development, pores connect the heterocysts with the vegetative cells, but later the pores are closed by the gradual thickening of the walls of the heterocyst. The function of the heterocyst is in doubt, but it separates the threads into sections, called **hormogonia** (singular, hormogonium). It has been suggested that the heterocysts serve as receptacles of stored food materials. Another theory advanced suggests that they are the remnants of reproductive organs which, for some reason or other, have degenerated as the processes of evolution have progressed. In the genus *Oscillatoria* there are no true heterocysts, but the hormogonia are delimited by dead cells known as **concave cells** or separation disks (Fig. 15-37*A*). Further differentiation of cells occurs in the reproductive processes of some species when heavy-walled **resting cells** (erroneously called spores) are formed (Fig. 15-35*C*).

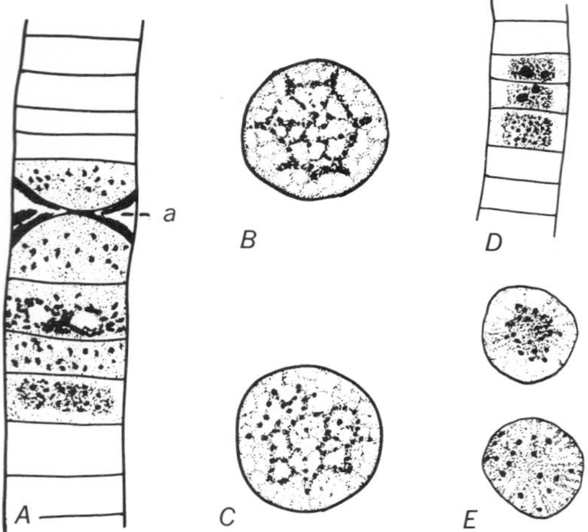

Fig. 15-37 Oscillatoria. *A, cells of a filament, showing granular contents; also a dead or concave cell at a; B, C, E, cross-sectional views, showing central body and peripheral cytoplasm; D, cells of a filament, showing central body.*

Reproduction The methods of multiplication and of reproduction are asexual by means of nonmotile structures. In all unicellular plants cell division is the chief method by which increase in number takes place. In many species of the blue-green algae, the exact mechanism of cell division is in dispute. In some electron-microscope studies, it has been established that elaboration of the plasma membrane takes place where cross walls occur and, in the case of *Symploca muscorum,* a filamentous form, the inward, simultaneous development of several cross walls occurs.

Little is known about the mechanism of division of the nuclear material, and those divisions attributed to amitosis leave much to be desired. Chromosomes have not been identified, and results of investigations previously published, wherein spindle formation associated with chromosomes or mitotic figures has been discussed, may be in doubt. There is some evidence, similar to that discovered in certain bacteria, for genetic recombination as shown in changes in the resistance of some blue-green algae to the antibiotics penicillin and streptomycin.

Fragmentation of the colony plays an important part in dissemination. In the filamentous species, the filaments usually break at the heterocysts. The hormogonia are thus set free, and each of them will again become a mature colony. Resting cells may be developed, under certain conditions, from the ordinary vegetative cells, which increase in size and become thick walled and resistant.

RHODOPHYTA

General characteristics The class Rhodophyta, or red algae, a large group composed mostly of marine algae, are of worldwide distribution, with the greatest display of species in the warm waters of the tropical, subtropical, and lower temperate zones. Some of the largest forms, however, grow in the colder oceans. The characteristic colors of the Rhodophyta come from two pigments accessory to chlorophyll. Red algae generally grow attached mostly to rocks near the shore. Sometimes they live as epiphytes on larger water plants, and a few are parasitic on other plants. Though most of the red algae are marine plants, a few grow in fresh water. A favorite habitat of the freshwater forms is in mountain streams, the outflow of springs, and other locations where there is good aeration of the water.

Form and structure Their bright coloration and delicate structure lend great beauty to many of the red algae (Fig. 15-38). On the whole they are larger than the members of the Chlorophyta but never attain the proportions of such brown algae as the kelps. In form Rhodophyta range from filamentous plants, often heterotrichous, to those with expanded thalli which may be 3 to 4 ft in length. Some kinds are ribbon-shaped and attain considerable differentiation of tissues. Small cells in an outer layer of the thallus contain chloroplasts and are therefore the seat of photosynthetic activity. A central portion is composed of larger cells. Some slender kinds are

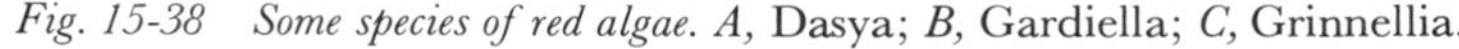

Fig. 15-38 Some species of red algae. A, Dasya; *B,* Gardiella; *C,* Grinnellia.

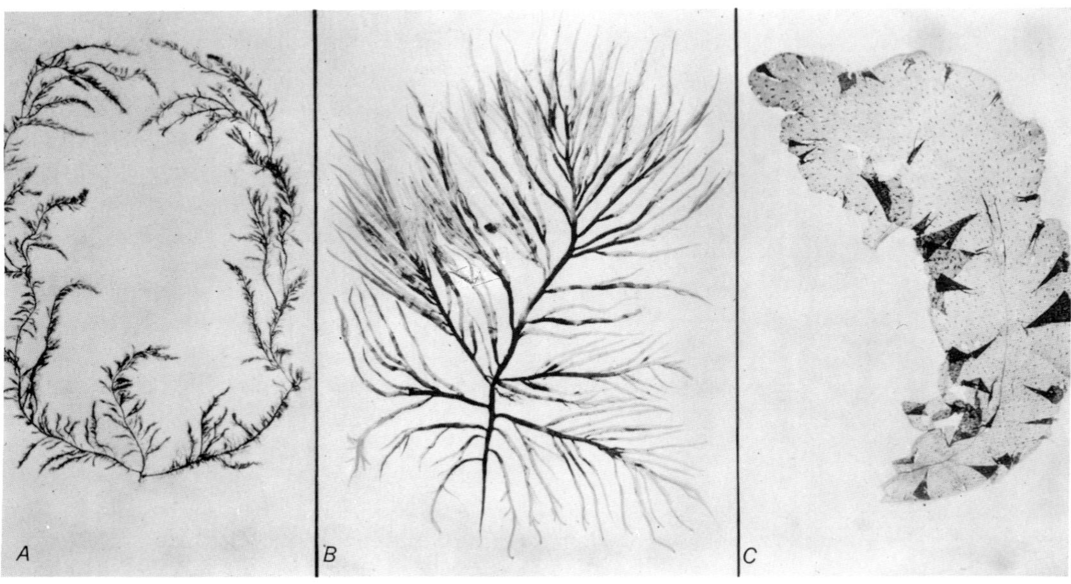

composed of several rows of cells appearing as tubes or siphons. The presence of gelatinous material in the thallus is characteristic of most red algae. In the simple branching forms, the filaments are covered with a sheath of a gelatinous nature which makes the whole plant slippery to the touch.

Agar, a gelatinous substance extensively used as a medium upon which fungi and bacteria are cultured, is prepared from the thalli of some species of the red algae. This substance is also used in articles of diet for invalids and in the preparation of certain medicines. Additional phycocolloids are **funori** and **carrageenin.** The first is water soluble and has been used as an adhesive and in the preparation of water-base paints. Carrageenin is a mucilaginous extract derived chiefly from *Chondrus crispus* (Fig. 15-39). This alga, growing along the North American Atlantic Coast, is locally of significant economic importance. It is dried and baled by the coastal residents. The extract carrageenin is mucilaginous. It is sometimes allowed to jell and is eaten as a dessert. But of greater economic use is its incorporation into products as a stabilizing agent. It may be used in the preparation of ice cream, cheese, cosmetics, water-base paints, and insect sprays. Red algae are extensively used as food by the people of Oriental countries and they are frequently on sale in the markets of coastal cities of the United States. When they are cooked, the red algae tend to retain both their color and their gelatinous nature.

The red algae have a definite cell structure of high organization. Generally there is a single nucleus although some cells are multinucleate.

Fig. 15-39 Baling dried, bleached Irish moss, Chondrus crispus, *at Yarmouth, Nova Scotia. (Courtesy of Dr. C. J. Hillson.)*

One or several plastids, which may contain pyrenoids, float in the cytoplasm. Because the rather simple plastid is similar to that of the blue-green algae, some workers have suggested a phylogenetic link between the two. The plastids of the red algae contain chlorophyll *a*, various yellow pigments, and accessory phycobilin pigments which account for the rather wide color range of these algae. The phycobilins are similar, but not identical, to those found in the blue-green algae and include both red and blue pigments. The red and blue pigments often obscure the green color of the chlorophyll. The products of photosynthesis are sugars. Food reserves in the red algae may occur as alcohols, but they are chiefly stored in the form of a polysaccharide, called **floridean starch.** This polysaccharide is stored as small granules free in the cytoplasm rather than in plastids. Upon treatment with iodine, floridean starch turns yellow, later changing to reddish purple. Conspicuous **protoplasmic connections** (Fig. 15-41*C*) that pass through openings, or pores, in the end walls of the cells are generally found in the tissues of the red algae.

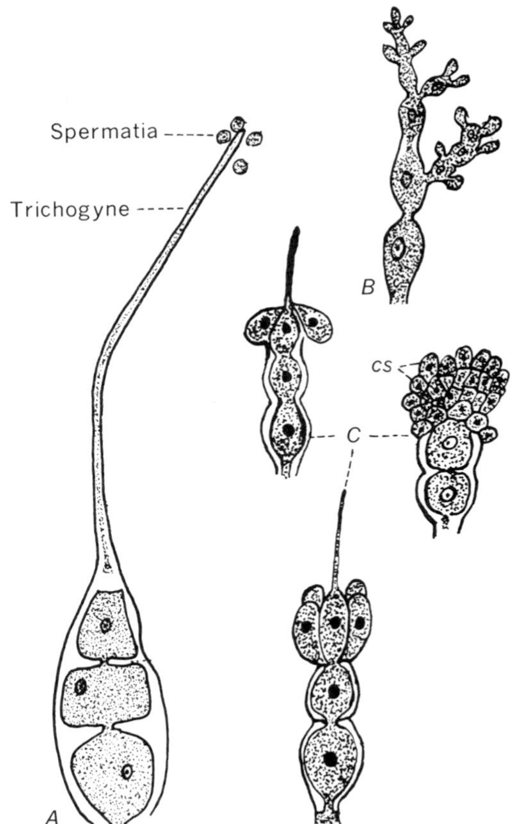

Fig. 15-40 Nemalion, *showing, A, carpogonial branch, with carpogonium and trichogyne with spermatia adhering; B, antheridial branch; C, stages in development of cystocarp and carpospores, cs, following fertilization. (Drawn by Helen D. Hill.)*

Sexual reproduction A unique characteristic of the Rhodophyta is a total lack of cilia, or flagella, and consequently of motility in all types of reproductive structures. Oögamous sexual reproduction is general and is achieved by small male cells, or microgametes, and larger female cells, or macrogametes, generally produced on separate plants. The male gametangia are simple, but the female gametangia are unusual structures that are associated with complex accessory tissue. The male gametangia, the spermatangia, are simple, single-celled structures produced abundantly in clusters on branches of the vegetative plant body. Each antheridium contains a single male gamete, or microgamete, called a **spermatium,** which escapes at maturity. The spermatia usually contain a single nucleus, but in a few genera this nucleus divides and forms two nuclei.

The female gametangium in the red algae is called the **carpogonium** (Figs. 15-40*A*, 15-41*D*). It consists of an enlarged or swollen basal portion and an elongated projection termed the **trichogyne.** There is considerable diversity in the form and size of the trichogyne in the various genera. The trichogyne may be cut off from the carpogonium proper by a cell wall or it may be merely a projection of the carpogonium. In those cases where the trichogyne is cut off as a separate cell, it has its own nucleus; in others it may

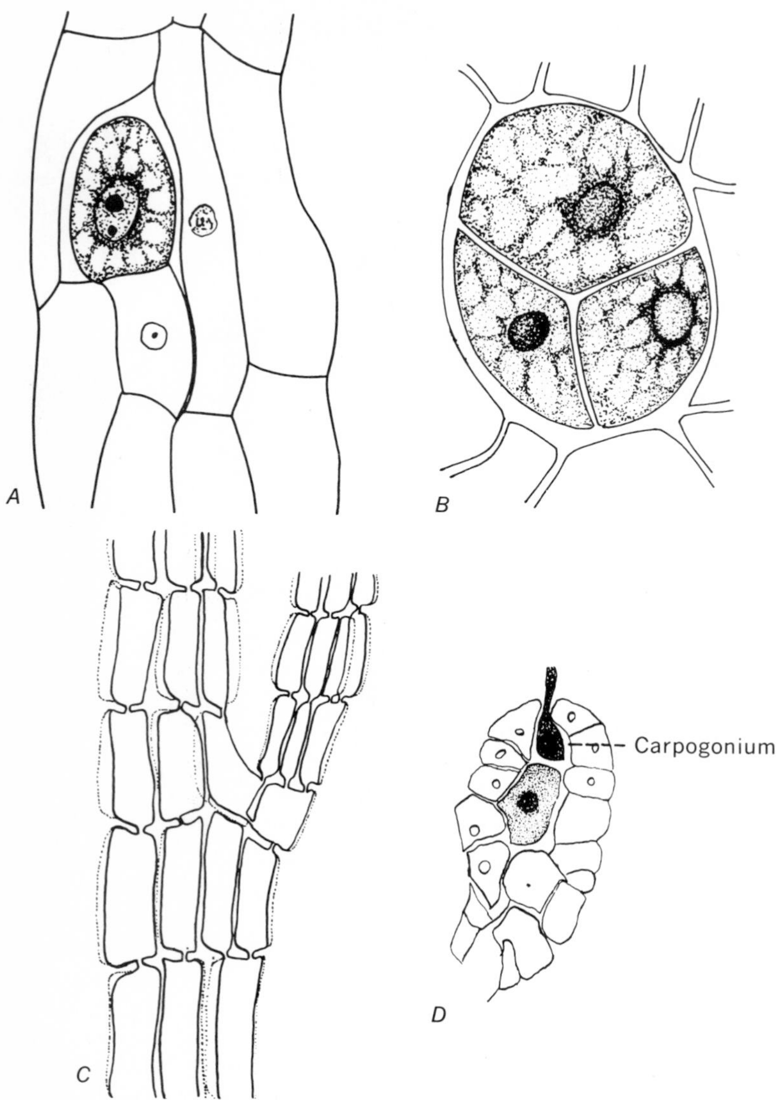

Fig. 15-41 Polysiphonia, *showing, A, young sporangium; B, tetraspores; C, portion of plant to illustrate the organization of filaments of several "siphons" and cytoplasmic connections between cells; D, young procarp showing carpogonium with trichogyne. (B–C drawn by Helen D. Hill.)*

not have a nucleus. The carpogonial cell corresponds in every essential with the oögonium of the green algae. It contains a protoplast, consisting of nucleus and cytoplasm, which comprises the female gamete. The carpogonium is borne upon a branch of the filament called the carpogonial branch, which is composed of several enlarged, specialized cells. When the gametes are mature, the spermatium floats to the female gametangium and adheres to the trichogyne. After the adjoining portions of the cell walls have been dissolved, the nucleus passes through the trichogyne down to the carpogonium, where it fuses with the nucleus of the female gamete. This constitutes fertilization.

Asexual reproduction Besides their diverse methods of sexual reproduction most red algae also have methods of asexual multiplication. One of the most common methods is the formation of a single-celled structure called a **monospore,** produced in a sporangium without change in chromosome numbers. When it is shed, the monospore germinates and grows into a new plant. Monospores are generally produced on the haploid plants and serve to multiply the individuals in this phase. Besides monospores there are **bispores,** produced two in a sporangium, and **polyspores,** with many in a sporangium. Another term, **paraspore,** is also applied to some of the asexual reproductive cells. Since diploid plants normally producing tetraspores (meiospores) sometimes form polyspores or paraspores, some investigators have considered these asexual reproductive cells to be derivatives of the tetraspores.

Nemalionales

Batrachospermum *Batrachospermum* is a freshwater genus found in clear cold water where there is some shade. The thallus has a predominant central axis with smaller lateral branches. These lateral branches are conspicuously forked, and from the basal cells of older branches downward-growing, rhizoidlike extensions enclose the single row of axial cells, causing the axis to appear multicellular. This growth habit results in a somewhat more complicated thallus than in many other freshwater forms.

Spermatangia (antheridia) and **carpogonia** (oögonia) may be formed on the same or on different thalli. Spermatangia are formed at the ends of filaments, and each spermatangium contains a single spermatium (male gamete). Carpogonia are borne on carpogonial filaments, and the carpogonium regularly has a conspicuous **trichogyne.**

Following fertilization, derivatives of the zygote nucleus migrate into several filaments, formed directly from the base of the carpogonium. Although not definitely established for *Batrachospermum,* most probably the zygote nucleus divides meiotically and haploid carpospores (meiospores) are produced. Under suitable environmental conditions, carpospores germinate to form new gametophytes.

Nemalion *Nemalion* grows as an annual attached to rocks or other suitable substrata in the intertidal zone. It is somewhat limited in its distribution, but it is abundant in places where it does occur. The mature plant body is made up of a central colorless core of intertwined filaments of elongated cells and a pigmented, enclosing layer of short, branched, lateral filaments.

Nemalion is homothallic, and both carpogonia (oögonia) and spermatangia (antheridia) are formed on the same thallus, although they are produced at different times. Spermatangial branches are produced, bearing many spermatangia, and a single nonmotile spermatium is formed within each spermatangium (Fig. 15-40*B*). Carpogonial filaments are formed from the division of a single carpogonial initial, and they consist of from three to five cells. The terminal cell of the carpogonial filament with its elongated protuberance, the trichogyne, is the carpogonium (oögonium) (Fig. 15-40*A*).

Nonmotile spermatia may lodge against the

trichogyne (Fig. 15-40*A*). Following the division of the single nucleus of the spermatium, both nuclei pass into the trichogyne and migrate toward the base of the carpogonium and the single carpogonial nucleus. One of the two male nuclei fuses with the female (carpogonial) nucleus to form the zygote nucleus. Division of the zygote nucleus is meiotic. Continued divisions of daughter nuclei and subsequent cell-wall formation result in the production of lateral filaments, each of which is called a **gonimoblast filament,** in which the terminal cell enlarges and becomes a carposporangium (Fig. 15-40*C*). The haploid carpospore is liberated from the carposporangium. At about the same time it is attached to some substratum, this naked spore becomes invested with a thickened wall. Sometime later, germination results in a protonemalike stage that soon develops into the adult gametophytic plant body.

Fig. 15-42 Polysiphonia. *Spermatangial plant.*

Ceramiales

Polysiphonia *Polysiphonia* is a relatively common red alga along both coasts of the United States. Its plant body is composed of tiers of cells, each tier consisting of a central cell surrounded by a number of pericentral cells; this organization is said to be polysiphonous. These filaments exhibit the protoplasmic connections between cells which are characteristic of red algae. The plants are small, only a few to several inches in length.

The gametophytes of most, if not all, species of *Polysiphonia* are heterothallic, and carpogonia and spermatangia are borne on female and male plants, respectively. Spermatangia are borne in clusters (Fig. 15-42) on fertile branches, and a single spermatium is liberated from each spermatangium by rupture of its wall.

A carpogonium, female gametangium, is borne at the end of a short filament of cells, called the carpogonial filament. The carpogonium, similar to that of *Nemalion,* has an elongated protuberance, the trichogyne (Fig. 15-41*D*). In a fashion similar to that described for *Nemalion,* spermatia liberated from the spermatangia lodge against the wall of the trichogyne, and, following the partial dissolution of their walls, the nucleus of a spermatium reaches the nucleus of the carpogonium and syngamy occurs. Unlike *Nemalion* (and other red algae not described in this textbook), the diploid nucleus does not undergo meiosis, and, at the tips of gonimoblast filaments, the carposporangia which are formed contain a single diploid carpospore (Fig. 15-43). After liberation of the nonmotile carpospores, germination results in the development of a diploid (tetrasporic) plant. Some of the cells of this diploid plant function as sporangia (tetrasporangia) in which the protoplast divides meiotically to form meiospores

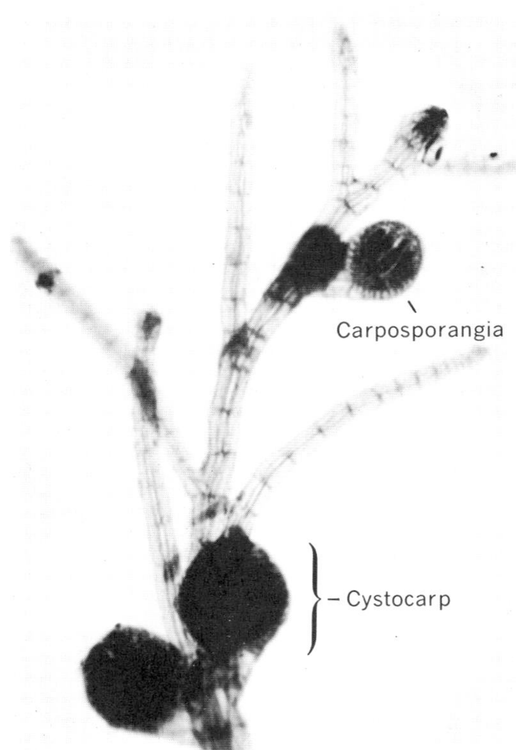

Fig. 15-43 Polysiphonia. *Carpogonial plant.*

(tetraspores). Following liberation, the meiospores germinate to produce gametophytic plants.

At about the same time as syngamy, additional filaments of sterile cells are produced in the area of the carpogonium. The more or less simultaneous development of the gonimoblast filaments, carposporangia, and sterile cells produces a rather complex structure called the **cystocarp** (Fig. 15-43).

QUESTIONS

1. What structural and functional features of algae can be considered to be common features?

2. Prepare a list of such features as kind of stored food material, motility, pigments present, etc., and behind each category assemble the algae showing the feature.

3. Of what value is a knowledge of the life cycle of an alga?

4. Explain the masking effect of certain pigments and relate this to the process of photosynthesis.

5. Propose or diagram synthetic life histories of algae involving isomorphic plants, heteromorphic plants, sexual reproduction, gametic reproduction, isogametes, heterogametes, oögonia, and antheridia.

6. Compare the economic importance of blue-green algae and red algae.

7. List other possible economic uses of algae, or their products, in industry, in public health, or as experimental organisms for biological research.

REFERENCES

Blum, J. L. 1956. The Ecology of River Algae. *Botan. Rev.,* **22:**291–341.

Chase, Florence M. 1941. Useful Algae. Smithsonian Inst. Ann. Rept. 1941, pp. 401–452. Excellent review of economic uses of algae.

Fogg, G. E. 1965. Algal Cultures and Phytoplankton Ecology. The University of Wisconsin Press, Madison, Wis. The importance of phytoplankton, its place in the food chain, photosynthetic output, growth of algae in culture.

Fritsch, F. E. 1935. The Structure and Reproduction of the Algae, vol. 1. Cambridge University Press, New York. A standard textbook.

Fritsch, F. E. 1945. The Structure and Reproduction of the Algae, vol. 2. Cambridge University Press, New York. A standard textbook.

Jackson, Daniel F. (ed.). 1964. Algae and Man. Based on lectures at the NATO Advanced Study Institute held at Louisville, Kentucky, July 22–August 11, 1962. Plenum Press, New York. Examination of current concepts of the biology of algae. Recent advances in a variety of fields including taxonomy, cytogenetics, physiology, and ecology.

PALMER, C. MERVIN. 1959. Algae in Water Supplies. U.S. Public Health Serv. Publ. 657. Interesting illustrations, practical approach.

PRINGSHEIM, E. D. 1964. (reprinted). Pure Cultures of Algae. Originally published in 1949. Hafner Publishing Company, Inc., New York. The preparation and maintenance of algal cultures.

SCAGEL, ROBERT F., et al. 1965. An Evolutionary Survey of the Plant Kingdom. Wadsworth Publishing Company, Belmont, Calif. Emphasis on comparative morphology.

SCHWIMMER, M. 1955. The Role of Algae and Plankton in Medicine. Grune & Stratton, Inc., New York. A survey of the medicinal uses of algae.

SMITH, G. M. 1955. Cryptogamic Botany, 2d ed., vol. 1: Algae and Fungi. McGraw-Hill Book Company, New York. Shorter treatment of both freshwater and marine algae.

SMITH, G. M. 1950. The Freshwater Algae of the United States, 2d ed. McGraw-Hill Book Company, New York. Considers only algae commonly found in fresh water. A standard textbook.

TYRON, C. A., Jr., and R. T. HARTMAN (eds.). 1960. The Ecology of Algae. Special publication No. 2 of the Pymatuming Laboratory of Field Biology, University of Pittsburgh. Edwards Brothers, Inc., Ann Arbor, Mich. A technical treatment of algal ecology.

16

BACTERIA, VIRUSES, ACTINOMYCETES, SLIME MOLDS

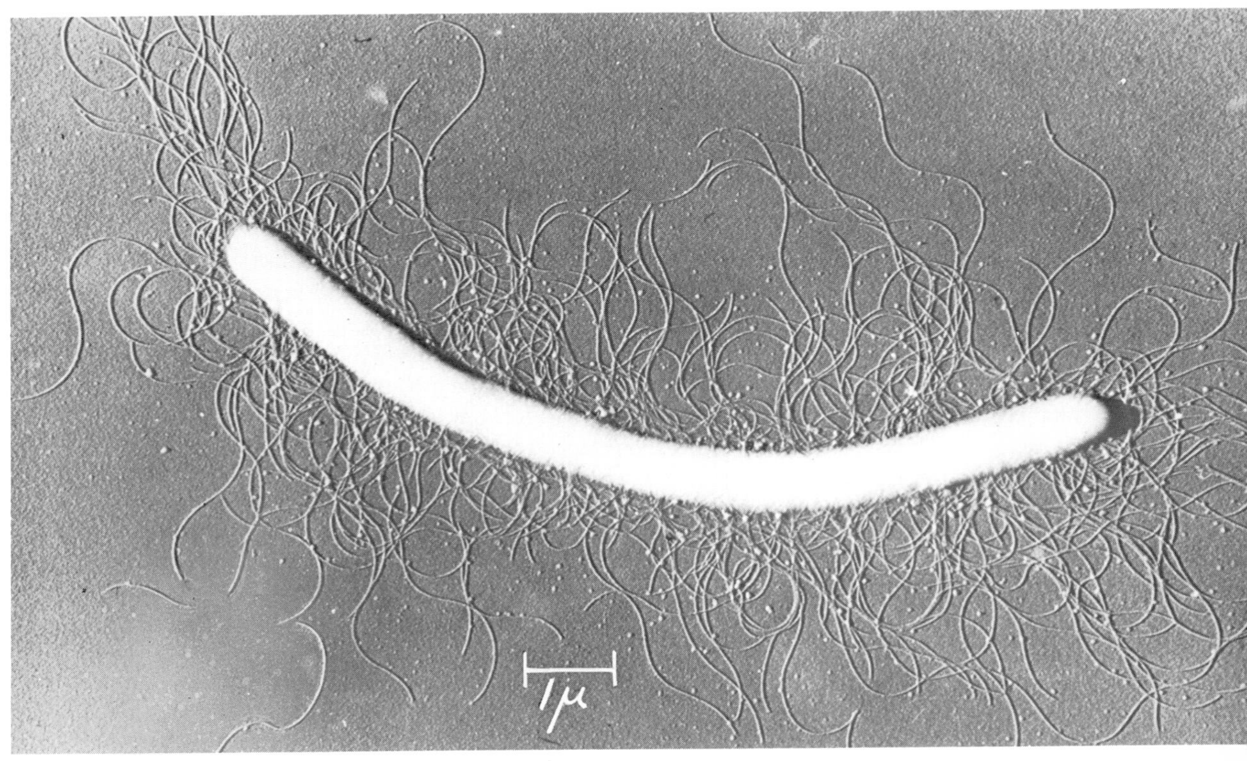

Electron micrograph of the bacterium Proteus vulgaris. *Grown on agar, washed in 5% formaldehyde, and washed in distilled water by centrifugation. Specimens prepared by Dr. C. F. Robinow, shadowed with chromium, and photographed by Dr. James Hillier. (Courtesy of Dr. C. F. Robinow and the Society of American Microbiologists.)*

This chapter deals with the principal characteristics of the bacteria, viruses, actinomycetes, and slime molds; the size, distribution, shape and form, culture and identification, and economic importance of the bacteria, bacterial photosynthesis and chemosynthesis, the nature and importance of viruses and bacteriophages, and a brief account of the slime molds. The grouping of these forms in this chapter does not imply that they are closely related.

BACTERIA

General characteristics Bacteria are microscopic unicellular organisms, which are often known as "germs" and "microbes." Their relationship to other living things is very obscure, but they are ancient organisms, being found as fossils in Precambrian rock more than 600 million years old. In their unicellular structure, in the absence of a definite nucleus and of a nuclear membrane, and in their asexual methods of reproduction, they show similarities to the blue-green algae. They differ from the blue-green algae in that the cells of only a few species contain bacteriochlorophyll and only these can carry on photosynthesis. Most bacteria are heterotrophic and in this respect they are like the fungi. Some species are **parasites,** attacking the living cells of other plants or of animals and securing their foods from that source. Most bacteria grow on the dead remains or the products of plant and animal life without a direct relationship with living cells and are therefore **saprophytes.** Parasitic bacteria are responsible for some of the diseases of plants and animals, whereas the saprophytic kinds may be very beneficial.

While most of the bacteria are heterotrophic (dependent) plants, a few species are autotrophic; i.e., they are capable of synthesizing carbohydrates out of carbon dioxide and water and hence can make all their own foods. Some of these, such as the purple bacteria, contain green pigments (bacteriochlorophyll) and can carry on a type of photosynthesis, while others, such as the hydrogen bacteria, iron bacteria, nitrifying bacteria, and some of the sulfur bacteria, lack green pigments but can manufacture carbohydrates by chemosynthesis.

Our knowledge of bacteria dates back only to the middle of the nineteenth century. It was initiated by the brilliant work of the French chemist and biologist Louis Pasteur, who was the first to demonstrate the importance of sterilization and aseptic conditions in preventing infection by these organisms. His work revolutionized medicine and surgery and paved the way for the development of the study of bacteria into the independent science of bacteriology.

Size and distribution Bacteria are the smallest of all known living things. The largest of them measure about 1⁄10 mm (1⁄250 in.) in greatest dimension and are nearly large enough to be visible to the unaided eye. The smallest known are 1⁄10 μ (1/10,000 mm or 1/250,000 in.) in length. This means that, if they were placed end to end, it would require 250 of the largest kind and 250,000 of the smallest kind to make a chain 1 in. long. The rod-shaped forms average about 2 μ in length and ½ μ in width (1 μ = 0.001 mm). Thousands of bacteria might be contained in a single drop of water and not be at all crowded for space. A cubic centimeter (20 to 30 drops) of sour milk contains many millions of them. Not all substrata are suitable for the growth of bacteria. They may be present on any exposed surface that does not possess properties fatal to their existence. In many such situations they are unable to multiply and hence their numbers do not increase. On or in a substratum suitable for their growth and development they may become extremely numerous. Decaying vegetable and animal materials and solutions rich in organic matter are usually excellent places for the growth of saprophytic species.

Shape and form In shape or form two principal types of bacteria may be differentiated, viz., the spherical, or globose, forms and the cylindrical (Fig. 16-1). The spherical forms are called **cocci** (singular, coccus) and are classified into several groups based mainly on the manner in which they remain together after dividing. If they divide in only one plane, they may remain in pairs and are called **diplococci,** or they may cling together in chains. The latter are called **streptococci.** Some forms divide in two planes and cling together in irregular masses shaped like bunches

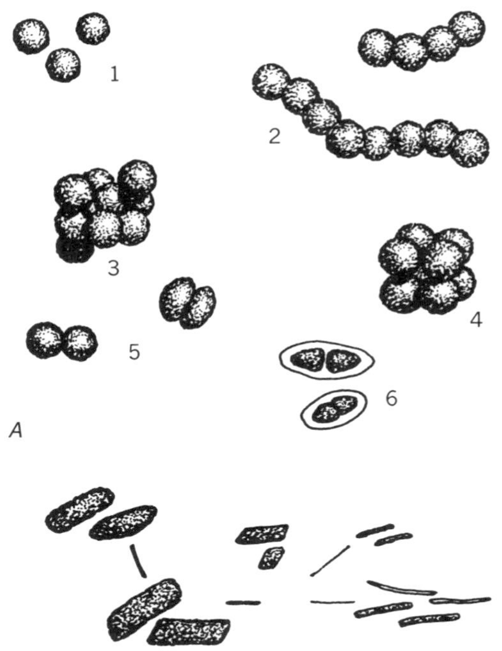

Fig. 16-1 Forms of bacteria. A, spherical forms; 1, coccus; 2, streptococcus; 3, staphylococcus; 4, Sarcina; *5, diplococcus; 6, encapsulated diplococci; B, nonflagellated bacilli; types, shapes, and groupings of bacilli. (Drawn by Dr. R. D. Reid.)*

of grapes. Such forms are called **staphylococci,** or **micrococci.** Still others divide in three planes and form cubical groups. These are placed in the genus *Sarcina.* Coccus forms are often distorted into oval or elliptical shapes.

The cylindrical types may consist of straight, rodlike forms collectively called **bacilli,** or the rods may be curved, as in the genus *Vibrio,* or spirally twisted. Those with rigid spirals consisting of one or more complete turns are placed in the genus *Spirillum,* while those with flexible spirals are called **spirochaetes** (Fig. 16-2).

Under certain conditions not well understood, a bacterial cell of one of the foregoing types may alter its shape to some unusual or abnormal form. Such alterations are not uncommon in some species and the altered cell shapes are known as **involution forms.**

Structure of the bacterial cell Because of the very minute size of the bacterial cell, its structural features have been very difficult to determine. Studies with the electron microscope have thrown some light on these features and have provided a better understanding of the cytology of bacteria. Cellulose has not been found in the cell walls of bacteria, but other polysaccharides and other substances may be present. The wall, in some species, is surrounded by a thin gelatinous sheath, or **capsule,** somewhat of the nature of the sheath present in many of the blue-green algae and consisting usually of carbohydrates and other substances. This sheath swells in aqueous solutions and makes the solutions slimy in character.

Just beneath the cell wall there is a plasma membrane (consisting of fats and proteins) which is differentially permeable. The cytoplasm contains granules of various sorts. Among them are what appear to be chromatin granules or filaments consisting of deoxyribonucleic acid (DNA). These represent the nuclear matter of the cell (Fig. 16-3). No definite nucleus with a nuclear membrane is present and there are no

Fig. 16-2 A, types of spirilla; B, spirochaetes. (Drawn by Dr. R. D. Reid.)

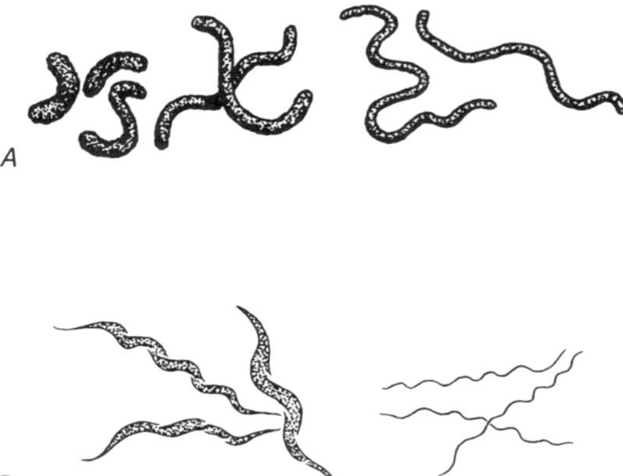

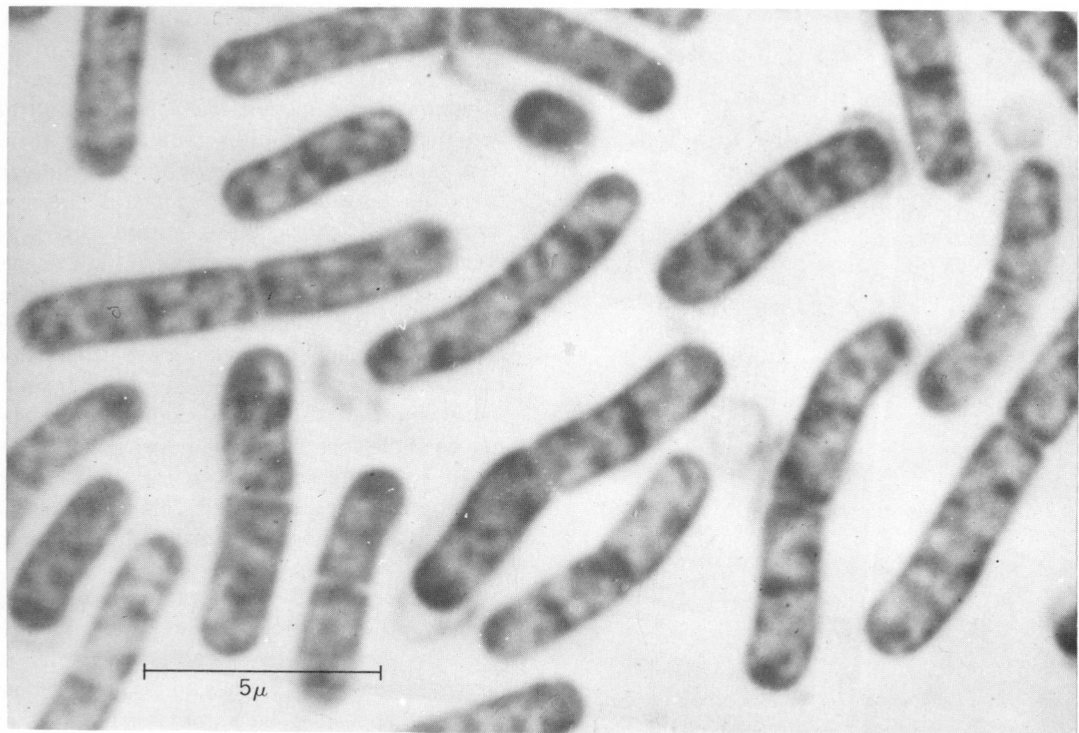

Fig. 16-3 Bacillus megatherium. *Cultivated from spores for 2 hr at 37° C, fixed with bouin through the agar and stained with 0.01 percent aqueous basic fuchsin for 20 sec. The uneven staining of the cell contents is due to the greater affinity of the basic fuchsin for the cytoplasm than for nuclear material. Empty spore cases are visible. In some cases a clear band is seen in the bacteria which represents the unstained transverse cell wall. (Courtesy of Dr. C. F. Robinow and the Society of American Bacteriologists.)*

mitochondria. Ribosomes involved in protein synthesis are present and membranous structures, suggestive of the endoplasmic reticulum of the cells of higher plants, can be seen. Except in a few species the protoplast contains no chlorophyll and the bacterial cell is, in most cases, colorless. A few species contain special chlorophylls, as well as carotenoids which give them green or purple hues.

Some bacteria are motile by means of flagella (Fig. 16-4*B*). Sometimes a single terminal flagellum is present; sometimes there are two or more at one or both ends, and in some species they are numerous on all sides of the bacterial cell. The coccus forms, with the exception of one or two species, have no flagella; most of the spirillum types are flagellated; of the bacillus types, some have flagella and some do not.

CULTURE AND IDENTIFICATION Many different kinds of methods are used in culturing and identifying bacteria. Broths are used for culturing bacteria in liquid media. These include meat broth, extracts of plant origin, inorganic and organic chemicals, and other substances. Sugar, starch, vitamins, and blood are added to the broth in special cases. Broths are first sterilized by heating in an autoclave at high temperatures, such as 121°C, for 20 min or longer. A solid medium is obtained by adding gelatin or agar. Agar cultures are often made in petri dishes in which colonies can be counted.

An examination of the cell content sometimes assists in the identification of different species, as, for example, the presence of certain granules, such as those found in the diphtheria bacillus, or the presence of certain fat globules and other substances. Some can be identified only by physiological tests involving temperature requirements, food requirements, gas production, or fermentation studies. Others can be identified only by their ability to cause diseases in plants or animals.

THE GRAM REACTION Two different kinds of bacteria are commonly differentiated on the basis of their response to a differential stain originally devised by the Danish investigator Gram. The stain consists of a crystal-violet solution and an iodine solution. Those species of bacteria which retain the stain are **gram-positive,** while those which do not are **gram-negative.** This differential staining is widely used by bacteriologists in the study and identification of bacteria. Furthermore gram-positiveness and gram-negativeness are associated with other properties of the bacteria, such as sensitivity to dyes and antibacterial substances, a knowledge of which is important in developing methods of control.

Reproduction Multiplication of bacterial cells is accomplished by the process of cell division. It is a rapid process, occupying usually about 30 min; but when in vigorous condition, some species are said to divide as often as every 20 min. It has been calculated that if division occurs only every hour, the descendants of a single cell after 24 hr would number 17 million individuals and that in two days the number would reach 281 billions of individuals. Fortunately such multiplication does not last long because of the lack of food and other favorable conditions.

A type of sexual reproduction, by conjugation, has now been shown to occur in certain bacteria. This was demonstrated experimentally in the laboratory in 1947 with two strains (mutants) of the bacterium *Escherichia coli.* More than 100 mutants are recognized in *E. coli,* most of them physiological in their expression. If two strains of *E. coli,* differing in two mutations, are grown together in one colony, the progeny not only exhibits the original mutations but also two new combinations. The occurrence of the new combinations is suggestive of Mendelian segregation and recombination. It definitely indicates sexual mating and exchange of hereditary material. *E. coli* is an inhabitant of the human intestine and is found in great numbers in sewage. Its presence in water indicates the pollution of water, and its absence is widely used as a test for the purity of water for human consumption.

Spore formation Under certain conditions changes take place in some bacterial cells which result in the formation of bodies called spores (Fig. 16-4*A*), in which the vital activities of the protoplasm are retained. Only two genera of

Fig. 16-4 A, bacilli showing positions of endospores; B, flagellated bacteria. (Drawn by Dr. R. D. Reid.)

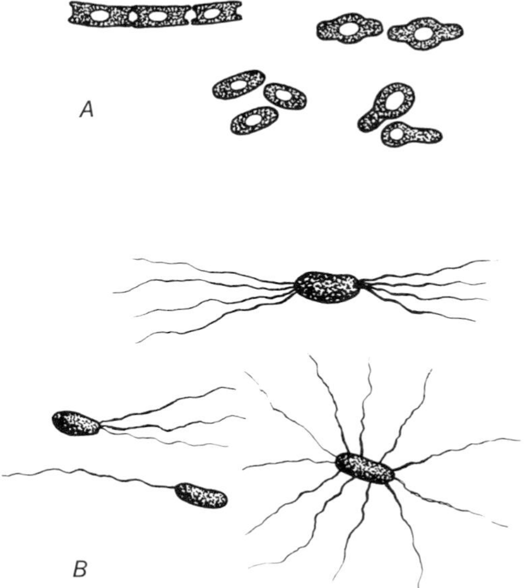

bacteria, *Bacillus* and *Clostridium,* form spores. Both are rod forms. Only one spore is formed in each bacterial cell. These spores are always nonmotile, but they are so small as to be carried easily in air currents. They are widely distributed in air and on all objects and are practically omnipresent. They differ from the usual bacterial cell in their capacity to resist prolonged unfavorable external conditions, such as drought, high or low temperatures, and disinfecting chemicals, and in their reaction with certain staining solutions. Some spores withstand as much as 16 hr of constant boiling. On the return of favorable conditions, these spores will germinate, assume the original form of the bacterial cell, and soon grow to normal size for the species. Spore formation is not to be considered a method of reproduction or of multiplication, since but one spore is formed in each bacterial cell and but one new bacterium comes from each spore. It may be regarded as a stage in the life cycle of the organism.

Bacterial photosynthesis and chemosynthesis As previously mentioned, a few species of bacteria are able to carry on photosynthesis. Among these are the **purple sulfur bacteria** and the **green sulfur bacteria.** Both are found in sulfur springs where hydrogen sulfide (H_2S) is available. Both utilize H_2S instead of water as a hydrogen source and liberate sulfur instead of oxygen. Both utilize light and CO_2 in making carbohydrates. Both are autotrophs. The purple sulfur bacteria contain a special variety of chlorophyll known as **bacteriochlorophyll,** as well as red and yellow carotenoids. The green sulfur bacteria have a special type of chlorophyll which differs from bacteriochlorophyll. The overall equation for photosynthesis of both these types of bacteria may be written as follows:

$$\mathbf{CO_2} + \mathbf{H_2S} + \text{light} + \text{bacterial chlorophyll} \longrightarrow \underset{\text{carbohydrate}}{(\mathbf{CH_2O})} + \underset{\text{elemental sulfur}}{\mathbf{S}}$$

The sulfur is stored inside the cells of the purple sulfur bacteria and is given off as a waste product in the green sulfur bacteria.

Another group of bacteria, the **purple bacteria,** i.e., purple *nonsulfur* bacteria, also use light as a source of energy for photosynthesis. They contain bacteriochlorophyll, as well as red and yellow carotenoids, but they require organic raw materials as a source of hydrogen in the process. Photosynthesis is carried on only in the presence of light and in the absence of oxygen. These forms do not use the absorbed organic raw materials directly as foods. In some species of purple bacteria absorbed organic raw materials are used directly as foods in the dark and in the presence of oxygen. Since both forms require organic raw materials in the environment, they are heterotrophs.

The **iron bacteria,** the **sulfur bacteria,** the **hydrogen bacteria,** and the **nitrifying bacteria** are chemosynthetic autotrophs; i.e., they can synthesize carbohydrates, utilizing inorganic substances as a source of energy instead of light. The iron bacteria use soluble iron compounds which they combine with oxygen to form insoluble iron compounds. The sulfur bacteria utilize either H_2S or molecular sulfur. These are converted into sulfur or sulfate ions respectively. The hydrogen bacteria use molecular hydrogen, which is combined with oxygen to form water. The nitrifying bacteria are of two types: the nitrite type, utilizing ammonia (NH_3) and converting it to nitrite (NO_2) ions; and the nitrate type, using nitrite ions and converting them to nitrate (NO_3) ions. The energy released in all these conversions is used in converting water and CO_2 into carbohydrates with the released oxygen used in respiration. This may be summarized as follows:

$$\mathbf{H_2O} + \mathbf{CO_2} + \text{inorganic energy source} \longrightarrow \underset{\text{carbohydrate}}{(\mathbf{CH_2O})} + \mathbf{O_2} + \text{inorganic end products}$$

The equations given above for bacterial photosynthesis and for chemosynthesis are highly simplified expressions of the processes. Actually these processes are much more complicated, involving many hydrogen- and electron-transfer mechanisms, such as those involved in photosynthesis in the higher green plants.

Economic importance of bacteria

FERMENTATION AND DECAY Many saprophytic species of bacteria are capable of producing profound chemical changes in the substrata on which they grow. The decay of plant and animal bodies and the process known as fermentation are changes of this type. Decay is the more comprehensive term and includes the decomposition of organic bodies into their constituents or into such simple compounds as water, carbon dioxide, ammonia, and hydrogen sulfide. When an abundant supply of oxygen is not available, as in the decay of the bodies of the larger animals in which nitrogenous materials are abundant, unpleasant odors are often developed, and the process is known as **putrefaction.** The processes of decay are important to man from at least two standpoints. They prevent the accumulation of organic matter, both plant and animal, on the earth, and they result in the formation of simple compounds or set free elements that are returned to the soil to be used again by plants.

Many of the fermentation processes that are carried on by bacteria are of household or commercial importance. Thus species of *Acetobacter* produce acetic acid (vinegar) from the alcohol formed from the juice of ripe fruits, especially apples and grapes. The alcohol itself is formed in a similar fermentation process by different organisms (yeast) acting on the carbohydrates in the fruit juices. Processes of this type have already been discussed in Chap. 12. The souring of milk is another fermentation process that involves the conversion of sugar of milk (lactose) into lactic acid. The making of dill pickles, the manufacture of cheeses, the making of sauerkraut, the retting of flax, and the tanning of leather are other examples of fermentation processes of commercial importance.

Most of the methods of sewage disposal are dependent for their efficiency on the activities of bacteria. This is especially true of those methods involving the use of septic tanks and cesspools. Certain bacteria rapidly break down the solid organic materials into soluble compounds. Likewise the use of sand filters for sewage disposal involves the action of bacteria, mostly of the aerobic types.

BACTERIA AND SOIL FERTILITY Of all the living organisms found in soils, bacteria are among the most abundant. It is not unusual to find a hundred million or more individual bacteria per gram of soil. They are especially abundant in the surface layers of soil, decreasing in numbers with depth of soil. These bacteria, along with other soil organisms, play a dominant role in soil fertility. In general they succeed in converting insoluble or unavailable materials into forms that can be used by higher plants. Among the simpler compounds they produce are carbon dioxide, ammonia, nitrates, and sulfates. The processes involved in nitrogen transformations, including ammonification, nitrification, denitrification, and nitrogen fixation, have already been considered in Chap. 7. The bacteria that bring about decomposition of cellulose and similar compounds are important in the production of humus. Among the soil bacteria there are anaerobic and aerobic forms, autotrophic as well as heterotrophic species.

BACTERIA AS DISEASE-PRODUCING AGENTS As stated previously, only the parasitic (pathogenic) species of bacteria are capable of causing disease in other organisms. Not all diseases of either plants or animals are caused by bacteria. Many plant diseases are caused by fungi. Fire blight of apple and pear trees (Fig. 16-5*A*), crown gall of various fruit trees and ornamental plants (Fig. 16-5*B*), and wildfire of tobacco (Fig. 16-5*C*) are examples of bacterial diseases of plants. Certain types of decay of parts of such herbaceous plants as cabbage,

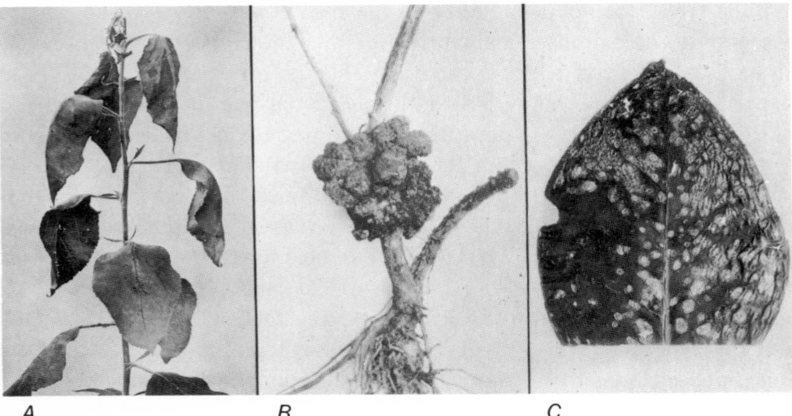

Fig. 16-5 Bacterial diseases of plants. A, fire blight of apple; leaves and tips of twigs suddenly wilt and die, the terminal portion of the twig usually curling over; B, crown gall on rose; woody gall-like growths are formed at the base of the stem; C, wildfire of tobacco; small circular dead areas are formed on the leaves.

celery, and various members of the root crops often become serious under storage conditions owing to the presence of bacteria.

Some of the worst diseases of man and of other animals are caused by bacteria. Examples are tetanus, diphtheria, tuberculosis, pneumonia, and typhoid fever. Botulism is a fatal form of food poisoning caused by a very potent toxin produced usually in spoiled foods by the saprophytic, anaerobic, spore-forming bacterium *Clostridium botulinum*. Tetanus (Fig. 16-6) and diphtheria are caused by parasitic bacilli which produce toxins in the tissues which they parasitize. The tuberculosis bacilli and the pneumonia cocci, as well as many other infectious bacteria, do not produce toxins but cause serious diseases in other ways.

Typhus fever, transmitted by body lice and rat fleas, and Rocky Mountain spotted fever, transmitted by a wood tick, are caused by a special group of spherical or rod-shaped organisms called the **Rickettsiae.** In shape and form they resemble bacteria, but their size is less than that of the smallest bacteria and near the size of large viruses. They are barely visible under the higher powers of a light microscope. Unlike the viruses, they do not pass through bacterial filters. All of them are obligate intracellular parasites. Like the viruses, they cannot be grown in nonliving media.

VIRUSES

There are many diseases of plants that can be transmitted from one plant to another by means of sap from diseased plants that has been passed through a porcelain filter so fine that it removes completely even the smallest bacteria. The causal agents of such infectious diseases are so small as to be invisible even under the highest-power magnifications of compound microscopes. They can, however, be seen with an electron microscope. Because they are able to pass through bacterial filters, they were formerly called **filterable viruses.** Viruses are now recognized as the causal agents in one of the largest and most important classes of plant and animal diseases, causing millions of dollars of damage annually to crop plants, serious losses of animals, and death of humans.

Virus diseases and infections in plants are

recognized and described on the basis of symptoms and transmissibility. The ones most readily recognized are the so-called yellows, mosaics, and ring-spot types. More than 50 virus diseases of such stone fruits as cherries, peaches, and almonds are known. Some typical diseases are peach yellows, necrotic ring spot of sour cherry, peach mosaic, and tomato mosaic.

While some plant viruses have a restricted host range, i.e., infect only a few plant species, others can be introduced into many different and unrelated plant species and will multiply (Fig. 16-7). For example, cucumber-mosaic virus has been shown to cause a mosaic disease in celery and in other plants, as well as in cucumbers. The fact that some viruses will infect a wide range of different plant species has led to complications in the methods of control. It is not uncommon for perennial weeds and ornamental plants growing near fields used for vegetables and other crops to become infected with viruses. These plants may then serve as reservoirs of viruses that are transmitted by insects or other means to crop plants. Serious outbreaks of disease in cultivated crops have been traced to such sources of infection. In general the plant viruses are transmissible by sap, by grafting, or by insects.

Virus diseases of animals include such maladies as foot-and-mouth disease of cattle, hog cholera, distemper of dogs, Newcastle disease of chickens, and psittacosis of birds. In man viruses cause rabies, smallpox, chicken pox, influenza, yellow fever, encephalitis, infectious hepatitis, poliomyelitis, measles, and common colds.

Fig. 16-6 Tetanus bacteria showing location of spore. (Courtesy of General Biological Supply House Incorporated, Chicago.)

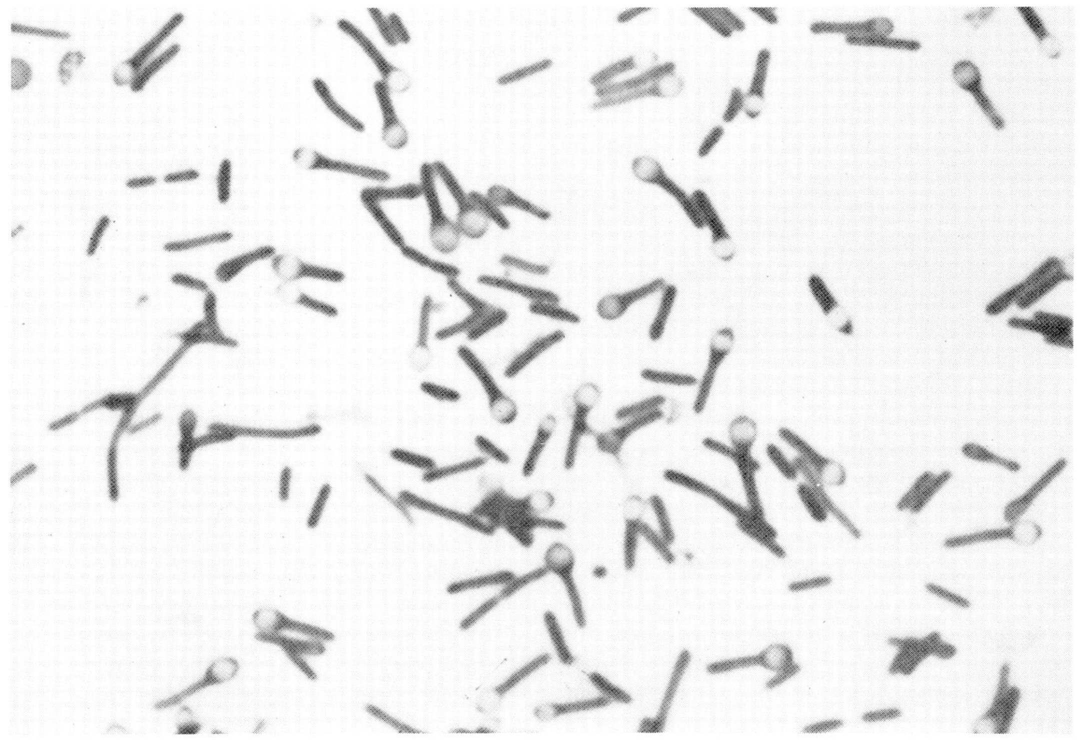

Fig. 16-7 Three cucumber seedlings inoculated with different isolates of necrotic ring-spot virus of sour cherry; uninoculated control on extreme right. Note mottling and dwarfing of infected plants. (Courtesy of Dr. J. S. Boyle.)

Viruses cannot be propagated on artificial media but multiply readily in living cells. By means of tissue-culture techniques much new knowledge has been acquired concerning the etiology of some diseases, particularly those attacking animals and humans. The preparation of poliomyelitis vaccine became possible only after it was learned how to propagate the virus in tissue culture, and its success has in turn paved the way for the production of a vaccine to immunize against measles.

Some viruses can be precipitated out of suspensions by chemical means without losing their activity, but the ultracentrifuge has now provided a method for the purification of the less stable forms. Each year many additional plant and animal viruses are purified and many have been critically studied chemically and structurally. All thus far isolated have been found to be nucleoproteins of large molecular size and weight.

Electron-microscope studies and X-ray diffraction studies show the single particles of tobacco-mosaic virus (Fig. 16-8) to be rigid rods near 15 mμ wide and 300 mμ long. The outer shell is composed of 2,200 subunits of protein which are arranged in a helix around the core of nucleic acid (RNA). Recently each subunit of tobacco-mosaic virus was found to consist of 158 amino acids and their sequence was established. An exciting discovery, which has had considerable influence on studies related to protein synthesis, was made when the nucleic acid of tobacco-mosaic virus was determined to be the infectious part of the virus particle.

There are many different kinds of viruses (Fig. 16-9) and apparently different races or strains within a type. Whether the viruses are living organisms, as originally believed, cannot be stated with certainty. In their ability to reproduce themselves in living tissues they resemble microorganisms. No one has ever succeeded, however, in demonstrating that they carry on respiration or have a metabolism. Certainly, those viruses which have been prepared in pure crystalline form and have been found to be nucleoproteins are not living organisms in the ordinary sense. They are considered by some to be degraded from primitive living organisms.

A special group of viruses known as **bacteriophages** cause the destruction of bacteria. They have most of the attributes of other viruses and exist in various strains, some of which are characterized by their ability to attack only certain types of bacteria. Some of the virulent types cause the complete disintegration of the bacterial cells they enter. Others, after entering the bacterial cells, do not prevent the bacteria from

Fig. 16-8 Electron micrograph of tobacco-mosiac virus particles. Actual size of particles, 15 by 300 mμ (Courtesy of Dr. J. S. Boyle.)

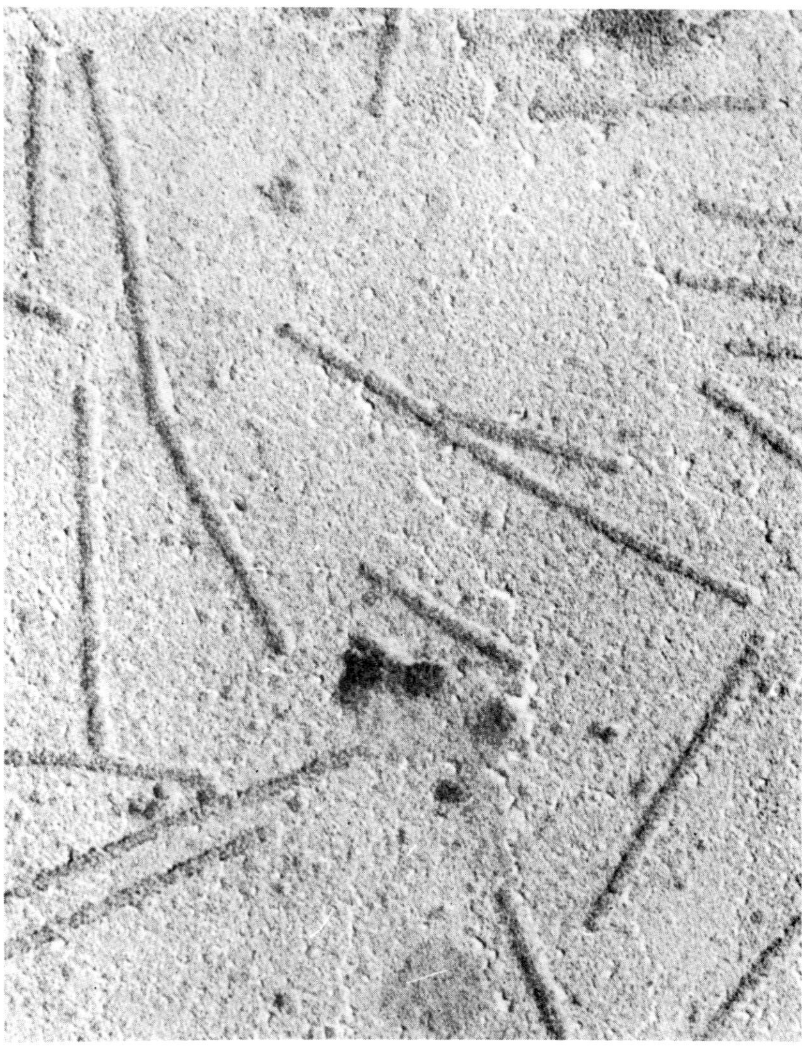

Fig. 16-9 Electron micrograph of lettuce-mosiac virus particles; shadow cast with palladium; polystyrene sphere, 514 mμ in diameter, used for calibration. (Courtesy of Dr. H. B. Couch and Dr. A. H. Gold.)

carrying on their normal physiological functions.

Electron micrographs of bacteriophages show them as small irregular spheres with a rodlike appendage. They consist of a central mass of deoxyribonucleic acid (DNA) covered by protein which protrudes as the rodlike appendage. In attacking a bacterium, the rodlike appendage attaches itself to the bacterial cell and the core of DNA is injected into the cell, becoming attached to the chromatin material of the bacterial cell. Following this, only bacteriophage particles are produced in the virulent species within the bacterial cell. The bacterial cell disintegrates and the new bacteriophages escape and may invade new bacteria.

ACTINOMYCETES

The actinomycetes are a group of microorganisms, some of which are moldlike and others bacterialike. For this and other reasons they are difficult to classify. Some authors place them with the Fungi Imperfecti; others consider them as an independent group of fungi; still others classify them with the bacteria in a separate order, the Actinomycetales. Only a brief mention can be given them here.

The actinomycetes differ from nearly all true fungi in the extreme fineness of their mycelia, the hyphae or individual threads of which are commonly only about 1 μ in diameter. The mycelium, or vegetative plant body, in those forms in which it is well differentiated, is branched and sometimes twisted. The protoplasm of the young hyphae appears to be undifferentiated, but the older parts of the mycelium show definite granules and vacuoles. Nuclei have been reported to be present in some species. The mycelium in some species commonly breaks up into small fragments called **arthrospores,** which often look like bacterial cells and which might easily be mistaken for the latter. Many species also produce asexual spores, called conidia, on aerial hyphae, which appear as a fine powdery coat on the surface of cultures. The conidia-bearing filaments are often spirally twisted (Fig. 16-10).

The spores of actinomycetes seem to be widely distributed in the atmosphere. An agar plate exposed to the air will often yield small, round, flat colonies of these organisms closely adherent to the medium, often highly colored and emitting a penetrating musty odor. The production of pigments is one of the most striking cultural characters of the group. The spores are often white or gray; the mycelium may be nearly colorless or may be colored red, orange, yellow,

Fig. 16-10 A spirally twisted conidiophore of Streptomyces *sp., bearing a chain of conidia. (Drawn by Dr. C. L. Fergus.)*

green, or blue. In addition, soluble pigments may diffuse out into the medium. These may be of the same color as the mycelium or different. Many species produce a characteristic brown discoloration of agar.

Like the fungi and most of the bacteria, the actinomycetes are all either saprophytes, obtaining their food from dead organic matter, or parasites, living on other living organisms. The saprophytic species are widespread in soil and take an active part in the decomposition of complex organic matter, breaking down such compounds as proteins, starch, cellulose, chitin, and perhaps lignin. Most of them are able to reduce nitrates to nitrites but not to ammonia or free nitrogen. Several species are parasitic on higher plants, causing such diseases as the common potato scab. Others cause spoilage of nuts and of other food and dairy products. Still others cause diseases of animals and man. The term actinomycosis is commonly applied to such diseases. Actinomycosis occurs in many wild and domestic animals as well as in man. One of the commonest of these diseases is lumpy jaw of cattle, caused by *Actinomyces bovis*. In man actinomycosis may involve all parts of the body.

From what has been said in the preceding paragraph it is obvious that the actinomycetes are of considerable economic importance. The importance of the group has increased since the discovery that some of the species produce antibiotics. Thus streptomycin, which is now being used in medicine to combat certain bacterial diseases, is obtained from *Streptomyces griseus*.

MYXOMYCETES

General characteristics The myxomycetes, commonly known as slime molds, or slime fungi, constitute a group of doubtful taxonomic position. Combining characteristics of both plants and animals, they were at one time named Mycetozoa, which means "fungous animals." Because they produce spores which have cellulose walls, and for other reasons, they are usually considered as members of the plant kingdom and are usually classified with the fungi. The vegetative body consists of a naked, slimy mass of protoplasm with many nuclei, called a **plasmodium,** which creeps over the surface of, or within, the substratum on which it grows, by an amoeboid movement engulfing solid food particles as it moves forward. All of them are saprophytes,[1] growing in moist, dark places on wood, leaves, rocks, in grass, and on other objects. (Some slime molds, now in the class **Plasmodiophoromycetes** and formerly classified as myxomycetes, are parasitic in higher plants, algae, and fungi. The so-called club root of cabbage and other crucifers is caused by one of them.) The plasmodia of the myxomycetes are excellent material for protoplasm studies.

Reproduction The vegetative stage is most often found in dark, moist places. Just prior to the reproductive stage, the plasmodium is usually found in drier, better-lighted situations. It may ascend the side of a stump (Fig. 16-11*B*), or, emerging from dead leaves or other debris on the ground, it may ascend the stems of grasses, small trees, or other objects (Fig. 16-11*A*, *C*) and produce one or more sporangia. In some species the sporangia are colorless; in others they are conspicuously colored. The sporangia of the different species vary greatly in size and form. In some they are small, round, puffball-like structures. In others they are borne on long or short stalks. A well-developed, though fragile, membranelike wall is usually formed over the surface of the sporangia. Eventually the protoplasm within the sporangia breaks up into numerous uninucleate, nonmotile spores with cellulose walls. These spores are usually borne within a network of tubular, simple or branched threads called the **capillitium.** Finally the sporangial wall breaks and the spores are

[1] The term **saprobe** has been suggested for an organism, plant or animal, that obtains its food from dead or decaying organic matter. The term saprophyte would then be restricted to plants obtaining their food in this manner.

Fig. 16-11 Developmental stages of three different species of myxomycetes. A, sporangium produced in the forks of a chrysanthemum plant which it has climbed; B, on side of stump, a plasmodium just beginning to form sporangia; C, fruiting bodies (aethalia) of Lycogala.

disseminated by the wind or fall to the ground.

On reaching a favorable moist substratum, the spores germinate, each producing one to four **swarm spores** (Fig. 16-12). The swarm spores consist of a naked mass of protoplasm containing a nucleus and two flagella, one of which is shorter than the other. They soon begin to divide to form more swarm spores. This process may continue for some time. If unfavorable conditions are met, each swarm spore may enter an encysted or resting stage, from which, on the return of favorable conditions, it may emerge again as a swarm spore. Sooner or later, the swarm spores lose their flagella and begin to

Fig. 16-12 A germinating spore of a myxomycete, showing how the protoplasmic mass escapes from the spore wall and differentiates, at one end, two long slender flagella, one of which is longer than the other. (After Frank A. Gilbert, Am. J. Botany, vol 15*(6): 345–352, pl 21, figs. 8, 9, 10.)*

fuse in pairs. These fusions involve also a nuclear fusion, and hence the resulting body is a zygote, the swarm spores thus behaving as gametes. Each zygote may develop directly into a plasmodium, or a number of zygotes may coalesce to form a plasmodium.

QUESTIONS

1. How would you prove that bacteria are present in laboratory air?

2. Why do bacteria fail to develop on properly prepared jellies?

3. Why are bacteria likely to grow in an open wound?

4. Why are vaccines not used with plants as they are with animals?

5. Why is bacterial photosynthesis less important in man's economy than the photosynthesis of higher green plants?

6. Why do foods keep better in a refrigerator?

7. Are viruses living organisms? Explain.

8. How could you determine whether a plant disease is caused by a virus?

9. Why are the slime molds considered to be plants rather than animals?

10. What characteristics of slime molds are similar to those of some lower animal forms?

REFERENCES

ALEXOPOULOS, CONSTANTINE J. 1962. Introductory Mycology, 2d ed. John Wiley & Sons, Inc., New York. Contains sections on myxomycetes.

ALEXOPOULOS, CONSTANTINE J. 1963. The Myxomycetes, II. *Botan. Rev.,* **29**(1):1–78. An excellent review including structure, physiology, reproduction, ecology, and classification of this group.

BAWDEN, F. C. 1964. Plant Viruses and Virus Diseases, 4th ed. The Ronald Press Company, New York. One of the best general accounts of plant viruses and the diseases they cause.

GEST, HOWARD, A. SAN PIETRO, and L. P. VERNON (eds.). 1963. Bacterial Photosynthesis. The Antioch Press, Yellow Springs, Ohio. Papers presented at the first Kettering Symposium, 1963, devoted entirely to bacterial photosynthesis.

PELCZAR, MICHAEL J., JR., and ROGER D. REID. 1965. Microbiology, 2d ed. McGraw-Hill Book Company, New York. A good general textbook for bacteria, viruses, etc.

17

FUNGI

Basidiocarps of Panaeolus sphinctrinus *from Mexico. (Photograph by Robert S. Beese, courtesy of Dr. Leon R. Kneebone.)*

General characteristics, classification, and economic importance of fungi, followed by detailed descriptions of the principal groups with life histories of representative species of each group, are discussed in this chapter.

General characteristics The group of plants known as the Eumycophyta, or true fungi, is an extremely heterogeneous one, comprising approximately 70,000 described species, with many more yet to be described. Such kinds as mushrooms, toadstools, molds, mildews, rusts, and smuts are more or less familiar to everyone, but thousands of others are so minute or so evanescent or grow in such obscure situations that they are to be found only by those trained in the search for them. Their habitats are also extremely diverse. Some are entirely subterranean; others are epiphytic on various other types of plants; one or more species are to be found as parasites on practically every species of higher plants as well as on insects and other animals, including man. They grow in drinking water, on food, feathers, leather, cloth, and even on optical equipment. The economic losses they cause are out of all proportion to their size and importance as plants.

These diversities render difficult an exact statement of the characteristics of the group. One feature they have in common is their universal lack of chlorophyll and hence their inability to manufacture their own carbohydrate foods through the agency of sunlight. They are therefore all either **saprophytes,** obtaining their food from dead organic matter, or **parasites,** living on other living organisms. The organism on which a parasite lives is known as the **host.** The presence of a parasite may cause abnormal physiological activity within the host that is correctly designated as a **disease.** The parasite in this case is commonly referred to as a **pathogen** and the host as a **suscept.** While the fungi are unable to make carbohydrates out of carbon dioxide and water, they can synthesize fats, proteins, and many other organic substances. In their inability to carry on photosynthesis they contrast strongly with the algae and agree with the bacteria. The bacteria differ, however, in the lack of a well-defined nucleus in the cell, in methods of growth, in reproduction, and in other ways.

The plant body The life history of most fungi consists of two phases, a vegetative stage, in which the fungus grows through the substratum, and a reproductive stage, in which it produces spores or other structures by which the plants are multiplied. To make clear this essential difference between these two phases of the life history, we may examine the ordinary mushroom of commerce. The conspicuous part of this fungus, i.e., the mushroom itself, is the reproductive body. Examination of the interior of the bed from which this mushroom grows reveals a multitude of fine white threads that collectively constitute the **mycelium,** or **spawn,** as it is commercially known. This is the vegetative stage of the fungus. Pieces of the spawn planted in the bed grow rapidly, permeate the entire bed, and finally emerge at definite points to form the mushrooms. The latter are the fruiting structures of the plant and, in fact, are usually spoken of as the fruit bodies or **sporophores** or, more commonly, the mushrooms. The mycelium is the plant body, and the mushroom originates from it.

The vegetative phase; mycelium

ORIGIN, STRUCTURE, AND GROWTH
With the exception of a few unicellular fungi, all species have a more or less extensive mycelium (Fig. 17-1). The mycelium originates through the germination of a spore. The simplest spores are unicellular bodies of microscopic size, containing a nucleus and cytoplasm. They may or may not be motile. A spore germinates by pushing out, from a thin place in its wall, a tubelike or filament-like structure known as a **germ tube** (Fig. 17-4*B*). By rapid elongation, accompanied by branching,

Fig. 17-1 Small portion of a typical mycelium, showing branching, septation, and the granular cytoplasm. (Drawn by Edna S. Fox.)

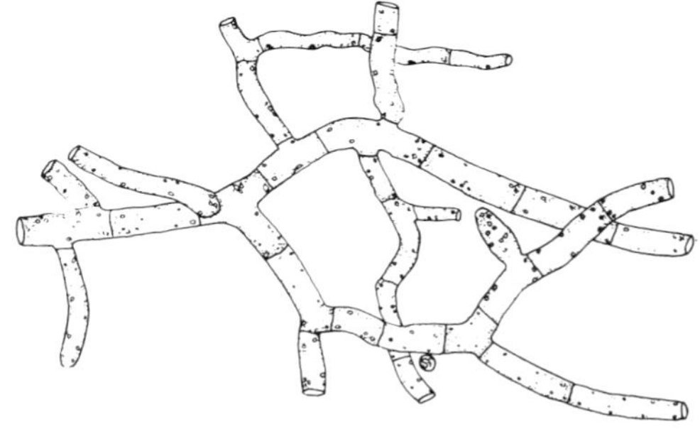

the mycelium is formed. Growth is more or less localized at the tips of the mycelial branches and is made possible by the absorption of foods from the substratum directly through the cell walls. Cross walls may or may not be formed. Their absence in one large group of the fungi results in a coenocytic type of plant body.

Structurally the mycelium consists of a complicated mass of interwoven, branched filaments (Fig. 17-2), each of which is termed a **hypha** (plural, hyphae). Hyphae, then, are the individual filaments or branches, and collectively the hyphae composing the vegetative stage of a given fungus plant are the **mycelium.** The medium upon which the mycelium grows is termed the **substratum.**

The hyphae, with the exception of the coenocytic condition noted above, are composed of cells, the walls of which are made up, in some cases, apparently of pure cellulose. In others little cellulose is present, the greater part of the wall substance being a fatty-acid complex with a chitin base. Within the cells the usual cell parts are present, consisting of one or more nuclei and a mass of vacuolated cytoplasm. Sugars and glycogen, but not starch, represent the carbohydrate type of food present, but fatty and oily materials, proteins, organic acids, and other substances also occur. Various types of pigments, but not chlorophyll, may also be present, either in the cell wall or in the cell content. These pigments sometimes diffuse into the surrounding medium.

Mycelium is usually abundant and easily obtained from any supply of decaying organic matter. In turning over piles of decaying leaves in late autumn or early spring, one may find it occurring as fine white strands running through them. On the lower sides of boards in damp situations, it develops rapidly (Fig. 17-2) or it may often be found forming large sheets or mats in the crevices of decaying logs. The presence of a reproductive body is evidence of the extensive development of mycelium in that region of the substratum. If a piece of bread or other food material rich in carbohydrates is kept moist for a few days and exposed to air, a white cottony growth, the mycelium of a fungus, usually develops over it.

Growing in such situations as have been enumerated, the tips of the mycelium are often able to penetrate directly through the cell walls of the tissue concerned, even though it be of the hardest wood. They do this by means of en-

Fig. 17-2 A, mycelium of a fungus as grown in a petri dish; B, white mycelial strands on the underside of an old log.

zymes, which they secrete, some of which have the power of dissolving the substances composing the cell walls. For example, the cellulose of the cell wall is ordinarily insoluble and would offer considerable resistance to mechanical penetration, but it is broken down by enzymatic hydrolysis and, moreover, yields glucose, which is assimilable by the fungus. The fungus then, in bringing about decay or disease, is only preparing, by an external digestive process, the food materials for its use. In the course of time this may result in the complete transformation of a sound log into a mass of rotten, useless material representing that portion of the tissue which the fungus is unable to use; or a perfectly sound apple, a potato, or an orange will, in the course of a few days, be reduced to a wet, rotten, pulpy mass. In the case of a mushroom or toadstool growing on a grassy lawn or in a pasture, the result of these activities is not so conspicuous, but they are continually taking place on the dead organic matter in the soil. Such activities of soil fungi, like those of soil bacteria, are important in breaking down organic matter in the soil, thereby providing materials for the growth of higher plants.

RESTING STAGES In some fungi the mycelium may pass into a dormant or resting stage by the formation of definite bodies of closely compacted hyphae known as **sclerotia.** These vary in size from half the size of a pinhead to several inches in diameter. When of sufficient size, some of them form a palatable food material for man and have been known under the names of Indian bread, tuckahoe, and other descriptive terms. Usually, however, they are quite small or are developed within the substratum, so that they are not easily found. On the return of favorable conditions, these sclerotia may grow out into a new mycelium or they may produce a sporophore of some sort. They are more frequently found in species that are parasitic upon annual plants or plant parts. In the case of parasites on the perennial parts of plants, the fungus usually hibernates as a dormant mycelium in the host tissue. In many other cases the mycelium in its ordinary condition is, if well protected, perennial in nature.

MYCORRHIZA A close relationship sometimes exists between the mycelium of certain fungi and the roots of certain higher plants, particularly forest trees. In this association the mycelium may form an encircling mantle around the finer rootlets or may be largely internal in the cortex of the root. Such infested rootlets are much enlarged and usually show an abnormal amount of branching (Fig. 17-3). This combination of fungus and host is known as a **mycorrhiza** and is a phase of those physiological relationships between organisms grouped under the term **symbiosis.** Mycorrhizas are probably essential for the proper growth of many species of plants with which they are associated.

The reproductive phase

THE SPOROPHORES The most effective method of reproduction in most fungi is by means of spores. The structure which produces these spores is called a **sporophore** and is frequently well differentiated from the purely vegetative structures. The reproductive organs are often aerial while the mycelium is usually within the substratum.

The variation in types of sporophores produced is almost endless. Some are extremely small, so minute that the naked eye cannot discern them. At the other extreme are such large sporophores as the common mushrooms or toadstools and the puffballs.

THE SPORES Motile spores, known as zoospores or swarm spores, are present in most aquatic fungi and represent the typical asexual method of reproduction of such forms. They are formed in zoosporangia, from which they usually emerge ("swarm") at maturity and, after a brief period of activity, germinate into a mycelium. The majority of fungi are more terrestrial

Fig. 17-3 Mycorrhiza of white pine. Note that the finer rootlets, as at a, are abnormally swollen and show an abnormal amount of branching.

in habitat and produce nonmotile, wind-disseminated spores (Fig. 17-4*A*). These spores may be formed in any one of several different ways.

Often a given fungus may have more than one kind of spore. Usually each species has one spore type of major importance and one or more types of minor importance or for different functions. Thus many species have a spore stage in which they pass the winter or other unfavorable environmental conditions, and thus the spore becomes a resting spore. The same species may have one or more other spore forms, allowing rapid dissemination through the growing season. These spores may be quite different in structure, color, thickness of wall, and other characters. Some species of fungi are known to have as many as five different types of spores in the life cycle.

Spores in general are unicellular, but there are many instances of multicellular spores. In such cases each cell behaves as a unicellular spore, and the structure might be better regarded as a compound spore.

The number of spores produced by fungi is enormous. A single mushroom sporophore is estimated to give off 40 million spores per hour during the period of active spore discharge, which covers several hours. Other fungi, such as the large puffballs, produce many more than this. When one considers the total number of spores produced during a growing season, one ceases to wonder at the universal presence of these plants and is surprised that there are not more of them. As a result of their presence, the air about us is always more or less contaminated with spores. To prove this, one has but to uncover a dish of sterilized culture medium for a minute or two. Spores from the air settle quickly on this substratum and begin development (Fig. 17-5).

After dissemination, which is accomplished by wind, air currents, or splashing rain, the spores may germinate if conditions are favorable. Some resting spores are so constructed that they must remain dormant for a time. An adequate moisture supply is essential for germination. Some spores require an actual film of water, and others are able to germinate in a moist atmosphere. A favorable temperature is also of great importance. In general, low temperatures as well as unusually high temperatures tend to suppress germination. The optimum for many spores is between 15 and 20°C. Spores usually germinate by extruding, through some point on the spore wall, a protoplasmic sac, which rapidly elongates into a narrow tube known as a germ tube (Fig. 17-4*B*). This tube usually continues to grow and elongate and eventually becomes

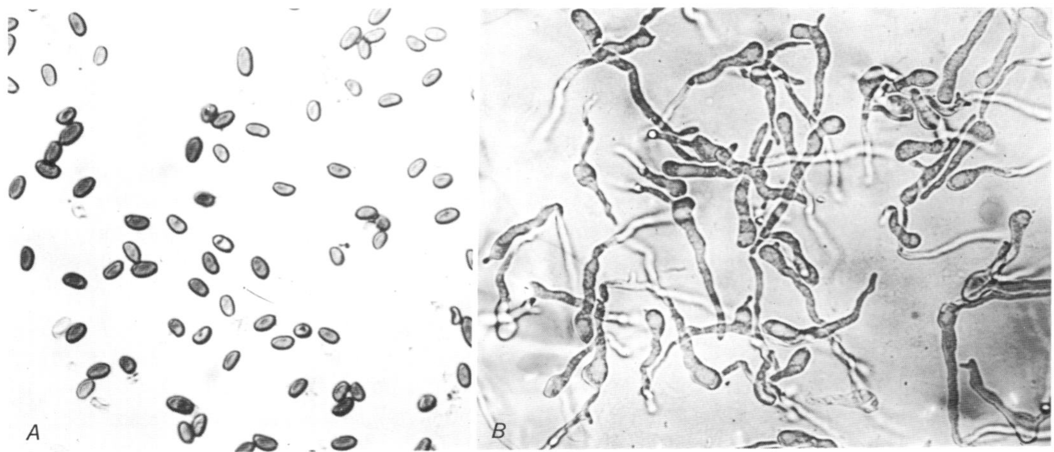

Fig. 17-4 Spores and spore germination. A, photomicrograph of spores of a common mushroom; B, photomicrograph of spores germinating. The bulblike swellings are the original spore bodies.

Fig. 17-5 A, dish of agar medium uncovered for 5 min in a crowded laboratory and then incubated for three days. During this 5-min interval of exposure, nearly 50 spores settled on the agar and developed into colonies. B, a similar dish of agar exposed for 5 min on a stump in the woods near the laboratory, at the same time as foregoing exposure. Only three colonies developed in the second dish.

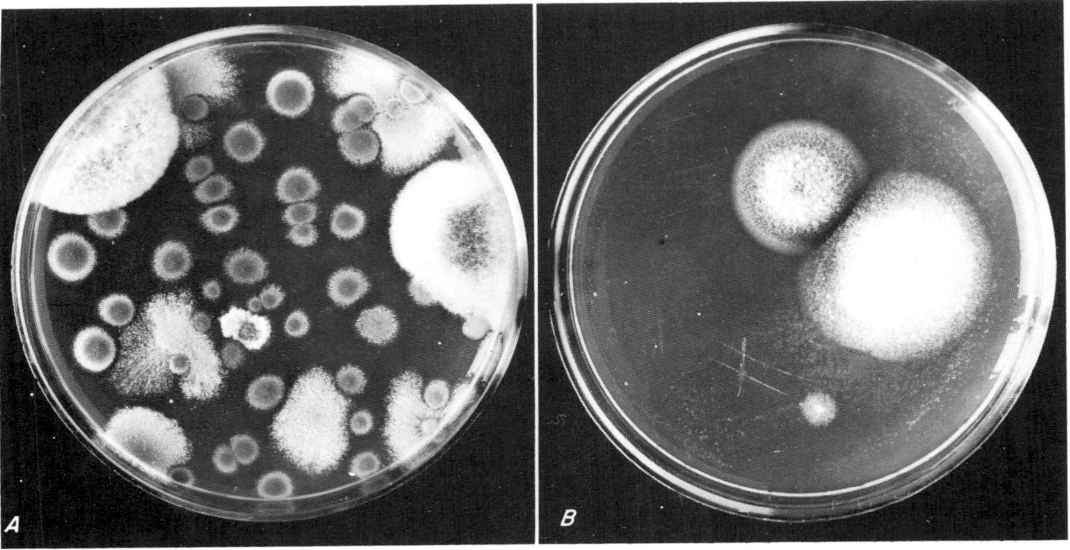

the mycelium. With its spores in the proper condition for germination on the proper host or substratum and with proper atmospheric conditions, a fungus may spread extremely rapidly. Such coincidences are the direct cause of the rapid and extensive spread of various plant diseases over a period of a few damp, sultry days during the growing season. Spores of many fungi can be readily germinated in tap water and are interesting objects for microscopic study (Fig. 17-4*B*).

Economic importance of fungi The economic importance of the different kinds of fungi is considered in connection with the discussion of the separate groups in the following pages. A few general matters may, however, be pointed out here. While many of the fungi are decidedly detrimental to man, others are beneficial and even indispensable. Some, like the mushrooms and truffles, are used as food. Others are used as sources of drugs. The yeasts are universally used in breadmaking, in the making of alcoholic beverages, and in various other fermentation processes, and are also a source of vitamins. Some fungi are used in cheese making and others in other manufacturing processes.

The pink bread mold, *Neurospora sitophila,* and other fungi are used as tools in genetic studies because they reproduce sexually and complete their life cycles in a few days. Individual ascospores can be isolated and studied in their development. In addition, mutants can be produced by means of radiation and other treatments and the mutant characters can be followed readily. Such studies have contributed much to our knowledge of the action of genes in inheritance of biochemical as well as morphological characteristics.

SOIL FUNGI The importance of soil fungi is often minimized in discussions of soil fertility. From studies at the Rothamsted Experiment Station in England, it has been estimated that living fungus cell material was present in fertilized soil to the extent of 1,700 lb per acre, which is about twice as much as the material from bacteria and all other soil microorganisms combined. These fungi are important in maintaining soil fertility, not only because they break down complex organic substances, such as cellulose, proteins, and other compounds, thereby transforming such substances into forms available to crop plants, but also because they utilize many inorganic substances which are thus prevented from being lost from the soil by leaching. Some of the soil fungi are more active in producing ammonia from proteins than are the ammonifying bacteria. Not all the soil fungi are beneficial, however. Some of them cause damping off of seedlings and others cause serious diseases of more mature crops. While in some cases only the spores or resting forms of these pathogens may be present in the soil, in other cases these fungi lead a saprophytic existence in the soil, later becoming parasites.

ANTIBIOTIC SUBSTANCES Many of the fungi produce certain chemical substances which are able to inhibit the growth of, or to destroy, bacteria and other microorganisms. Such substances are called **antibiotics.** The best-known of these substances is **penicillin,** which, since about 1941, has been widely used as a therapeutic agent against infections produced by certain bacteria, particularly those of the gram-positive type. Penicillin is a metabolic product of the fungus *Penicillium notatum,* from which it is produced in commercial quantities (Fig. 17-6).

Another antibiotic, **streptomycin,** so named because it was originally isolated from *Streptomyces griseus,* one of the actinomycetes, has been found to be effective against some of the bacteria that are unaffected by penicillin, particularly the gram-negative organisms. **Aureomycin,** also obtained from one of the actinomycetes, has been found effective in the treatment of many serious diseases of animals and man that do not respond readily to other antibiotics. Many

Fig. 17-6 Battery of fermentors used in the production of antibiotics at the Cherokee plant of Merck and Co., Inc. (Courtesy of Merck and Co., Inc., Danville, Pa.)

other antibiotics have been reported from fungi and from other organisms, but many of them will probably never be used as therapeutic agents because of their toxic properties.

DISEASES OF ANIMALS AND MAN; MEDICAL MYCOLOGY Many different fungi infest the bodies of animals and man. A disease caused by such infestation is called a **mycosis.** Examples of these diseases are ringworm (Fig. 17-7), athlete's foot, valley fever, histoplasmosis, and aspergillosis. There are many others. Some of the fungi causing mycoses affect only the skin, while others may cause generalized infection of a more serious nature. While fatal fungous diseases in man are less common than those caused by bacteria, they are nevertheless important. According to the vital statistics of the United States, of all deaths caused by parasitic diseases, about 3 per 1,000 are caused by fungi. Mycoses are more prevalent in tropical regions. The nonfatal types are probably as common as are some bacterial diseases. **Medical mycology,** which is the study of fungi causing mycoses, is a prominent branch of the general study of fungi.

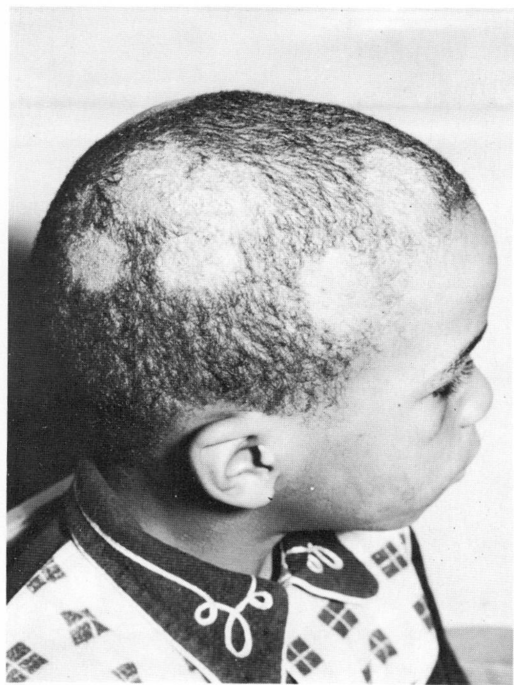

Fig. 17-7 Tinea capitis, *or ringworm of the scalp, a mycosis caused by* Microsporum audouini. *(Photograph courtesy of Dr. A. M. Kligman, Duhring Laboratories, Hospital of the University of Pennsylvania, Philadelphia.)*

PLANT DISEASES; PLANT PATHOLOGY

Diseases of plants may result from many different causes, such as mineral deficiencies or excesses; unfavorable environmental conditions of light, temperature, moisture or oxygen supply; toxic gases and other chemicals; nutritional unbalance or other internal physiological disturbances; viruses; the attacks of eelworms, insects, mites, spiders, and other animals, bacteria, actinomycetes, and fungi. Plant pathology deals with all these diseases. Of all parasitic diseases of plants, those caused by fungi are the most common. Fungi causing plant diseases are found in all the main groups of fungi, and there is hardly a plant that is not subject to the attack of one or more of them. Not only do they cause millions of dollars of loss annually in crop plants, but many molds, mildews, and other fungi also cause spoilage of foods, destruction of fabrics, fibers, paper, and leather, and rotting of timber. It is the problem of the plant pathologist to prevent and to control such ravages. A knowledge of the fungi is therefore basic to him.

In attempts to prevent and to control diseases of plants, several methods of attack are used. Among these may be mentioned the following:

1. *Exclusion and quarantine,* by which the transportation of diseased plants from one region to another is prohibited. In 1912 the United States Congress enacted the first quarantine act for the control of plant diseases. Since then 50 or more additional Federal disease quarantines have been put into operation for preventing the spread of particular plant parasites. Almost all foreign countries as well as the individual states of the Union now have similar laws.

2. *Sanitation,* or the maintenance of sanitary conditions in the vicinity of growing plants. This involves the destruction, by burning or other means, of diseased plants or plant refuse which may harbor disease-producing organisms, or the elimination of breeding places of such organisms.

3. *Development of resistance or immunity to disease.* Some varieties of plants are more resistant than others and may be wholly immune to particular pathogens. Such resistance is often a heritable character. The selection and breeding of such varieties sometimes offers the most effective or the only method of controlling certain diseases. Particularly is this true of some virus and bacterial diseases of plants, but the method has been applied to many fungous diseases as well.

4. *Selection of disease-free seed and propagating stocks.* Some diseases, like the anthracnose of beans, are carried over from crop to crop by means of diseased seed. Similarly potato tubers and other propagating stock may harbor disease organisms. Care in the selection of disease-free seed and propagating stock may prevent the spread of certain diseases.

5. *Crop rotation.* The spores or mycelia of some fungi are able to live over from one year to another in the soil. The length of time that such fungi can remain alive in the soil varies, but the ravages of these organisms can be reduced by not planting the same crop in the same soil year after year. In fact, one of the chief reasons for rotating crops is to reduce plant diseases.

6. *Correcting mineral deficiencies or excesses in soil.* Diseases caused by deficiencies of certain minerals in the soil often manifest themselves by yellowing, mottling, and other discolorations of leaves and other parts of plants and by reduced growth. Examples of such deficiency diseases are sand drown of tobacco caused by magnesium deficiency, heartrot of beets resulting from boron deficiency, little leaf or rosette of fruit trees caused by zinc deficiency, frenching of the leaves of various plants, sometimes caused by copper deficiency, chlorosis of spinach due to manganese deficiency, and pineapple chlorosis caused by iron deficiency. Plants also become chlorotic when there is an inadequate supply of nitrogen. In some cases these diseases are overcome by spraying the plants with salts of the deficient element and in others by applying such salts to the soil. An excess of salts of boron, zinc, copper, or other elements may also be toxic to plants. An excess of soluble nitrogen salts has frequently been reported to render plants more susceptible to infectious diseases, such as fire blight, a bacterial disease of apples and pears. It is likely that there is a relationship between the relative amounts of various inorganic substances absorbed by plants and their susceptibility to invasion by parasites.

7. *Sterilization of soil or other media* by means of heat or chemicals. Partial sterilization of soils or soil pasteurization by means of electric heat or steam is a common method of disease control in greenhouses and seedbeds. It is possible to select a temperature that will inactivate many destructive organisms without impairing most of the beneficial ones or otherwise adversely affecting the soil as a growth medium. Formaldehyde and other chemicals are also sometimes used for soil sterilization.

8. *Seed treatments* by means of heat or chemicals. The object of seed treatments is to kill pathogenic organisms found in or on the seeds themselves and to prevent their invasion from the soil when the seeds are planted. The treatment may involve the use of heat. Thus a hot-water treatment is used to overcome loose smut of wheat, which is perpetuated by an internal infection of the grain. More commonly the seeds are treated with chemicals, either in solution or as dusts. Formaldehyde or compounds of mercury, copper, or sulfur may be used. Many seedsmen thus treat their seeds before putting them on the market. Organic mercury compounds are widely used for this purpose. Some seeds, however, cannot be treated in this manner. Care must be exercised to avoid seed injury by the chemicals.

9. *Spraying or dusting plants with fungicides* (Fig. 17-8) chiefly to prevent the invasion of pathogens but also to check the spread of those already present. The active ingredients of sprays and dusts are often copper or sulfur. Bordeaux mixture, which consists of lime and copper sulfate, and lime sulfur, which is a mixture of lime and sulfur, were formerly used extensively for this purpose. Many organic spray materials are now being used. Usually an insecticide is also added to the spray material. Care must be exercised to avoid spray injury, and the application of the spray or dust must be timed carefully to be effective. In the case of spray materials that may be toxic to humans, the government regulates the tolerance of spray residues permitted on food crops.

The manufacture of fungicides has become a thriving business enterprise. While their use has increased the cost of crop production, it has likewise increased the value and quality of the crops produced. In fact, without the use of proper fungicides many crops could not be produced at all.

10. *Eradication.* There are some rusts of plants

Fig. 17-8 Spraying dormant apple trees with a Speed Sprayer to control diseases. (Photograph by R. S. Beese, courtesy of Dr. H. C. Fink.)

that require two separate kinds of hosts to complete their life cycles. Thus currants and gooseberries are the alternate hosts of the white-pine blister rust; red cedar is the alternate host of the apple rust; and the common barberry is the alternate host of the black stem rust of wheat. It has been possible in some regions to eradicate these diseases by destroying the less important host, thereby preventing the fungus from completing its life cycle. An extensive program is now being carried out in various parts of the United States to destroy the common barberry in an attempt to control the black stem rust of wheat. A considerable downward trend in average annual losses of wheat has been achieved by it.

Other methods besides the ones enumerated have been used in special cases, as, for example, the injection of chemicals into trees to overcome mineral deficiencies or to combat parasites. Antibiotics are used in the form of sprays against such bacterial diseases as the fire blight of pears, apples, and quinces. The use of vaccines, antitoxins, and serums, which are very effective in certain diseases of animals and man, has not met with much success with plants. This is partly because the plant pathologist must deal usually with great numbers of plants rather than with individuals and partly because plants do not have a circulatory system comparable with that of the higher animals and man. As to control measures in general, preventive measures are usually much more effective with plant diseases than are attempts to combat a disease already present.

Classification The Eumycophyta, or true fungi, were formerly divided into the following four classes:

1. **Phycomycetes,** or algalike fungi, in which spores are usually produced in sporangia, and hyphae are either lacking or, when present, are destitute of cross walls. All the following groups have septate hyphae.
2. **Ascomycetes,** or sac fungi, in which the spores are produced in asci.
3. **Basidiomycetes,** or basidium fungi, in which the spores are produced, usually in fours, on a specialized structure known as a basidium.
4. **Fungi Imperfecti** (Deuteromycetes), or imperfect fungi, a grouping of the many species the life histories of which are not entirely known but which do not fall readily into any of the foregoing classes.

At the present time the Phycomycetes are divided into six separate classes and with the Myxomycetes constitute the **Lower Fungi.** The Ascomycetes, Basidiomycetes, and Deuteromycetes are then considered the **Higher Fungi.** We here consider the lower fungi as a group with details on a few representative orders.

THE LOWER FUNGI

General characteristics The vegetative plant body is either a simple protoplasmic mass or an extensive mycelium, but in neither case are internal cell walls usually formed. The plant body is therefore coenocytic in structure, although, in an old mycelium and during the formation of reproductive organs, cross walls may be laid down. Zoospores are present in the aquatic members of the group and usually nonmotile wind-disseminated spores in the terrestrial and epiphytic species. Some species are isogamous, others heterogamous, and in some sexual reproduction does not occur. Four orders are considered here, namely, the *Chytridiales,* the *Saprolegniales,* the *Mucorales,* and the *Peronosporales.*

Chytrids (Chytridiales) Most of the members of this order are aquatic single-celled parasites, or saprophytes, without a true mycelium (Fig. 17-9*A*). Some are parasitic on vascular land plants, others on algae, microscopic aquatic animals, and aquatic fungi. They may parasitize the host externally or internally. The cell walls consist of fungus chitin and sometimes contain cellulose.

Asexual reproduction (Fig. 17-9*B*) is by means of uniflagellate zoospores produced in masses

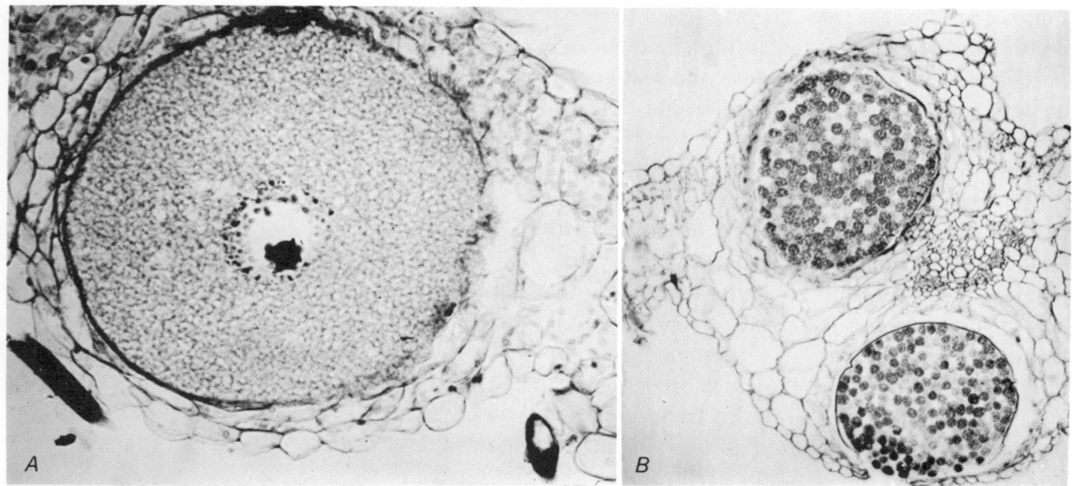

Fig. 17-9 Synchytrium decipiens, *a parasitic chytrid on* Amphicarpa monoica. *A, a uninucleate, single-celled thallus inside a swollen host cell; B, two sori of sporangia, each of which contains many zoospores which are freed by rupture of the host tissue.*

within the thallus. In some genera the entire protoplast divides into spores, or gametes, and in others the fertile portion is distinct from the vegetative portion. The zoospores often have an amoeboid movement when they swarm out of the parent cell. Each one is capable of developing into a new plant.

Sexual reproduction is by isogamous union of zoogametes formed usually at the end of the growing season in a manner similar to that of the formation of zoospores. In appearance the zoogametes resemble zoospores. They fuse in pairs. The resulting biflagellate zygotes swim about for a time, then come to rest and penetrate the host-cell walls. After entering the host cell, the zygote enlarges and becomes a thick-walled, resting cell. In the following spring, by repeated nuclear division, a multinucleate protoplast is formed which ultimately breaks up into zoospores, each of which may infect the host.

Some of the chytrids cause serious diseases of crop plants. Among these are the crown wart of alfalfa, the brown spot of corn, and the black wart of potato (Fig. 17-10).

Water molds (Saprolegniales) This order is composed almost entirely of aquatic fungi living as saprophytes on decaying organic matter or more rarely as parasites, particularly of fish and amphibians. They sometimes cause serious losses of fish in hatcheries and in aquaria. Some are found in soil. They usually have a well-developed, white, moldlike mycelium which grows both on the surface and within the substratum. The cell walls are reported to consist of cellulose. The multinucleate colorless cytoplasm is highly vacuolated and often very granular.

Asexual reproduction is by biflagellate zoospores produced in club-shaped, multinucleate zoosporangia formed at the tips of hyphae or sometimes in a globose form at intervals along a filament, from which they are cut off by a cross wall (Fig. 17-11*A–C*). The protoplasm of the sporangium forms a parietal layer surrounding a large vacuole. Irregular furrows appear extending outward from the vacuole, ultimately dividing the protoplast into numerous uninucleate zoospores (Fig. 17-11*C*). In other cases the undivided protoplast is discharged from the sporangium as a naked mass that becomes invested with a membrane. The protoplasm then divides to form the zoospores. The zoospores are terminally biflagellate and pear shaped. After they have escaped (by swarming) from the sporangium, they swim about for a time and then go into a resting stage, from which, after several hours, they emerge in a bean-shaped, laterally biflagellate form. Ultimately the zoospores germinate by each producing a germ tube that develops into a mycelium.

Fig. 17-10 Potato wart caused by Synchytrium endobioticum.

In several genera the protoplasmic content of the sporangium becomes separated into spore-like bodies without flagella. These bodies germinate in place, each producing a coenocytic germ tube that grows out through the sporangial wall (Fig. 17-11*D*). Some species produce gemmae in which a filament becomes divided into short cells that swell into ellipsoid or globose form (Fig. 17-11*E, F*). After a short rest period, each of these may germinate and develop into a mycelium.

Sexual reproduction, in those species in which it is known to exist, is by heterogamy, with well-developed oögonia and antheridia. The unicellular oögonia (Fig. 17-11*G, H*) originate as globular swellings on terminal hyphae cut off by cross walls. Many of the nuclei of the oögonium degenerate, but one or several remain and become the nuclei of eggs. The antheridium (Fig. 17-11*G, H*) is developed from a hypha below the oögonium or from another hypha. It contains several nuclei and grows up until it comes into contact with the oögonium, whereupon it puts out one or more lateral fertilization tubes which pierce the wall of the oögonium. The tips of these tubes come into contact with the eggs. No sperms are developed, but male nuclei fertilize the eggs. The fertilized eggs develop into thick-walled oöspores which, after several months of dormancy, germinate by producing germ tubes which may soon develop zoosporangia and zoospores. In some cases the oöspores are produced parthenogenetically, i.e., without fertilization.

Black molds (Mucorales) The fungi of this group are mostly saprophytic and very abundant on decaying organic matter. They often cause rotting of fruits and vegetables in storage. Some are parasitic on other fungi and on higher plants. The term black molds is derived from the black color of the asexual spores produced in profu-

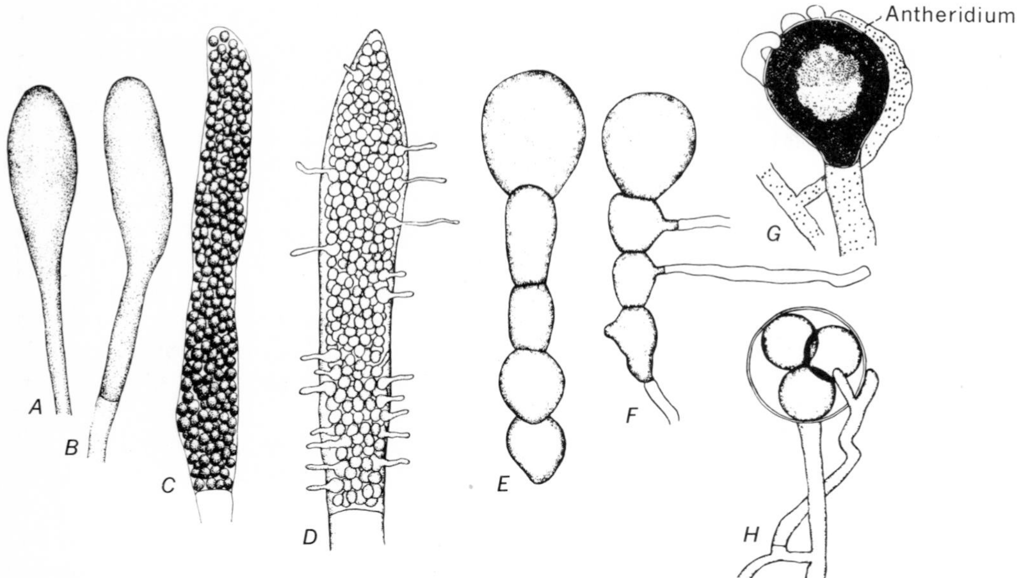

Fig. 17-11 A–C, Saprolegnia ferax; *A, B, young zoosporangia forming at the tips of the hyphae and becoming dense with cytoplasm; the sporangium is then cut off from the rest of the hypha by a cross wall, as in B; C, mature sporangium packed with zoospores ready to emerge; D–F, asexual reproduction in Saprolegniales; D, tip of filament containing zoosporelike bodies, some of which have germinated, the germ tubes penetrating the hyphal wall; E, F, formation of gemmae and their germination into hyphae, in a species of* Pythiopsis; *G, H, sexual reproduction in Saprolegniales; G, a pear-shaped oögonium of a species of* Saprolegnia *to which are attached four antheridia; H, a mature oögonium of a species of* Achlya, *containing three eggs, with a fertilization tube penetrating the oögonial cavity. (A–C drawn by Christian Hildebrandt;D–H drawn by Edna S. Fox.)*

sion in the mature stage. The spores of these fungi are universally present in the air. In order to obtain a luxuriant growth of mycelium, it is only necessary to expose a piece of bread in a damp atmosphere for a few days (Fig. 17-12*A*). From the preceding order, the Mucorales differ chiefly in the more terrestrial habitat, the absence of zoospores in the life history, and the presence of a special type of sexual reproduction known as gametangial copulation.

The mycelium is colorless or slightly brownish and covers the substratum with a thick growth of tangled hyphae. A portion of it is embedded in the substratum. At intervals on the superficial hyphae, rootlike projections known as rhizoids, which penetrate the substratum, are developed in a few species. The mycelium consists of branched, nonseptate filaments containing abundant, often streaming cytoplasm and numerous minute nuclei. Cross walls are frequently found in the older hyphae, and reproductive parts are usually separated by septa from the vegetative portions.

Asexual reproduction (Fig. 17-13*E–I*) is by nonmotile, wind-disseminated spores produced in sporangia formed at the apex of erect hyphae, termed **sporangiophores.** The sporangia are visible as small round black bodies on the mycelium (Fig. 17-12*B*). The sporangia develop as globose, terminal swellings filled with multinucleate protoplasm which eventually becomes transformed into uninucleate spores. In *Rhizopus stolonifer,* the common bread mold, the spore-bearing portion of the sporangium is separated by a wall from a central, dome-shaped portion called the **columella.** The spores, on reaching a suitable substratum, germinate and develop into new plants.

Sexual reproduction (Fig. 17-13*A–D*) results when two gametangia fuse. In *Rhizopus stolonifer,* which may serve to illustrate the group, two hyphae lying in proximity each put out a short, multinucleate, club-shaped branch. When these branches meet, end to end, the tip of each is cut off by a cross wall from the remainder of the branch and becomes a multinucleate gametangium. The remainder of the branch is called a **suspensor.** The double wall between the two gametangia breaks down, and the protoplasm of the two gametangia mingles to form a multinucleate zygospore. The zygospore develops a dark thick, often spiny, wall and, after resting for a considerable period, germinates by producing a single sporangiophore and sporangium, called a germ sporangium, with nonmotile spores.

Within the Mucorales as well as in other groups of fungi it has been found that there are two distinct types of species, namely, **homothallic** types and **heterothallic** types. In the former, gametangial copulation will take place between hyphae originating from a single spore;

Fig. 17-12 A common black mold, Rhizopus stolonifer. *A, half a loaf of bread kept in a moist container for several days; the cottony growth is the mycelium; the minute dots are sporangia; B, photograph of part of A, enlarged to show black sporangia.*

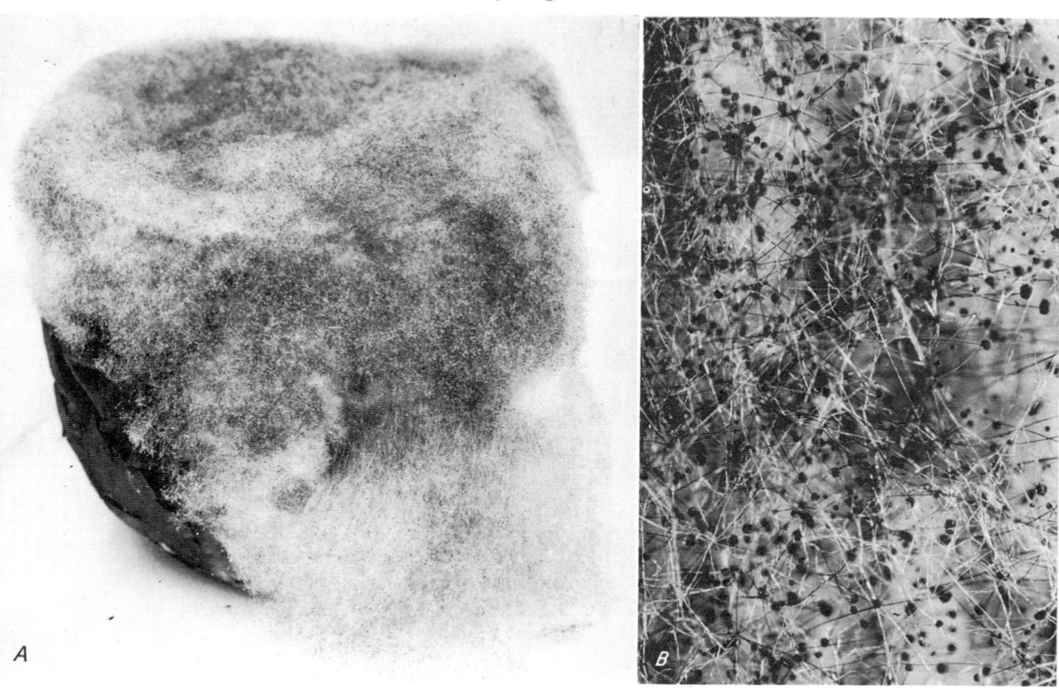

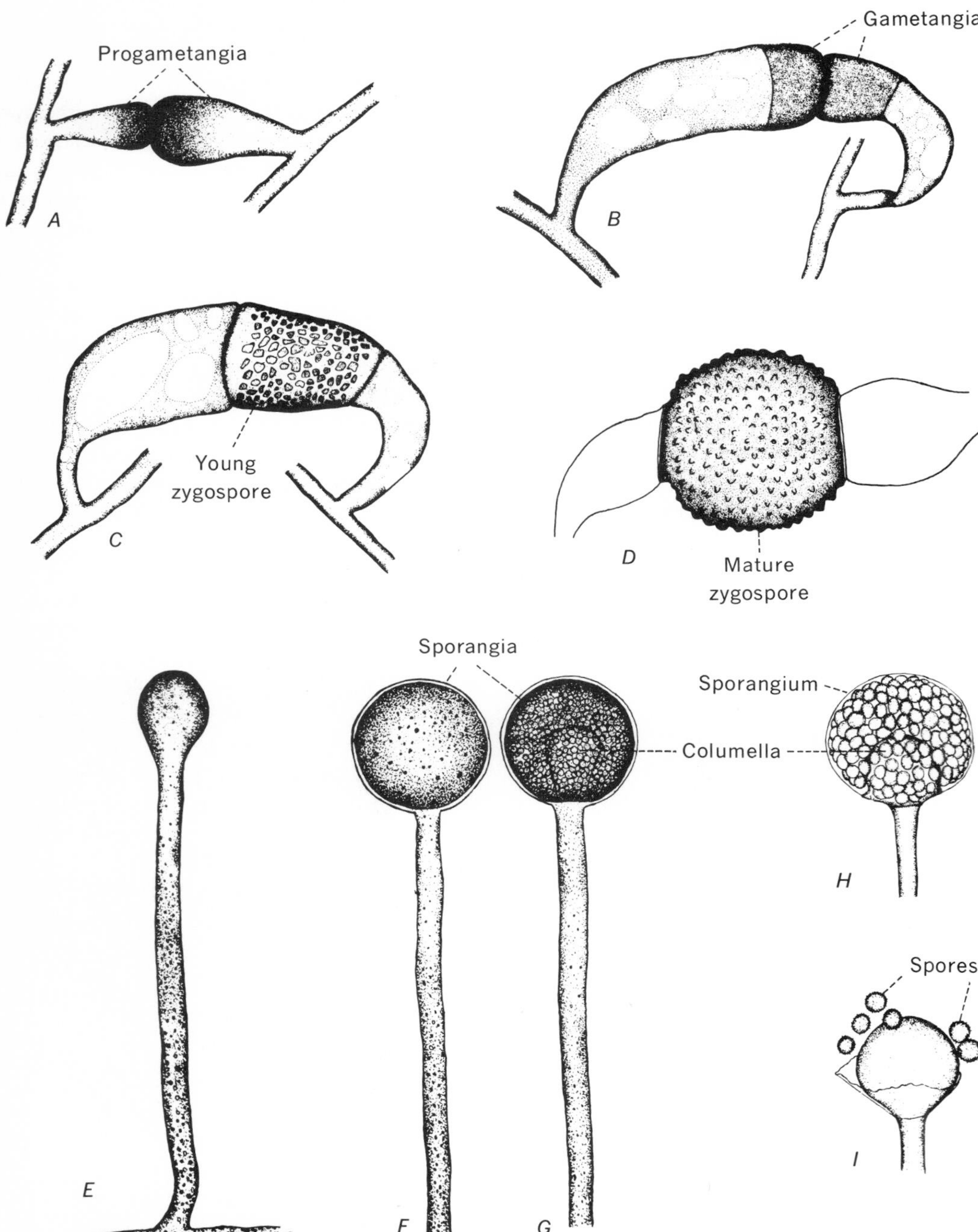
Progametangia
A
Gametangia
B
Young
zygospore
C
Mature
zygospore
D
Sporangia
Sporangium
Columella
E
F
G
H
Spores
I

i.e., two hyphae from the same mycelium may fuse. In heterothallic species, fusion is dependent upon the presence of mycelia which have arisen from two inherently different spores. They have been designated as plus strains and minus strains. In this case zygospores will be formed only when a plus mycelium and a minus mycelium come into contact.

Homothallic species always produce, in the germ sporangium, spores that are alike, i.e., spores that will always produce homothallic mycelia. In some heterothallic species, however, the germ sporangium will contain either all plus spores or all minus spores. In other heterothallic species the germ sporangium contains both plus spores and minus spores. This indicates that during the development of the zygospores and their subsequent germination there has first been a fusion of nuclei (fertilization), probably in pairs, and a further reduction of chromosomes (meiosis) to form the spores.

Downy mildews (Peronosporales) and related forms are of considerable economic importance because most of them are parasitic on vascular plants, including important crop plants. Thus the downy mildew of grapes is caused by one of these fungi. It was in connection with this disease that the fungicidal properties of bordeaux mixture were discovered. The late blight of potatoes and tomatoes is also caused by a member of this group. The ravages of this disease caused the Irish famine of 1845–1846 and brought many immigrants from Ireland to America, where they have had a pronounced effect on American history. The damping off of seedlings in greenhouses and nurseries is often caused by fungi in this group.

The mycelium is located within the tissues of the host, usually completely filling the intercellular spaces or occasionally, by means of haustoria, penetrating to and developing profusely within the cell cavity. There is little or no mycelium developed external to the substratum, but in some cases, after the death of the host, the mycelium becomes somewhat external on the decaying portions. A few species are able to live as saprophytes in moist soil.

Asexual reproduction is usually by wind-disseminated sporangia. In one genus, *Albugo*, a common parasite of members of the mustard family, these bodies are cut off in chains from the tips of special hyphae, the sporangiophores, that are grouped in white blisterlike patches under the epidermis of the host (Fig. 17-14*A*). On this account the fungus is sometimes called a white rust. On reaching maturity, the sporangia burst through the epidermis and are disseminated by the wind. In most other genera these bodies are produced on erect, aerial, simple, or branched hyphal stalks that arise from the internal mycelium, frequently emerging to the outside by way of the stomata of the host plant (Fig. 17-14*B–D*). In a few species the sporangia germinate while still attached to the parent hyphae. The sporan-

Fig. 17-13 A–D, stages in the sexual reproduction of a black mold; A, two lateral hyphal branches in contact, the protoplasm accumulating in the ends of each branch; B, tip of each branch cut off by a cross wall, two gametangia in contact; C, coalescence of the two gametangia has occurred, and pads of thickening are beginning to appear on the wall of the young zygospore; D, mature zygospore with black-warted wall; E–I, sporangium development in a black mold; E, hyphal tip enlarging to form a sporangium; F, sporangium before spore formation; G, sporangium with spores cut out and columella visible; H, sporangium with mature spores and a distinct columella; I, sporangium with wall broken, leaving a ragged fringe at the base of the globose columella. (Drawn by Ernest Geisweite.)

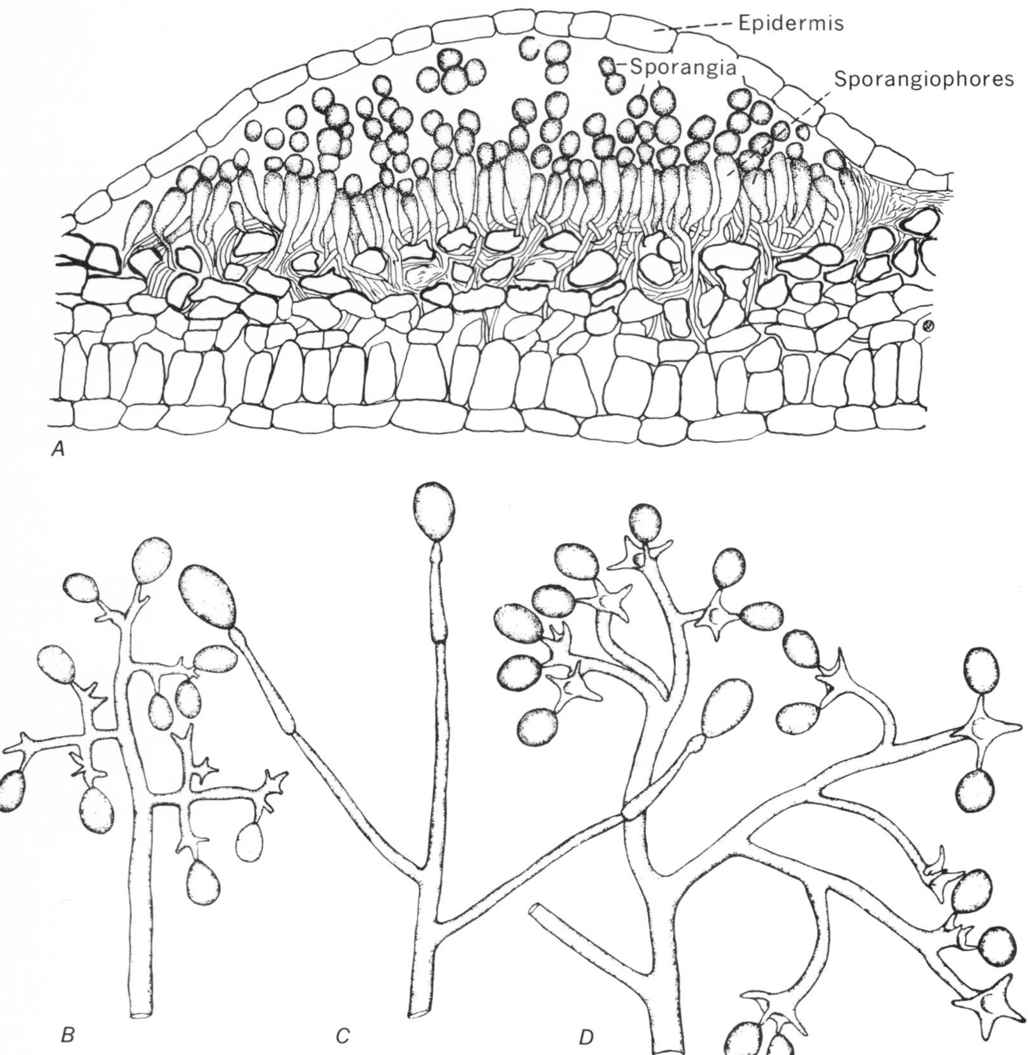

Fig. 17-14 Asexual reproduction in Peronosporales. A, section through a leaf of Capsella, *bearing a sporangial pustule of* Albugo candida; *the sporangia are formed in chains from a basal palisadelike layer of club-shaped sporangiophores; the pustule is covered by the leaf epidermis; B, sporangiophore and sporangia of the fungus* (Plasmopara viticola), *causing downy mildew of grapes; C, sporangiophores and sporangia of the fungus* (Phytophthora infestans), *causing the late blight of potatoes; D, sporangiophores and sporangia of the fungus* (Bremia lactucae), *causing the downy mildew of lettuce; the sporangiophores in B–D are exserted through the stomata on the lower leaf surface. (A drawn by Edna S. Fox; B–D by Ernest Geisweite.)*

gia are deciduous and usually germinate by producing a number of bean-shaped zoospores, each with two lateral flagella. In a few genera the sporangia germinate by forming germ tubes. When the sporangia come to rest on leaves of susceptible plants, the germ tubes or the zoospores, under suitable conditions, infect the host tissue. It has been reported that the zoospores in some genera, under unfavorable conditions, encyst and then swarm a second time, as in the Saprolegniales, but with this difference, that the second zoospore is like the first one but perhaps smaller. These fungi are called downy mildews because the abundant production of the zoosporangia on the surface of the host presents a white powdery or cottony appearance.

Sexual reproduction is heterogamous; i.e., oögonia and antheridia are formed. The oögonia arise as globose swellings on the hyphae in the intercellular spaces of the host plants. The antheridia usually arise near the bases of the oögonia as short, lateral, club-shaped branches which curve up and over the oögonia. Both of these structures are at first multinucleate (Fig. 17-15). In nearly all species, all but one of the nuclei in the oögonium disintegrate, the remaining one becoming the nucleus of the single egg that is always matured. The tip of the antheridium comes into contact with the tip of the oögonium and a passageway is dissolved between the two. A fertilization tube enters the oögonium, and a single male nucleus passes into the oögonium and fuses with the egg nucleus. The resulting zygote (oöspore) becomes a heavy-walled resting body that persists within the tissues of the host plant over winter. In the spring the decay and disintegration of the host tissue set the oöspores free, and under favorable conditions they germinate, in some genera forming a mycelium, in others giving rise to zoospores.

ASCOMYCETES

General characteristics The most significant characteristics of the fungi in this class are the formation of septate mycelia, lack of motility, the production of **asci** (singular, ascus), or sacs, each with a definite number of ascospores, usually eight, and the production in most species

Fig. 17-15 Sexual reproduction in Albugo. *A, section through an oögonium in which the single central egg nucleus has just been fertilized and a wall is forming about the oöspore, separating that body from a peripheral zone of cytoplasm containing degenerating nuclei; B, section through an oögonium and an antheridium, each with numerous nuclei.*

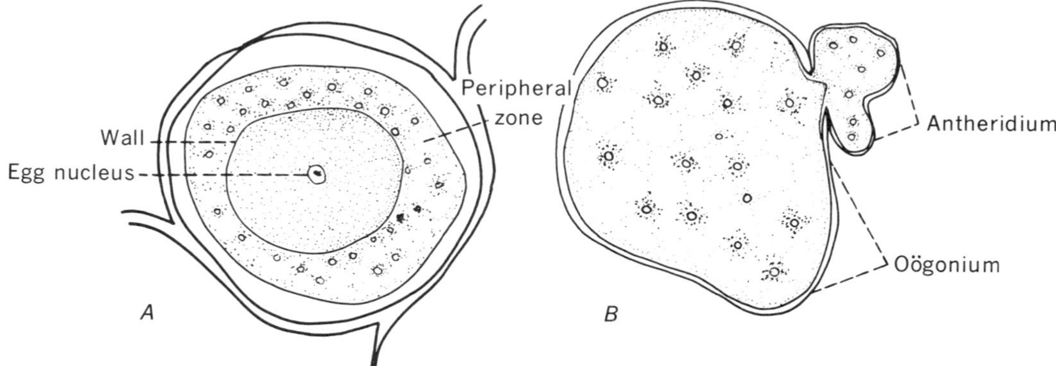

of a multicellular fruiting body. The ascus is the product of sexual reproduction, and the ascospores are meiospores. In addition, a variety of spores are produced in asexual reproduction.

The mycelium is well branched and divided by septa into uninucleate or multinucleate cells. The septa are usually perforated by a small central opening through which cytoplasm may flow from cell to cell. The mycelium may be external or internal and with or without special hyphal branches called haustoria, which penetrate the living cells. In the yeasts, which are unicellular, no mycelium is produced.

Sexual reproduction varies but always results in the formation of asci, which are actually the zygotes. The ascus stage has been called the **perfect stage.** Typically the asci are cylindrical in shape and the ascospores are produced in one or two rows in them (Fig. 17-16*E, F*). Less frequently they are more or less globose and the spores lack a definite arrangement (Fig. 17-16*B*). Often the asci are arranged in a layer called the **hymenium** (Fig. 17-17*B*), but the asci may stand singly (Fig. 17-16*A*) or show an irregular arrangement. In some of the unicellular yeasts the entire plant becomes an ascus. Sexual reproduction consists of the following processes:

Fig. 17-16 Asci, paraphyses, ascospores, and spore arrangements in Ascomycetes. A–C, broader types of asci; A with eight spores, B with five spores, and C with two spores, in Physalospora, Podosphaera, *and* Phyllactinia, *respectively; D, antlered paraphyses of* Propolis faginea; *E, eight-spored cylindrical ascus of* Propolis faginea *with spores in one row; F, cylindrical ascus and a cylindrical septate paraphysis of* Leotia stipitata *with spores two- to four-celled; G, ascus of* Ophiobolus fulgidus *with eight elongate several-celled spores. (Drawn by Edna S. Fox.)*

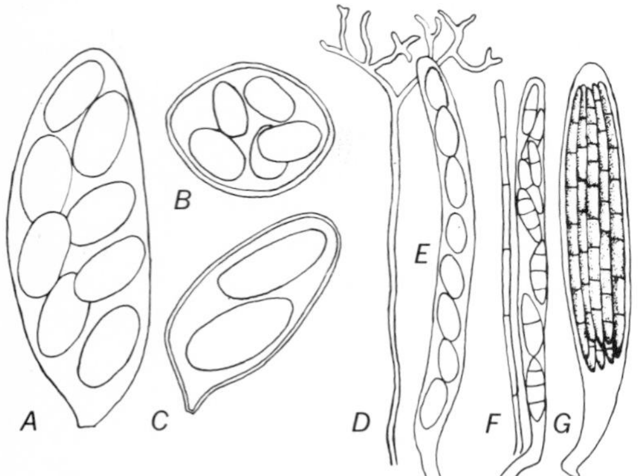

1. The copulation of two similar gametangia, with the ascus developing out of the fusion cell. This is true of many of the yeasts in which the yeast cells themselves act as gametangia and copulate.

2. The heterogamous fusion of an antheridium with an ascogonium, as in *Pyronema omphalodes.* The ascogonia arise as clusters of erect hyphae (Fig. 17-18*A*). The individual ascogonium is at first club shaped and multinucleate. A curved, tubelike trichogyne is produced at the tip. The basal part, i.e., the ascogonium proper, soon becomes balloon shaped. The multinucleate antheridia arise from another hypha and, growing upward, come in contact with the trichogyne. A passageway is dissolved at the point of contact, and the contents of the antheridium flow through the trichogyne into the ascogonium. The male nuclei pair with the female nuclei, but do not fuse, thus originating a **dicaryon.** Ascogenous hyphae grow out from the wall of the ascogonium, and the paired nuclei migrate into them. The ascogenous hyphae (Fig. 17-18*B*) elongate, and the dicaryons divide repeatedly. Eventually the binucleate tip of each ascogenous hypha forms an ascus, in which the nuclei fuse to form a diploid nucleus. The fusion nucleus immediately undergoes meiosis and mitosis to supply the eight haploid nuclei for each of the eight ascospores (Fig. 17-19). Usually a certain amount of cytoplasm is left over after the ascospores are formed; this is known as epiplasm.

3. Spermatization of an ascogonium as in *Neurospora sitophila.* No antheridium is formed; instead, the male nuclei reach the ascogonium by means of spermatia (microconidia) or conidia

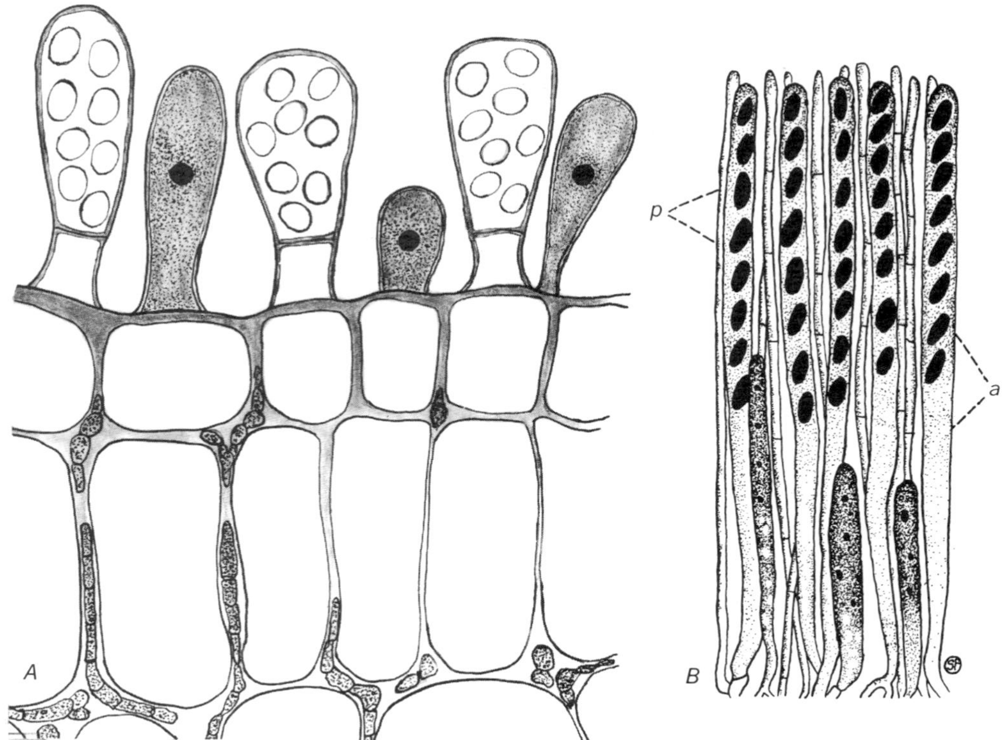

Fig. 17-17 A, three mature asci of Taphrina deformans, *the peach-leaf-curl fungus, protruding singly from the epidermal cells of the leaf, each ascus with eight small spherical spores; the three darker structures are immature asci; B, asci and paraphyses of* Peziza *standing closely united to form a hymenial surface, each ascus with eight spores; a, ascus; p, paraphysis. (B drawn by Edna S. Fox.)*

(macroconidia). Spermatia are small uninucleate nonmotile cells produced by spermatiophores. The spermatia may be wind blown, rain splashed, or insect spread to the trichogyne. The walls break down after fusion, and the spermatial nucleus migrates to the ascogonium to pair with the ascogonial nucleus. Conidia may also function as spermatia. Ascogenous hyphae and ascus development are as described previously.

4. No gametangia being produced, as in some species, such as *Helvella elastica.* The vegetative hyphae fuse, and the nuclei migrate throughout. This process is called **somatogamy.** Ascogenous hyphae arise from certain cells after hyphal fusion.

COMPATIBILITY In many Ascomycetes any two fusing elements are compatible and sexual reproduction is completed from hyphae and branches arising from a single spore. However, there are species which are composed of two compatibility types which are self sterile but cross fertile. In these species a plus male (antheridium or spermatium) can unite only with a minus female (ascogonium) and a minus male only with a plus female. This has been called heterothallism.

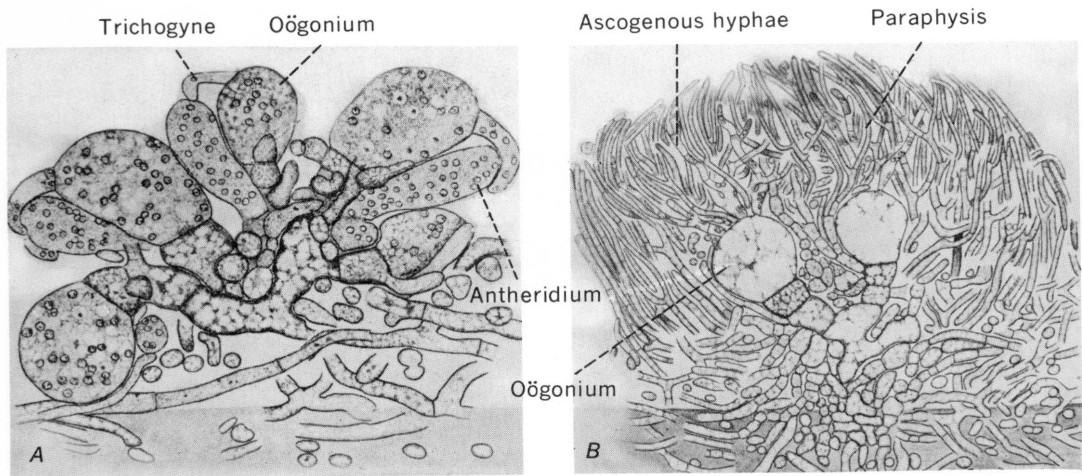

Fig. 17-18 Sexual reproduction in Pyronema. *A, development of antheridia and ascogonia; B, development of ascogenous hyphae from the ascogonia. (After Claussen.)*

Fruiting structures, generally known as ascocarps, are produced by many Ascomycetes. The simpler Ascomycetes possess no definite ascocarps, and the asci are produced in a scattered unprotected manner, as in the yeasts and *Taphrina* (Fig. 17-17*A*). In the more advanced Ascomycetes the ascocarp consists of the hymenium (asci and paraphyses), together with sterile protective hyphae. In the **apothecium** the hymenium is exposed as the inner lining of a cup- or saucer-shaped ascocarp (Fig. 17-31). If the ascocarp is rounded or flask shaped, with or without an opening, it is called a **perithecium** (Figs. 17-20, 17-21). A perithecium without an opening is called a **cleistothecium.** There are some ascocarps which are not readily classified as either apothecia or perithecia (Fig. 17-33).

Ascospores are usually discharged from the asci by means of an internal force that projects them into the air, where they are caught up by

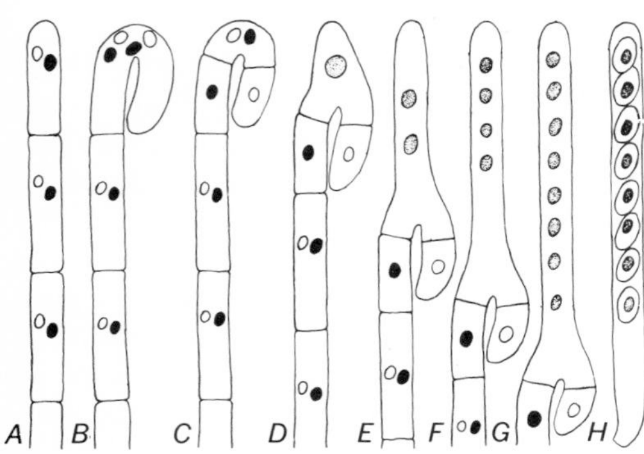

Fig. 17-19 Diagrammatic representation of the usual mode of ascus formation. A, an ascogenous hypha with binucleate cells, the nuclei shown unlike to indicate their sex difference; the tip of such an ascogenous hypha turns over to form the apical hook illustrated in B; the two nuclei of this ascus hook divide, forming four nuclei, as in B; transverse walls are laid down in this hook, forming a uninucleate terminal cell, a binucleate penultimate cell, and a uninucleate antepenultimate cell, as in C; the two nuclei in the dome cell fuse, and that cell begins to elongate upward, as in D; as it elongates, a series of three nuclear divisions results in the appearance of eight nuclei, as in E–G; walls are laid down cutting out the eight spores, as in H. (Drawn by Edna S. Fox.)

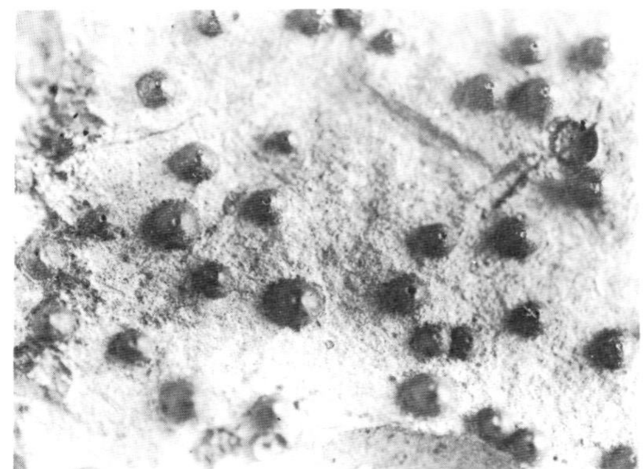

Fig. 17-20 Perithecia of the peach-pit fungus (Caryospora putaminum), *enlarged to show the apical ostioles.*

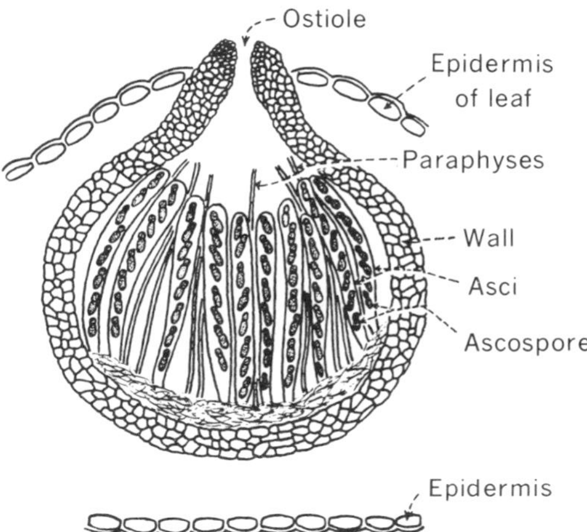

Fig. 17-21 A perithecium of the apple-scab fungus, Venturia inaequalis, *as seen in the cross-sectional view of an apple leaf. (Drawn by Otto E. Schultz.)*

Fig. 17-22 Different types of conidial or imperfect stages in fungi. A, a pycnidium with dense hyphal wall lined with conidiophores and producing globose conidia that escape through the ostiole; this pycnidium is buried in the leaf tissue of its host; B, septate conidia produced on erect clustered conidiophores that protrude from the leaf tissue; C, an acervulus type of fruiting body with conidia on conidiophores arranged in a definite layer below the leaf cuticle; D, E, conidia produced at irregular intervals on branches of the mycelium. (Drawn by Edna S. Fox.)

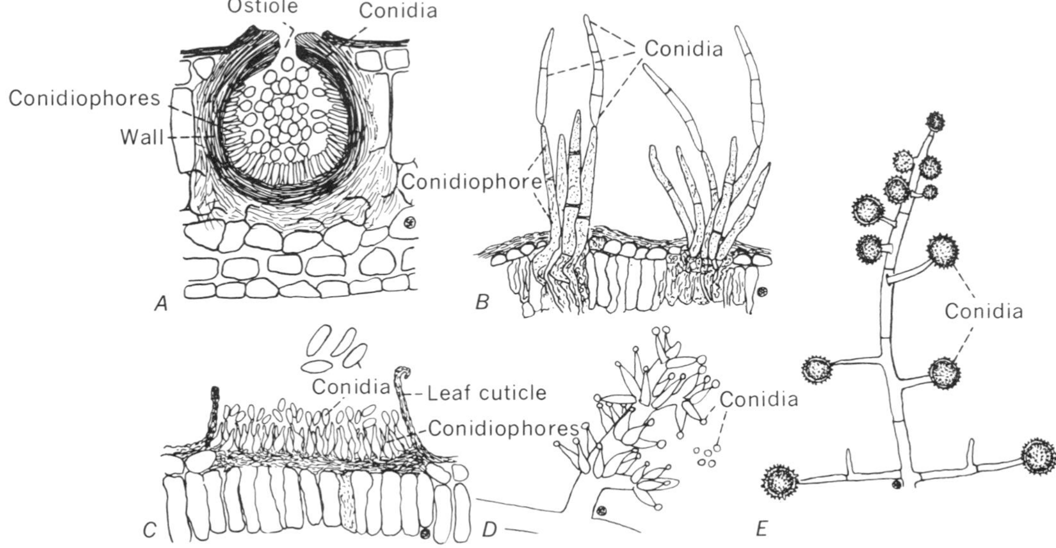

air currents and disseminated. In some apothecial forms a slight jar or the blowing of the breath will cause a distinctly visible cloud of ascospores to be thrown out. In some Ascomycetes the asci dissolve and the ascospores are pushed out of the ascocarp like toothpaste out of a tube. These sticky spores are readily spread by insects or rain splash.

Asexual reproduction is accomplished by budding (yeasts) (Fig. 17-23*B–D*) or by conidial formation. This stage is called the **imperfect stage.** Conidia are produced at the tips of special simple or branched hyphae known as conidiophores. Conidiophores may be scattered irregularly (Fig. 17-22*B, D, E*), aggregated into exposed flat or saucer-shaped acervuli (Fig. 17-22*C*), or grouped into rounded or flask-shaped pycnidia (Fig. 17-22*A*).

The spores produced in this stage serve to disseminate the fungus rapidly throughout the growing season. They are usually short lived and germinate by producing a germ tube which elongates and branches to form a mycelium similar to that produced by the germination of ascospores. The structures produced in the perfect stage, especially the perithecia, often serve to tide the fungus over the winter. Other Ascomycetes overwinter as hyphae in the host plant or as resistant hyphal structures, such as **sclerotia** (Fig. 17-27*A*).

No attempt is made to present a complete classification of this class of approximately 40,000 species. Only those groups of more or less biological interest are described to indicate the different kinds.

Yeasts (Saccharomycetales) are unicellular fungi that reproduce asexually by budding (Fig. 17-23*B–D*). Following fusion of two yeast cells (Fig. 17-23*E*), the fused cell buds, but finally becomes an ascus containing one to eight ascospores (Fig. 17-23*F*). Meiosis occurs as the ascospores are formed. Thus the ascospore nucleus is haploid. The ascospore buds and many haploid yeast cells form. Fusion of two such cells is followed by nuclear fusion to form the diploid zygote, which may in turn bud for a while, but meiosis occurs with the formation of the haploid ascospores.

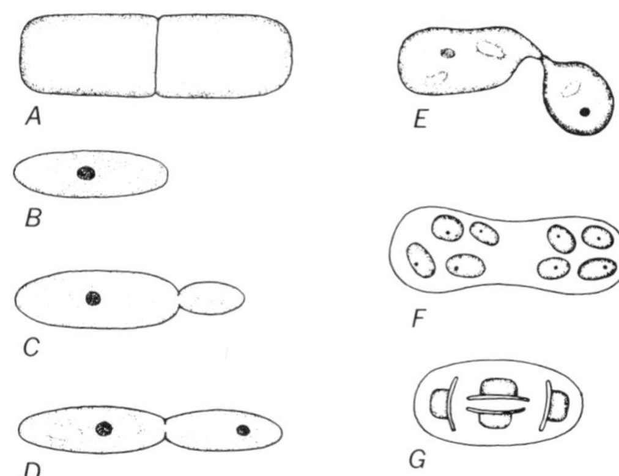

Fig. 17-23 Asexual and sexual reproduction in yeast. A, a single-celled plant of yeast (Saccharomyces *sp.*) *in the process of dividing into two cells in asexual reproduction; B–D, stages in the process of budding (asexual reproduction) of a cell of yeast* (Saccharomyes *sp.*); *E, fusion of two cells of yeast* (Saccharomyces octosporus) *in sexual reproduction; F, ascus containing eight ascospores and resulting from the fusion of the two yeast cells in E; G, ascus containing four hat-shaped ascospores of the yeast* Hansenula anomala. *(Drawn by Otto E. Schultz.)*

Yeasts are of considerable economic importance, being capable of fermenting carbohydrate solutions with the production of ethyl alcohol and carbon dioxide. In this way wines from grapes and berries, fermented cider from apples, and beers from sprouted barley are obtained. It is not always necessary to add yeast to the solutions containing the fruit sugars since desirable wild yeasts are present in nature on the surfaces of the fruits. The carbon dioxide released in fermenting dough raises the bread in making its escape. Yeasts are also important as a commercial source of vitamins.

Taphrina species (Taphrinales) are parasites of the leaves, stems, and fruits of living plants. Perhaps the most important of these is the fungus that causes leaf curl on peach trees. The parasitized leaves become much curled, thickened, and distorted. The asci are produced externally in an unprotected layer on the leaf surface (Fig. 17-17*A*). The ascospores bud like yeast cells and, lodged in bark crevices or bud scales, survive the winter. Suitable fungicides applied two weeks before the buds begin to unfold give effective control.

Blue and green molds (Aspergillales) are a heterogeneous group best known because of the genera *Aspergillus* and *Penicillium,* which are found as molds on citrus fruits, jellies, leather, paper, and ensilage. Several species are responsible for the flavors of some cheeses. Some cause allergy diseases. Some are used for the commercial production of chemicals, including the antibiotic penicillin. The color of these molds is due to the conidia which are produced in chains on special conidiophores (Fig. 17-24*B, C*). Thin-walled perithecia (Fig. 17-24*A*) are formed by

Fig. 17-24 A, cleistothecium of Aspergillus *sp., showing eight asci, each with eight ascospores; B, conidiophore of* Aspergillus *sp., showing chains of conidia arising from a globose vesicle at the terminal end of a conidiophore; C, brushlike conidiophore of* Penicillium *sp., with chains of conidia. (Drawn by Otto E. Schultz.)*

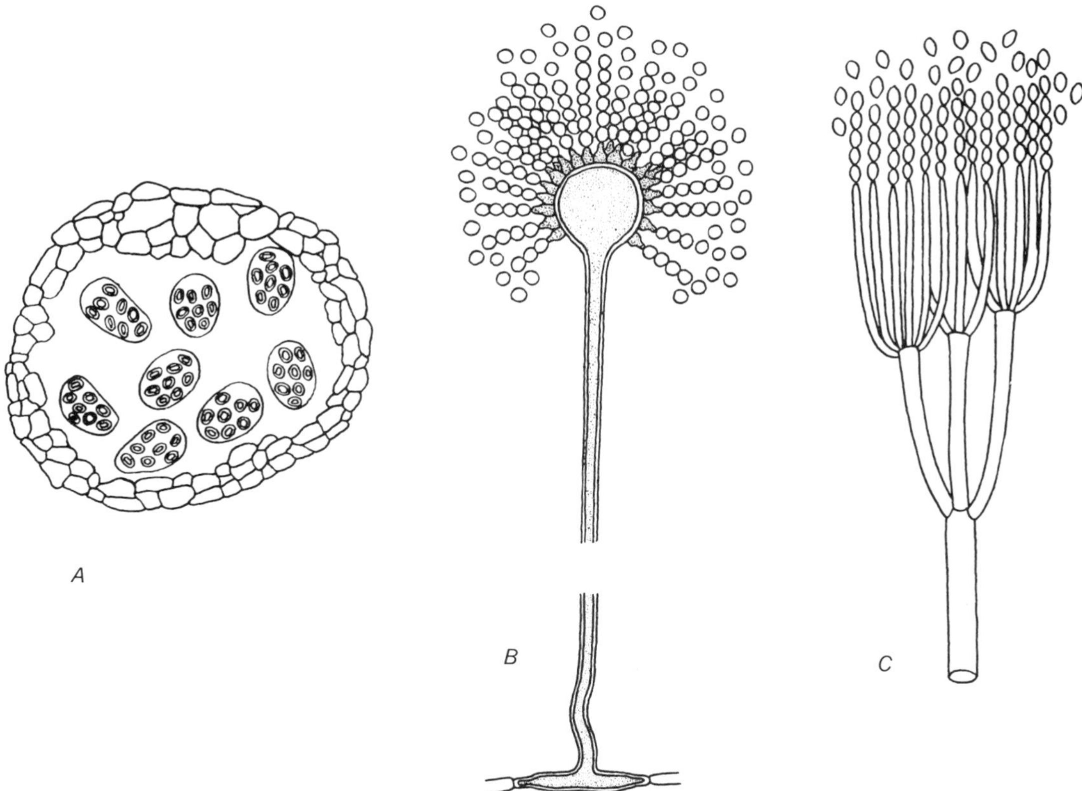

only a few of the many species of *Penicillium* and *Aspergillus*. In most species perithecia have never been seen, and thus many authorities refer these genera to the Fungi Imperfecti.

Powdery mildews (Erysiphales) are unique in developing a mycelium that is largely superficial on the surface of leaves, flowers, or fruits. All the powdery mildews are obligate parasites and have never been cultured on artificial culture media. They obtain their food by forcing special absorbing organs called haustoria into the living cells of the host, often in such a manner that the host cells are not killed. During the growing season the powdery mildews form a gray or white coating on the surfaces of the host they attack (Fig. 17-25*A*). The appearance may at times be powdery, because of the formation of a large number of chains of conidia (Fig. 17-26*E, F*). The conidia spread the fungus from plant to plant. Late in the summer or fall cleistothecia are formed on the same mycelium and appear as small black dots scarcely visible to the unaided eye (Figs. 17-25*B;* 17-26*A, B*). Hyphal appendages of peculiar and characteristic form are present. The enclosed asci (Fig. 17-26*C*) are liberated the following spring when the perithecial wall disintegrates. The ascospores are disseminated and invade the foliage. Dusting infected plants with flowers of sulfur has proved successful in combating these dis-

Fig. 17-25 Powdery mildew of Phlox. *A, leaf of phlox, showing white mildew patches on its upper surface; B, photograph of small area of infected leaf surface, magnified to show the globose black cleistothecia.*

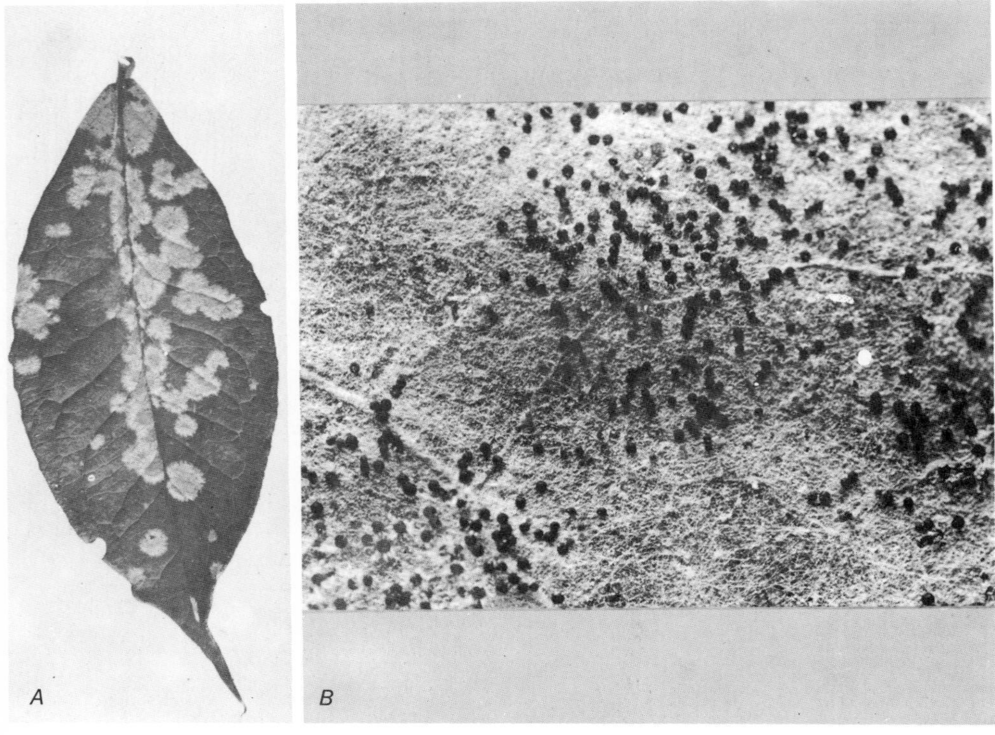

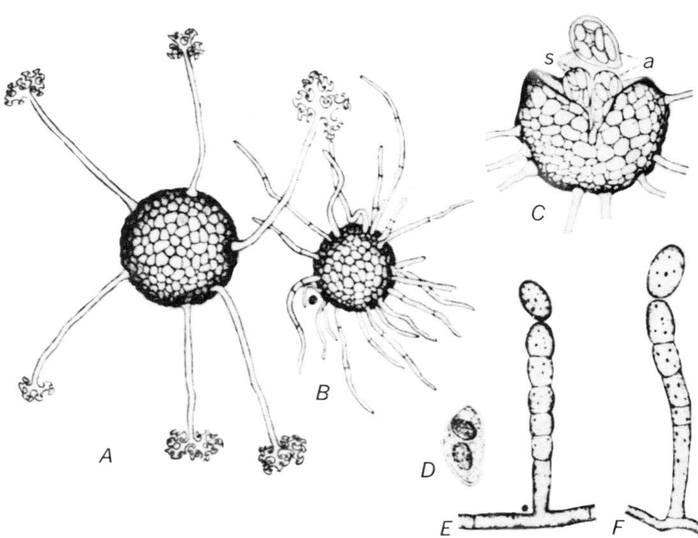

Fig. 17-26 Reproduction in powdery mildews. A, B, perithecia of two different species of mildews; note cellular walls of the perithecia and the elongated appendages with dichotomously divided tips in A; C, perithecium partially burst open, showing the contained asci, a, with their ascospores, s; D, a single ascus with two spores; E, F, conidiophores with chains of conidia. (Drawn by Edna S. Fox.)

eases, some of which, such as the mildew of rose, apple, and wheat, are of considerable economic importance.

Other perithecial Ascomycetes (Hypocreales and Sphaeriales) The characteristic feature of the Hypocreales is the bright color (white, yellow, or red) and the waxy consistency of the perithecia. Many are parasitic on other fungi or on insects, such as the pupae of beetles. A few are of some economic importance.

The species of *Claviceps* are parasitic on grasses, including rye and barley and occasionally wheat. The hard, black, cylindrical sclerotium of this fungus, known as an **ergot,** entirely replaces a grain in the fruiting head (Fig. 17-27*A*). These sclerotia are resting stages, and the next spring they give rise to small stalked knobs in which are embedded numerous perithecia which produce the spores in asci (Fig. 17-27*B*). The ascospores infect the ovaries of grasses, producing a mycelium which gives rise to a type of conidium capable of spreading the fungus to other hosts. Eventually the mycelium entirely destroys the ovary and organizes in its place the sclerotium. The sclerotia are poisonous to livestock, often causing disease and abortion. Improper methods of cleaning grain have led to death of human beings from eating bread made from ergot-infected rye. As recently as 1951 many people in a small village in France were poisoned in this manner, five died, and thirty were driven temporarily insane with hallucinations of fire, demons, and snakes. Extracts of ergot are used in medicine.

The order Sphaeriales is probably the largest of the Ascomycetes, and its species are probably the most common of all the members of the

Fig. 17-27 Claviceps purpurea. *A, sclerotia replacing grains in head of rye; B, photomicrograph of vertical section through the stalked knobs which arise from sclerotia; the circle of elongated cavities toward the periphery of the head consists of perithecia that contain asci.*

class. They are found on many dead sticks in the woods, as well as on a variety of other plant materials, usually appearing as black spots or crusts, more rarely as fingerlike projections from the substratum (Fig. 17-28*A*). Most of them produce both conidia and ascospores. The perithecia are usually black and carbonous.

Many of the species of this order cause serious plant diseases, such as oak wilt, the black rot of grapes, black knot of cherry and plum (Fig. 17-28*B*), a leaf spot of strawberries, twig blight of sycamore, a black rot and canker of apples, and apple scab (Fig. 17-29). The fungus *Endothia parasitica* causes the blight of the chestnut tree (Fig. 17-30). This disease, apparently introduced from China about 1900, has virtually destroyed all the native chestnut trees of the United States. The fungus is parasitic on the cambium and the living cortical cells of the tree.

Cup fungi (Pezizales) are so called because the ascocarp is an apothecium, or cuplike body, often mounted on a distinct stalk (Fig. 17-31). *Peziza* is a common genus with many species occurring as saprophytes on decaying vegetable matter or on the ground. Many have rather large fruiting bodies, often as much as 4 in. in diameter, and more frequently saucer shaped than cup shaped. Some are brilliantly colored. The hymenial layer of asci and paraphyses lines the

Fig. 17-28 *A, dead-man's-fingers, the fruiting structure of* Xylaria polymorpha; *B, black knot of plum caused by* Dibotryon morbosum. *(B courtesy of R. S. Kirby.)*

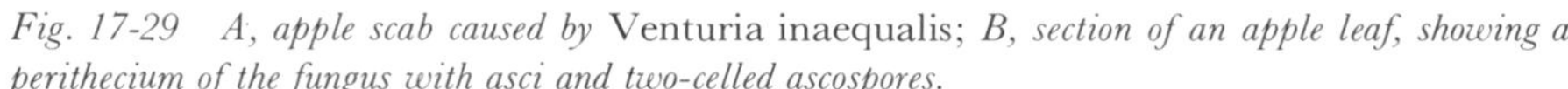

Fig. 17-29 *A, apple scab caused by* Venturia inaequalis; *B, section of an apple leaf, showing a perithecium of the fungus with asci and two-celled ascospores.*

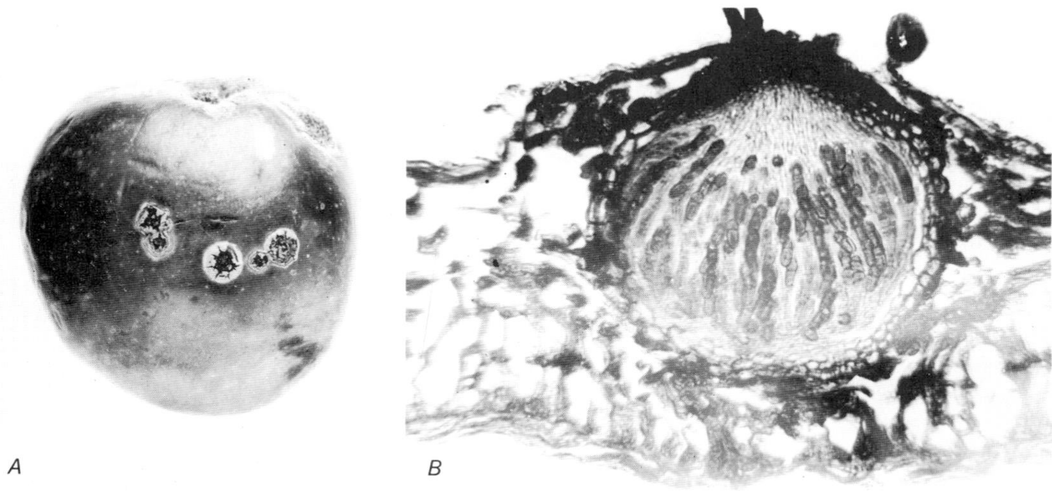

Fig. 17-30 Chestnut blight caused by Endothia parasitica.

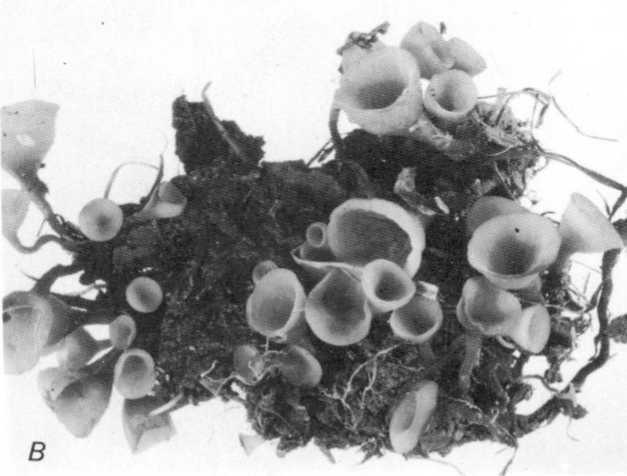

Fig. 17-32 Brown rot of peach caused by Monilinia fructicola; *A, rotting peach fruit; B, apothecia developing on a "mummied" fruit.*

cup on the inside. The asci are cylindrical or club shaped and nearly always contain eight spores.

A few species, such as *Monilinia fructicola*, are of economic importance. This fungus causes the brown rot of cherries, plums, and peaches (Fig. 17-32). Particularly in wet seasons these fruits, when attacked, become brown and soft and a moldlike growth develops over them. Sometimes as much as 50 percent of the crop is destroyed. The rot also develops on the fruits in storage and in transit. On the tree the fungus is spread rapidly by conidia produced on the rotting fruits. The fruits of the peach and plum, so attacked, often dry up on the trees and fall to the ground. In these "mummies" the fungus hibernates over winter. In the spring the mycelium grows out from these fallen mummies to produce apothecia (Fig. 17-32*B*), in which are produced the ascospores that may infect the next crop. The fungus also attacks the flowers

Fig. 17-31 Cup-shaped apothecia of the genus Peziza.

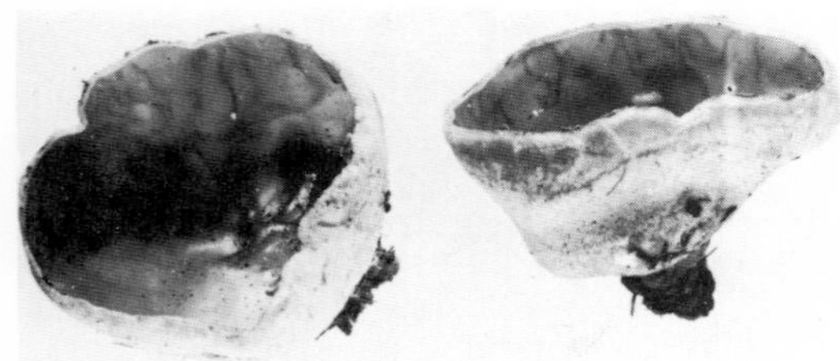

and gains entrance to the twigs, where it is perennial and may form conspicuous cankers. In the control of this disease it is advisable to destroy all mummies, especially since the mycelium in this stage may survive for several years.

Many of the cup fungi are involved in the association of fungi and algae, known as lichens.

Morels and truffles (Helvellales and Tuberales) One genus, *Morchella*, of the Helvellales is rather well known as an edible mushroom under such names as "morel" and "sponge mushroom." To the mushroom lover, no species surpasses these in their culinary delicacy. The ascocarp consists of a hollow stem, bearing a somewhat conical, enlarged upper portion that is marked with ridges and furrows (Fig. 17-33), in which the hymenium develops. These mushrooms may be found in old apple orchards about the time the trees are in bloom, along fence rows, in sandy soil along creeks and rivers, and sometimes in open oak and hickory woods. In Europe morels are sold, in season, on the city markets.

The members of the order Tuberales are entirely subterranean fungi, and the ascocarps are commonly known as truffles, which are edible. They are rounded in form, usually brown or black in color, and rather rough on the outside. The asci and spores are borne internally, the hymenium lining internal passageways that usually open to the exterior by one or more pores. There are fewer than 100 species of this order, and about half of them are in the single genus *Tuber*. In size they vary from the size of a pea to 4 in. in diameter and weigh as much as a pound. They grow apparently in close association with the roots of certain trees, particularly oaks and chestnuts, on which they are probably parasitic. The ascocarps are found just beneath the surface of the soil or under leaves. In France, Spain, and Italy they form an important article of commerce. They are gathered usually with the assistance of trained dogs or trained pigs. These animals locate them by the odor of the truffles. In France they are canned for export. In America they occur mainly on the Pacific Coast, but some have been found in the East. All American species, numbering about 25, are said to be of yellowish or white color.

Lichens Many species of Ascomycetes and a few species of Basidiomycetes are found growing in close association with one of several genera of green or blue-green algae. Such com-

Fig. 17-33 Morels (Morchella esculenta).

Fig. 17-34 Diagrammatic representation of a cross section through an apothecium on a foliose-lichen thallus. (Drawn by Ernest Geisweite.)

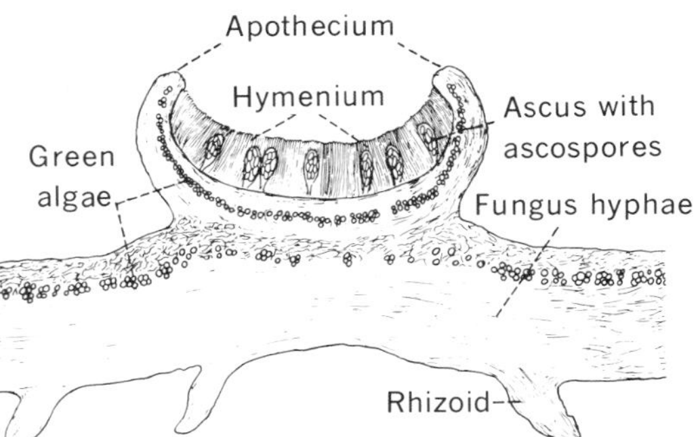

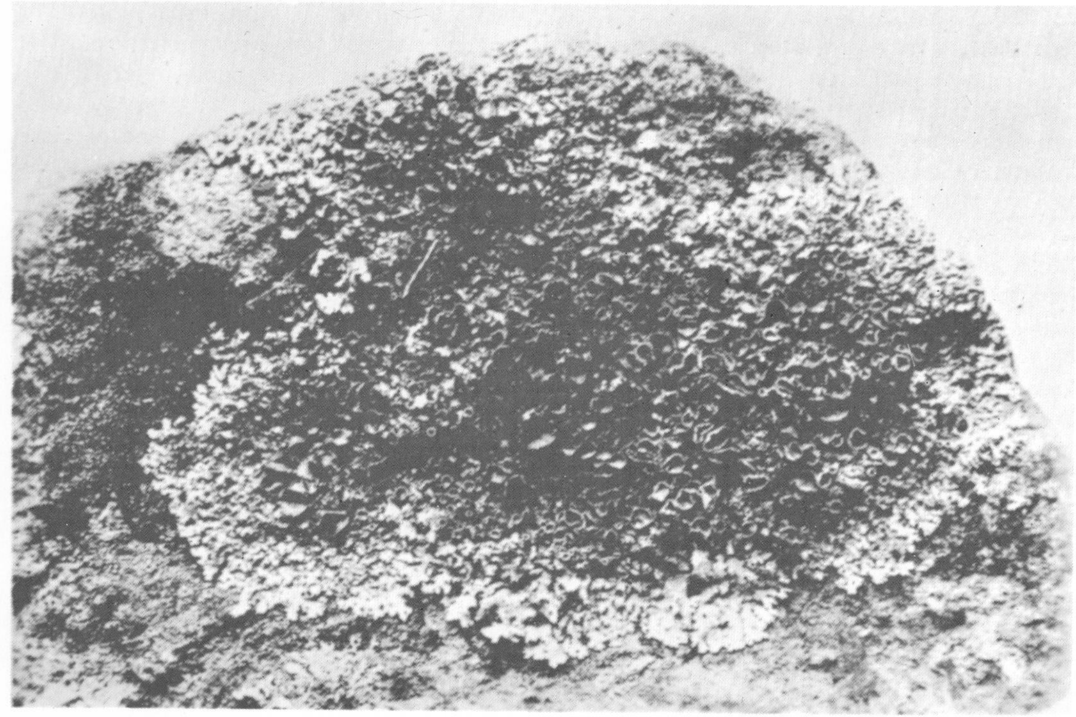

Fig. 17-35 Parmelia *sp., a foliose lichen growing on a rock and bearing apothecia.*

posite plants are called lichens. By some authors this association is regarded as symbiosis, in which both of the participants are benefited. Others regard the relationship as one of parasitism, the fungus being parasitic on the alga, as evidenced in some cases by the presence of fungous haustoria in the algal cells. *Protococcus* and *Cystococcus* are common green algae involved, and *Nostoc, Gloeocapsa,* and *Rivularia* are among the blue-green algae found in this relationship. Both the algae and the fungi reproduce, the ascomycetous fungi by producing apothecia or perithecia containing asci and ascospores (Figs. 17-34, 17-35, 17-37). Special reproductive bodies known as **soredia** are also formed. These consist of definite masses of fungous hyphae containing embedded algal cells that are cut out and break off to continue growth in a new location.

In form, three different types of lichens are recognized, the **foliose** type, resembling a thallose liverwort (Fig. 17-35), the erect, branched, **fruticose** type (Fig. 17-36), and the **crustose** type (Fig. 17-37), in which the body is a very thin crust entirely adnate to the substratum.

Lichens are found in a wide variety of habitats. Many of them are able to grow in situations where no other plants could survive. Consequently they are usually the pioneer plants in a succession of plants beginning in a habitat of bare rocks or a severely burnt-over area. They occur on exposed rocks (Fig. 17-37), on the sides of trees (Fig. 17-38), on old logs, and on the ground in woods. In damp woods of warm

Fig. 17-36 Cladonia rangiferina, *together with a little* Cladonia alpestris, *both fruticose lichens. (Photograph by A. A. Hansen.)*

regions they often hang in long festoons from trees. Their growth on bare rocks initiates the weathering away of such rocks. In arctic regions they develop as cushionlike masses on the ground, and because of the lack of competition of other plants they become an important part of the flora, replacing as food for such animals as the reindeer the grasses of more southern regions. To a slight extent they are also used as food by man in arctic regions.

Dyes of different colors have been obtained from lichens but seldom in commercial quantities. Litmus solution is made by grinding the lichen, *Roccella tinctoria,* and extracting the color-

Fig. 17-37 Lecidia platycarpa, *a crustose lichen inseparable from its rock substratum and bearing small black perithecia.*

ing matter, after which paper is soaked in the neutralized solution and is then known as litmus paper.

BASIDIOMYCETES

This is the class of the more conspicuous fungi, including puffballs, mushrooms, toadstools, rusts, and smuts. All possess one feature in common: all species produce four external basidiospores. Like the ascus, the basidium is the modified terminal cell of a hypha in which two nuclei fuse to form a diploid nucleus which immediately undergoes meiosis. No specialized sex organs are formed, but all reproduce sexually. Fusion can occur between ordinary hyphae, between two special cells (oidia), or between spermatia and receptive hyphae. Some species are heterothallic, fusion occurring between plus and minus strains. Others are homothallic, fusion occurring between any two cells of any two hyphae.

Two subclasses of Basidiomycetes may be recognized: the **Heterobasidiomycetes,** including the parasitic smuts and rusts with basidia transversely divided into four cells (Fig. 17-39*A*) and the jelly fungi with forked (Fig. 17-39*C*) or longitudinally septate basidia (Fig. 17-39*B*); and the **Homobasidiomycetes,** including the saprophytic fleshy and woody forms, the former usually known as mushrooms and toadstools, with one-celled basidia (Fig. 17-39*D*).

HETEROBASIDIOMYCETES

Smuts (Ustilaginales) The smuts are parasites of higher plants that cause millions of dollars worth of damage to crops. They receive their name from the black, sootlike masses of **telio-**

Fig. 17-38 Lichens growing on the north side of a tree. (Photograph by A. A. Hansen.)

spores (formerly incorrectly called chlamydospores) that are usually produced externally on the living host. Hosts are seldom killed outright, but their vitality is lowered and often the plants are stunted. Two families are recognized, based upon the germination of the teliospores (Fig. 17-40*B, C*). Briefly, the life histories consist of the following points: the teliospores germinate in the spring, producing a basidium, composed of one or four cells. Two nuclei in the young teliospore fuse, meiosis occurs, and the basidiospores are uninucleate and haploid. Basidiospores, or hyphae coming from their germination, fuse, but the nuclei do not usually fuse at this time. Consequently the cells of the mycelium that form from the fusion remain dicaryotic until the early stages of teliospore formation, when the nuclei fuse. The mycelium is intercellular, and often haustoria develop.

Three types of smuts are recognized on the basis of time and place of infection. In the smut of corn (*Ustilago zeae*), the smut tumors may appear on many localized parts of the host. They are masses of black dusty teliospores (Fig. 17-41). The teliospores mature in the summer or autumn and lie dormant on the stalks or in the soil. In the spring they germinate, producing many basidiospores which are windblown. The basidiospore germinates to form a germ tube that may infect any growing part of the corn plant. *Ustilago zeae* is heterothallic. Two plus and two minus basidiospores are produced from one teliospore. The germ tube develops into a mycelium which is either plus or minus. If the plus mycelium develops near a minus mycelium, fusion occurs, but the nuclei do not immediately fuse. After the mycelial fusion the dicaryotic mycelium forms a gall. Most of the hyphal cells in

Fig. 17-39 Basidia and mycelium of Basidiomycetes. A–C, basidia of Heterobasidiomycetes; A, cylindric, four-celled type; B, longitudinally septate type; C, one-celled and two-spored forked type; D, one-celled, four-spored basidium of Homobasidiomycetes; E, mycelium with clamp connections.

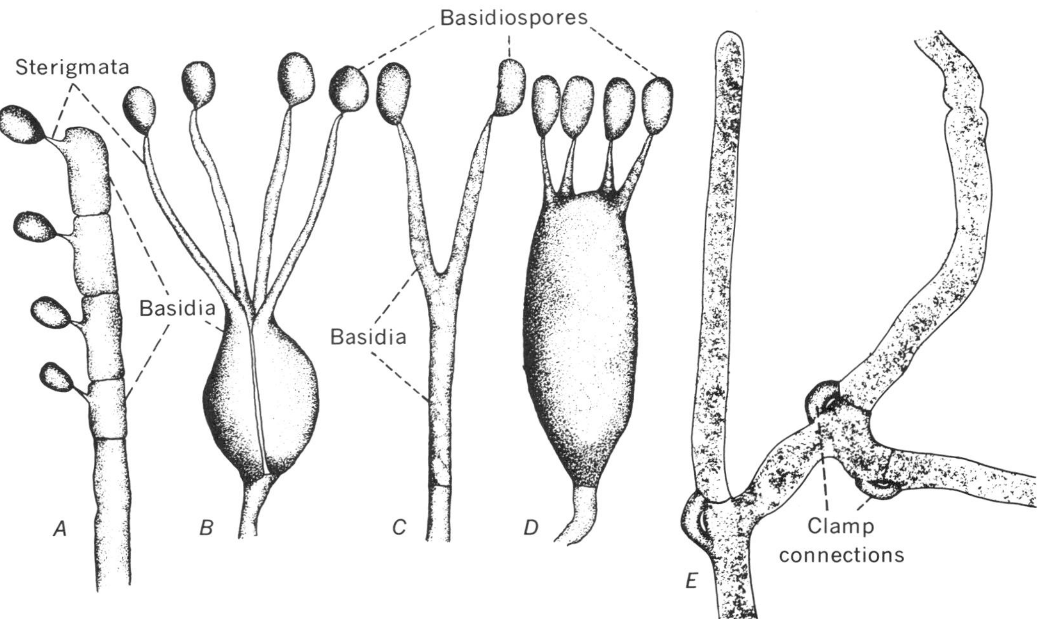

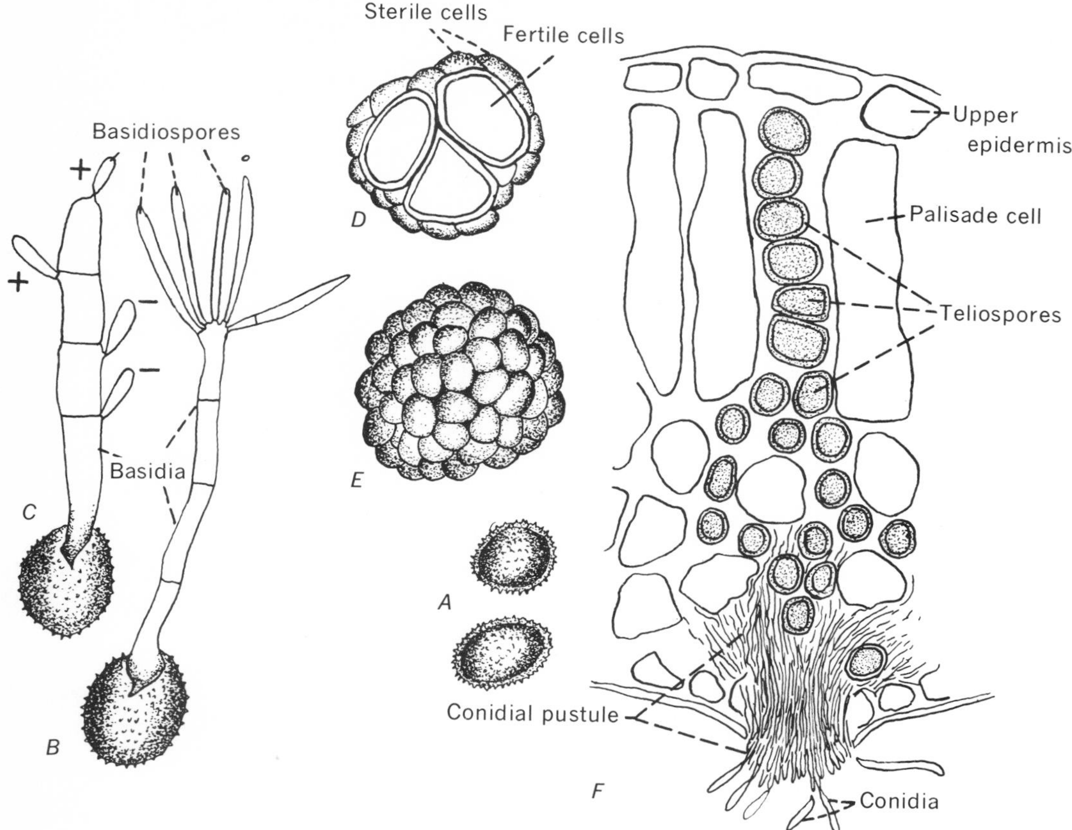

Fig. 17-40 Spores and teliospore germination in smut fungi. A, teliospores of corn smut; B, C, germination of smut teliospores and the formation of basidia and basidiospores; D, section through a smut ball of Urocystis agropyri, *showing the three fertile central cells and the surrounding layer of sterile cells; E, surface view of smut ball of* Sorosporium everhartii; *F, section through a leaf of* Physalis *infected by one of the internal smuts,* Entyloma australe.

the gall change into teliospores. No adequate and practical method of control is known. Since the teliospores overwinter in the soil and in fodder, rotation of crops is important in preventing damage due to corn smut.

In the loose smut of wheat (*Ustilago tritici*) the teliospores produced on the smutted heads lodge on the healthy heads in flower. Here they germinate (Fig. 17-42*B*), the germ tube infecting the ovaries without producing basidiospores. There is no external evidence of this infection, except sometimes in the shriveled condition of the grain. The mycelium lies dormant in the seed until planting time. When the embryo of the seed starts growth, the mycelium again becomes active, growing up through the seedling as new tissues are formed. The parasite causes infected plants to grow faster than uninfected, and so

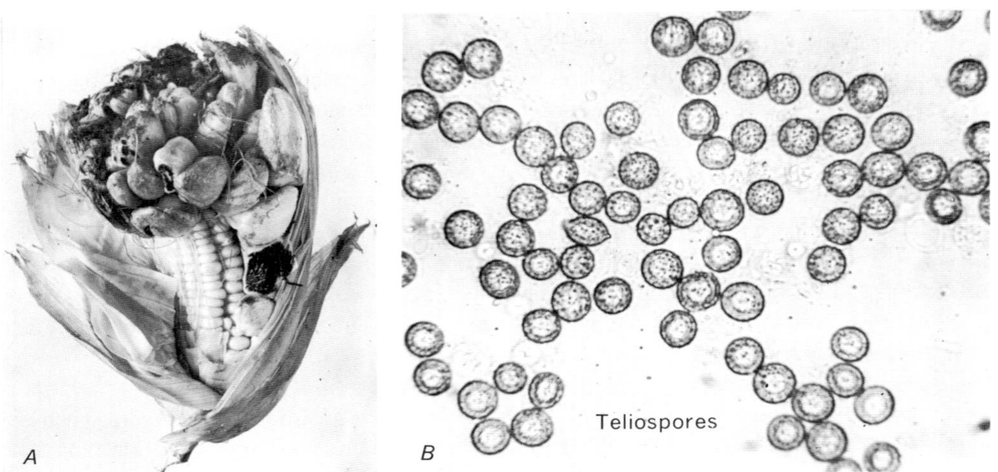

Fig. 17-41 *Corn smut. A, ear of corn with tumors of smut disseminating spores; B, view of teliospores enlarged.*

Fig. 17-42 *A, stinking smut of wheat: left, infected head and the resulting grains, right, healthy head and grains; B, loose smut of wheat: left, two diseased heads, the entire head, except the rachis, being destroyed, right, healthy head; C, loose smut of oats: left, healthy head, right, diseased head; D, E, photomicrographs of teliospores of the stinking smuts of wheat* (Tilletia foetens *and* T. tritici, *respectively).*

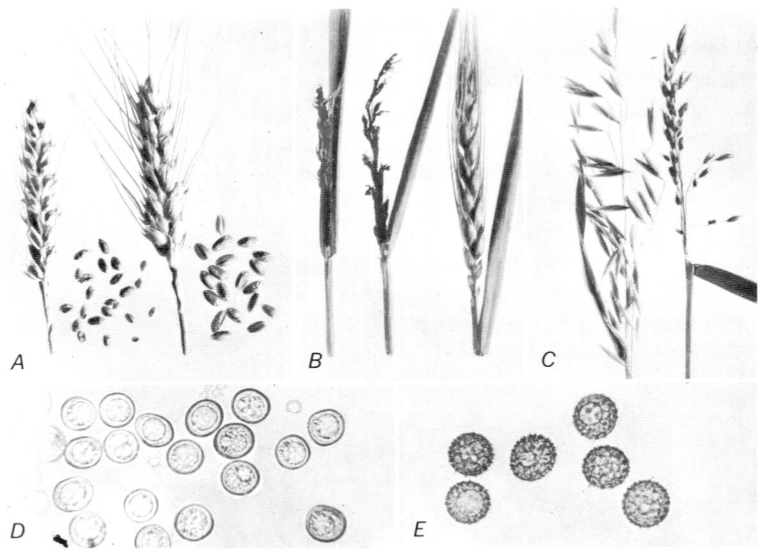

they produce flowers earlier. However, the flower heads become masses of black teliospores just as healthy plants are blossoming and their infection follows. An efficient method of control is to immerse the seeds in water hot enough (125 to 129°F for 10 min) to kill the mycelium but not the embryo.

The stinking smuts of wheat (*Tilletia foetens* and *Tilletia tritici*) and loose smut of oats (*Ustilago avenae*) produce teliospores in the infected grains (Fig. 17-42*A, C*). These are freed from the coats of the infected grains at the time of harvest and are mixed with the healthy grains to which they adhere. Masses of teliospores smell like rotting fish; hence the common name. In the spring when the healthy grains are planted, each of the teliospores which are able to survive on the surface of the stored grain germinates to produce the basidium which produces basidiospores. These immediately infect the very young seedlings. The mycelium grows up with the plant, invading the new tissues as they are formed. Serious dwarfing and a failure to flower are often the result of the infection. Some do flower, but the ovary tissues are replaced by hyphae which form large numbers of teliospores. The loss in oats is estimated to run as high as 30 percent of the crop in some instances. Practically all losses from both of these diseases are preventable except in extreme cases of heavy soil infestation, since it is possible to treat the seed with chemicals that will kill the smut teliospores but not harm the wheat grain.

Rusts (Uredinales) The term rusts has been used to designate a great variety of obligate parasites of seed plants and ferns that cause brown or rusty spots on living plants. However, in different stages of development they may produce brighter colors (yellow, orange, red). There are many different rusts, and they show a wide variety of life histories. However, most of them produce teliospores in which two nuclei fuse and meiosis occurs with the formation of basidia with four haploid basidiospores. Compatibility factors (self-sterility) or heterothallism has been discovered in many of them, so that two basidiospores are of a plus strain and two are minus. Some rusts, like the common bean rust, are **autoecious,** i.e., complete their entire life cycle on a single host, while others are **heteroecious,** requiring two separate unrelated hosts to complete a life cycle.

Stem rust of wheat (*Puccinia graminis*) is economically one of the most important plant diseases known. Its absence in some years and abundance in others are controlled to a large extent by certain environmental factors, such as temperature and moisture. This rust (Fig. 17-43) is heteroecious; i.e., it requires two unlike hosts, wild barberry and wheat, to complete its total developmental pattern. In the spring small yellow or reddish spots containing mycelium and spermagonia appear on the upper leaf surface of barberry. *Puccinia graminis* is heterothallic, with both plus and minus **spermagonia** (pycnia) containing **spermatia** (pycniospores) (Fig. 17-44*A*). These are developed on mycelium that originated from the germination of plus and minus basidiospores, respectively. Insects, attracted by the sweet odor and the color of the spermatial masses extruded from the spermagonia, carry the sticky spermatia from one spermagonium to the receptive hyphae of other spermagonia. If a plus spermatium is carried to a minus receptive hypha, or vice versa, fusion occurs and the nucleus of the spermatium moves into the receptive hypha, and about two days later, on the lower surface of the leaf, aecia are formed. The **aecia** are cup shaped (Fig. 17-44*B*) and enclose chains of bright orange or yellow dicaryotic ($N + N$) **aeciospores.** The aecium ruptures, freeing the aeciospores, which are wind disseminated. Although produced on barberry, the aeciospores cannot infect barberry but can infect wheat. It has now been demonstrated that **somatogamy** (fusion) of a plus mycelium with a minus mycelium can occur inside the

Fig. 17-43 Puccinia graminis, *black stem rust of wheat. A, telial stage on stem of wheat; B, enlarged photograph of two aecial pustules on lower surface of barberry leaf; C, barberry leaf enlarged to show aecial pustules on lower surface.*

barberry leaf, resulting in aecia and aeciospore formation.

The germ tubes of germinating aeciospores grow through the stomata of the wheat plant, developing into an extensive branched dicaryotic ($N + N$) intercellular mycelium with haustoria. After a lapse of two weeks, a third sorus, the **uredinium,** forms on wheat. From the tips of hyphae in the uredinium, one-celled **uredospores** (Fig. 17-44*C*), which are dicaryotic ($N + N$) and brown or reddish in color, are produced, which burst the epidermis and are wind disseminated. The spores in this stage spread the rust to other wheat plants. Weather conditions at this time determine whether or not a severe epidemic occurs. After ten days or two weeks a new crop of uredospores is formed. Uredospores are not capable of infecting barberry plants.

Within a short time further formation of uredospores ceases, and from the mycelium that has been producing uredospores, or from mycelia originating from uredospores, a fourth kind of spore, the **teliospore,** is formed. These are heavy-walled, two-celled spores on a conspicuous stalk, or pedicel (Fig. 17-44*D*), and are

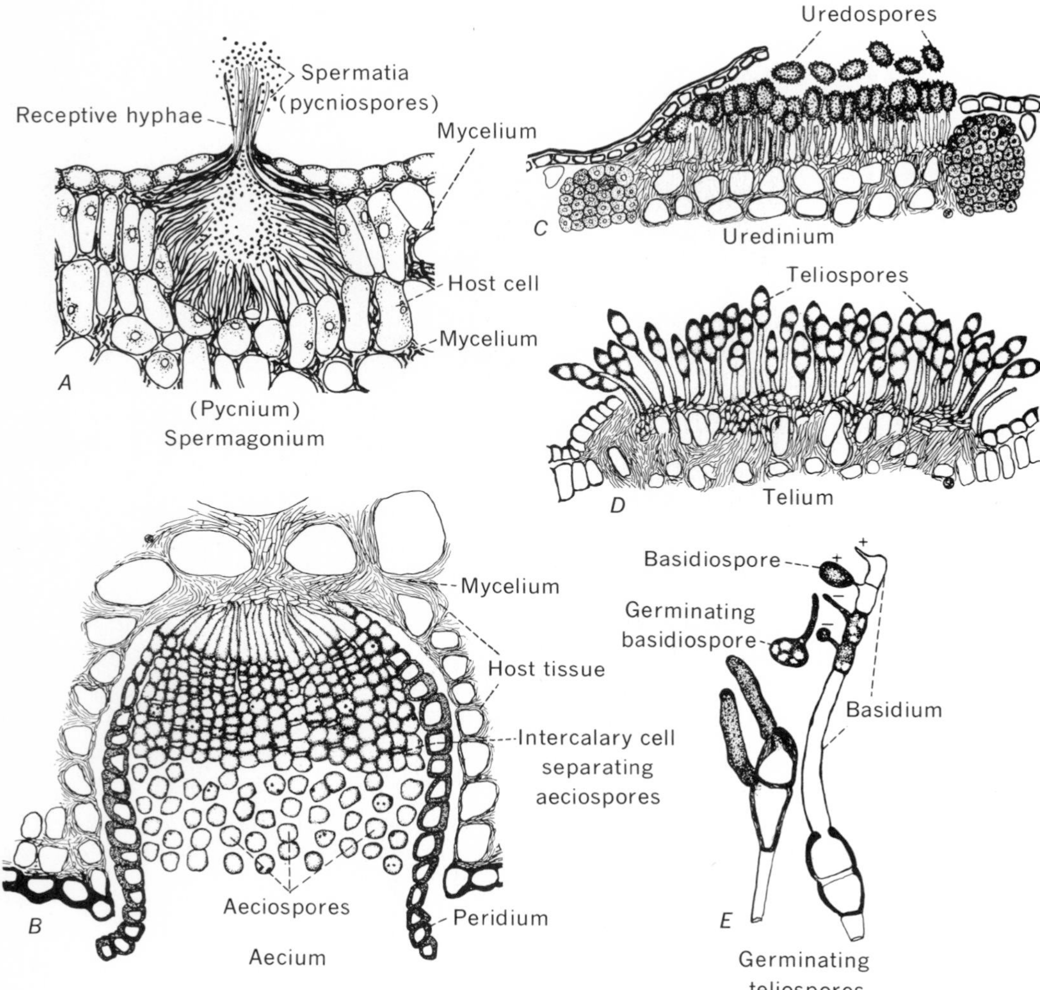

Fig. 17-44 Puccinia graminis. *A, section through a spermagonium (pycnium) with receptive hyphae and spermatia (pycniospores); B, section through an aecium embedded in the tissue of a barberry leaf; C, section through a uredinial sorus, with the one-celled, rough-walled uredospores borne on slender stalks that arise from a subepidermal mycelial complex; D, section through a telial sorus showing the two-celled teliospores on long pedicels arising from an internal mycelial complex; E, germinating teliospores; left, both cells of the spore germinating, right, only the apical cell germinating; the germ tube has been transformed into a basidium, from each cell of which a basidiospore has been or is being formed.*

black in mass. By this time the wheat plant is mature, and further growth of the fungus ceases. The teliospores remain in place on the wheat stalk in the field and tide the fungus over the winter.

With the return of warm weather and rains in April or May, the teliospores start to germinate. Each cell of the young teliospore contains two nuclei ($N + N$), which fuse to form a diploid nucleus. As the teliospore germinates, meiosis occurs with the formation of four haploid nuclei, each of which migrates into a **basidiospore** (Fig. 17-44*E*). Two basidiospores are plus, and two are minus. These are wind disseminated, but it is not possible for them to infect a wheat plant. However, if one falls on the leaf of barberry, the germ tube that it produces may enter the leaf and, after a short period of growth, the mycelium forms the spermagonia as already described.

In northern climates the black stem rust of wheat cannot perpetuate itself more than a single season without both the barberry and the wheat plants. Consequently, eradicating all barberry bushes near wheat fields will aid in controlling disease losses. Apparently, in portions of southern Texas, the winters are mild enough to allow the uredospores to winter over in viable condition and so reinfect wheat again the next spring. Furthermore some of the uredospores may be carried by the wind to states farther north where they infect wheat. However, eradication of barberry is still advantageous because not only does it reduce the number of spores in a region, but it also tends to reduce hybridization of the rust and the production of new strains which are able to infect varieties of wheat resistant to the usual wheat-rust races. New wheat varieties have been bred that are highly resistant or immune to the rust. However, these occasionally become severely infected by the rust, revealing the occurrence of **physiological races,** morphologically indistinguishable but different in ability of infecting different wheat varieties.

Another rust (*Gymnosporangium juniperi-virginianae*), of considerable economic importance, attacks the leaves and fruits of the apple. It is heteroecious, producing the spermagonia and aecia on the apple (Fig. 17-45*B*) and the telia (uredia are not produced) on the red cedar or juniper (Fig. 17-45*A*).

White-pine blister rust (*Cronartium ribicola*), introduced from Europe prior to 1900, has become a serious disease of the white pine. Spermagonia and aecia are produced on pine (Fig. 17-45*C*), and uredia and telia on the leaves of wild and cultivated currants and gooseberries (Fig. 17-45*D*, *E*). The method of control consists in the eradication of currants and gooseberries from the vicinity of white-pine plantations or forests.

Many other rusts are of considerable economic importance. All our cereal plants have one or more such diseases. They attack forage crops such as clover and timothy; fruit trees including apple, pear, peach, and quince; ornamental plants such as carnation, hollyhock, roses, and chrysanthemums. Beans and asparagus are attacked in the gardens; in the forests willows, poplars, pines, hemlocks, and spruces are attacked.

HOMOBASIDIOMYCETES

This subclass includes such common fungi as mushrooms, toadstools, and puffballs. They have conspicuous fruiting bodies **(basidiocarps)** with basidia which do not arise directly as germ tubes from spores, as in the rusts and smuts, but are terminal cells of an extensive mycelium. In this character the basidium corresponds to the ascus of the Ascomycetes but produces its spores externally rather than internally. Also, there is no such variety of spores produced

Fig. 17-45 *A, B, cedar-apple rust; A, a cedar apple on juniper; this is their appearance in April or May, when the telial horns bearing the teliospores are prominent; B, apple leaf, lower surface view, showing the aecial stage; C–E, white-pine blister rust; C, aecial stage on main stem of seedling white pine; the orange bladdery pustules are a conspicuous feature of the disease on white pine in May; D, telial stage on lower surface of wild gooseberry leaf; the telial columns appear as brown hairs, as shown in the enlarged photograph in E.*

within this group as occurs in the rust fungi, the basidiospores being, in most cases, the only type of spore produced. The basidia are usually grouped into a definite hymenial layer (Fig. 17-46*A*). They are one celled, usually producing four basidiospores on as many short, apically situated stalks, the **sterigmata.**

The mycelium of these rather uniformly saprophytic fungi is usually widely dispersed within the substratum and in many cases is perennial.

Frequently it shows peculiar and characteristic structures known as **clamp connections** at the cross walls (Fig. 17-39*E*). Arising from germination of a spore, the mycelium eventually produces a sporophore with spores, and apparently the life cycle in most cases is not more complicated than this. Sporophores are initiated by the massing of the mycelium at definite points near the surface of the substratum. This mass ultimately takes on a definite form and emerges to the surface, where it develops into the characteristic fruiting body of the species.

The simplest type of sporophore is a flat or more or less cushion-shaped body (Fig. 17-47*A*) in which the hymenium is likely to be spread over the entire free surface of the sporophore. Other types include the bracketlike sporophore with the hymenium restricted to the lower surface (Fig. 17-47*B*), the mushroom type, consisting of a stalk and cap (Fig. 17-48), and the puffball type (Fig. 17-52), in which the spores are completely covered until the surrounding tissue collapses and sets them free.

Sexuality Definite sex organs are not present in any species of this group; yet they have the equivalent of a sexual process. The cells of the mycelium, which is produced by the germination of a spore, eventually become binucleate. Apparently, at times, this is brought about through a fusion between two cells of adjacent hyphae, but variations of this method are known. As a result, the terminal cell of a hypha, like the terminal cell of the ascogenous hyphae of the Ascomycetes and like the teliospore of the rusts, is binucleate. This terminal cell (Fig. 17-46*B*) becomes the basidium. In the young basidium the two nuclei fuse (Fig. 17-46*C*). This fusion nucleus then divides twice (meiosis) to form four haploid nuclei (Fig. 17-46*D, E*). From the

Fig. 17-46 A, small portion of a hymenial layer; basidium a has matured and discharged its spores, leaving the four slender sterigmata; basidium b shows immature spores; basidium c shows four spores ready for discharge; the slightly smaller structures between the maturing basidia are immature basidia that may or may not produce spores; B, young binucleate basidium on the end of a hypha; C, basidium in which the two nuclei have fused; D, E, meiosis in the young basidium; D, basidium again binucleate through the meiotic division of the fusion nucleus; E, a four-nucleate basidium showing the beginning of sterigmata; F, basidium with the spores forming on the sterigmata and the beginning of nuclear migration through the sterigmata; G, basidium with nearly mature spores; H, basidium that has discharged its spores.

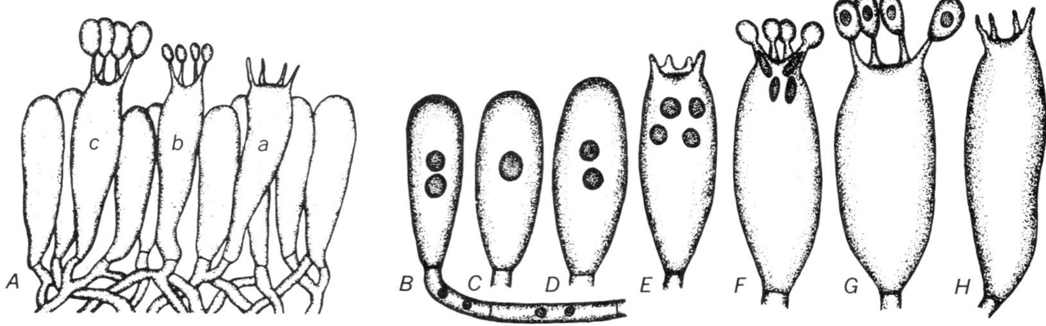

Fig. 17-47 Sporophores (basidiocarps) of Homobasidiomycetes. A, Poria albobrunnea, *a type which lies flat on the substratum; B,* Fomes applanatus, *a bracket-type sporophore on a dead beech tree.*

tip of the basidium, four (usually) minute tubelike processes, the young sterigmata, arise and a small enlargement (Fig. 17-46*E, F*) appears on the tip of each. Into each of these enlargements there passes, through the hollow sterigma, a small amount of cytoplasm, together with one of the nuclei from the basidium (Fig. 17-46*F, G*). These terminal enlargements of the sterigmata become the spores.

Different kinds of Homobasidiomycetes The gill fungi (family Agaricaceae), commonly known as mushrooms and toadstools, are probably the best-known members of the group. The basidiocarp usually consists of a stalk, or **stipe,** and a cap (the **pileus**) (Figs. 17-48, 17-49), with the basidia covering the surfaces or leaflike plates or **gills** on the lower surface of the cap. The gills vary in color in different species because of the various colors of the spores. Sometimes the gills are covered for some time previous to the maturity of the spores by a **veil** (Fig. 17-48). This veil breaks as the basidiocarp increases in size and may remain on the stipe in the form of a distinct membrane, or **annulus** (Fig. 17-48). In addition, there is sometimes a veil in which the entire basidiocarp is enclosed during the early stages of development. This veil usually splits and allows the basidiocarp to emerge, itself remaining at the base of the stipe as a cup, or **volva** (Fig. 17-49). There are several hundred species in this family in the eastern United States, of which less than 25 are known to be poisonous. Most of the poisonous species belong to the genus *Amanita.*

Another group (family Polyporaceae) contains the more conspicuous bracket fungi (Fig. 17-50), in which the basidia line numerous tubes that open downward in an even layer on

Fig. 17-48 Mushroom type of sporophore (Agaricus campestris, *the common mushroom of commerce). Note the stipe and pileus with gills on lower surface and well-developed veil tearing away from the margin of the cap and remaining on the stipe as an annulus.*

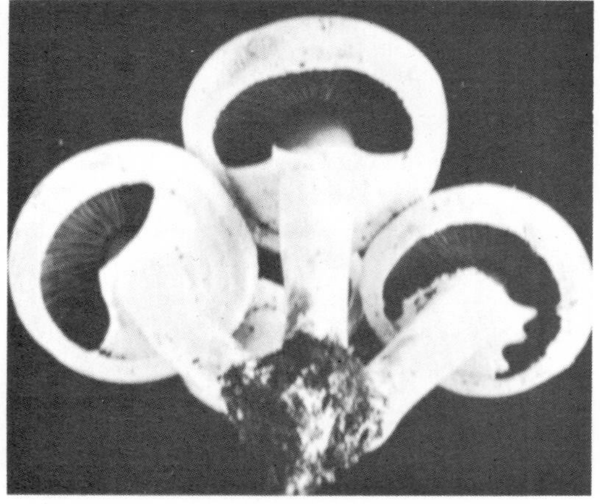

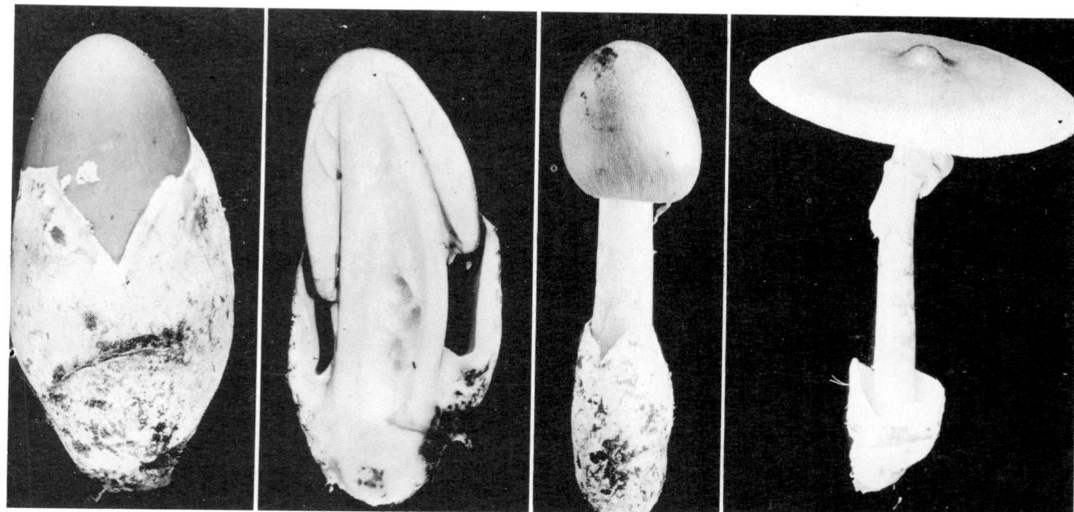

Fig. 17-49 Stages in the development of the basidiocarp of Caesar's agaric (Amanita caesaria), *one of the most beautiful of the gill fungi and one of the few members of the genus* Amanita *that are edible. The stipe, pileus, gills, and annulus are orange to yellow in color, and the open, saclike volva at the base is white. Although it has been prized as an article of food since the time of the Caesars, care must be taken not to confuse it with the poisonous fly agaric* (Amanita muscaria) *and other amanitas.*

the lower surface of the bracket. Some of this group are edible, some poisonous. Most are of a tough or woody consistency. The special importance of the group lies in their ability to bring about timber decays.

In the toothed fungi (family Hydnaceae) the hymenium covers the surfaces of downward-projecting teeth, or spines (Fig. 17-51*A*). The coral fungi (family Clavariaceae) often have beautiful, delicate fruiting bodies (Fig. 17-51*B*). In the puffballs (Fig. 17-52) the spores are retained internally until fully matured, when they can be forced out in the form of a fine dust or powder. One species, the giant puffball, sometimes reaches a diameter of more than 2 ft. All puffballs are edible when young.

The basidiocarp of the bird's-nest fungi is in the form of a cup in which are contained small egglike bodies which hold the spores (Fig. 17-53). In the stinkhorns the basidiocarp is typically a cylindrical body, at the apex of which the spores are produced (Fig. 17-54). These fungi are readily recognized by their bizarre shapes and their peculiar and unpleasant odors.

Economic importance Mushrooms have been used by man as food since ancient times, but not until the seventeenth century were they cultivated as a crop, and then not extensively. The early cultivation of mushrooms took place in France. Soon many other European countries started to grow them. In the United States they were not grown before the late part of the nineteenth century, and the method of growing them was a guarded secret for a long time. Now they are produced in many places all over the world. They are grown in caves, cellars, and greenhouses, but especially in houses built for this

Fig. 17-50 Bracket type of basidiocarp. A number of basidiocarps of Polyporus tsugi *protruding from the trunk of a dead hemlock tree. (Photograph by W. A. Campbell.)*

Fig. 17-52 A group of puffballs (Lycoperdon gemmatum).

purpose. The annual production in the United States alone is about 175 million lb a year, with a wholesale value of about 50 million dollars. They are sold as fresh mushrooms or are canned and made into soups, sauces, and other food products.

In the United States *Agaricus campestris* (Fig. 17-55) and perhaps three other related species are practically the only ones grown and sold. *A. campestris* is also found as a wild species. There

Fig 17-51 A, Hydnum erinaceum, *one of the toothed fungi in which the hymenium covers the downward-projecting teeth; B,* Clavaria *sp., one of the coral fungi.*

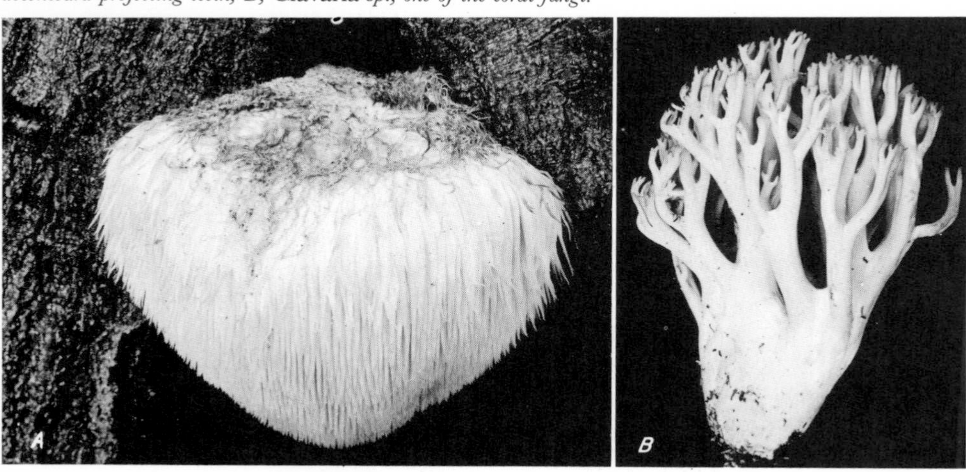

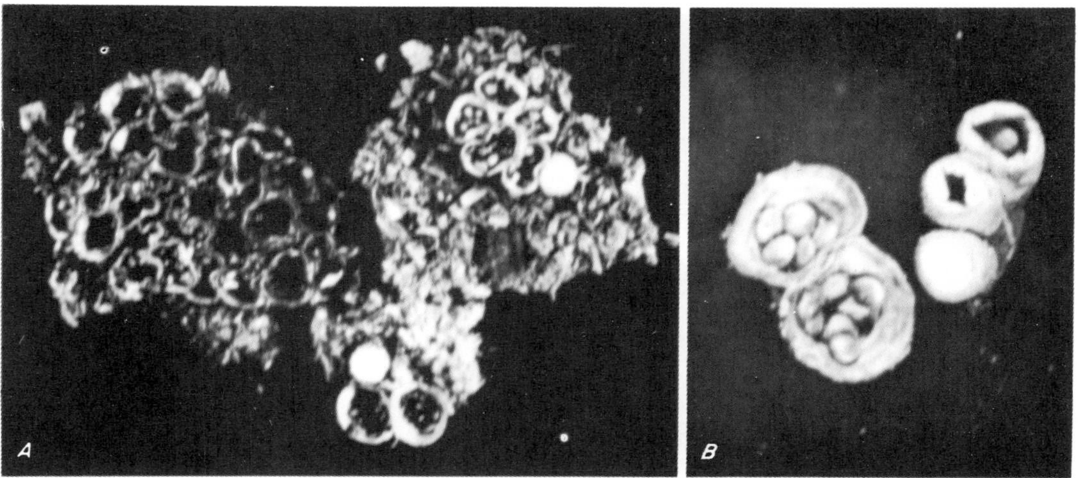

Fig. 17-53 Bird's-nest fungus. A, Cyathus stercorarius; *B, enlarged view of* Cyathus vernicosus.

are more than 500 other edible wild species (Fig. 17-56) found in the eastern United States. There is no botanical distinction between mushrooms and toadstools, edible and poisonous species occurring in the same family and often in the same genus (Fig. 17-57). There is no simple test that can be applied to determine whether they are edible or poisonous. The only way of distinguishing between them is to learn to recognize the different species much as one recognizes other plants. On the markets of the large European cities, as many as 50 or 60 different species of wild mushrooms are for sale in their season, indicating a much more extensive and intimate acquaintance with these plants.

Several species of *Psilocybe* (Fig. 17-58), a genus of Agaricaceae, when eaten, produce hallucinations. They are considered to be sacred mushrooms by the natives of southern Mexico and are used during secret ceremonies.

Although most members of the Homobasidiomycetes are saprophytic, some of them cause tree diseases and timber decays. Many species grow in the heartwood of living trees (Fig. 17-59*B*), breaking down the tissues and rendering the wood useless for lumber. When lumber is kept dry or when it is treated with preservatives such as creosote and asphaltum, the fungi are unable to live. However, these wood-decaying fungi must be regarded also as a benefit to man. If it were not for them, the trees and shrubs that die and fall in the forest

Fig. 17-54 Stinkhorns (Phallus ravenelii).

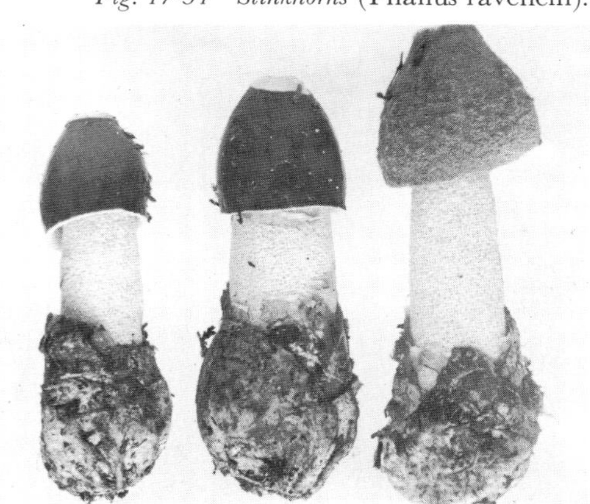

Fig. 17-55 A break of mushrooms in a commercial mushroom house. (Photograph courtesy of Dr. Leon Kneebone.)

Fig. 17-56 Edible mushrooms. A, Pleurotus ostreatus, *the oyster agaric, with white gills and spores; it grows on tree trunks and branches; B,* Coprinus comatus, *the prized shaggy-mane mushroom found in lawns and other grassy places; C,* Boletus edulis, *one of the fungi found on the ground in open woods.*

Fig. 17-57 Poisonous mushrooms. A, Amanita muscaria, *the fly agaric, with orange to yellow cap spotted with white scales and white to yellowish gills; the stem and volva are white; B,* Clitocybe illudens, *the jack-o'-lantern toadstool, so named because of its strong luminescence at night; all parts are orange to saffron yellow in color; C,* Panaeolus retirugis, *a small, dark, smoky-gray to white toadstool with black spores, commonly found in lawns.*

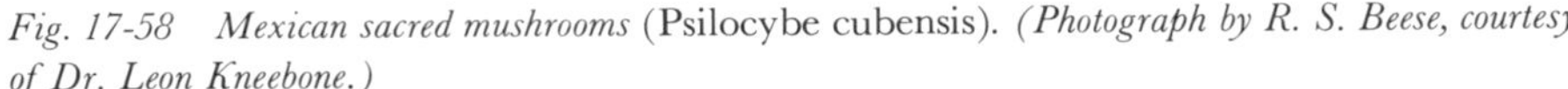

Fig. 17-58 Mexican sacred mushrooms (Psilocybe cubensis). *(Photograph by R. S. Beese, courtesy of Dr. Leon Kneebone.)*

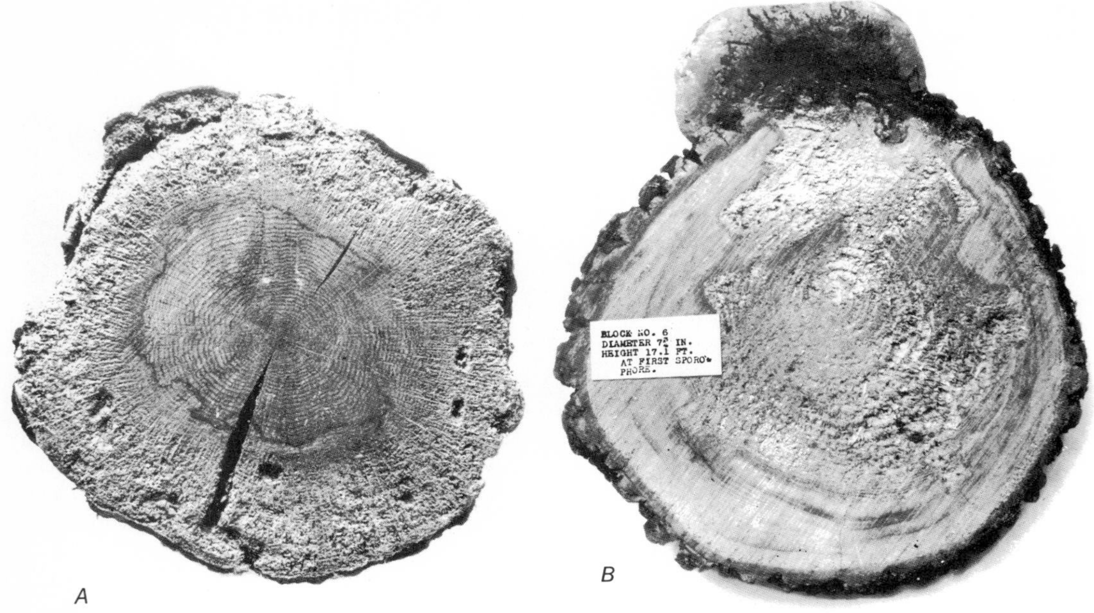

Fig. 17-59 Fungi and wood decay. A, a block of oak wood with the sapwood entirely decayed; B, a block of ash wood with the heartwood entirely decayed. The basidiocarp of the fungus is shown on the upper side of the block.

would accumulate until the woods would be impenetrable. At the same time, the decay continually adds new mineral elements and organic matter to the soil and CO_2 to the atmosphere.

IMPERFECT FUNGI (DEUTEROMYCETES)

In addition to those species of fungi falling properly into the main classes of fungi, there exists a large number of species the life histories of which are in doubt but which are known to possess characteristics that would place them in one or another of the previously described classes. Some of these consist only, as far as known, of sterile mycelium, but in most of them asexual spores comparable with those produced by many of the Ascomycetes are present. A few such species, previously referred to this group, are now known to be conidial stages of Basidiomycetes, and many more to be similar stages in the life histories of Ascomycetes. Perhaps some have lost all connection with other stages, reproducing entirely by asexual means. The life histories, as far as known, are therefore quite simple. The different methods of producing spores are shown in Fig. 17-22.

The majority of the species of the group are saprophytes, but large numbers are parasitic, particularly on the leaves and fruits of plants. In addition, most of the fungal pathogens of man belong here. Several of the imperfect fungi are able to trap nematodes (microscopic roundworms that often infest the roots of crop plants) by forming hyphal rings which constrict about the nematode when stimulated by contact. Many of the saprophytic forms are important soil organisms. Some of the parasitic members of the group cause the formation of cankers on branches of trees or spots and rots on fruits. Still others cause degradation of fabrics. Because of the importance of the imperfect fungi, many species previously referred to this group have been thoroughly studied and now are known to be asexual stages of Ascomycetes and are so classified.

QUESTIONS

1. What evidence have we that mold spores are universally present in the atmosphere?

2. If you were making a collection of polypores, where would you look for them?

3. Would you recommend spraying corn plants with a fungicide to control corn smut?

4. How would you prove that wheat rust is heteroecious?

5. What method of control would you recommend for the following: *a.* loose smut of wheat; *b.* apple scab; *c.* black stem rust of wheat; *d.* damping off of seedlings?

6. Why have we been unable to control chestnut blight?

7. Why do the fallen trunks of trees decay more rapidly in the forest than does lumber cut from trees and stored in a lumberyard?

8. Why do fence posts usually decay more rapidly near the surface of the ground than above the ground or deeper in the ground?

9. Why is it advantageous to spray dormant apple trees in early spring?

10. How does an ascus differ from a basidium?

REFERENCES

ALEXOPOULIS, CONSTANTINE J. 1962. Introductory Mycology, 2d ed. John Wiley & Sons, Inc., New York. A good general textbook on the fungi.

ATKINSON, G. F. 1900. Mushrooms, Edible, Poisonous, etc. Andrus and Church, Ithaca, N.Y. One of the best books on native mushrooms containing descriptions and excellent photographs, many colored, of most of the common genera of the United States, with chapters on the chemistry and toxicology of mushrooms and recipes for cooking. The book has long been out of print but may be found in most good libraries.

CHRISTENSEN, C. M. 1961. The Molds and Men, rev. ed. The University of Minnesota Press, Minneapolis.

FERGUS, C. L. 1960. Some Common Edible and Poisonous Mushrooms of Pennsylvania. Penn. State Univ. Agr. Expt. Sta. Bull. 667. Descriptions of 44 of the most common mushrooms of Pennsylvania and of the northeastern part of the United States.

KRIEGER, L. C. C. 1936. The Mushroom Handbook. The Macmillan Company, New York.

18

LIVERWORTS AND MOSSES

Antheridial plants of the moss Polytrichum. *(Copyrighted photograph, courtesy of General Biological Supply House Incorporated, Chicago, Illinois.)*

This chapter includes a brief discussion of the liverworts and mosses including their numbers, general characters, physiology, development, ecology, and distribution. Some attention is devoted to a comparative consideration of the characteristics of several of the liverworts and the mosses. The basis is provided for an understanding of useful information necessary for the identification of members of this relatively small group of plants.

The division Bryophyta is a small group of terrestrial plants numbering 20,000 to 25,000 species. As a group, they have worldwide distribution, occurring in all climates of the earth from the tropics to the subarctic and subantarctic regions, wherever there is moisture enough to sustain plant life. Although the bryophytes are regarded as terrestrial plants, the range of habitat is diverse. At one extreme they grow in situations where they are almost constantly submerged in water; at the other, on dry rocks, where the soil is scant and moisture occasional. They are also found in all situations between these extremes. In general, however, the plants can be said to occur in moist situations, as on damp rocks in ravines or sheltered cliffs and moist soil, and on decaying logs in forests.

The bryophytes are often early invaders in ecological succession, especially after fire. Their persistence is in part the result of their ability to withstand periodic drying (pollacauophyte). Some tolerate salt in their environment and occur along seacoasts in the so-called splash zone. Both liverworts and mosses occur as fossils in the Tertiary period of the Cenozoic era and in the Mesozoic and Paleozoic eras (Fig. 19-1).

The plant body of the members of the bryophytes is generally small, usually attaining a length of a few inches. However, the gametophores of an Australasian genus regularly reach a height of 2 ft or more and one specimen of *Polytrichum commune*, growing in water in New Zealand, was reported to be 6 ft in length. Like that of the algae and fungi, the plant body is a thallus. The plant usually grows attached to the soil or other substratum by branches of the plant body called rhizoids, which are single-celled or multicellular filamentous structures. Bryologists generally agree that those members of the Bryophyta exhibiting considerable external dissection of the thallus present a "leafy" appearance and are primitive, whereas those with a less-dissected appearance are more highly specialized and advanced.

In many bryophytes specialization of the thallus has resulted in the formation of stemlike and leaflike structures. In most discussions of the bryophytes these structures are referred to as stems and leaves, but it must be recognized that the terms are used for convenience and imply a physiological similarity to the stems and leaves of vascular plants. It has been suggested by some botanists that the term **caulidium** or **caulid** be substituted for stem and **phyllidium** or **phyllid** for leaf. The bryophyte plant body is regarded as a highly differentiated thallus (Figs. 18-1, 18-12, 18-21, 18-27).

In some orders the sporophyte is a simple structure with little differentiation of tissue, but in others the sporophytic structures are more highly developed. The capsule, or spore case, of the moss sporophyte, with its complex differentiation of tissues and elaborate mechanism

Fig. 18-1 Conocephalum conicum, *a liverwort, showing habit of growth on a rock.*

of dehiscence, is a structural advance over the sporophytes of the algae. The sporophyte is always dependent upon the gametophyte for at least a portion of its food, water, and minerals.

The gametophyte is the independent structure in the life history and often shows remarkable development in the matter of asexual reproduction. Many of the bryophytes have developed such efficient methods of asexual reproduction that they approach independence of gametic reproduction as a means of increase and distribution. The formation of **multicellular gametangia,** antheridia, and archegonia, surrounded by a jacket of sterile cells, is an important feature distinguishing them from the algae. The Bryophyta are a well-defined group with no recognizable link to algae or vascular plants. Their evolution is largely speculative.

Three classes of Bryophyta are recognized, namely, the **Hepaticae,** made up of thallose and leafy liverworts; the **Anthocerotae,** or horned liverworts; the **Musci,** or mosses.

HEPATICAE—THALLOSE AND LEAFY LIVERWORTS

The Hepaticae are the simplest of the Bryophyta. They are found inhabiting damp or wet situations to a much greater extent than are the mosses, but only in rare cases are they entirely aquatic. The plant body is prostrate, or practically so, and varies from a flat, dichotomously branched thallus in the thallose hepatics to a plant in which the thallus is specialized into a creeping caulid (stem) and expansions or divergences called phyllids (leaves). These are referred to as the leafy liverworts. In all prostrate forms the symmetry is distinctly dorsiventral; i.e., there are an upper or dorsal surface and a ventral or lower surface. This is in contrast to

Fig. 18-2 Bazzania, *one of the Jungermanniales, showing habit of growth.*

the radial symmetry of mosses. Rhizoids are developed on the lower surface of the plant, except for those cases in which the plant is aquatic. These rhizoids are unicellular structures in contrast to those found in mosses which have cross walls and are multicellular.

The gametangia are either embedded in the dorsal tissue of the plant or are produced on special branches known as **gametophores.** Following fertilization of the egg, a sporophytic plant body develops which varies in complexity but is always attached to the gametophyte, is devoid of chlorophyll, and is dependent upon the gametophyte for its food. Fragmentation of the gametophyte, with subsequent growth of the tips of each of the fragments, serves effectively in vegetative reproduction.

The class Hepaticae is sometimes described as including seven orders, but if the Anthocerotae are retained as a class, then only these six orders are included: Calobryales, Sphaerocarpales, Jungermanniales, Monocleales, Metzgeriales, and Marchantiales.

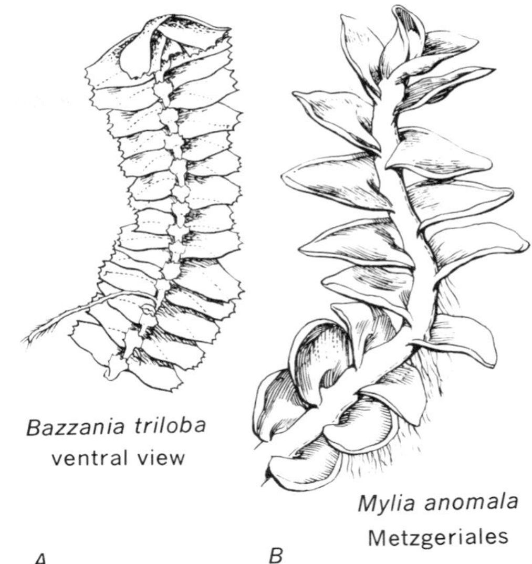

Fig. 18-3 A, Bazzania triloba, *a member of the Jungermanniales with two ranks of lateral phyllids and a ventral row of amphigastria; B,* Mylia anomala, *a member of the Metzgeriales. (Redrawn from R. M. Schuster, "Boreal Hepaticae." Am. Midland Naturalist,* **49:** *257–684.)*

Jungermanniales

The members of the order Jungermanniales may be found growing on damp soil and rock or on decaying logs. Some forms are epiphytic and others aquatic (Fig. 18-2). The gametophyte is generally dorsiventrally flattened with two lateral ranks of phyllids and a single rank (amphigastria) on the ventral surface (Fig. 18-3*A*). The phyllids found on the ventral surface of the plant usually are simpler in shape than the lateral phyllids. Rhizoids are absent or scarce in the primitive Jungermanniales but in more advanced forms, where they occur, they are smooth walled. Branching of the thallus is irregular and the structure of the caulid (axis) is simple with only the cortical region generally containing chlorophyll. Gametophytes of the Jungermanniales are unisexual or bisexual. They are **acrogynous** and are characterized by the development of the archegonia at the apex of the thallus. The archegonia (Fig. 18-4*A*) are typically flask shaped but usually more massive than the corresponding structures of the Marchantiales. The antheridia are usually spherical or ovoid and are borne singly or in groups on multicellular stalks. The mature antheridia resemble those of the Marchantiales.

The sporophyte is usually more elaborate than that of the Marchantiales and at maturity consists of a foot, the long flexuous stalk or seta, and a capsule which is usually ovoid or spherical in shape (Fig. 18-5*B, C*). The single outer layer of the capsule, the **amphithecium,** differentiates into a variable number of cell layers that form the sterile jacket. A single inner layer of cells, the **endothecium,** is the central tissue, some of which forms sterile cells and some of which becomes the archesporial cells.

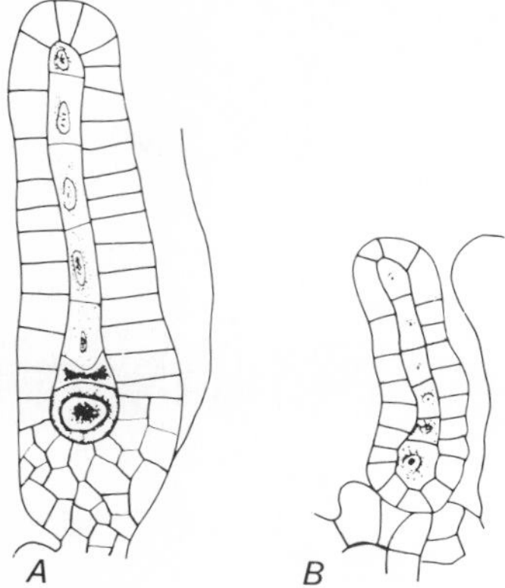

Fig. 18-4 Archegonia of liverworts. A, Porella, *a member of the Jungermanniales; B,* Pellia, *a member of the Metzgeriales.*

The sterile cells specialize to form **elaters,** which are single cells, cylindrical in shape and with a number of wall thickenings (Fig. 18-5). The elaters are distributed between the meiospores, and, through contraction and sudden extension, they aid in the dispersal of spores from the capsule. Spore mother cells, derived from the archesporium, divide meiotically to produce tetrads of meiospores. The wall of a meiospore is distinctively sculptured and the protoplasmic content includes chlorophyll and stored food.

The structure of the capsule and the manner of its dehiscence are characteristic of most genera of the order. The capsule wall regularly opens along four longitudinal lines and at maturity breaks into four valves (Fig. 18-5*B*). This regular method of dehiscence is in contrast to the capsule of the Marchantiales, which opens irregularly.

Vegetative reproduction occurs when undifferentiated masses of vegetative cells, formed on both the caulid and phyllid, fall to the substratum and grow. Vegetative reproduction may occur with the growth of new plants from phyllids while they are attached to the parent thallus or after they have dropped to the substratum.

Metzgeriales

Members of the Metzgeriales have sometimes been included within the order Jungermanniales and were previously designated as the **anacrogynous** members of this order. Many characters are similar to those described for the Jungermanniales, but bryologists consider the differences between the two orders to be significant and reasonable for the establishment of the Metzgeriales.

The vegetative form of the Metzgeriales is generally thallose. The rhizoids, like those of the Jungermanniales, are unicellular and originate from epidermal cells on the lower (ventral) surface of the thallus. Lateral ranks of phyllids are present but no row of ventral phyllids develops (Fig. 18-3*B*). The archegonia are structurally similar to those of the Jungermanniales (Fig. 18-4*B*), but they are borne on special dorsal branches and not at the apex of the thallus. The growth of the thallus is indeterminate. Antheridia are structurally similar to those found in the Jungermanniales and occur on short branches that develop from the dorsal surface of the thallus. The sporophyte resembles that of the Jungermanniales (Fig. 18-5*B*).

Marchantiales—thallose liverworts

Members of the order Marchantiales, usually considered to be the most highly evolved hepatics, constitute a group of widely distributed plants generally inhabiting moist locations. Some species, however, grow in exposed situations that are relatively dry at certain seasons of the year although they are moist at others. Liverworts of this order usually develop a flat, branching thallus, or gametophyte, 1 to 4 in. long, often with conspicuous lobes. Because of the fancied resemblance of the thallus to the mammalian liver, these plants have been called liverworts, or liver plants. The Marchantiales exhibit variation in size and in structural development of the thalli. They are usually simple in form and structure with very little cell differentiation. In some the thallus is differentiated into an epidermis, chlorophyll-bearing cells, and storage tissue. Although they are not universally found, air chambers may be present in the thallus and often are well developed with pores in the epidermis that function in the exchange of oxygen and carbon dioxide. The sporophyte of the Marchantiales is a comparatively small body that is relatively simple in structure. Within the group, sporophytes vary in size and differentiation.

Species of *Riccia* are small thallus plants growing in aquatic habitats, such as quiet, shallow ponds, swamps, and bogs. A part of their life is spent floating on the surface of the water and often a part on the mud at the bottom of the pond after the water has dried up during the summer months. Structurally the thallus of *Riccia* is adapted to existence in a moist situation where the mineral nutrients are derived from the aqueous medium in which it grows. The carbon dioxide is probably derived from the water rather than from the air, much as in the case of the algae. This is especially true in *Riccia fluitans,* since much of the thallus is submerged in water.

The mature *Riccia* thallus is a small, green, dichotomously branched mass of tissue. In one species the thallus is very thin and of simple

Fig. 18-5 Structural features of Cephalozia, *a member of the Jungermanniales. A, plant showing phyllids, caulid, rhizoids, and capsule arising above involucre, on a branch; B, capsule showing valves after dehiscence; C, detail of capsule, the upper figure indicating spore dispersal.*

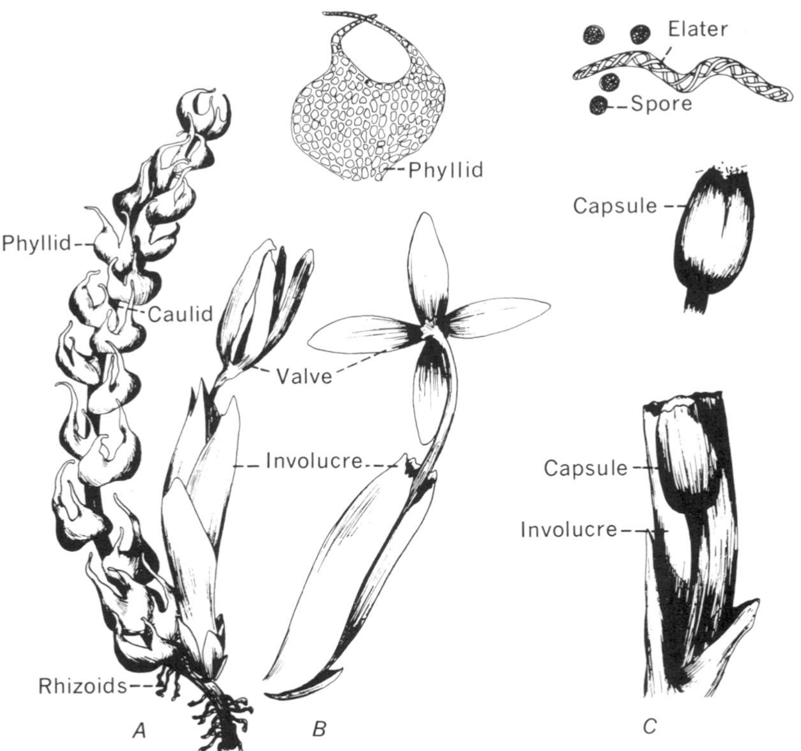

structure, almost algalike (Fig. 18-6*A*). It attains a length of possibly 2 or 3 in. and has the form of a long, narrow, branched ribbon. The thallus of *Ricciocarpus* is heart shaped, thicker, and more compact, the larger specimens attaining a maximum diameter of possibly ¾ to 1 in. (Fig. 18-6*B*). The upper surface consists of a single layer of epidermal cells. Some cells of the lower surface bear rhizoids and thin scales, the latter composed mostly of a single layer of cells. The upper portion of the thallus is composed of chlorophyllose cells arranged in vertical rows or columns, with broad chambers of air spaces between them. No well-developed pores for gaseous exchange are found in *Riccia.* The thalli in some species have rudimentary pores, but in others, owing to the peculiarities of development of the epidermal cells, pores are entirely lacking.

The thallus of *Riccia* develops from a meiospore. *Riccia* is without specialized methods of asexual reproduction but may be vegetatively multiplied by fragmentation. As growth progresses and the older portions of the thallus decay or are broken, the younger branches may grow as separate plants.

THE LIFE CYCLE Structurally the antheridia and archegonia are alike in all species. Their positions on the thallus, however, are different. The gametangia of *Riccia* develop from segments of the apical cell located in the notch of the thallus. The antheridia usually develop first and archegonia follow later in the same thallus. As the plant grows, new vegetative tissue forms at its apex in advance of the development of the young gametangia. As growth progresses, the gametangia appear distributed along the notch, with the antheridia toward the base and the archegonia nearer the apex of the thallus. At maturity the gametangia are found deeply sunken in the thallus, each one in a separate cavity (Fig. 18-8*A, B*).

The sporophyte develops from the zygote and at maturity is a spherical body consisting of a central mass of spores. The entire sporophyte, completely dependent upon the gametophyte,

Fig. 18-6 Vegetative features of A, Riccia fluitans; *B,* Ricciocarpus natans. *(A, drawn by Elsie M. McDougle.)*

A

B

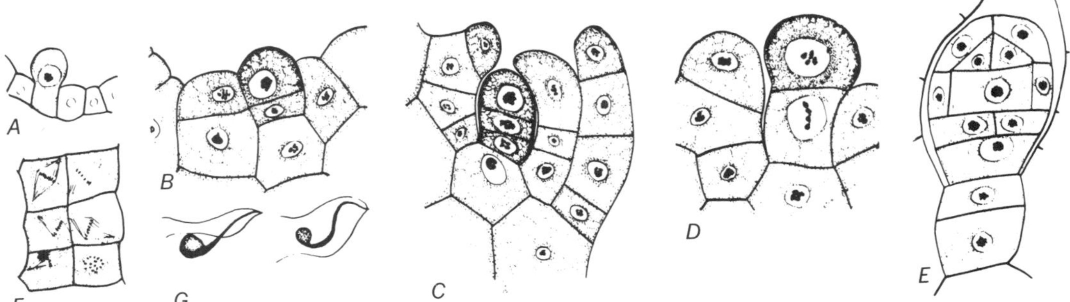

Fig. 18-7 Development of antheridium of Riccia. *A–E, young stages in antheridial development; F, final division to form sperm cells; G, mature sperms.*

is enclosed in the enlarged venter of the old archegonium, now called the **calyptra.** The combined structures, sporophyte and calyptra, are embedded in the thallus (Fig. 18-8*C*). Spores are released from the sporophyte by decay of the thallus tissue and the wall of the sporophyte. Upon germination the spores produce new gametophytes.

Marchantia and related forms; family Marchantiaceae The members of this family are terrestrial plants of thallose form. The flat, dichotomously branched thallus growing prostrate, with the ventral side attached to the substratum, is one of the characteristics of this group of liverworts. The gametangia are generally produced on gametophores, which are highly specialized branches of the thalli and which are differentiated as to sex. In related plants that do not have gametophores, the gametangia are developed in groups on slightly raised portions of the thalli.

The members of the family Marchantiaceae differ from other liverworts in the structural

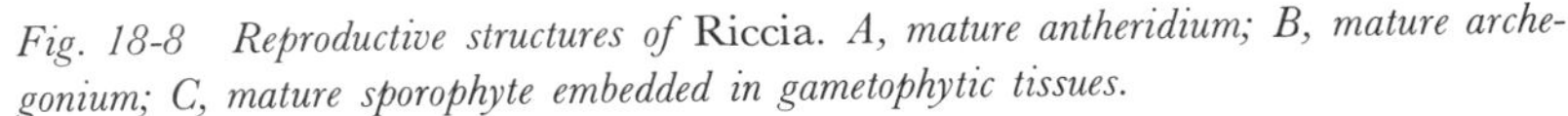

Fig. 18-8 Reproductive structures of Riccia. *A, mature antheridium; B, mature archegonium; C, mature sporophyte embedded in gametophytic tissues.*

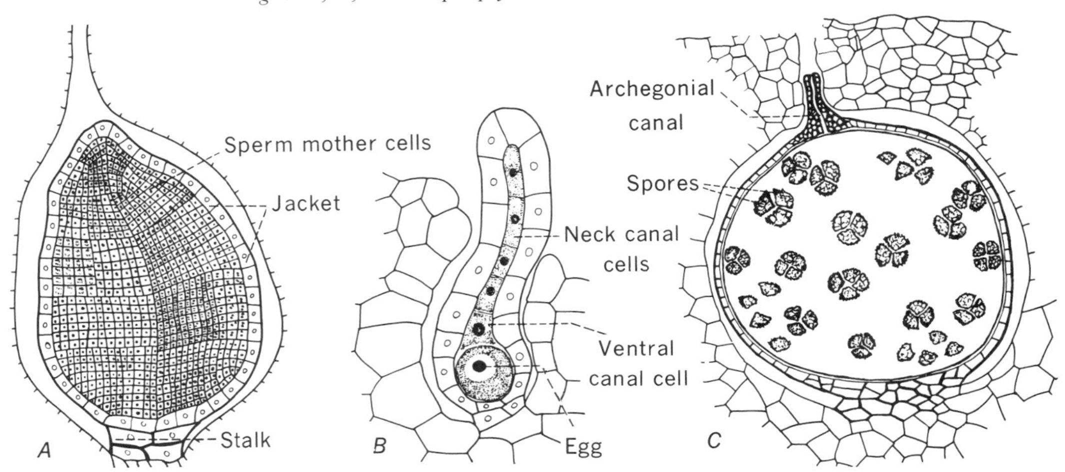

development of the sporophyte. A well-developed **foot,** a short **stalk,** and a **capsule** characterize the sporophyte of the Marchantiaceae. **Elaters** are produced among the spore-forming cells, or **sporocytes.**

THE GAMETOPHYTIC PHASE The gametophytic phase has its origin in the maturation of the meiospores. Upon germination each spore develops a germ tube, a short filament of cells which, by growth from an apical cell, soon broadens into the thallus.

The thallus is a dichotomously branched structure with dorsiventral symmetry (Fig. 18-10). It increases in size by growth from one or more apical cells. The apex of each branch of the thallus is more or less notched and at the base of each notch an apical cell is located. In *Marchantia* the mature thallus attains a length of 2 or 3 in., and in other genera of the Marchantiaceae it may be even larger. Rhizoids attach the thallus to the substratum. These are single-celled, hairlike structures that originate from the outer cells of the lower surface of the thallus. In many of these structures the inner surface of the wall is smooth, but in others it is roughened by wartlike or fingerlike projections into the cell cavity. These projections may extend nearly across the cavity, but they do not form complete partitions. Thin leaflike scales also occur on the underside of the thallus (Fig. 18-9*A*).

The dorsal epidermal cells usually contain chlorophyll, and the tissue is provided with pores or openings. Each pore is surrounded by a group of specialized epidermal cells. The number and arrangement of these cells vary considerably in the different genera of the family (Fig. 18-9). In some they are almost suppressed, but in others there is elaborate development of these cells. The genus *Marchantia* occupies an intermediate position in this respect, having several tiers of these cells arranged in the form of a chimney.

Internally the thallus body is divided into two

Fig. 18-9 Structural features of the thallus of members of the Marchantiales. A, cross section through thallus of Marchantia; *B, apical region of an archegonial head, showing origin of air chambers; C, D, air pores, C, in* Asterella, *and D, in* Plagiochasma.

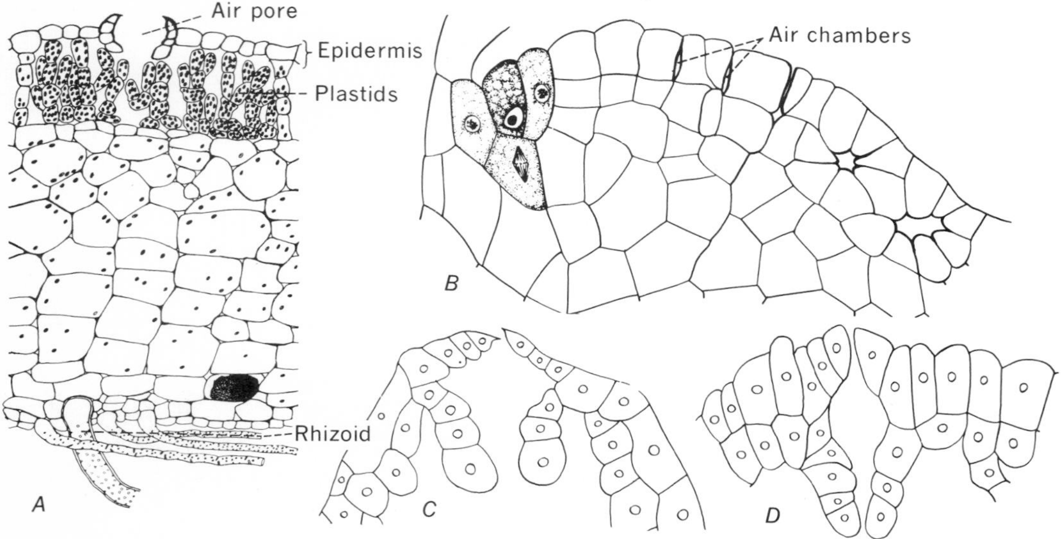

regions. The lower half or more of the thallus consists of compactly arranged, thin-walled parenchyma cells which probably serve as storage cells. These cells have little chlorophyll. The space between this region and the upper epidermal layer is divided into definite compartments separated by erect columns of cells supporting the epidermis. Each compartment opens to the outside through a single pore described above. Within these compartments are produced loosely arranged chlorophyll-containing cells. In *Marchantia* these cells are in the form of branched filaments originating from the floor of these cavities. The loose arrangement of these cells and the presence of pores facilitate gaseous exchange for respiration and photosynthesis (Fig. 18-9*A*).

Asexual multiplication may occur by fragmentation. The growing tips of the branching thalli continue development, while the central, basal portion decays, finally separating the branches. In certain genera, as in *Marchantia* and *Lunularia,* special asexual reproductive bodies, called gemmae, are produced. These **gemmae** are formed in small **cupules** on the upper surface of the thallus (Fig. 18-10). Each cupule has an elevated margin and the gemmae are produced on short stalks attached to the bottom of the cupule. The gemmae originate from single cells arising from the superficial layer of the bottom of the cupule. By a series of transverse walls, followed by vertical walls, a thin platelike structure is formed. The gemma at maturity is a few layers of cells in thickness and notched on two opposite sides. In each notch there is found an apical cell which is capable of continuing the growth of the gemma when it is suitably placed for growth. These gemmae are green in color and are held erect by their short stalks (Fig. 18-11).

SEXUAL REPRODUCTION A distinguishing feature of the members of the Marchantiaceae is the tendency to produce the gametangia in groups. In some genera of the Marchantiaceae the gametangia are merely grouped on the thallus. In others they are located in specialized structures called **receptacles.** The receptacles are, according to the genera, either sessile or stalked. The stalked receptacles are called **gametophores** (Figs. 18-12, 18-13). A rather clearly defined evolutionary series of forms can be arranged from those which have the gametangia merely grouped on the thallus to those with stalked gametophores.

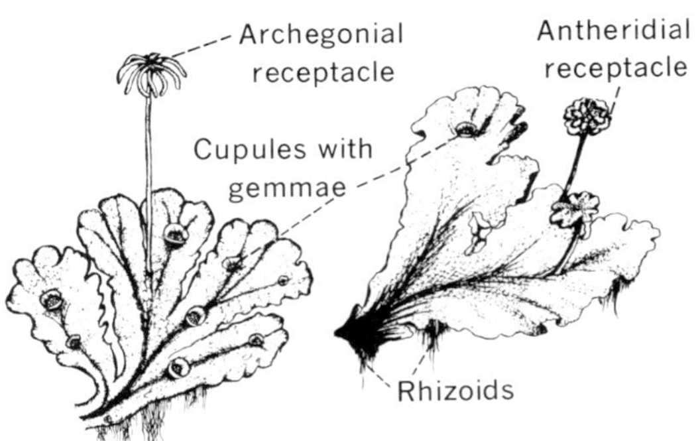

Fig. 18-10 Archegonial and antheridial plants of Marchantia. *(Drawn by Edna S. Fox and Elsie M. McDougle.)*

Marchantia produces highly specialized archegonial and antheridial gametophores on distinct thalli. The male and female plants may be distinguished by the characteristics of their gametophores (Figs. 18-12, 18-13). The gametophore consists of a stalk bearing an expanded, lobed, disklike upper part, the receptacle. The receptacles are provided with several apical cells. In the archegonial receptacle these are located between the long fingerlike projections (Fig. 18-12); in the antheridial receptacle they are in the notches of the margins of the characteristic disk (Fig. 18-13). The gametangia are developed in each case from the segments of the apical cells produced on the upper or dorsal surface of the receptacle. The antheridia remain on the upper side, but in the female receptacle the growing points are pushed underneath by the rapidly enlarging cells of the upper part of the structures, and therefore the archegonia are

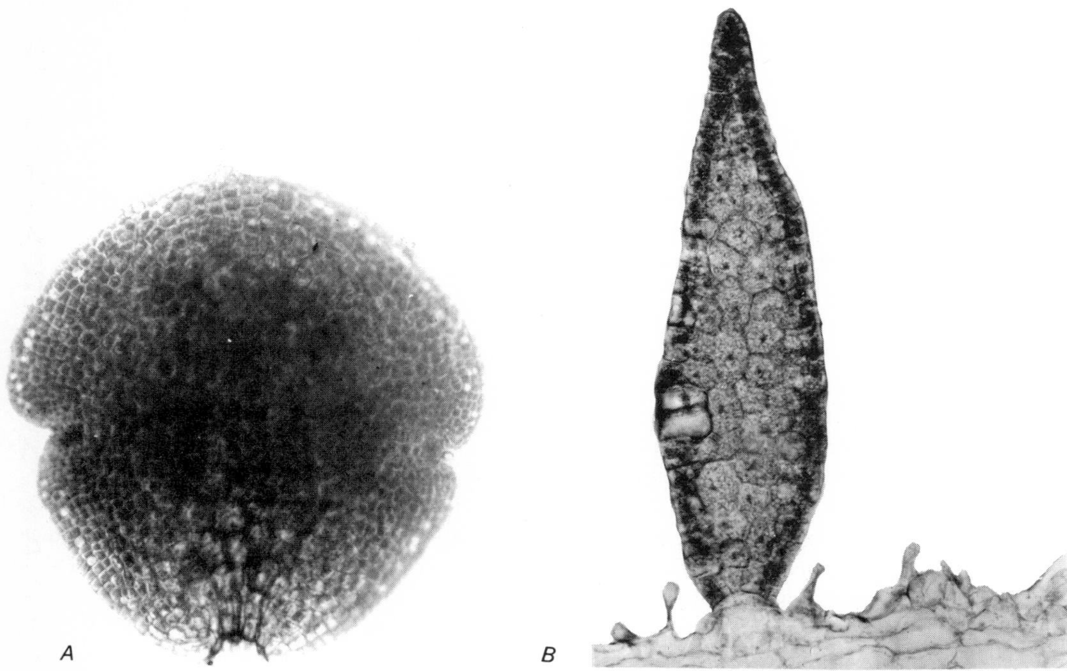

Fig. 18-11 Gemmae of Marchantia. *A, surface view of a gemma; B, lengthwise section of a gemma.*

forced to the underside of the receptacle. The archegonia are protected from excessive evaporation by the presence of thin layers of protective tissues or scales hanging from the underside of the branches of the receptacles and covering the young archegonia. At maturity the archegonia are pendulous on the underside of the receptacles.

THE GAMETANGIA (SEX ORGANS) The gametangia of the Marchantiales are essentially alike in all genera. Both antheridia and archegonia originate from superficial cells derived from segments of the apical cell near the tips of the thallus or, in the higher forms, near the growing points of modified branches (Figs. 18-7, 18-14).

At maturity an antheridium, or microgametangium, is an ovoid structure supported by a short multicellular stalk. In transverse section it appears circular. Structurally an antheridium consists of an outer wall of sterile cells and an inner fertile mass of small cubical cells known as the spermatogenous tissue (Fig. 18-8*A*). Each cell of this tissue divides diagonally forming two cells, from each of which a minute biflagellate sperm, or microgamete, is developed. At maturity the wall of the antheridium ruptures, and the sperms are free to swim to the archegonia. In *Marchantia* the antheridia are produced in large numbers, the older ones toward the center and the younger ones toward the margins of the disk-shaped receptacles. The surrounding vegetative tissue on the surface of the antheridial receptacle develops an overgrowth which makes the antheridia appear to occur in depressions.

At maturity the archegonium, or macrogametangium, is a flask-shaped structure. It has an enlarged basal portion, the **venter,** a long **neck,** and a short **stalk** attaching the organ to the thallus. The venter of the archegonium contains the egg, or macrogamete, and a **ventral canal cell.** The neck consists of a central row of cells and the surrounding neck wall cells. The upper part of the neck is made up of cap cells (Fig. 18-14*G*). Both venter and neck are circular in transverse section. When the archegonium is mature, the neck canal cells and the ventral canal cell disintegrate and a canal is opened from the tip of the neck to the egg cell in the venter. The sperms move through the liquid formed from the disintegration of the neck canal cells.

Mature archegonia are frequently found in young receptacles before much elongation has occurred in the stalks of the gametophores. This location, together with the habitat of these plants, explains the ease with which the sperms reach the archegonia after their release from the antheridia. The liverworts all grow in situations where at times there is sufficient water to serve as a medium through which the sperms reach the eggs. Under greenhouse conditions,

Fig. 18-12 Archegonial plants of Marchantia.

Fig. 18-13 Antheridial plants of Marchantia.

with just the ordinary watering of the plants, fertilization generally occurs.

SYNGAMY, OR FERTILIZATION The sperms, or microgametes, are attracted by chemicals given off by the archegonium. Frequently several of these microgametes penetrate the neck of the archegonium and reach the venter, but normally only one of them unites with the egg, or macrogamete (Fig. 18-15*A*). Each gamete carries the 1*N* number, or haploid complement, of 8 chromosomes. Their union brings the 2*N* number, or diploid complement, of 16 chromosomes into the nucleus of the zygote.

THE SPOROPHYTIC PHASE Fertilization initiates the diploid phase of the life cycle. With union of gametes, the **zygote,** or first cell of the sporophyte, is formed. It also results in an activation or physiological stimulation that expresses itself in the growth and further development of that structure.

The development of the young sporophyte, the **embryo,** proceeds from the growth of the single-celled zygote. The first division of the zygote is transverse to the vertical axis of the archegonium and the resulting cells give rise to different parts of the sporophyte. The outer of the two cells develops the capsule, containing the spore-bearing tissue, and the inner one forms the foot and stalk. Further development of the embryo proceeds by mitotic divisions, which first form four and then eight cells. Later divisions produce cells in irregular formation. As the young embryo develops, it elongates vertically (Fig. 18-15).

The sporophyte developing from the zygote is retained in the venter of the archegonium during its embryonic stages. The foot portion of the sporophyte is embedded in the archegonial receptacle, and as the young sporophyte develops, the venter of the archegonium enlarges and for a considerable time entirely envelops the embryo. The old archegonium is at this stage called the **calyptra** (Fig. 18-15*E, F*). Eventually, by elongation of the stalk, the capsule is pushed through the calpytra, which remains at the base, around the foot.

THE MATURE SPOROPHYTE The mature sporophyte is only a few millimeters long. Structurally it consists of an expanded foot, a short, relatively thick stalk, and a spore case, or capsule. The sporophyte is pendulous from the female receptacle to which it is attached by the anchorlike foot. The foot and stalk show only slight differentiation of tissues. The capsule contains the spore-bearing tissue and numerous sterile elongated cells, the **elaters.** For a while, after emerging from the calpytra, the sporophyte may be slightly green, but at maturity it is yellowish in color and is dependent upon the old gametophytic thallus for its nutrition.

The capsule (Fig. 18-15*F*) is an ovoid structure and has a wall consisting of a single layer of cells, surrounding the spore-bearing tissue, and is the most highly differentiated portion of the sporophyte. Within the spore-bearing tissue, or **archesporium,** are produced isodiametric cells that will produce meiospores; for this reason the archesporial cells are called **spore mother cells.** Besides these there are slightly elongated cells which are the young elaters. The elaters develop by further elongation and by the secondary deposition of material on their walls. At maturity they appear as long narrow cells with spirally thickened walls. The twisting of the hy-

Fig. 18-14 Development of archegonium in Marchantia. *A–E, young stages; F, older archegonium; G, mature archegonium.*

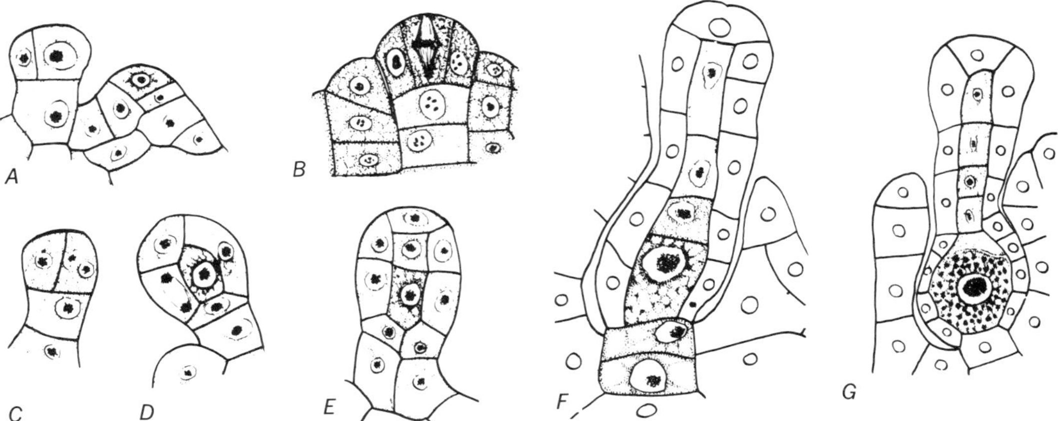

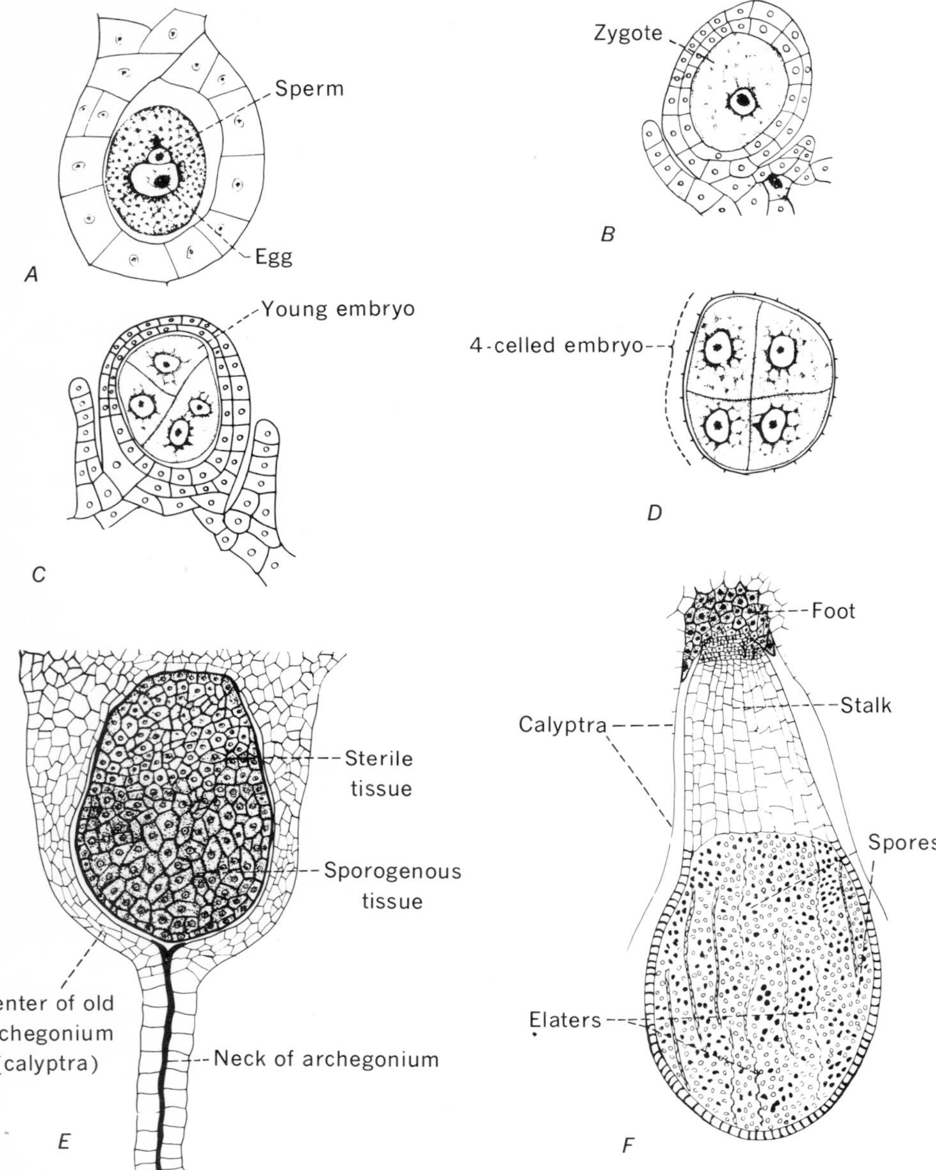

Fig. 18-15 Stages in development of sporophyte of Marchantia. *A, fertilization of egg by sperm; B, young zygote; C, D, early stages in division of young sporophyte; E, multicellular sporophyte, showing differentiation of potential sporogenous and sterile tissue; F, mature sporophyte.*

groscopic elaters assists in the dispersal of spores produced in the spore case. As the capsule develops, the spore mother cells increase in size, separate slightly, and become spherical in form. As a result of meiosis, each spore mother cell, or meiocyte, forms a quartet of meiospores. During the process the diploid number of 16 chromosomes in each meiocyte nucleus is reduced to 8 in the nucleus of each meiospore. The meiospores are thus $1N$ and are the first cells of the gametophytic phase. At maturity, when the capsule dehisces in an irregular manner, the meiospores and elaters are released.

DEVELOPMENT OF THE NEW GAMETOPHYTE When the meiospores are released from the capsule, they fall to the ground, where, if conditions are favorable, they may germinate. Upon germination a meiospore produces a short filament of several cells. Growth takes place by division of an apical cell located in the tip of the filament. As growth proceeds, the thallus develops and eventually forms gametangia and gametes (Fig. 18-16).

ANTHOCEROTAE—HORNED LIVERWORTS

The class Anthocerotae represents a very small group of liverworts consisting of five genera. *Anthoceros*, found in temperate regions, usually occurs on very moist clay banks, frequently along hillside roads or along streams.

The noteworthy features of the members of the Anthocerotae are the very simple form and structure of the thallus, the sunken gametangia, and, most prominent of all, the unusual development of sporophytic structures (Fig. 18-17). The thallus, in comparison with that of other liverworts, is simple in structure. It lacks differentiation in external form and internal structure. An unusual feature of the cells of the gametophyte is the presence of a single chloroplast which contains a pyrenoid. Pyrenoids are characteristically present in the chloroplasts of the cells of algae but are unknown elsewhere, with the exception of *Anthoceros* (Fig. 18-18*H*).

Both archegonia and antheridia are developed within the tissues of the thallus rather than superficially as in other liverworts. Although the gametangia may appear sunken in other forms, they actually develop superficially and come to lie in pockets formed by the more rapid growth of thallus tissue immediately surrounding the gametangia. In *Anthoceros* the antheridia are borne in groups sunken in the thallus. The archegonia are not so completely embedded as the antheridia, but the wall of the venter and part of the neck are confluent with the tissues of the thallus (Fig. 18-18).

The structure of the sporophyte is the feature of greatest interest in the group (Fig. 18-19). It is a slender cylindrical structure of fairly uniform thickness throughout its length, consisting of a foot and capsule. It stands erect upon the thallus to which it is attached by means of the foot. There is a central tissue composed of sterile cells, called the columella. In transverse section the columella shows 16 cells in rectangular arrangement. Surrounding the columella is a cylinder of sporogenous tissue, which in turn is surrounded by a cylinder of sterile tissue and finally by the epidermis. In the cylinder of sporogenous tissue, alternating layers or bands of sterile and spore-producing cells occur. In the spore-producing bands, sporocytes, or spore mother cells, eventually round off, undergo meiosis, and give rise to meiospores. The outer tissues of the capsule are abundantly supplied with chlorophyll, which enables them to synthesize carbohydrates. The ability of the sporophyte to carry on photosynthesis is more highly developed in this group than in any other of the liverworts. Stomata are present in the epidermis of the sporophyte.

A feature not shown at all by the sporophytes of any other liverworts is the presence in the *Anthoceros* capsule of a mass of meristematic tissue located in the lower portion of the capsule,

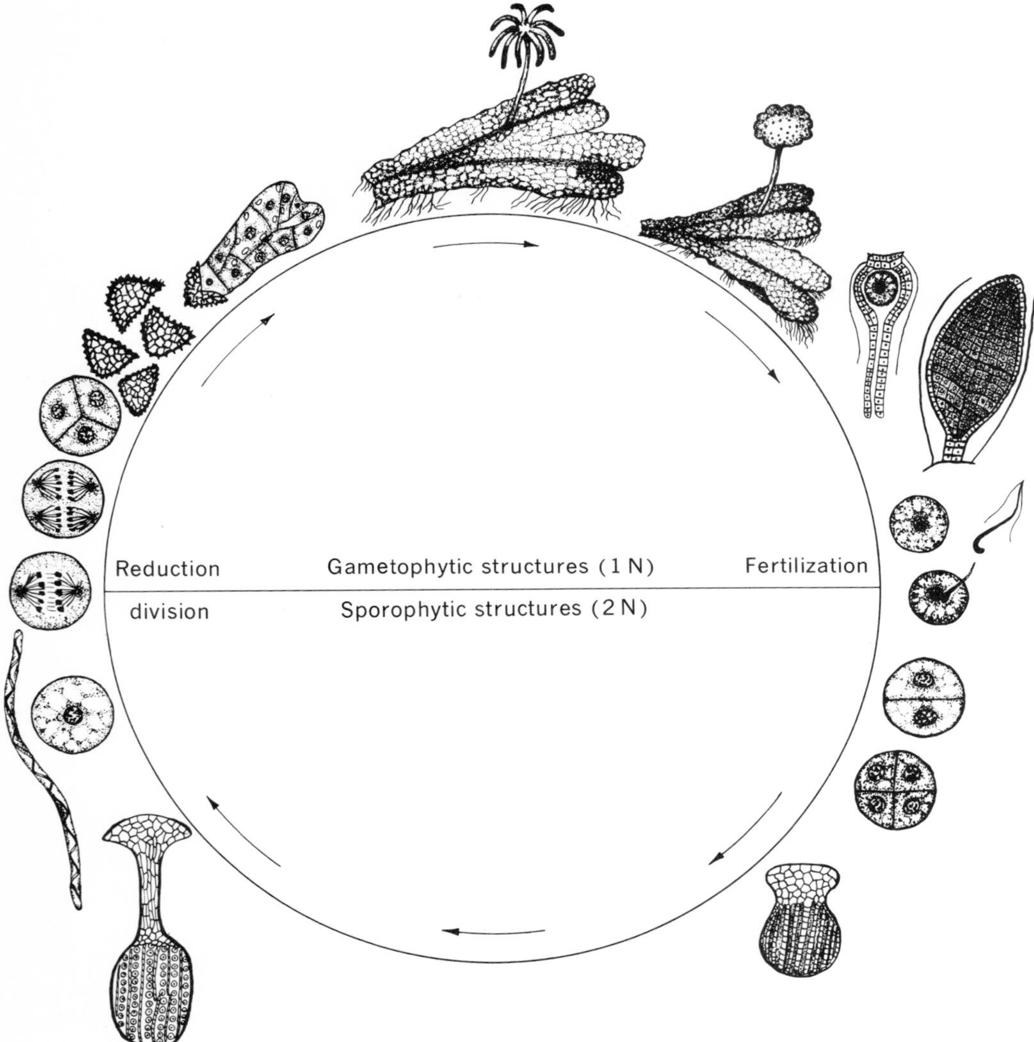

Fig. 18-16 Diagrammatic representation of the life cycle of Marchantia.

which provides for continued growth of the capsule over an extended period and prolongs the period of meiospore production. Mature meiospores may be found in the tip of the capsule while the meristematic tissue at the base is still giving rise to sporocytes. At maturity the capsule dehisces into two valves and its appearance at this time justifies the common name horned liverworts often applied to them.

The *Anthoceros* sporophyte, with its alternating layers of sterile and spore-producing tissue, is highly specialized. Bower, a British botanist, made the arrangement of sterile and spore-producing tissue in the capsule of the Anthocerotae the basis of a theory for the origin and evolution of leaves and sporangia of the higher plants.

MUSCI—MOSSES

The mosses, a large group of plants, constitute the higher bryophytes. They are worldwide in distribution, growing in every kind of habitat from hydrophytic to xerophytic. Most species of mosses, however, grow in moist situations, such as moist woodlands.

The conspicuous part of the moss plant consists of the more or less erect caulid which bears expanded phyllids. Rootlike strands, known as rhizoids, anchor it to the substratum. The "leafy stems" are known as the **gametophores** because they bear the gametangia and the gametes. After fertilization the sporophyte is developed at the apex or on the side of the gametophore, depending on the position of the archegonium.

The outstanding structural features which distinguish mosses from liverworts are the algalike **protonema** which develops from the meiospore, the radial symmetry of the "leafy stems," and the elaborate capsule of the mature sporophyte. The class Musci is subdivided into six subclasses, and many families are assembled into about fifteen orders. Recognition of the six subclasses is based largely on the characters of the highly developed capsule and its method of dehiscence. The six subclasses are Tetraphidae, Polytrichidae, Sphagnidae, Andreaeidae, Buxbaumiidae, and Bryidae. In this limited discussion *Sphagnum,* the single genus representing the subclass Sphagnidae, and a generalized moss plant are considered.

Sphagnum

General features The single genus *Sphagnum,* the bog or peat moss, is represented by several hundred species of worldwide distribution. These mosses grow only in moist situations, especially in ponds or bogs, where, in many cases, growth of the *Sphagnum* is the primary cause of the filling up of such bodies of water. Aquatic species of *Sphagnum,* growing both along the shoreline and as extensive mats on the surface of the water, contribute to the organic material which, as it accumulates, fills both the shallow and deeper portions of the bog. The increased acidity of the water, contributed to by *Sphagnum,* assists in this process by delaying the decay of organic material. As the water is replaced with these deposits, other plants unable to establish themselves in an aquatic habitat may invade the area. In a somewhat opposite role, *Sphagnum* often invades drier areas some distance from the shoreline of the bog, and because of the plant's ability to hold large quantities of water, the environment becomes semiaquatic and too wet sometimes for the growth of trees that normally would occupy such areas.

Sphagnum is of considerable commercial importance. Peat, used for fuel in some regions, is composed of carbonized *Sphagnum* and its bog-plant associates. *Sphagnum* is gathered and dried and, upon remoistening, is extensively used for packing living plants for shipment.

Fig. 18-17 Anthoceros *thalli with sporophytes. (Drawn by Elsie M. McDougle.)*

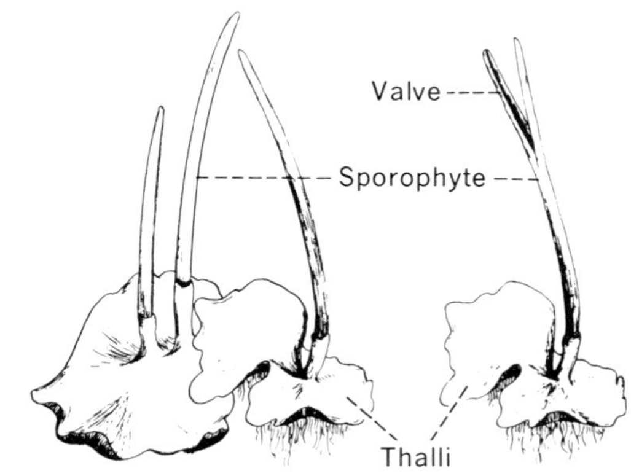

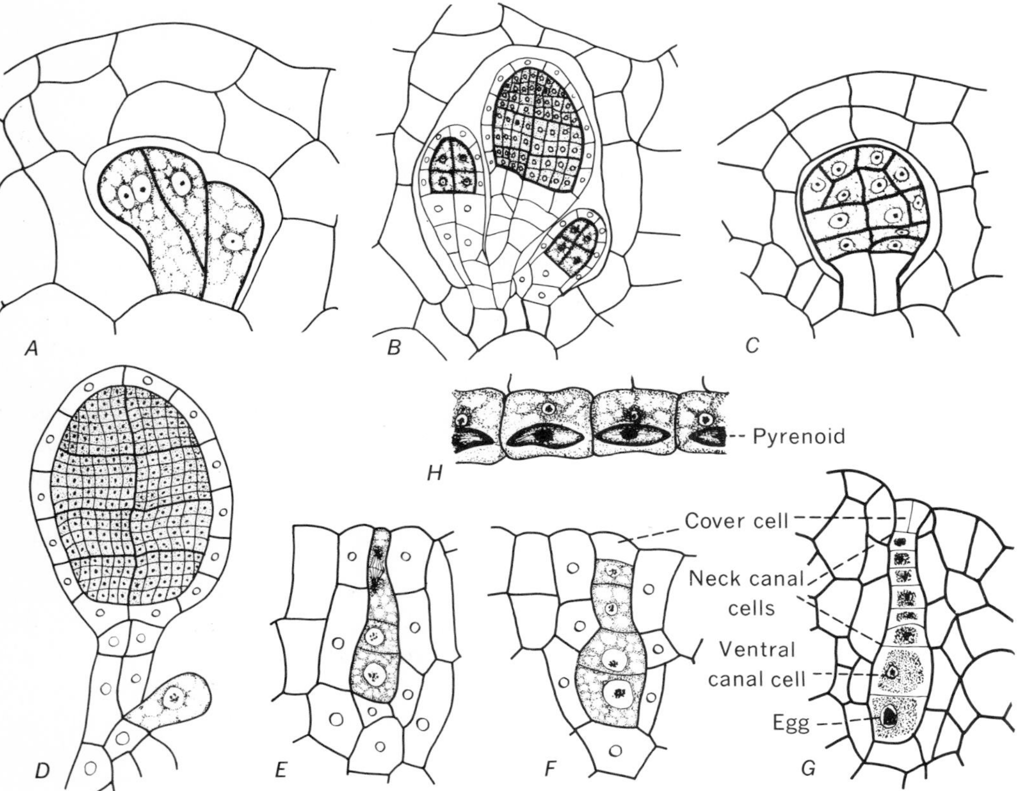

Fig. 18-18 Structural details of Anthoceros. *A–D, stages in development of antheridia; A, B show development by budding of several antheridia in a chamber; D, mature antheridium; E–G, stages in development of archegonium; E shows spindle in cell dividing to form cover cell and primary neck canal cell; G, mature archegonium; H, detail of vegetative cells from thallus with the characteristic pyrenoids.*

The structural features of *Sphagnum* indicate that it occupies a position in the same evolutionary line as other Musci, but it is somewhat isolated from them.

The gametophytic structures Upon germination the meiospore produces a very short filament which eventually forms a branched thallus body resembling that formed by some of the liverworts (Fig. 18-20*A*). The thallus body forms erect branching structures which, upon maturity, become the gametophores and are the recognized *Sphagnum* plants (Fig. 18-21). The gametophores consist of a central caulid with numerous branches which are covered with phyllids. The branches of the gametophores are of two kinds, short erect branches, which, clustered near the apex, form a sort of **rosette,** and long pendant structures, which fall gracefully around the caulid (Fig. 18-21). The short erect branches produce the gametangia. *Sphagnum* plants are either monoecious or dioecious; but even when both kinds of gametangia are on the same gametophore, the antheridia and archegonia are produced on separate branches.

The caulids and phyllids of the gametophore have a relatively complex structure which varies with the different species (Figs. 18-22*A*; 18-23*I*, *J*). The caulids consist of a central mass of tissue and an outer region. The cells of these two regions are differentiated, those of the central portion being elongated in the vertical direction. The phyllids are made up of two kinds of cells, large hyaline cells, often with conspicuous thickenings and large pores, and continuous chains of small elongated cells which contain the chloroplasts (Fig. 18-22*B*). When a portion of the phyllid is magnified, the small cells appear in a reticulate formation with the large hyaline cells occupying the "meshes" of the network. The absorptive properties of *Sphagnum* are probably to be attributed to the water-holding power of the hyaline cells of the phyllids.

The antheridia at maturity are spherical bodies borne on long stalks and located on the short erect caulids in the axils of the phyllids (Fig. 18-23*C*–*G*). The sperms are elongated, spirally curved structures and are biflagellate. The archegonia, which in general resemble those of *Marchantia*, are produced on the ends of the erect caulids; at maturity they are relatively large structures with stalk, long neck, and massive

Fig. 18-19 Development of the sporophyte in Anthocerotae. A–C, Anthoceros; *A, B show old archegonial canal; B, C show sporogenous tissue; D, somewhat diagrammatic drawing of sporophyte of* Notothylas *with tetrads in sporogenous region.*

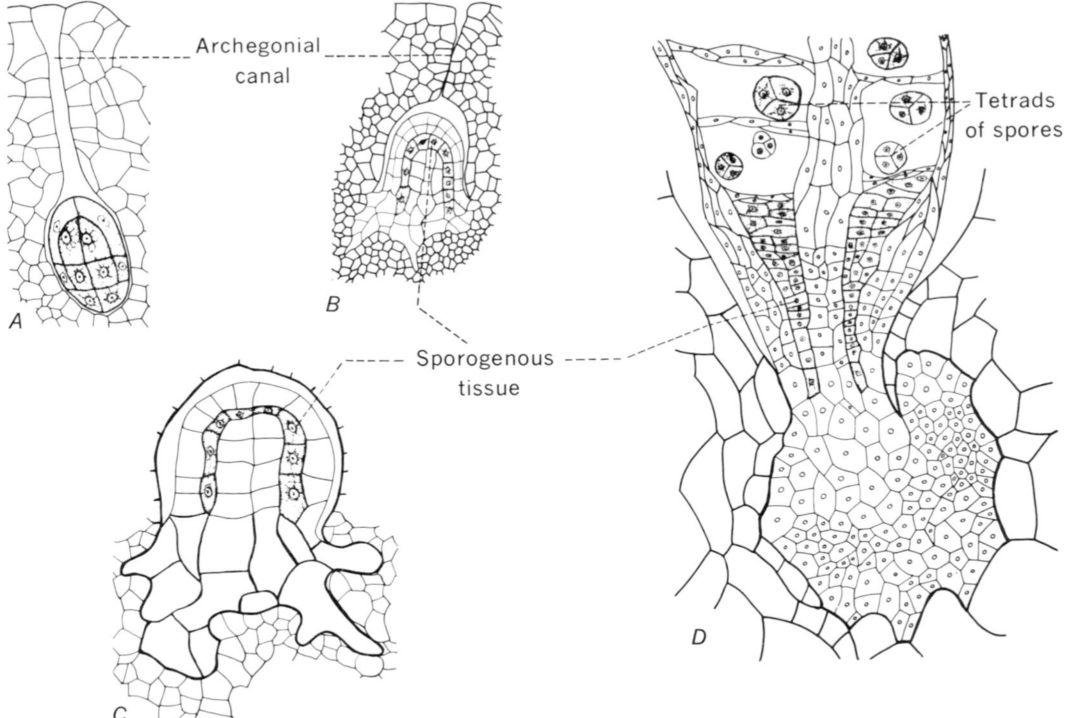

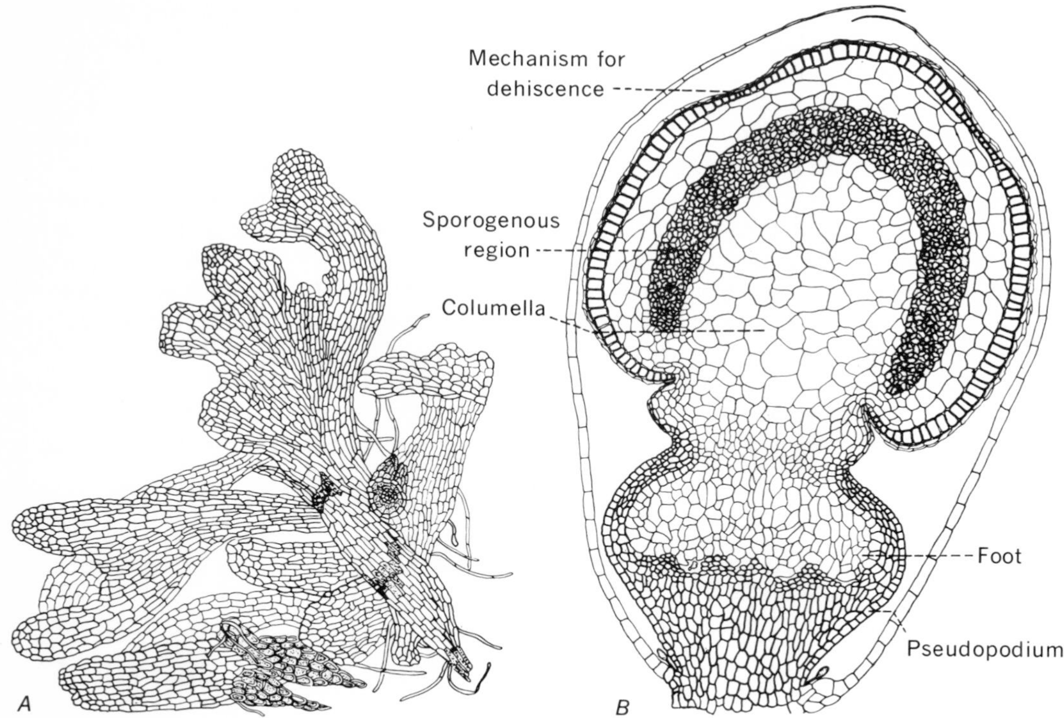

Fig. 18-20 A, young thallose gametophyte of Sphagnum *with rhizoids and bud from which the erect gametophore may arise; B, sporophyte. (Drawn by Laura K. Wilde.)*

venter. The egg cell is relatively small and of about the same size as the ventral canal cell, which it greatly resembles (Fig. 18-23*H*).

The sporophytic structures The sporophytes of *Sphagnum* are generally regarded as of infrequent occurrence. Their scarcity probably results from the infrequency of fertilization, which in turn appears to be correlated with the water level in the bogs where *Sphagnum* grows. When they are found, the sporophytes are generally located on plants growing in clumps upon little knolls or hummocks in the bog or, perhaps, in depressions. Investigators have secured abundant sporophytes by artificially applying the sperms to the female plants at the time when the eggs are mature within the archegonia.

The zygote, formed by the union of the sperm and egg, develops into the sporophyte, which eventually produces meiospores. The mature sporophyte is small, consisting of a spherical capsule, dark brown or black in color, a very short seta or stalk, and an expanded foot, which serves to attach the sporophyte to the gametophyte (Figs. 18-20*B*, 18-21). The growth of the **pseudopodium,** a gametophytic structure, elevates the sporophyte above the apex of the branch which bore the archegonium containing the egg. The pseudopodium is an outgrowth from the apex of the caulid and is a portion of the gametophyte. The dome-shaped spore layer, or **archesporium** (Fig. 18-20*B*), formed within the capsule resembles the spore layer in the capsule of the liverwort *Anthoceros*.

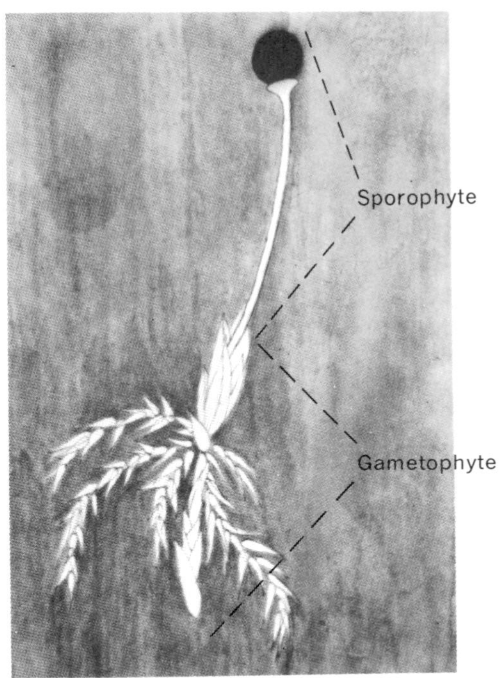

Fig. 18-21 Gametophyte and sporophyte of Sphagnum.

OTHER MOSSES

Mosses in general comprise a group of small, green, mostly terrestrial plants of wide distribution. Some of them grow in exposed and rather dry situations as on the surface of rocks and on tree trunks. Most of them, however, are found in moist and, frequently, in shaded locations. A few grow submerged in water. Multicellular rhizoids attach the plant to the substratum. Moss plants vary in size from specimens scarcely visible without the aid of a hand lens to large forms, like the water moss *Fontinalis*. In general they are more slender and delicate than the thallose liverworts, but at the same time they are more complex. Some grow as unbranched erect structures, and others are branched. The branched types may have a prostrate stem growing horizontally on the surface of the substratum from which arise erect branches. Moss plants are elaborate in form, resembling the leafy liverworts in this respect. The main axis of a moss plant is called a caulid and its expanded parts phyllids. These structures belong to the $1N$, or gametophytic, phase of the life cycle and are not homologous with the true stems and leaves of the flowering plants, which are $2N$ and are borne on the sporophyte. True leaves and stems have a well-differentiated epidermis with stomata and guard cells. There is a vascular system which serves as conducting tissue and which provides support in the leaves and stems of all higher plants. The "stems" (caulids) and "leaves" (phyllids) of mosses lack these well-developed structures.

Fig. 18-22 Structural features of Sphagnum. *A, surface view, and B, transverse view, of cell structure in leaf-like organs. (Drawn by John Shuman.)*

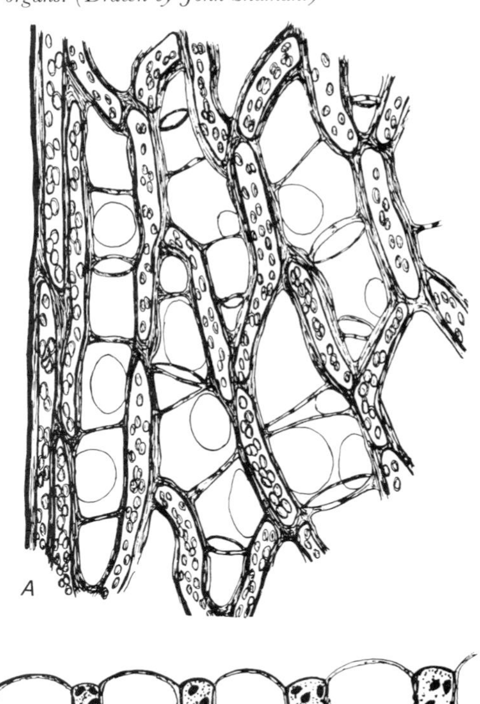

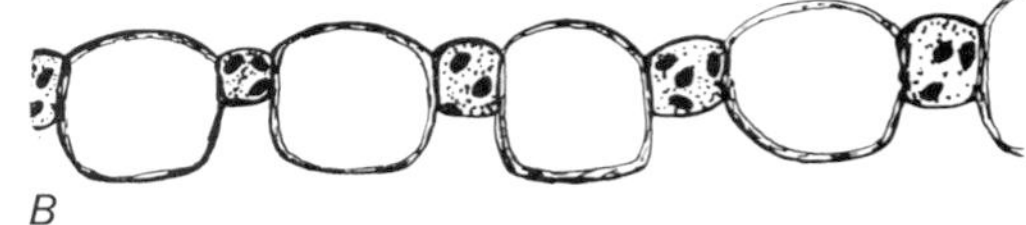

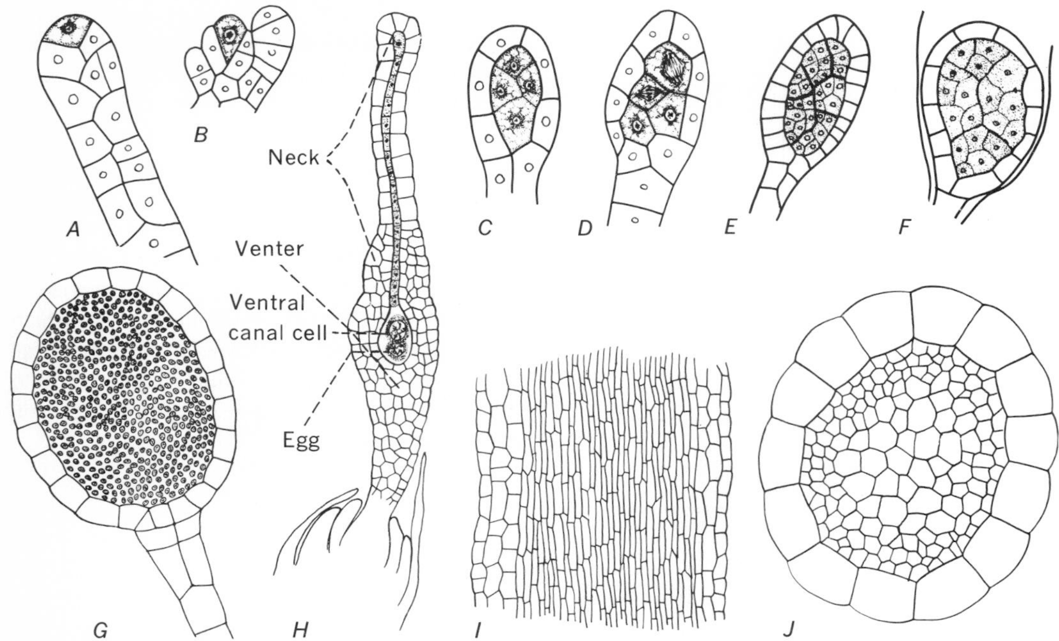

Fig. 18-23 Structural features of Sphagnum. *A, B, apical cells; C–F, stages in development of young antheridia; G, mature antheridium; H, mature archegonium; I, longitudinal, and J, transverse section of a caulid showing slight differentiation in tissue. (G, I, and J drawn by Helen D. Hill.)*

The caulids of mosses are differentiated into an epidermal layer, a cortical region of mostly isodiametric, thin-walled cells, and a central cylinder composed of thick-walled cells that are of small diameter and somewhat elongated. The epidermis of moss caulids lacks stomata. Epidermal cells and cells in the younger cortex are generally chlorophyllose. It is thought that the central core functions as supporting rather than conducting tissue. Caulids of the moss plants, which may be either branched or unbranched, increase in length through the activity of a single apical cell which rarely may have two but more frequently has three cutting faces. Similar apical cells are also present in the growing points of branches. In their young stages the position of the phyllids is determined by the type of apical cell in the apex of the caulid. An apical cell with two cutting faces determines that the phyllids will occur in two ranks. The three-faced apical cell gives rise to a radial arrangement of phyllids which occur in three ranks.

As the plant develops, the phyllids tend to lose their definite positions and may no longer appear to be in vertical rows. All three-ranked types, however, maintain their radial arrangement. The structure and complexity of moss phyllids vary in the different kinds of mosses. In most mosses the phyllids have only one layer of similar cells making up the blade. Usually there is a midrib (costa) of several layers of cells. Mosses with delicate phyllids of simple structure usually grow in moist, shaded locations. Other mosses, often growing in dry localities, have phyllids composed of many cell layers.

The gametophytic phase The adult vegetative moss plant, the most conspicuous part of the life cycle, is gametophytic, but it does not represent the entire gametophyte. The meiospore is the first cell of the gametophyte, and structures produced during the gametophytic phase include (1) the protonema which grows from a germinating meiospore; (2) the erect gametophores which grow from the protonema; (3) the antheridia and archegonia; (4) their gametes, sperms and eggs, which are produced on the gametophore. The **protonema** is an inconspicuous filamentous algalike structure. A protonema may develop in a prostrate position upon the surface of the substratum or at times in a slightly aerial position. It is composed of short cylindrical cells, each with a nucleus, chloroplasts, and cytoplasm. A protonema may, in its growth, penetrate the substratum; the cells then tend to lose their chlorophyll and the cell walls become brownish in color (Fig. 18-24*B*). The leafy moss plants are erect, highly developed branches of the protonema. Several buds may develop on a single protonema. This accounts in part for the occurrence of moss plants in clumps.

After the gametophores are established, the protonemal structures may disappear or persist, depending on the species. Septate rhizoids grow from the bases of the gametophores and serve to attach the plant to the soil and to absorb moisture and minerals. The gametangia develop at the apex of the caulids or their branches (Fig. 18-23*A–H*). This feature and the fact that they arise from the prostrate protonema suggest that the erect leafy moss plants are in reality gametophores, homologous with the erect gametophores growing on the thalli of *Marchantia*.

THE GAMETANGIA The gametangia (Figs. 18-25*A, B;* 18-26) are produced terminally, either on the main axis of the gametophore or on its side branches. Some species are monoecious, with both kinds of gametangia borne on the same plant, either in the same cluster or in different clusters. Other species are dioecious with the antheridia and archegonia borne on separate plants. These may be called male, or antheridial, plants and female, or archegonial, plants. In some species male and female plants may be strikingly different and may be recognized at a glance (Fig. 18-27*A, B*).

The mechanism of dehiscence of an antheridium generally consists of one or two large cells that form a cap at its apex. When a mature antheridium is wet, the wall cells absorb moisture and swell. Pressure within the wall cells, including the larger cap cells, may become sufficient to burst the cap cells or to break their connections with adjacent cells, thus freeing the mass of sperms (Fig. 18-25*B, C*).

Fertilization Fertilization, or syngamy, in the mosses occurs when a haploid sperm, or microgamete, fuses with a haploid egg, or megagamete, and forms a diploid zygote. Since this union marks the beginning of the diploid or sporo-

Fig. 18-24 Moss protonema. A, with bulbils; B, general habit of growth.

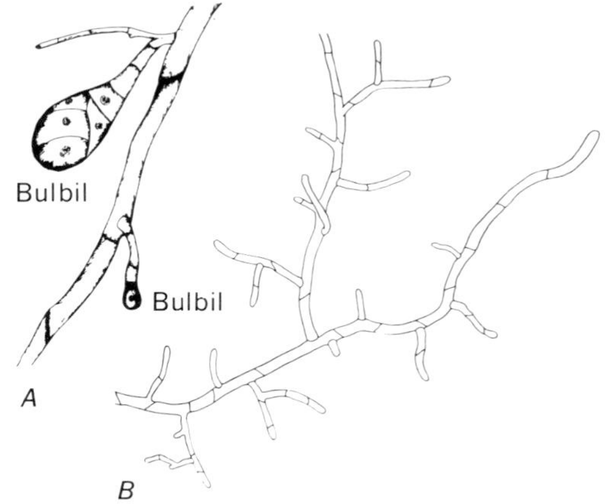

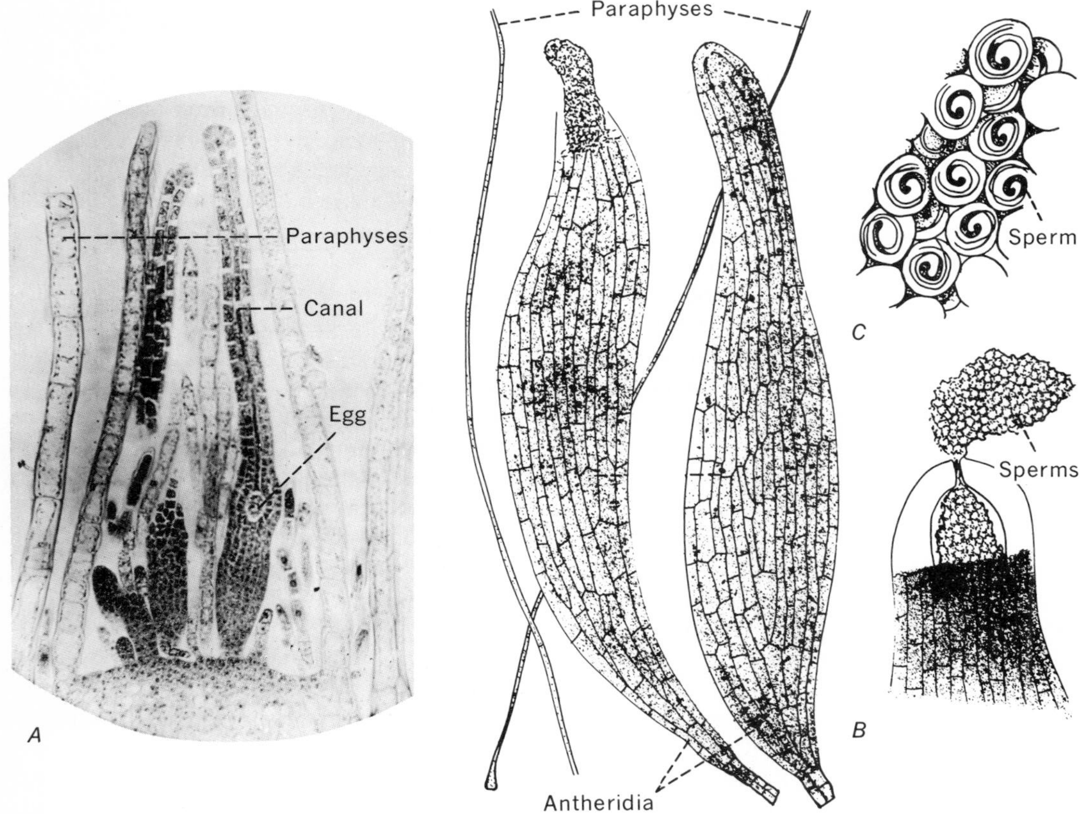

Fig. 18-25 Gametangia of a moss. A, showing archegonium with egg and long canal; B, left, two antheridia and paraphyses, right, tip of antheridium enlarged to show sperms emerging; C, biflagellate sperms enlarged.

phytic phase, it is one of the two critical points in the reproductive cycle. A stimulus to growth following fertilization is expressed in the development of the embryo, or young stage of the sporophyte (Fig. 18-28).

The sporophytic phase The zygote formed at fertilization is the first cell of the diploid or sporophytic structure. The first division of the zygote is transverse to the long axis of the archegonium. This is followed by other divisions which finally form apical cells in both the upper and lower ends of the elongating embryo (Fig. 18-28*A–F*). The early development of the sporophyte is largely a sequence of cell divisions and growth of the resulting cells. Since the zygote is located in the archegonium, all the early stages of embryonic development take place within the venter of the archegonium. As the embryo increases in size, the tissues of the archegonium are also stimulated to growth and there is a corresponding enlargement of the venter.

THE MATURE SPOROPHYTE Later development of the embryo results in the production of the mature sporophyte with its elaborate structural and cellular differentiation. The mature sporophyte consists of: (1) a **foot** embedded in, and attaching the sporophyte to, the apex of the old gametophore; (2) an elongated stalk, or **seta,** connecting the foot; and (3) the **capsule,** or spore case (Fig. 18-27*D*). Only slight tissue differentiation is shown by the foot. Water and minerals in solution, necessary for the life and growth of the sporophyte, are absorbed from the gametophytic tissues. The seta shows more differentiation than the foot. A well-defined central strand of small thin-walled cells, similar to those found in the caulid, may conduct water. A layer of thick-walled cells that may serve as supporting tissue surrounds the central strand.

The distal end of the sporophyte is differentiated into an elaborate sporangium, or capsule. Moss capsules are of various shapes; most often they are almost spherical or pear shaped. Many moss capsules are circular in cross section. The capsule may be erect or pendant from the apex of the stalk. Structurally the capsule consists of an upper urn-shaped portion, the **theca,** where the spores are produced, and often an enlarged, sterile portion at the base, the **apophysis.** An apical covering, a cap, is called the **operculum.** This structure, a part of the mechanism of dehiscence, takes its name from a kind of cap worn by some of the Roman soldiers. At the base of the operculum there is also a ring of specialized epidermal cells called the **annulus.** These cells are often enlarged and, with drying, break to form a line of separation at the time the operculum is released.

Beneath the operculum there is a ring of pointed multicellular structures called teeth. Collectively the teeth constitute the **peristome,** so named because of its position surrounding the mouth of the capsule. The peristome is made up of one or more rows of teeth of varying number. The number and structure of the teeth are used as characters in the determination of the genera and species of the mosses. During periods of dampness the teeth curl inward, tending to close the mouth of the capsule. When they are exposed to dry conditions, the teeth curl upward, opening the capsule, and spores are likely to be shed in dry weather when conditions for dissemination are favorable.

In the mosses, generally, there is an additional covering called the **calyptra,** carried over the apex of the capsule and external to the operculum. The calyptra is not a part of the sporophyte but is a structure developed through the enlargement of the archegonium. As the zygote in the venter of the archegonium starts growth, the tissues of the archegonium also grow. During the early stages of development the growth of the calpytra parallels that of the sporophyte, and for a time the sporophyte and the archegonium enlarge proportionally. Eventually, however, the sporophytic tissues outgrow those of the archegonium. The tension is relieved by the rupture of the archegonium at its base. Thus freed, the archegonium, now called the calyptra, is carried as an apical covering of

Fig. 18-26 Stages in development of gametangia of moss. A, B, young archegonia; C, mature archegonium; D–G, early stages in the development of antheridia.

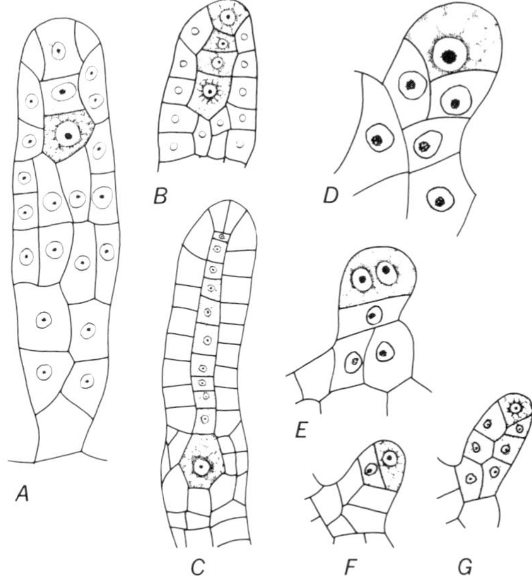

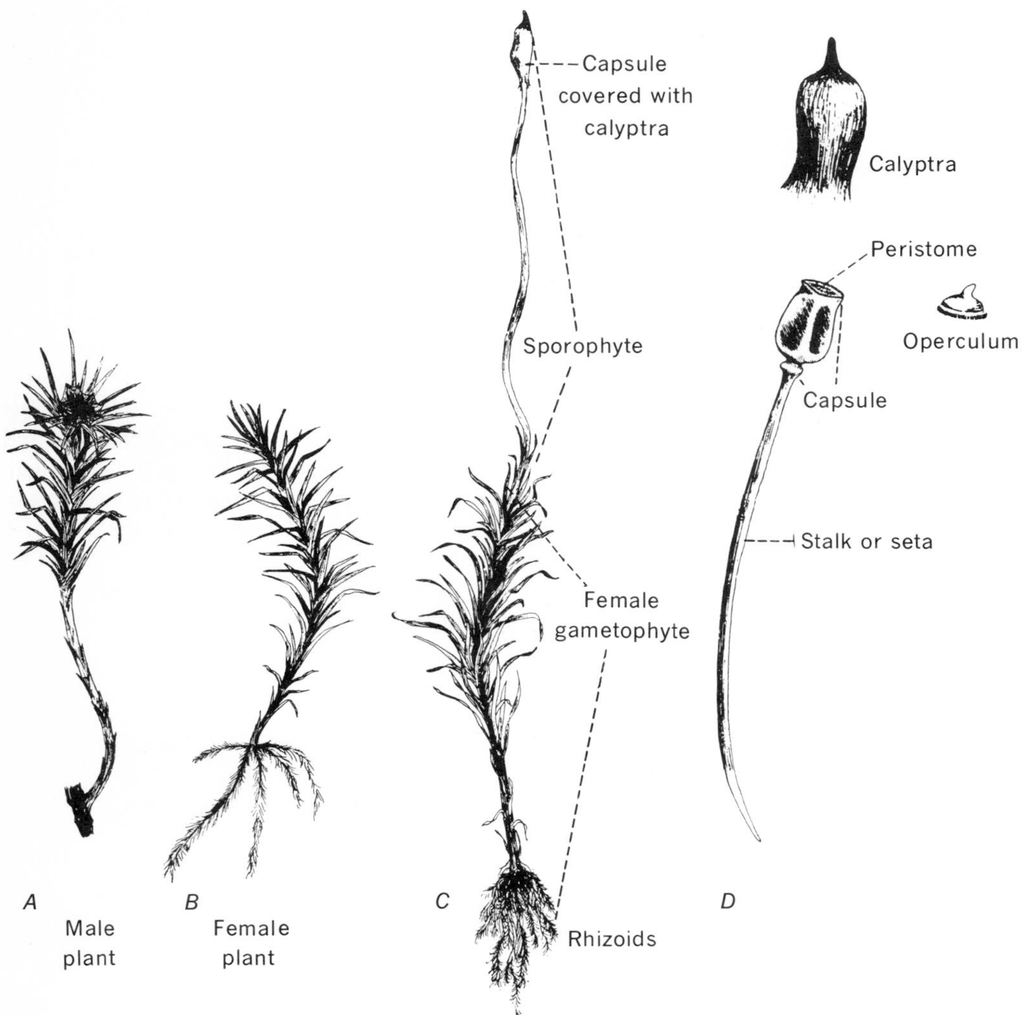

Fig. 18-27 Polytrichum commune, *a conspicuous moss. A, male gametophyte; B, female gametophyte; C, female gametophyte with sporophyte; D, sporophyte parts enlarged.*

the sporophytic capsule (Fig. 18-27*D*). Although the entire sporophyte is diploid tissue, the calyptra, originating as the archegonium, is haploid tissue.

DIFFERENTIATION OF TISSUES IN THE CAPSULE There is considerable differentiation of tissues in the capsule. A well-defined epidermal layer with stomata and guard cells, resembling those of the higher plants, covers the capsule. The stomata open into large intercellular cavities in the capsule, as in the higher plants. Just beneath the epidermis is a region, two or more layers of cells in thickness, that forms the wall of the capsule. The apophysis is a mass of sterile tissue often present as a bulbous enlargement in the lower portion of the capsule. Chlorophyllose tissue and stomata are abundant in this part of the capsule, although not restricted to this region. In the capsules of some mosses there are radial columns of chlorophyllose cells beginning in the apophysis and extending upward into the capsule. Within the capsule are two portions, an inner group of tissues, the **endothecium,** and an outer group, the **amphithecium.** These portions are differentiated during the development of the embryo. The endothecium consists of a central mass or core of tissue, the **columella,** and a cylindrical sporogenous layer, the **archesporium,** surrounding it. The amphithecium forms the outer layers of the capsule (Fig. 18-28*G*).

Meiosis and the formation of meiospores As the capsule develops, large numbers of sporocytes are formed in the sporogenous layer. Each of these sporocytes undergoes meiosis, forming a quartet of meiospores. Thick walls develop during the maturation of the spores, which are the first cells of a new gametophytic phase. At maturity of the capsule very large numbers of spores are found in a cylindrical or barrel-shaped mass surrounding the remnants of the central columella.

Vegetative, or somatic, multiplication The gametophytes of mosses have several methods of multiplication. The protonemata branch extensively and form numerous buds which de-

Fig. 18-28 Development of sporophyte of a moss. A, early stage just prior to fertilization; B–F, stages in growth of the sporophyte; G, transverse section of a young capsule.

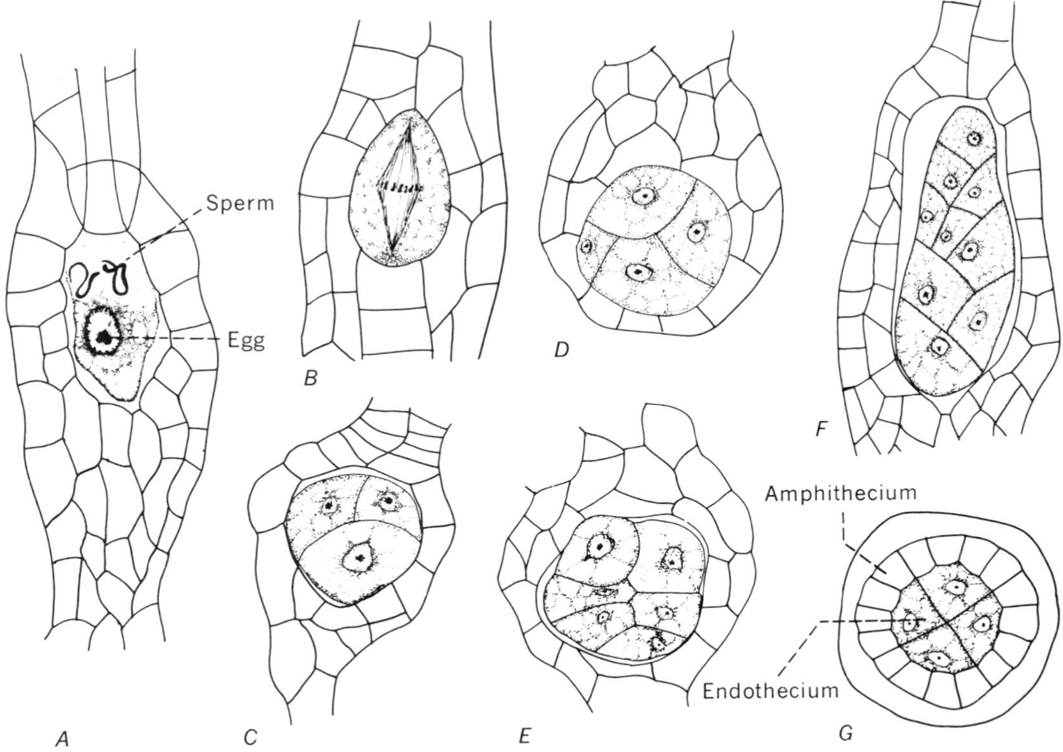

velop the leafy moss plants. Decay of the older parts of the protonemata leads to the separation of these plants as independent entities. Small bulbils may also develop on the protonemata. These become resting structures, which tide the organism over periods generally unfavorable to vegetative growth (Fig. 18-24*A*). Small, green, multicellular gemmae, resembling the gemmae produced on the thalli of *Marchantia*, are developed on many moss plants. These may grow into new plants. Secondary protonemal structures may develop from almost any part of a moss plant. They develop from the caulids, the rhizoids, and other structures. It has been possible in experimental material to induce artificially the production of protonemata from the diploid or sporophytic structures. These protonemata had the diploid number of chromosomes and eventually developed diploid gametophores, bearing gametangia with diploid gametes. Fusion of these diploid, or 2*N*, gametes produced 4*N* sporophytes.

QUESTIONS

1. What morphological characters are useful in determining the possible evolutionary pattern of the liverworts?

2. What information can be learned from a study of the bryophytes that may assist in our understanding of the invasion of a terrestrial habitat by aquatic plants?

3. Summarize anatomical features that might, if present, contribute to the adaptation and subsequent survival of plants in a terrestrial environment.

4. What facts are important in summarizing the nutritional relationships between gametophyte and sporophyte?

5. What characteristics of the bryophytes are most useful in identification? In what phase of the life cycle are these characters recognized?

6. What relationship, if any, does the effectiveness of spore dispersal have to survival of the species?

REFERENCES

BOWER, F. O. 1935. Primitive Land Plants. Macmillan & Co., Ltd., London. A part of the classical morphological literature; especially interesting for its discussion of the invasion of the land environment by aquatic plants.

CAMPBELL, D. H. 1895. The Structure and Development of Mosses and Ferns. Macmillan & Co., Ltd., London. Detailed discussion of structural features; a part of the classical literature.

CONRAD, H. S. 1956. How to Know the Mosses and Liverworts. H. E. Jaques, Mt. Pleasant, Iowa. A manual useful for identification of common species of mosses and liverworts.

GROUT, A. J. (ed.). 1928–1940. Moss Flora of North America North of Mexico. Published by the editor, Newfane, Vt. A three-volume comprehensive treatment of mosses.

INGOLD, C. T. 1965. Spore Liberation. Oxford University Press, Fair Lawn, N.J. Chapter 8 deals with the mechanisms of spore release in bryophytes.

SCAGEL, ROBERT F., et al. 1965. An Evolutionary Survey of the Plant Kingdom. Wadsworth Publishing Company, Belmont, Calif. Modern treatment, comparative in approach.

SCHUSTER, R. M. 1953. Boreal Hepaticae, a Manual of Liverworts of Minnesota and Adjacent Regions. *Am. Midland Naturalist*, **49**:257–684. Good illustrations.

SMITH, G. M. 1955. Cryptogamic Botany, 2d ed., vol. 2, Bryophytes and Pteridophytes. McGraw-Hill Book Company, New York. Relatively modern treatment of the structure of mosses and liverworts.

WARDLAW, C. W. 1955. Embryogenesis in Plants. John Wiley & Sons, Inc., New York. Plants from all divisions of the kingdom.

WATSON, E. V. 1964. The Structure and Life of Bryophytes. Hutchinson University Library, London. A brief, interesting treatment of the bryophytes.

19

VASCULAR PLANTS; THE LOWER GROUPS

Impression of fossil leaves of Sphenophyllum *and one of the Cycadofilicales from the Paleozoic era.*

This chapter contains a brief review of the classification of the lower groups of the vascular plants and a short comparative study of *Lycopodium, Selaginella, Equisetum,* and the ferns, including several life histories. An examination of these plants, together with several extinct genera, provides the opportunity for a brief discussion of vascular plants as inhabitants of terrestrial habitats, the origin of leaves, cauline and foliar sporangia, and the geologic timetable.

The development of a vascular system is regarded as one of the very important structural advances in the evolution of the plant kingdom, and it is also physiologically and ecologically significant. It is correlated with the ability of plants to develop into organisms of greater size in a terrestrial habitat. The Tracheophyta include both fossil and living vascular plants of the subdivisions Psilopsida, Lycopsida, Sphenopsida, and Pteropsida.

THE LOWER VASCULAR PLANTS

Subdivision **Psilopsida.**

Order Psilophytales. The oldest known vascular plants from the Silurian and Devonian periods. All forms now extinct.

Order Psilotales. Two living genera resembling the fossil Psilophytales.

Subdivision **Lycopsida.** The lycopods.

Order Lepidodendrales. Flourished as tree forms in the Carboniferous forests. All forms now extinct.

Order Lycopodiales. The living members are small plants in only two genera.

Order Selaginellales. A group of relatively small living plants included in a single genus with numerous tropical, subtropical, and temperate species. In many ways these plants resemble the fossil Lepidodendrales.

Order Isoetales. Contains the single living genus *Isoetes.*

Subdivision **Sphenopsida.** The plants with wedge-shaped leaves.

Order Calamitales. A group of giant tree forms that were prominent features of the Carboniferous forests. All forms extinct.

Order Sphenophyllales. Small fossil plants with wedge-shaped leaves from the Paleozoic era.

Order Equisetales. A group of generally small living plants, also represented in fossils by very large forms.

Subdivision **Pteropsida.** The ferns.

Class Filicinae.

Order Ophioglossales. Living fernlike plants.

Order Filicales. The common ferns with many living genera with ancestral types found as fossils in the Mesozoic era and extending back into the Carboniferous period of the Paleozoic era.

The Pteropsida also include the higher vascular plants (gymnosperms and angiosperms), which are taken up in the following chapter.

The relative size and importance of the gametophyte and the sporophyte In all members of the Tracheophyta the structures developed during the sporophytic phase are more prominent and live longer than those of the gametophytic stage. The sporophyte is the recognized plant body of all Tracheophyta, with the gametophyte relatively small, of simple structure, and, in the higher forms, restricted to microscopic proportions. The sporophyte of the Tracheophyta generally is differentiated into true roots, stems, and leaves with well-developed vascular tissue throughout all plant parts. Some primitive members of the group may lack roots and well-developed leaves. In all cases the sporophyte produces sporangia and meiospores as a result of meiosis.

Sporangia and meiospores The sporangium of the vascular plants is the structure in which meiosis takes place. More specifically, it is the place where sporocytes, or spore mother cells, divide meiotically to form groups of four meiospores, called tetrads. In the most primitive vascular plants and even in some advanced forms, sporangia are borne on stems; i.e., they are **cauline** in origin. In other, usually more advanced, vascular plants, sporangia are associated with leaves **(foliar),** and these sporangia-bearing leaves are called **sporophylls.**

To identify sporangia as foliar or cauline is useful to the botanist. Primitive vascular plants (living and fossil) frequently have no leaves; if sporangia are formed, they develop from stem tissue and are called cauline sporangia (Fig. 19-3). Other plants, which, it is believed, have evolved more recently, have leaves with the sporangia borne on them (Figs. 19-39, 19-40). The leaf, in this case, is referred to as a sporophyll and the sporangia are foliar in origin. Developmental morphological and paleobotanical evidence suggests that leaves evolved from a system of dichotomously branched stems and the involvement of sporangia-bearing stems in this evolutionary development resulted in sporangia-bearing leaves.

Sporangia may be rather widely dispersed on a plant, or they may be organized into a structure called a **strobilus,** or **cone.** In addition to the fact that sporangia may be borne on stems or leaves, there are two fundamental types of sporangia important in the vascular plants, the **eusporangium** and the **leptosporangium.** The eusporangium is found in most vascular plants and it develops from several sporangial initials. By maturity it has formed a sporangial jacket several cell layers in thickness. The leptosporangium develops from a single initial, and at maturity its jacket, or sporangial wall, is a single cell layer in thickness. The leptosporangium is found only in some of the more advanced ferns, commonly referred to as the leptosporangiate ferns. Plants in which only one type of meiospore is produced are **homosporous;** if two kinds of meiospores are formed, the plant is **heterosporous.**

Gametangia and gametes In vascular plants the gametangia are unisexual and produce one kind of gamete. Male gametes (microgametes, or sperms) are produced in antheridia. In the lower vascular plants female gametes (megagametes, or eggs) are produced within archegonia. Gametophytes that develop from homospores are independent plants and are not retained within the original spore case. Gametophytes (both microgametophytes and megagametophytes) that develop from heterospores are, for the most part, retained within the original spore case. Typically the gametophytic plant formed from the homospore is **monoecious** and bears both antheridia and archegonia, whereas the gametophytic plants formed from the heterospores are **dioecious** and the male and female gametangia are produced on separate male and female plants.

Sporophyte → homosporous meiospores → monoecious gametophyte →
megagametangia and microgametangia

Sporophyte → heterosporous meiospores → dioecious gametophytes →
megagametangia or microgametangia

PSILOPSIDA

The **Psilopsida,** the oldest and most primitive group of vascular plants, include the order Psilotales, represented by living plants, and the extinct order Psilophytales, found as Silurian and Devonian fossils (Fig. 19-1). In general the Psilopsida are primitive types of plants without roots and true leaves, although expanded appendages of the stem function as photosynthetic organs. Protosteles are characteristic of the vascular structures, but some living species are siphonostelic in the aerial stems. Leaf gaps, typical of the higher vascular plants, are lacking in the Psilopsida. In these primitive plants sporangia are eusporangiate and are borne terminally on stems.

Psilotales

The order Psilotales has two living genera, *Psilotum* (Fig. 19-3) and *Tmesipteris* (Fig. 19-4). The Psilotales are of special interest because they resemble more closely than any other living plants the extinct Paleozoic members of the Psilophytales. *Psilotum* is native in the tropical and subtropical regions of both the Northern and Southern Hemispheres and is small (usually

GEOLOGICAL PERIODS IN THE EARTH'S HISTORY
(Oldest period at bottom, figures represent estimated duration of periods in millions of years)

Era	Period	Epoch	Predominating Animals and Plants	Miscellaneous Events etc.
Cenozoic 75	Quaternary 1	Recent	recent animals; man recent plants	origin of crop plants; rise of man
		Pleistocene 1	mammals; birds; herbs	extinction of many plants and animals; glaciation
	Tertiary 74	Pliocene 9	mammals; birds; herbaceous plants	appearance of man; cooling climate; restriction of plant distribution
		Miocene 20	mammals; birds; angiosperms	retreat of polar floras; carnivores abundant
		Oligocene 10	mammals; forests	1st anthropoids; world-wide forests
		Eocene 20	herbivores; angiosperms	forest browsing herbivores; tropical plants in polar regions
		Paleocene 15	reptiles; mammals; angiosperms	modern birds and marine mammals appear; angiosperms ascending
Mesozoic 130	Cretaceous	60	reptiles; gymnosperms	ancestors of modern mammals appear; gymnosperms decreasing; angiosperms increasing
	Jurassic	30	dinosaurs; cycads	modern fishes; first birds; 1st angiosperms; conifers common; *Ginkgo*
	Triassic	40	dinosaurs; cycadophytes	1st mammals; plant types fewer; some conifers
Paleozoic 295	Permian	25	reptiles	much glaciation; land vertebrates rise; fewer plant types; early cycads and conifers
	Pennsylvanian (upper carboniferous)	25	coal flora	primitive insects; early reptiles; great coal forests and swamps; primitive gymnosperms
	Mississippian (lower carboniferous)	25	amphibians; lycopods; horsetails; seed ferns	marine life abundant; rank vegetation
	Devonian	45	fishes; early land plants	the age of fishes; earliest amphibians; extensive land flora of vascular plants
	Silurian	35	marine animals; algae	marine life; first fossils of land plants
	Ordovician	65	marine animals; marine algae	many fossil corals; brachiopods; starfishes; first vertebrates
	Cambrian	75	invertebrates; marine algae	trilobites dominant
Proterozoic 900			marine invertebrates; bacteria; algae	earliest fossil algae; first animal fossils; worms; crustacea
Archeozoic 500			unicellular organisms	no fossils; igneous rocks; few sedimentary rocks

Fig. 19-1 Geologic timetable.

Fig. 19-2 Late Devonian landscape. The smaller trees with whorled branches are Calamites; *the large tree in the foreground is* Archaeosigillaria; *and the tree with leafy fronds on the right is a seed fern. (Courtesy of Chicago Natural History Museum.)*

4 in. to 4 ft), herbaceous, perennial, and devoid of roots and true leaves. It grows both as an epiphyte on plants, such as palms and tree ferns, and as a terrestrial plant. The sporophytic plant body consists of an irregularly branching rhizome and erect, aerial, dichotomously branched, green shoots (Fig. 19-3*A*). Mycorrhizal fungi usually invade the cortical area of the rhizome, and the aerial stems are ridged. Anatomically the stem consists of an epidermis, a broad cortex, some cells of which contain starch granules, an endodermis, and an exarch star-shaped stele. The radial xylem and phloem surround a central parenchymatous region, which in older stems becomes heavy, thick-walled sclerenchyma. The stem of *Psilotum* is regarded as being protostelic in the rhizome portion, tending to be siphonostelic in the erect aerial parts. Stomata are present, usually between the longitudinal ridges of the aerial axis.

The aerial stems of *Psilotum* have short branches, each of which bears a terminal spo-

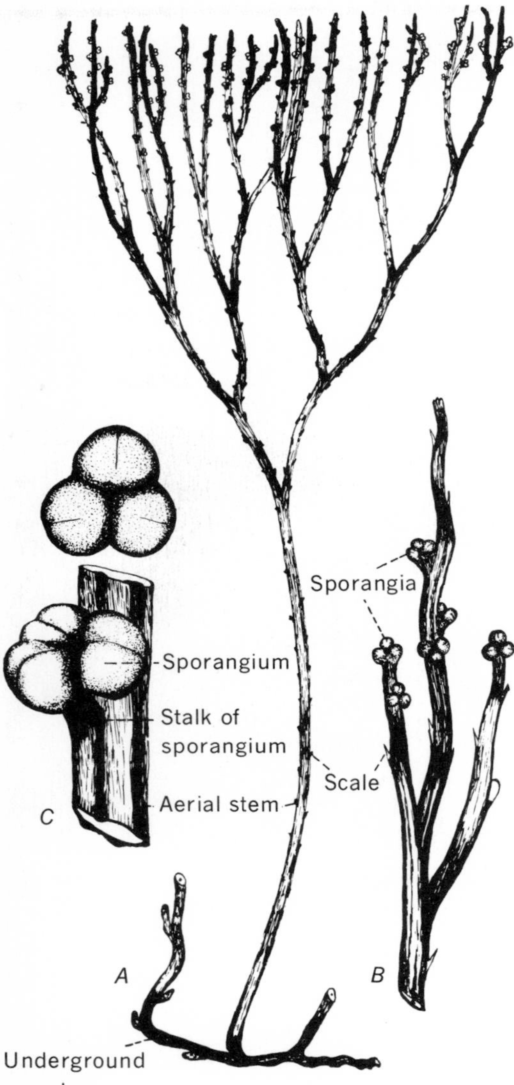

Fig. 19-3 Psilotum, *a primitive living vascular plant. A, habit of plant, with rootless, underground stem and aerial, dichotomously branched stem with numerous scale leaves and three-lobed sporangia; B, enlarged portion of branch tips, showing scale leaves and sporangia; C, three-lobed sporangia attached to the stem; upper figure shows line of dehiscence in each lobe.*

rangium. The sporangia, sometimes called synangia, are eusporangiate and generally three lobed, and they are subtended by the sterile branch of the appendage (Fig. 19-3*B, C*). *Psilotum* is homosporous. It multiplies vegetatively by buds, or gemmae, formed on the rhizome, or underground stem.

Tmesipteris (Fig. 19-4) is found in Australia, New Zealand, and other islands of the South Pacific Ocean. Generally it grows epiphytically although one species has been described as

Fig. 19-4 Tmesipteris parva, *an Australian species. (Photograph by Dr. Samuel Chan.)*

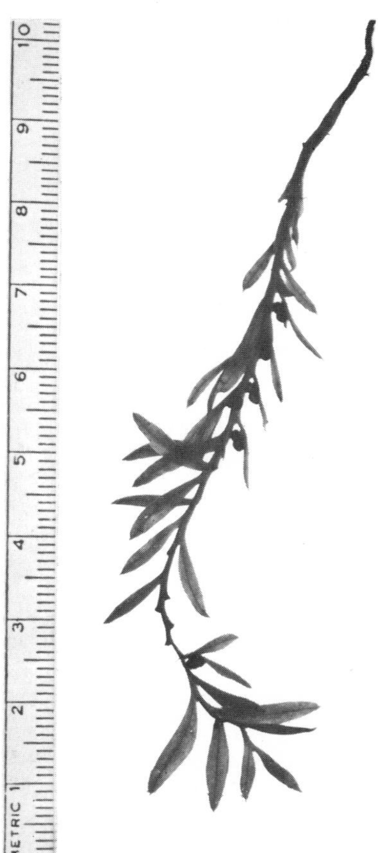

being terrestrial. The stem consists of a rhizome and aerial portions, which are flexuous and most often pendant. The aerial stems of *Tmesipteris* bear leaflike appendages that are expanded and green. The sporangia of *Tmesipteris* are two-lobed, elongated structures that are borne terminally on short side branches of the stem appendages. This plant is also homosporous.

The gametophytes of the Psilotales are small structures growing on the ground or underground and are cylindrical and branching. Like those of *Lycopodium* and the Ophioglossales, the gametophytes of the Psilotales are inhabited by a mycorrhizal fungus. The gametophyte of both *Psilotum* and *Tmesipteris* is described as resembling a piece of the rhizome of the sporophyte. Lacking chlorophyll, it is brown and covered with rhizoids. Gametangia, both antheridia and archegonia, are produced over all surfaces of the same thallus and are intermixed. Antheridia are superficial structures resembling those of the common leptosporangiate fern. The sperms are multiflagellate. Archegonia are developed with their basal portions, or venters, sunken but with the short necks protruding from the surface of the thallus tissue. There is evidence that the axial row of the archegonium may be reduced to two neck canal cells and a single ventral canal

Fig. 19-5 Sporophytes of three species of Lycopodium. *A,* L. complanatum, *with sporophylls in stalked strobili; leaves scalelike; B,* L. obscurum, *with sessile strobili and reduced scalelike leaves; C,* L. lucidulum, *with large bulbils and sporangia in axils of upper leaves. (Photographs A and B by Homer Grove.)*

cell. There are some indications that the central cell may function as the egg.

After the entry of the multiflagellate sperm into the canal of the archegonium and fertilization of the egg to form the zygote, the first division of the zygote results in the formation of two cells. The hypobasal cell, the one toward the body of the thallus, develops into a foot that penetrates the gametophyte. The other cell, at the apex, gives rise to the main portion of the sporophyte through repeated cellular divisions.

Instances have been reported of tracheids, phloem, and endodermis being present in the gametophyte of *Psilotum.* Cytological investigations suggest that the gametophyte of *Psilotum* is diploid and that the sporophyte is a tetraploid. Gross similarities between the sporophytic rhizome and the gametophyte and the presence of vascular tissue in the gametophyte are considered by some to support the homologous theory of alternations (see Chap. 14).

LYCOPSIDA

The Lycopsida include both extinct forms, known only from their fossil remains, and living plants. Among the plants classified as Lycopsida, there is diversity in size, form, and methods of reproduction, implying a lack of close relationships. The fossil order Lepidodendrales, with its large dendroid types, and the living orders Lycopodiales and Selaginellales are representative of the Lycopsida. Members of the Lycopsida line have roots and leaves, representing an advance over the Psilopsida. Although leaves are present in the Lycopsida, they are usually small and there are no associated leaf gaps or breaks in the vascular cylinder. Protosteles are characteristic of the Lycopsida, but siphonosteles have been found in some fossil types and in living members. Sporangia are developed either on the upper, or adaxial, surface of sporophylls or, in some cases, from the stem near the axil of the leaf. Strobili, or cones, consisting of a stem and aggregated sporophylls, are characteristic of both living and fossil Lycopsida.

Members of the Lycopsida were the dominant plants of the coal flora of the Carboniferous period of the Paleozoic era (Fig. 19-1). The two most important genera of this period were *Lepidodendron* and *Sigillaria* (Fig. 19-2). *Lepidodendron* was an erect dichotomously branched tree; specimens that were 100 ft in height have been found. Its leaves were linear, or needle shaped, 6 to 7 in. long, densely and spirally arranged on the stem. The characteristic scars left by these leaves, from which the name *Lepidodendron* was derived (*lepidos,* meaning "scale," and *dendron* meaning "tree"), may still be seen on the preserved fossils. Unlike the living *Lycopodium,* which they somewhat resembled in form, some of the species of the Lepidodendrales were heterosporous. They were more like the living *Selaginella* in this respect.

Sigillaria differed from *Lepidodendron* in general appearance by a usual scarcity of branches, by a shorter, more tapering trunk, and by the manner in which the sporophylls were borne. The name *Sigillaria* is derived from the Latin *sigillum,* meaning "seal," which the leaf scar is thought to resemble. The *Sigillaria* species were probably among the largest of the trees of the Carboniferous period. Their trunks reached 6 ft in diameter. These trunks, like those of *Lepidodendron,* had an unusually thick corky layer. Erect rigid leaves completely covered the young portions of the stem. In contrast to the spiral arrangement of the leaves of *Lepidodendron,* the leaves of some species of *Sigillaria* occurred in vertical rows.

CLASSIFICATION OF THE LYCOPSIDA

Class Lycopodinae:

Order Lepidodendrales	Fossil forms
Order Lycopodiales	Living forms
Order Selaginellales	Living forms
Order Isoetales	Living forms

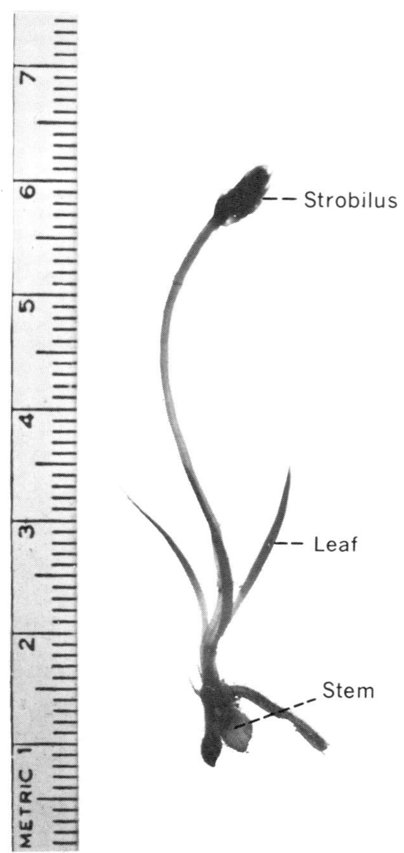

Fig. 19-6 Phylloglossum Drummondii. *(Photograph by Dr. Samuel Chan.)*

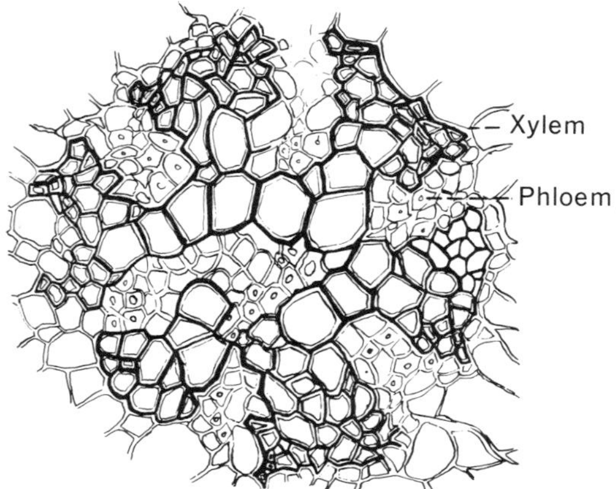

Fig. 19-7 Lycopodium, *cross section of center of stem showing xylem and phloem.*

Lycopodiales—club mosses

The order Lycopodiales is small; it consists of two living genera, *Lycopodium* (Fig. 19-5) and *Phylloglossum* (Fig. 19-6). The species of *Lycopodium* are widespread throughout the subtropical and temperate zones, and the single species of *Phylloglossum* is native to Australia and New Zealand. The plants of *Lycopodium* are evergreen perennials of relatively small size and in the temperate regions are generally low growing or trailing. In the warmer subtropical regions many species of *Lycopodium* are epiphytes, growing on trees. Because of their form, structure, and low-growing habit, *Lycopodium* species are commonly known in the United States by such names as ground pine, trailing pine, ground hemlock, club moss, and Christmas greens because of their extensive use at the holiday season. The terms club moss, ground pine, and ground hemlock are misleading since these sporophytic structures are in no way comparable to mosses and they have no close relationship with the pines and hemlocks, which are seed plants. In parts of Mexico where some species of *Lycopodium* grow as epiphytes on trees, they are commonly called riscos.

Fig. 19-8 Stages in development of sporangium in Lycopodium. *A, sporangial initial; B, C, developing sporangium; D, tetrad of meiospores.*

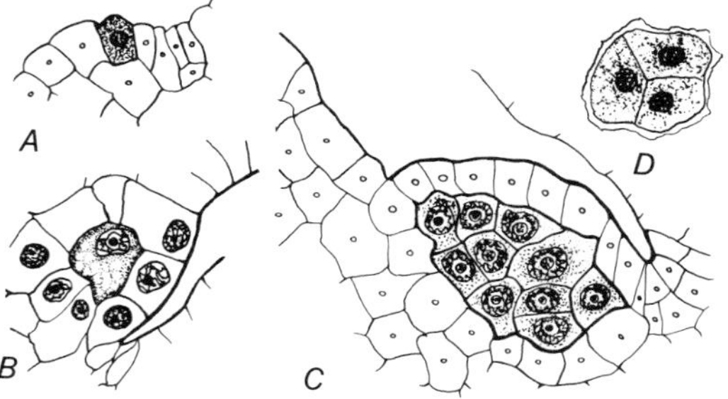

Lycopodium

SPOROPHYTIC STRUCTURES The sporophyte of *Lycopodium* always has a much-branched stem that in some species is erect and in other species is creeping. The stem is covered with small elliptical leaves attached by a broad base. In some species most leaves of the year's growth bear single, rather large sporangia in their axils (Fig. 19-5*C*). In other species the sporophylls are smaller, lack the green color of the foliage leaves, and are closely compacted in an elongated **strobilus,** or cone (Fig. 19-5*A, B*). Each sporangium originates from a row of several cells occupying a transverse position near the base of the leaf (Fig. 19-8). At maturity the sporangium consists of a stalk and a capsule. Numerous spore mother cells, or sporocytes, divide meiotically to form tetrads of meiospores within the capsule. A large mass of sterile tissue occupies the lower central portion of the capsule. This tissue is known as the **subarchesporial pad.**

The vascular system of the stem varies somewhat with the age of the plant. In the young stages it is a radial protostele (Fig. 19-7). In some older stems the stele consists of isolated bundles with an irregular arrangement. Both of these types of steles are primitive in organization. A suspensor develops from one half of the fertilized egg and is considered to be an advanced character comparable to the suspensor found in seed plants. *Selaginella* is the only other member of the lower vascular plants that develops a suspensor.

GAMETOPHYTIC STRUCTURES *Lycopodium* is homosporous. The gametophytes of *Lycopodium* are very inconspicuous and are rare. These gametophytes are small, being only a few millimeters in length. In some species they are entirely underground and lack chlorophyll; in other species they are partly buried but develop a crown of green tissue (Fig. 19-9). An endophytic fungus usually is found associated with the tissues of the gametophyte.

Antheridia and archegonia develop at the apex of the gametophyte (Fig. 19-10). The antheridia develop in the same way as the eusporangia. At maturity the antheridia of *Lycopodium* are relatively large structures, sunken in the tissues of the thallus, and they produce many coiled biflag-

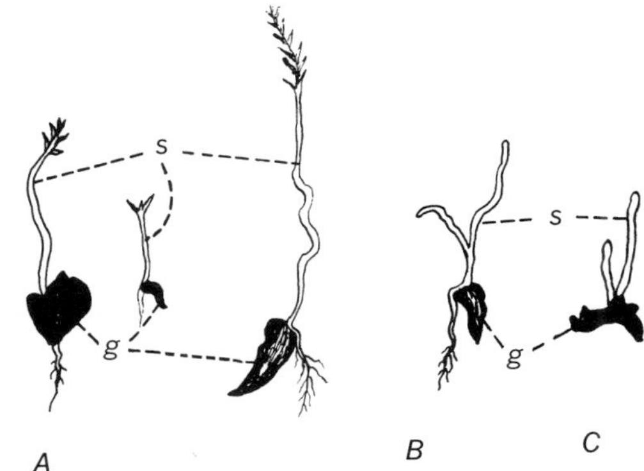

Fig. 19-9 Lycopodium *gametophytes, g, with young attached sporophytes, s. A,* L. selago; *B,* L. complanatum; *C,* L. clavatum.

Fig. 19-10 Details of gametangia of Lycopodium. *A–C, antheridia; D, archegonium.*

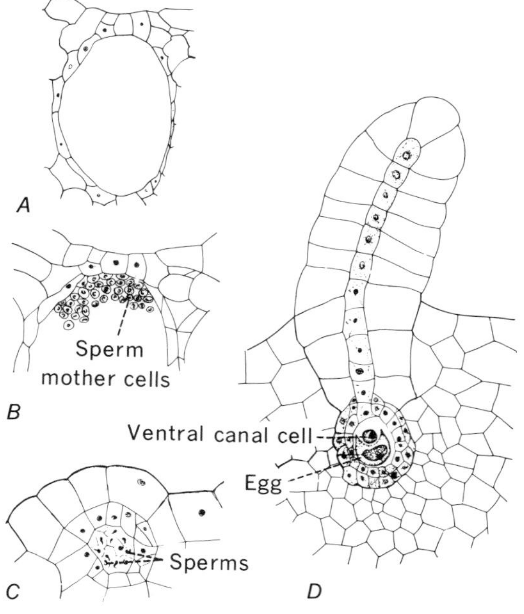

ellate sperms instead of the multiflagellate type characteristic of *Equisetum* and the ferns. The archegonia are generally large with long necks with as many as 10 to 14 or more neck canal cells. The gametophytes of some species of *Lycopodium* develop short-necked archegonia with only one, or a few, neck canal cells. At maturity the basal portions of the archegonia are sunken in the thallus, but the necks protrude slightly above the surface. A characteristic feature of the gametophytes of *Lycopodium* is their slow development after a long delayed germination of the spores. Spores may take three years or more for germination.

THE EMBRYO The embryo of *Lycopodium* develops from the fertilized egg cell, or zygote, in the venter of the archegonium. The first division of the zygote is transverse to the long axis of the archegonium, forming an upper, or outer, cell and a lower, or inner, cell. The embryo proper develops from the lower of these two cells, whereas the upper one, called the **suspensor,** remains functionless. The young sporophyte of *Lycopodium* remains attached to the old gametophyte for a considerable time (Fig. 19-9). Eventually the gametophyte disintegrates.

Phylloglossum The plant, combining some primitive and some fairly advanced features, consists of a tuberous stem that bears simple leaves. A strobilus, or cone, is produced at the apex of the stem (Fig. 19-6). The plant resembles *Lycopodium* in being homosporous. The gametophytes produced by the spores are a primitive type, resembling those of certain spe-

Fig. 19-11 Selaginella *colony.*

cies of *Lycopodium. Phylloglossum,* with an amphiphloic siphonostele, shows a higher anatomical development than *Lycopodium.* Primitive features of *Phylloglossum* include the tuberous stem, homospory, and tuberous gametophytes. Advanced features are the siphonostelic condition and the production of a strobilus.

Selaginellales—little club mosses

The living members of the Selaginellales all belong to a single genus, *Selaginella,* with possibly 600 species.

Selaginella Numerous species of *Selaginella* occur in the tropics, but only a few extend the range of the genus into the colder temperate regions. *Selaginella* is related to *Lycopodium,* differing chiefly in being heterosporous.

THE SPOROPHYTIC STRUCTURES The mature sporophyte of *Selaginella,* although it is similar to that of *Lycopodium,* is usually more delicate in appearance (Fig. 19-11). The stems are, in many cases, prostrate or only semierect but are sometimes climbing. The vascular system of the stem is usually a protostele, though in some species it is a siphonostele or a dictyostele. There is no cambium and the bundles are closed (Fig. 19-12). Spiral and annular tracheids make up the protoxylem. The metaxylem consists largely of scalariform elements. Of small size and simple structure, the leaves approximate the structure of the leaves of seed plants. They have an epidermis, with stomata mostly on the lower surface of the leaf. Sometimes the mesophyll is differentiated into a palisade region and a spongy mesophyll. There is usually a single vascular bundle in the mesophyll. In some species there are two rows of small leaves on the upper side of the prostrate stem and two rows of larger leaves, one on each of the lateral faces of the stem (Fig. 19-13). At the base of each leaf there is a small membranous outgrowth called a **ligule** (Fig. 19-15*C*). The presence of the ligule can be traced to Paleozoic forms that are thought to be related to and possibly ancestral to *Selaginella.* The presence of a ligule helps to distinguish *Selaginella* from *Lycopodium,* which lacks this structure. The roots of *Selaginella* form no extensive system but grow from unique structures, the **rhizomorphs,** or **rhizophores,** that are probably more stemlike than rootlike. These rhizophores develop adventitiously on the underside of the prostrate stem and grow downward into the soil or substratum. The roots proper, which are small and fibrous, develop at the tips of the rhizophores.

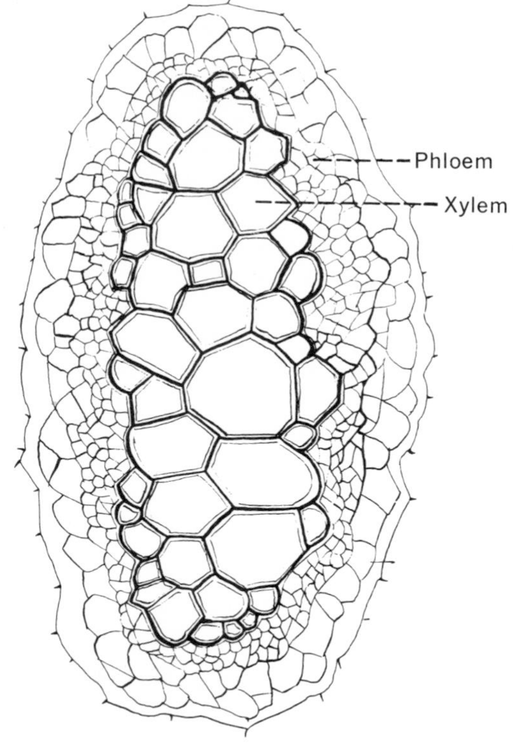

Fig. 19-12 Selaginella. *Transverse section through the vascular tissue in the stem, a protostele.*

Sporophytic structures grow as a result of divisions of apical cells that occupy the tips of the stems and roots (Fig. 19-15*D*). The apical cells are the primordial meristem. The sporophylls are grouped into more or less definite cones, or strobili, at the apices of the branches (Figs. 19-13*A*, 19-14). These cones bear two kinds of sporangia—microsporangia and megasporangia—producing, respectively, numerous small red meiospores, called microspores, and four large yellow meiospores, called megaspores. The sporophylls, therefore, are distinguishable as microsporophylls and megasporophylls. The megasporangia and microsporangia may occur randomly in the cone or they may be arranged in a more definite pattern.

Development of microsporangia is eusporangiate. They contain a large number of microspore mother cells, all of which form tetrads of microspores through the usual meiotic divisions.

The development of the megasporangium is identical with that of the microsporangium up to the spore-mother-cell stage, but many of the megaspore mother cells fail to produce meiospores. In fact, usually only one forms a tetrad of meiospores. Consequently, following the maturation processes, not more than four megaspores are produced in each sporangium and sometimes one or two or three of these may not develop; therefore often only one megaspore is produced.

GAMETOPHYTIC STRUCTURES Production of two kinds of spores in heterosporous

Fig. 19-13 Selaginella. *A,* S. mortensii, *stems, leaves, and cones; B,* S. krausiana, *stems, leaves. and cones.*

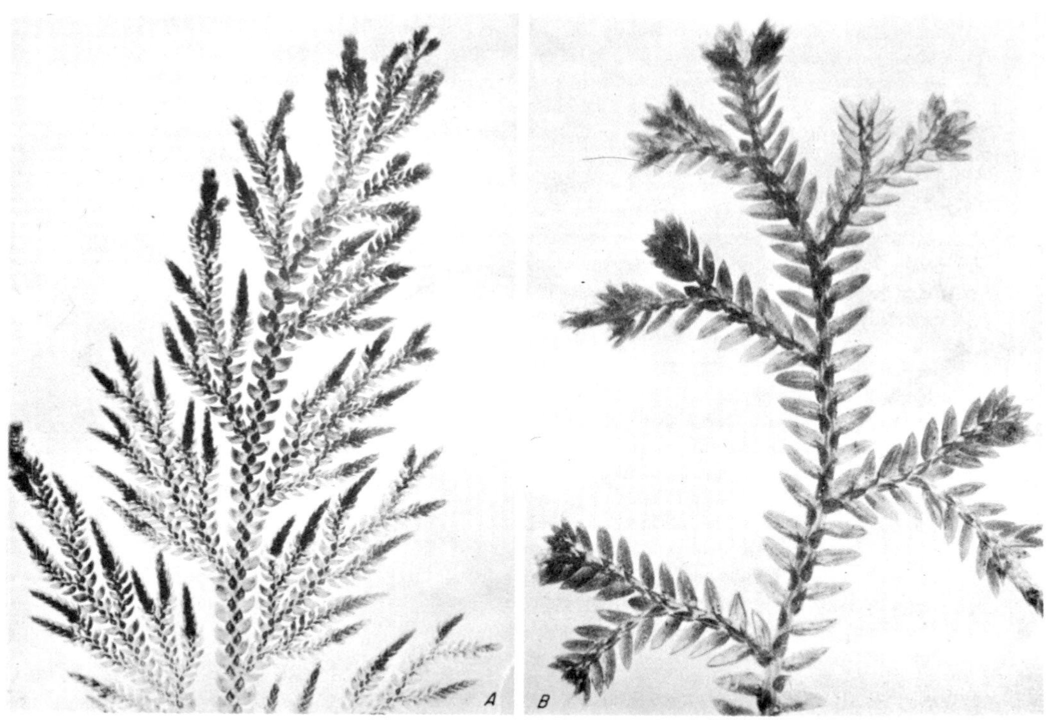

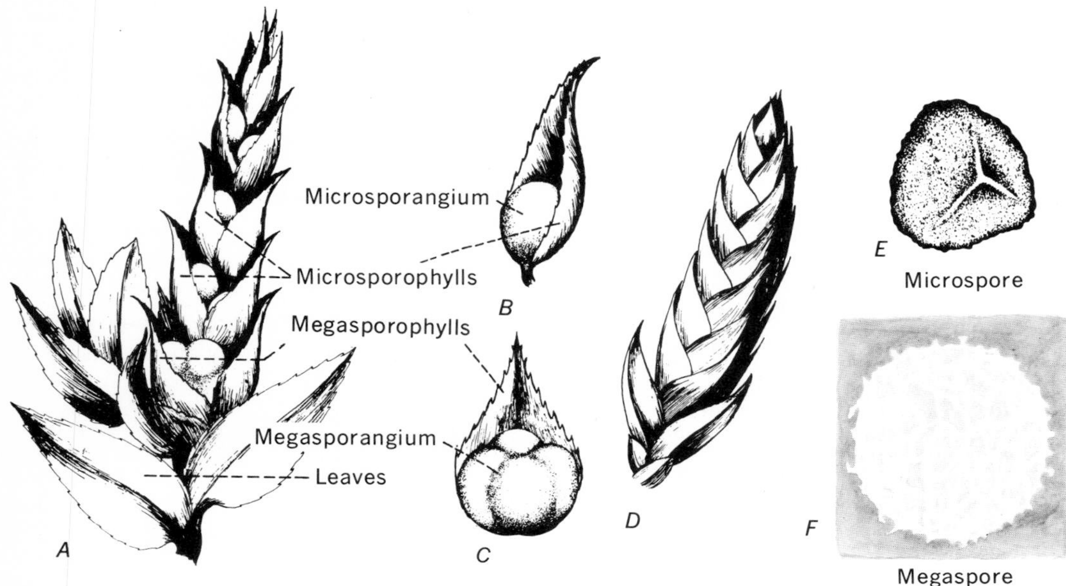

Fig. 19-14 Strobili, sporangia, and spores of Selaginella. *A, branch showing vegetative leaves and strobilus with megasporophyll below and microsporophylls above; B, microsporophyll with microsporangium; C, megasporophyll with four-lobed megasporangium; D, strobilus previous to shedding stage; E, microspore, and F, megaspore, both enlarged but not to same scale. (Drawing by Elsie M. McDougle.)*

plants results in separate male and female gametophytes that show conspicuous differences. In the heterosporous Lycopsida the male gametophytes, or microgametophytes, which develop from the microspores, are small. They lack chlorophyll and are not physiologically independent plants. The female gametophytes, or megagametophytes, develop from the larger of the two kinds of spores, the megaspores. Although the female gametophytes are small structures, they are larger than the male gametophytes. They contain chlorophyll and are, to some extent, independent plants.

Germination and development of the microspore of *Selaginella* occur entirely within the microspore wall, which does not rupture until the male gametophyte is mature. In the process of germination the microspore divides to form a very small vegetative cell and a larger antheridial cell. The single vegetative cell is homologous with the thallus of other plants, including the haploid thallus of the algae and of the bryophytes. The antheridial cell, by a series of divisions, produces a small mass of cells, the outer layer of which becomes differentiated into the wall, or jacket, of the antheridium, whereas the internal cells produce the sperms. Up to this point the development takes place entirely within the microspores. Before the male gametophytes mature, the red-colored microspores are set free by the bursting of the sporangial wall. Some of them lodge near the megasporangium or even within the cleft of the megasporangial wall, where they are in the vicinity of the female gametophyte and the archegonia. As the sperms mature in the antheridium within the micro-

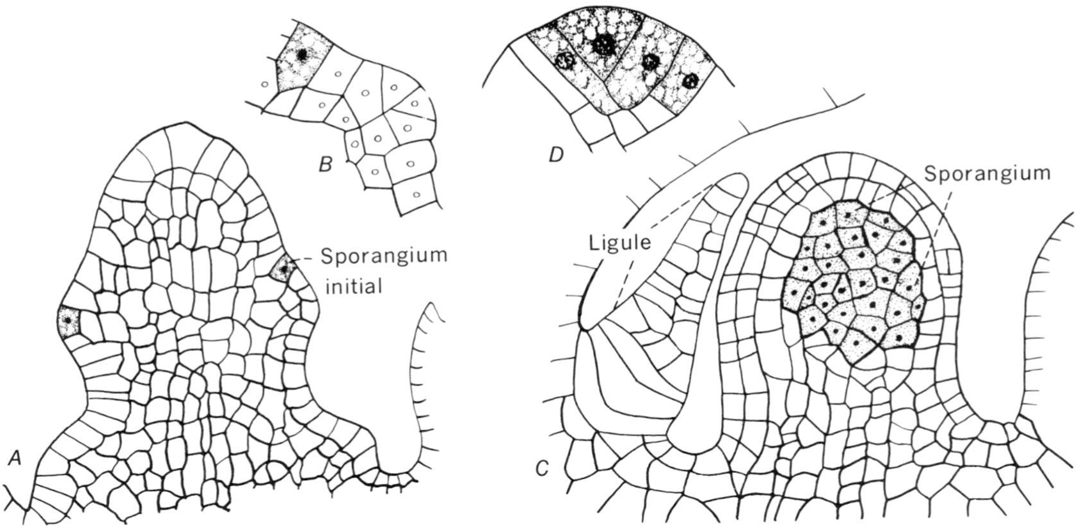

Fig. 19-15 Structural details of Selaginella. *A, leaf primordium, with sporangium initial; B, sporangium initial enlarged; C, ligule and young sporangium; D, apical cell.*

Fig. 19-16 Stages in development of female gametophyte of Selaginella. *A, megaspore with single nucleus surrounded by inner wall; B, free nuclei developed by division from A; C, parietal placing of free nuclei; D, E, development of tissue at "beak," enlarged in F to show archegonial initial; G, archegonium with single neck canal cell, ventral canal cell, and egg; H, I, mature archegonium with egg cell.*

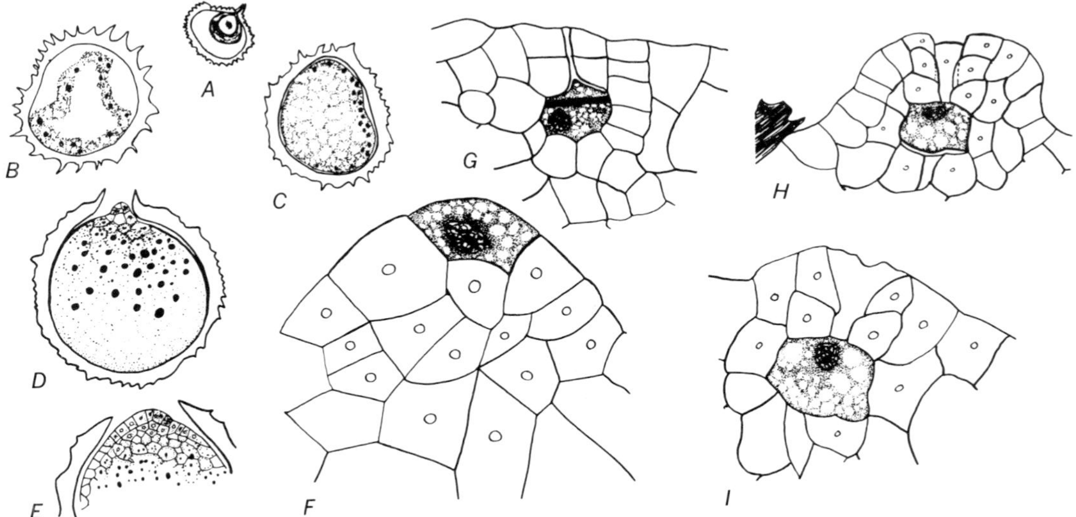

spore, the wall of the microspore ruptures and the few biflagellate sperms produced in each antheridium are set free.

In the development of the female gametophyte, the nucleus of the megaspore divides many times, forming a mass of free nuclei that later become surrounded by walls. At an early stage in this development, which takes place at one end of the spore cavity, a large vacuole, said to be filled with an oily reserve food material, appears at the opposite end of the spore (Fig. 19-16). This eventually becomes cellular. Up to this point the growth of the multicellular female gametophyte takes place entirely within the megaspore wall. Increasing pressure within the spore causes the wall to burst at the apex, where the gametophyte then protrudes slightly. The archegonia are formed in considerable numbers on this protruding portion. The archegonia are reduced structures consisting of an egg cell, a ventral canal cell, and one neck canal cell, all of which are surrounded by a few indefinite wall cells (Fig. 19-16). Fertilization usually occurs while the megaspores are still within the sporangium.

FERTILIZATION AND THE DEVELOPMENT OF THE EMBRYO When water is available, the sperms freed from the antheridium swim to the archegonium and descend the neck to the egg, with which one of them fuses, thus accomplishing fertilization. Union of sperm and egg initiates development of the sporophytic phase, and the embryo begins development immediately. A suspensor is developed as in *Lycopodium,* but it is larger, and its growth, by pushing the embryo downward, aids in bringing the young sporophyte into contact with the storage tissue in the lower part of the gametophyte. Although an embryonic stem and leaf are formed early, the later appearance of the root may be correlated with the large amount of food available in the female gametophyte. Sometime after the embryo begins to develop, the mega-

Fig. 19-17 Habit of growth of Isoetes. *(Photograph by Dr. D. A. Kribs.)*

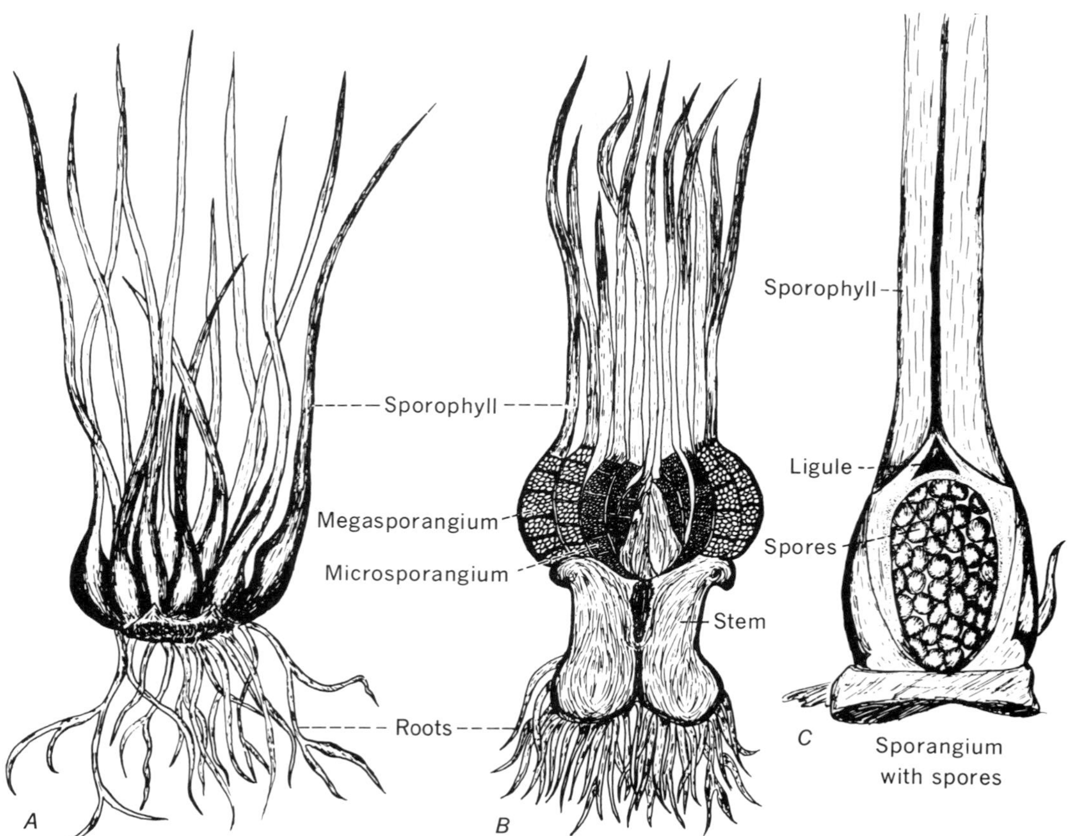

Fig. 19-18 Leaves and sporangia of Isoetes. *A, entire plant, general form; B, longitudinal section through plant; C, longitudinal section of one sporophyll with sporangium at the base, showing spores. (Drawings by Clyde Shipman.)*

spores and megagametophytes enclosed by the megaspore wall drop out of the old shriveled sporangium. At this stage the embryo plant has the general appearance of a seedling of the seed plants. The roots of the young sporophyte grow into the soil, and the plant begins an independent existence.

Isoetales—quillworts

The order Isoetales contains the single living genus *Isoetes* (Fig. 19-18*A*), with more than 60 species and some extinct fossil forms known as *Isoetites*. In some features the members of this order resemble *Selaginella*. *Isoetes* is common in cooler climates and grows immersed in shallow water or in swampy locations (Fig. 19-17). The axis of the plant is extremely short and rather broad with crowded appendages; the axis is referred to as a corm (Fig. 19-18*B*). The upper part of the axis is stem and bears leaves; the lower, or root-producing portion of the axis, is usually referred to as the rhizophore (Fig. 19-18*B*). Each leaf is ligulate (Fig. 19-18*C*) and potentially a microsporophyll or megasporophyll (Fig. 19-18*B, C*). Both megaspores and microspores are liberated with the decomposition of the sporangial wall and their germination is usually delayed. The male and female gametophytes resemble those of *Selaginella* and the sperms are multiflagellate.

SPHENOPSIDA

The Sphenopsida include the extinct order Sphenophyllales, a group of small plants bearing wedge-shaped leaves, several other orders of fossil plants, and the order Equisetales with both fossil forms and the single living genus *Equisetum*. Among fossil Sphenopsida were the giant *Calamites,* of the order Calamitales, that lived during the Carboniferous period. These plants were generally slender tree forms, frequently 20 to 40 ft in height but usually not exceeding 10 to 15 in. in diameter; sometimes, however, they were larger (Fig. 19-2). Whatever their size, they had underground rhizomes and jointed, erect, aerial stems, such as those of the present-day *Equisetum*. Fossil Sphenopsida are found in the Devonian and extend through the Carboniferous and Permian periods of the

Fig. 19-19 Colony of Equisetum.

Paleozoic era (Fig. 19-1). Forms resembling the modern *Equisetum* have been found as fossils in more recent geological periods.

Jointed stems are characteristic of all Sphenopsida. Both protosteles and siphonosteles are found in the group, which is characteristically without leaf gaps. The leaves are generally small or reduced but are large in some fossil forms. Leaves in the Sphenopsida are usually developed in whorls at the nodes of the aerial stem or its branches. The sporangia are borne on specialized structures or stem appendages, called sporangiophores, and are commonly aggregated into strobili, or cones.

CLASSIFICATION OF THE SPHENOPSIDA

Class Equisetinae:

Order Sphenophyllales	Fossil forms
Order Calamitales	Fossil forms
Order Equisetales	Both fossil and living forms

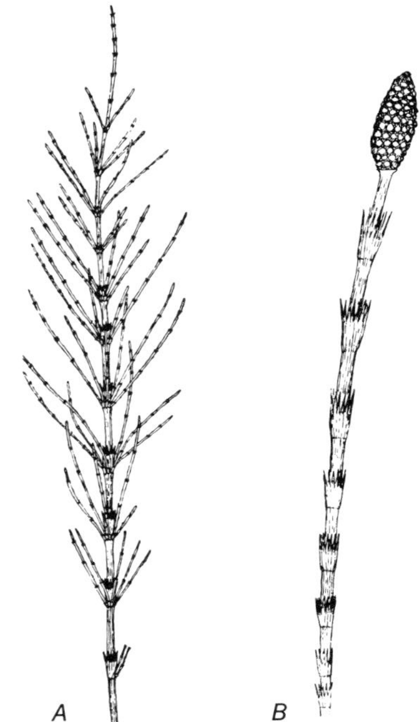

Fig. 19-20 Equisetum arvense. *A, sterile vegetative stem; B, fertile stem bearing cone at apex. (Drawings by Edna S. Fox.)*

Equisetales—horsetails, or scouring rushes

The living members of the order Equisetales, comprising about 25 species of the genus *Equisetum,* are only a remnant of a group of plants that, during an earlier geological period, constituted a prominent part of the vegetation of the world. Species of *Equisetum* are commonly called scouring rushes or horsetails. The former designation came from their use in scouring household utensils. The term horsetail, particularly applicable to the branching forms, is equivalent to *Equisetum,* from the Latin *equus,* "horse," and *saeta,* "bristle." Most plants of *Equisetum* now living are small in size with slender stems less than a yard high; the members of one species, however, are said to have stems 25 to 30 ft high, although they are not much more than 1 in. in diameter.

Sporophytic structures In some species of *Equisetum* the mature sporophytes produce aerial stems of two distinct types. One is short lived, unbranched, without chlorophyll and produces spores in April or early May (Fig. 19-20*B*); the other is sterile and green and persists throughout the growing season (Figs. 19-19, 19-20*A*). In other species there is only one type of aerial stem, a green structure bearing a cone at its apex (Fig. 19-21). Some of the aerial branches are annual, while some persist for more than one year. The stems may have slender lateral branches produced in whorls at the nodes, as in *E. arvense,* the common horsetail, or they may be unbranched, erect, and wandlike. All species have an extensive perennial development of

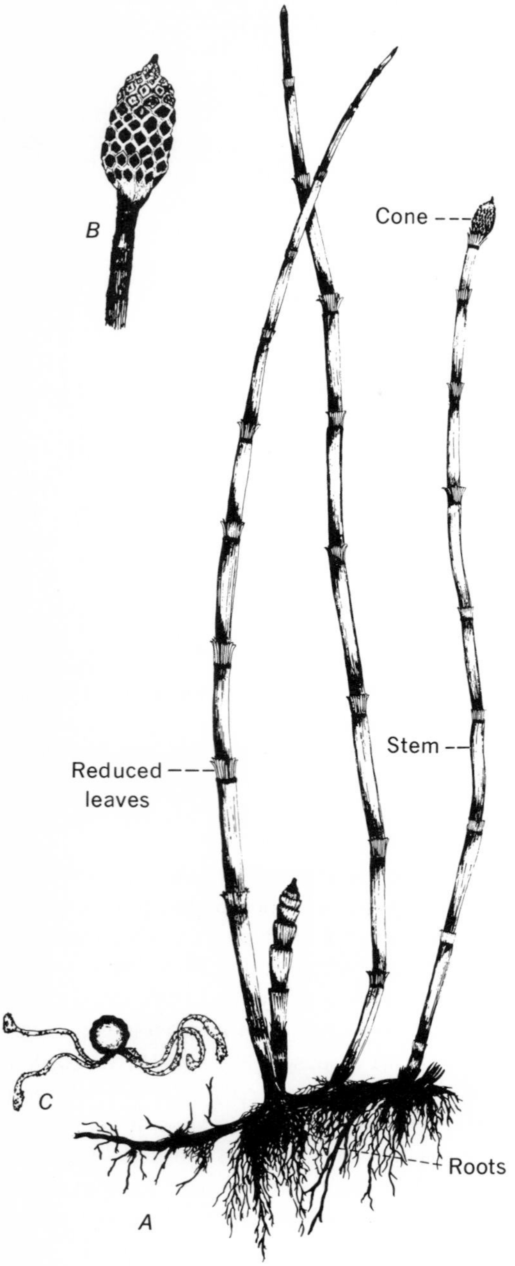

Fig. 19-21 Equisetum prealtum. *A, sporophyte; B, detail of cone, showing sporangiophores; C, one spore with elaters, greatly enlarged. (Drawings by Elsie M. McDougle.)*

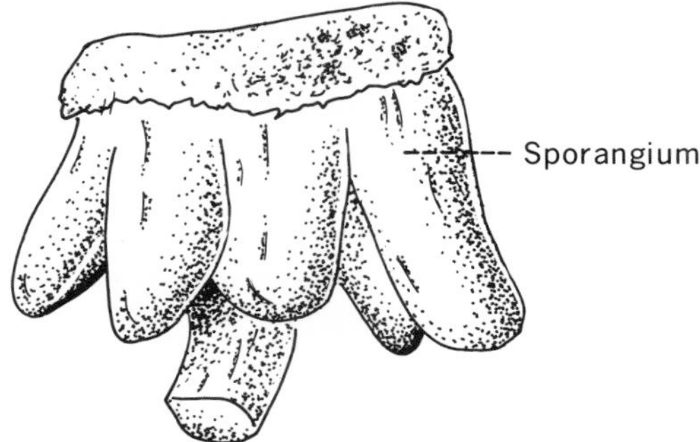

Fig. 19-22 Sporangiophore of Equisetum.

underground stems, or rhizomes. The vegetative stems are green and photosynthetic. Stems of *Equisetum* are longitudinally grooved or furrowed. In most species silica is present in the epidermis of the stems giving them their value as scouring rushes. The stems are conspicuously jointed and may easily be pulled apart at the nodes.

The center of the stem is at first occupied

Fig. 19-23 Equisetum. *A, transverse section of a vascular bundle from the stem; cavity in bundle toward center of stem; B, section through stoma from stem of* Equisetum.

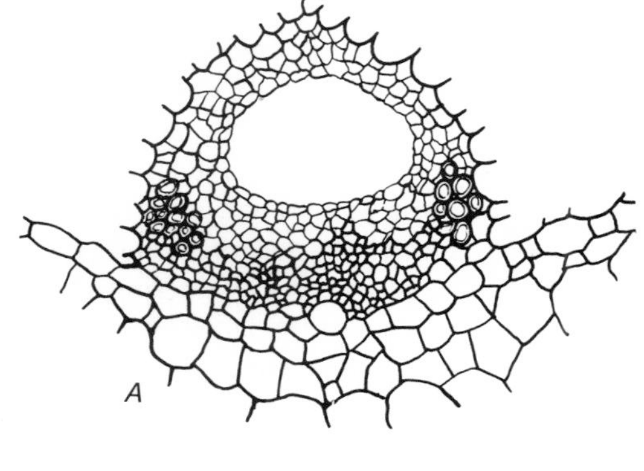

by pith, which soon disappears, leaving most of the older stems hollow. The permanent tissues of the stem, epidermis, cortex, and vascular strands form a thin cylinder surrounding the central cavity. In addition to the central lacuna, or canal, two other types of longitudinal canal are regularly present in the *Equisetum* stem. These are (1) **vallecular canals,** which are located in the cortex, each of which is associated with a longitudinal groove or furrow of the stem; (2) **carinal canals,** which are located deeper in the tissue on a radius with one of the longitudinal ridges of the stem and each of which is associated with a vascular bundle.

The vascular system is made up of collateral bundles arranged as a siphonostele. In the internodes the bundles are distinct and widely spaced, with each bundle located beneath one of the superficial ridges of the stem. In the nodal regions the bundles form a continuous ring of tissue. Each bundle consists of primary xylem and phloem. Although the phloem is well developed, the xylem is sparse and poorly developed (Fig. 19-23*A*). Generally there is neither cambium nor secondary tissue in the living species of *Equisetum*. Endodermal tissues are regularly present and are well developed in the stems. They are distributed in a variety of ways. There may be a single endodermis external to the vascular system; there may be both an external and an internal endodermis; in the absence of a continuous endodermis, each individual bundle is surrounded by this tissue.

The branches in *Equisetum* arise from adventitious buds developed at the nodes. The leaves develop as whorls at the nodes and are always small scalelike structures that coalesce to form a toothed sheath around the stem. The roots are small and wiry, having a single vascular strand with the tissues arranged radially. Roots are thought to arise from the basal portion of the branch primordia rather than directly from the nodal tissues of the main stem. Growth in the sporophyte is by means of a pyramid-shaped apical cell in the tip of the stems and roots. These apical cells are the primordial meristems and, by their division, give rise to new cells that mature into the tissue of these organs (Fig. 19-24*B, C*).

The spore-bearing organs of *Equisetum* are aggregated in definite cones at the apex of the stem (Figs. 19-20*B*, 19-21). The cone consists of a main central axis with specialized sporangia-bearing structures, called **sporangiophores** (Fig. 19-22), developed in whorls. Each sporangiophore consists of a hexagonal plate, attached at a right angle to the cone axis by a short stalk. Some botanists have regarded the sporangiophore as a sporophyll, i.e., specialized spore-bearing leaf, whereas others believe the sporangiophore is a specialized stem structure or a composite of both leaf and stem. Whatever

Fig. 19-24 Equisetum. *A, sporangial initial; B, apical cell in longitudinal section; and C, the same in transverse section.*

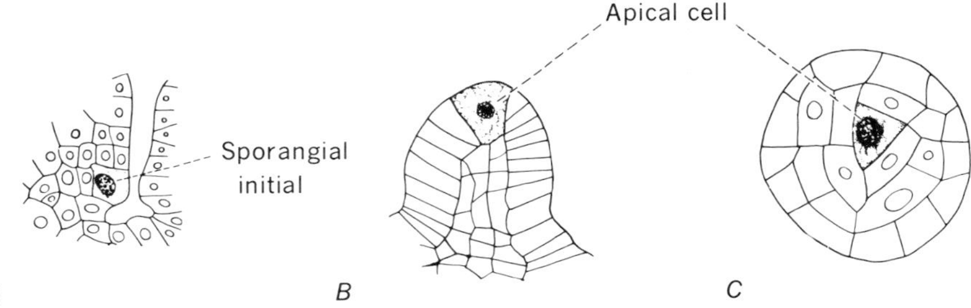

may be the exact morphological interpretation of the sporangiophore, diploid mother cells in the sporangia undergo meiosis and form tetrads of haploid spores that are all alike.

Gametophytic structures Meiospores are the products of meiosis and are the first cells of the gametophyte. The spores contain chlorophyll, are dark green, and, as soon as they are shed, the spores germinate to produce the gametophytic thallus. Although a few fossil relatives of *Equisetum* were heterosporous, all living plants of this group are homosporous. Four ribbonlike **elaters** are formed from the outer layer of the wall of the spores, i.e., deposited in four strips which are broken at maturity. The hygroscopic nature of the elaters results in their curling and uncurling under varying conditions of moisture (Fig. 19-21*C*).

The mature gametophytic thallus of *Equisetum* consists of a heavy cushionlike central or basal disk that is several layers of cells in thickness. Green, irregularly shaped lobes grow from the upper surface and from the marginal regions of the disk. The cells of the upper portion contain chlorophyll, but those in the lower part are colorless. Rhizoids attach the dorsiventral thallus to the ground and are developed from the underside of the central portion of the gametophyte. The gametophytes grow in a prostrate position on the surface of the soil and, in this respect, resemble those of the leptosporangiate ferns (Filicales) and differ from those of all other lower vascular plants. The

Fig. 19-25 Structure of gametangia of Equisetum. *A–C, antheridia; A, young stage; B, intermediate stage; C, young antheridia and older antheridium with sperms; D–F, series of stages in development of archegonia; F, a mature archegonium.*

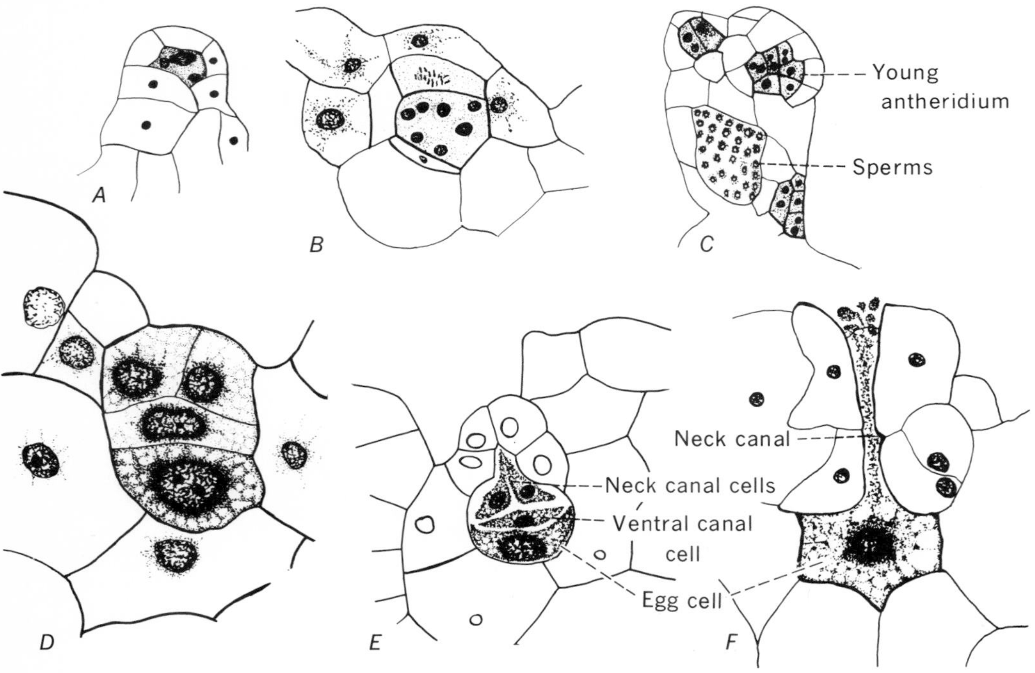

margins of the thalli are meristematic and continue an irregular growth sometimes for as long as two years. At maturity the gametophytes are small, usually less than ⅓ in. in diameter, although some species are said to produce larger ones of a conspicuous reddish color.

Under the usual conditions of growth both antheridia and archegonia develop on the same thallus. When gametophytes are grown in culture or under crowded conditions, plants may produce only male gametangia. Archegonial development normally precedes that of the antheridia. The archegonia are located in the cells of the central disk between the upright, leafy lobes and continue to form from certain cells derived from the marginal meristem of the central or basal disk. The initial cell of the archegonium of *Equisetum* is a superficial cell. This initial divides with a wall parallel to the surface of the thallus. The outer of the two cells formed develops the neck cells, and the inner cell forms the axial row of the archegonium (Fig. 19-25*D–F*). At maturity the archegonium is mostly sunken in the thallus but the end of the neck may protrude slightly (Fig. 19-25*E, F*).

Antheridia are usually formed after some archegonia are present. Antheridia develop from the periphery of the central disk or from the tissue of upright lobes. An antheridial initial divides in a plane parallel to the surface of the thallus, thus forming an outer and an inner cell. Of these, the outer cell develops into the wall of the antheridium and the inner cell produces the sperm-bearing tissue. At maturity the antheridium, containing a large number of sperms, appears sunken beneath the surface

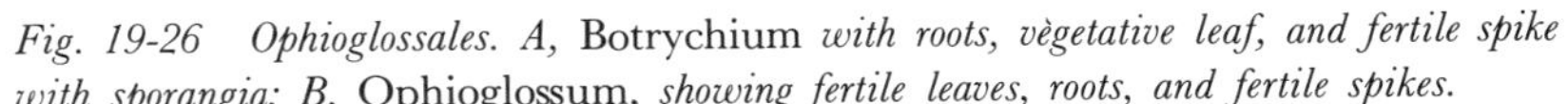
Fig. 19-26 Ophioglossales. A, Botrychium *with roots, vègetative leaf, and fertile spike with sporangia; B,* Ophioglossum, *showing fertile leaves, roots, and fertile spikes.*

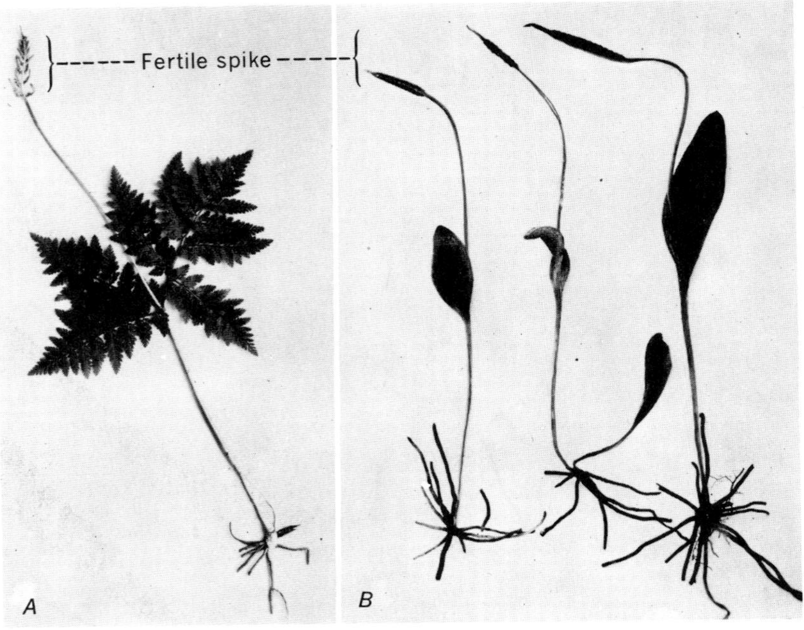

of the thallus (Fig. 19-25*A*–*C*). The large sperms are spirally coiled, motile structures with numerous flagella.

Syngamy and the development of the embryo. Several of the motile sperms may penetrate a single archegonium by way of the neck canal, but only one unites with the egg to produce a zygote. The development of the embryo is initiated by the enlargement of the zygote and its division in a plane transverse to the long axis of the archegonium. In some species, at least, the upper cell (epibasal cell) of the two thus formed develops into the embryonic stem and leaf. The root appears late in development and is formed from the lower (hypobasal) cell. A foot may or may not form. Several eggs may be fertilized and may develop embryos in the numerous archegonia of a single thallus.

THE PTEROPSIDA

The Pteropsida include all living and fossil ferns and the seed plants. The pteropsid line is distinguished, as are the other lines, by specific morphological and anatomical features. In general the Pteropsida have large leaves and leaf gaps, or breaks in the vascular cylinder.

Class Filicinae. The ferns.

Order Ophioglossales. The adder's-tongue and grape ferns. Genera *Ophioglossum* and *Botrychium.*

Fig. 19-27 Botrychium. *A, transverse section of root, showing protostele surrounded by endodermis, cortex, and epidermis; B, enlarged view of stele.*

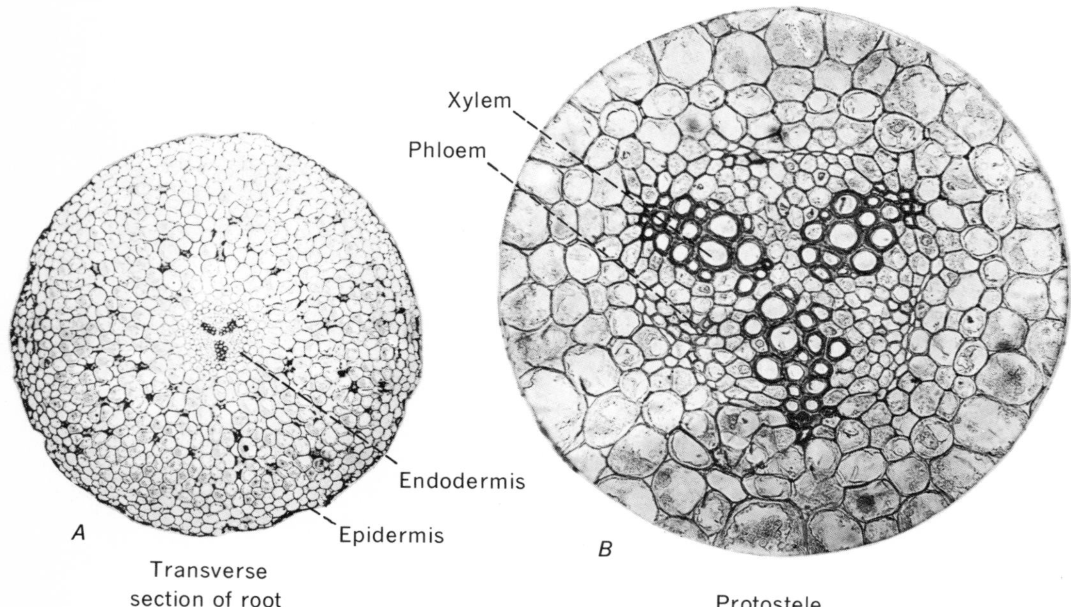

Order Filicales. The true ferns. Genera *Polypodium, Pteris, Marsilea,* etc.
Fossil Forms. The fossilized remains of many extinct types of ferns are known.

FILICINAE—THE FERNS

Ophioglossales—adder's-tongue ferns

The Ophioglossales form a small order of interesting ferns consisting of three genera, two American and one Asiatic. Both of the American genera, *Ophioglossum,* the adder's-tongue fern, and *Botrychium,* the rattlesnake fern, are widely distributed (Figs. 19-26, 19-27).

The members of this order of ferns form a distinct group, showing advanced evolutionary structures combined with primitive features. In the sporophytes the presence of a cambium and the development of secondary tissues are evidences of an advanced evolutionary position. Simplicity of form and structure in stem and leaf is regarded as a reduction rather than a primitive feature. The development of tuberous gametophytes is a primitive character, although the form and structure of the sperms are of an

Fig. 19-28 Diagrammatic representation of the life cycle of a common fern.

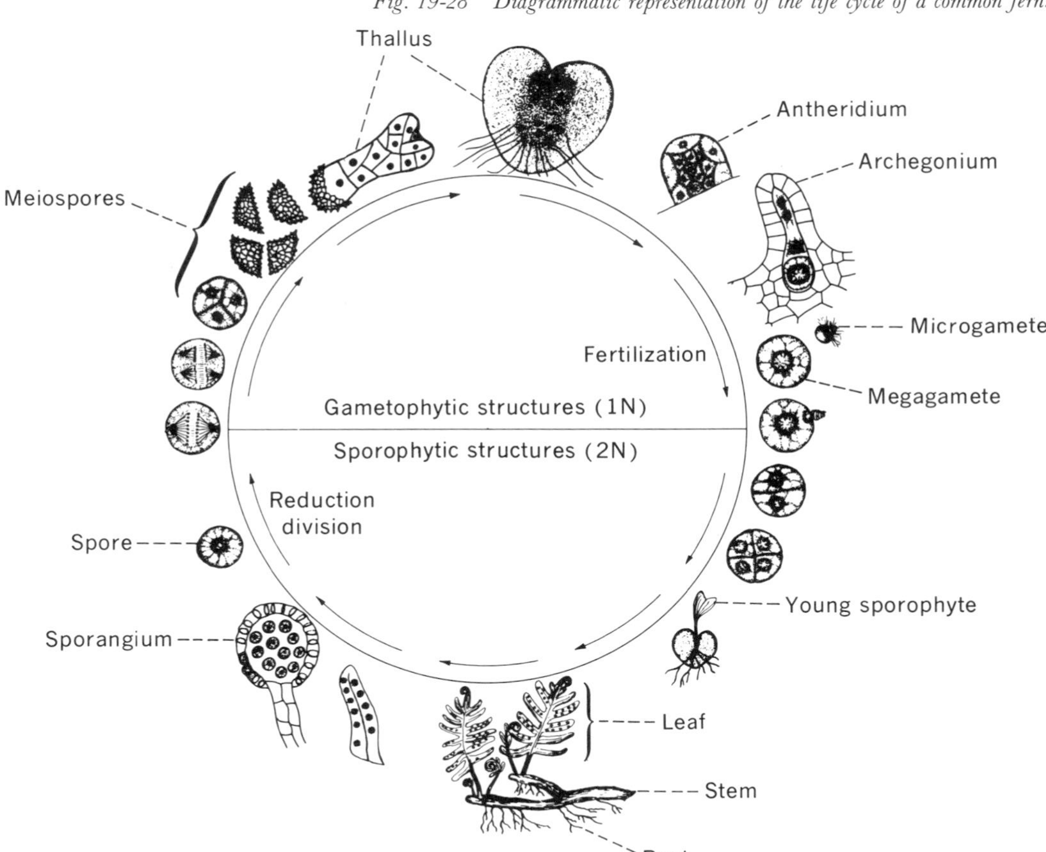

advanced nature. The presence of mycorrhizal fungi in the underground tissues of both sporophyte and gametophyte and the lack of chlorophyll in some of the gametophytes are regarded as evidences of degeneration.

Filicales—ferns

The order Filicales includes the common ferns which are widely distributed and are referred to as the true ferns. They are used extensively by florists in bouquets, and in greenhouses orchids are commonly raised on pieces of the trunk of the tree ferns or on fibrous material secured from the fern *Osmunda*. The tips of young fern leaves are eaten by natives of the Malaysian area and by gourmets in the finest restaurants. Since the days of the Romans the fern *Dryopteris filix-mas* has been used for its medicinal value in the cure of tapeworm.

The sporophyte of the fern, the conspicuous phase of the life cycle, is, in part at least, a perennial plant. In contrast, the gametophyte, a small thallus structure, lives only for a short time. The common ferns are homosporous (Fig. 19-28). There is another group of ferns called the water ferns that produce two kinds of spores and are heterosporous.

The ferns are widely distributed, occurring in both temperate and tropical regions. Ferns are generally small or medium-sized plants making up a part of the ground flora in moist forests (Fig. 1-5). The largest living representatives are the tree ferns of the tropics that may attain heights as great as 80 ft, making them a conspicuous feature of the flora (Fig. 19-29). Some ferns grow as floating aquatics (Fig. 19-48), some are rooted in mud and may be partially covered with water (Fig. 19-49), others are epiphytes on tree trunks in moist forests (Fig. 19-30), and still others grow under dry conditions. Some of these xerophytic forms may be found in the crevices of rocks on the sides of vertical cliffs.

The sporophyte In its early, physiologically dependent stages, the sporophyte consists of the embryonic root, stem, leaf, and a foot that is attached to the independent gametophyte and functions as an absorbing organ (Fig. 19-31). The mature sporophyte with roots, stem, and leaves bearing sporangia is the recognized fern plant (Fig. 19-32).

THE STEM The fern stem is generally a small, creeping, underground structure, but sometimes it is larger and erect. The stems of tree ferns are relatively large structures, attaining heights of 30 to 40 ft or more and diameters as great as 12 in. (Fig. 19-29). The stems of ferns are perennial and each year new leaves are produced. Conspicuous features of the fern stem include the persistent leaf bases and a

Fig. 19-29 Tree fern.

Fig. 19-30 Polypodium polypodioides, *an epiphytic fern growing on a large rock. (Courtesy Carolina Biological Supply Company.)*

hairy outgrowth called the **ramentum** (Fig. 19-32).

THE ROOT The roots are generally small, wiry, adventitious structures which grow from the stem and are frequently abundant enough to form a mass around the underground rhizomes (Fig. 19-32).

THE LEAF The leaves (fronds) of common ferns vary in size and form, but all are characterized by circinate vernation (Fig. 19-33). Simple leaves are present in some species, such as the walking fern (*Camptosorus*) (Fig. 19-35), but in most species the leaves are compound—most often once compound, sometimes twice or three times compound (Fig. 19-34). The leaflets are known as **pinnae,** and if the pinnae are also dissected, the ultimate divisions are known as **pinnules.** The leaf of the walking fern is only a few inches long (Fig. 19-35), but leaves on some tree ferns are as much as 6 ft in length (Fig. 19-29).

ANATOMY In most instances the anatomy of the Filicales is similar to that of other members of the Pteropsida. However, no cambium develops in the so-called true ferns, and no secondary vascular tissue forms.

The anatomical features of fern **roots** are similar to those seen in the primary roots of other vascular plants. The three primary meristems—protoderm, ground meristem, and pro-

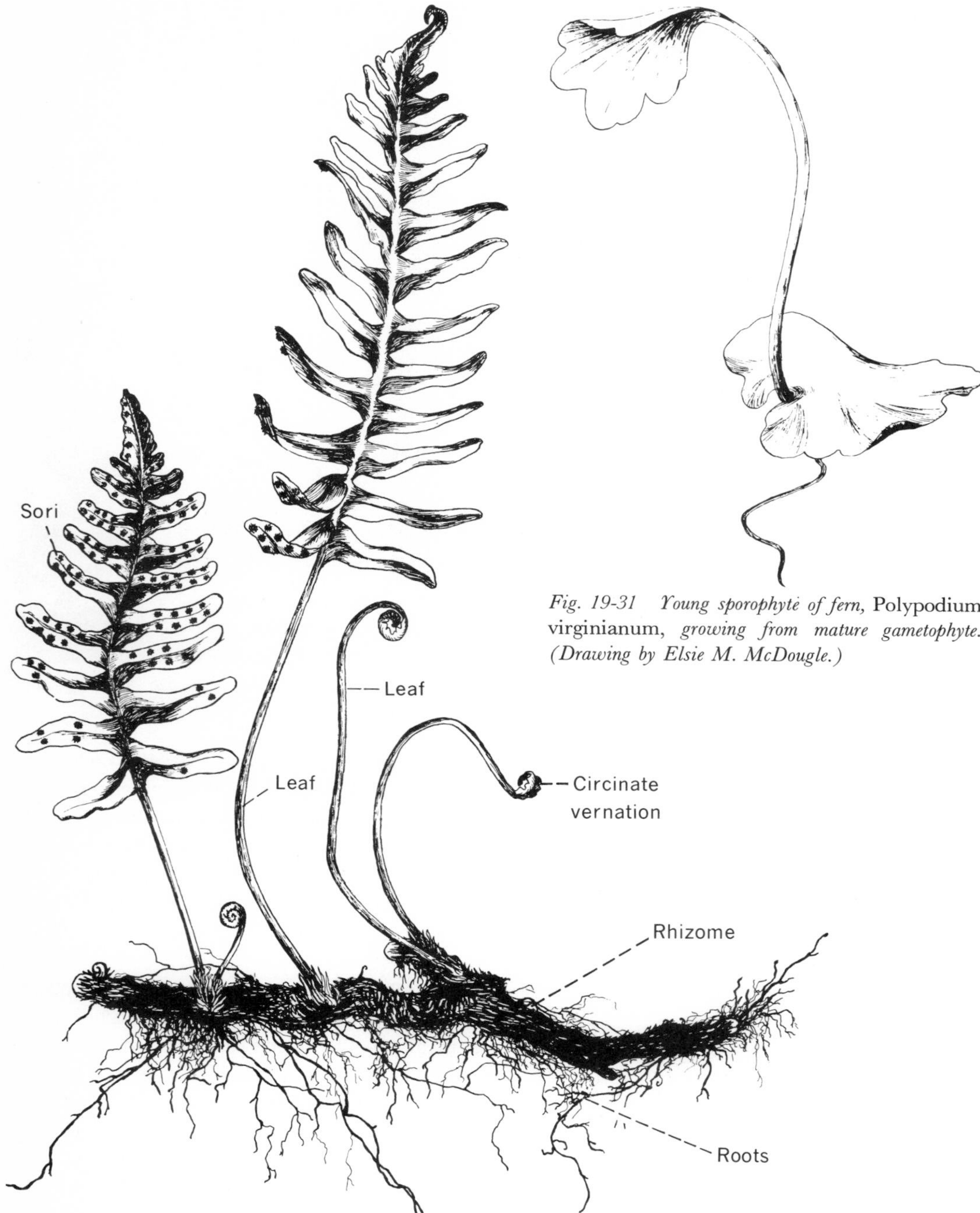

Fig. 19-31 Young sporophyte of fern, Polypodium virginianum, *growing from mature gametophyte. (Drawing by Elsie M. McDougle.)*

Fig. 19-32 Habit of fern. Mature sporophyte of fern, Polypodium virginianum. *(Drawing by Elsie M. McDougle.)*

Fig. 19-33 Portion of Schiede's tree fern, Cibotum schiedei, *showing circinate vernation. (Photograph courtesy of Conservatories, New York Botanical Garden.)*

Fig. 19-34 Fern-leaf types. A, Dryopteris spinulosa *with sori; B,* Osmunda claytoniana, *vegetative leaf interrupted by groups of fertile leaflets; C,* Osmunda regalis, *terminal, fertile leaflets.*

cambium—originate from cells derived from the pyramid-shaped apical cell. The stele is a protostele with a radial exarch arrangement of xylem and phloem (Fig. 8-33*A*). The epidermal cells are usually thin walled and they cover the cortical parenchyma. A variable amount of the inner cortical region may consist of sclerenchymatous cells. An endodermis is always present and well defined; each endodermal cell has the familiar Casparian strip. It is reported that in ferns lateral roots originate in the endodermis rather than in the pericycle.

Although no secondary vascular tissues develop in the **stems** of ferns, a considerable amount of sclerenchyma may differentiate in both the cortical region and in the pith. In some instances this sclerenchyma is very extensive (Fig. 19-36). The radial maturation of xylem is

Fig. 19-35 Camptosorus rhizophyllus, *walking fern, showing vegetative reproduction by turned-down leaf tips, which give rise to new plants, p; sori on leaves, s; roots, r. (Drawing by Elsie M. McDougle.)*

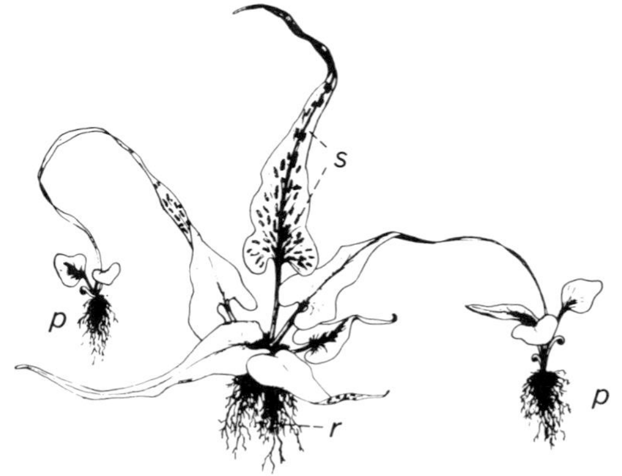

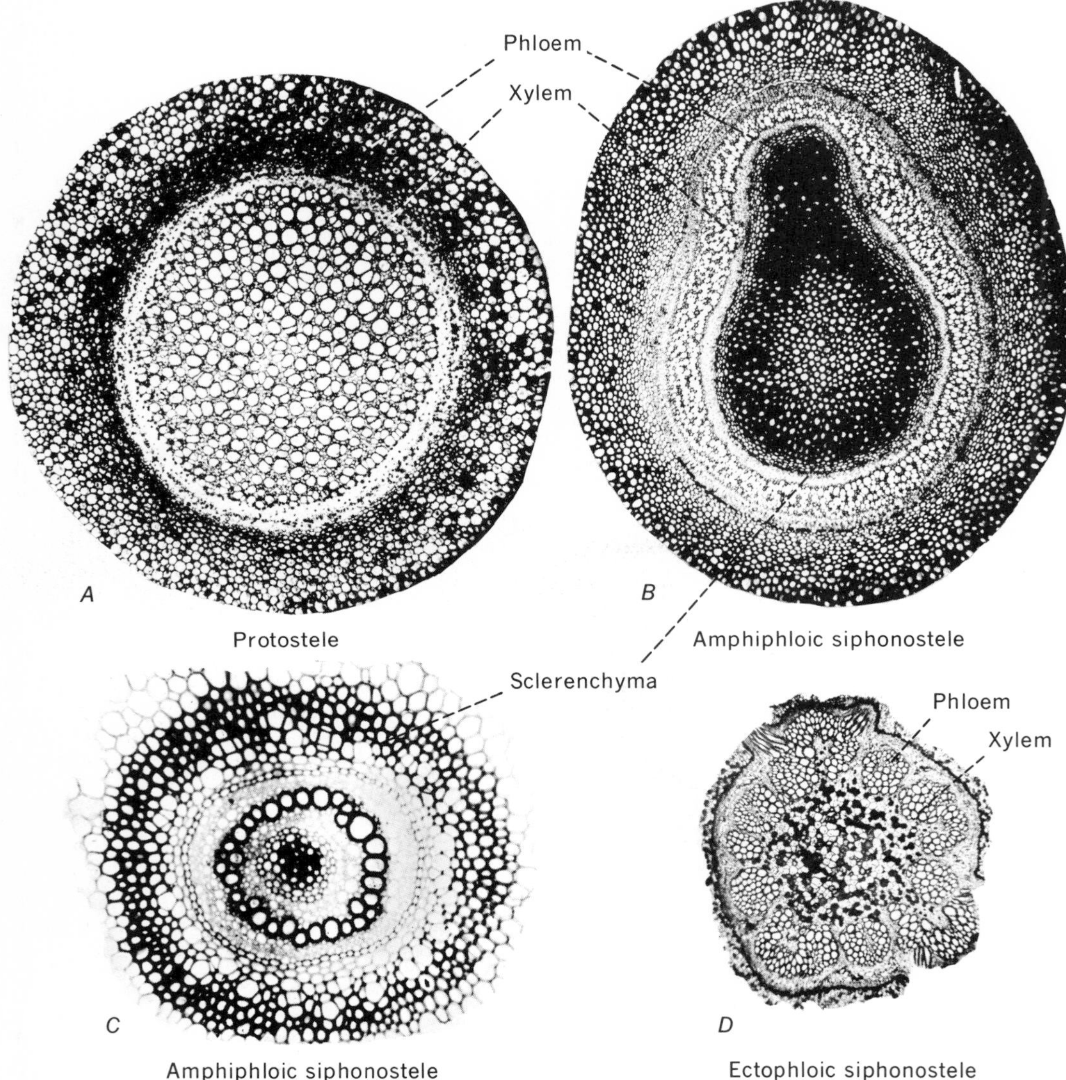

Fig. 19-36 Anatomy of ferns. A, transverse section of protostele of Gleichenia, *a tropical fern; B, transverse section of amphiphloic siphonostele of* Adiantum; *C, transverse section of the amphiphloic siphonostele of* Marsilea; *D, transverse section of the ectophloic siphonostele of* Osmunda.

typically mesarch. The epidermis of the fern stem is relatively permanent.

The vascular tissue is generally bounded by a pericycle and a well-defined endodermis. The xylem is made up of pitted or scalariform tracheids, and the phloem is made up of sieve tubes with sieve plates on the side walls. The vascular system of the ferns shows great diversity, and the stelar types include protostele, siphonostele, and dictyostele (Figs. 19-36, 19-37). Another anatomical feature of phylogenetic significance is the presence of leaf gaps in the vascular cylinder of those genera having a siphonostele.

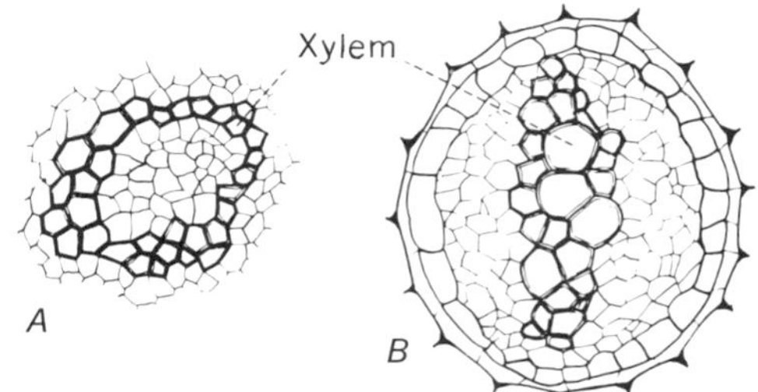

Fig. 19-37 Vascular structure of ferns. A, transverse section of portion of stem of young Pteris *sporophyte, showing stage in development of a siphonostele; B, transverse section of portion of stem of* Polypodium, *showing a single vascular bundle.*

The structure of fern **leaves** is similar to that of leaves of higher plants. There is a well-developed epidermal layer with stomata that are usually confined to the lower surface. Guard cells with chlorophyll occur as in higher plants (Fig. 19-38). The epidermal cells of the common mesophytic fern leaves are irregular in outline and contain chloroplasts. Internally the differentiation of the mesophyll into a palisade region and a spongy region is usually evident, but these regions are not sharply marked in ferns that live mostly in shaded habitats. The vascular system of the leaves is not materially different from that of the stems. The venation, however, is peculiar in that the larger veins are, in many instances, dichotomously branched. Generally the vascular bundles of the larger veins are concentric or bicollateral, whereas those of the smaller veins are collateral.

Fig. 19-38 Detail of leaf structure of ferns. A, Pteris, *cross section showing structure; B,* Nephrolepis, *surface view of lower epidermis showing stomata.*

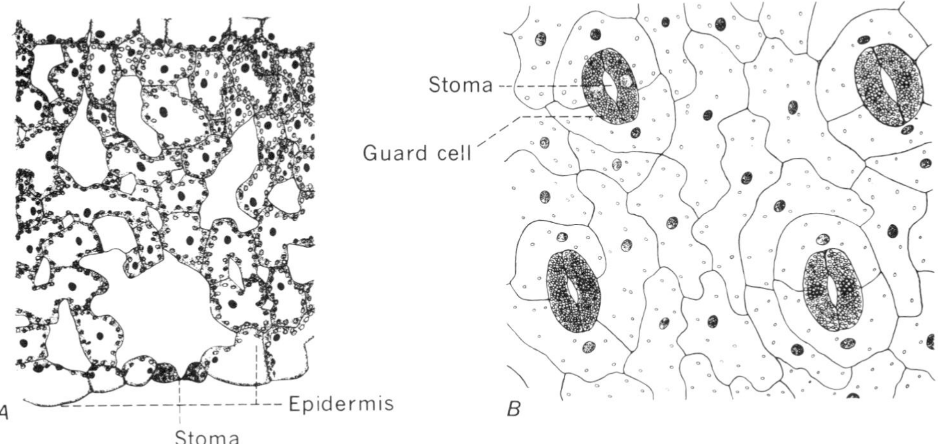

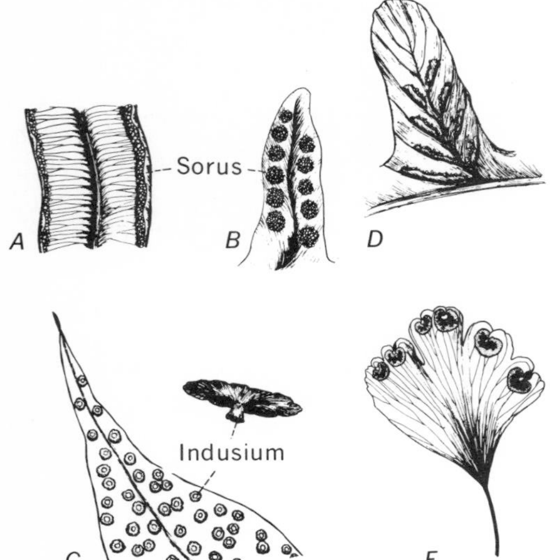

Fig. 19-39 Fern sori. A, portion of pinna of a species of Pteris, *the brake fern, showing the false indusium covering the marginal sporangia; B, sori of* Polypodium vulgare, *the polypody fern, without indusia; C, portion of pinna and a single detached indusium of* Cyrtomium falcatum, *the holly fern; D, pinna of* Asplenium *sp., one of the spleenwort ferns, showing the indusia attached along the margins of the sori; E, pinnule of* Adiantum pedatum, *the maidenhair fern, with marginal sori covered by false indusia. (Drawing by Elsie M. McDougle.)*

VEGETATIVE LEAVES AND SPOROPHYLLS In many of the ferns the sporangia are produced on the underside of vegetative leaves (Figs. 19-39, 19-40). In some genera, such as the royal fern and the Clayton fern, sporangia are borne on specialized portions of the foliage leaves (Fig. 19-34*B, C*). In the Clayton fern several fertile segments near the middle of the leaf, differing from the sterile segments in that they contain no chlorophyll and are contracted in size, bear all the sporangia. In other genera, such as the sensitive fern, the sporangia are produced on fertile leaves that are distinctly different in appearance. They usually lack chlorophyll and are not photosynthetic. In cases in which there are two kinds of leaves, sterile vegetative leaves and fertile spore-bearing leaves, the plant is said to show dimorphism in its leaves. Leaves that bear sporangia are called sporophylls regardless of the degree of structural specialization.

THE SPORANGIA AND THE SORI The sporangia of ferns show almost every conceivable manner of distribution over the surfaces of the leaves (Fig. 19-39); this is a character used in identifying families and genera. Usually the sporangia occur on the underside of the leaf surface. Their distribution varies from an arrangement in which a sporangium occurs singly, to an irregular arrangement in which the sporangia cover patches of the undersurface of the leaf, to an arrangement in which they sometimes entirely cover the undersurface of the sporophylls. When the sporangia are grouped, the group is termed a **sorus** (plural, sori). The sori occupy definite positions on the leaf surface, usually being found on the veins or at vein endings. The sori may be covered by a thin membranous tissue attached to the leaf surface, called the **indusium** (Figs. 19-39*C*, 19-40). In other genera the sporangia are partially covered by the reflexed margin of the leaf; this is called a false indusium (Fig. 19-39*A*).

DEVELOPMENT OF THE SPORANGIUM AND MEIOSPORES The sporangium of the Filicales follows a pattern of development referred to as being leptosporangiate. The sporangium develops from a single outer initial formed by the division of a superficial cell of the sporophyll (Fig. 19-41). By a series of divisions, the initial produces the sporangium, including the stalk, the sterile wall, and the inner spore-bearing tissue (Fig. 19-42*B*).

In the young developing sporangium there is a large triangular **central cell.** By a series of divisions in the central cell, there is differen-

Fig. 19-40 A pinna of the fern Dryopteris. *(Photograph by C. Neidorf, courtesy of Smithsonian Institution.)*

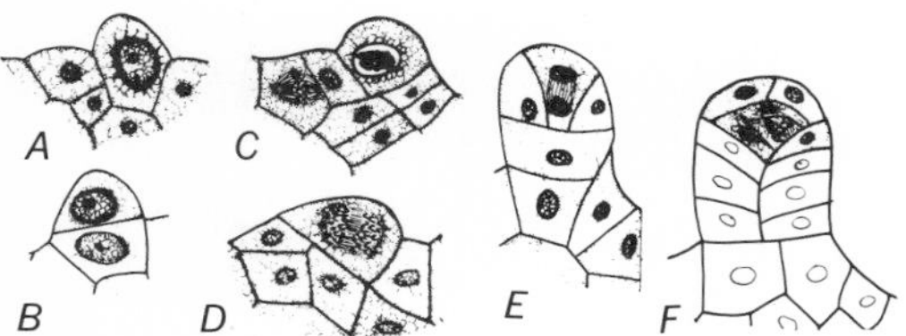

Fig. 19-41 Early stages in development of sporangia of ferns. A, initial cell of the sporangium just protruding above the surface; B, the basal cell has been cut off; C, D, stages in division; D, showing anaphase of division which will result in the development of a vertical wall as shown in E; E, young sporangium with basal cell and three curved vertical walls, two of which are shown in drawing and one of which is in the plane of the page; the central cell, which is pyramidal in shape, is now known as the apical cell of the sporangium; the division figure in the apical cell is the division to form a cell from which is developed the upper part of the wall as shown in F; F, the central cell dividing to complete the formation of the three primary tapetal cells.

tiated a layer of nutritive tissue, called the **tapetum,** surrounding the primary sporogenous cell (Fig. 19-41*F*). By four successive mitotic divisions the primary sporogenous cell forms typically 16 spore mother cells, or sporocytes. Each of the diploid sporocytes eventually undergoes meiosis and forms a tetrad of haploid meiospores, which are the first cells of the new gametophytic phase of the fern. As the sporangium enlarges and develops, the walls of the tapetal cells disintegrate and their cytoplasmic contents and nuclei form a nourishing medium for the sporocytes and spores.

While the spores are maturing, the rest of the sporangium continues its development. The mature sporangium consists of a slender **stalk,** attaching it to the tissues of the sporophyll, and a **capsule,** or spore case, that contains the spores (Fig. 19-42). The wall of the sporangium is one cell layer in thickness. A portion of the wall is specialized to form the **annulus,** or structure of dehiscence, and the **stomium,** a group of specialized cells, sometimes called the **lip cells** (Fig. 19-42*B*). This arrangement of cells permits an easy cleavage when the sporangium begins

Fig. 19-42 Fern spores and sporangia. A, tetrad of meiospores; B, a single sporangium. (Drawings by Elsie M. McDougle.)

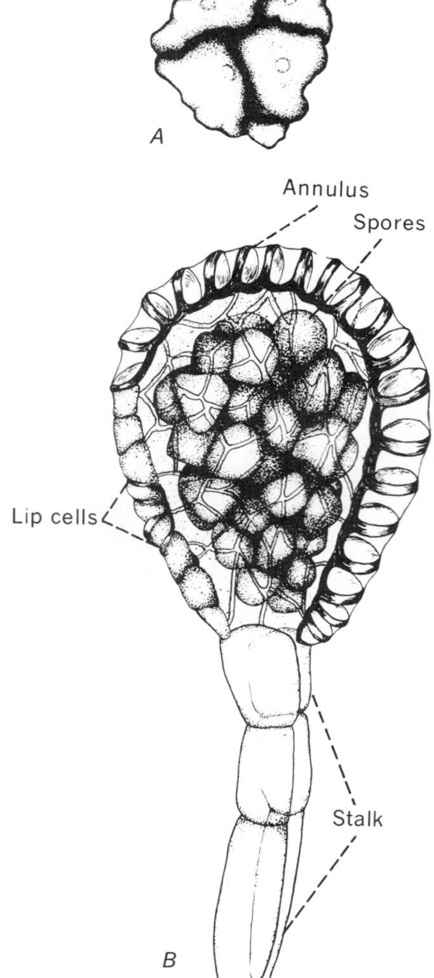

to dehisce. The opening of the capsule can be demonstrated by placing unopened ripe sporangia in a drop of glycerin and observing under a microscope.

The gametophyte Upon germination the spore produces a short filament of cells resembling that of a green alga or the protonema of a moss. Some fern gametophytes continue in the filamentous condition, when they are grown in poor light, even to maturity, but the typical mature fern gametophyte is a small, flat, heart-shaped thallus, ⅛ to ⅓ in. in diameter (Figs. 19-43, 19-44).

The thallus is developed by the repeated divisions of a wedge-shaped apical cell which is formed at the end of the filament. As a result of the rapid growth of the two lobes of the thallus, the apical cell occupies a position in the notch of the heart-shaped thallus. At maturity the thallus is a thin, flat mass of tissue. The lobes consist of a single layer of cells, whereas the central part is several cell layers in thickness. Rhizoids are produced on the undersurface (ventral) of the thallus and serve as organs of attachment (Fig. 19-44).

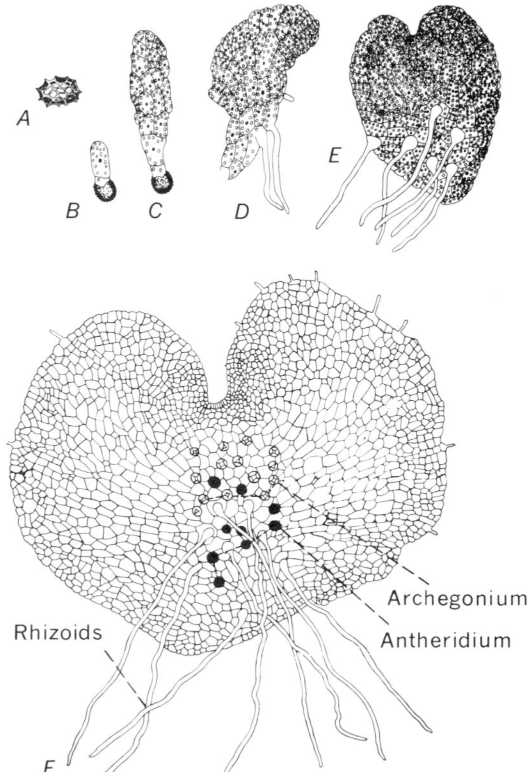

Fig. 19-44 Stages in development of fern gametophyte. A, mature spore; B, germinating spore; C–E, developing thallus; F, mature thallus.

Fig. 19-43 Fern prothallia on soil.

THE GAMETANGIA The antheridia and the archegonia are produced on the ventral surface of the thallus. Young thalli produce only antheridia, but older thalli produce both antheridia and archegonia. The antheridia are located on the thinner portion of the thallus, often among the rhizoids. The archegonia are produced on the thickened portion of the thallus, just back of the notch in which the apical cell is located.

An outstanding feature of the gametangia of

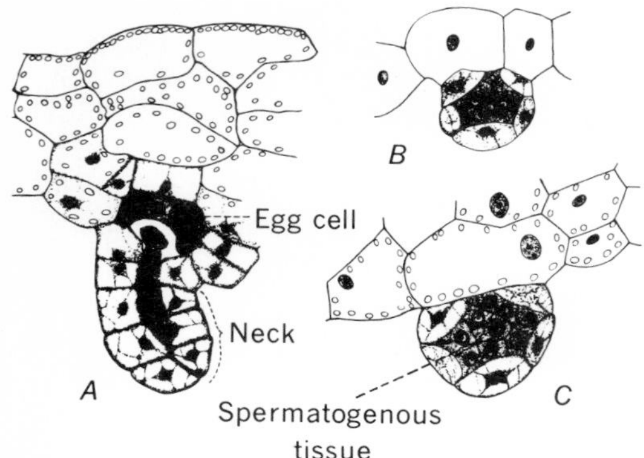

Fig. 19-45 Sex organs of fern. A, mature archegonium with ventral canal cell and neck canal cells disintegrating; B, C, stages in development of an antheridium. (From slides prepared by Dr. H. A. Wahl.)

ferns is the reduction in the amount of the sterile tissue. This is seen in the few cells of the wall of the antheridium and in the reduction of the number of neck cells of the archegonium. The wall of the mature antheridium consists of only three cells (Fig. 19-45*C*). The enclosed central portion of the antheridium originally is comprised of a central cell, which is the primary spermatogenous cell. The central cell divides until about 32 spermatogenous cells, or more in some primitive genera, are formed. At maturity the sperm consists of a nucleus within a spirally coiled body to which are attached many flagella. The antheridium breaks open and the sperms escape. When it is shed, the sperm is within a thin membrane that soon dissolves in water and the sperm is freed.

The mature archegonium of the common fern consists of an enlarged venter which is deeply sunken in the tissues of the thallus, a very short neck, curved backward because of the unequal growth of the neck cells, and an axial row of normally two neck canal cells, a ventral canal cell, and the egg in the venter (Fig. 19-45*A*). When the egg is ready for fertilization, the neck canal cells and the ventral canal cell disintegrate, the neck of the archegonium opens, and a portion of the disintegrating neck canal cells and the ventral canal cell are extruded. The remainder of the material forms a fluid through which the sperm may pass to fertilize the egg. The egg is a single, large, nearly spherical cell with a depression on the side toward the neck of the archegonium. The depressed portion becomes the "receptive spot" which admits the sperm at fertilization. It has been demonstrated that the sperm is caused to swim to the archegonium by the chemical attraction of substances secreted by the archegonium. With fertilization, the diploid number of the chromosomes is reestablished and the new sporophytic phase begins.

Fig. 19-46 Early stages in development of the embryo of a fern (Pteris). *A, fertilized egg; B–E, early stages in growth by cell division; F, somewhat older stage, with apical cell of root differentiated. (From slides prepared by Dr. H. A. Wahl.)*

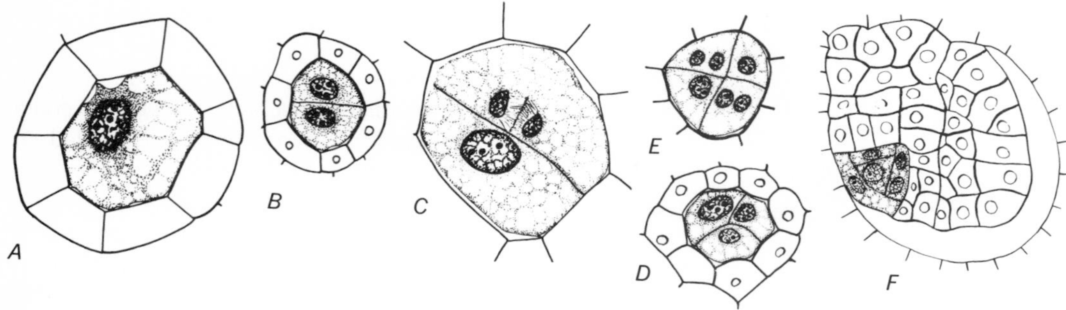

The embryo and its development As soon as fertilization is accomplished, the zygote develops a membrane. Mitotic division takes place, dividing the zygote by a vertical wall into an anterior (front) half and a posterior (back) half, and then a transverse division forms four cells. The two anterior cells produce the **stem** and the first **leaf,** and the two posterior cells produce the **root** and the **foot** (Fig. 19-46). At maturity the foot is a short, projecting, hemispherical mass of cells in close contact with the cells of the thallus. Soon the root and the leaf of the embryo emerge from the archegonial wall, and, with their development, the young sporophyte begins to manufacture its own carbohydrate food, absorbs water and minerals from the soil, and soon becomes independent of the gametophyte.

The water ferns The water ferns (Hydropteridineae), so named because of their aquatic habit, are a small group consisting of four

Fig. 19-47 Azolla, *growing on wet soil in greenhouse. (Much enlarged.)*

Fig. 19-48 Habit of the water fern, Salvinia. *(Much enlarged.)*

genera, all of which are heterosporous. These genera are *Azolla* (Fig. 19-47) and *Salvinia* (Fig. 19-48), which usually float on the surface of the water, and *Marsilea* (Fig. 19-49) and *Pilularia,* which grow rooted in the mud in quiet water.

There is considerable diversity among the four genera. *Marsilea* will be discussed briefly. The stem is an amphiphloic siphonostele (Fig. 19-36*C*), and the four pinnae of the leaf are so arranged that they resemble a four-leaf clover (Fig. 19-49). Sporocarps are laminar in origin and are produced on highly specialized branches of the leaf petioles (Fig. 19-49). Sori occur on the inner side of each half of the sporocarp, and each sorus consists of a group of either megasporangia or microsporangia covered with a saclike membranous indusium (Fig. 19-50*B*). Following meiosis in a megasporangium, only one megaspore remains functional but many functional microspores are produced within each microsporangium.

Sporocarps are long lived and, when they are placed in water, they gradually swell and open. Following imbibition of water by the tissues within the sporocarp, a translucent gelatinous ring, to which the sori are attached, is gently extruded (Fig. 19-50*A*). At the time of spore germination, the megaspore (Fig. 19-51*A*) produces a small green thallus, the female gametophyte, which usually protrudes from one end of the megaspore (Fig. 19-51*C*). On germination of a microspore a much-reduced microgametophyte is formed.

The female gametophyte produces a single reduced archegonium with a relatively large egg (Fig. 19-52*A, B*). The single archegonium consists of a few neck cells, a neck canal cell, a ventral canal cell, and an egg. The male gametophyte, growing entirely within the wall of the microspore, consists of only a few cells. Two central spermatogenous cells divide to produce 16 sperms which escape through the ruptured

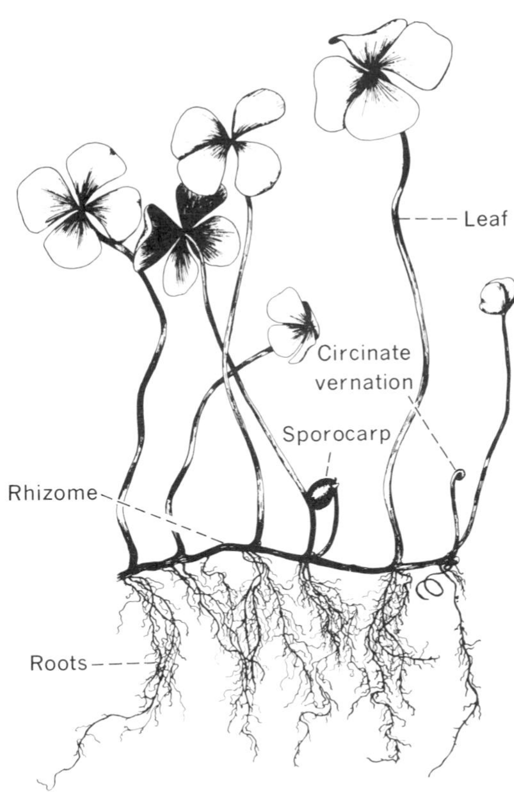

Fig. 19-49 Marsilea, *one of the heterosporous water ferns. (Drawing by Elsie M. McDougle.)*

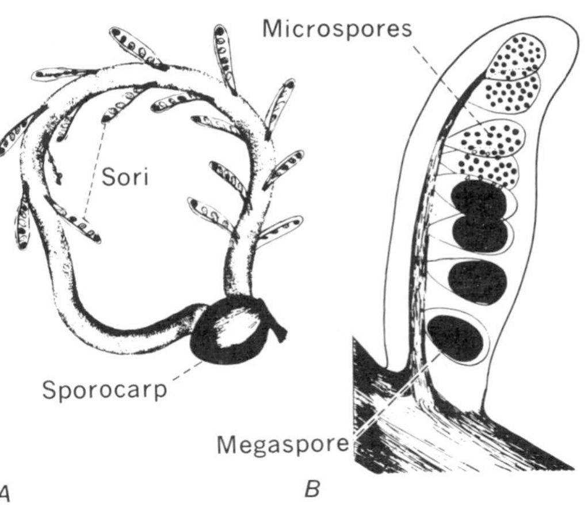

Fig. 19-50 Marsilea. *A, sporocarp with protruding gelatinous ring carrying sori; B, a sorus with megaspores and microspores.*

microspore wall. The sperms swim to the archegonium, to which they are attracted, and one penetrates the egg and fuses with it to produce the zygote. Following fertilization, the zygote divides to form a quadrant (Fig. 19-52*C*), the

Fig. 19-51 Marsilea. *A, megaspore; B, microspores; C, a megaspore germinating with protruding female gametophyte.*

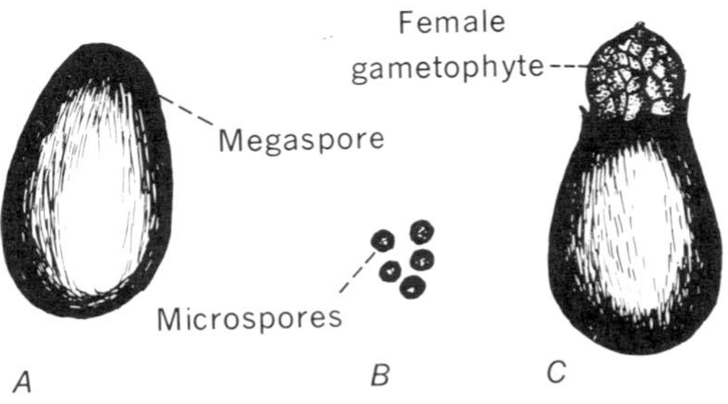

Fig. 19-52 Marsilea. *A, detail of female gametophyte, showing archegonium with egg cell and indications of nutritive tissue beneath it; B, fertilization of egg by spirally coiled sperm; C, D, detail of stages in development of embryo still contained in the old female gametophyte; C, four-celled stage; D, number of cells increased by division.*

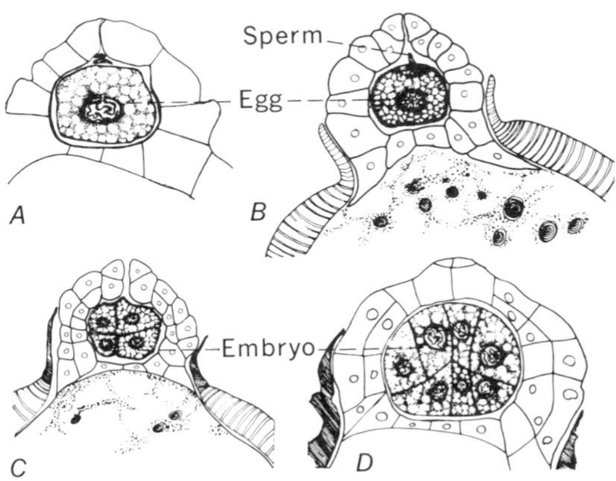

outer two cells of which develop into the leaf and root and the inner two cells of which develop into the foot and stem.

QUESTIONS

1. In what forms are fossil plant materials of considerable economic value?

2. In what ways do students of botany make use of paleobotanical evidence?

3. What might the landscape be like on the campus if vascular tissue had not evolved?

4. What would be the effects on your daily life if all plants were leafless?

5. Do you think there are significant differences in the spores produced by *Lycopodium* and those produced by *Selaginella?* Why?

6. Explain why some closely related genera, such as *Tmesipteris* and *Psilotum,* are found in different areas very remote from each other.

7. Do the kinds of spores produced by the water ferns better equip these plants for survival than the kinds of spores produced by the common woodland ferns?

8. What is the morphological and physiological relationship between micorrhizal fungi and the perennial gametophytes of *Lycopodium?*

9. A chart summarizing the geological periods of the earth's history is presented in this chapter. Where would you go and what would you look for in an attempt to verify this information?

10. List the features of an environment in which epiphytes might persist.

REFERENCES

ANDREWS, H. N. 1961. Studies in Paleobotany. John Wiley & Sons, Inc., New York.

ARNOLD, C. A. 1947. An Introduction to Paleobotany. McGraw-Hill Book Company, New York.

BOWER, F. O. 1935. Primitive Land Plants. Macmillan & Co., Ltd., London. Excellent reference.

EAMES, A. J. 1936. Morphology of Vascular Plants, pp. 117–304. McGraw-Hill Book Company, New York. A standard work presenting basic features of vascular plants.

FOSTER, A. S., and E. M. GIFFORD, JR. 1959. Comparative Morphology of Vascular Plants, pp. 101–319. W. H. Freeman and Company, San Francisco. A detailed discussion of topics in plant morphology.

SCAGEL, ROBERT F., et al. 1965. An Evolutionary Survey of the Plant Kingdom, pp. 355–465. Wadsworth Publishing Company, Belmont, Calif. Modern comparative treatment, excellent figures.

SINNOTT, E. W. 1960. Plant Morphogenesis. McGraw-Hill Book Company, New York. Comprehensive treatment of plant development, types of meristems, polarity, etc.

SMITH, G. M. 1955. Cryptogamic Botany, 2d ed., vol. 2, Bryophytes and Pteridophytes. McGraw-Hill Book Company, New York. A standard textbook presenting pertinent information.

20

VASCULAR PLANTS; THE HIGHER GROUPS; SEED PLANTS

El Gran Arbol de Tule, *the Big Tree of Tule,* Taxodium mucronatum Ten. *This tree is the national tree of Mexico and grows native or as a cultivated plant in most of the states of the Republic of Mexico. Many of them were planted by the Aztec kings more than 500 years ago. This specimen grows in a churchyard in the village of Santa Maria del Tule, near Oaxaca, and its age has been estimated to be about 2,000 years. (Photograph by Dr. Charles J. Chamberlain, previous to 1910, courtesy of Chicago National History Museum, Chicago, Illinois.)*

The emphasis in this chapter is placed on a review of the Gymnospermae, including their distribution in time and their importance today as living plants. The nature and evolution of the carpel and stamen, the possible origin of the seed, the structure of the cone, and the unique features of flowers are discussed. A comparison of the gametophytes of gymnosperms and angiosperms, the formation of pollen, its distribution and germination, fertilization in seed plants, and the development of the embryos of gymnosperms and angiosperms are included.

The Gymnospermae and Angiospermae are classes of the subdivision Pteropsida and, with the class Filicinae (Chap. 19), comprise the more or less coordinate groups within this subdivision. The Pteropsida are characterized by the development of **megaphylls** (large leaves), in contrast to the **microphylls** (small leaves) of the Psilopsida, Lycopsida, and Sphenopsida, the other subdivisions of the Tracheophyta, or the vascular plants. The pteropsid leaf is a megaphyll, which is distinguished from the microphyll by its larger size, by its complex venation, and by having evolved from a major branch system.

The gymnosperms and angiosperms are seed-producing plants and, in the past, have been referred to the division Spermatophyta by some botanists. The seed plants are widely distributed and they constitute a large part of the vegetation of the earth today. They overshadow all other groups of plants in their economic importance. Directly or indirectly, most of the food of land animals, including man, comes from the seed plants, and man generally is dependent upon them for much of his clothing, shelter, and fuel.

Although ideas concerning the origin of the seed are largely speculative, it is obvious that heterospory, as well as the retention of the megaspore and its product, the megagametophyte, on the sporophytic plant, was prerequisite to the seed habit. The ovule, which develops to form a seed, consists of a centrally located megasporangium, or nucellus, surrounded by one or more integuments. A seed has been defined as a matured ovule, containing an embryo and stored food.

In the Gymnospermae the ovule most often is borne exposed, or naked, on a megasporophyll and food is stored in the cells of the large megagametophyte. In the Angiospermae the ovule develops into the seed within the cavity formed by a closed carpel and food is stored within the seed in the endosperm tissue or within the cotyledons (see also Chap. 11).

Many of the reproductive structures in the seed plants were named before it was known that they were homologous with other structures in the lower vascular plants. For this reason a comparison of the terms used may lead to a better understanding of the terminology in this chapter.

lower vascular plants	*seed plants*
Megasporophyll	Ovuliferous scale, carpel
Megasporangium	Nucellus
Megaspore	Megaspore
Megagametophyte	Megagametophyte, embryo sac
Megagamete, egg	Megagamete, egg
Microsporophyll	Microsporophyll, stamen
Microsporangium	Microsporangium, anther
Microspore	Microspore
Microgametophyte	Microgametophyte, pollen
Microgamete, sperm	Microgamete, sperm

THE GYMNOSPERMS

General features The gymnosperms include ancient lines of plants, and the long evolutionary history of the gymnosperms includes many extinct forms. In some cases their geological age may be computed in millions of years (Fig. 19-1). In some instances both macroscopic and microscopic features of these fossil plants have been perfectly preserved (Fig. 20-3*A*). A knowledge of fossil plants is essential to a complete conception of the plant kingdom because some of these plants were the ancestors of present-day plants.

Some gymnosperms are known both as fossil and as living forms. Today these groups are represented by only a remnant of a once vast assemblage of species and genera (Figs. 20-15, 20-16). Among the living forms of gymnosperms are short, stubby, almost herblike plants, shrubby and even vinelike forms, and many tree types.

Although some gymnosperms are restricted

in their geographical range, as a group, they are found throughout the temperate and tropical zones and even in arctic regions. The greatest development of living gymnosperms has been attained by the coniferous forms in the temperate climates of both hemispheres. In general, gymnospermous plants are found in fairly dry situations and, in some instances, even in semidesert regions.

Classification

Class Gymnospermae. Plants with seeds usually borne on the upper side of open scales that are often produced in cones. In some forms ovules were developed on the surface of or at the tips of vegetative leaves. Woody tissues are mostly composed of single-celled elements, the tracheids. Two subclasses and seven orders of Gymnospermae are recognized:

Subclass Cycadophytae The cycadophyte line.	
Order Cycadofilicales or Pteridospermae	Fossil forms
Order Bennettitales	Fossil forms
Order Cycadales	Fossil forms and living plants
Subclass Coniferophytae The coniferophyte line.	
Order Cordaitales	Fossil forms
Order Ginkgoales	Fossil forms and living plants
Order Coniferales	Fossil forms and living plants
Order Gnetales	Living plants

Cycadofilicales

The earliest fossil remains of seed plants are found in the strata of the Paleozoic era and, from various calculations, are estimated to be from 100 million to 250 million years old (Fig. 19-1). Among the oldest recognized fossil seed plants are some that greatly resemble ferns, especially in their leaf forms. So great is this resemblance that they were at first thought to be ferns, but the discovery of seeds attached to the fernlike foliage identified them as seed plants. Because of these features they are also called **pteridosperms** (Figs. 19-2, 20-1).

The nature of the epidermal tissue indicates closer similarities to gymnosperms than to ferns. Fossil remains show that the Cycadofilicales usually had slender stems bearing large fernlike leaves that were widely spaced in a spiral arrangement. The slender stem and weakly developed vascular system indicate that some were probably vinelike. Others were sturdier and probably resembled modern tree ferns (Fig. 19-2).

Seeds were borne singly on the leaves (Fig. 20-1). In some cases, possibly the most primitive, the seeds were produced on branched stalks reminiscent of the sporangia-bearing stalks of the Psilopsida. In genera regarded as possibly more advanced, the seeds were produced on the surfaces of the leaves.

Bennettitales

This group, often designated as the cycadeoids, is thought to have originated, like the cycads, from the Cycadofilicales. Long of worldwide distribution, they became extinct in the Mesozoic era (Fig. 19-1). Some had short stubby stems from a few inches to a yard or more in length that bore large fernlike or palmlike leaves (Fig. 20-2). Others had slender branching stems with small-bladed leaves, and some may have been trees. The most significant feature of these plants was their production of both microsporophylls and megasporophylls in the same spiral structure (Fig. 20-3*B*). This bisporangiate structure differentiates them from both groups of plants to which they may be related, i.e., the Cycadofilicales, which produced staminate and ovulate strobili on the same plant, and the Cycadales, all of which produce staminate and ovulate cones on separate plants.

Fig. 20-1 A carboniferous landscape. The fernlike, foliage-bearing seeds are Cycadofilicales; the large stems and cones on the left are those of lepidodendrids; on the right a calamite; the small plants in the foreground are sphenophylls. (Courtesy of Chicago Natural History Museum.)

During the Mesozoic era, when giant dinosaurs roamed the earth, the cycadeoids were very widespread. They have been found on all continents. A famous collecting ground in the United States is found in the Black Hills of South Dakota. These fossils had the attention of botanists in both field and laboratory, especially of Wieland, whose investigations provide most of the information about them.

Cycadales

The Cycadales are represented in the fossil flora of the early Mesozoic era (Fig. 19-1) and they extend to the present. Living cycads, represented by nine genera and perhaps sixty species, are only a remnant of a former abundant flora. They are the most primitive living seed plants and live to a great age. They are usually

Fig. 20-2 Monanthesia magnifica. *A cauliflorous cycadeoid from the Mesa Verde Upper Cretaceous of the San Juan Basin, northwestern New Mexico. Tangent section through armor of old frond bases with small flower stalk in each axilla, except at A, where a frond base fails of later growth. The relation is nearly as persistent as that of bract and scale in a pine cone. The bundle patterns of the frond bases appear; likewise the bundle cylinder of the peduncles. The lower bracts are cut. At R is indicated the enveloping ramentum, or chaff. (Figure and explanation by Dr. George R. Wieland.)*

Fig. 20-3 A, photomicrograph of a stem section of a fossil gymnosperm; B, Cycadeoidea. *Diagrammatic sketch of longitudinal section through the bisporangiate strobilus. At the center is the apical cone closely invested by a zone of short-stalked abortive (?) ovules and interseminal scales. On the left is a single frond of the hypogynous staminate disk, with much-reduced pinnules bearing densely crowded sporangia. On the right a similar fertile frond is arbitrarily shown in an expanded position. Exterior to the fronds are the hairy bracts. About three-fourths natural size. (Redrawn from sketch by Dr. George R. Wieland, Am. J. Sci., **11**:424, 1901.)*

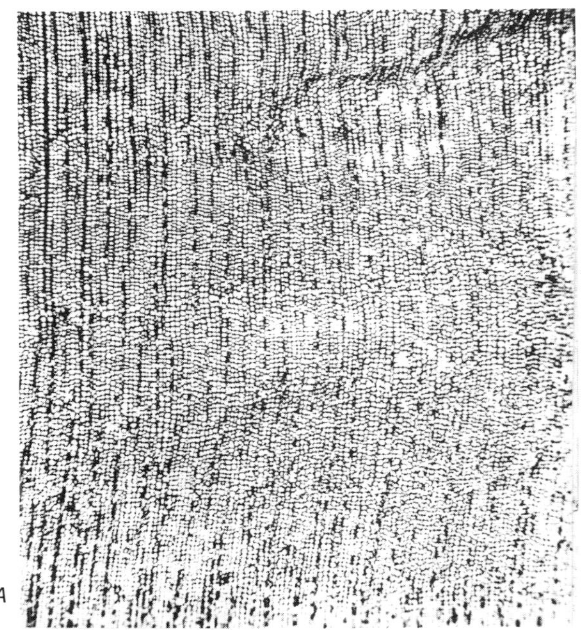

short, thick-stemmed plants, producing large fernlike or palmlike leaves (Fig. 20-4). Some species attain heights of 50 or 60 ft and are treelike in form. The leaves of the genus *Cycas* (Fig. 20-5), known as palm leaves to the floral trade, are used extensively in decoration. Cycads are indigenous to tropical and subtropical zones of both hemispheres and are grown as ornamentals in those regions.

Anatomical features of the cycad stem include a large pith, a scanty xylem region, and a large cortex. The leaf bases remain on the stem and closely cover it. Although some cycads are

Fig. 20-4 Dioön edule, *a cycad growing at Chavarrillo, Mexico. (From C. J. Chamberlain, "Elements of Plant Science," McGraw-Hill Book Company, New York, 1930.)*

branched, unbranched stems are characteristic of the group.

Living cycads are dioecious, the staminate and ovulate cones being produced on separate plants. The microsporophylls are aggregated into relatively large staminate strobili, or cones (Figs. 20-7, 20-9). The megasporophylls range from structures somewhat resembling leaves, which are aggregated into a loose strobilus, as in the genus *Cycas* (Fig. 20-6), to highly reduced structures aggregated into a compact cone, as in *Zamia* (Figs. 20-8, 20-9). On germination the microspores give rise to male gametophytes and the megaspores to female gametophytes. The mature female gametophyte consists of a mass of tissue within the nucellus and the integuments of the ovule. Four or five archegonia are formed, each consisting of two neck cells, a ventral canal cell, which usually disintegrates early, and an egg (Fig. 20-10*A–D*). The developing male gametophyte is surrounded by the microspore wall. Two successive mitotic divisions result in the formation of three cells: a basal prothallial cell, a generative cell, and a tube cell. The prothallial cell persists throughout the life of the male gametophyte.

The male gametophyte (pollen) is shed in the three-celled stage, is wingless, and is distributed by the wind (Fig. 20-11). When the pollen reaches an ovule on the ovulate cone, a pollen tube grows through the nucellus. The pollen tube does not carry the male cells to the archegonium, but it serves as a haustorium and is presumed to be important in securing nutrient material for the developing microgametophyte. The growth of many pollen tubes destroys the cells of the nucellus, and the tissue is often completely disorganized above the region of the female gametophyte containing the archegonia.

Further development of the male gametophyte consists of the division of the generative cell to form a stalk cell and a body cell (Fig. 20-11*I, J*). The body cell, by division, forms two male cells, or sperms, each with a coiled struc-

Fig. 20-5 *Cycads* (Cycas revoluta). *(Photograph furnished by Conservatories, New York Botanical Garden.)*

Fig. 20-6 Cycas revoluta. *Ovulate cone.*

Fig. 20-7 Cycas revoluta. *Staminate cone.*

Fig. 20-8 Zamia florida. *Ovulate cone. (Photograph by Chicago Natural History Museum from negative by Dr. Charles J. Chamberlain.)*

ture, called a blepharoplast, bearing hundreds of flagella (Fig. 20-10*E, F*). The motile sperm of the cycads is one of their most primitive features. The sperms swim about in the liquid within the pollen chamber and penetrate the neck of the archegonium, and one fertilizes the egg nucleus. The embryo is produced by repeated divisions of the zygote, and, at maturity, it develops two cotyledons.

The covering of the seeds of cycads consists of three layers, a middle stony one and inner and outer fleshy ones, both of which have vascular bundles. The outer fleshy layer dries and adheres tightly to the middle layer, forming a hard type of seed.

Ginkgoales

The *Gingko* tree, native in China, is the only living representative of the order Ginkgoales, which flourished with numerous genera and species throughout the Mesozoic era (Fig. 19-1). This tree, now widely cultivated as an ornamental, has an unbroken history far back into past geological ages. Scarcely any other living plant so completely fulfills the name of living fossil as does the *Ginkgo* (Fig. 20-12). The two-lobed leaves have dichotomously branched veins (Fig. 20-13*C*). The common name maidenhair tree refers to the resemblance of the leaves to those of *Adiantum,* the maidenhair fern.

Long branches bear scattered leaves, and short branches bear clustered leaves and the reproductive structures. *Ginkgo* is dioecious; one tree bears staminate strobili, and the other tree ovulate strobili. The staminate strobili are loose

Fig. 20-9 Staminate and ovulate cones of Zamia.

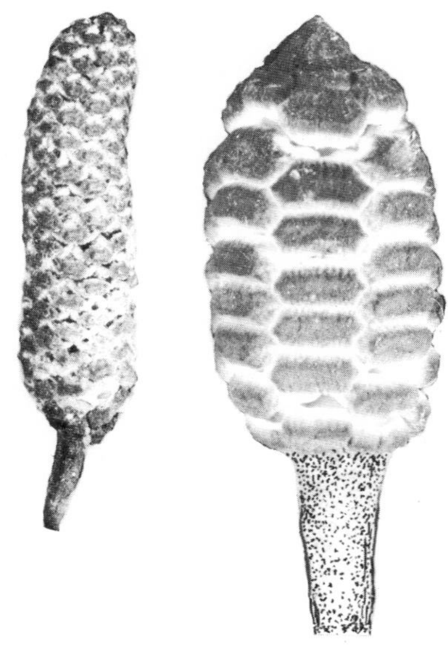

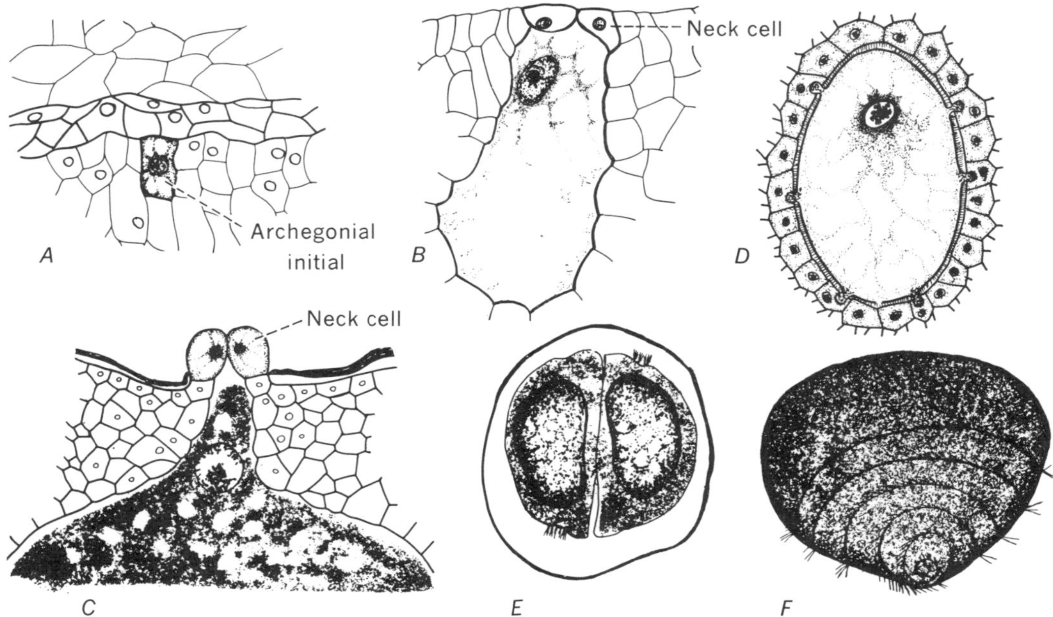

Fig. 20-10 Development of archegonium and sperms in cycads. A–D, archegonia; A, archegonial initial, below a nutritive layer; B, young archegonium, showing central cell and two neck cells; C, older stage with nucleus preceding division into egg and ventral nuclei; D, mature archegonium, cut to one side, showing egg with nucleus and haustoria extending into surrounding jacket cells; E, sperms within body cell; F, flagellate sperm.

catkinlike structures, and the microsporophylls and sporangia are primitive in type. The ovulate strobili take the form of long, slender, fused stalks bearing single ovules (Fig. 20-13*A*, *B*).

The gametophytes of *Ginkgo* follow the same general lines of development as the cycads (Fig. 20-14). The seed of *Ginkgo* superficially resembles a stone fruit, but the two are not homologous (Fig. 20-13*C*).

Coniferales

General features The order Coniferales was well represented in past geological ages and is represented in the modern flora by the families Pinaceae, with 29 genera and 245 species, and the Taxaceae, with 11 genera and 105 species. The Pinaceae are characterized by distinct ovulate cones, whereas in the Taxaceae solitary ovules are exposed and the seeds develop a fleshy covering. In the Pinaceae some of the outstanding genera are *Pinus,* pine; *Larix,* larch; *Tsuga,* hemlock; *Pseudotsuga,* Douglas fir; *Abies,* fir; *Sequoia,* redwood; *Taxodium,* bald cypress; *Juniperus,* cedar; *Thuja,* arborvitae (Figs. 20-15–20-18); *Araucaria,* a Southern Hemisphere conifer; *Agathis,* a broad-leaved tropical form (Fig. 20-19). The most prominent genera in the Taxaceae are *Podocarpus,* native to the Southern Hemisphere, and *Taxus,* the northern yew (Fig. 20-20).

In form and size the Coniferales vary from small, creeping plants, such as *Juniperus horizontalis* (Fig. 20-21), and shrubby, bushlike plants,

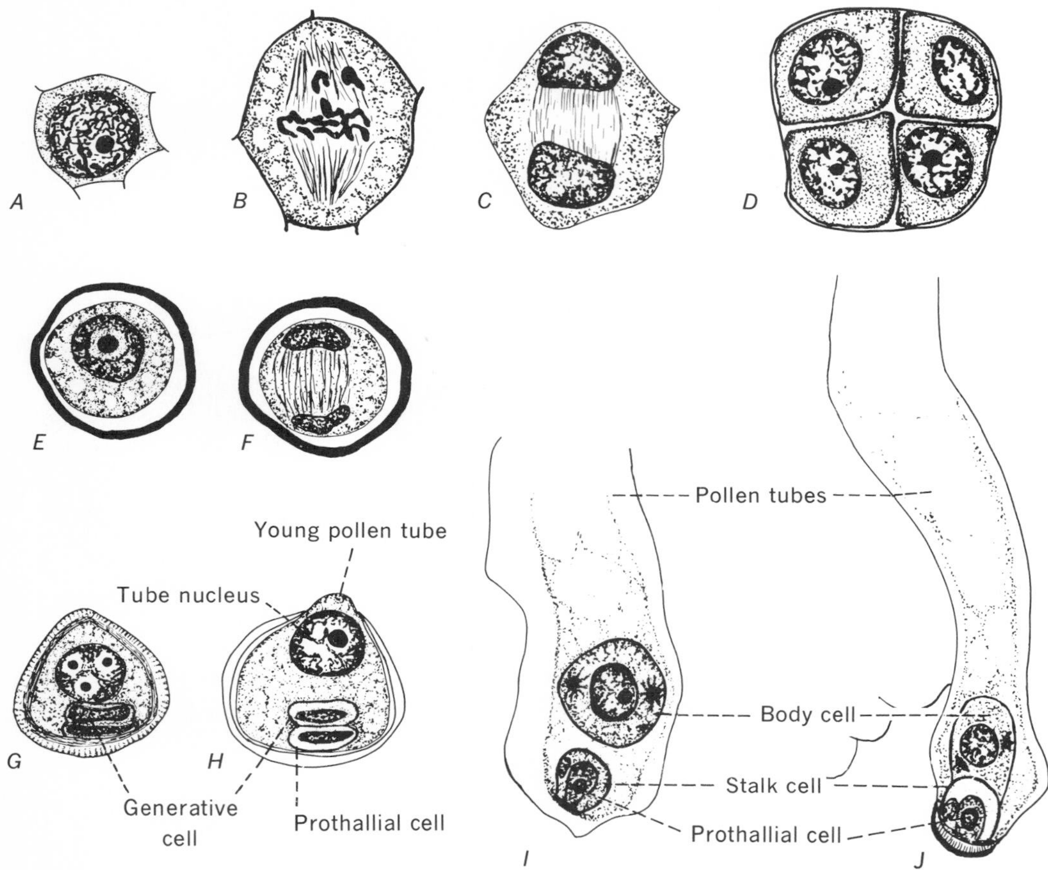

Fig. 20-11 Development of microspores and male gametophyte of a cycad. A, microspore mother cell; B, metaphase, and C, telophase, of first maturation division; D, tetrad of microspores; E, microspore; F–H, stages in development of the microgametophyte from a microspore; I, J, nearly mature pollen tubes.

such as *Taxus canadensis,* one of the yews of the north temperate regions, to the big trees of California, *Sequoia* (Figs. 20-16, 20-22). The majority of the living members of the group are tall evergreen trees of the temperate forests of both hemispheres.

Structurally the conifers are diverse. Resin ducts are abundant in certain genera and entirely absent in others. Wood parenchyma may be a conspicuous feature in the xylem of certain genera and rare in others. Diversity is found in both male and female gametophytes, in the embryo and in its mode of development, and in the number of cotyledons, which varies from 2 to 15.

The great interest and sentiment that have always been associated with the cone-bearing trees, their wide distribution, and their great economic importance make this coniferophyte group deserving of the extended literature about them. The evergreen character of most of the members has made them favored as ornamentals. Many of the conifers are of great value as Christmas trees and they are used as lumber in construction. Some of them are important for their specialized products, such as pitch, turpentine, tar, and resin. The wood of the yew was famous as fine material in making the crossbow, as well as the longbow, and was thus of historic importance in the battles where these weapons were used.

Form and structure of the sporophyte The conspicuous plant body of the conifers is the sporophyte, consisting of roots, stems, and leaves. In form, nearly all these trees are tall and stately, with excurrent branching. Side branches are characteristically longer on the lower part of the stem, becoming gradually shorter toward the apex. This arrangement of branches gives the tree a graceful, tapering appearance (Fig. 1-6). The branches in some genera, such as *Pinus*, are of two types. The long branches, when they are young, bear only scale leaves, and the short branches bear the needle-shaped leaves (Fig. 20-24).

In the pines the leaves are produced in clusters of two to five leaves attached to a short branch. These short branches develop at intervals on the sides of the longer stems. Close observation of the short branch, with its fascicles of leaves, reveals that each of the short leaf-

Fig. 20-12 Ginkgo *tree on the campus of The Pennsylvania State University, University Park, Pa. (Photograph by S. E. Northsen. The Pennsylvania State University, College of Agriculture, Photographic Service.)*

Fig. 20-13 Reproductive structures of Ginkgo biloba. *A, the staminate strobili at about the time pollen is shed; B, young ovules as they appear at the time the staminate strobili are in the stage shown at A; C, mature ovules developing seeds. (Photographs of A and B by D. S. Wright.)*

bearing branches is subtended by a thin scale leaf. The position of the short branch is that of a lateral branch growing from the node of the long branch.

The leaves of the conifers exhibit marked variation in form. There may be differences in form between the juvenile leaves and the leaves of the adult plant. In some genera of the family Cupressaceae the juvenile leaves are needlelike and the leaves of the adult plant are scalelike. There is also variation exhibited by the leaves of the adult plants of conifers, and several types are recognized. Univeined leaves, needlelike in shape and tetragonal in cross section, are found commonly among the Pinaceae, Araucariaceae, Podocarpaceae, and Taxodiaceae (Fig. 20-23*A*). A second general type is represented by a univeined leaf that is more linear in shape and is bifacially flattened. This is considered to be the most common type found among living conifers (Figs. 20-20, 20-23*C*). Scalelike leaves are represented by the adult leaves of the Cupressaceae, and the leaf of *Agathis* (Fig. 20-19) represents the broad, multiveined leaf type. Leaves are often retained from three to five years but in *Larix,* commonly called larch or tamarack, the leaves are shed each year.

Anatomically the conifers show a considerable advance over the living representatives of the lower Tracheophyta. For a discussion of the anatomy, the reader is referred to Chap. 8.

The spore-producing structures—cones The sporophylls of the conifers are organized into cones, and this suggests the name conifer (Fig. 20-24). The cones are of two kinds, the **staminate** (microsporangiate) and the **ovulate** (megasporangiate), the latter at maturity being the structure commonly recognized as the cone. The staminate cones consist of a central axis, or stem, and the attached microsporophylls, each of which normally bears two microsporangia on its under, or abaxial, surface (Figs. 20-25, 20-26, 20-31*D*). The ovulate cone most often consists of a central axis and the closely overlapping, attached bracts, or scales, some of which are megasporophylls (ovuliferous scales), each normally bearing two ovules attached to its upper, or adaxial, surface (Fig. 20-24*C*). The ovules, each of which consists of a megasporangium surrounded by a single large integument (Fig. 20-31*C*), are borne exposed on the surface of the sporophyll; this is characteristic of the gymnosperms. In some genera referred to the family Taxaceae, the ovules do not occur in well-defined cones but are exposed and at least partially are surrounded by a fleshy covering called the **aril.**

Most conifers are monoecious, i.e., bear both kinds of cones on the same tree. Each cone is

Fig. 20-14 Gametophytic and embryonic structures of Ginkgo. *A–E, early stages in development of male gametophyte; A, microspore; B, C, early stages in development; D, three-celled stage; E, mature pollen grain; F, micropylar end of female gametophyte; G, older stage; H, free nucleate stage of young embryo; I, proembryo with small cells at apex. (Drawings except E and G by Helen D. Hill.)*

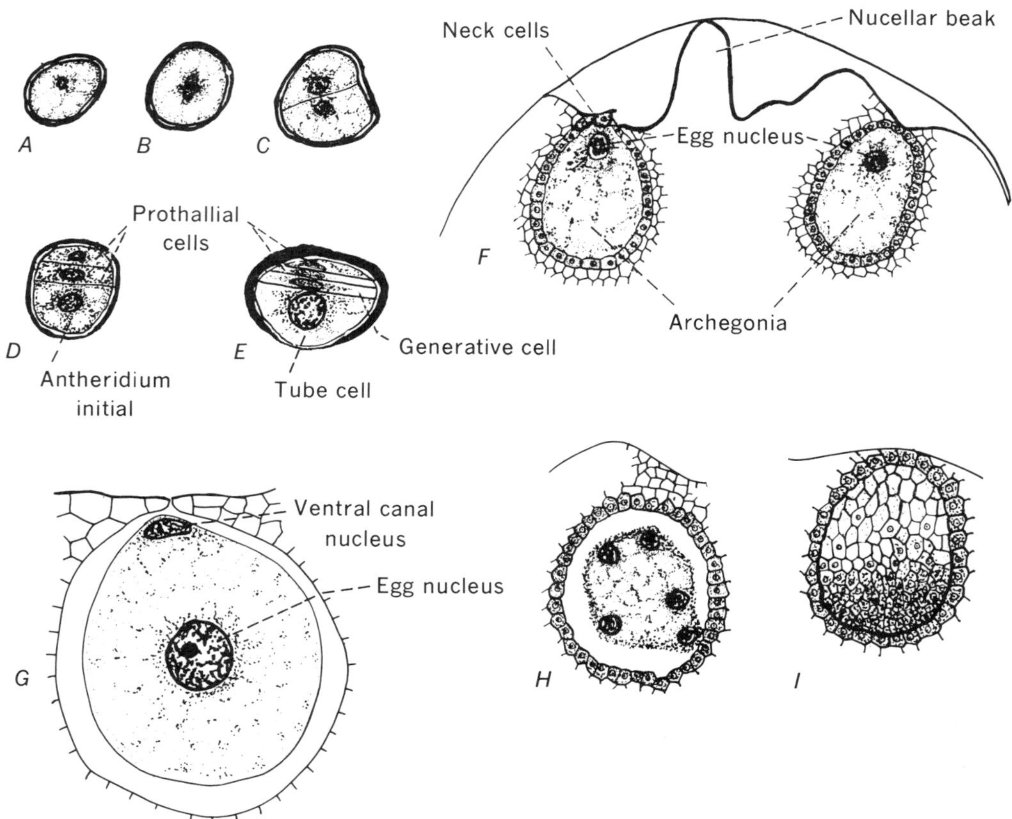

Fig. 20-15 Fossil tree identified as Sequoia, *on Specimen Ridge in Yellowstone National Park, Wyo. (Photograph courtesy of National Park Service, Yellowstone Park.)*

Fig. 20-16 Sequoia gigantea. *General Sherman tree in Giant Forest, Sequoia National Park, Calif., enters the contest for one of the oldest living things with estimates of 3,500 years. It is approximately 272 ft in height, 101.6 ft in girth. (Photograph by Eddy and Mains, courtesy of U.S.D.I., National Park Service.)*

Fig 20-17 Bristlecone pine (Pinus aristata), *growing in the White Mountains of Inyo National Forest, Calif. Some of these trees have been estimated by Dr. Edmund Schulman, University of Arizona, to be more than 4,000 years old. (Photograph by Daniel Todd, courtesy U.S. Department of Interior.)*

Fig. 20-18 Taxodium distichum, *bald cypress.*

Fig. 20-19 Leaves of Agathis. *(Drawn by Helen D. Hill.)*

Fig. 20-20 Vegetative leaves and staminate cones of Taxus, *a single cone at right and a single sporophyll from a cone at the left. (Drawn by Helen D. Hill.)*

usually separate from the other kind, and generally they occur on separate branches. Exceptions to this condition are found. The plants of some species are prevailingly dioecious, i.e., bear only one type of cone on a tree. Rarely, on monoecious plants, cones occur which are bisporangiate, i.e., bear both megasporangia and microsporangia on different parts of the same cone.

The staminate cones are small, rather inconspicuous structures, enduring in most genera only a few weeks in the spring. Within each microsporangium the microspore mother cells

Fig. 20-21 Juniperus horizontalis, *the dark, low-growing shrub, a conifer growing prostrate on sand dunes and lake shores. (Photograph taken at Lake Bluff, Ill., and furnished by Dr. George D. Fuller.)*

Fig. 20-22 Sequoia gigantea *at Mariposa Big Tree Grove, Calif. (American Forests Magazine, photograph courtesy of U.S. Forest Service.)*

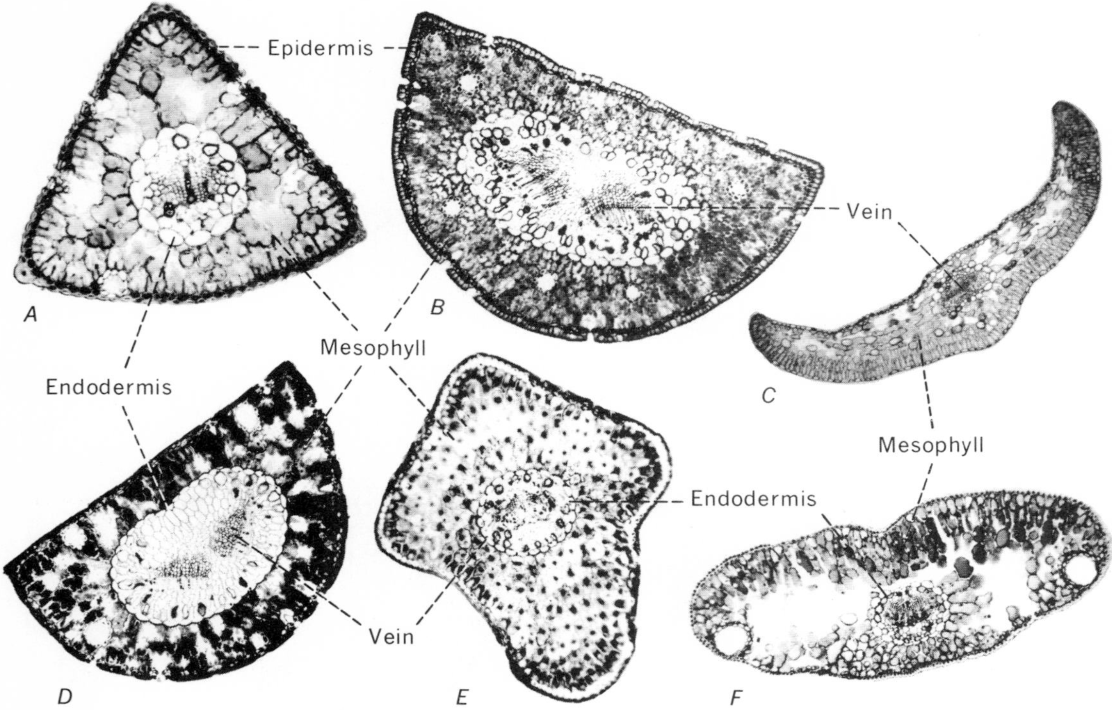

Fig. 20-23 Photomicrographs of transverse sections of various conifer leaves. A, Pinus strobus; *B,* P. laricio; *C,* Taxus *sp.; D,* Pinus *sp.; E,* Picea excelsa; *F,* Pseudotsuga *sp.*

are produced. These, by the processes of maturation (Fig. 20-27), involving meiosis, give rise to the microspores. Each microspore mother cell forms a spherical tetrad of four microspores, which are produced in tremendous numbers and appear at maturity as a yellow dust or powder. The microspores have the haploid number of chromosomes and initiate the male gametophytic phase.

The ovulate cones remain on the tree longer than the staminate cones (Figs. 20-24, 20-28). In spruce (Fig. 20-29*D*) the ovulate cones develop in one year; in pine in two years. After maturity the cones may remain on the trees, in some cases, for several years. In the early stages the ovulate cones of most genera are very inconspicuous and they may escape observation (Figs. 20-24, 20-30).

The ovules, each consisting of the megasporangium surrounded by the integument, are located on the upper surface of the ovuliferous scale. In the pines the megasporangium is an ovate body and the integument, which later becomes the seed coat, covers it as a cuplike overgrowth. At the tip of the ovule, the encircling parts of the integument do not quite come together, leaving an opening called the micropyle (little gate). This opening allows the pollen grains to enter and come into contact with the megasporangium (Fig. 20-31*A*).

The megasporangium, now called the nucellus, is generally undifferentiated in the young stages, but sooner or later one or several megaspore mother cells appear toward the center of the nucellus. In most genera a single spore mother cell is formed; in some genera, and under certain conditions, from one to five or six may be formed. Each megaspore mother cell undergoes meiosis and gives rise to a group of four megaspores (only three megaspores in some species of pine), each with the haploid number of chromosomes. The four megaspores are usually arranged in a vertical row in the sporangium, forming a linear tetrad. The linear arrangement of the megaspores is characteristic of many seed plants, but it is not universal.

Since the sporophyte of the Coniferales is perennial, its life does not end with the forma-

Fig. 20-24 Leaf, branch, and cone characters of conifers. A, branch of pine with leaves and ovulate cones; B, ovulate cone of spruce cut longitudinally to show scales with ovules; C, detail of one scale with two ovules. (Drawings by Elsie M. McDougle.)

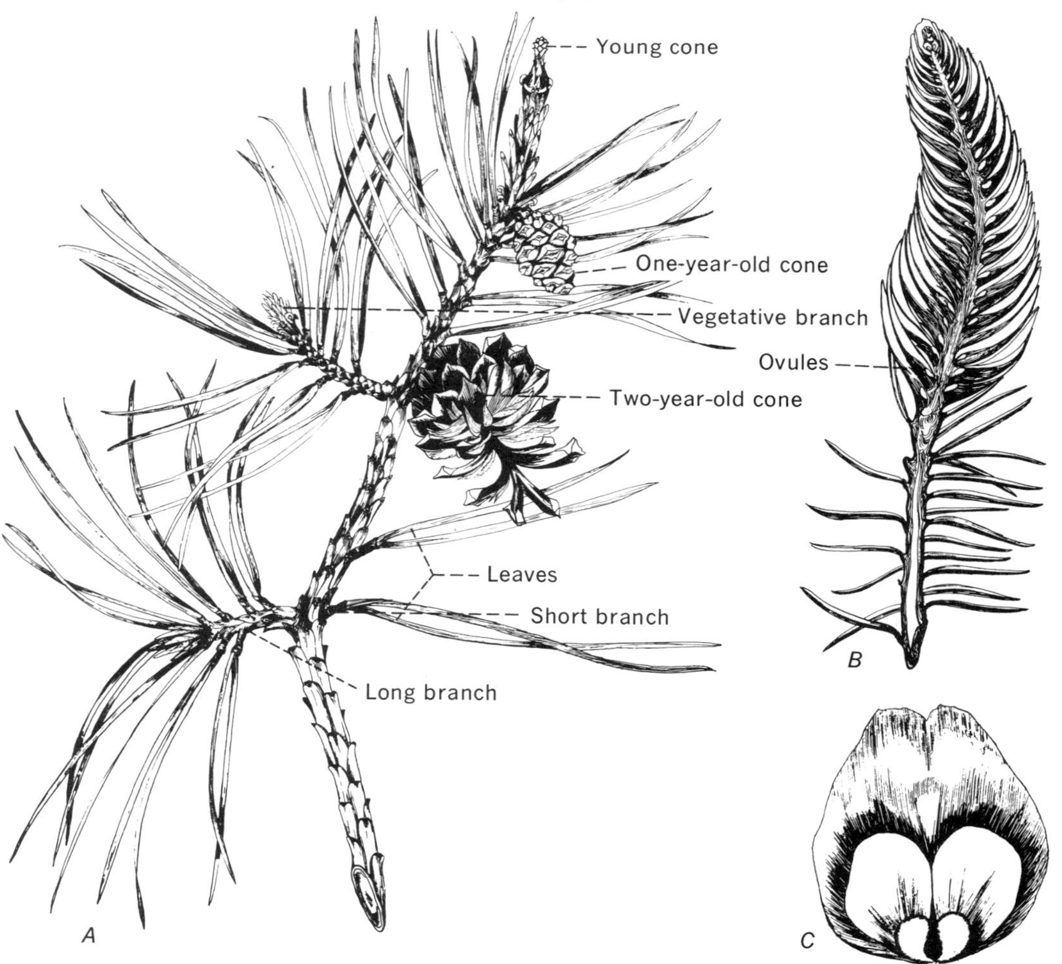

Fig. 20-25 A group of mature staminate cones of Pinus pungens.

Fig. 20-26 Staminate cones of pine.

tion of megaspores and microspores. The maturation processes initiate the gametophytic phase, resulting in the development of the male and female gametophytes, but the leaves, roots, and stems persist from year to year.

The gametophytic structures Heterospory, found occasionally in the lower vascular plants, is established as the universal condition in the gymnosperms. The tendency toward reduction in the gametophytic structures, first observed in other members of the Pteropsida, is carried much further in the gymnosperms. In the male gametophyte of the Coniferales, the sperms are reduced to nonmotile cells, a feature found for the first time in the evolution of plants. This lack of motility in the Coniferales is considered to be another step in the reduction of the gametophytic structures. It should be noted, however, that swimming sperms still occur in two orders

Fig. 20-27 Maturation divisions in microspore mother cells of conifers. A, prophase; B, metaphase of first division; C, diad stage; D, anaphase of second division. As a result of these divisions, four microspores are formed in each microspore mother cell, as shown in Fig. 20-32A.

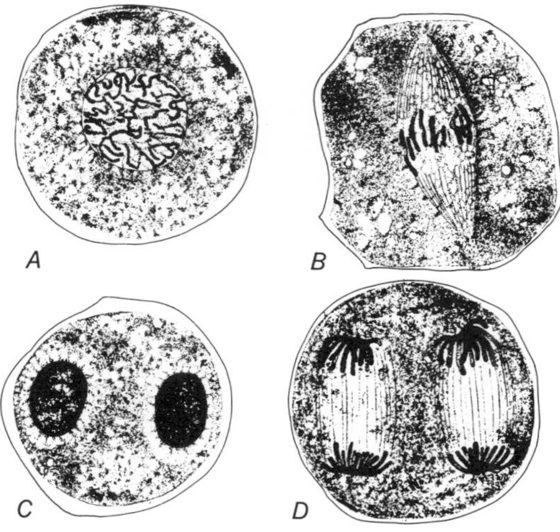

Fig. 20-28 Cone types of conifers. A, B, ovulate cones of Pseudotsuga taxifolia; *A, young, and B, mature (one year old); C, mature ovulate cone of* Pinus strobus *(two years old).*

of the gymnosperms, the Cycadales and the Ginkgoales.

THE FEMALE GAMETOPHYTE The development of the female gametophyte takes place entirely within the megasporangium, or nucellus, of the ovule (Fig. 20-31*B, C*). The gametophyte is entirely dependent for nutrition upon the old sporophytic structures. In the pine the basal megaspore of the linear tetrad develops into a female gametophyte. The functional megaspore remains in a resting condition for a time and begins to develop only after a considerable portion of the nucellus has been digested by the surrounding nutritive, tapetumlike tissue. The megaspore, occupying a central position in a vacuole resulting from the digestion, then undergoes a series of divisions, which result in the production of about two thousand nuclei. These nuclei occupy a central position in a common mass of cytoplasm and later become separated by walls, the formation of which begins at the micropylar end and eventually extends to the chalazal end of the gametophyte, which becomes a solid ovate mass of tissue.

Archegonia are generally developed at the micropylar end of the ovate female gametophyte (Fig. 20-31*C*). Each archegonium develops from a single archegonial initial, a superficial cell of the gametophyte, and consists of a very short neck, a ventral canal cell, and an egg. There are no neck canal cells, the neck consisting only of the wall cells, which vary in number from 2 to 12, with 8 the most common number. The ventral canal cell is usually a smaller cell than the egg cell. It soon disintegrates, leaving the egg cell the sole content of the archegonium. The wall of the archegonium is made up of the adjacent sterile vegetative cells of the female gametophyte, called the

Fig. 20-29 Cone types of spruce, Picea excelsa. *A, young staminate cone; B, mature staminate cone; C, ovulate cone at time of pollination; D, mature ovulate cone. The ovulate cones of the spruce reach maturity in one season.*

archegonial jacket, which, in addition to serving as the archegonial jacket, seem also to act as nourishing cells for the archegonium.

At maturity the archegonium is an oval structure rather deeply sunken in the female gametophyte, and the egg, with its nucleus and abundant cytoplasm, fills the entire structure. Generally the archegonia are grouped together in the micropylar region of the female gametophyte, and the number of archegonia varies greatly in the different genera. In pine from 2 to 5 archegonia are developed, whereas in the giant redwood (*Sequoia*) the number is reported to be sometimes as high as 60. Other genera have archegonia varying in number from 2 to 100.

THE MALE GAMETOPHYTE The early developmental stages of the male gametophyte (Fig. 20-32) occur entirely within the microspore wall and within the microsporangium. The male gametophyte (pollen) is usually shed as a four-celled structure from the staminate cone. After the pollen reaches the pollen chamber at the micropylar end of an ovule, the pollen tube and male gamete develop (Fig. 20-32*E, F*).

In pine the first divisions within the microspore result in the production generally of two sterile **vegetative cells** (Fig. 20-32*B*). These are small lens-shaped cells usually occupying one side of the developing pollen grain and disintegrating very soon. They are regarded as being homologous with the gametophytic thallus developed in the nonvascular plants and in the more primitive members of the Tracheophyta. In other genera of the Coniferales the number of vegetative cells may vary from none to as many as 48. Where they occur, they are formed by successive divisions of the **central cell** of the microspore.

Following the formation of the vegetative (thallus) cells, the central cell becomes the **antheridial cell** and is considered to be homologous with the antheridium of lower plants. This cell divides to form the **tube cell** and the **generative cell.** The tube cell is related to the subsequent development of the pollen tube. The generative cell is a small spherical cell which consists of a large nucleus and a relatively small amount of rather dense cytoplasm surrounded by a delimiting membrane but there is no cellulose wall. The generative cell divides to produce the **sperms,** or male gametes. Typically in the pine the young male gametophyte (pollen) consists of four cells: the two sterile vegetative cells, the tube cell, and the generative cell (Fig. 20-32*D*). At about this stage of development, the pollen grains are shed from the microsporangia and are transported by the wind (pollination) to the female cones where they come into contact with the ovules (Fig. 20-31*A*).

POLLINATION AND FERTILIZATION The pollen of some genera is winged, whereas in other genera it is wingless. When they are present, the wings are formed by the extension or inflation of the outer covering of the spores. These extensions occur on two sides of the spore, forming a pollen grain with two wings. At the time of pollination the young female cones stand erect at the ends of branches (Fig. 20-30). The pollen grains sift down among the scales of the young ovulate cone, reaching the ovules at the base of the ovuliferous scales, or megasporophylls. The ovules secrete a drop of resinous material into which the pollen grains fall, and as this resinous material dries, the pollen grains are drawn through the micropyle, bringing them into contact with the tissues of the megasporangium, or nucellus. Following pollination, the male and female gametophytes complete their development.

After a variable period of time in different genera, the pollen tube begins to grow and penetrates the tissues of the megasporangium. Early in the period of pollen-tube growth, the tube nucleus migrates from the pollen grain to the tip of the pollen tube. Later the generative cell divides into two cells, one termed the **stalk cell** and the other the **body cell** (Fig. 20-32*E*). The

Fig. 20-30 Young ovulate cone of pine at time of pollination.

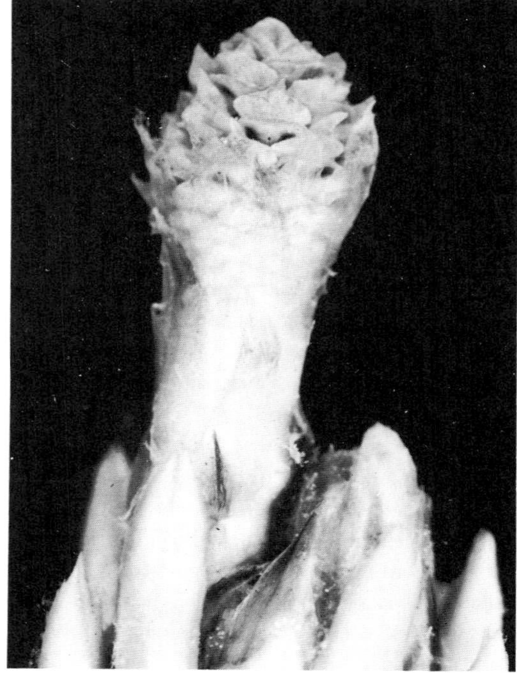

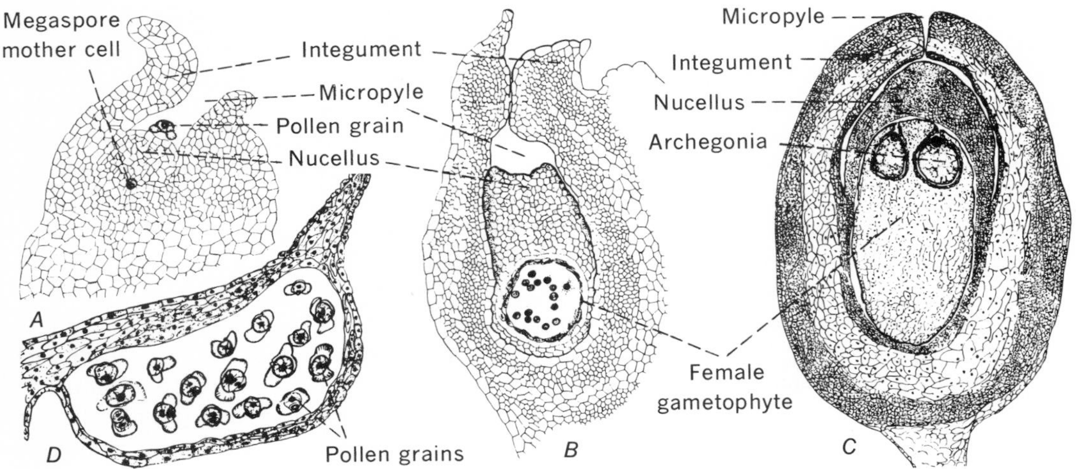

Fig. 20-31 Development of the reproductive structures in the Coniferales. A–C, longitudinal sections through ovules; A, young stage with pollen grain shortly after pollination; B, free nucleate condition of the female gametophyte; C, older stage with mature female gametophyte, showing two archegonia; D, section through microsporangium of staminate cone of spruce, showing the mature pollen grains. (A and B drawn by Helen D. Hill; C and D by Jeannette Ritter.)

stalk cell is sterile and takes no part in reproduction. As the pollen tube elongates, the stalk cell passes down into the tip of the tube. The final division occurring in the male gametophyte is the division of the body cell to form the two male gametes (Fig. 20-32*F*). The male gametes in the Coniferales are nonmotile.

The pollen tube with its two male cells, or sperms, grows through the cone-shaped nucellar tissue (nucellar beak) between the micropyle and the female gametophyte. A pollen tube grows into the neck of an archegonium, crushing the neck cells, and penetrates the archegonium. When the pollen tube penetrates the archegonium, its contents are discharged. Fertilization is accomplished by the fusion of the nucleus of one sperm (the other disintegrating) with the nucleus of the egg (Fig. 20-33*A*). Fusion is accomplished slowly, with the chromosomes of each gamete remaining distinct for some time. The time elapsing between pollination and fertilization varies considerably. In some genera pollination and fertilization occur within a few weeks. This is true of the spruce, hemlock, arborvitae, and others. In other genera, including the pine, pollination occurs in the spring of one year and fertilization in the spring of the year following, a period of about 11 months.

The embryo, the seed, and the seedling The embryo of the Coniferales differs considerably in the various genera. The development of the embryo in the pine is considered in this discussion.

After fertilization the nucleus of the zygote divides. The two resulting nuclei also divide, thus forming four nuclei (Fig. 20-33), which migrate to the base of the archegonium and arrange themselves in a plane. A simultaneous division of the four nuclei results in the production of two tiers, each consisting of four nuclei

(Fig. 20-33*C*). Following this, walls begin to develop, separating these nuclei. The outer tier of nuclei does not have complete walls, but the cells are open on the upper side and the nuclei are therefore connected with the general cytoplasm of the original egg. Divisions in the upper and lower tiers follow, and, as a result of these divisions, the proembryo consists of four tiers of four cells each (Fig. 20-33*E*). The lowest of these four tiers of cells is the **apical** tier. The tier adjacent to it forms the **suspensor,** an elongated structure which pushes the developing embryo deep into the old female gametophyte, which is rich in stored food. The tier above the suspensor is the so-called **rosette,** and the uppermost tier most often is referred to as the **open** tier.

The development of several embryos **(polyembryony)** from the proembryo is of rather regular occurrence in pine and in many other

Fig. 20-32 Microspores and development of the male gametophyte. A, tetrad of microspores; B, pollen grain with two prothallial, or vegetative, cells, central cell, and wings; C, division of central cell to form generative cell and tube cell; D, stage showing generative cell, tube cell, and two prothallial cells; E, diagrammatic representation of germinating pollen grain, showing the development of the pollen tube; the two prothallial cells are shown degenerated; the generative cell has divided to form a stalk cell and a body cell; F, diagrammatic representation of mature pollen tube with contents; the body cell has divided to form the two male cells, or sperms; the stalk cell, the tube cell, and the prothallial cells are still visible. (Drawings by Helen D. Hill.)

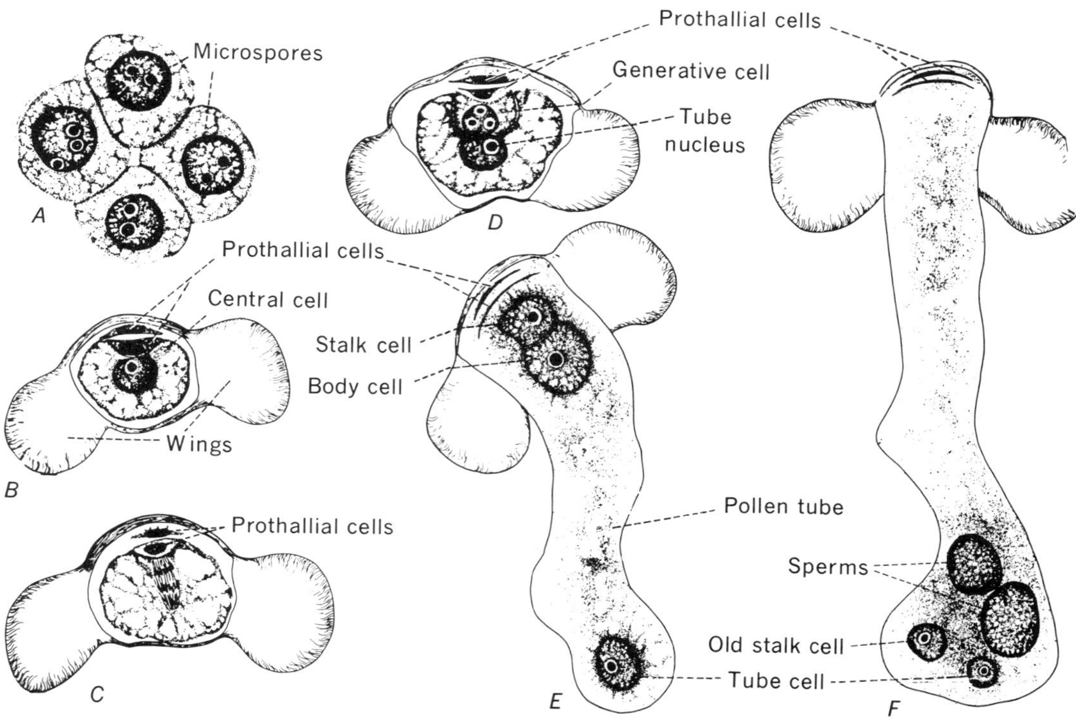

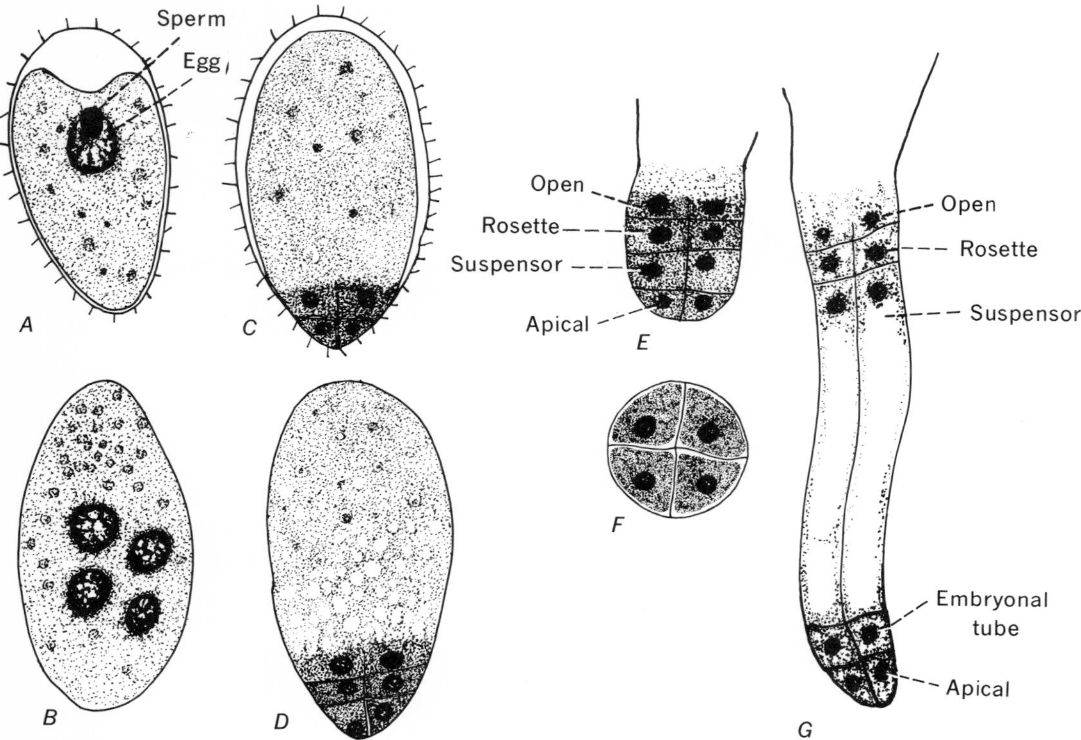

Fig. 20-33 Fertilization and development of the embryo in conifers. A, syngamy; B, early development of the embryo; C, two tiers of four cells each; D, E, continued development of the embryo; F, cross-sectional view of E; G, formation of suspensor cells. (Drawings by Helen D. Hill.)

conifers. In the proembryo, consisting of sixteen cells, four tiers of four cells each, the uppermost tier is open, and the cells of this tier soon disintegrate. The next tier of four cells is made up of the rosette cells, and on occasion one or more of these cells may give rise to rather ephemeral, small rosette embryos that soon abort. A much more important aspect of multiple embryo production involves the longitudinal separation (cleavage) of the lower tiers of cells and their subsequent development into embryos.

During the elongation of the tier of suspensor cells, the tier of four apical cells divides to form additional cells (Fig. 20-33*G*) that elongate like the suspensor cells above them and are called **embryonal tubes.** Additional divisions may give rise to other embryonal tubes. Cleavage results in four independently developing embryos, each consisting of an apical cell, one or more embryonal tube cells, and a primary suspensor cell. Competition between the simultaneously developing embryos usually results in the survival of one. In the mature seed no recognizable remains of additional embryos can be found.

The mature embryo of pine consists of a primary root and shoot, the apex of which is surrounded by a whorl of cotyledons. The **seed** is made up of the embryo embedded in the remains of the massive megagametophyte, which is in turn covered by the seed coat. The seed coat consists essentially of a harder outer portion derived from the stony layer of the integument and the remains of the inner fleshy part of the integument as a thin papery layer.

There is considerable variation of size in the seeds of conifers. The piñon pine in the southwestern portion of the United States bears seeds, called piñon nuts, that are eaten by the residents of this region.

The seeds generally germinate slowly after a prolonged resting period. Some seeds, however, are reported to germinate before they are shed from the cone. Upon germination the embryo emerges, usually pushing the seed coat up through the ground ahead of the cotyledons. Finally the several cotyledons are freed from the seed coat and the young seedling conifer begins a long period of growth to develop a mature sporophyte (Fig. 20-34).

Fig. 20-34 Pine seedlings.

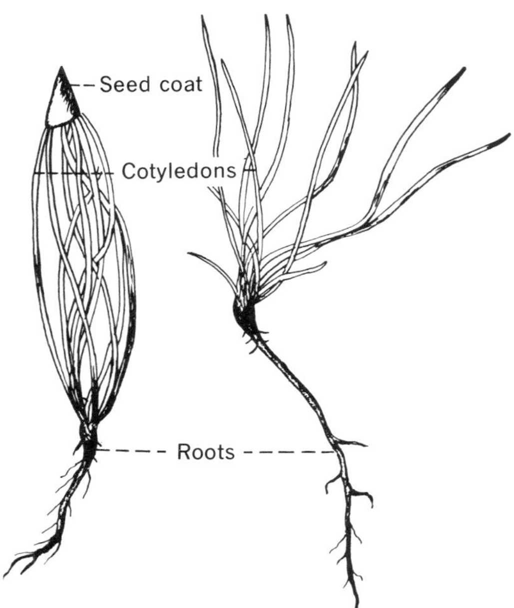

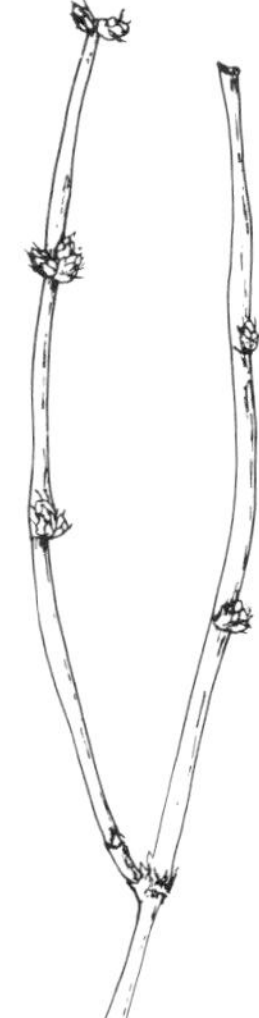

Fig. 20-35 Ephedra *showing branch with staminate cones.*

Gnetales

No ancestors of the Gnetales are known. Compared with the other orders of gymnosperms, Gnetales are a modern group, consisting of three genera—*Ephedra, Welwitschia,* and *Gnetum*. These genera comprise a group of plants of uncertain relationship but undoubtedly gymnospermous. The plants are diverse in every respect, and it has been suggested that each genus should be placed in a separate order. *Ephedra,* a bushy or trailing plant (Fig. 20-41), developing only scale leaves, grows in the dry regions of both hemispheres (Fig. 20-35). Some of the species grow in Mexico and in the southwestern region of the United States. *Welwitschia* is restricted to the dry regions of South Africa. The stem of *Welwitschia* is a blunt, tuberous structure, tapering into a long taproot. Only two leaves are produced. These leaves, renewed by a basal meristematic region, are described as being split into long leathery ribbons by the force of the wind (Fig. 20-36). The species of *Gnetum* are mostly vines (Fig. 20-37), producing large netted-veined leaves resembling those of

Fig. 20-36 Welwitschia mirabilis, *Brandberg, South-West Africa. Male plant, showing condition of the leaves after exposure to the elements for a number of years. (Photograph courtesy of Dr. R. J. Rodin.)*

the dicotyledons. Some species of *Gnetum* are tree forms. *Gnetum* is native in the moist tropical forests of Asia, Africa, and South America. Anatomically the Gnetales have a combination of the typical gymnospermous tracheid with bordered pits and a type of vessel.

The Gnetales are dioecious, although there is a tendency to produce both microsporangia and megasporangia together, in some instances. Superficially the strobili resemble simple types of angiospermous flowers. An important feature of the strobili of the Gnetales is that both staminate and ovulate strobili are compound. The ovulate strobilus (cone) of the conifers is compound, but in the Gnetales the staminate strobilus, as well as the ovulate strobilus, is composed of bracts and sporophylls. All the Gnetales produce naked ovules, a feature characteristic of gymnosperms.

The details of reproduction in the Gnetales vary with each of the genera. *Ephedra* does not advance beyond the type of reproduction characteristic of other living gymnosperms (Fig. 20-38). It has a female gametophyte, producing archegonia. However, *Welwitschia* and *Gnetum* do show advances over other gymnosperms in their reproductive structures. Neither of these genera develops archegonia. In the case of *Gnetum* the mature female gametophyte is partially in the free nuclear condition. In this genus one of the free nuclei in the female gametophyte functions as an egg.

In *Ephedra* and *Welwitschia* the male gametophytes are typically gymnospermous. This is indicated by the development of prothallial cells. In *Ephedra* the male gametophyte contains one cell and one free nucleus. Apparently, in *Welwitschia,* a single prothallial nucleus without a cell wall is produced. However, *Gnetum* produces no prothallial cells or free nuclei. The pollen grains have three nuclei, and the generative cell forms a stalk cell and a body cell, the latter forming two male gametes.

Many theories have been proposed to show that the Gnetales are a connecting link between the gymnosperms and angiosperms and consequently the ancestor of the angiosperms. However, the evidence really suggests that the Gnetales are a very highly specialized group (or groups) of gymnosperms and any relationship to the angiosperms is remote.

THE ANGIOSPERMS

General features The angiosperms comprise a large group of plants, growing in many kinds of ecological sites and under many climatic con-

Fig. 20-37 Gnetum scandens. *A, staminate plant growing on bamboo; B, staminate strobili. (Photographs by Dr. Samuel Chan.)*

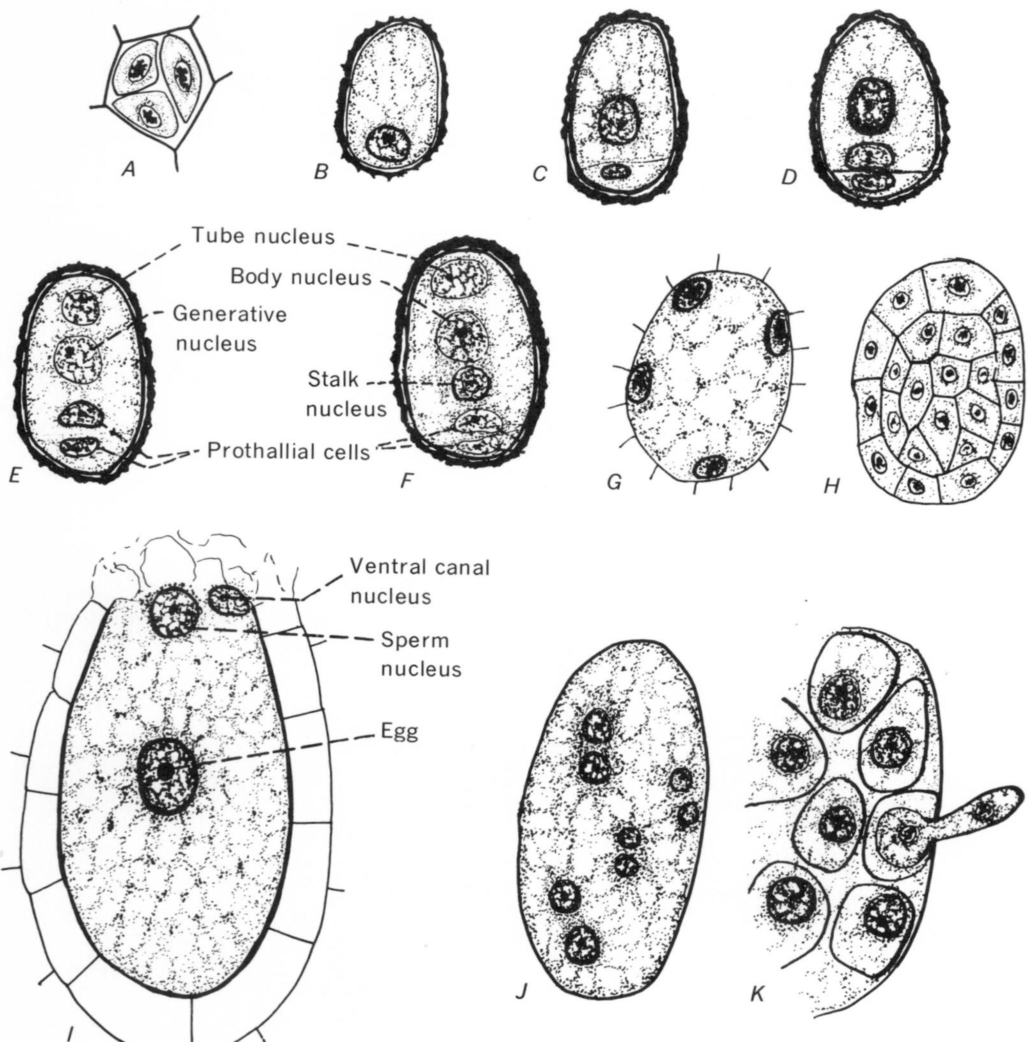

Fig. 20-38 Gametophytic and embryonic structures in Ephedra. *A–F, stages in development of male gametophyte beginning with microspores, A; B, single microspore, or pollen grain; C–F, successive stages resulting finally in F, the male gametophyte; stage E shows the generative nucleus before it has divided to form the body nucleus and the stalk nucleus; the final stage, in which the body cell divides to form the sperms, is not shown; G, young female gametophyte with four free nuclei, parietally placed; H, female gametophyte somewhat older; I, archegonium with egg and ventral canal nucleus; a sperm nucleus is entering the archegonium; J, eight free nuclei in embryo; K, tubular elongation of one of proembryonal cells with divided nucleus. (Drawings by Helen D. Hill from slides prepared by Dr. W. J. G. Land.)*

ditions and exhibiting great variation in size and form. Angiosperms range in size from minute forms, such as *Wolffia,* which is only a fraction of an inch in diameter, to giant trees. Every structural feature varies, leaves and flowers being especially diverse.

The angiosperms are the flowering plants and are recognized as the higher seed plants, in contrast to the gymnosperms, or lower seed plants. The flower characterizes the group (Fig. 20-40). In the angiosperms the ovule, which matures into a seed, is produced within the space enclosed by a carpel, and the angiosperms are thus said to produce enclosed seeds. The matured ovary, or fruit, exhibits a wide variety of forms (see also Chap. 11). It may be a simple or a compound fruit, a dry or a fleshy fruit, and may or may not open at maturity. In many cases additional parts of the flower are incorporated with the developing carpels to form the mature fruit. Whatever the exact composition of the fruit, the seeds are enclosed and this is a condition accepted as being more advanced in evolution than the exposed seeds of the gymnosperms (Fig. 20-24*C*).

The leaves of the angiosperms and the gymnosperms are megaphylls and have evolved from a major branch system. The megaphyll is distinguished from the microphyll by its complex venation and the relationship of the vascular tissue of the leaf to an easily identified leaf gap in the stele of the axis. Although the leaves of both the angiosperms and the gymnosperms are megaphylls, those of the angiosperms are often large and broad in contrast to the frequently small needlelike leaves of many gymnosperms. The structure of the xylem tissue of the angiosperms and the gymnosperms differs in ana-

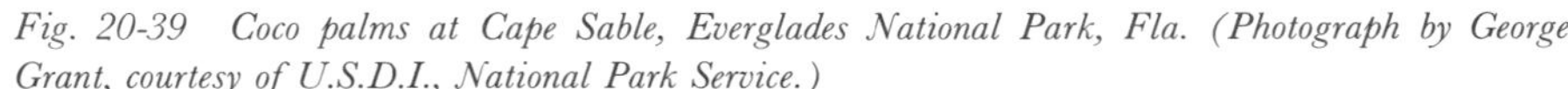

Fig. 20-39 Coco palms at Cape Sable, Everglades National Park, Fla. (Photograph by George Grant, courtesy of U.S.D.I., National Park Service.)

tomical detail (see Chap. 8). The xylem of gymnosperms is characterized by tracheids (Fig. 2-25*B*) and the xylem of angiosperms by vessels (Fig. 2-26).

More than 200,000 species of angiosperms have been recognized and the class Angiospermae is divided into two subclasses—the Monocotyledoneae, or monocotyledons, and the Dicotyledoneae, or dicotyledons. The monocotyledons are a group of plants of great economic importance to mankind. Many prominent food plants, such as the banana and all the cereals and grasses including corn, wheat, rice, barley, and oats, belong to this group. In addition lilies, tulips, orchids, and many other common flowering plants are monocotyledons. The embryo of the monocotyledons has a single cotyledon which bears slight resemblance to the more familiar cotyledons of dicotyledonous plants (Figs. 20-45 to 20-47). The major veins in the leaves of some monocotyledons, such as the grasses, are parallel or nearly so, but in other carefully investigated species the major veins converge and join near their distal ends. Interconnections by small veinlets are common and in some instances a complex reticulum is developed. The floral members are usually in threes or some multiple of three, as three sepals, three petals, three or six stamens, and three carpels (Fig. 20-40). Although a few monocotyledonous plants, such as some of the palms (Fig. 20-39), are fairly large trees, many of the monocotyledons are grasslike or are small herbaceous plants (Fig. 20-41).

Fig. 20-40 Flower of the lily, one of the monocotyledons. A, general view of the flower, showing the parts; B, longitudinal section of the flower, showing parts of the perianth, stamens, and pistil. (Drawings by Elsie M. McDougle.)

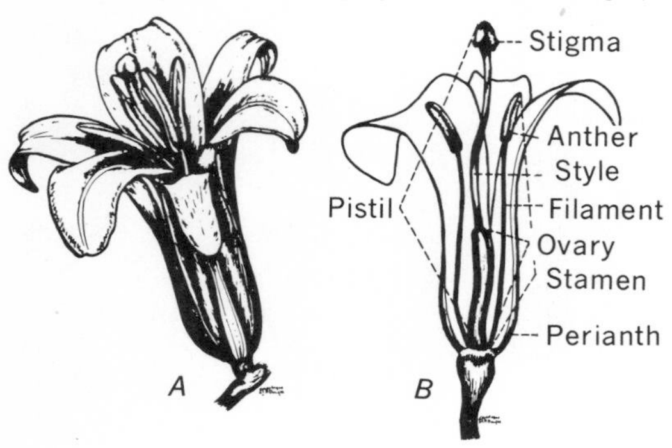

The dicotyledons constitute a large group of seed plants. The characteristics of the group include the development of two cotyledons in the embryo, generally a complex reticulum of veins and veinlets with numerous branched veins, ending free in the mesophyll of the leaf, and the floral members usually in fours or fives or multiples of these numbers. Many forest trees, such as the oak, the hickory, the walnut, and the maple, as well as most of the fruit trees including apple, orange, pear, peach, and plum, are dicotyledons. Beans, peas, potatoes, and peanuts constitute some of the principal dicotyledonous food crops. A number of the ornamentals, such as roses, petunias, asters, lilacs, and snapdragons, and fiber crops, such as cotton, hemp, and flax, are dicotyledons.

The sporophyte and gametophyte In the angiosperms, as in most of the vascular plants, the sporophytic structures attain greater size than the gametophytic structures. The roots, stems, leaves, flowers with their sporophylls and sporangia, the mature fruit, and the seed constitute the sporophyte, or diploid structure.

The gametophytic structures in the angiosperms attain the greatest degree of reduction found anywhere in the plant kingdom. The haploid gametophytes are microscopic structures (Figs. 20-42, 20-43).

The spore-bearing structures—flowers Because of the general absence of the flower in the fossil record, we are largely ignorant of its evolutionary history, its origin, its development, and its relationship to similar structures in other vascular plants to which it may be related. Some botanists have defined a flower as a stem beset

Fig. 20-41 A group of xerophytic plants, mostly monocotyledons. In the center foreground and to the right are three specimens of the Agave, *or century plant. The background in front of the building is occupied by palms. An* Ephedra, *one of the Gnetales, occupies the left center. (Photograph furnished by Dr. A. F. Hemenway.)*

with sporophylls (sporangia-bearing leaves) and have compared it with the strobilus of *Selaginella* (Fig. 19-13) or the cone of pine. Although the flower is a sporangia-bearing structure, it is at the same time quite distinct from the cone of the gymnosperm or other vascular plants. The flower is distinctive of angiosperms and is a unique structure. The classical concept, which is widely accepted, is that the appendages of the flower are foliar; i.e., the flower is an axis with highly specialized leaves. Marked resemblances can be demonstrated between the vegetative leaves of plants and the sporangia-bearing, as well as the non-sporangia-bearing, appendages of the flower. Similarities include the origin and early stages of development, the form, and the presence of chloroplasts and stomata. Studies have shown that the initiation and early stages of development of sepals, petals, stamens, and carpels of the flower are essentially the same as the early developmental stages of vegetative leaves (Figs. 9-1, 10-11).

Recognition of the fundamental similarities in origin and development of the sporangia-producing appendages has led to the conclusion that the microsporophyll of pine and the stamen of the angiosperm, as well as the megasporophyll of pine and the carpel of the angiosperm, are comparable morphological structures. Comparisons can also be made between the stamen and the carpel of the angiosperm and the sporangia-bearing structures of other heterosporous vascular plants (*Selaginella, Isoetes, Salvinia,* etc.).

The pollen sac, or microsporangium, is the anther, and the young anther contains four narrow, elongated masses of sporogenous tissue which in transverse section appear as four lobes (Fig. 10-1*D*). Following disintegration of the sterile tissue between pairs of adjacent lobes, the sacs are reduced to two in number (Fig. 10-3*A*). Dehiscence of the anther occurs with its opening at the lip cells (Fig. 10-3*A*), as a result of pressure exerted by the movement of the hygroscopic bands, or wall thickenings, of the endothecium.

The ovule, a megasporangium (nucellus) enclosed by one or more integuments, is produced from nucellar initials which are cells of the carpel (Fig. 10-14*A*). These initials usually are associated with the carpellary margin, although this is not always the case, and they are limited to the basal, expanded region of the carpel, called the ovary (Fig. 10-3*B*). The place or tissue from which the ovules develop is called the placenta, and the location of the several ovules within the ovarian cavity (carpellary cavity) is called placentation (Fig. 10-4*E–H*). The ovule is attached to the placenta by the funiculus (Fig. 10-4*D*).

The integuments of the ovule arise as collarlike rings of tissue from the young megasporangium (Fig. 10-14*B*) and overgrow it except for a small pore, the micropyle, at the distal end of the ovule (Fig. 10-14*C*). The region where the funiculus, integuments, and the nucellus merge is the chalaza (Fig. 10-14*C*). Later in development, pollen tubes may enter through the micropyle **(porogamy)** and penetrate the megagametophyte, or they may enter through the outer tissue of the ovule nearer the chalaza **(chalazogamy).**

Although it is comparatively easy to describe the static nature of a stamen as consisting of a filament with attached anther (Fig. 10-1) and a carpel with margins fused to form a closed structure, called the pistil, consisting of stigma, style, and ovary (Fig. 10-3*B*), it is more difficult to reach a final conclusion concerning the origin and evolution of these structures. The evolved stamen is considered to be a highly specialized structure, largely as a result of reduction. Although it is stemlike in general appearance, the stamen basically is of leaf rank and is comparable to other parts (organs) of the flower. The primitive stamen of such angiosperms as the Ranales is leaflike (laminar) in appearance and lacks much of the clear distinction between sterile and fertile portions characteristic of the evolved, reduced, and specialized stamen.

The ancestral form of the carpel is a stalked, laminar, dorsiventral organ with slight distinction into fertile and sterile parts. The elaboration of the style and the stigma is considered to be an evolved character. In some genera the folded carpel is not completely closed at the time of pollination; in others the margins of the carpel are not fused histologically but are only appressed or interlocked by small projections.

Regardless of the origin and evolution of the microsporophyll and the megasporophyll, each bears sporangia in which sporogenesis takes place and microspores and megaspores are produced as products of meiosis (Figs. 10-13*A–F*, 10-15*A–H*, 10-18*A–E*, 10-19*A–C*).

The gametophytes The gametophytes of the angiosperms have their origin with the megaspores and microspores. The prefixes *mega-* and *micro-* applied to these spores imply that there is a difference in size of the spores. In some instances there is a variation in size of the spores in angiosperms; in many cases, however, there is actually very little difference. Rather, the names indicate comparisons with the large and small spores of other heterosporous plants. The gametophytes produced by germination and

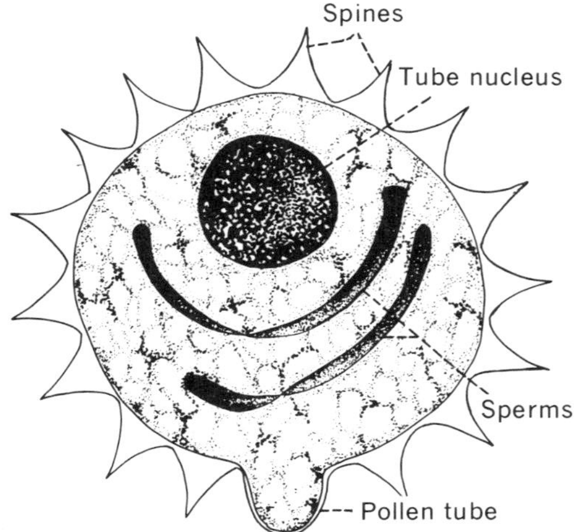

Fig. 20-42 Development of male gametophyte (germinating pollen grain) of Silphium.

growth of the megaspores and microspores are extremely reduced structures compared with those of the lower plants. The production of gametes capable of entering into the process of fertilization remains the same (Figs. 10-13*G, H;* 10-16*A–G;* 10-19*D–F;* 10-20*A–C*). The mature megagametophyte of angiosperms, rather typically, consists of an egg cell, two synergid cells, two polar nuclei, and three antipodal cells (Figs. 10-16*G,* 10-20*C*). The description or definition of the male microgametophyte varies among different authors, and the confusion seems to be related to the use of the terms microgametophyte and pollen grain as synonyms. Pollen may be either two or three celled, depending on the time of the division of the generative cell to form two sperm cells, as related to the time it is shed from the anther. However, if the microgametophyte is defined in terms of being mature, or ready to effect fertilization, then it consists of three cells—the tube cell and two sperm cells (Fig. 20-42). Generally it is believed that neither the tube cell nor its nucleus controls the development of the pollen tube and the pollen tube is not a part of the microgametophyte although it may enclose most of it.

Pollination, fertilization, and the embryo Unlike that of gymnosperms, the pollen of the angiosperms does not come into intimate contact with the ovule but is deposited by various

Fig. 20-43 Mature female gametophyte of Silphium *at time of fertilization.*

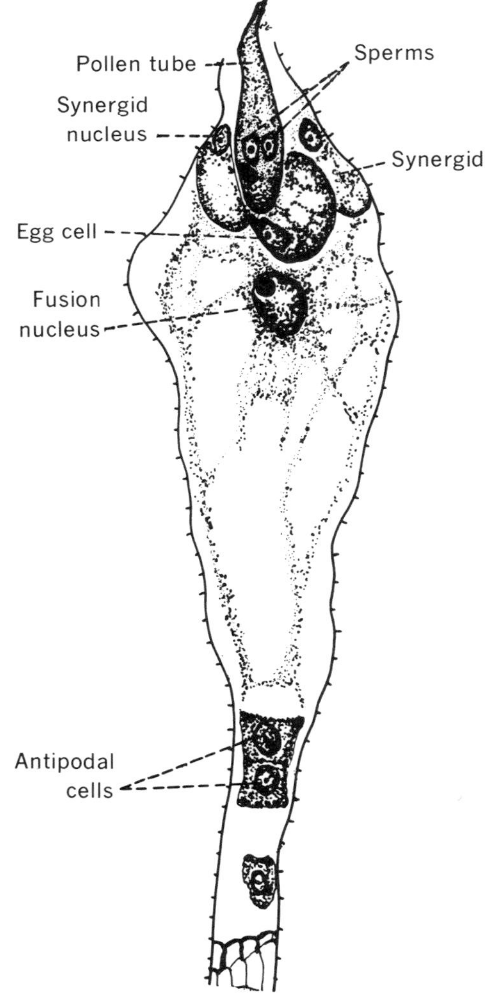

means (wind, insects, etc.) on a receptive surface of the pistil, called the stigma. Most often this surface is limited to the tip of the pistil, but in some instances it may be continuous, essentially along the length of the style, or, in the case of several investigated primitive genera, the more or less open margins of a **conduplicate carpel** serve as the receptive surface. After the location of the pollen grain on the receptive or stigmatic surface, it germinates with the production of a pollen tube that grows to the ovule and penetrates the megagametophyte (Fig. 20-44).

The two male gametes deposited in the female gametophyte are both normally involved in the process of fertilization, in contrast to only one gamete (of the two deposited) in the gymnosperms. One fuses with the nucleus of the egg cell **(syngamy)** to form the zygote, and the second male gamete fuses with the two polar nuclei, sometimes after they have already fused into a single nucleus, to produce the primary endosperm nucleus (see Chap. 10; Figs. 10-20*C, D;* 20-43; 20-45). Subsequent division of the primary endosperm nucleus produces the endosperm tissue (Fig. 20-45*J*). In many angiosperms the early development of the endosperm precedes the development of the embryo. As the embryo develops, some of the food material stored in the endosperm cells is utilized by the developing embryo (see also Chaps. 10 and 11). Endosperm tissue is not produced in the gymnosperms, but food material is often stored in the cells of the massive megagametophyte (Fig. 20-31*C*). The zygote develops to form the embryo.

Although the development of the embryo presents considerable diversity in the angiosperms, certain features may be regarded as being characteristic. The zygote, at first, divides into two cells—a basal cell and an apical cell (Fig. 20-45*A–D*). Other divisions follow, and an elongated structure, consisting of four or five cells, is soon formed (Figs. 20-45*E,* 20-46*A*). This structure is termed the **proembryo** and quite early becomes differentiated into a suspensor and an enlarged terminal portion, the embryo proper (Fig. 20-45*F–J*). Growth of the embryo results in the production of an axis with

Fig. 20-44 Development of the pollen tube. A, Cypripedium, *or lady's slipper, one of the orchids; pollen tube with two male cells, or sperms, entering the embryo sac; B,* Cypripedium, *pollen tube entering the embryo sac; C, pollen tube of* Lilium *entering embryo sac; nuclei of embryo sac not shown.*

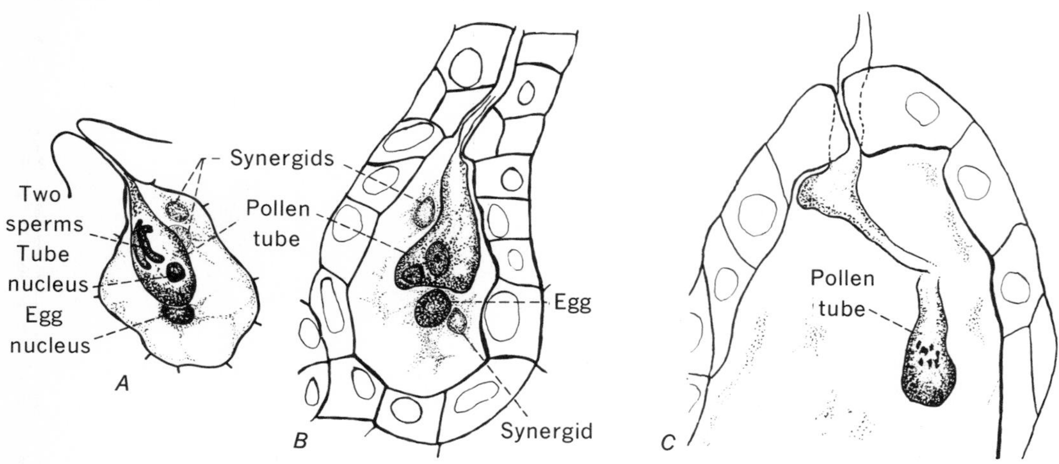

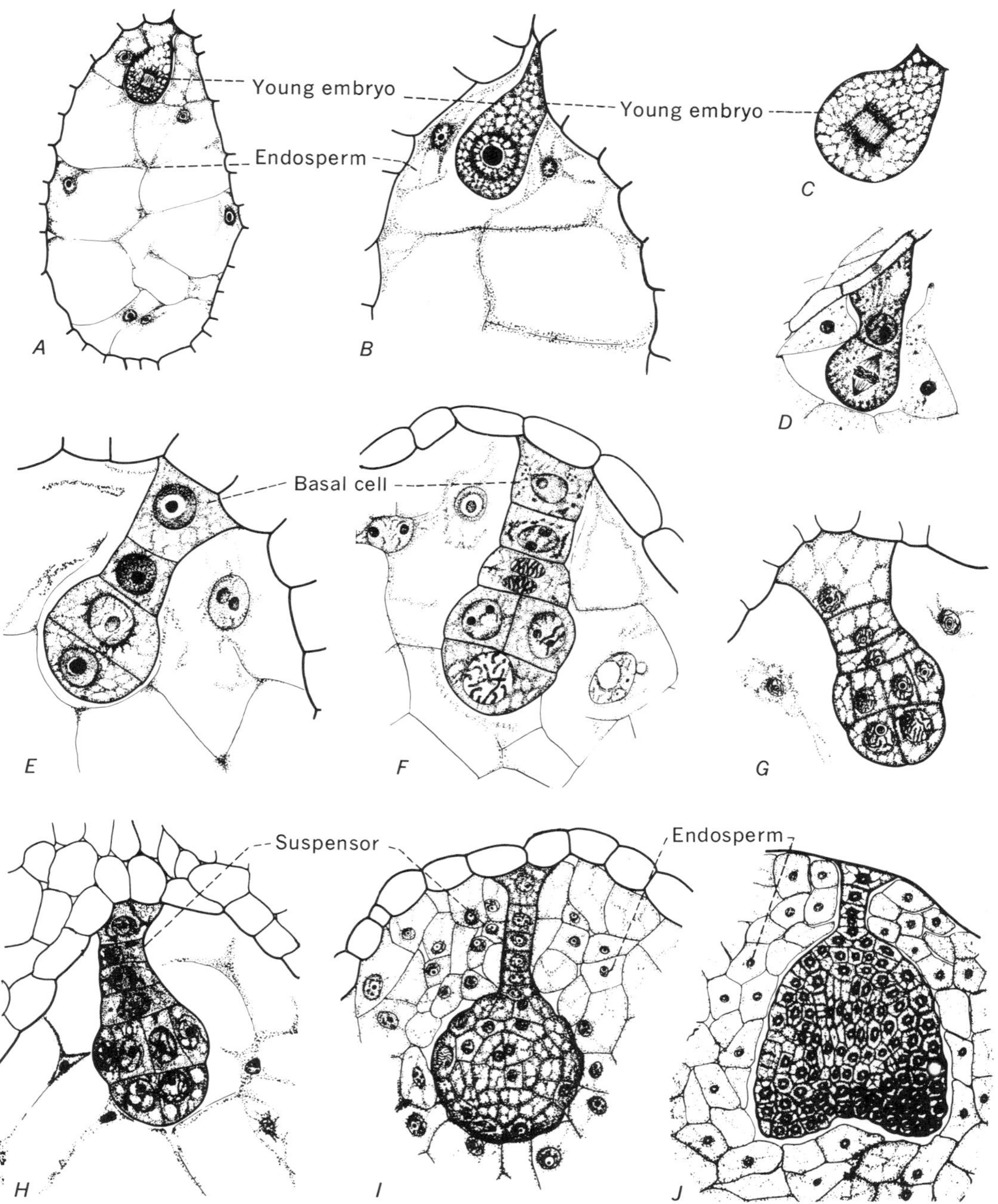

Fig. 20-45 Development of the embryo and endosperm in tobacco (Nicotiana), *a dicotyledon. A, young embryo and developing endosperm; B, detail of A; C–J, stages in development of embryo and surrounding endosperm by cell multiplication and differentiation, endosperm in later stages becoming more compact; J, cotyledons beginning to be differentiated. (J drawn by Paul Sacco.)*

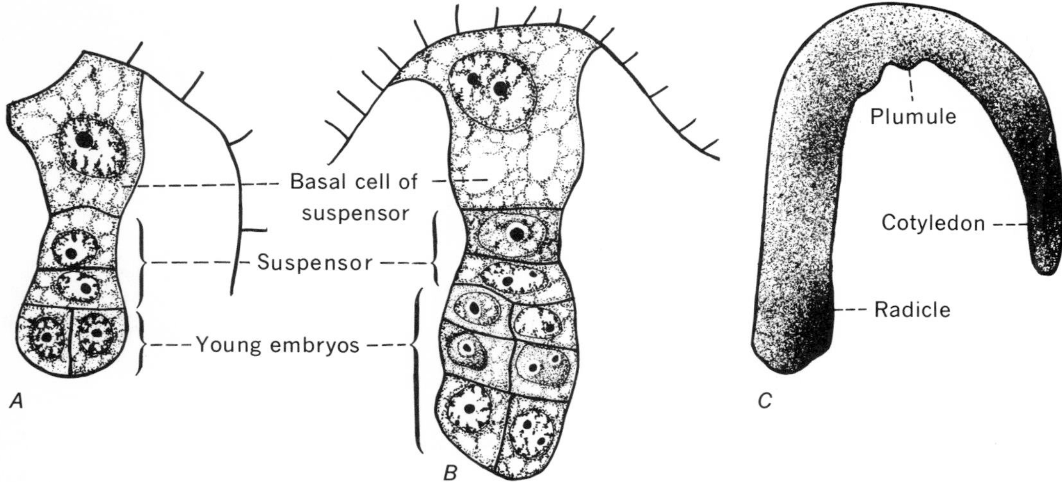

Fig. 20-46 Embryo development in Sagittaria, *a monocotyledon. A, B, young stages showing basal cell, suspensor, and the embryo proper; C, older embryo.*

Fig. 20-47 Embryo of wheat.

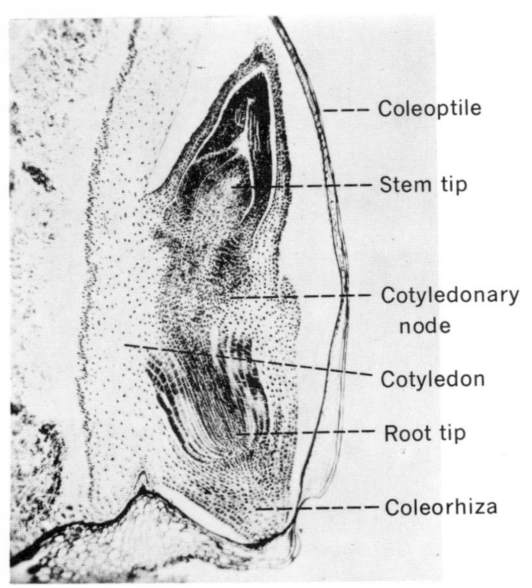

root, stem, and cotyledons (Figs. 10-21, 20-45*J*, 20-46*C*, 20-47). In some monocotyledonous embryos, illustrated by members of the grass family, an additional mass of tissue, called the **coleorhiza,** covers the root tip and a specialized leaf, called the **coleoptile,** covers the stem tip. The single massive cotyledon of some monocotyledonous plants, although it is a lateral divergence of the plant axis, is so large that it appears to be the axis with a laterally attached root and stem (Fig. 20-47).

QUESTIONS

1. List features of the seed plant interpreted to indicate specialization and evolution in what you consider to be their order of importance.

2. Explain what is meant by parallel evolution by citing specific plants or lines of plants.

3. Trace the reduction of the gametophytes in vascular plants and interpret the significance of such reduction.

4. What is the seed habit?

5. What has been the significance of the evolution of vascular tissue and the seed habit in relation to the distribution of plants?

6. If the concept is rejected that floral parts are foliar in nature, what interpretation of sepals, petals, stamens, and carpels can be made?

7. What is the pollen tube? How does it differ in location and function in gymnosperms and angiosperms?

8. What significance is attached to the disappearance of motile sperms in the angiosperms? What other plants or animals have nonmotile male gametes?

9. What is the significance of microsporogenesis and megasporogenesis in the gymnosperms and angiosperms?

10. Do you think it is very significant for the paleobotanists to find the so-called missing links of the plant kingdom to substantiate a theory of evolutionary direction?

REFERENCES

ARNOLD, C. A. 1947. An Introduction to Paleobotany. McGraw-Hill Book Company, New York.

CHAMBERLAIN, C. J. 1919. The Living Cycads. The University of Chicago Press, Chicago.

CHAMBERLAIN, C. J. 1935. Gymnosperms: Structure and Evolution. The University of Chicago Press, Chicago.

EAMES, ARTHUR J. 1961. Morphology of the Angiosperms. McGraw-Hill Book Company, New York.

FOSTER, ADRIANCE S., and ERNEST M. GIFFORD, JR. 1959. Comparative Morphology of Vascular Plants. W. H. Freeman and Company, San Francisco.

JOHANSEN, D. A. 1950. Plant Embryology. Chronica Botanica Company, Waltham, Mass.

MAHESHWARI, P. 1950. An Introduction to the Embryology of the Angiosperms. McGraw-Hill Book Company, New York.

MEEUSE, A. D. J. 1966. Fundamentals of Phytomorphology. The Ronald Press Company, New York. The new phytomorphology—the phylogeny of fruits and seeds, the stamen, carpel, and ovule of the angiosperms. A modern treatment of significant problems in angiosperm morphology, providing an alternate to the classical concepts presented in an older manner.

WARDLAW, C. W. 1955. Embryogenesis in Plants. John Wiley & Sons, Inc., New York.

21

ECOLOGY; THE PLANT AND ITS ENVIRONMENT

Maroon Lake near Aspen, Colorado.

The following discussion introduces concepts that include the biosphere, the biomass, and the ecosystem, all of which are important to an understanding of ecology. Edaphic, atmospheric, and biological aspects of the environment are examined briefly. Biological succession, a review of plant communities, and a discussion of adaptation conclude the chapter.

A detailed or even significant treatment of the effects of the environment on the development of a plant or plant community or of the effects of such a community on the environment lies beyond the intent of this textbook. In many ways it is difficult to define the environment or simultaneously to recognize all of its parts because it must include not only the molecular environment but also the cosmos. A discussion of the environment must include an interpretation of the effects of all of the factors in terms of action, as well as interaction or feedback.

Ecology (Greek *oikos,* meaning "place to live") was, at first, largely confined to the recording and description of the kinds of plants and their distribution. Growth form, i.e., trees, shrubs, and herbs, is the level of recognition in **plant geography.** This approach to ecology provides an overall and usually worldwide view of predominant plant communities and is still an active area of study. Investigation of groups of organisms and their interdependency is called **synecology.** When the research emphasis is on an organism, it is identified as **autecology.**

Life exists in a relatively narrow band where the land, water, and atmosphere meet; this relatively narrow, hospitable zone in which life can exist is the **biosphere.** The amount of living material produced within different parts of the biosphere is the **biomass,** which varies and is expressed as weight of organisms per unit area of habitat, e.g., grams of dry weight of organic matter per day per square meter of surface. Grobstein states that in moist forests, moist grasslands, and most agricultural areas from 3.0 to 10.0 g are produced per square meter of surface; in the shallow oceans above the continental shelf, 3.0 g; on grasslands and mountains, from 0.5 to 3.0 g; in deserts and in deep oceans, 0.5 g. Net production of organic material in selected ecosystems has also been expressed as thousands of kilograms per hectare per year, and in a summary prepared by Smith included among others were: old field in South Carolina one year after abandonment, 5; coral reef, Eniwetok Atoll, 81; corn in Wisconsin, 15; an alpine meadow (exclusive of roots), 0.0002 to 0.0011. The most easily recognized factors limiting the production of living material include solar energy or a converted form of the sun's energy, together with raw materials, such as carbon dioxide, water, and minerals (see also Chaps. 5 and 7).

Although the use of solar energy by photoautotrophic plants (primary producers) is the most important of the energy conversions (see also Chap. 5), many plants (see also Chaps. 16 and 17) are heterotrophic and are dependent upon their ability to use a fixed form of energy (food) for their existence. However, the statement that there is a cycling of energy within the biomass, from solar energy to autotrophic organism to heterotrophic organism, with some energy loss from the system through nonrecoverable heat energy, oversimplifies the complexities involved.

Groups or assemblages of functioning units, i.e., the groups that exist in an organization in which the individuals bear a characteristic association with each other and with the environment, form the **ecosystem.** The dynamic nature of the total ecosystem is influenced by its smaller units, and the adjustive change within the biomass is, in large measure, dependent upon its heterogeneity. Chemotrophic bacteria, which are neither photoautotrophic nor heterotrophic but which produce organic complexes from simple raw material (see also Chap. 16), and many food chains, each distinct and recognizable but obviously a part of the total dynamics of the ecosystem, delay or accelerate the rate of the ecosystem metabolism. Both conceivably and practically, it is convenient to select smaller ecosystems within a large ecosystem for study and analysis. It may be desirable to study the dynamics of a pond, a field, an orchard, or a space capsule. In such units it is possible to determine the energy input, the direction of its flow, and the products in terms of energy and materials.

A plant, which is a part of the biomass, is influenced by its environment both in growth and in development (see also Chap. 9). The response of a plant, or of groups of plants, to environmental factors has generated great interest not only among ecologists but also among morphologists, physiologists, geneticists, sociologists, taxonomists, agriculturalists, and other scientists to study such response under controlled conditions. It has been suggested that the final presentation of physiological (functional), as well as structural (morphological), facts concerning plants is most meaningful when it is expressed within the concept of the ecosystem(s).

An enzyme produced as a result of a message carried by RNA (see also Chaps. 2, 12, and 13) into the cytoplasm of a cell makes it possible for a certain pattern of cellular differentiation to occur (see also Chap. 2), but the absence of the enzyme may result in some other sequence of differentiation. The expression of the message may be through a feedback system which turns off a particular enzymatic reaction when there is, for example, an accumulation of certain end products, thus resulting in a minimal change in the environment or behavior. This tendency of biological systems to be self-regulating and to minimize change is called **homeostasis** and is operative within the ecosystem at the subcellular or cellular level as well as at the biomass level. The synthesis of a protein that combines with fat to produce the lipoprotein of a selectively permeable membrane may determine what molecules diffuse into the cell and influence its metabolism. An examination of the dynamics of the ecosystem at the molecular level differs markedly from an examination of the annual rainfall in Death Valley or the average mean temperature on the steppes of central Asia. Both—the enzyme (molecular) and the rainfall and the temperature (atmospheric)—are parts of the environment in which plants or groups of plants develop. Obviously it is the total influence of all factors of the environment, including those modified by the organism, that determines the final product. The biomass influences the biosphere and, as a result of such modification, life may occupy new niches. The generally inhospitable part of the biosphere that follows after fires, volcanic eruptions, or atomic explosions is soon modified by the already existing biomass, usually in adjacent occupied areas, so that invasion of new life soon occurs. Wind-borne spores of ferns, the soredia of lichens, lighter, often winged seeds of angiosperms, heavier seeds carried in the digestive tracts of birds, and perhaps such massive seeds as those of the coconut palm, which are carried by ocean currents, find their way to the temporarily uninhabited part of the biosphere. The establishment of a rather unusual assemblage of plants is followed by the invasion of other plants, e.g., seed plants, the seeds of which now find a more hospitable environment in which to germinate and grow.

An extension of a part of the biomass into the biosphere can be illustrated by the seed of the epiphytic plant that germinates in a compatible environment only after the tree fern or palm is established or by the maple seed that germinates in the humus accumulated around crustose lichens on the surface of a rock. An extension of a part of the biomass into the biosphere is illustrated also by the bird that, through the building of a nest, creates a slight but significant change so that its eggs can be kept warm enough to hatch.

Some overriding factors of the environment are readily recognized (see also Chap. 9). If no water is available, plants will not live; if a small amount is available, some plants will grow but they may be highly specialized (adapted) for the resulting dry climate. If a large quantity of water is available, the plants may be succulent, floating, or anchored to the ground in some specialized fashion, as in the case of cypress (Fig. 20-18) or water lilies (Fig. 4-10). Not only is the amount of water available to a plant important, but the temperature of the water, the salts dissolved in it, and the acidity of the water are also factors of the environment and contribute to the dynamics of the ecosystem.

Because biological research is usually performed under conditions that hold all factors constant except the variable being investigated, the results of many investigations have become facts of ecological value. Consequently the literature has become voluminous and a discipline that, for many years, was largely static and descriptive has now become highly experimental and dynamic.

There are three large categories of environmental factors influencing plant growth, development, and distribution. Atmospheric factors include, among others, the gases of the atmosphere, movement of the atmosphere, moisture found in the atmosphere, and the energy passing through the atmosphere. Consideration of energy must necessarily involve the total electromagnetic spectrum, not merely the small portion which is visible light (see also Chap. 5). Edaphic, or soil, factors include water content and water-holding capacity, soil temperature, particle size and arrangement, porosity, organic content, minerals, and particularly the presence or absence of carbonates (see also Chap. 7). Biological considerations include, among others, the microorganisms found in the soil; competition of growing plants for space, light, minerals, and water; nutritive relationships that may be established between host and parasite (see also Chap. 5); the effects of animals, including man.

Man interrupts the ecosystem, and he has played an important role in altering, often in a deleterious manner, the atmospheric, edaphic, and biotic factors of the plant environment. He has polluted the atmosphere with the toxic gases of industry, killing plants within a radius of many miles of the center of pollution. He has destroyed soil texture and structure through farm practices favoring wind and water erosion and through the physical removal of the earth's crust by mining. He has, in some places, eradicated all aquatic life from thousands of miles of streams with acid mine drainage, with organic pollution from the untreated wastes of metropolitan areas, and with the introduction of harmful industrial chemicals. Man, through farming, fire, and lumbering, has eradicated plants from their original habitat and has introduced others, including many undesirable species called weeds.

Succession The environment is complex and is made up of a great variety of factors in a state of flux which, in turn, elicit from plants and plant communities a wide range of reactions. Plant **communities** are never completely static, and although the changes that take place may be slight or pronounced, they do occur. In most instances these are predictable within the limits and effects of the known environment. In abrupt changes, as in the denuding of land by fire, there is a radical change of the environment. The cessation of all rainfall or a permanent alteration in temperature would produce equally dramatic differences in the environment. The gradual filling of open water with vegetation or inorganic deposits (see discussion under *Sphagnum*, Chap. 18), the failure of certain seeds to germinate in the shade of an upper canopy of vegetation, and the accumulation of certain plant parts, with a resulting change in the acidity of the soil, are more imperceptible changes, and the reaction of existing plant communities is more gradual and much less noticeable.

Very often moisture relationships determine the ability of plants to become established in an environment. A dry habitat is termed **xeric,** an intermediate habitat **mesic,** and a wet habitat **hydric.** The gradual changes reflected in the plant communities that occupy such areas are, in turn, referred to as xerarch, mesarch, or hydrarch succession. Succession tends toward stability and the establishment of a community that succeeds itself. Usually this single community represents the most mesophytic type that can be sustained under the existing climatic conditions, and it is called **climax.**

The living (biotic) community is a group of plants in the same environment, interacting with each other. Within the community one or several species are likely to be abundant and dominant and an established biotic community tends to be homogeneous. The lateral or horizontal extent of a community may be strictly defined, e.g., as a clearing in a wooded area, a pond, a burn, or a mountaintop. Its boundary with another community may be distinct. Or its boundary may not be well defined, and it is then characterized by considerable overlapping with an adjoining community. Some representatives of one or both of the contiguous communities may be numerous or they may be present only in the overlapping portion called the **ecotone.** The ecotone is sometimes more heterogeneous, less stable, and more dynamic than the adjacent communities. These edges or borders, as they are sometimes called, often provide an environment in which are found more kinds and greater numbers of birds, mammals, insects, and plants than are found outside the ecotone.

In addition to horizontal stratification of a community, vertical layering often occurs. The upper canopy of trees may overtop lower layers of shrubs and herbs.

Some plant communities and their distribution
A few types of plant forms make up the principal vegetation of the earth; they are **tundra, grassland, shrubland,** and **forest.** These are broad designations, based principally on growth forms, but within each there are smaller ecological units of considerable interest, such as the aquatic habitat of a mountain lake enclosed

Fig. 21-1 Bog at Bear Meadows in central Pennsylvania. (Photograph by Wallace R. Bell.)

Fig. 21-2 Alpine tundra and timberline in Rocky Mountain National Park, Colo.

by trees, the swamp surrounded by a forest (Fig. 21-1), or the pothole encircled by a grassland.

The major worldwide types of vegetation are largely determined by climate, the main ingredients of which are rainfall, radiation, and temperature. The average annual rainfall, together with its distribution, is important. Different types of vegetation are supported by rain that falls in a short rainy season rather than in each month of the year. Moisture that falls mostly as snow will exert its influence in a way entirely different from that of rain that falls in a warm climate. Extremes of temperature might be just as important as the average temperature, especially in limiting the growth of certain kinds of vegetation. It becomes apparent that even when only two factors, rainfall and temperature, are considered, there are many variables.

TUNDRA The tundras of the world are cold, moist places where low-growing plants predominate. Essentially, such places are without trees and lie between the limit of trees and the areas of perpetual snow and ice. Vegetation consists mostly of grasses, sedges, mosses, and lichens, with some low-growing herbs or shrubs occurring in more favorable locations. **Arctic tundra** is typically associated with the Aleutian Islands, northern Canada, Greenland, Iceland, and northern Russia. **Alpine tundra** occurs at extremely high elevations, at the tops of mountains, as a smaller ecological unit within other major vegetation types of the world (Fig. 21-2).

DESERTS Deserts are hot and dry. Precipitation is low and erratic, and the air temperature usually is high during the day, with a sudden drop at night. The days are sunny, with low relative humidity. Although the essential features of deserts are the same, the plants occurring there vary considerably, depending upon local conditions. The giant saguaro (Fig. 8-45), for example, is quite limited in its distribution and is found in the uplands of the Sonoran Desert (Fig. 21-3). Mesquite is characteristic of the **desert scrub formation** found in southern New Mexico and western Texas, extending southward into Mexico. Sagebrush is abundant in the Great Basin of Nevada and Utah. Although they are similar in some respects, the lower elevations of such deserts as the Mojave Desert, including Death Valley, present more extreme conditions where such plants as the Joshua tree and the sagebrush give way to creosote bush and burroweed.

The vegetation characteristic of deserts consists of plants able to withstand extreme desiccation but, in addition, there are also many annuals that grow rapidly and flower as a result of a brief rainy season. At times the deserts of southwestern United States are a flowering paradise. Deserts are a conspicuous part of the world's vegetative picture. Not only are they relatively extensive in western and southwestern United States and northern Mexico, but they are also predominant both in northern and southern Africa, central Asia, and central Australia.

GRASSLANDS Tundra and deserts are essentially inhospitable to both plants and man and consequently they support only a meager part of the world's population. On the other hand, the grasslands have become our most important agricultural areas. Annual rainfall and temperatures are such that the tall grasslands, or tall-grass prairies, of central United States have become, through cultivation, the corn belt. Similar land provides soil to grow fruit crops and vegetables. The somewhat poorer short-grass country (rainfall about 12 in. a year) to the west of the prairies, called the Great Plains in the United States, furnishes natural grazing and supports much of the cattle and sheep industry. The rather sparse native grasses hold the soil in place and prevent erosion. From time to time, however, these grasses have been destroyed by overgrazing, or man has removed them and planted wheat. In good years, with adequate rainfall and little wind, the wheat crop may succeed, but often the winds blow and erode the land, creating disastrous dust bowls. In the United States grasslands are most often called the prairies and plains; grasslands in Argentina are called the **pampas** and in the Ukraine they are called **steppes.**

Primitive methods of agriculture are not sufficient to utilize grasslands efficiently. Only with the development of modern methods of farming were peoples of the world able to cope with the grasslands and, in a sense, to extract from these areas staple crops necessary for the establishment of towns and cities. In this country and in many areas of the tropical grasslands, the struggle of man against the grasslands has been only partially successful. In short-grass areas of low rainfall, existence is always perilous, and historically this has stimulated the migrations that have played such an important role in the conquest and development of China, India, and western Europe.

SHRUBLANDS At a time when much less was known about the world than today, Aristotle set forth his theories about climates. He suggested that to the north of the area of the Mediterranean world it was too cold except for barbarians and to the south of the temperate land in which he lived it was too hot for civilized man. The intermediate, or temperate, area of the Mediterranean is characterized by a scanty but sufficient supply of rainfall with abundant sunshine. There is too little rainfall to produce the forest characteristic of northern Europe and too much for the development of a desert such as the Sahara to the south. The Mediterranean area is typified by the develop-

Fig. 21-3 Sonoran Desert, Ariz. Some of the plants shown are bur sage, ocotillo, paloverde, desert ironwood, organ-pipe cactus, and saguaro cactus. (Photograph by National Park Service, courtesy of Dr. F. D. Kern from "The Essentials of Plant Biology," Harper and Row, Publishers, Incorporated, New York, 1947.)

ment of shrublands, with plants somewhat intermediate between those of the forest and those of the grassland or desert.

In the United States there is a rather extensive area from southern Oregon southward into Lower California and on the lower slopes of the Rocky Mountains where shrubs constitute the dominant type of vegetation. They are resistant to drought and often thorny; they are frequently referred to as **chaparral.**

There is also considerable justification to include under shrublands some of the formations discussed under "Deserts." Thus the sagebrush of the northern part and the creosote bush of the southern part of the Great Basin are often referred to as desert scrub.

FORESTS The world's forests occupy much less area today than they have in the past. In nearly all countries of the world, they have been cut as a resource and to make living space. Forests have almost entirely disappeared in some of the countries that nurtured early civilizations, and in the United States little more than half of the land mass is still covered with trees.

The terms used to designate forest types reflect at least two things. One term most often refers to the climate (temperature and rainfall), and the second describes the characteristics of the trees themselves. The trees may be conifers or softwoods (pine, spruce, fir) and most often evergreen, or they may be deciduous hardwoods (oak and hickory), which drop their leaves during some part of the year. They may exhibit a rich, exuberant growth, or they may reflect minimal conditions necessary for existence.

TROPICAL FORESTS Primarily as a result of variations in rainfall, the forests of the tropics differ in their makeup. The high temperatures of the tropics and the rapid rate of water evaporation may produce a comparatively dry habitat even with relatively large amounts of rainfall. In the Amazon basin there exists the most extensive and, perhaps, most luxuriant forest growth in the world. Similar but smaller expanses of tropical rain forests (Fig. 21-4) occur elsewhere and characteristically grow in areas where the rainfall is as much as 150 in. a year. The tropical rain forest contains many species of plants, most of which are broad leaved. The forest canopy is high and dense, and there may be little underbrush. The development of **lianas** and **epiphytes** is striking, and the forest is evergreen.

Not all forests of the tropics are of this type, however. In areas where there is less rainfall, the canopy may be less dense. The understory may be thick and nearly impenetrable (jungle), with the trees semideciduous rather than evergreen. In still other instances smaller amounts of rainfall may result in the formation of a relatively sparse (savanna-type) forest, the understory of which is usually grass. Many of the trees develop as thorny shrubs producing the **tropical scrub forest.**

TEMPERATE FORESTS The temperate forests of the world are better known than any other forest. In the United States two basic forest types exist—the **coniferous forest** and the broad-leaved, or **deciduous, forest.** In places there is intermingling to produce a mixed forest growth.

We have already seen that the climatic distribution of plants is the result of a series of complicated and interacting factors. In mountainous areas temperature is of striking importance. Although most students of plant science recognize that one can find cooler temperatures by walking toward the poles, fewer realize that more sudden changes in temperature are encountered in vertical ascent. The coniferous forest of more northern latitudes, with lower temperatures, often extends southward at high elevations. Similarly other vegetation may reach north if climatic conditions support its development.

BOREAL FOREST The boreal forest formation is a coniferous forest that spans the continent in a band beginning roughly at the northern border of the United States and extending northward into Canada until it meets the arctic tundra. On the east coast it extends from Newfoundland southward to the New England states, and in the West it reaches northward to the Pacific Coast of Alaska.

The growing season of the boreal forest is short. Temperatures generally are cool, and the winters are very cold. Rainfall is approximately 20 in. per year, although it is somewhat higher on the western coast. White spruce and balsam fir are best developed in the central and southern part of the zone. Balsam fir disappears toward the northern limits of the forest formation, and black spruce, tamarack, paper birch, and jack pine become dominant.

Other coniferous forests are the southeastern coniferous forest and two western associations, those of the Rocky Mountains and of the Pacific Coast. The southeastern coniferous forest extends along the Atlantic coastal plain westward as far as Texas and is characterized by longleaf and shortleaf pines, loblolly pine, cypress, and

Fig. 21-4 Tropical rain forest, Belém, Brazil. (Courtesy of Dr. Thomas R. Soderstrom, U.S. National Museum, Smithsonian Institution.)

southern white cedar. Intermixed with the conifers are such broad-leaved species as magnolia, red gum, and live oak.

The Rocky Mountain coniferous forest (Fig. 21-5) is characterized by western yellow pine, lodgepole pine, Engelmann spruce, Douglas fir, and white alpine fir. It extends from British Columbia to Central America and produces a great quantity of the commercial timber used in the United States. The Pacific Coast coniferous forest (Fig. 21-6) stretches southward to about the center of California and is character-

Fig. 21-5 Pinus ponderosa, *yellow-pine forest in the Rocky Mountains of New Mexico.*

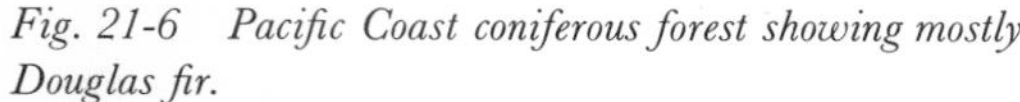

Fig. 21-6 Pacific Coast coniferous forest showing mostly Douglas fir.

ized by Douglas fir, sugar pine, and the redwoods (Fig. 21-7).

DECIDUOUS FOREST FORMATION Most of the eastern United States, with the exception of southern Florida, is occupied by a deciduous forest (Fig. 21-8). In the north there is a transition to conifers and, in the west, to grasslands. The forest is very diversified, reflecting age, topography, soil types, glaciation, temperature, and many other ecological factors. Some of the better-known associations within this mixed deciduous forest are the beech-maple association, the hemlock-hardwood association, and the oak-hickory association.

Adaptation Adaptation is the ability (inherited potential) of an organism to adjust to its environment. The inference is, of course, that the environment exerts a stimulus to which the plant reacts. Unfortunately the impression is sometimes created that the reaction of the organism, its adaptability, is purposive. Xerophytic plants do not develop water-storage tissue or heavily cutinized plant surfaces in order to live in a dry climate, but those individuals in which an adaptation results in the more efficient storage of water can and do, in fact, exist in dry climates.

The plants we see today exist in a variety of environments. Whatever their characteristics, they are present as a result of inheritance and environment and their presence is proof that they are adapted to their environment. Although it may be easier to relate adaptation to an individual, it is obvious that the flora around us is not just one individual after another but rather groups of individuals or **populations.** Populations exhibit adaptive changes. Not all members of a population produce offspring. The kinds of offspring are not produced in equal numbers, and so the succeeding generation or population reflects adaptive changes from the previous one. Although there are exceptions, it is biologically true that parents best able to cope with their environment will probably contribute a larger number of offspring to succeeding populations than relatively unsuccessful parents.

Fig. 21-7 Pacific Coast coniferous forest of redwoods. (Photograph by George A. Grant, courtesy of U.S.D.I., National Park Service.)

Fig. 21-8 A beech-maple-cherry-hemlock forest of Pennsylvania. (U.S. Forest Service photograph, courtesy of Dr. F. D. Kern from "The Essentials of Plant Biology," Harper and Row, Publishers, Incorporated, New York, 1947.)

Not all changes, from one population to the next, are adaptive. Certain changes may make a negative contribution to survival of the population (or individuals of the population), other changes have no recognizable influence on survival, and some may enhance the chances of survival in the environment under consideration. Those changes that enhance survival are adaptive changes and play an important role in evolution.

QUESTIONS

1. How do you think the most recent change in the weather might affect the plants growing on your campus?

2. What is the most obvious difference between environments described as xeric and those described as hydric?

3. What more subtle differences exist in two environments, both of which are recognized as being hydric?

4. Rank the relative importance of ecological factors that determine the distribution of plants in your area.

5. What is meant by teleology? Evaluate the concept from a biological position.

6. Discuss the role of solar energy in the establishment of a plant community.

7. Explain when an environment that appears to be hydric may be, in fact, xeric in its effect on plant distribution.

8. Diagram a model of a pond and insert

emerging aquatic plants, algae, aquatic insects, minnows, and black bass. Indicate the source and flow of energy in this ecosystem and list the material products.

9. What is an ecological niche? Explain in terms of biotic succession how the bare rock face of a mountain might provide the habitat for a developing plant community.

10. What visible changes occurred in the ecosystem that formerly existed in the place where the latest four-lane highway has been built in your neighborhood? Was man an important biotic factor in this change?

REFERENCES

FRYXELL, PAUL A. 1957. Mode of Reproduction of Higher Plants. *Botan. Rev.,* **23:**35–233. Much of the article is devoted to a listing of plants to show their mode of reproduction. In addition to other categories, the listing of plants as being cross-fertile, self-fertile, self-incompatible, etc., is of value in understanding the dynamics of pollination and the role it plays in the establishment of populations.

GARB, SOLOMON. 1961. Differential Growth-Inhibitors Produced by Plants. *Botan. Rev.,* **27:** 422–443. A review of plant growth inhibitors showing that they are widespread in the plant kingdom and are found in various plant parts. They are of significance in ecology. Lists many additional references.

GROBSTEIN, CLIFFORD. 1965. The Strategy of Life. W. H. Freeman and Company, San Francisco.

OOSTING, H. J. 1956. The Study of Plant Communities. W. H. Freeman and Company, San Francisco. A basic text considering the most important features of a plant's environment.

SMITH, ROBERT L. 1966. Ecology and Field Biology. Harper & Row, Publishers, Incorporated, New York. The approach reflects recent concepts of ecology blended with the useful and practical information required for an understanding of ecology. Very understandable.

WALLACE, B., and A. M. SRB. 1961. Adaptation. Foundations of Modern Biology Series. Prentice-Hall, Inc., Englewood Cliffs, N.J.

WENT, F. W. 1955. The Ecology of Desert Plants. *Sci. Am.,* **192:**20, 68–75. An interpretation of adaptation as it affects survival in a dry climate.

WENT, F. W. 1957. Plant Life, pp. 45–61, 137–145, 146–155. Scientific American Book, published by Simon and Schuster, Inc., New York. A series of essays on various ecological topics.

GLOSSARY

In studying any science, one must become familiar with its vocabulary. This involves acquiring a clear understanding of the technical terms of the science. In general these terms have been devised to enable scientists the world over to understand each other when discussing a structure, a process, a theory, or any other scientific matter. Furthermore the use of a scientific term often makes it possible to avoid lengthy descriptions and explanations. It is important, therefore, for the beginning student to make every effort to learn a new term the first time it is presented. If he does this, he will have made a major stride in learning the subject.

The following glossary defines terms according to the way they are used in this book. In some cases the terms have additional meanings. For these a standard botanical dictionary should be consulted.

Although most of the terms in the glossary have already been defined where they are introduced in the text, a special effort has been made to include all terms that are used again in the text without definition.

abscission layer a zone of cells, the breakdown of which results in the separation of a leaf or other structure from the stem.

accessory bud an additional bud; a bud in addition to the axillary bud.

accessory fruit a fruit formed not only from the ovary but also from other or accessory parts.

achene a simple, dry, and usually one-seeded indehiscent fruit with the seed attached to the fruit wall at only one point.

acrogynous referring to a condition in the Jungermanniales in which the apical cell of the thallus produces a terminal gametophore.

acropetal developing in a direction of differentiation or origin from some place below the tip toward the tip.

actinomorphic radially symmetrical; usually used to mean a flower of regular shape or a star pattern.

actinostele a star-shaped protostele in which protoxylem and protophloem cells are located on different and alternating radii.

adaptation the ability of an organism to adjust to its environment.

adenine a purine component of nucleic acids and nucleotides.

adenosine phosphates a group of organic phosphates including adenosine monophosphate (AMP), adenosine diphosphate (ADP), and adenosine triphosphate (ATP). They function in phosphate transfer in the cell, particularly in the transfer of the high-energy phosphate bonds of ADP and ATP. ATP is the most directly utilizable source of energy of the living cell.

adnation in flowers, the congenital growing together of unlike parts, such as two or more whorls of floral parts.

ADP (see adenosine phosphates).

adsorption the adhesion, in a very thin layer, of the molecules of gases, dissolved substances, or liquids to the surfaces of solid bodies with which they are in contact.

adventitious referring to structures arising from places other than the usual or expected, as of roots growing from leaves, buds developing at locations other than in leaf axils.

adventitious bud a bud located at some place other than in the leaf axil or the apex of a stem.

aeciospore a binucleate spore formed in an aecium by rust fungi.

aecium a sorus or cuplike structure in which aeciospores are produced by the rust fungi.

aerobe an organism requiring free oxygen for respiration and life.

aerobic respiration oxygen respiration, or respiration in the presence of free oxygen.

agar a gelatinous product used extensively as a culture medium for the growth of bacteria and fungi and derived from certain red algae.

agaric any of a group of fungi belonging to the family Agaricaceae; a gill fungus.

aggregate fruit a fruit consisting of a number of similar small fruitlets all derived from the ovaries of a single flower, as in blackberries.

aglycon the nonsugar part of a glycoside.

akinetes single thick-walled vegetative cells most often produced by certain algae.

aleurone proteinaceous material in the form of small grains found in the outer layer of the endosperm of many seeds.

algae relatively simple plants that contain chlorophyll and are photosynthetic.

algin (alginic acid) a phycocolloid found in the Phaeophyta; salts of alginic acid (alginates) of economic use as stabilizers.

algology (phycology) the science dealing with algae.

alkaloid an organic substance of strong basic properties occurring in some plants and usually poisonous to animals.

allele either of a pair of contrasting Mendelian characters such as round peas and wrinkled peas; also applied to genes located at comparable positions in the two members of a pair of homologous chromosomes.

allelic genes genes located at comparable positions in the two members of a pair of homologous chromosomes.

alpine designating plants growing at high elevations; properly denotes plants of the Alps.

alternate phyllotaxy a type of phyllotaxy in which the leaves, one at a node, occur in two rows on opposite sides of the stem; sometimes used for any spiral arrangement with one leaf at a node.

alternation of generations the alternation of plant generations designated as the gametophyte and the sporophyte. The former develops from meiospores, following meiosis, and the latter from the fusion of gametes.

ament a catkin (which see).

amino acid an organic carboxylic acid containing one or more NH_2 groups and having the general formula $R \cdot CHNH_2 \cdot COOH$. They are the principal structural units of proteins.

amitosis a division of a nucleus or nuclear material without the usual recognizable stages of mitosis; without mitosis.

ammonification the decomposition of amino acids with the formation of ammonia, especially by the action of bacteria on nitrogenous organic matter.

AMP (see adenosine phosphates).

amphicribral bundle a vascular bundle in which the xylem tissue is surrounded by the phloem tissue.

amphiphloic siphonostele a siphonostele in which phloem occurs both internally and externally with respect to the xylem.

amphithecium the outer and surrounding tissues of the moss capsule.

amphivasal bundle a vascular bundle in which the xylem tissue surrounds the centrally located phloem tissue.

amylase an enzyme which hydrolyzes starch to maltose.

amylopectin one of the components of starch, consisting of a branched chain of glucose units.

amyloplast a starch-storing (or perhaps starch-producing) plastid; a type of leucoplast.

amylose one of the components of starch, consisting of an unbranched chain of glucose units.

anabolism the building-up or synthetic phase of metabolism.

anaerobe an organism capable of living in the absence of free oxygen.

anaerobic respiration respiration in the absence of free oxygen.

anaphase that stage in mitosis when half chromosomes (chromatids) move to opposite ends of the cell.

anatomy a branch of morphology dealing with the structure of organisms.

androecium a collective term for the stamens of a flower.

androspore a special kind of spore formed by some algae which, on germination or growth, produces a special filament of cells.

angiosperm one of the flowering plants.

anisogamy gametic reproduction in which syngamy is between gametes similar in form but different in size.

annual plant a plant that completes its life cycle and dies within a year.

annual ring a layer of secondary xylem (wood) produced in the stem of a plant in one year.

annular vessel a ringed vessel; one with internal lignified rings.

annulus in agarics, the remains of the veil, attached to the pileus; in mosses, the cells along the rim of the capsule to which the peristome is attached; in ferns, a row of specialized cells in a sporangium involved in the opening of the sporangium.

antagonism the influence of one ion on the absorption of another ion by plants.

anther the pollen-bearing portion of a stamen.

antheridium a male gametangium in which male gametes, or microgametes, are formed.

antherozoid a male cell, usually motile.

anthocyanidin the aglycon or nonsugar part of the anthocyanins.

anthocyanin one of a class of water-soluble pigments, usually causing red, blue, or violet colors in plant parts.

antibiotic a product formed by a living organism that has properties detrimental to the growth of another organism.

antipodals the nuclei or cells found in the female gametophyte (embryo sac) of angiosperms and located at the end opposite that containing the megagamete (egg) and synergid nuclei.

antithetic theory a theory of the alternation of generations that pictures a gradual decrease in gametophyte and increase in sporophyte as parts of the life cycle of most plants; an increase in importance of one generation at the expense of another.

apical meristem the meristematic cell or cells located at the tips of the plant axis.

aplanospore a thick-walled resting spore formed by some algae; a nonmotile spore.

apoenzyme the basic or protein part of a holoenzyme.

apophysis the usually enlarged and sterile basal portion of a moss capsule.

apothecium a cuplike or saucer-shaped ascocarp.

archegonium the multicellular female gametangium of Bryophyta and some vascular plants.

Archeozoic the geological era preceding the Proterozoic and lasting approximately 500 million years.

archesporium the spore-bearing tissue of bryophytes and other plants.

Archichlamydeae (Choripetalae) one of the subdivisions of the dicotyledons, the plants of which have flowers in which the individual members of the calyx and corolla are entirely separate from each other or in which the perianth as a whole is poorly developed.

arthrospore a jointed spore; a spore resulting from the fragmentation of a hypha, as in some actinomycetes.

ascocarp the sporocarp or ascospore-bearing structure of the Ascomycetes.

ascogenous hyphae hyphae producing asci and ascospores and arising from an ascogonium.

ascogonium the oögonium or female gametangium of the Ascomycetes.

Ascomycetes the sac fungi; a class of fungi in which the meiospores, called ascospores, are produced in asci.

ascorbic acid vitamin C, the water-soluble antiscorbutic vitamin.

ascospore a meiospore developed within an ascus.

ascus the characteristic reproductive structure of the Ascomycetes consisting of a saclike cell usually having eight ascospores.

asexual reproduction vegetative reproduction without benefit of gametes or their fusion.

assimilation the transformation of food into the living substance, protoplasm.

aster a cytoplasmic radiation found at the poles of certain cells during mitosis.

atactostele a stele consisting of many separate vascular bundles which lack any apparent order of arrangement but appear scattered through the fundamental tissue, as in many monocotyledons.

ATP (see adenosine phosphates).

aureomycin an antibiotic obtained from one of the actinomycetes.

autecology the study of an organism in relation to its environment.

autoclave a strong metallic vessel, gastight when closed, using steam under pressure for sterilization.

autoecious referring to a parasitic fungus completing its entire life cycle on a single host.

autonomic movements movements independent of changes in the external environment; movements that are the result of internal growth changes.

autosomes chromosomes other than sex chromosomes.

autotrophic plant a self-nourishing plant; one capable of making its own food.

auxin a plant hormone effective in regulating cell elongation.

auxospore a spore formed by the diatoms, resulting from the fusion of two cells or their contents.

axil the upper angle between a leaf petiole or twig and the stem from which it grows.

axile placentation a type of placentation in which the placentae are borne on the central axis of the ovary.

axillary bud a bud located in the axil of a leaf.

bacillus a rod-shaped bacterium.

bacteriology the study of bacteria.

bacteriophage a virus that invades bacterial cells and causes their destruction.

bark all the tissues of a stem or root from the cambium outward.

basidiocarp a fruiting structure of the Basidiomycetes producing basidiospores.

basidiospore the meiospore of the Basidiomycetes, formed exogenously on basidia.

basidium the structure, often saclike or stalklike, from which basidiospores are formed.

berry a simple, fleshy fruit in which all parts of the pericarp, except in some cases the skinlike exocarp, are fleshy or pulpy; usually referred to as superior berries if formed from a superior ovary, or inferior berries if formed from an inferior ovary with surrounding parts.

bicollateral bundle a vascular bundle in which phloem is located both externally and internally with respect to the xylem, all tissues lying on the same radius, with the internal phloem lying next to the pith.

biennial a plant that completes its life cycle in two years and then dies; usually it produces a rosette of leaves the first year and flowers, fruit, and seed the second year.

biflagellate bearing two elongated cytoplasmic extensions, the flagella.

bilateral capable of being divided into two equal parts, one of which is the mirror image of the other; a kind of symmetry.

binomial a name consisting of two terms; applied to the scientific names of plants, the genus name and species name always given together.

biology the science dealing with living organisms.

bioluminescence the emission of light from living organisms as a result of internal oxidative changes.

biomass the living material within the biosphere.

biosphere the narrow zone, where land, water, and atmosphere meet, that supports life.

biotic living.

biotin one of the vitamins of the vitamin B complex.

blade the expanded portion of a leaf; the lamina.

blepharoplast a structure within the cell from which cytoplasmic extensions, cilia or flagella, originate; especially in male cells of plants.

bordered pit a thin place in the wall of a xylem cell, surrounded with secondary wall thickening in such a way as to form a border.

boreal northern.

botany the science dealing with plants.

bract a specialized leaf from the axil of which a flower or floral axis arises; or a leaf borne on the floral axis itself.

bromelin a protease obtained from ripe pineapples.

bryology the study of liverworts and mosses.

Bryophyta one of the divisions of the plant kingdom comprising the mosses and liverworts.

bud an unelongated shoot or stem; a small lateral protuberance on some leaves and roots that contains rudimentary leaves or floral parts, or both.

bud scale one of the specialized leaves covering a bud.

bulb a specialized bud, often underground, consisting of a greatly shortened stem surrounded by fleshy leaves or leaf bases.

bulbil an unelongated axis, often with thickened phyllids, that serves in vegetative propagation in moss.

bundle scars scars left on leaf scars by broken vascular bundles that extended from the stem into the petioles of the leaves before the leaves fell.

calciferol vitamin D_2, the fat-soluble antirachitic vitamin.

callus a parenchymatous tissue formed as an overgrowth of a wound or in tissue culture.

calorie a small calorie (cal) is the amount of heat energy required to raise the temperature of 1 g of water 1°C; a large calorie (Cal) is the amount of heat energy required to raise the temperature of 1 kg of water 1°C.

calyptra the hood or cap covering the capsule of a moss plant.

calyx a collective term embracing all the sepals of an angiosperm flower.

calyx tube a tube formed by the fusion of the lateral edges of a number of sepals.

cambium a lateral meristem that gives rise to secondary tissues, the vascular cambium giving rise to secondary xylem and secondary phloem, the cork cambium (phellogen) giving rise to the periderm.

capillary water that part of soil water which is free to move from one soil particle to another.

capillitium a fine threadlike growth among the spores in a sporogenous body, often forming a net, as in the Myxomycetes.

capsule an enclosing sheath of some bacteria; the spore-bearing or spore-producing structure of mosses and liverworts; a simple, dry, dehiscent fruit developed from a compound pistil.

carbohydrase any enzyme involved in the hydrolysis of carbohydrates.

carbohydrate an organic compound consisting of hydrogen, oxygen, and carbon, in which the hydrogen and oxygen are in the same proportion as they are in water.

carboxylase an enzyme capable of splitting off carbon dioxide from the carboxyl group (COOH) of certain organic acids.

carotene any of a class of yellow, orange, or red hydrocarbon pigments found in the plastids of plants and in animal tissues.

carotenols (xanthophylls) any of a class of yellow plastid carotenoid pigments differing from the carotenes by having oxygen in the molecules.

carpel a highly specialized leaf that makes up the structural unit of the pistil of an angiosperm flower.

carpogonium the female gametangium of some red algae, e.g., *Polysiphonia.*

carposporangium a sporangium of some red algae in which carpospores are formed, as in *Nemalion.*

carospore a spore produced in the carposporangium in red algae; may be either $1N$ or $2N$.

carrageenin a phycocolloid found in some red algae; of economic importance as a gel, used as a food and as a stabilizer in ice cream and other products.

caruncle a spongy, water-absorbing organ found on a few seeds, such as castor bean.

caryopsis (grain) a simple, dry, one-seeded, indehiscent fruit in which the seed coat is fused to the pericarp over its entire surface as in the grasses.

Casparian strip thickening, usually on the radial and end walls of the endodermis.

catabolism the breaking-down phase of metabolism, including such processes as digestion and respiration.

catalase an enzyme of wide distribution in plants and animals that accelerates the decomposition of hydrogen peroxide with the liberation of water and free oxygen.

catalyst a substance that changes the rate of a reaction, usually accelerating it, appearing unchanged at the end of the reaction.

catkin (ament) a pendulous spike, sometimes scaly, consisting of either staminate or pistillate flowers, and characteristic of oaks, willows, poplars, birches, and others.

caulid (caulidium) the vegetative shoot of a bryophyte, stemlike in appearance.

cell the biological unit of structure and function.

cell membrane (plasma membrane) the outer membrane of the protoplast, adjacent to the cell wall.

cell sap the contents of a cell found within the vacuoles. It may include a variety of substances such as water, water-soluble pigments, sugars, and inorganic substances.

cellobiase an enzyme facilitating the hydrolysis of cellobiose into glucose; one of the enzymes active in the digestion of cellulose.

cellobiose ($C_{12}H_{22}O_{11}$) a disaccharide consisting of two molecules of glucose; a hydrolytic product of cellulose.

cellulase an enzyme facilitating the hydrolysis of cellulose to cellobiose.

cellulose a complex carbohydrate (polysaccharide) found in greatest abundance in the cell walls of plants.

Cenozoic the geological era following the Mesozoic and extending approximately from 75 million years ago to the present.

centripetal developing from the outside toward the center; said of an inflorescence in which the lower or outer flowers bloom first.

centromere the region or place of attachment of a spindle fiber to a chromosome.

centroplasm the central region of the cell of a blue-green alga; contains chromatin; shows areas of low electron density and crystalline granules.

centrosome a small, deep-staining body found near the nucleus in many animal cells and in some lower plant cells. Some authors use this term as a synonym of centriole.

chalaza the portion of the ovule opposite to the micropylar region; the area common to the base of the integuments and nucellus.

chalazogamy the entrance of a pollen tube and microgametes into the female gametophyte near the chalazal portion of the ovule.

chaparral a type of vegetation consisting of small trees or shrubs, most often growing in semiarid regions, with the plants frequently bearing evergreen leaves.

chemosynthesis a process by which a chemical source of energy instead of light is used by certain bacteria, such as the hydrogen bacteria and the nitrifying bacteria, in making carbohydrates out of CO_2 and water.

chemotropism a growth movement or response to a chemical stimulus.

chitin a horny substance common in the exoskeleton of insects but occurring also in some of the fungi in connection with cellulose.

chlamydospore a thick-walled, asexual, resting spore developed from hyphae but not on conidiophores.

chlorenchyma a tissue containing chlorophyll.

chlorophyll one of a group of green pigments found in chloroplasts and important in photosynthesis.

chloroplast a cytoplasmic structure containing chlorophyll; a green plastid.

chlorosis a condition of plants resulting from the loss of chlorophyll; often associated with absence of light, lack of essential minerals, or other causes, in which case the plant is referred to as being chlorotic.

chondriosomes (mitochondria) small grains or bodies found within the cytoplasm, important in respiration.

chromatids the two, paired parts of the chromosome at the time of its longitudinal replication.

chromatin a part of the chemical content of the nucleus; the chemical material of the chromosome.

chromatography a technique used to separate various chemical substances, including pigments, based on differential solubility and adsorption by an inert carrier such as filter paper, a column of a finely powdered substance, or gel.

chromatophore a pigment-containing body.

chromonemata the long slender threads of the chromosomes becoming visible during the prophase of mitosis.

chromoplasm the peripheral region of the cell of a blue-green alga; contains photosynthetic lamellae, pigments, and crystals.

chromoplast a colored plastid; a plastid containing only carotenoid pigments, as distinguished from a chloroplast, which contains green pigments as well.

chromosome one of the nuclear structures of definite number, consisting of chromatin and bearing hereditary units or genes.

chromosomes, sex chromosomes that carry the genes predominantly associated with the determination of sex; X, Y chromosomes.

chrysolaminarin a storage polysaccharide in some algae (leucosin).

chytrid a member of the fungus order Chytridiales.

circinate referring to a kind of vernation in which the leaf is rolled from the tip downward toward the base, as in ferns.

circumnutation the bending or nodding of a growing stem tip from one side to another, caused by unequal growth rate of different sections around the stem.

circumscissile referring to a type of dehiscence of capsules involving the development of a circular line of opening, as in *Portulaca*.

citric acid cycle see Krebs cycle.

cladophyll a leaflike stem.

clamp connection in the Basidiomycetes, a lateral connection formed between two adjoining cells of a filament and arching over the septum between them, by means of which a type of pseudosexual process takes place.

class in classification, a group of plants made up of orders.

cleistothecium a closed ascocarp, opening by rupture.

climax association a group of plants judged to be the permanent vegetation occupying any habitat under the existing environmental conditions.

clone the aggregate of the vegetatively produced progeny of an individual.

coalescence state of growing together; in flowers, the union of parts of the same kind, as united sepals (synsepaly).

cocarboxylase the coenzyme of carboxylase.

coccus a sphere-shaped bacterial cell.

coenobium a colony of independent cells or individuals such as found in the green alga *Volvox*.

coenocyte a multinucleate plant body or filament in which there is no division by walls into separate protoplasts, as in some algae and fungi.

coenzyme a substance necessary for the action of certain enzymes, particularly oxidative enzymes.

coenzyme A a derivative of the B-complex vitamin, pantothenic acid. It functions in the acceptance and release of the acetyl group (CH_3CO—).

cohesion theory a theory explaining the ascent of sap in stems, based primarily upon the cohesive power of water molecules, that is, the ability of water molecules to stick together when a pull caused by transpiration is applied at the upper end of the column of water.

colchicine an alkaloid derived from *Colchicum autoumnale* that interferes with spindle formation during nuclear and cell division.

cold rigor a coagulation or setting of protoplasm resulting from the application of a subminimal temperature.

coleoptile in monocotyledons of the grass type, the leaf ensheathing the other young foliar leaves in the growing stem tip of the seedling.

coleorhiza a part of the grass embryo that sheathes the root tip.

collateral bundle a vascular bundle in which the phloem and xylem lie on the same radius, with the phloem toward the periphery of the stem and the xylem toward the center.

collenchyma living cells in which secondary wall thickening is more or less limited to the angles of the cell.

colloid a substance composed of particles ranging in size from about $\frac{1}{1,000,000}$ to $\frac{100}{1,000,000}$ mm in diameter, dispersed through some medium; a liquid colloid is called a sol, and a solid colloid, a gel; a state of fine division.

columella in mosses, the central sterile axis of the capsule or sporangium; also used to designate the central axis of certain fruits; in fungi, the sterile axial body within a sporangium.

community a group or assemblage of plants living in the same environment and interacting with each other.

companion cell a parenchyma cell found in phloem tissue and associated with sieve-tube cells.

complete flower a flower that has all parts present, i.e., sepals, petals, stamens, and pistils.

compound leaf a leaf, the blade of which is made up of a number of leaflets, either palmately or pinnately compound.

compound pistil a pistil made up of two or more united carpels.

concave cell (see separation disk).

concentric bundle a vascular bundle in which one type of vascular tissue surrounds the other, xylem around phloem or phloem around xylem.

conceptacle in some algae (*Fucus*), a part of the vegetative growth surrounding reproductive organs; sometimes used to designate the moss capsule or a peridium of fungi.

conduplicate referring to a type of vernation in which a leaf is folded along a center line or midrib and the two halves of the face of the blade are brought together; also used in phylogeny to indicate the evolution of a leaf to form a carpel.

conduplicate carpel a carpel formed by the folding of a leaf along a center line; a phylogenetic development interpreted to be primitive and simple.

cone a strobilus; the sporangia-bearing structures of conifers and other vascular plants.

conidiophore a specialized hypha or sporophore bearing conidia.

conidium a thin-walled, asexual, one-to-many-celled reproductive spore of fungi, usually formed by the cutting off of terminal or lateral cells of conidiophores or by the breaking up of hyphae into separate units.

conifer a common name applied to the members of the order Coniferales, within the Gymnospermae; a cone-bearing gymnosperm.

conjugation the fusion of similar gametes (isogametes).

convolute referring to a type of vernation in which the leaf is rolled like a scroll, as in roses.

cordate heart-shaped.

cork a suberized tissue, the phellem, developed from the phellogen or cork cambium of stems and roots; the commercial product secured from cork oak trees grown extensively in southern Europe.

corm a short, thickened, upright, underground stem with scaly or papery leaves.

corolla a collective term to designate all the petals of a flower.

cortex the primary parenchyma tissue lying between the epidermis and the vascular tissue of roots and stems.

corymb an inflorescence like a raceme except that the pedicels of the flowers become shorter from base toward the apex, resulting in a flat-topped or convex flower cluster, as in candytuft.

cotyledon a seed leaf; the first leaf or leaves of an embryo.

crenate having the margin cut into rounded scallops; a kind of leaf margin.

crossing-over an exchange of corresponding segments between chromatids of homologous chromosomes.

cross-pollination the transfer of pollen from the stamens of one plant to the stigmas of another plant; or from the staminate cones of one plant to the ovulate cones of another plant.

crustose lichen a lichen in which the habit of growth is such that it is flat and closely appressed to the substratum.

cryptogam an older term used to designate that part of the plant kingdom not producing flowers or seeds.

cultivar a kind or type of organism originating and persistent under cultivation.

cupule a cuplike structure in which vegetative reproductive bodies, gemmae, are produced on the thallus of several liverworts.

cuticle a waxy layer found on the outer walls of epidermal cells.

cutin a waxy substance found on the outer walls of epidermal cells, especially of leaves, flower parts, and fruits.

cycad a member of the order Cycadales of the Gymnospermae, commonly used for decorative purposes. Some are often mistaken for palms.

cyme an inflorescence in which the flowers arise from terminal buds, sometimes forming a flat-topped or convex cluster in which the central flowers bloom first, as in some members of the pink family.

cystocarp the rather complicated structure formed around and involving the carpogonia of red algae such as *Polysiphonia.*

cytase an enzyme or enzyme system involved in the digestion of hemicelluloses.

cytochrome oxidase an enzyme system involved in the final stages of aerobic respiration in which hydrogen is united with oxygen to form water.

cytochromes metalloprotein respiratory pigments in which the effective group is an iron-containing hematin; usually designated as cytochromes a, b, and c. They are part of the enzyme system, cytochrome oxidase, and act as hydrogen carriers in oxidation-reduction reactions.

cytokinesis cell division as distinguished from mitosis or nuclear division.

cytology the science dealing with the cell.

cytoplasm the complicated living material of a cell exclusive of the nucleus.

cytosine a pyrimidine constituent of nucleic acids.

cytosome the cytoplasm of a cell.

2,4-D (2,4-dichlorophenoxyacetic acid) a growth substance widely used as a herbicide.

deciduous referring to trees and shrubs that lose all their leaves at the end of the season.

dehiscence splitting open of a capsule or other plant part at maturity.

dehiscent fruit a fruit that splits open at maturity.

dehydrogenase an oxidizing enzyme effective in removing hydrogen from certain organic compounds.

deliquescent literally, melting away; applied to trees having trunks which rise for some distance above ground and then divide into branches which, in turn, branch again and again.

denitrification the breaking down of nitrates to nitrites and ultimately to free gaseous nitrogen by soil organisms, especially denitrifying bacteria.

dentate toothed, with the sharp points directed outward, as in some leaf margins.

deoxyribonucleic acid (DNA) a nucleic acid consisting of the sugar deoxyribose, together with phosphate, adenine, guanine, cytosine, and thymine.

deoxyribose a 5-carbon sugar with one oxygen atom less than the related sugar ribose; a component of deoxyribonucleic acid (DNA).

desmid any of a group of microscopic, unicellular green algae belonging to the family Desmidiaceae.

determinate referring to an inflorescence with a single terminal flower opening before those below, as in a cyme.

development differentiation accompanying growth.

dextrose (glucose, $C_6H_{12}O_6$) a monosaccharide sugar of wide occurrence in plants.

diadelphous referring to stamens, combined in two, often unequal, parts.

diarch having two protoxylem points or groups.

diastase an enzyme complex involved in the hydrolysis of starch to sugar.

diatom any of a group of unicellular or colonial golden-brown algae with silicious cell walls which fit together like halves of a pillbox.

dicaryon (dikaryon) the binucleate condition of a mycelium following a fusion between two sexual cells and before nuclear fusion occurs, as found in Ascomycetes prior to the formation of ascospores.

dichotomy a system of branching in liverworts and other plants in which the main axis forks repeatedly into two branches.

dicotyledon any of a class of plants with embryos which have two cotyledons.

dictyostele a dissected siphonostele; one consisting of a network of separate vascular units resulting from leaf gaps.

differentiation a progressive change from a formative to a mature state as when a meristematic cell changes into a xylem cell or a phloem cell.

diffuse-porous wood wood in which the vessels are more or less uniform in diameter and in distribution throughout the annual rings.

diffusion the movement of molecules resulting from their kinetic energy and tending to cause them to distribute themselves equally throughout a medium.

diffusion pressure deficit (suction force) the term applied to the net force causing water to enter a cell; the water-absorbing power of the cell.

digestion the enzymatic processes involved in rendering substances soluble and diffusible, thus making it possible for them to be transported or utilized in the general metabolism of the plant.

dihybrid an individual heterozygous for two genetic factors.

dioecious literally, two households; having staminate and pistillate flowers on separate plants or having male and female organs on separate plants.

dipeptide an organic compound consisting of two amino acids, the NH_2 group of one being united with the carboxyl group of the other.

diphosphopyridine nucleotide (DPN) a coenzyme derived from the vitamin nicotinic acid and functioning in dehydrogenation reactions. Now usually referred to as nicotinamide adenine dinucleotide (NAD).

diplococcus a form of bacterium consisting of two spherical cells (cocci) remaining together.

diploid having a double set of chromosomes per cell as in the sporophytic generation; twice the haploid number.

disaccharide a sugar consisting of two units of a monosaccharide.

DNA abbreviation for deoxyribonucleic acid (which see).

dominant gene that gene which, when present in a hybrid with a contrasting gene, completely dominates in the development of the character it controls. Thus in peas the character roundness is dominant over wrinkledness.

DPN abbreviation for diphosphopyridine nucleotide (which see).

drupe (stone fruit) a simple, fleshy, mostly one-seeded fruit in which the exocarp is usually skinlike, the mesocarp fleshy, and the endocarp hard and stony; examples: peach and plum.

ecology the study of plants in relation to their environment.

ecosystem assemblages of functional units that exist in an organization in which individuals bear a characteristic association with each other and with the environment.

ecotone the overlapping area of two contiguous plant communities.

ectophloic siphonostele a siphonostele in which the

xylem cylinder lies next to the pith and is surrounded by the phloem cylinder.

ectoplast (plasma membrane) the peripheral cytoplasm of a protoplast, lying next to the cell wall.

egg a female gamete.

elater in liverworts, an elastic, spirally twisted filament occurring among the spores in the sporangium; in *Equisetum,* four hygroscopic bands attached to the spores and serving for dispersal.

electrolyte a substance which dissociates into ions in water solutions.

embryo a young sporophyte plant still attached to the plant or part of a plant producing it; the young plant within a seed.

endarch literally, inner origin; referring to the origin or position of the protoxylem relative to the metaxylem, the metaxylem developing outwardly from the protoxylem.

endocarp the inner layer of a pericarp (fruit wall).

endodermis a tissue of roots and of some stems, consisting of a layer of cells, often with Casparian strips on the radial and transverse walls, lying on the inner side of the cortex.

endoplast the central cytoplasm of the protoplast, as opposed to the ectoplast.

endosperm the nutritive tissue developed around an embryo in the ovule. In some plants the endosperm is consumed by the embryo before the seed matures, as in beans and peas.

endosperm nucleus the nucleus resulting from the fusion of one of the sperms with the two polar nuclei in the female gametophyte of angiosperms. It gives rise to endosperm.

endothecium a tissue in an anther just under the epidermis, involved in the dehiscence of the anther and the inner or central tissue of a moss capsule.

endothermic referring to a chemical reaction in which heat or energy is absorbed or stored, as in photosynthesis.

enzyme any of a class of organic substances, protein in nature, produced by living organisms and functioning as catalysts in metabolic reactions.

epicotyl the part of the axis of an embryo above the region of attachment of the cotyledons.

epidermis the outer, usually cutinized layer of cells of all parts of a young plant and of some parts of older plants, such as leaves and fruits.

epigyny a condition of a flower in which a floral tube, made up of the bases of the other floral parts, is united with (adnate to) the wall of the ovary, thereby making the other parts appear to arise from the top of the ovary.

epiphyte a plant growing attached to another plant or object but not parasitic; an air plant.

ergot the disease of cereals and other grasses caused by species of *Claviceps;* also the fungus itself or the sclerotia of the fungus.

esterases a class of enzymes involved in the hydrolysis (digestion) of fats and other esters.

etiolation the blanched, elongated, spindly condition of plants resulting from lack of light and other causes.

euphyll an ordinary foliage leaf.

eusporangium a sporangium developed from several initial cells and found in all vascular plants except the leptosporangiate ferns.

eustele a stele in which there are separate bundles of vascular tissue alternating with fundamental tissue throughout the axis.

evergreen a plant that does not shed all its leaves at the end of the growing season.

exarch literally, outer origin; referring to the origin or position of the protoxylem relative to the metaxylem, the metaxylem developing centripetally, or toward the center, as in all roots.

excurrent referring to a type of branching in trees in which there is one main vertical stem or trunk tapering from base to summit from which lateral branches radiate outward, giving the tree a conical shape.

exocarp the outer layer of a pericarp (fruit wall).

exothermic referring to a chemical reaction in which heat or energy is released, as in respiration.

extracellular digestion digestion outside the cell.

eyespot a small, pigmented body sensitive to light and present in motile algae and in the reproductive cells of certain other algae.

F_1 first generation resulting from mating parents, plant or animal, differing in specific traits; first generation following hybridization.

F_2 second generation following hybridization.

FAD abbreviation for flavin adenine dinucleotide (which see).

family in classification, a group of genera.

fascicular cambium the cambium within a vascular bundle.

fat an organic compound used as a food and consisting of carbon, hydrogen, and oxygen, with relatively more carbon and hydrogen and less oxygen than a carbohydrate; glycerol esters of fatty acids, insoluble in water but soluble in ether and chloroform.

fatty acid any of a group of organic acids, the higher members of which occur in natural fats; examples: stearic acid and palmitic acid. The lower members include formic acid and acetic acid.

fermentation an oxidative process occurring in the absence of free oxygen; anaerobic respiration.

ferredoxin according to Arnon, an iron-containing protein which functions as an electron carrier on the "hydrogen side" of pyridine nucleotides and plays a key role in the energy-transfer mechanisms of photosynthesis.

fertilization fusion of gametes in sexual reproduction.

fertilizer an organic or inorganic material added to the soil to provide essential inorganic substances for plants or to improve the physical condition of the soil.

fiber an elongated, thick-walled, strengthening cell found in many parts of plants.

field capacity the condition of a soil with respect to its water content after all the gravitational or free water has drained away.

filament the stemlike portion of a stamen bearing the anther; any threadlike body, as in algae; a slender row of cells.

flagellum a long, slender, whiplike extrusion of protoplasm by which unicellular organisms, swarm spores, and zoospores move about in liquid media; of two types, smooth and tinsel.

flavin adenine dinucleotide (FAD) a coenzyme, derivative of vitamin B_2 (riboflavin), functioning as an electron or hydrogen carrier in oxidation-reduction reactions involved in aerobic respiration.

flavin mononucleotide (FMN) a derivative of vitamin B_2 (riboflavin) functioning in electron transfers in photosynthesis and respiration.

flavone a water-soluble yellowish pigment related to the anthocyanins.

floridean starch a type of starch or polysaccharide found in the red algae. In the iodine-starch reaction it gives a wine-red or reddish-violet color instead of a deep blue-violet characteristic of ordinary starch.

florigen a hormone effective in flowering of plants.

flower the structure involved in the sexual reproductive processes of angiosperms. It consists of a short stem bearing specialized leaves, some of which are sporophylls.

flower bud a bud that develops into a flower.

fluorescence the emission of light (radiation) by a substance or body while exposed to the action of certain rays of the spectrum and distinct from the reflected or transmitted radiation of the source radiation.

FMN abbreviation for flavin mononucleotide (which see).

foliose lichen a leafy lichen.

follicle a simple, dry, dehiscent fruit derived from a simple pistil and splitting open at maturity along one suture, as in milkweed.

food an organic substance (carbohydrate, fat, or protein) furnishing energy or building material for protoplasm.

food material the raw material out of which foods are made.

foot in ferns, mosses, and liverworts, the basal part of the young sporophyte which attaches it to the gametophyte.

fossil an impression or the remains of a plant or animal of a past geological age changed to a stony consistency.

free central placentation a type of placentation in which the ovules are all attached to a central axis free from the ovary wall.

frond the leaf of a fern and of some other plants with large leaves.

fructose ($C_6H_{12}O_6$) fruit sugar, a monosaccharide found in many plants.

fruit a structure consisting of one or more matured ovaries, together with any accessory structures adhering to them.

fruit bud a bud giving rise to a flower or flowers and fruit.

fruiting body a term applied to any reproductive structure of plants.

fruticose lichen a lichen that is erect and highly branched.

fucoxanthol (fucoxanthin) a brown xanthophyll pigment; one of the principal pigments of the brown algae.

fungicide a substance that destroys fungi or inhibits the growth of the spores or hyphae.

fungus any of a group of thallus plants lacking chlorophyll and hence subsisting on other plants or animals or on organic matter. Its vegetative body is called a mycelium.

funiculus the stalk of an ovule.

funori a phycocolloid found in some red algae; of economic importance, as in the preparation of adhesives.

gametangial copulation the fusion of two gametangia or their protoplasts, giving rise to a zygote.

gametangium a structure bearing gametes.

gamete a cell capable of uniting with another in sexual reproduction to produce a zygote.

gametophore a structure in bryophytes on which the sex organs are borne.

gametophyte the gamete-producing plant or generation.

gamopetaly (see sympetaly).
gamosepaly (see synsepaly).
gel a jellylike colloid.
gemma an asexual, vegetative structure arising from a thallus, as in liverworts, and capable of developing into a new plant.
gene material carried in chromosomes, determining one or more hereditary characters; a part of a DNA molecule.
generative nucleus the nucleus of pollen grains, which by division forms the sperms.
genetics the science of heredity.
genome the group or "set" of chromosomes present in a gamete.
genotype the genetic makeup of an organism, determined by the assemblage of genes it possesses.
genus a group of closely related species.
geotropism a growth curvature induced by gravity.
germination the beginning of growth of a spore, seed, or other structure.
germ tube a tubular process from a germinating spore, developing into a hypha.
gibberellic acid a growth substance obtained from the fungus *Gibberella fujikuroi.*
gills the platelike structures on the undersurface of the pileus of agarics.
globule the spherical antheridium in the Charales.
glucose (dextrose or grape sugar, $C_6H_{12}O_6$) a monosaccharide found in many plants.
glucoside an organic compound yielding glucose on hydrolysis.
glume an outer, lowermost bract of a grass spikelet.
glutathione an autooxidizable substance involved in certain oxidations in the cell.
glycerol a sweet, sirupy trihydroxyalcohol, colorless and odorless, formed by the digestion of fats and fixed oils in the plant.
glycogen a white, amorphous, tasteless carbohydrate related to starch and found in the liver of most animals and in some plants, particularly some algae and fungi.
glycolysis the enzymatic oxidative breakdown of sugar to simpler compounds without involving free oxygen; the early stages of respiration.
glycoside an organic compound yielding a sugar on hydrolysis.
Golgi bodies small structures in cell cytoplasm, possibly functioning in the formation of vesicles in which secretory products are stored.
gonidium a propagative cell produced asexually and separating from the parent; in lichens, an algal cell of the thallus.
gonimoblast filaments filaments which are often clustered and which arise from the fertilized carpogonium of certain algae.
grain (caryopsis) a simple, indehiscent, dry, single-seeded fruit in which the seed coat is fused to the pericarp over its entire surface, as in the cereal grasses.
gram-negative (see Gram reaction).
gram-positive (see Gram reaction).
Gram reaction a method of differential staining of bacteria by treating them with a special iodine solution after they have been stained with gentian violet. Certain species (gram-positive) retain the purple dye and others (gram-negative) are decolorized, thus affording a basis for classification.
grand period of growth the total period of enlargement of an organ or a plant.
granum (pl. grana) a small unit of a chloroplast functioning in photosynthesis.
gravitational water (free water) that part of the water falling on a soil that sinks in response to gravity.
ground meristem a primary meristem which gives rise to pith, pith rays, and cortex.
growth an increase in mass or volume of an organism, accompanied by an irreversible change in form and structure, all resulting from the activities of protoplasm.
growth by accretion growth by the synthesis of new substances within the protoplasm.
growth by distension growth by the absorption of materials from the exterior of the protoplast.
growth substance an artificial or natural substance which in minute quantities affects the growth and often other physiological activity of plants.
guanine a purine constituent of nucleic acids.
guard cells the two specialized chlorophyllous epidermal cells enclosing a stoma.
gum any of a class of colloidal, water-soluble substances, gluey when wet but hardening on drying, exuded by or extracted from plants.
guttation the giving off of water in liquid form by plants, usually through special structures called hydathodes.
gymnosperm a plant that bears naked seeds without an ovary.
gynoecium the pistil or collective pistils of a flower.

halophyte a plant which grows naturally in soil impregnated with salt, particularly on seacoasts and alkaline deserts.
haploid (monoploid) having a single complete set of chromosomes per cell or half the number found in the somatic cells of the sporophytic generation;

usually abbreviated as *N*, and referring to the cells of the gametophytic generation.

hardening the treatment of plants in such a way as to increase their resistance to low temperature or drought.

hardwood trees trees belonging to the angiosperms; a commercial term.

haustorium a special branch of a hypha penetrating plant cells and serving as an organ of attachment or absorption; in vascular plants, a parasitic root.

head an inflorescence consisting of a dense cluster of nearly sessile flowers on a very short axis, as in dandelion or sunflower.

heartwood the older, central part of the xylem of a tree, usually darker in color than the sapwood.

heat rigor a coagulation or gelation of the protoplasm of cells resulting from their being subjected to supramaximal temperatures.

helicoid cyme a cyme in which there is only one bractlet to a pedicel and the position of this occurs spirally at the base of successive pedicels; found in monocots.

hemicellulose any of a group of polysaccharides less complex than cellulose and easily hydrolyzable to simple sugars.

herb a plant with no persistent woody stem aboveground; also a plant used in medicine or in seasoning.

herbaceous having the characteristics of an herb and contrasting with woody.

herbaceous perennial a herbaceous plant with perennial underground parts which develop aerial shoots each year.

herbarium a collection of dried, pressed plants; also the room or building in which such a collection is kept.

heredity the transmission of genetic morphological, physiological, or mental characteristics from parent to offspring.

hesperidium a berry with an outer, leathery, separable rind, e.g., an orange.

Heterobasidiomycetes a group of parasitic Basidiomycetes to which the rusts and smuts belong.

heterocyst a large, inert cell in the filaments of blue-green algae that separates contiguous hormogonia.

heteroecious requiring two separate hosts to complete a life cycle, as in some rusts.

heterogametes gametes that are dissimilar.

heterogamy fusion of gametes that are dissimilar.

heterosis the diverse effects following a cross between heterozygous parents, particularly increased vigor (hybrid vigor) or capacity for growth displayed by crossbred animals or plants.

heterosporous producing microspores and megaspores.

heterothallic referring to plants (chiefly algae and fungi) in which the male gametangia are produced on one filament or plant and the female gametangia on another.

heterotrichous in some algae, having a plant body divided into a prostrate portion and an upright or erect portion.

heterotrophic plant a plant that is incapable of synthesizing carbohydrates out of CO_2 and water and hence is either a saprophyte or a parasite.

heterozygous having contrasting genes of a gene pair present in the same organism, e.g., a yellow-seeded pea plant with genes for yellowness (*Y*) as well as for greenness (*y*); yellowness being dominant (compare with homozygous).

hexose a 6-carbon sugar.

hilum the scar left on a seed at the place where the seed has broken away from the funiculus.

hip the fruit of the rose, consisting of a fleshy floral tube surrounding a group of achenes.

holdfast a specialized rootlike structure in some algae; serves in attachment of the alga to a substratum.

holoenzyme a complete or intact enzyme, consisting of apoenzyme and coenzyme.

homologous chromosomes a pair of chromosomes essentially alike in size and shape, each derived from a different parent and carrying genes for the same characters.

homologous theory a theory of alternation of generations which postulates that plants have evolved from ancestral algal stock in which both gametophytic and sporophytic phases were already differentiated.

homosporous producing meiospores of one type only, as compared with a heterosporous condition.

homothallic referring to a thallus (plant) producing compatible gametes that fuse.

homozygous having identical genes of a gene pair present in the same organism, e.g., a yellow-seeded pea plant with genes (*YY*) for only yellowness (compare with heterozygous).

hormogonium a section of a filament of the blue-green algae separated from another section by a heterocyst.

hormone a substance naturally produced in one part of an organism and transported to another part where, in extremely minute quantities, it is capable of producing marked physiological effects.

host the plant or animal on or in which a parasite exists.

humus decomposing organic matter in the soil.

hybrid the offspring of parents that differ genetically.

hybridization producing hybrids.

hydathode (water pore) a structure through which liquids are exuded by guttation.

hydrarch referring to plant successions from ponds, lakes, or other bodies of water.

hydration union of a chemical substance with water without chemical decomposition.

hydrolase an enzyme capable of accelerating hydrolysis.

hydrolysis a change or splitting of compounds into simpler compounds by the chemical addition of water.

hydrophilic water-loving, applied to certain colloids.

hydrophyte a plant which grows in water or in a saturated soil.

hydroponics the growing of plants in water solutions of inorganic substances.

hydrotropism a growth movement or bending resulting from the stimulus of water or water vapor.

hygroscopic moisture the very fine film of water adhering to soil particles in an air-dry soil.

hymenium the fruiting surface or layer of an ascocarp or basidiocarp.

hypanthium an enlargement or development of the torus under the calyx; a floral tube.

hypha a single filament of the mycelium of a fungus.

hypocotyl the axis of an embryo below the cotyledons.

hypogyny a condition of flowers in which stamens, petals, and sepals arise from the receptacle below, and entirely free from, the ovaries of the pistils.

imbibition the absorption of water by colloidal substances; one of the methods by which root hairs and other plant parts obtain water.

imperfect flower a flower that lacks either stamens or pistils.

incomplete flower a flower lacking one or more of the four sets of parts (calyx, corolla, stamens, or pistils).

indehiscent fruit a dry fruit that does not split open at maturity.

indeterminate inflorescence a type of inflorescence in which the apex continues to grow, the flowers being borne laterally in the axils of leaves or bracts, with the youngest nearest the apex.

indusium in ferns, a membrane which covers or invests the sorus.

inferior ovary an ovary in which a floral tube is united with the ovary wall, the ovary thus appearing to lie below the other floral parts.

inflorescence the disposition of the flowers on the floral axis; a flower cluster.

insectivorous plant a plant that captures and digests insects, e.g., a pitcher plant.

integument(s) the outer covering of an ovule, which becomes the seed coat.

intercalary meristem a meristem usually occurring at the bases of the internodes of stems of grasses and other plants, being parts of the apical meristem which have become separated from the apex by differentiated tissues.

interfascicular cambium the cambium that develops between vascular bundles.

internode the part of a stem between two successive nodes.

intracellular digestion digestion occurring within a cell.

inulin a white, tasteless polysaccharide, yielding fructose on hydrolysis and found especially in the roots, tubers, and rhizomes of some plants.

invertase (also called saccharase, sucrase, or *β-d*-fructosidase) an enzyme which accelerates the hydrolysis of sucrose to glucose and fructose.

involucre a ring of bracts surrounding an inflorescence, as in some composites, or surrounding the nut type of fruit, as in hazelnuts.

involute referring to a type of vernation in which both edges of the leaf are inrolled lengthwise on the upper surface, toward the midrib, as in violets.

ion an electrically charged atom or group of atoms.

irregular flower a bilaterally symmetrical or zygomorphic flower.

irritability ability to respond to a stimulus, a property of living protoplasm.

isodiametric having equal diameters in all directions.

isogametes gametes that are similar in size and structure.

isogamy the fusion of gametes that are similar in size and structure.

isomorphic literally, like form, as when the plants of the haploid and diploid generations are similar in size and appearance.

isotope one of several possible forms of a chemical element differing from other forms in atomic weight but not in chemical properties.

isthmus the narrowed connection between half cells of desmids.

karyolymph the nuclear gel or nuclear sap.

kelp a large brown alga.

kinetin a growth substance prepared from deoxyribonucleic acid, which promotes cell division in plants.

kinin any substance that promotes cell division.

kombu a food made from kelp and usually eaten with fish or other meat.

Krebs (citric acid) **cycle** a series of oxidation-reduction reaction whereby, step by step, hydrogen and CO_2 are liberated from a series of organic acids through the action of various enzymes in the aerobic oxidation of pyruvic acid in respiration. The Krebs cycle is involved also in other metabolic activities.

lactiferous duct a tube or duct consisting of either latex vessels or latex cells, carrying latex.

lamina the blade of a leaf.

lanceolate referring to leaves shaped like a lance head, that is, several times longer than wide, broadest toward the base, and narrowed to the apex.

latex a milky, usually white, fluid found in the lactiferous ducts of certain plants and used commercially in the production of rubber, chicle, and gutta-percha.

latex cells living, coenocytic cells of lactiferous ducts.

latex vessels elongated cells joined end to end that form one type of lactiferous duct.

leaf one of the usually green, expanded, lateral outgrowths of a stem often consisting of a blade, petiole, and sometimes stipules.

leaf gap a gap or break in a vascular cylinder above a leaf trace caused by the diversion of the vascular tissue from the stem into a leaf.

leaflet one of the separate divisions of the blade of a compound leaf.

leaf mosaic an arrangement of leaves caused by one-sided illumination in which the blades all face toward the illuminated side.

leaf primordium a lateral divergence from an apical meristem which develops into a leaf.

leaf scar the scar left on a stem by a fallen leaf.

leaf trace a vascular strand passing from the vascular cylinder of a stem into a leaf.

legume a simple, dry, dehiscent fruit developed from a simple pistil and splitting at maturity along two sutures; also used for any member of the family Leguminosae.

lemma in the grasses, the lower of the two bracts enclosing the flower.

lenticels multicellular structures found in the outer corky layer of stems, appearing as dots or ridges and functioning in gaseous exchange.

leptosporangium a sporangium derived from one superficial cell, as in the common ferns.

leucoplast a colorless plastid.

liana any climbing plant that roots in the ground and depends upon a tree or other plant for mechanical support, as a woody liana characteristic of tropical rain forests.

lichen any of a group of thallus plants growing usually as epiphytes on rocks, bark, or other objects and consisting of a fungus and an alga living together.

lignin a complex organic constituent of the walls of some cells, especially those of the xylem, combining with cellulose to form lignocellulose.

ligule a projection from the summit of the leaf sheath in grasses; also the flattened spreading limb of some marginal or ray flowers of the Compositae.

linkage the tendency of certain genes (or the corresponding characters) to remain associated in inheritance because of their location in the same chromosome.

lipase an enzyme that hydrolyzes fats to fatty acids and glycerol. Under certain conditions it can reverse this action and synthesize fats out of glycerol and fatty acids.

liverwort any plant belonging to the class Hepaticae.

locule a cavity in an ovary in which the ovules lie.

loculicidal dehiscent into the cavity of a locule of an ovary through the dorsal suture, as applied to the dehiscence of capsules.

lodicule one of two or more small parenchymatous scales at the base of the flower of a grass.

loment a segmented legume as in tick trefoil.

longevity length of life.

luteol a xanthophyll pigment commonly found in leaves.

lycopene a carotenoid pigment found in the fruit of the tomato and of other members of the nightshade family.

Lycopsida a subdivision of the Tracheophyta to which the club mosses and quillworts belong.

macrandrous in algae, having large or long male plants; applied to sexual reproduction when the antheridia arise on long or regular-sized filaments.

macro- a combining form meaning unusually large or long.

macroconidium a large conidium produced usually at a different period or in a different sporocarp from the one producing microconidia.

maltase an enzyme which hydrolyzes maltose to glucose.

maltose ($C_{12}H_{22}O_{11}$) malt sugar, a disaccharide made up of two molecules of glucose.

medullary ray (pith ray) an extension of groups of pith cells between the vascular bundles.

mega- a combining form meaning great.

megagametophyte the female gametophyte.

megaphyll a leaf with complex venation evolved from a major branch system.

megasporangium a sporangium which produces megaspores.

megaspore the meiospore of vascular plants from which develops the female gametophyte.

megaspore mother cell (megasporocyte) a diploid cell which by meiosis produces four megaspores.

megasporocyte a megaspore mother cell, from which megaspores are developed by meiosis.

megasporophyll a specialized leaf bearing megasporangia.

meiocyte a cell capable of undergoing meiotic division.

meiosis a process consisting of two nuclear divisions by which the number of chromosomes is reduced to one half; associated with production of gametes in animals and meiospores in plants.

meiospore a spore resulting from meiosis.

mericarp one of the two or more units of a schizocarp.

meristem undifferentiated tissue made up of parenchyma cells capable of active division and differentiation into other types of cells.

meristoderm outer (epidermal) meristematic layer in some Phaeophyta.

mesarch a condition of the primary xylem in which the metaxylem differentiates in all directions from the protoxylem points, as found in ferns.

mesocarp the middle layer of a pericarp.

mesophyll the interior of a leaf between the upper and lower epidermal layers.

mesophyte a plant that grows under medium conditions of moisture, neither too dry nor too wet.

Mesozoic the geological era between the Paleozoic and the Cenozoic and lasting approximately 130 million years.

messenger RNA RNA produced in the nucleus. It moves from the nucleus to the ribosomes of the cytoplasm and determines the sequence of amino acids in the protein molecule.

metabolism the sum of the processes involved in the building up and tearing down of protoplasm; the synthetic, or building-up, processes constitute the anabolic phase of metabolism; the tearing-down processes, the catabolic phase.

metaphase the stage of mitosis in which the chromosomes are aggregated at the equator of the spindle.

metaphloem primary phloem formed after the protophloem has been differentiated.

metaxylem primary xylem formed after the protoxylem has been differentiated.

micro- a combining form meaning small.

microconidium a small conidium produced usually at a different period or in a different sporocarp from the one producing macroconidia.

microelement a trace element; an inorganic substance that is essential to a plant in extremely minute amounts, such as copper, molybdenum, and zinc.

microgametophyte a male gametophyte.

micron (symbol μ) a unit of measurement of length or distance; $1\ \mu = 0.001$ mm or 0.000039 in.

microphyll a leaf with simple vascularization evolved from a minor branch system.

micropyle the small pore or opening in the integuments of an ovule through which the pollen tube enters to reach the female gametophyte.

microsporangium a sporangium that produces microspores.

microspore a spore which gives rise to the male gametophyte in vascular plants; in seed plants, the young pollen grain.

microspore mother cell (microsporocyte) a diploid cell which by meiosis produces four microspores.

microsporocyte the microspore mother cell which by meiosis gives rise to four microspores.

microsporophyll a specialized leaf bearing microsporangia.

middle lamella a layer of pectic material, between two adjacent protoplasts and remaining as a cementing layer between adjacent cell walls.

midrib the central and principal vein of a pinnately veined leaf.

millimicron (symbol mμ) a unit of measurement commonly used in measuring wavelength of radiation; $1\ \text{m}\mu = 0.001\ \mu = \frac{1}{1{,}000{,}000}$ mm.

mitochondria (chondriosomes) small, often rod-shaped, cytoplasmic particles containing respiratory enzymes and coenzymes.

mitosis nuclear division, involving replication of chromosomes and their separation into two groups of equal numbers to form two daughter nuclei.

mixed bud a bud containing both rudimentary leaves and rudimentary flowers.

mole (mol) a molecular weight in grams.

molecule a compound made up of atoms held together by covalent bonds.

monadelphous referring to stamens, united by their filaments into a tube or column.

monocotyledon a plant, the embryo of which has only one cotyledon.

monoecious with stamens and pistils in separate flowers on the same plant; or having microsporangia and megasporangia borne on the same plant.

monohybrid an individual heterozygous for a single genetic factor.

monoploid literally, one fold, having one set (the basic number) of chromosomes per cell; haploid.

monopodial branching a type of growth of stems with a main axis which continues its line of growth because of annual additions from terminal buds and which gives off lateral branches, as in horse chestnut (compare sympodial branching).

monosaccharide a simple sugar, containing five or six or fewer carbon atoms.

monospore an asexual spore produced on the haploid plants of some of the red algae.

morel any of the edible fungi belonging to the genus *Morchella.*

morphogenesis development of size, form, and other differentiation of organisms.

morphology the science dealing with form and structure and their development in organisms.

moss any member of the bryophytic class Musci; applied loosely to other plants resembling mosses.

multiple fruit one of a class of fruits consisting of the ripened ovaries of an entire flower cluster, together with any accessory structures, all adhering together as a single unit, as in pineapple and mulberry.

mutation a sudden heritable change in an individual resulting from a change in genes or chromosomes.

mycelium a group or mass of hyphae constituting the vegetative body of a fungus.

mycology the science dealing with fungi.

mycorrhiza the association of fungi with the roots of plants. The fungi may cover or penetrate the rootlets.

mycosis a disease of animal tissues caused by fungi.

myxomycete a slime mold.

NAD abbreviation for nicotinamide adenine dinucleotide (which see).

NADP abbreviation for nicotinamide adenine dinucleotide phosphate (which see).

nannandrous referring to algae which have a type of sexual reproduction in which the antheridia are produced on special dwarf filaments, as in some species of *Oedogonium.*

nasties movements of bilaterally symmetrical or flat organs, such as leaves and flower petals, in which the movement is independent of the direction from which the stimulus is applied.

nicotinamide adenine dinucleotide (NAD) a coenzyme derived from the vitamin nicotinic acid and functioning in dehydrogenation reactions; formerly referred to as diphosphopyridine nucleotide (DPN).

nicotinamide adenine dinucleotide phosphate (NADP) the phosphate of NAD, functioning in dehydrogenation reactions; formerly referred to as triphosphopyridine nucleotide (TPN).

nicotinic acid a member of the water-soluble B group of vitamins.

nitrification the oxidation of ammonia and ammonium compounds into nitrites and nitrates through the action of nitrifying bacteria in the soil.

nitrogen fixation the conversion of free gaseous nitrogen into combined form.

node the part of a stem where leaves and axillary buds arise.

nodule an enlargement on the roots of certain plants in which nitrogen-fixing bacteria live.

nomenclature literally, a list of names; the usage of scientific names in taxonomy.

nuclear gel (or karyolymph; also called nuclear sap) the liquid or jellylike part of the nucleus.

nucleic acid one of a class of immense macromolecules often having molecular weights of several million and composed of joined nucleotide complexes; the principal types are deoxyribonucleic acid (DNA) and ribonucleic acid (RNA).

nucleolus a small, often rounded body, found within the nucleus. A nucleus may contain one or more nucleoli.

nucleoside a molecule consisting of a 5-carbon sugar combined with a purine or pyrimidine, i.e., a combination of sugar and a nitrogen base.

nucleotide a molecule consisting of a 5-carbon sugar (either ribose or deoxyribose), phosphate, and a purine or pyrimidine, i.e., a combination of sugar, nitrogen base, and phosphate.

nucleus a protoplasmic structure consisting of an outer membrane, nuclear gel, nuclear reticulum made up of chromatin, and one or more nucleoli; it is the center of much of the physiological activity of the cell and functions in the transmission of hereditary characteristics.

nucule the female sexual structure of *Chara.*

nut an indehiscent, mostly one-seeded, dry fruit in which the pericarp is hard or crustaceous throughout. It is usually derived from a compound pistil, only one carpel of which develops, and is usually surrounded by an involucre, e.g., hazelnut.

nyctinasties nasties brought about by the alternation of night and day; the so-called sleep movements of plants.

oidia short, cylindrical, thin-walled parts of hyphae of some Basidiomycetes, serving as reproductive units or as diploidizing agents.

oil a fat that is liquid at ordinary temperatures.

ontogeny the life history or development of an individual organism.

oögamy (heterogamy) sexual reproduction involving unlike gametes or sperm and egg. The egg is usually immobile.

oögonium the unicellular female sex structure of certain algae and fungi, containing one or more eggs.

oöspore a spore developed from a zygote resulting from the fertilization of an egg by a sperm.

open bundle a vascular bundle with cambium.

operculum in mosses, the lid or cap of the capsule.

order in plant classification, a group made up of families.

organelle part of a cell, subcellular in size, such as chloroplast or mitochondria.

osmosis the phenomenon which results in a difference in rate of movement (or passage through the membrane) of solvent molecules from opposite sides of a selectively permeable membrane separating two solutions of different concentration or separating a solution from its pure solvent.

osmotic pressure of a solution, the maximum hydrostatic pressure that this solution could develop when separated from its pure solvent by a membrane permeable only to the solvent. A molar solution of any nonelectrolyte has an osmotic pressure of 22.4 atm.

ovary the enlarged basal portion of a pistil, containing the ovules and developing into the fruit.

ovate having the shape of the longitudinal section of an egg with the broader end basal.

ovule the immature seed in the ovary, containing before fertilization the female gametophyte surrounded by nucellus and integuments. At maturity the ovule becomes the seed.

oxidase any of a group of enzymes that promote the oxidation of various substances.

oxygen respiration respiration requiring the presence of free oxygen.

palea (or palet) in grasses, the upper bract that subtends the flower.

paleobotany the study of plants of past geological periods.

Paleozoic the geological era between the Proterozoic and the Mesozoic, beginning approximately 500 million years ago and lasting approximately 295 million years.

palisade mesophyll the elongated chlorophyllous cells lying just beneath the upper epidermis of leaves.

palmate leaves leaves having lobes radiating from a common point.

palmate venation a type of venation in which the principal veins branch out from the apex of the petiole like fingers on the hand.

pampas vast treeless plains, especially those of Argentina.

panicle an inflorescence somewhat like a raceme in which the pedicels have branched, the branching being somewhat irregular, as in *Yucca* and many grasses.

pantothenic acid one of the water-soluble B group of vitamins.

papain a proteolytic enzyme obtained from the juice of the green fruit of the papaya or papaw tree.

pappus tufts of hairs on achenes and other fruits; the limb of the calyx of composite florets.

parallel venation a type of venation in which the larger veins of the leaf blade appear to be roughly parallel to each other.

paraphyses sterile filaments occurring in the fruiting bodies of certain fungi and in other lower plants.

parasite an organism that obtains its food from another living organism.

paraspore a type of asexual spore produced by some species of red algae.

paratonic movement movement of parts of plants caused by external stimuli such as gravity, chemicals, heat, light, and electricity.

parenchyma a simple tissue consisting of thin-walled, roughly isodiametric cells.

parietal placentation a type of placentation in which the ovules appear to be attached to the outer wall of the ovarian cavity.

parthenocarpy the development of fruit without fertilization.

parthenogenesis the development of a gamete into a new individual without fertilization.

parthenospore (azygospore) a body resembling a zygospore but not developed from the fusion of the contents of two sexually different cells.

pathogen an organism or virus causing a disease.

pathology the science of diseases, their nature, causes, effects, and treatment.

pectic material chemical substances derived from pectin, a jellylike substance found in fruits and other parts of plants; the chief cementing material of plant cells.

pedicel one of the ultimate divisions of a common peduncle, bearing a single flower.

peduncle the stalk or stem of a flower borne singly, or the main stem of an inflorescence.

penicillin an antibiotic obtained from the fungus *Penicillium.*

pentarch referring to the condition of a stele with five protoxylem points.

pentose a 5-carbon sugar.
pepo an inferior berry with a leathery, inseparable rind, such as a pumpkin.
pepsin a proteolytic enzyme.
peptidase an enzyme involved in the hydrolysis of peptides to amino acids.
peptide a combination of amino acids in which the amino group of one is united with the carboxyl group of another.
peptone any of a class of initial, water-soluble decomposition products of protein digestion, noncoagulable by heat and not capable of being precipitated with saturated ammonium sulfate.
perennial a plant continuing to live from year to year; one that lives more than two years.
perfect flower a flower that bears both stamens and pistils.
perfect stage of Ascomycetes and Basidiomycetes, the stage that produces the organs in which meiosis occurs, i.e., the stage that produces the asci and the basidia.
perianth the external envelope of a flower, i.e., the calyx and corolla together.
pericarp the wall of the ripened ovary or fruit; in red algae, the outer, urn-shaped jacket of the cystocarp.
pericycle a tissue in roots and stems lying between the endodermis and the phloem, and in roots giving rise to branch roots and to the phellogen.
periderm a tissue consisting of cork (phellem), cork cambium (phellogen), and phelloderm.
peridium the outer covering of the sporangia or sporophores of many fungi.
perigyny a condition in flowers in which the bases of the petals, stamens, and sepals are fused, forming a cup or floral tube surrounding the free ovaries of the pistils.
perisperm nutritive tissue of the seed derived from the nucellus.
peristome a ring of teeth surrounding the opening of a moss capsule.
perithecium a spherical, oval, or flask-shaped ascocarp usually with a small opening.
permanent-wilting percentage (wilting coefficient) the percentage of water left in a soil in which plants permanently wilt.
permeable membrane a membrane that permits both solvents and solutes to pass through it.
peroxidase an enzyme involved in the transfer of oxygen from peroxides to substances to be oxidized.
petal one of the individual parts of a corolla.
petiole a leaf stalk.
pH a symbol denoting the relative concentration of hydrogen ions in a solution; pH values extend from 0 to 14; the lower the value, the higher the acidity or the more hydrogen ions the solution contains. pH is actually the negative logarithm of the hydrogen-ion concentration. Water at 25°C has a concentration of H ion of 10^{-7}; the pH therefore is 7.
phanerogam an older term applied to a seed plant or flowering plant when all plants were divided into the two classes, cryptogams and phanerogams.
phellem cork.
phelloderm a secondary tissue formed by cork cambium on the inner side toward the phloem.
phellogen (cork cambium) the tissue that gives rise to the periderm.
phenotype the external or visible appearance of an organism.
phloem a food-conducting tissue consisting usually of sieve tubes, companion cells, phloem parenchyma, and often fibers.
photoautotrophic of organisms, capable of converting solar energy into potential energy.
photonasties nasties caused by changes in light intensity.
photoperiodism growth, development, and flowering of plants in response to the length of their exposure to light.
photosynthesis a process by which chlorophyll-containing cells store radiant energy and liberate oxygen in making carbohydrates out of water and carbon dioxide.
phototropism a growth movement of plants induced by the stimulus of light.
phycobilin (biliprotein) a class of pigments found in both the blue-green and red algae. Although not identical, they produce red and blue colors in both algal divisions.
phycocyanin a blue pigment found in the blue-green algae.
phycoerythrin a red pigment found in the red algae.
phycology (algology) the science dealing with algae.
Phycomycetes a group of fungi formerly considered to be a class but now divided into six classes as follows: Chytridiomycetes, Hyphochytridiomycetes, Oömycetes, Plasmodiophoromycetes, Zygomycetes, Trichomycetes.
phyletic of or pertaining to a phylum or line of descent.
phyllid (phyllidium) "leaf" of moss.
phyllome a general term for all organs which are morphologically leaves, as bracts, scales, and petals.

phyllotaxy the system or order of leaf arrangement.

phylogeny the race history of an animal or vegetable type.

physiology the science dealing with life processes and the functions of the different organs and tissues of living organisms.

phytochrome a blue-green photoreceptive pigment through which light acts to control flowering and other plant-growth phenomena.

phytohormones plant hormones.

phytoplankton a variety of small plants suspended in water.

pileus the umbrellalike cap of a mushroom or similar fungus fruiting body.

pinna leaflet of a fern leaf.

pinnate literally, featherlike; having parts arranged along two sides of an axis.

pinnately compound leaves leaves with a central axis, the rachis, to which all leaflets are attached.

pinnately veined leaves leaves with one main central vein from which all other veins branch off.

pinnatifid pinnately cut, referring to leaf margins.

pistil one of the main parts of a flower consisting usually of stigma, style, and ovary.

pistillate flower a flower that bears pistils but no stamens.

pit a thin place in a cell wall.

pith a parenchymatous tissue occupying the center of stems.

pith ray (medullary ray) an extension of pith into the vascular tissue of the stem between the vascular bundles.

pitted vessel a vessel with pits in the side walls.

placenta the ovarian tissue to which the ovules are attached.

placentation the disposition of the placentae and ovules in the ovary.

plain-sawed wood wood sawed along a tangential plane.

plant food an organic substance (carbohydrate, fat, or protein) furnishing energy or building material for protoplasm.

plant pathology the science treating of plant diseases, their nature, causes, effects, and treatment.

plant succession the replacement of one community of plants by another under changing conditions.

plasma membrane (ectoplast) the peripheral membrane of a protoplast, lying next to the cell wall.

plasmodesmus (*pl.* plasmodesmi; in many textbooks, *sing.* plasmodesma, *pl.* plasmodesmata) protoplasmic connection between adjacent cells.

plasmodium a slimy, naked mass of protoplasm with many free nuclei; the vegetative body of the Myxomycetes.

plasmolysis shrinking of protoplasm away from the cell wall as a result of water loss.

plastid a cytoplasmic body, colorless or pigmented, with definite physiological functions; the types are leucoplasts, chloroplasts, and chromoplasts.

plicate literally folded like a fan; referring to a type of vernation found in palmately veined leaves.

ploidy the degree of replication of chromosomes, e.g., diploidy, polyploidy, etc.

plumule the primary bud of an embryo.

plurilocular sporangia the many-loculed or many-celled sporangia of some of the brown algae.

pollacauophyte a plant that is able to withstand periodic drying.

pollen (pollen grains) the partially germinated microspores or developing male gametophytes shed from the staminate organ of seed plants.

pollen sac the microsporangium of seed plants.

pollen tube the tube, formed by the pollen grain, which grows through stigma, style, ovary, and micropyle to the female gametophyte.

pollination the transfer of pollen from anther to stigma or from a staminate cone to an ovulate cone.

polyadelphous of stamens, united by their filaments into several sets or bundles.

polypeptide a compound consisting of three or more amino acids, the amino group of one being united to the carboxyl group of another.

polypetalous having separate (not united) petals.

polyploid having more than two complete sets of chromosomes per cell.

polypore any of a group of Basidiomycetes in which the basidia line numerous tubes or pores of the basidiocarp.

polysaccharide any carbohydrate consisting of many monosaccharide units.

polysepalous having separate (not united) sepals.

polysiphonous referring to the structure of the plant body of *Polysiphonia,* a red alga; with a central cell surrounded by a number of pericentral cells.

polysomes (polyribosomes) a group of ribosomes with a strand of messenger RNA involved in protein synthesis.

polyspores asexual spores of the red algae, with many produced in each sporangium.

pome a simple, fleshy, accessory fruit in which the outer portion is developed from the floral tube that surrounds the ovary, as in apples and pears.

population a group of interbreeding plants.

poricidal of capsules, opening by means of pores.

porogamy the entrance of the pollen tube through the micropyle of the ovule.

potometer an instrument for measuring the loss of water by transpiration.

primary meristem meristem of the shoot or root tip, giving rise to the primary tissues.

primary root the main root which develops directly from the embryo.

primary tissue tissue developed directly from apical meristem.

primordium the beginning or rudimentary structure of a plant part, as a leaf primordium.

procambium a primary meristem which gives rise to primary vascular and cambial tissue.

procarp the female reproductive organ of the gametophyte in certain red algae.

proembryo a group of cells developed from the division of the zygote before the embryo proper becomes differentiated.

promeristem (primordial meristem) the apical meristem of roots and stems.

prophase an early stage in mitosis in which the chromosomes become distinct prior to their migration to the center of the cell.

prosthetic group an additional substance necessary for the action of certain enzymes, particularly oxidative enzymes.

protease (proteinase) any enzyme capable of hydrolyzing a protein or such protein hydrolytic products as proteoses, peptones, and polypeptides.

proteins complex organic constituents of all living cells, consisting of carbon, hydrogen, oxygen, nitrogen, and usually sulfur and phosphorus. They are made up of chains of amino acids and may have other complex substances attached to them.

proteose any of a class of initial, water-soluble decomposition products of protein digestion, noncoagulable by heat and capable of being precipitated with saturated ammonium sulfate.

Proterozoic the geological era preceding the Paleozoic, beginning approximately 1,400 million years ago and lasting approximately 900 million years.

prothallium the gametophyte of the ferns, sometimes used to designate the early stages of development of any gametophyte.

Protista a name proposed for a third kingdom to which simple, usually unicellular, flagellated organisms are referred.

protoderm the primary meristem that gives rise to the epidermis and, in some instances, to subepidermal cells.

protonema (*pl.* protonemata) the algalike, filamentous structure in the early development of moss gametophytes.

protophloem the first primary phloem developed from the procambium.

protoplasm the living substance of which cells are composed.

protoplast the organized living unit of a single cell.

protostele a stele with a solid central core of xylem, surrounded by phloem.

protoxylem the first primary xylem developed from the procambium in which cell elongation and differentiation most often overlap.

pseudopodium in Myxomycetes, a projection from the plasmodium by which the plant moves from place to place; a gametophytic outgrowth in *Anthoceros* elevating the sporophyte above the location of the archegonium.

Psilopsida a subdivisiion of the Tracheophyta consisting of primitive vascular plants.

Pteridophyta an older term for one of the main divisions of the plant kingdom including the ferns, horsetails, club mosses, and *Selaginella.*

pteridophyte a member of the Pteridophyta.

Pteropsida a subdivision of the Tracheophyta consisting of ferns and seed plants.

puffball a usually rounded basidiocarp in which the spores are retained internally until fully matured, when they can be forced out in the form of a fine dust or powder.

purine a nitrogen base consisting of a double carbon-nitrogen ring as in adenine and guanine.

putrefaction the anaerobic decomposition of organic substances, particularly proteins, brought about by bacteria and other microorganisms.

pycnidium a variously shaped cavity common in some of the imperfect fungi and bearing pycnidiospores which are more often called conidia (not to be confused with pycnium).

pycniospore a spore produced in a pycnium; in the rusts, a sperm cell of either a plus or a minus strain; a spermatium.

pycnium (spermagonium) a globose or flask-shaped structure of the rust fungi, containing the pycniospores (spermatia).

pyrenoids small, rounded, granular bodies occurring within the chloroplasts of certain algae and liverworts and associated with starch formation.

pyridoxine the B_6 vitamin.

pyrimidine a nitrogen base consisting of a single carbon-nitrogen ring as in cytosine, thymine, and uracil.

pyxis a capsule with circumscissile dehiscence.

quarantine a forced stoppage of the transportation of diseased plants from one region to another.

quarter-sawed wood wood sawed along a radial plane so as to expose the radial surface; cf. plain-sawed wood, which is cut along a tangential plane.

raceme a type of inflorescence in which the flowers are borne on an elongated axis on pedicels more or less equal in length.

rachilla the shortened axis of a spikelet.

rachis in compound leaves, the extension of the petiole from which the leaflets arise; in some inflorescences, the main axis from which the pedicels arise.

radial symmetry (actinomorphy) a type of floral symmetry in which the flower may be cut in half in any plane passing through the center to obtain two similar halves, as in tulips.

radicle the part of the embryo that develops into the primary root.

raffinose a trisaccharide found in barley, cottonseeds, beet roots, and other parts of plants.

raphe in seeds in which the funiculus is bent so as to bring the micropyle close to the hilum, the ridge along the testa formed by the fusion of the seed stalk with the testa; in diatoms, the median line or rib of a valve.

receptacle the apex of a pedicel or peduncle, from which the flower parts grow.

recessive character that member of a pair of contrasting characters which, when both are present, is masked by the other, or dominant, character.

reclinate (inflexed) referring to a type of vernation in which the upper part of the leaf is bent down on the lower part, as in the tulip tree.

rennin a proteolytic enzyme.

respiration biological oxidation of materials resulting in the liberation of energy; all catabolic changes involving substantial energy release and any gaseous exchange accompanying this.

reticulum a netlike structure or network.

revolute referring to a type of vernation in which both edges of a leaf are inrolled lengthwise on the lower surface toward the midrib, as in dock.

rhizoid rootlike filaments of some fungi, mosses, liverworts, and ferns that attach the gametophyte to the substratum and absorb water and inorganic substances.

rhizome (rootstalk) a horizontal underground stem.

rhizomorph a rootlike structure; in fungi, a rootlike branched strand of mycelial hyphae.

rhizophore in club mosses, a leafless branch which gives rise to roots when it comes in contact with soil.

riboflavin the B_2 or water-soluble growth vitamin.

ribonucleic acid (RNA) a nucleic acid consisting of the sugar ribose, together with phosphate, adenine, guanine, cytosine, and uracil.

ribosome a submicroscopic particle found in cytoplasm; contains RNA and is the site of protein synthesis.

rickettsia an organism intermediate in size between a virus and a bacterium; parasitizes insects and ticks.

ringed vessel a vessel strengthened by internal rings of lignocellulose.

ring-porous wood wood in which the vessels of one part of an annual ring are greater in diameter than those of the other part of the ring, causing them to stand out as rings of pores.

RNA abbreviation for ribonucleic acid (which see).

root a major organ of a vascular plant. It is devoid of leaves and reproductive organs but is provided with a growing point and functions as an organ of absorption of water and inorganic substances, as a food reservoir, or as a means of support.

root cap a thimblelike mass of cells covering and protecting the apical meristem of a root.

root hairs slender outgrowths from the epidermal cells of young roots that function in the absorption of water and inorganic substances.

root pressure pressure developed in a root by osmotic forces and causing exudation from the cut stump of a severed main stem.

rootstock (rhizome) a horizontal underground stem; in horticulture, a rooted stem base used in grafting.

rosette arrangement of parts (structures) in a circular form, e.g., the phyllids of a moss plant.

runner a stem growing horizontally along the surface of the ground, as in strawberry plants.

rust any of an order of parasitic fungi (the Uredinales) causing discolorations on leaves, stems, and other organs of higher plants; also the diseases caused by these fungi.

saccharase (sucrase or invertase) an enzyme capable of hydrolyzing sucrose to glucose and fructose.

sagittate shaped like an arrowhead, the basal lobes directed downward or backward.

samara a simple, dry, indehiscent, winged fruit, usually with only one seed.

saprobe (saprotroph) an organism (plant or animal) that obtains its nourishment from dead or decaying organic matter.

saprophyte a plant that obtains its food from dead or decaying organic matter.

sapwood the comparatively young xylem of a tree comprising the several annual rings most recently formed, which are usually more active physiologically and lighter in color than the heartwood.

Sarcina spherical bacteria which divide in three planes, forming cubical groups.

savanna a tropical or subtropical grassland containing scattered trees and xerophytic undergrowth.

scalariform having markings suggestive of a ladder, as in scalariform vessels.

schizocarp a simple, dry, indehiscent fruit of the carrot and mallow families, consisting of two or more carpels, which separate at maturity into mericarps, each usually single-seeded.

scion a detached shoot used in grafting.

sclereid a strongly thickened or lignified, usually roughly isodiametric cell.

sclerenchyma a strengthening tissue consisting of thick-walled fibers or stone cells.

sclerotium in fungi, a compact mass of hardened mycelium from which fruiting bodies may develop.

scorpioid cyme a cyme in which the terminal flowers have a single bractlet or none; when the bractlet is present, it occurs alternately to right and to left at the base of successive pedicels so that the structure bears flowers alternately to right and to left along one side. In some cases it coils (Boraginaceae); in others it does not (Portulacaceae).

scouring rush the common horsetail, *Equisetum*.

scutellum the single cotyledon of the grass embryo.

secondary meristem a meristem developed from living parenchyma tissue already differentiated.

secondary root a branch of a primary root.

secondary tissues tissues developed from cambiums or from tissues already differentiated.

seed a ripened ovule consisting of an embryo and stored food enclosed by seed coats.

seedling the young plant developing from a germinated seed.

segregation of genes separation of allelic genes during meiosis.

self-pollination transfer of pollen from anthers to stigmas of the same flower or flowers on the same plant or from the staminate cones to the ovulate cones of the same plant.

semicell one half of a desmid.

semipermeable membrane a membrane that permits a solvent to pass through it but prohibits the passage of many solutes.

sepal one of the individual parts of the calyx.

separation disk dead cells, also called concave cells, that delimit the hormogonia in the filament of *Oscillatoria*.

septicidal of capsules, dehiscing through the partitions and between the locules, thus dividing the capsule into its component carpels.

septifragal of capsules, designating a type of dehiscence by which the valves (lengthwise sections) break from the partitions, leaving these partitions attached to the axis of the fruit.

serrate having sharp teeth pointing forward.

sessile lacking a petiole, as in some leaves, or a pedicel, as in some flowers and fruits.

seta a bristle; in the mosses and liverworts, the stalk which supports the capsule.

sex chromosomes chromosomes that determine sex; X, Y chromosomes.

sex linkage the presence in sex chromosomes of genes that determine characters other than sex; or the association of characteristics with one or the other of the two sexes.

sexual reproduction reproduction that requires meiosis and fertilization to complete a life cycle.

shoot a young branch with leaves; or the aboveground part of a young plant.

shrub a relatively low-growing perennial woody plant with several main stems; a bush.

sieve plate the perforated wall area of a sieve-tube member.

sieve tube a series of specialized cells with sieve plates and joined end to end, forming a tube through which chiefly foods are conducted.

silicle a short, broad silique.

silique the elongated, two-loculed capsule of the mustard family.

simple fruit a fruit developed from a single (simple or compound) pistil.

simple leaf a leaf in which the blade is not divided into leaflets.

simple pistil a pistil consisting of only one carpel.

simple pit a pit not surrounded by an overarching border.

sinus in desmids, the recess between the half cells.

siphonostele a stele with a central pith.

smut a fungus belonging to the order Ustilaginales; also, a disease caused by these fungi.

softwood wood of the conifers.

soil solution soil water containing dissolved substances.

solute a dissolved substance.

solution a liquid consisting of the molecules or ions of one or more solutes homogeneously dispersed among the molecules of a solvent.

solvent the medium, usually liquid, in which another

substance is dissolved; the part of a solution through which the solutes are dispersed.

somatogamy in fungi, the fusion of a plus mycelium with a minus mycelium.

soredium in lichens, a specialized reproductive body consisting of algal cells surrounded by fungous hyphae.

sorus in ferns, a cluster of sporangia; in fungi, a heap of spores or an erumpent spore mass, as in the rusts and smuts.

spadix a fleshy spike or head with small, often imperfect flowers of both types, commonly surrounded by an enveloping sheath called the spathe.

spathe the enveloping sheath of a spadix.

spawn mycelium of mushrooms prepared for propagating purposes.

species a category in classification between a genus and a variety; a group of plants or animals having in common one or more distinctive characters and being capable of interbreeding and reproducing their characters in their offspring, thus remaining relatively stable in nature.

spectrum an image formed when radiant energy is dispersed so that its rays are arranged in the order of their wavelengths, e.g., the visible spectrum, a rainbow, the electromagnetic spectrum.

sperm a male gamete.

spermagonium (pycnium) a flask-shaped structure which produces spermatia (pycniospores).

spermatangium the male sex organ or antheridium of red algae.

spermatid mother cell of antherozoids or motile sperms.

spermatium the male gamete of the red algae; in rusts, a pycniospore or cell borne at the tip of a hypha in the spermagonia (pycnia) which fuses with a cell of opposite strain, following which aecia develop.

Spermatophyta an older term for a main division of the plant kingdom; the seed plants.

spermatophyte a member of the Spermatophyta; a seed plant.

Sphenopsida a subdivision of the Tracheophyta which includes the horsetails and related fossil forms.

spike an inflorescence in which the flowers are sessile on a more or less elongated axis.

spikelet a cluster of one or more flowers subtended by a common pair of glumes, as in grasses.

spine a sharp, pointed structure, usually a specialized leaf or leaf part.

spiral vessel a vessel with a spiral internal band of lignocellulose.

Spirillum a genus of bacteria consisting of long, rigidly spiral or curved rods.

spirochaete a slender, spirally undulating bacterium.

sporangiophore a sporophore bearing a sporangium.

sporangium a spore case.

spore a unicellular or few-celled structure of many types and forms, which is often produced in sporangia and which, with the exception of bacterial spores and meiospores, is usually involved in asexual reproduction.

sporocarp a many-celled body serving for the formation of spores.

sporocyte a cell capable of undergoing meiotic division.

sporogenesis production of spores.

sporophore a spore-bearing structure.

sporophyll a spore-bearing leaf.

sporophyte in alternation of generations, the diploid plant in which meiosis occurs, resulting in the production of meiospores.

springwood the part of an annual ring formed during the early part of the growing season.

stalk any lengthened support of an organ.

stalk cell in conifers, one of the cells formed by the division of the generative cell.

stamen the pollen-producing organ of a flower, usually consisting of a filament and an anther.

staminate flower a flower bearing stamens but no pistils.

staphylococcus any of a group of gram-positive bacteria that form grapelike clusters.

starch a white, odorless, tasteless, water-insoluble polysaccharide made up of glucose molecules; the commonest storage carbohydrate of plants.

stele the central cylinder in the stems and roots of vascular plants; originally applied only to the primary tissues but often considered as the vascular skeleton of the plant.

stem the ascending, usually branched axis of a plant, consisting of nodes and internodes and usually having leaves and buds at the nodes.

steppes vast tracts in southeastern Europe and in Asia, generally level and without forests.

sterigma (*pl.* sterigmata) a small stalk or spiculelike extension of the apex of a basidium, on which the basidiospore develops.

sterilization of a medium, the act or process of killing all living organisms in the medium; of organisms, loss of sexual structures.

stigma the receptive part of a pistil, which receives the pollen and on which the pollen germinates.

stimulus an environmental agent or change that induces a reaction in a living organism.

stipe a stalk, in some algae and fungi, lacking vascular tissue.

stipule one of a pair of usually small, leaflike structures borne at the base of a leaf in many plants.

stock the stem or plant on which a scion is grafted.

stolon a stem that grows horizontally along the surface of the ground; also formerly applied to a stem or branch that bends down to the ground and takes root at the tip; in fungi, a horizontal hypha which sprouts where it touches the substratum and forms haustoria and rhizoids in the substratum and aerial mycelium or sporophores above it, as in the Mucorales.

stoma a small pore bordered by guard cells found in the epidermis of leaves, stems, and other plant parts and through which gaseous exchange occurs.

stomium an opening on the side of fern sporangia, between the lip cells, at which dehiscence occurs.

stone cell a more or less isodiametric sclerenchyma cell with very thick, hard walls.

stone fruit a drupe, such as a cherry or plum.

streptococcus any of a group of gram-positive bacteria, occurring in chains and dividing in one plane only.

streptomycin an antibiotic obtained from *Streptomyces,* one of the actinomycetes.

strobilus a conelike group of sporophylls borne on a stem axis.

style the slender part of many pistils between the stigma and the ovary and through which the pollen tube grows.

subarchesporial pad a cushionlike mass of sterile cells found at the base of the capsule portion of the sporangia of *Lycopodium.*

suberin a waxlike substance found in the walls of the cells of cork.

succession (see plant succession).

succulent a plant with watery or juicy tissues.

sucrase (see saccharase).

sucrose ($C_{12}H_{22}O_{11}$) a disaccharide made up of a molecule of glucose and one of fructose; cane sugar.

suction force (diffusion pressure deficit) a term applied to the net force causing water to enter a cell; the water-absorbing power of the cell.

summerwood the part of an annual ring formed during the later part of the growing season.

superior ovary an ovary situated above, and free from, the calyx and corolla.

suscept a plant or organism subject to a given disease brought about by a given causal complex.

suspensor a cell or group of cells developed from a zygote in higher plants, which attaches the embryo to the embryo sac; in the Mucorales, the cell which supports the conjugating cell.

suture the line of junction of contiguous parts, as in capsules of seed plants.

swarm spore a motile naked protoplasmic body, as in Myxomycetes; a zoospore.

syconium the multiple fruit of the fig, in which the upper part of the peduncle becomes fleshy and completely envelops the fruits.

symbiosis an association of two or more kinds of organisms in which both are benefited.

Sympetalae (Metachlamydeae) a subdivision of the dicotyledons consisting of plants in which the petals of the flowers are united into a gamopetalous corolla.

sympetaly (gamopetaly) coalescence of petals.

sympodial branching a type of branching brought about by the death of terminal portions of the branch or stem, the new growth proceeding from the development of lateral buds and resulting in a series of short branches, each one attached to the next preceding one, forming an irregular or broken line, as in *Catalpa.*

synapsis the pairing of homologous chromosomes in meiosis.

synecology the study of a group of organisms and their interactions with the environment and with each other.

synergids the two nuclei which, with the egg, form the egg apparatus of the female gametophyte in the micropylar end of the ovule.

syngamy fertilization, or union, of gametes.

syngenesious consisting of united anthers.

synsepaly (gamosepaly) a condition in which sepals are united.

2,4,5-T an abbreviation for 2,4,5-trichlorophenoxyacetic acid, a growth substance widely used as a herbicide to kill woody species.

tannin a bitter, astringent organic substance found in the bark and other parts of some plants.

tapetum nutritive tissue in a sporangium, especially in anthers.

taproot a primary root which grows vertically downward and gives off smaller lateral roots.

taxies paratonic movements of motile organisms or motile structures of plants such as zoospores in response to unilateral stimulation.

taxon a group of organisms recognized as a formal unit.

taxonomy the science dealing with the description and classification of organisms.

teliospore a spore, usually a resting spore, characteristic of the Heterobasidiomycetes, in which fusion of nuclei and meiosis occur and from which a basidium develops.

telium (*pl.* telia) a sorus in which teliospores are borne.

telophase the final stage of mitosis during which daughter nuclei are developed.

tendril a slender, spirally coiling organ which serves as a means of attachment to a supporting body or surface.

testa the outer seed coat.

tetrad a group of four, usually referring to the four meiospores formed by meiosis.

tetraploid with four sets of chromosomes per nucleus, i.e., four times the haploid number.

tetrarch referring to the condition of a stele that consists of four protoxylem points.

tetraspore one of the four spores of a tetrad; nonmotile spores of the red algae produced in fours.

tetrasporophyte a plant bearing tetraspores, as found in *Polysiphonia*.

Thallophyta an older term for a main division of the plant kingdom consisting of bacteria, slime molds, algae, and fungi.

thallus a relatively simple plant body without true roots, stems, and leaves.

thermonasties nasties caused by changes in temperature.

thermotropism growth movement in response to the stimulus of temperature.

thiamin vitamin B_1, the water-soluble antineuritic or antiberiberi vitamin.

thorn a short, sharp-pointed branch.

thymine a pyrimidine constituent of nucleic acids.

tinsel a flagellum covered with small projections.

tissue a group of cells performing a special function; simple tissues are made up of similar cells, complex tissues of different kinds of cells.

α-tocopherol vitamin E, the fat-soluble antisterility vitamin.

tonoplast the cytoplasmic membrane bordering a vacuole.

torus the thickening of the closing membrane in bordered pits.

toxin a poisonous substance or secretion of a plant or animal.

TPN abbreviation for triphosphopyridine nucleotide (which see).

trace element (microelement) an element that is essential for plants in extremely minute quantities.

trachea a vessel.

tracheid an elongated, thick-walled cell of the xylem, having tapering ends and functioning in water conduction and support.

Tracheophyta a main division of the plant kingdom to which all vascular plants are referred.

transfer RNA the RNA specific for individual amino acids involved in protein synthesis.

translocation the transfer or movement of foods and other products of metabolism.

transpiration the giving off of water vapor from living plants.

tree a woody, perennial plant with a single main stem or trunk.

triarch referring to the condition of a stele consisting of three protoxylem points.

trichogyne the receptive filament or tube of the female gametangia of certain algae and fungi.

triphosphopyridine nucleotide (TPN) a coenzyme derived from the vitamin nicotinic acid and functioning in dehydrogenation reactions. Now usually referred to as nicotinamide adenine dinucleotide phosphate (NADP).

triploid with three sets of chromosomes per nucleus, i.e., three times the haploid number.

tropism a bending or growth movement of radially symmetrical organs in response to a stimulus, caused by greater growth rate on one side.

truffles the edible ascocarps of the fungus order Tuberales.

trypsin a proteolytic enzyme.

tube nucleus one of the nuclei of a pollen grain thought to influence the growth and development of the pollen tube.

tuber a short, thick, fleshy underground stem.

tundra one of the level or undulating treeless plains of northern arctic regions.

turgor the distension or turgidity of a living cell resulting from its fluid content.

turgor pressure the internal hydrostatic pressure exerted by the cell content against the cell wall.

tylose or **tylosis** (*pl.* tyloses) growth of one cell into the cavity of another; more specifically, the intrusion of a parenchyma cell into the cavity of a xylem vessel or tracheid.

tyrosinase an oxidizing enzyme of the copper oxidase type, attacking the amino acid tyrosine.

umbel an inflorescence in which the pedicels all arise from a common point, as in onion.

unilocular sporangia the single-celled sporangia of *Ectocarpus* in which meiosis takes place.

uracil a pyrimidine constituent of nucleic acids.

urease an enzyme which hydrolyzes urea to ammonia and CO_2.

uredinium a sorus containing uredospores.

uredospores the red, unicellular summer spores of the rust fungi.

vaccine a suspension of dead or weakened bacteria or other pathogens which may be injected into an organism to immunize against the same species or kind of pathogen or its toxins.

vacuole a cavity or vesicle in the protoplasm of a cell, containing a watery solution of various substances called cell sap.

valve one of the pieces into which a capsule splits; in diatoms, each half of the silicified wall.

vascular of or pertaining to tissue for the conduction of fluids.

vascular bundle a strand of conducting and strengthening tissue consisting of primary xylem, primary phloem, and often cambium, sometimes enclosed by a sheath of parenchyma or sclerenchyma.

vascular cambium cambium that gives rise to secondary xylem and phloem.

vascular tissue tissue involved in conduction and strengthening, chiefly xylem and phloem.

veil a structure found in the mushroom type of fruiting body of many agarics as a thin membrane covering the gills on the lower side of the cap and breaking as the sporophore enlarges, part of it remaining on the stem as an annulus. In some cases the entire sporophore in its early stages is covered by a veil, which is broken through as the sporophore enlarges, the veil remaining at the base of the stem as a cup or volva.

vein one of the vascular bundles forming the framework of fibrous tissue of a leaf.

velamen a parchmentlike sheath or layer of spiral-coated air cells on the roots of some tropical epiphytic orchids and aroids.

venation the arrangement and disposition of the veins in the tissue of a leaf blade.

venter the enlarged basal portion of an archegonium in which an egg cell develops.

vernalization treatment of germinating seeds or seedlings with low (or high) temperatures to induce flowering at maturity.

vernation the disposition or method of arrangement of foliage leaves within a bud.

versatile anther an anther attached to the filament at or near the middle of the anther, so as to swing freely.

vessel a xylem tube or duct for the conduction of water and inorganic substances.

vessel segment an individual cell or unit of a vessel.

viability the ability to live and grow, as applied to seeds and spores.

virus an ultramicroscopic pathogen capable of passing through bacteriological filters and reproducing only in living tissue. All viruses thus far isolated have been found to be proteins.

vitamin any of a group of organic substances naturally occurring in plants and necessary in minute quantities for metabolic activity and growth of plants and animals.

volva a cuplike structure found at the base of the sporophores of some fleshy fungi, originally covering the entire sporophore as a universal veil.

water requirement the weight of the water lost by transpiration divided by the dry weight of the plant, or the number of units of water transpired in the production of one unit of dry matter.

weed any plant growing in cultivated or otherwise utilized ground to the detriment of the crop or desired vegetation or to the disfigurement of the area; an economically useless, unsightly, or undesired plant.

whorl a circle of flower parts or leaves.

whorled referring to leaf arrangement in which three or more leaves occur at a node; also referring to flowers and other organs having several parts arranged in a circle at the same level.

wilting coefficient (permanent-wilting percentage) the percentage of water left in a soil in which plants permanently wilt.

wood the xylem, especially of trees and shrubs.

wood ray a radial row of parenchyma cells in secondary xylem.

xanthophyll a yellow, carotenoid, plastid pigment.

xerarch succession a succession originating in a dry habitat.

xerophyte a plant living in a dry habitat; a drought-resisting plant.

xylem a complex vascular tissue of plants consisting of such components as vessels, tracheids, wood fibers, parenchyma, and ray cells; wood.

zeaxanthol (zeaxanthin) the chief yellow, carotenoid pigment of corn.

zoosporangium a sporangium bearing zoospores.

zoospore a motile spore.

zygomorphic bilaterally symmetrical.

zygospore a thick-walled resting spore resulting from the conjugation of isogametes.

zygote a fertilized egg; a cell resulting from the fusion of gametes.

zymase an enzyme complex involved in alcoholic fermentation of sugar.

INDEX